METHODS in
MICROBIOLOGY

METHODS in
MICROBIOLOGY

Edited by

J. R. NORRIS
Borden Microbiological Laboratory,
Shell Research Limited,
Sittingbourne, Kent, England

D. W. RIBBONS
Department of Biochemistry,
University of Miami School of Medicine,
and Howard Hughes Medical Institute,
Miami, Florida, U.S.A.

Volume 5 B

 1971

ACADEMIC PRESS
London and New York

ACADEMIC PRESS INC. (LONDON) LTD
Berkeley Square House
Berkeley Square
London, W1X 6BA

U.S. Edition published by
ACADEMIC PRESS INC.
111 Fifth Avenue
New York, New York 10003

Library of Congress Catalog Card Number: 68–57745
ISBN: 0–12–521545–2

PRINTED IN GREAT BRITAIN BY
ADLARD AND SON LIMITED
DORKING, SURREY

LIST OF CONTRIBUTORS

PER-ÅKE ALBERTSSON, *Department of Biochemistry, University of Umeå, Umeå, Sweden*

K. E. COOKSEY, *Shell Research Limited, Milstead Laboratory of Chemical Enzymology, Sittingbourne, Kent, England*

E. L. DUGGAN, *Deceased April, 1968*

K. HANNIG, *Max-Planck-Institut für Eiweiss-und Lederforschung, Munich*

OSAMU HAYAISHI, *Department of Medical Chemistry, Kyoto University Faculty of Medicine, Kyoto, Japan*

D. HERBERT, *Microbiological Research Establishment, Porton, Nr. Salisbury, Wilts., England*

D. E. HUGHES, *Microbiology Department, University College, Cardiff, Wales*

D. LLOYD, *Microbiology Department, University College, Cardiff, Wales*

W. MANSON, *Hannah Dairy Research Institute, Ayr, Scotland*

F. J. MOSS, *School of Biological Technology, The University of New South Wales, Kensington, N.S.W., Australia*

MITSUHIRO NOZAKI, *Department of Medical Chemistry, Kyoto University Faculty of Medicine, Kyoto, Japan*

P. J. PHIPPS, *Microbiological Research Establishment, Porton, Nr. Salisbury, Wilts., England*

PAMELA A. D. RICKARD, *School of Biological Technology, The University of New South Wales, Kensington, N.S.W., Australia*

G. H. ROPER, *Department of Biological Process Engineering, School of Chemical Engineering, The University of New South Wales, Kensington, N.S.W., Australia*

J. R. SARGENT, *Department of Biochemistry, Marischal College, University of Aberdeen, Scotland*

W. D. SKIDMORE, *Armed Forces Radiobiology Research Institute, Bethesda, Md. 20014, U.S.A.*

R. E. STRANGE, *Microbiological Research Establishment, Porton, Nr. Salisbury, Wilts., England*

I. W. SUTHERLAND, *Department of General Microbiology, University of Edinburgh, Edinburgh, Scotland*

J. SYKES, *Department of Biochemistry, University of Sheffield, Sheffield, S10 2TN, England*

J. F. WILKINSON, *Department of General Microbiology, University of Edinburgh, Edinburgh, Scotland*

J. W. T. WIMPENNY, *Microbiology Department, University College, Cardiff, Wales*

O. VESTERBERG, *Karolinska Institutet, Stockholm, Sweden*

ACKNOWLEDGMENTS

For permission to reproduce, in whole or in part, certain figures and diagrams we are grateful to the following—

The American Society of Biological Chemists; Aminco Inc., Silver Springs, Maryland, U.S.A.; Beckman Instruments Inc., Fullerton, California, U.S.A.; Elsevier Publishing Co., Amsterdam; Macmillan & Co., Ltd.

Detailed acknowledgments are given in the legends to figures.

PREFACE

Volume 5 of "Methods in Microbiology" is concerned with the microbial cell—methods of observing it, of studying its properties and behaviour, of analysing it chemically and immunologically, and of purifying and characterizing its various "organelles" and macro-molecular components. Wherever possible, the emphasis has been placed on quantitative methods.

We have tried to cover relatively new techniques such as reflectance spectrophotometry, isoelectric focusing and polyacrylamide gel electrophoresis which appear to us to have considerable future potential in microbiology in addition to more generally used techniques such as those for cell disintegration and hybridization of nucleic acids which are not fully described in a concise form elsewhere.

As with earlier Volumes in the Series we have left the treatment of the different topics largely to the individual authors, restricting our editorial activity to ensuring consistency and avoiding overlaps and gaps between the contributions.

As contributions accumulated it became obvious that there was too much material for a single Volume and the content was divided. Volume 5A contains Chapters concerned with the direct observation or study of whole cells or organelles while Volume 5B is concerned with the disintegration of cells, their chemical analysis and the techniques used to separate and characterize their components.

Our thanks are due to the pleasant way in which our authors have co-operated with us and particularly to those who agreed to update their contributions when delay in the publication process made it necessary.

J. R. NORRIS
D. W. RIBBONS

April, 1971

CONTENTS

LIST OF CONTRIBUTORS v

ACKNOWLEDGMENTS vi

PREFACE vii

CONTENTS OF PUBLISHED VOLUMES xi

Chapter I. The Disintegration of Micro-organisms—D. E. HUGHES, J. W. T. WIMPENNY AND D. LLOYD 1

Chapter II. Centrifugal Techniques for the Isolation and Characterization of Sub-Cellular Components from Bacteria—J. SYKES 55

Chapter III. Chemical Analysis of Microbial Cells—D. HERBERT, P. J. PHIPPS AND R. E. STRANGE 209

Chapter IV. Chemical Extraction Methods of Microbial Cells— I. W. SUTHERLAND AND J. F. WILKINSON 345

Chapter V. Biphasic Separation of Microbial Particles—PER-ÅKE ALBERTSSON 385

Chapter VI. Separation and Purification of Proteins—MITSUHIRO NOZAKI AND OSAMU HAYAISHI 425

Chapter VII. Zone Electrophoresis of the Separation of Microbial Cell Components—J. R. SARGENT 455

Chapter VIII. Free-flow Electrophoresis—K. HANNIG . . . 513

Chapter IX. Preparative Zonal Electrophoresis—W. MANSON . 549

Chapter X. Disc Electrophoresis—K. E. COOKSEY . . . 573

Chapter XI. Isoelectric Focusing and Separation of Proteins— O. VESTERBERG 595

Chapter XII. Reflectance Spectrophotometry—F. J. MOSS, PAMELA A. D. RICKARD AND G. H. ROPER 615

Chapter XIII. Base Composition of Nucleic Acids—W. D. SKIDMORE AND E. L. DUGGAN 631

AUTHOR INDEX 641

SUBJECT INDEX 665

CONTENTS OF PUBLISHED VOLUMES

Volume 1

E. C. ELLIOTT AND D. L. GEORGALA. Sources, Handling and Storage of Media and Equipment

R. BROOKES. Properties of Materials Suitable for the Cultivation and Handling of Micro-organisms

G. SYKES. Methods and Equipment for Sterilization of Laboratory Apparatus and Media

R. ELSWORTH. Treatment of Process Air for Deep Culture

J. J. McDADE, G. B. PHILLIPS, H. D. SIVINSKI AND W. J. WHITFIELD. Principles and Applications of Laminar-flow Devices

H. M. DARLOW. Safety in the Microbiological Laboratory

J. G. MULVANY. Membrane Filter Techniques in Microbiology

C. T. CALAM. The Culture of Micro-organisms in Liquid Medium

CHARLES E. HELMSTETTER. Methods for Studying the Microbial Division Cycle

LOUIS B. QUESNEL. Methods of Microculture

R. C. CODNER. Solid and Solidified Growth Media in Microbiology

K. I. JOHNSTONE. The Isolation and Cultivation of Single Organisms

N. BLAKEBROUGH. Design of Laboratory Fermenters

K. SARGEANT. The Deep Culture of Bacteriophage

M. F. MALLETTE. Evaluation of Growth by Physical and Chemical Means

C. T. CALAM. The Evaluation of Mycelial Growth

H. E. KUBITSCHEK. Counting and Sizing Micro-organisms with the Coulter Counter

J. R. POSTGATE. Viable Counts and Viability

A. H. STOUTHAMER. Determination and Significance of Molar Growth Yields

Volume 2

D. G. MACLENNAN. Principles of Automatic Measurement and Control of Fermentation Growth Parameters

J. W. PATCHING AND A. H. ROSE. The Effects and Control of Temperature

A. L. S. MUNRO. Measurement and Control of pH Values

H.-E. JACOB. Redox Potential

D. E. BROWN. Aeration in the Submerged Culture of Micro-organisms

D. FREEDMAN. The Shaker in Bioengineering

J. BRYANT. Anti-foam Agents

N. G. CARR. Production and Measurement of Photosynthetically Useable Light

R. ELSWORTH. The Measurement of Oxygen Absorption and Carbon Dioxide Evolution in Stirred Deep Cultures

G. A. PLATON. Flow Measurement and Control

RICHARD Y. MORITA. Application of Hydrostatic Pressure to Microbial Cultures

D. W. TEMPEST. The Continuous Cultivation of Micro-organisms: 1. Theory of the Chemostat

C. G. T. EVANS, D. HERBERT AND D. W. TEMPEST. The Continuous Cultivation of Micro-organisms: 2. Construction of a Chemostat

J. ŘíČICA. Multi-stage Systems

R. J. MUNSON. Turbidostats

R. O. THOMSON AND W. H. FOSTER. Harvesting and Clarification of Cultures—Storage of Harvest

Volume 3A

S. P. LAPAGE, JEAN E. SHELTON AND T. G. MITCHELL. Media for the Maintenance and Preservation of Bacteria

S. P. LAPAGE, JEAN E. SHELTON, T. G. MITCHELL AND A. R. MACKENZIE. Culture Collections and the Preservation of Bacteria

E. Y. BRIDSON AND A. BRECKER. Design and Formulation of Microbial Culture Media

D. W. RIBBONS. Quantitative Relationships Between Growth Media Constituents and Cellular Yields and Composition

H. VELDKAMP. Enrichment Cultures of Prokaryotic Organisms

DAVID A. HOPWOOD. The Isolation of Mutants

C. T. CALAM. Improvement of Micro-organisms by Mutation, Hybridization and Selection

Volume 3B

VERA G. COLLINS. Isolation, Cultivation and Maintenance of Autotrophs

N. G. CARR. Growth of Phototrophic Bacteria and Blue-Green Algae

A. T. WILLIS. Techniques for the Study of Anaerobic, Spore-forming Bacteria

R. E. HUNGATE. A Roll Tube Method for Cultivation of Strict Anaerobes

P. N. HOBSON. Rumen Bacteria

ELLA M. BARNES. Methods for the Gram-negative Non-sporing Anaerobes

T. D. BROCK AND A. H. ROSE. Psychrophiles and Thermophiles

N. E. GIBBONS. Isolation, Growth and Requirements of Halophilic Bacteria

JOHN E. PETERSON. Isolation, Cultivation and Maintenance of the Myxobacteria

R. J. FALLON AND P. WHITTLESTONE. Isolation, Cultivation and Maintenance of Mycoplasmas

M. R. DROOP. Algae

EVE BILLING. Isolation, Growth and Preservation of Bacteriophages

Volume 4

C. BOOTH. Introduction to General Methods

C. BOOTH. Fungal Culture Media

D. M. DRING. Techniques for Microscopic Preparation

AGNES H. S. ONIONS. Preservation of Fungi

F. W. BEECH AND R. R. DAVENPORT. Isolation, Purification and Maintenance of Yeasts

MISS G. M. WATERHOUSE. Phycomycetes

E. PUNITHALINGHAM. Basidiomycetes: Heterobasidiomycetidae

ROY WATLING. Basidiomycetes : Homobasidiomycetidae

M. J. CARLILE. Myxomycetes and other Slime Moulds

D. H. S. RICHARDSON. Lichens

S. T. WILLIAMS AND T. CROSS. Actinomycetes

E. B. GARETH JONES. Aquatic Fungi

R. R. DAVIES. Air Sampling for Fungi, Pollens and Bacteria

GEORGE L. BARRON. Soil Fungi

PHYLLIS M. STOCKDALE. Fungi Pathogenic for Man and Animals: 1. Diseases of the Keratinized Tissues

HELEN R. BUCKLEY. Fungi Pathogenic for Man and Animals: 2. The Subcutaneous and Deep-seated Mycoses

J. L. JINKS AND J. CROFT. Methods Used for Genetical Studies in Mycology

R. L. LUCAS. Autoradiographic Techniques in Mycology

T. F. PREECE. Fluorescent Techniques in Mycology

G. N. GREENHALGH AND L. V. EVANS. Electron Microscopy

ROY WATLING. Chemical Tests in Agaricology

T. F. PREECE. Immunological Techniques in Mycology

CHARLES M. LEACH. A Practical Guide to the Effects of Visible and Ultraviolet Light on Fungi

JULIO R. VILLANUEVA AND ISABEL GARCIA ACHA. Production and Use of Fungal Protoplasts

CHAPTER I

The Disintegration of Micro-organisms

D. E. HUGHES, J. W. T. WIMPENNY AND D. LLOYD

Microbiology Department, University College, Cardiff, Glam.

I.	Introduction	1
	A. History	1
	B. Methods of assessment	3
	C. Enzyme location	5
II.	Physical Methods of Breakage	6
	A. Mechanism of mechanical cell disintegration	6
	B. Liquid or hydrodynamic shear	13
	C. Solid shear	30
III.	Chemical Methods	40
	A. Osmosis	40
	B. Drying and extraction	41
	C. Autolysis	42
	D Inhibition of cell wall synthesis	42
	E. Enzymic attack on cell walls	43
	F. Bacteriophage and other lytic factors	47
	G. Ionizing radiation	47
IV.	Choice of Methods	47
	References	48

I. INTRODUCTION

A. History

Although enzymes released from micro-organisms have long played a part in industrial process such as retting and tanning, the deliberate development of methods for extracting microbial enzymes marks the beginning of the modern phase not only of microbiology but also of biochemistry.

The experiments of Buchner (1897) which led to the isolation of cell free fermenting extracts of yeast are well known and often referred to as one of the basic archetypal experiments of modern biochemistry ranking with that of Lavoisier in chemistry. It is often forgotten however that it was in the main the criticisms of this work which led to its further extension. For instance, many critics were loath to believe that the relatively low rates of fermentation achieved by the extract could in any way represent the massive effects of a few live yeast cells, and attempts to rectify this later led to the discovery of the role of phosphates and of coenzymes (cf. Harden and Young, 1906). Attempts to improve the yield of the intermediate enzymes in glycolysis led to the development of many individual methods of enzyme preparations particularly by Lebedev and others.

In addition to extracting enzymes from disintegrated cells much early work was concerned with the extraction of immunologically active material. Methods evolved around 1900 (cf. MacFayden and Rowland, 1903) included grinding in liquid air in ball mills etc., which were later applied to microbial enzymology. It is somewhat remarkable that the more elegant methods today used for extracting enzymes are little used in immunology.

From 1900 for the next 30 to 40 years microbial biochemistry was largely concerned with the characterization of metabolic pathways and the enzymes which catalyzed them. This is reflected in the increasing sophistication of methods for enzyme extraction. Most of these methods are aimed at increasing speed of extraction, maintaining a low temperature in order to increase the yield of the more unstable and delicate enzymes from the most refractory organisms: these include for example the enzymes of oxidative phosphorylation from Gram-positive cocci, nocardia and mycobacteria, a process which still leaves much to be desired.

In recent years, the art of enzyme extraction has centred less upon the extraction of enzymes *per se* but rather on studies of the relationship between structure and function and thus the localization of enzymes, first between so-called "soluble" and "bound fraction", and later in identifiable intracellular organelles. This has led to a reappraisal and modification of a wide range of available methods most of which had been developed previously. An additional need for this reappraisal is the growing use of extraction methods in the study of enzyme regulation and control. Interpretation of these phenomena based on the quantitation of enzyme activities in extracts often pays far too little attention to possible changes in enzyme properties and kinetics introduced by the very act of extraction.

In reflecting on these later developments, the late Dr. Majory Stephenson's analogue of Microbial Biochemistry as a person observing a house and noting food delivery and analysing dustbin contents is brought to mind. The further step of bombing the house and interpretation of the relationship between architecture, function of the inhabitants and their regulation and control would extend this analogue to our present studies, especially if the interpretation was made with almost a complete disregard for the random effect of the explosion.

The following material is an attempt to rationalize the position as far as is possible by describing the available methods for disintegrating microbes for preparing biologically active fractions. Some brief advice on using the techniques themselves is also given. In addition a summary is made of the theory of the methods, their mode of action as well as some guidance as to choice of method and the assessment of efficacy.

B. Methods of assessment

1. *Cell disintegration.*

(a) Direct measurement of cell disintegration especially when one of the chemical methods of extraction is used, can sometimes be dispensed with and the criteria of yield of the desired enzyme activity be relied on. Nevertheless, in any comparative work and where control or location is studied by means of a physical method of disruption or with spheroplasts, some method of assessing disintegration is essential. This may include classical methods of direct counting either of numbers or mass (this Series, Vol. 1) but these often have to be adapted to take account of the effects of material released from the cells. This is especially so where for instance DNA or other polymeric material interferes with dilution prior to counting. Some methods, e.g. Coulter counter are rendered almost useless by such material as well as by the presence of cell envelopes or other large particles. In these cases an approximate estimation of the degree of disintegration may be made by proportional counts of stained smears. Here advantage may be taken of differences between whole and damaged cells: for instance damaged Gram-positive bacteria often become Gram-negative, damaged yeast cells take up basic stains more densely than undamaged cells. In the electron microscope the absence of electron dense material in cells examined in unstained drops and the shape of cells in shadowed specimens is diagnostic and can often be used quantitatively: total counts can be made in shadowed specimens if a known number of polystyrene beads (1–2 μm dia.) are mixed thoroughly with the control and disintegrated cells prior to preparing grids by the blot-dry or spraying method.

(b) *Indirect.* Estimation of cell disintegration by measuring the release of typical cell components can be made quantitatively more precise than is direct counting. However there is some difficulty in interpretation and in obtaining a normal zero time and 100% standards. For instance the measurement of the release of soluble protein is often used. This depends partly on the degree of disruption but also on the method used to separate soluble protein from insoluble debris. In the case of yeast we have adopted the standard method of separating cell debris by centrifugation for 10 min at 6000 g and estimating protein in the supernatant by the method of Lowry *et al.*, (1951) or with Biuret reagent (Layne, 1957).

These conditions are not entirely arbitrary but were arrived at by methods described in (a) where it was found that when ca. 100% of the yeast appeared as empty cell walls by staining and by electron microscopy, ca. 80% of the protein appeared in solution. Further comminution of the separated cell walls resulted in a slow release of protein into solution together with increasing amounts of cell wall carbohydrates. Under these conditions it was assumed that a value of 80% protein in solution represented ca. 100% cell disintegration. In adapting this method to bacteria, account must be taken of the higher centrifugal forces needed to remove both whole and partially disintegrated cells and higher concentration of DNA in bacteria and its tendency to precipitate with proteins especially as the pH is reduced below 6·0 must also be taken into account. It is essential to ensure that sufficient buffering capacity is present to counter the fall in pH which results in almost all bacterial suspensions upon disintegration. The use of DNA'ase and the addition of extra alkali, e.g. Na or KHCO$_3$ after disintegration counter these effects.

Other chemical estimations, e.g. 260 nm/280 nm ratio, DNA and RNA, have also been used but it must be emphasized that in each case further information especially by methods in (a) are necessary before degrees of cell disintegration can be estimated. It is worthwhile to recall too, in connection with studies on the intracellular location of enzymes and other components, that the study of the time constants of their release from cells may yield extremely valuable information (Hughes, 1961; Marr and Cota Robles, 1957).

2. *Measurement of enzyme activity*

The measurement of enzyme activity in crude cell or partially purified cell-free extracts particularly from bacteria presents pitfalls which appear not to be generally appreciated. These arise in the first place from the fact that the assay method is often one which is based on the kinetics of the purified enzyme and that it is assumed that these apply in cell-free extracts and intact cells with which they are often compared. There are however now sufficient cases where such assumptions are known to be wrong. Two such may be quoted from our own work. The first concerns the properties of the glutamate decarboxylase and glutaminase of *Clostridium welchii* whose K_m, pH curve and solubility altered considerably not only during the growth of the organism but also during purification in cell-free extracts. This was due to the presence of an unidentified competitive inhibitor which appeared in the cell towards the end of the log phase (Hughes, 1950). The other occurred in studies on polyphosphate synthesis in corynebacteria where crude cell-free extracts were found to be completely inactive until polynucleotides were removed by streptomycin and manganese treatment: the increase in specific enzyme activity suggested that changes in the turnover value of the enzyme might be occurring during purification (Muhhammed, 1959).

It is clear therefore that apart from questions of permeability which arise when comparing whole cells and extracts, changes in the kinetics of enzymes and their stability in cell-free extracts may play an important role in interpretation of questions of regulation and control based on a comparison of enzymic activity. A check of the apparent enzyme yield, kinetics and stability at all stages of isolation and purification therefore is essential, although this is rarely done. A comparison of these properties in extracts prepared by different methods of disintegration and extraction are also of great value in assessing the effects of the method of disintegration.

C. Enzyme location

The criteria already discussed for comparing enzyme activities in whole cells and cell free extracts become of paramount importance in studies on enzyme location by the analytical cytological method (Marr, 1960). Additional techniques are needed and these are briefly mentioned here as far as each is affected by the method of cell disruption.

1. Quantitative criteria

For the most part the location of a particular site of enzymic activity in a cell organelle has relied simply on the finding that a major part of the enzymic activity found in the intact cell is located in that fraction of a cell-free extract (Hogeboom, Kuff and Schneider, 1957). It will be evident that the amount of the enzymic activities found in any fraction depends largely on the criteria already discussed and on checking the stability both of the organelle and enzyme in the fractionation procedure. An additional factor is the production of artefacts especially due to the formation of vesicles from disrupted membranes or the absorption of more soluble enzymes on insoluble proteins which may or may not be produced by the method of disruption. For instance prolonged sonication may precipitate initially soluble enzymes (e.g. pyruvate oxidase from lactobacilli, Hughes unpublished).

2. Morphological criteria

Morphological studies with microbes inevitably involve the electron microscope with its attendant advantages and disadvantages. The limit of resolution (ca. 0.2 μm) of classical histochemical methods does not assist here (cf. Marr, 1960; Holt and Sullivan, 1958) but newer methods involving radioautography at high resolution (Reith et al., 1967) and the use of ferritin labelled antibodies (this Series, Vol. 5a) may improve this approach and assist interpretation of results obtained on isolated intracellular fractions. Few studies at high resolution of the processes of microbial disintegration have been made and associated with the intracellular location of enzymes. But recent work on the relationship between mesosomes and the protoplasmic membrane indicate how useful such an approach may be.

II. PHYSICAL METHODS OF BREAKAGE

A. Mechanism of mechanical cell disintegration

1. *Crushing and grinding*

The methods used for crushing and grinding dry microbial powders are similar to those used widely in industry and elsewhere but with the important difference that the particle size, especially of bacteria, is very much smaller than that usually dealt with. This imposes several problems not least of which is that the machines must be made to high precision and that this must be maintained in the face of constant wear. One other important difficulty is that the phenomenon of caking associated with fine powders (10–1 μm diam.) is met with almost immediately and not in the later stages of the process. However it is probably unnecessary to comminute each microbial cell into finer fragments because the aim of grinding is to facilitate enzyme extraction. Changes in permeability due to the introduction of cracks through the envelope is probably all that is necessary. This is an important consideration because both the power requirements and elaboration of equipment increase exponentially with the degree of fineness of the product. Dry grinders and crushers include ball and vibratory mills, roller mills, plate mills and powder attrition mills. No general theory is applicable to these. Three main theoretical treatments, those of Bond, Kick and Rittinger differ mainly in their definition of the measurement of comminution (cf. Marshall, 1966). The equation (1) expresses an overall balance,

$$\frac{dE}{dX} = -C\frac{1}{X}N \tag{1}$$

where E = net energy required/weight/unit of comminution. X = size factor N = an exponent, if we neglect differences in definition of size factor of feed and product. It follows that the energy requirements and the grinding index C, depend on the properties of the material and the degree of comminution required. The solid properties of dry or frozen microbial cells as yet escape any precise definition. Nevertheless it is likely that they will behave as a crystalline or semicrystalline material when frozen and an elastic or viscoelastic solid when dried. The comminution process proceeds by the introduction of cracks at points of strain introduced by imposing either a tensile or shear stress in each particle even in beater systems such as hammer mills and ball mills where the initial stress is compressive. This is shown diagrammatically where the compressive stress is transmitted in much the same way as in a beam test—in this model an additional bending moment will also be produced in flexible material, but this is weak compared to the other stresses.

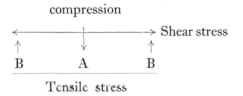

Tensile stress

Estimations of the force required to break bonds in ideal crystals have been made (Joffe, 1928) and show that the energy required (ca. 400 Kgcal/ml) is of the order to melt an ionic crystal. No such estimations, even approximate, can be made of the more complex structures dealt with here, but it is likely that malformations and imperfections such as reduce considerably their tensile strength will arise when microbes are dried or frozen. This reduces the energy requirements for fracture which will however remain quite large due to their small size (3–5 μm). Energy is expended producing stresses below that level resulting in cracks or breakages and this energy appears as heat and must be taken into account when working with labile material such as enzymes. Since the degree of fineness is comparatively great, surfaces in the mill such as the liners, balls or abrasives, appear rough in comparison: these too suffer attrition often quite rapidly and it is quite common to find a high degree of metal contamination in these systems. This is particularly undesirable when the metals contain toxic elements such as nickel or copper as for instance in stainless steels. There are no reports of machines made from relatively non toxic material such as titanium or the newer impact resistant ceramics.

2. Wet milling

Many mills used for dry grinding and crushing can also be used for wet milling microbes. Others designed specifically for this use include, the Booth Green Mill, Muys Mill, Kalnitsky Mill, Mickle, Nossal, Braun shakers. The common feature of all these devices is that breakage is probably produced directly and predominantly by hydraulic or liquid shear rather than by an initial compressional stress as in dry milling. These methods are thus analogous in action to the Chaikoff and French presses as well as to that of ultrasonics. Fluid Energy Mills are also hydraulic shear systems, but have not yet been used in microbiology.

Wet grinding or attrition mills of use in microbiology fall into the general category of so-called "Colloid Mills" because the particle size of the product approaches that of colloidal particles, i.e. ca. 0.02 μm (Travis, 1929). In practice however it is difficult to reduce particle size below 1 μm. The rate of attrition is slow especially if no abrasive is added and wear on the moving parts with subsequent contamination is high. In addition, changes in the rheological properties consequent upon cell rupture may slow down attrition markedly especially where the process is by multiple by-flow (Booth Green and Muys Mills). The release of DNA and of wetting or foaming polymers is particularly troublesome in vibration mills and it is a common practice to add some antifoam agent: tributyl citrate is particularly useful. Although therefore the same generalized equation (1) will apply to these methods the grinding factor will in addition have to take account of the rheological properties of the released material. The importance of this may be realized when it is often found that the release of a gel material from the cells inhibits altogether further grinding either because of loss of contact between cells and grinding surfaces or the damping of shear stress due to high viscosity.

Vibratory mills used in microbiology generally use small glass beads as shearing surfaces (Ballotini Beads). A study of the mechanism of action of one such device the Sonomec Shaker, (Rodgers and Hughes, 1960; Rodgers, Nyborg and Hughes, 1961; Rodgers and Nyborg, 1967) suggest that the action is similar to that when cell suspensions are passed through sintered glass filters. In fact it was this consideration that led to this method first reported by Hogg and Kornberg (1963). The rate of disintegration was found to be very dependent on the air in the vessel which was shown to distribute between top and bottom due to bubble formation caused by the collapse of surface waves at high amplitudes. Studies with high speed films suggested that the glass beads which were essential for disintegration moved towards the centre of the vessel and remained almost stationary. Hydraulic shear is developed as the liquid passed rapidly through the spaces between the beads. Most other vibratory mills do not lend themselves so readily to the theoretical treatment developed for the Sonomec (Rodgers and Nyborg, 1967) but it is likely that their mechanism is similar to that described above. It is worthwhile to stress that considerable amplification of power input was due to the "air hammer" effect near resonance. Under steady state conditions the amplitude, and hence shear stress, of the water column exceed by a factor of four the peak-to-peak amplitude of the vessel itself, and this was very dependent on the initial height of the air column. This is important as it is a common finding that other vibratory mills are also dependent on the liquid load/air ratio. Studies of industrial vibratory mills are concerned mainly with power requirements and add little further to the understanding of the mechanism of disintegration except that it is a general conclusion that energy requirements are lower when the vibration frequency is close to the resonant frequency (Lissenden, 1965).

3. *Ultrasonics*

It has become accepted that cell disintegration by ultrasound is due to cavitation: disintegration is thus independent of frequency except so far as cavitation itself depends on frequency. (Esche, 1952). Most ultrasonic disintegrators in use today therefore work at lower frequencies than previously used, i.e. in the range of 15–25 K Hertz, because it is relatively cheaper and easier to produce power levels above the cavitation threshold at these lower frequencies. Piezo-crystalline transducers (lead zirconate, barium titanate) or magnetostrictors (Neppiras, 1960) have replaced the quartz plate oscillators for this purpose. Disintegration by cavitation has been considered as a "single hit" phenomenon. Thus with a constant cavitation level disintegration curves are exponential, and this explains why early attempts to use ultrasonics for sterilization were abandoned. It is however likely that this assumption is an oversimplification, since it does not account for the time dependence (cf. Hughes, 1961) of the release of cell components or cell death. It is probably more realistic to regard the phenomenon as one of fatigue of different biological cellular components particularly of those such as the mucopeptides which compose the high tensile rigid layer of the bacterial wall.

Other factors which may affect cavitation and sonic cell disintegration include nucleation or those which modify bubble activity such as viscosity, surface tension, or ambient pressure. The nature of the gas phase especially if it is CO_2 by reason of its solubility in water may also modify both disintegration and the stability of the product (Hughes, 1961; Neppiras and Hughes, 1964).

Cell disintegration by sonically induced cavitation may be due to shock waves, chemical attack by free radicals (Noltingk and Neppiras, 1950, 1951; El 'Piner, 1964) or, as is now becoming more generally accepted, by cavitation microstreaming (Elder, 1959; Hughes and Nyborg, 1962). In this latter case the physical force does not vary sinusoidally with time as do the various first order components of the acoustic field such as pressure, or velocity. Nor is it necessary to employ sound pressures above the cavitation threshold, to produce "transient cavities" (Flynn, 1964) since it has been shown experimentally that "stable cavities" can produce streaming velocities sufficiently great to break even bacteria (Hughes and Nyborg, 1962; Nyborg and Hughes, 1967). It is worthwhile to mention that especially high velocities of streaming can be generated by sound fields at phase boundaries, and this may account for some polymer degradation in the absence of cavities (Hawley, Macleod and Dunn, 1963). Recently Nyborg (1966) has calculated hydraulic shear gradients to be expected by streaming induced by a radially pulsating bubble and at sound pressures below those usually employed for cell disintegration. Under these conditions the velocity gradient may be considered as uniform in a thin layer proximate to the bubble, zero elsewhere and to take up forms shown in Fig. 1. The thickness of the boundary layer ϵ is of the order of $(2\,v/f)^{\frac{1}{2}}$ where v = kinematic viscosity, f = frequency. The velocity gradient G in the boundary layer is given by v_L/ϵ where $v_L = o/Ro$, o = oscillating amplitude

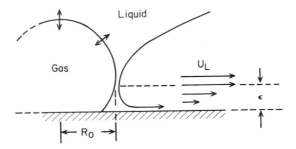

FIG. 1. Microstreaming near a radially vibrating bubble.

of the bubble, and Ro the stationary diameter of the bubble. Where o/Ro is 1/30 and 1/10 respectively, u_L is approximately 2·3 cm/sec and 21 cm/sec. At 10 K Herz, = ca. $5·7 \times 10^{-4}$ giving G as $4·1 \times 10^3$. The pressure amplitude of the sound to drive the bubble at resonance (P_{10}) is $7·7 \times 10^{-3}$ ats. This is well below the cavitation threshold of ca. 0·5 ats which is that of normal ultrasonic disintegrators. No account has been taken of surface waves which are known to be formed at higher amplitudes and which would enhance considerably the velocity of microstreaming. Likewise near the stage at which bubbles become unstable, i.e. the onset of transient cavitation, liquid jets may also form centres of high shear gradients.

It is noteworthy that solids which effectively speed up sonication probably do so by reason of the air which they carry in pores approaching resonant size (Hughes, 1961): diatoms were particularly effective.

4. *Hydraulic shear*

Shear in liquids may be defined as a frictional force set up during flow or deformation. In any liquid moving in parallel layers, that is in streamline flow, the friction force per unit area f is given by

$$f = \eta G \tag{2}$$

where G = velocity gradient and η = coefficient of viscosity. However, above a critical velocity streamline flow breaks down and flow becomes turbulent. The critical velocity (u) is given by

$$u = \frac{K\eta}{Pr} \tag{3}$$

where K is Reynold's number, and u is velocity, ρ = density of fluid; r = radius of channel through which the fluid flows.

The onset of turbulence creates vortices near which there may be very much higher shear rates than exist throughout the bulk of the liquid. Equations for this situation had been developed by Taylor (1935, 1936), but no simple treatment is available for estimating shear under these conditions. Most equipment in which microbial disintegration is assumed to be by hydraulic shear, operate at Reynolds numbers well above the critical value. Their efficiency probably depends largely on these local and intense shear gradients in Taylorian vortices rather than on simple shear stress in streamline conditions.

Several workers have attempted to apply these and more sophisticated rheological treatments to cell disintegration in streamline flow generally, by treating them as dilute emulsions (Taylor, 1935, 1936). Again the difficulty lies in the assumptions made about the solid/liquid state of individual cells. For instance Runscheidt and Mason (1961) discuss the behaviour of oil droplets in a liquid such as water, and show that at a certain value of G the droplet becomes unstable and then separates into two by a process of pinching and elongation. The critical value of G at which this occurs is given by ($T/2b$) where T = interfacial tension; b = spherical radius: and is generally much lower for cells than for oil droplets. At low flow rates, shear stress was transmitted across the interfacial boundary without diminution, and weak vortices were set up inside the droplet. This was decreased by the presence of a viscoelastic film at the interface such as probably occurs around most cells. Nevertheless deformations of the cell envelope and pinching of the small spheroblasts from red cells has been observed by us in low amplitude sound fields around resonant bubbles under the conditions described by Hughes and Nyborg (1962). Similar deformations have been predicted for red cells during laminar flow in tubes (Charm and Kurland, 1962).

Critical values for G also apply to the breakage of macromolecular threads such as DNA by hydraulic shear. Levinthal and Davison (1961) have considered flow through capillaries as exerting a stretching force upon a thread of DNA. Similarly Pritchard, Peacocke and Hughes (1966) have shown that shear gradients exerted by microstreaming around bubbles at very low sound amplitudes ($G = 10^4$ to 10^5 sec^{-1}) are sufficient to degrade

DNA. More recently work in this laboratory with homogenizers of the Potter-Elvejhem type and under conditions of streamline flow have yielded similar values of G, which cause polymer degradation and cell rupture (T. Coakley, personal communication).

Such deformations and bursting as occur in red cells or protozoa caused by weakly pulsating bubbles or in streamline flow do not readily affect the more rigid cells of bacteria. Degradation is therefore likely to depend on erosion of the outer cell wall polymers, particularly at weakened places such as division or budding scars. For bubbles to be effective at all against these cells it is likely that they must operate close to transient cavitation where it is impossible as yet to give a theoretical treatment. Under these conditions, streaming is likely to be more analogous to vortex streaming due to surface waves rather than to streamline flow. As already mentioned in liquid just below the level of transient cavitation the amplitude and velocity will approach that of liquid jets and G is likely to be very high indeed.

The finding that G values sufficiently high to degrade polymers and cells can occur in sonically induced stable cavities is of special significance because the deleterious effects of sonication are mostly due to the effects of transient cavitation, i.e. high temperature, shock waves and free radicals (cf. El 'Piner, 1964).

5. *Freeze pressure* (Hughes and x presses)

In discussing the mechanism of action of the Hughes press (1951) it was assumed by Hughes that the disintegrative effect was due largely to the abrasive action of ice crystals during compression and to re-gelation brought about by the high pressure (15 tons/in^2). More critical studies by Edebo (1961a, b) however suggested that a major contribution was played by a conformational change when ice crystals in phase 3 changed under pressure to the phase 2 form (Bridgman, 1937). More recently unpublished experiments in this laboratory support the idea that abrasion by ice crystals does play a role. The plunger of a Hughes press was given one blow sufficient to produce the cracking sound associated with the phase change, and to force about 2 mm of a 3 cm plug of frozen cells into the press reservoir. Disintegration was estimated in the material in the reservoir and at 3–5 mm intervals in the plug remaining in the cylinder. About 95–98% of yeast cells were disintegrated in the reservoir, 80% at both ends of the plug and almost no disintegration was found in the centre. This supports strongly the idea that attrition occurring while ice and cells move across the gap from cylinder to reservoir plays an important role. The nature of the rupture also supports this idea, since if this was due entirely to the volume change during the phase shift, much more fragmentation would occur than is normally the case in the Hughes press, when the majority of cells appear to have only a single, or at the most, two breaks through their outer envelopes. The appearance of cells broken in a Hughes press either with added abrasive at $+2°C$ or without abrasive at $-25°C$ is similar.

B. Liquid or hydrodynamic shear

1. *The French press*

The French press was first used by Milner, Lawrence and French (1950) to prepare extracts from the alga *Chlorella*, but since then this pressure cell has achieved deserved popularity as an effective method for breaking many species of micro-organisms.

(a) *Description of apparatus.* In its simplest form the French press consists of a steel cylinder having a small orifice and needle valve at its base, and a piston with some form of pressure tight washer. Pressure is applied to the piston by means of a laboratory hydraulic press capable of delivering 10–20 tons total pressure. Drawings and photographs of the French presses used in the authors laboratory are shown in Fig. 2.

The needle valve can be actuated by a simple capstan as in Fig. 2(b) or by a close tolerance anti-backlash reduction gear assembly such as we have fitted to the commercially available Aminco French press (American Instrument Company, Silver Springs, Maryland, U.S.A. (Fig. 2(a) and 2(b). In our experience reduction gear allows more precise control of the flow through the press when the needle valve is built into the cylinder, but seems not to be necessary in models with an external needle valve.

Various degrees of automation are available. Aminco market an automatic hydraulic press which allows any chosen pressure to be maintained whilst crushing cells. The most refined modification to the French press principle is the Sorvall-Ribi fractionator (Ivan Sorvall Inc., Connecticut, U.S.A.). In this model an external needle valve is chilled with precooled nitrogen. The press is automatic in operation incorporating a re-filling cycle and is refrigerated. Needle valve temperature is monitored and the whole assembly is enclosed in a sealed UV sterilizable container. Whilst the Ribi fractionator represents an advance in automation for most laboratories the simpler presses are adequate.

(b) *Operation.* The press is precooled to around $0°C$ and filled with the suspension of micro-organisms. Air should be excluded as it is highly compressed during operation and can squirt out explosively with the last of the suspension. This can be very dangerous at very high pressures should any component of the press fracture, although in practice we have never heard of any case of this type of failure. The Aminco press can be filled from the

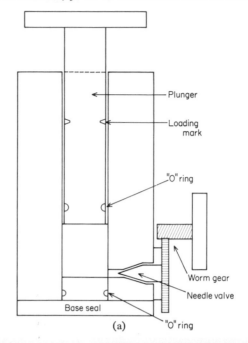

(a)

(b)

Fig. 2. (a) The commercially available "Aminco" French pressure cell fitted with anti back lash reduction gear to facilitate operation of the needle valve. (b) A larger pressure cell having an externally capstan controlled needle valve (kindly loaned by Aminco, Inc. Silver Springs, Maryland, U.S.A.).

bottom with the base removed. The manufacturers provide a stand to facilitate this procedure. Air should be forced out of the open needle valve. This is then closed and pressure applied; at the chosen pressure the valve is cautiously opened and pressure maintained by pumping. The broken suspension is collected on ice.

(c) *Experimental observations—effectiveness*. Protein release from a number of different micro-organisms has been compared at different pressures. At the pressures chosen only Gram-positive cocci and *Chlorella* proved hard to break. Mycobacteria which are resistant to breakage in other ways are easily broken in the French press (Table I). Viable counts show that highly clumped mycobacteria are easily dispersed at low pressures in the French press before rupturing as pressure increases (Fig. 3). The virtue of the press in its ability to operate over a large range of pressure and flow rates is at once clear.

TABLE I

Percentage of total nitrogen released from various organisms subjected to different pressures in French pressure cell

Organism	$1 \cdot 0 \times 10^4$ psi	$1 \cdot 33 \times 10^4$ psi	$1 \cdot 66 \times 10^4$ psi	$2 \cdot 9 \times 10^4$ psi
Escherichia coli	65	87	88	—
Baker's yeast	28	48	56	70
Mycobacterium tuberculosis BCG	52	57	61	67
Bacillus megatherium	26	35	46	57
Chlorella pyrenoidosa	—	27	32	35
Sarcina lutea	8	9	14	25 (37*)
Staphylococcus aureus	2	3	8 (12*)	31 (21*)

* Second passage.

(d) *Effect on structures*. The way in which increasing extrusion pressure breaks the cell has been investigated using *Escherichia coli*. Electron micrographs at increasing pressure demonstrate increasing comminution to a fairly uniform size fragment of cell wall-protoplast membrane at the highest pressures (Fig. 4).

(e) *Effect on labile molecules*. When pure alcohol dehydrogenase was subjected to increasing pressures in the French press no inactivation of this enzyme was observed.

Four enzymes of the TCA cycle were measured using *E. coli* as test organism: the activity in French press extracts was compared with data

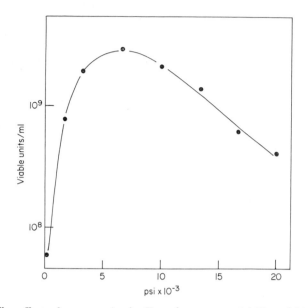

FIG. 3. The effect of pressure in the French press on viability of *Mycobacterium tuberculosis* var *bovis* B.C.G.

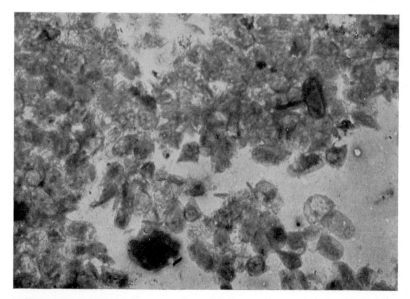

FIG. 4 (a)

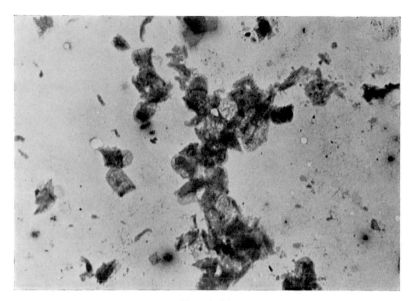

Fig. 4 (b)

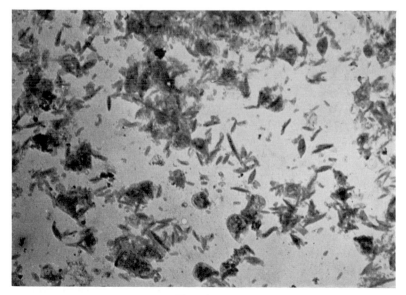

Fig. 4

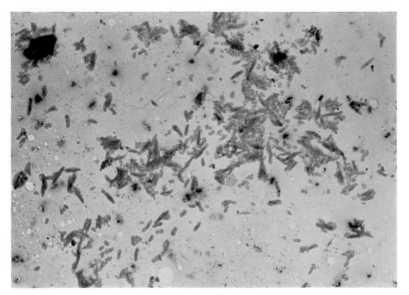

FIG. 4 (d)

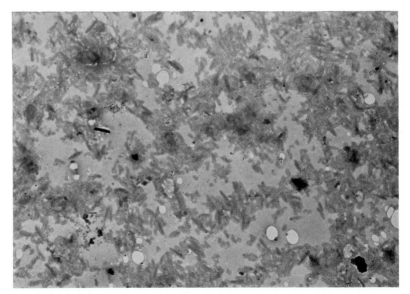

FIG. 4 (e)

FIG. 4. Cell wall-membrane fractions of *E. coli* broken in the French press at progressively higher pressures; Magnification : 10,000×. (a) $0\cdot4\times10^4$ psi; (b) $0\cdot8\times10^4$ psi; (c) $1\cdot2\times10^4$ psi; (d) $1\cdot8\times10^4$ psi; (e) $2\cdot4\times10^4$ psi. 1 cm = 1 μm.

from Hughes press and ultrasonic treatment (Table II). It is clear that within reasonable limits all methods are similarly useful.

TABLE II

Effect of breakage method on the activity of four enzymes from *Escherichia coli*. **Activity in French press extracts at 100.**

	French press‡	Hughes press‡	Ultrasonic treatment
Aconitase	100	75	109
Fumarase	100	93	86
Isocitrate dehydrogenase	100	97	115
Malate dehydrogenase	100	93	122

* 10,000 psi
† M.S.E. 60 W generator 4 min.
‡ Broken at −27°C

The effect of pressure on these enzymes is shown in Fig. 5. If a cell suspension is kept at the crushing pressure for some time before release through the needle valve inactivation of all enzymes occurred; however, aconitase and fumarase were most sensitive to prolonged pressure (Fig. 6).

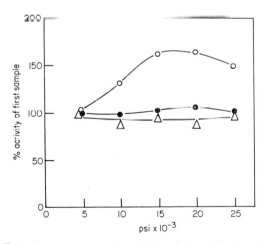

FIG. 5. The effect of pressure on the activity of three TCA cycle enzymes from *E. coli* disrupted in the French press.
△———△ Isocitrate dehydrogenase; ●———● Fumarase; ○———○ malate dehydrogenase.

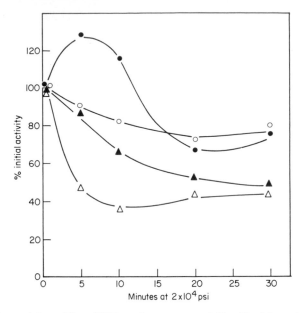

Fig. 6. The activity of four TCA cycle enzymes of *E. coli* subjected to 2×10^4 psi for various times before passing through the French press needle valve. △————△ Aconitase; ▲————▲ Fumarase; ○————○ Isocitrate dehydrogenase; ●————● Malate dehydrogenase.

Some experiments on the effect of cell concentration on enzyme activity, in which ultrasonic disintegration and extrusion pressure have been compared appear in Fig. 7. In all cases except with isocitrate dehydrogenase there appears to be a critical concentration below which enzyme inactivation occurs. Other experiments with yeast cells (Fig. 8) show also that breakage is more effective at high-cell concentrations.

A homogenizer for preparing emulsions on a large scale (Manton Goring 15M 18 BA) has recently been used for disintegrating yeasts and some bacteria. Volumes of several gallons/h have been successfully processed for enzyme extraction (Dr. M. D. Lilly, personal communication). This machine can be considered as a continuous " French press" since breakage takes place when the suspension is passed at high speed and pressure through a nozzle.

2. *Chaikoff press*

This instrument was originally designed as a homogenizer for the preparation of high yields of intact nuclei from animal tissue. It functions (Emanuel and Chaikoff, 1957) by extruding a liquid suspension through an

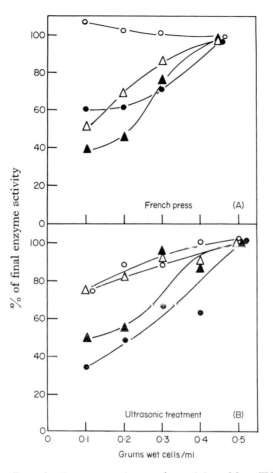

FIG. 7. The effect of cell concentration on the activity of four TCA cycle enzymes of *E. coli* broken in (A) the French press (B) by ultrasonic treatment. △———△ Aconitase; ▲———▲ Fumarase; ○———○ Isocitrate dehydrogenase; ●———● Malate dehydrogenase.

annular orifice around a steel rod (Fig. 9) according to Bernoulli's principle ($V\alpha^1$/pressure). The clearance through which the suspension is forced can be chosen by a system of interchangeable steel rods and spares between 0·00025 and 0·005 in. (6 and 130 μm) at room temperature, therefore slightly greater at working temperature) have been used. The pressures required to operate the instrument vary with the nature of the material treated and with the clearance chosen; pressures in excess of 2000 or 3000 psi are rarely attained when the annular orifice size is larger than 0·001 in., except

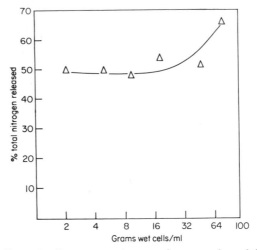

Fig. 8. The effect of cell concentration on nitrogen released from bakers yeast, suspensions treated at 2×10^4 psi in the French press.

with material containing fibrous tissue when pressures as high as 14,000 psi are developed. The buffer to tissue ratio is important and a ratio as high as 2·5 is effective.

Mitochondria obtained from rat liver after homogenization in a Chaikoff Press showed respiratory control (Klucis *et al.*, 1968). The Chaikoff press has not been extensively used for breaking micro-organisms, but fragile organisms such as ascites tumor cells, myxomycetes and *Tetrahymena* can be completely disrupted with an orifice size of 0·0015 in. (rod diam. 0·357 in.). We have obtained satisfactory results with the flagellate *Polytomella caeca* and the amoeba *Hartmanella castellanii*: a high degree of integrity of organelles of these fragile organisms was preserved. The Chaikoff press affords a method of great reproducibility of controlled cell disruption for such organisms although its usefulness is limited in microbiology by the relatively low shear gradients developed.

3. *Homogenizers*

The use of the Potter–Elvehjem homogenizer for preparing mammalian tissue extracts has been reviewed by Potter (1955); modifications of these are described by Hughes and Cunningham (1962). The application of this technique in microbiology is limited, although it has been used to disrupt the more fragile species of algae, protozoa and some fungi.

Various parameters have been examined in our laboratory by Mr. T. Coakley. In an homogenizer with a teflon pestle with a gap between pestle

TABLE III

Effectiveness of two different freeze pressing techniques. Breakage is expressed as a percentage and was estimated microscopically

Organism	Hughes Press		"X" Press −1 passage −25°C	"X" Press −25°C	No. of Passages
	Frozen paste −35°C	paste with equal volume pyrex glass powder −10°C −15°C			
Clostridium welchii	80–90	100	—	—	—
Lactobacillus arabinosus	80	100	—	—	—
Escherichia coli	90–95	100	70–90	100	3
Proteus morganii	80–90	100	—	—	—
Streptococcus faecalis	80	100	40–70	100	5
Mycobacterium phlei	80	100	40	70–90	5
Mycobacterium smegmatis	60–80	ca 90	—	—	—
Bacillus subtilis		100	—	—	—
Micrococcus aureus	—	—	40–70	70–90	3
Streptococcus pyogenes	—	—	—	70–90	3
Rhodospirillum rubrum	—	—	—	100	3
animal cells	—	—	—	100	3

and homogenizer wall of 100 μm (0·004 in.) various treatment times were tested. Homogenization for 30 sec with 4 rapid strokes gave complete breakage of the flagellate *Polytomella caeca* and of *Hartmanella castellanii* at 5000 rpm. Two 30 sec periods of treatment gave 20% disruption of *Euglena gracilis*. A total of 5 min homogenization time at 12,000 rpm with a 50 μm gap gave 30% breakage of the colourless alga *Prototheca zopfii* while 6 min treatment at this speed produced 20% disruption of baker's yeast. The average stress required to give 50% breakage of the flagellate *Polytomella* under conditions of turbulent flow was measured at about 1000 dynes/cm^2. Since flow was turbulent estimations of shear rate are not possible.

Rapid strokes of the pestle were found to be very necessary to minimize the effects of local heating. Under the conditions used (15 ml volume, 100 μm gap, 4 strokes, 30 sec homogenization time) the theoretical rise in temperature is 15°C; heat dissipation led to a measured temperature rise of 8°C in the bulk of the homogenate.

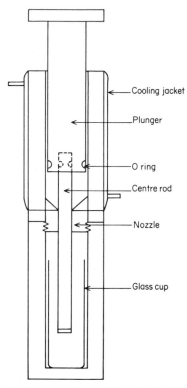

FIG. 9. Cross section of the Chaikoff Press.

There are many instances in the literature of reports of the use of homo-genizers under rather uncontrolled conditions. Difficulty is often experi-enced in exactly repeating methods using the homogenizer technique, for this reason. In cases where the intact nature of intracellular organelles is of importance it would seem better, in general, to avoid the high liquid shearing forces employed in this technique, although Watson and Smith (1967) have reported the isolation of intact mitochondria from *Aspergillus niger* after homogenization of mycelial masses. Suyama (1966) has described the use of a cream homogenizer for disrupting *Tetrahymena* prior to isolation of intact mitochondria.

4. *Wet mills*

The Booth and Green wet mill (Booth and Green, 1938) consists of a modified roller bearing driven by an electric motor. The suspension is passed between the roller and the race and recirculated by pumping through a chilled reservoir.

A relatively fragile bacterium such as *Escherichia coli* can be disrupted completely after 2 h treatment. After 4 h *Sarcinia lutea* has lost only a quarter of its weight into a centrifuged supernatant.

Clearly this is not very effective or convenient but it seems possible that the principle would be useful in the design of a continuous breakage method, if the effectiveness could be increased. Whilst the performance of the Booth and Green mill leaves much to be desired it has been used by many workers in the past to obtain enzymically active preparations. Some applications have been reviewed by Hugo (1954).

Barnard and Hewlett (1912) describe a wet ball mill which consists of a phosphor-bronze body containing five accurately fitting hardened steel balls located by baffles. The authors describe successful breakage of yeast, *E. coli* and *Staphylococcus aureus*.

The mill described by Muys (1949) consisted of a rotating glass ball joint through which a cell suspension was forced continuously by a syringe. More than a single passage was required for effective cell breakage and the instrument has not been widely used.

5. Vibration mills

Lamanna and Mallette (1954) have shown that the liquid shear gradients set up at the edges of the blade of high speed blendors are not sufficient to cause significant breakage of bacteria. However when glass beads were added to the suspension breakage was increased. This method was also used by Vitols and Linnane (1961) to prepare phosphorylating particles from baker's yeast. The effect of the glass beads is similar to that in the instruments in which micro-organisms are shaken violently with ballotini beads. Furness (1952) and Cooper (1953) used laboratory flask shakers, while Mickle and Ray (1948) and Nossal (1953) and Braun have developed high amplitude shakers; heat is generated by all these methods and its effect can be partly mediated by pulsing treatments of CO_2 cooling. These methods have been extensively used for the disruption of micro-organisms; the most efficient breakage is produced by the Nossal shaker.

Rodgers and Hughes (1960) have pointed out that both the Mickle and Nossal shakers suffer from the following disadvantages:—

(1) Small amounts (25 ml) of material only may be treated at one time.
(2) Lengthy treatment is necessary to obtain an acceptable percentage breakage.
(3) A high ratio of particles to micro-organisms is required.
(4) Mechanical unreliability.
(5) Inability to disintegrate micro-organisms in concentrated sucrose solutions.

(6) Dilute suspensions only can be used, depending on the organism and the foaming of extracts.

These authors investigated the use of a commercially available "Sonomec" Wave–Pulse Generator (Fig. 10) which does not present any of the above disadvantages. In this machine both amplitude and frequency of vibration of the shaker can be independently varied, and a variety of varying dimensions up to 250 ml were used. Factors affecting the rate of disruption of baker's yeast, *Corynebacterium xerosis* and *Bacillus megatherium* were studied and a comparison was made of the activities of several enzymes in fractions obtained by differential centrifugation of the cell dispersion with similar fractions obtained after breakage in the Hughes press. The instrument has been used since then for disintegrating a wide variety of cells including yeasts for preparing mitochondria.

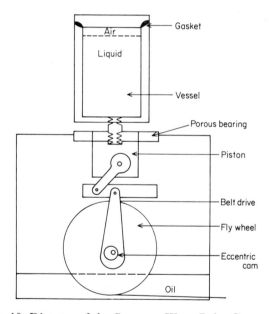

FIG. 10. Diagram of the Sonomec Wave-Pulse Generator.

A number of other commercially available vibration mills and vibrators have been tested in this laboratory but none so far has proved as generally useful as the Sonomec shaker, which is now no longer commercially available, but the Braun shaker has effectively replaced it for smaller volumes.

6. *Filters*

From investigation of cell disruption by shaking with glass beads in the Sonomec shaker Rodgers and Hughes (1960) suggested that the passage of cells through a stationary bed of beads in the form of a sintered glass funnel, might prove an effective way of disrupting fragile micro-organisms. The method was first applied by Hogg and Kornberg (1963) for *Tetrahymena pyriformis* and has been routinely used in our laboratory for the disruption of this protozoan and also for the flagellate *Polytomella caeca* and the amoeba *Hartmanella castellanii*. Cell suspensions are forced under reduced pressure (provided either by a filter pump or a vacuum pump) through a precooled sinter with a suitable pore size (15–40 μm) and the broken suspension collected in a cooled receiver. The method is a useful one for the preparation of extracts containing intact organelles such as mitochondria, but is not suitable where large quantities of cells have to be broken, due to progressive clogging of the pores of the sinter.

7. *Ultrasonic disintegration*

The use of ultrasound as a means of breaking micro-organisms was first suggested by experiments of Harvey and Loomis (1929) and by Chamber and Flosdorf (1936). Since this time the technique has been extensively used.

(a) *Description of apparatus.* The a.c. output from an electronic oscillator and amplifier is converted to mechanical waves by either a magneto-strictive or a piezo-electric transducer, of barium titanate or lead zirconate. The output from the transducer is coupled to the suspension undergoing treatment by a half wave metal probe, which oscillates at circuit frequency. Although soft metals such as brass have been used, extensive pitting of the end of the probe occurs and stainless steel, or preferably titanium, probes are more suitable. The amplitude of the wave generated is inversely related to the diameter of the probe tip, and the efficiency of coupling energy to the matching of acoustic impedance of probe and liquid. Choice of the probe diameters is governed to a certain extent by the volume of suspension undergoing treatment (Hughes, 1961).

(b) *Use of ultrasonic disintegrators.* The probe is fitted securely to the transducer. Hughes (1961) has demonstrated the importance of good coupling at this point and uses silicone stop-cock grease to make a tight joint. The machine is switched on and allowed to warm up. The tip of the probe is immersed a few millimeters beneath the liquid surface. Provision is made to cool the suspension either by packing the sonication cup in ice and salt or by circulating coolant through a jacketed cup. The generator is set at the required power setting or amplitude, if this facility is available, and tuned to the resonant frequency. At this point a sharp hissing sizzling sound is heard and to a great extent successful breakage is proportional to the intensity of the sound and can be judged by ear. However at the resonant frequency the current drawn by the transducer may also be used as a criterion. After the required length of time the instrument is detuned or power to the transducer reduced. The probe may be immersed in ice-water to keep cool in between operations. Some form of acoustic shielding around the transducer and probe is advisable as the "white noise" emitted can become distressing.

(c) *Efficiency of breakage.* A ranking system for relative ease of disruption has been provided by Hughes (1961) and Davies (1959). In general micro-organisms are more readily broken by ultrasound than by other methods. Exceptions to this are mycobacteria which break very easily in the French press and *Micrococcus aureus* which is more easily broken by freeze pressing techniques.

2. *Viscosity.* Cavitation is inhibited in viscous solutions. This situation occurs (a) during the course of disintegration as high molecular weight molecules are released from the cells (b) initially in some preparations which have been frozen and thawed before use; this necessitates treating at low power until the viscosity caused by DNA is reduced and then stepping up power. It is clear that concentrated suspensions of this type are most difficult to treat for this reason (this is in contrast to breakage in the Hughes press or the French press). Breakage in viscous solvents such as sucrose or glycerol or solvents with low surface tensions such as Tween "80" or ethanol also affects cavitation and reduces breakage.

(d) *Factors influencing cell disruption*
1. *Amplitude.* Above the power setting at which cavitation occurs significant free radicle formation takes place. This can easily be detected by measuring iodine released from a standard solution of KI (Hughes, 1961). Whilst free radicle formation probably does not affect breakage it certainly influences the subsequent biological activity of the preparation and free radicle scavengers such as gluthathione or AET can protect sensitive material to a certain degree: pregassing with H_2 can also be effective. At still higher amplitudes, the turbulence and the formation of large bubbles under the probe cause the probe to operate partially in air. This phenomenon is correlated with cavitation unloading and causes excessive heating.

3. *The effect of added nuclei.* Nuclei in treated solutions can act as triggers of foci of cavitation and hence cause increased breakage. Hughes has used

1% suspensions of Hyflo supercel, powdered glass, Embacel and norite during sonication and shown increased disruption of baker's yeast.

4. *Carbon dioxide.* The presence of carbon dioxide effectively suppresses cavitation and can be serious in suspensions of micro-organisms actively oxidizing a carbon source.

5. *Probe design.* The amplitude at the tip of the probe at constant power level is proportional to the end diameters. Where the tip is smaller the tip amplitude is increased and the output (Watts/cm^2) is therefore more intense. High amplitudes in small volumes will produce violent turbulence and overheating unless the power output can be reduced. As a general guide to the sound amplitude and volume should be adjusted so that many cavitation streamers with loud hissing occur and violent turbulence and large bubbles do not.

(e) *Continuous treatment.* Several designs for continuous flow ultrasound treatment vessels have been suggested, one of these is illustrated (Fig. 11) Neppiras and Hughes (1964) have described some experiments using

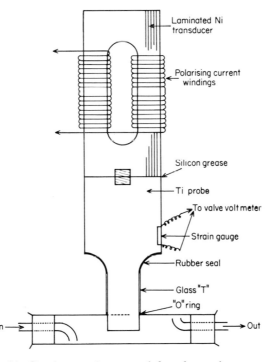

FIG. 11. Continuous flow vessel for ultrasonic generator.

continuous flow cells with baker's yeast. As might be expected breakage in this case depends on the mean residence time of the sample as well as the geometry of the vessel and the power output of the transducer. Of all mechanical breakage methods commonly used, ultrasonic treatment is most amenable to continuous operation. However very little experimental work has been done to critically examine the various parameters.

C. Solid shear

1. *Grinding*

Grinding techniques may be applied to cells prepared as (1) cell pastes (2) frozen cell pastes (3) dried cells and with or without added abrasive.

To grind with glass powder the cells are harvested, drained free from water, mixed with the abrasive until the paste becomes a free-running powder, and ground by hand or mechanically in a chilled pestle and mortar: the ratio of cell paste to abrasive can be critical and proportions of 1 part cells up to 5 parts of abrasive are used. McIlwain (1948) introduced commercial grades of aluminium oxide such as Griffin and Tatlock's "microid polishing alumina" washed and dried at 100°C. A ratio of 1–2 parts by weight of alumina to cell paste yielded more than 80% of the enzyme activity required after 2–5 min grinding. Cell disintegration is indicated by the powder changing to a clay like consistency. Other abrasives used here include diamond, glass, and aluminium.

Lloyd (1965) ground the colourless alga *Prototheca zopfii* with 0·15 mm glass beads for 5 min in a mortar chilled to 0°C by exerting considerable hand pressure. Mitochondria so prepared phosphorylated ADP and had good respiratory control. Buetow and Buchanan (1964) used similar techniques to obtain mitochondria from a bleached strain of *Euglena gracilis*.

A number of mechanical grinding methods have also been employed. Kalnitsky *et al.* (1945) used a rotating cone and stationary funnel system with cell paste mixed with pyrex powder. A device marketed by Gifford-Wood Co., Hudson, New York, called the MV 6–3 Micro mill is a colloid mill, and has been used recently to obtain intact phosphorylating mitochondria from yeast and from *Neurospora*. The Micro mill has an accurately calibrated gap between rotor and stator. Mattoon and Balcavage (1967) use 40 g wet weight of yeast, 150 ml of a special medium together with 70 g 0·2 mm glass beads. The chilled mill is operated at a gap setting of 35 (0·035 in.) for 30 sec with recycling of suspension. Greenawalt *et al* (1967) used a ratio of two parts 0·2 mm glass bead to one part by weight of a conidial or hyphal preparation of *Neurospora* and ground at maximum speed for 1 min at a setting of 0·030 in. In both cases mitochondria showed reasonably good phosphorylation properties.

Several groups of workers have used various mortars and pestles, cone and funnel mills or modified ball mills at sub zero temperatures. These have been reviewed by Hugo (1954).

Grinding techniques have been applied to lyophilized preparations either simply by grinding with alumina in a pestle and mortar (Korkes et al. (1951), Dolin (1950)) or in more sophisticated freezing ball mills (Mudd et al. (1937)).

The main virtue of grinding methods seems nowadays to lie in the production of intracellular organelles by very carefully regulated mills of the Micromill type or by limited grinding in a pestle and mortar. Most bacterial systems are more effectively disrupted by other methods.

2. The Hughes press

A most effective method for breaking micro-organisms was described by Hughes (1951) and is based on solid shear by compression of ice crystals or abrasive powder with cells. Since 1951 other presses based on this principle have been devised, and will be considered below.

(a) *Description of the apparatus.* The original press and the numerous modifications devised by Hughes and by other workers consist of devices that force a frozen suspension or paste of cells through a small gap into a receiving chamber, at very high pressures (10–80,000 psi). The original Hughes press had a split steel body bored to take a steel plunger and secured by bolts. The slit at the base of the body feeds into a reservoir.

A press at present used in our department avoids some of the problems of the split body and is both simpler to operate and trouble free (Fig. 12(b). This press has a cylindrical drilled body butting onto a base and retained by longitudinal bolts. In the base a central steel column, slightly larger than the base of the press, forms an annular slit during operation. The remainder of the base forms a reservoir.

An "X press" designed by Edebo (1960) has the advantages that (a) the cell suspension may be passed through a circular hole several times to increase the breakage rate and (b) that later models operate aseptically, as the press has been constructed with tight nylon gaskets. It can be seen (Fig. 12(c)) that this press can work in either direction, in each case pressure is applied in one cylinder forcing cell material through the central hole into the second which acts as a reservoir.

Raper and Hyatt (1963) described a modified press in which a series of slots operates between a cylindrical body and the reservoir. In this case no bolts are necessary to secure the body to the reservoir. Sutton and King (1966) have devised perhaps the simplest modification to the original Hughes press. In this case the slit is formed by bolting two pieces of steel

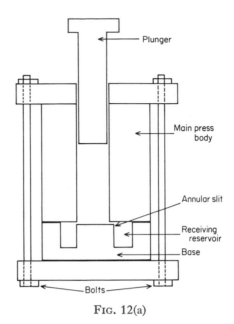

FIG. 12(a)

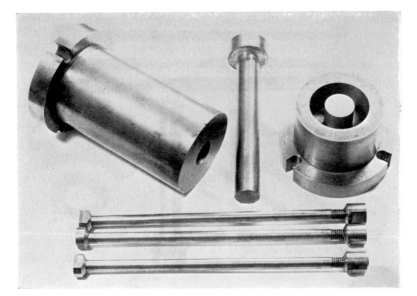

FIG. 12(b)

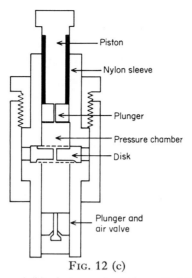

FIG. 12 (c)

FIG. 12. (a) Drawing and (b) photograph of the cylindrical Hughes press. N.B. The gap may be altered by using metal shims between the body and the reservoir (c) The "X" press.

to the base of a drilled solid block which acts as the cylinder. The reservoir is an ordinary glass beaker in a recess in the platform of the hydraulic press. A piston extractor for withdrawing the plunger from a frozen press to allow repeated pressings whilst it is still chilled is described by Mandell and Roberts (1966).

(b) *Operation.* Most modifications of the press are assembled and then chilled in dry ice to −25°C to −30°C. The cell suspension which may be as concentrated as possible is poured or pipetted, or small pieces of pre-frozen suspensions are placed in the cylinder. A simple method is to pre-freeze the preparation in plastic tubes or dialysis tubing of the same or smaller diameter than the bore of the press. The frozen core can then be placed into the press after removing the tubing. The press was originally activated using a fly press. However it is simpler and more convenient to use a 10 or 20 ton hydraulic press, such as those marketed by Apex Engineering Ltd., in this country or in the U.S.A. The pressure is built up to 5–10 tons and the preparation is forced through the slit or hole in a series of cracks. To activate the largest types of Hughes press a weight dropper proved most effective (Fig. 13).

(c) *Effectiveness.* Of all the breakage methods available freeze pressing appears to offer the most complete rupture of a large variety of cell material

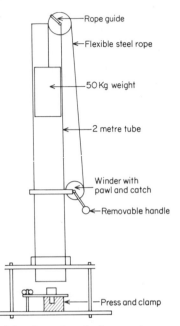

FIG. 13. Sketch of a weight dropping device used to apply pressure to a large Hughes press containing 80 g wet weight cells.

ranging from bacteriophage at one end of the scale to whole animal tissues at the other. Even the most robust organisms (staphylococci, green algae, yeast, fungal mycelia etc.) appear fairly easy to break. Table III includes some published comparative data from different sources.

The broken cells appear to be more intact using this method of breakage than with other mechanical methods, and it represents the method of choice for preparing large cell wall-protoplast membrane fractions from different bacteria (Fig. 14(b)). In the authors' experience the degree of comminution of cell fragments is a function both of the design of the press and more especially of the width or radius of the slot or hole through which the suspension passes. The major disadvantage of freeze-pressing is that a few enzymes or enzyme systems are sensitive to freezing and thawing, however most enzymes tested so far appear to be about as active when assayed after freeze pressing as after other breakage procedures see Table II for example).

In the press shown in Fig. 12(b) a slit width of 1/1000 has proved satisfactory. In the original press widths of 5/1000–15/1000 were used; efficiency fell abruptly above 15/1000. The relatively large diameter of the hole in the "X" press may explain the need for several passages of material in order to obtain breakages similar to the Hughes press.

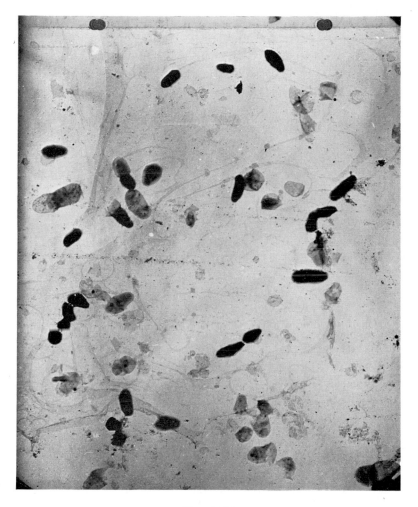

FIG. 14(a)

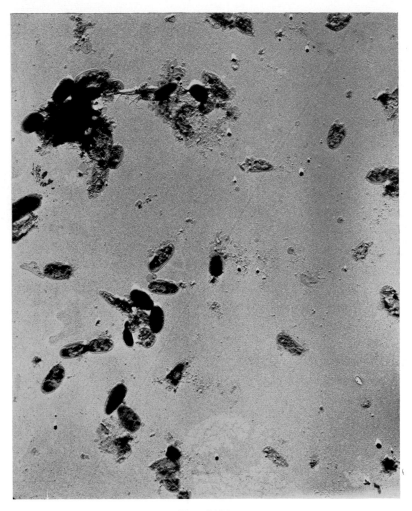

FIG. 14(b)

FIG. 14(c)

FIG. 14. Cell wall membrane fractions from cells disintegrated in a Hughes press. (a) Osmium fixed blot dry preparation. (b) Osmium fixed and shadowed. (c) Yeast cell wall membranes.

3. *Explosive decompression*

This method is based on the fact that at equilibrium the pressure of a gas in solution equals the external gas pressure. When a cell suspension is brought to a high pressure (approx. 1000 psi) under a gas, a relatively large amount of gas dissolves in the suspension medium and diffuses into the cells until equilibrium is achieved. On suddenly releasing the external pressure, the internal pressure is no longer balanced, and if it is large enough can rupture the walls and membranes of cells.

Explosive decompression was first tested by Fraser (1951) for disrupting *E. coli*. He forced nitrous oxide into a steel bomb to a pressure of 500 psi and allowed equilibration for 3 min with shaking at 37°C. The cell suspension was then expelled as quickly as possible into a receiver. One treatment gave 75% breakage as determined by total or by viable counts. N_2 at a pressure of 900 psi gave a similar degree of breakage; the higher pressure was necessary presumably because of its lower solubility compared with that of nitrous oxide. Carbon dioxide at 500 psi was also tested and in this case a large discrepancy was observed in results based on direct cell counts and viable counts, which Fraser attributed to cell damage by the acidity produced by this gas at high pressure. The most important factor in obtaining good disruption was the speed of pressure release; expulsion through narrow orifices (down to 0·004 in. diam.) gave no improvement in percentage disruption. The efficiency of breakage was also found to depend on the age of the culture (exponentially dividing cells were most easily disintegrated) and on the volume of suspension used; the method was less effective with volumes larger than 50 ml in a 350 ml container.

Hunter and Commerford (1961), described the construction and use of a pressure changer for treating mammalian tissues at 900 psi (Fig. 15). Their steel "bomb" was constructed by Artisan Metal Products, Waltham, Mass. and a similar device has been produced in Britain by Baskerville and Lindsay Ltd., Manchester. The homogenizer is in three parts all constructed from heavy stainless steel and bolted together by tightening 12 Allen head screws (C); this operation compresses the polyethylene gasket (B) between the lid and the cylinder. A container for the cell suspension (A) is placed inside the cylinder. One outlet which extends almost to the bottom of the cylinder is used for explusion of the suspension while a second outlet acts as an emergency outlet for reducing the pressure without expelling the contents. After tightening down the lid, nitrogen was slowly admitted and the pressure was maintained at 900 psi for 20 min during which time the apparatus was shaken at frequent intervals to aid equilibration between gas and aqueous phases. At the end of this period the suspension was expelled from the apparatus into a flask containing a trace of antifoam. All operations were performed at 1–5°C.

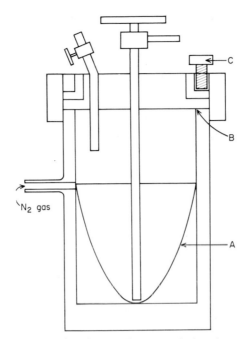

FIG. 15. Vessel for breaking cells by explosive decompression.

With this equipment Hunter and Commerford prepared broken cell suspensions containing intact nuclei from rat liver. From the nuclear preparation they were able to purify deoxyribonucleo-protein which was shown to be relatively undenatured (sedimentation constant (S_0) = 56S). The extract also contained mitochondria which were still capable of carrying out oxidative phosphorylation (P/O ratio of 2·4 with glutamate as substrate). More recently it has been observed (Klucis et al. (1968)) that mitochondria thus prepared also show an intact appearance in the electron microscope and respiratory control, probably the best criterion of mitochondrial intactness.

The original authors pointed out that although it is known that very high pressure can cause protein denaturation, pressures of the order of 1000 psi can actually stabilize protein to the action of denaturing agents. Also since adiabatic cooling occurs during rapid expansion of the gas, the effects of local heating at the time of cell breakage would be avoided.

Unfortunately the method does not appear to be very successful when applied to organisms with rigid cell walls. Yeast is highly refractile to pressure homogenization Foster, Cowan and Maag (1962) applied the method to the breakage of various bacteria using nitrogen cavitation. They obtained figures of 31–59% breakage of *Serratia marcescens* at 1740 psi and 10–25%

breakage of *Brucella abortus* and *Staphylococcus aureus* under similar conditions. *Mycoplasma gallinarum* and *Leptospira pomona* were "extensively broken into small fragments". To our knowledge no studies of enzyme activities in extracts of more fragile micro-organisms broken by this method have been made although theoretically it should be an excellent one for use with, for example, protozoa.

III. CHEMICAL METHODS

Section III describes methods of cell disintegration and enzyme extraction loosely considered as chemical methods. Most of these are aimed at modifying the cell-wall or cytoplasmic membrane, or both, so that the cells either become leaky or burst due to the effects of turgor pressure. The effect of bacteriophage infection is probably correctly included in this category although not yet greatly exploited: ionizing radiation is also included although the mechanisms of induced lysis used by radiation are not yet clear.

A. Osmosis

In this method cell disruption occurs as a result of hydrostatic pressure exerted against the membrane and as the membrane bursts, the force disappears. Thus osmosis affords the gentlest of mechanical methods of cell disruption. Robrish and Marr (1957) have reported that *Azotobacter agilis*, *Rhodospirillum rubrum* and *Serratia plymuthica* can be broken by increasing the solute concentration of the cytoplasm by brief exposure to 1M glycerol. Halophilic organisms are in osmotic equilibrium with their environment and can be disrupted by suddenly diluting the medium, Ingram (1957). In our laboratory we have used 1M sucrose for the osmotic lysis of fragile algae and protozoa.

Most micro-organisms cannot be disrupted by osmotic shock unless their cell walls are first weakened by enzymic attack or by growth under conditions of inhibited cell wall synthesis (see IIID, E).

It has recently been shown that osmotic shock can be used to release several hydrolytic enzymes from bacteria without loss of viability (Neu and Heppel, 1964, 1965; Nossal and Heppel, 1966; Heppel, 1967). The procedure is carried out for *E. coli* as follows. Well-washed cells in the exponential phase of growth are suspended in 80 parts of 0.5M sucrose containing dilute tris buffer and 10^{-3}M EDTA. The mixture is centrifuged and the supernatant solution is removed. The pellet of cells is then rapidly dispersed, by vigorous shaking, in 80 parts of cold 5×10^{-4}M MgCl$_2$. Once more the suspension is centrifuged and the supernatant solution ("shock fluid") is removed. This fluid contains alkaline phosphatase, ribonuclease I, deoxyribonuclease (endonuclease I), acid hexose phosphatase, non-specific acid phosphatase, cyclic phosphodiesterase, 5'-nucleotidase, UDPG-pyrophosphatase and ADPG-pyrophosphatase. A similarly selective release of enzymes has been noted with other Gram-negative bacteria, including *Aerobacter aerogenes, Salmonella typhimurium, Shigella sonnei, Serratia marcescens* and members of the paracolon and citrobacter groups (Glaser, Melo and Paul, 1967).

B. Drying and extraction

Where applicable this method is of immense use in preparing large quantities of relatively stable enzymes. It consists of drying the cells by one of a variety of methods and extracting the dried cells. There is no criterion by which to judge its application in any particular case and reference to the many instances in which it is used (cf. Hugo, 1954; Gunsalus, 1955) will give some idea of the range of applications. Particular points of the technique which may affect the yield of enzyme are (a) rate of drying, (b) the nature of the extractant, (c) the temperature of extraction. Rapid drying from the frozen state, i.e. freeze drying, is often less effective than slow or so-called "sloppy" drying either in air or after desiccation. This is presumably because the rate of drying affects damage to the cells and subsequent autolysis in the extracting fluid. When drying by acetone, ethanol, ethyl ether etc., it is usually necessary to use freshly redistilled solvents and to keep these below -5 to $-10°C$. Rapid handling to prevent water uptake when finally removing the solvent is also necessary.

The range of extractions is very large, and often a great degree of purification can be achieved by a proper selection after small scale experimentation at this point. Usually the more alkaline buffers such as borate at pH 8–9 will extract more material than buffers below pH 7·0. Thus an enzyme which is well extracted at the lower pH will often have a high initial specific activity. In some cases the addition of a membrane solvent such as ether during the extraction will speed it up. Non-aqueous solvents such as butanol have been used to release bound enzymes (R. K. Morton, 1950).

The temperature during extraction may greatly affect the rate and amount of activity in extracts. Where further disruption due to autolytic processes aids extraction, then a temperature as high as 40–45°C can be useful. On the other hand, low temperatures around 2–5°C are essential for labile enzymes.

The development of a suitable extraction method from dried cells is greatly aided by prior knowledge about enzyme solubility and stability. This is conveniently studied with extracts made by another method, e.g. Hughes or French press. (cf. Hughes and Williamson, 1952).

C. Autolysis

Stephenson (1928) used an autolytic method for the preparation of cell free extracts of *E. coli* and Weidel (1951) introduced a method using toluene for the preparation of cell walls of this organism. Since these early observations there have been a number of reports of methods of preparation of extracts of both Gram-positive and Gram-negative organisms by controlled autolysis.

Washed sporulating cells of *Bacillus* species produce lytic enzymes when incubated in buffer with toluene while undergoing autolysis, and the enzymes so produced are capable of dissolving cell walls of vegetative organisms. These enzymes are often used to lyse walls of other species. (Norris, 1957). Strange and Dark (1957a, b) have described a method for the preparation of lytic enzymes from *Bacillus cereus* and Strange and Dark (1957b) have prepared protoplasts by allowing the autolytic process to occur in the presence of sucrose as an osmotic stabilizer. These authors have evidence that two distinct enzymes are responsible and their method is quoted fully by Spizizen (1962). Richmond (1957) has described a lytic enzyme produced by growing cultures of *B. subtilis* and anaerobic lysis of this species has been studied by Nomura and Hosoda (1956a, b) and by Kaufmann and Bauer (1958). Kronish *et al.* (1960) have described a method for the autolytic production of osmotically sensitive forms of *B. cereus* and *E. coli* (cited by Spizizen, 1962). A number of clostridia also autolyse when suspended in cold buffers (Hughes and Galli, 1966).

Mitchell and Moyle (1956, 1957) have developed procedures for the autolytic production of osmotically sensitive forms of *Staphylococcus aureus* and *E. coli*. Their method for *S. aureus* consists of a 2 h incubation of log phase cells (5 mg dry weight/ml) at 25°C in a medium containing 1·2M sucrose and 0·33M sodium acetate at pH 5·8. For *E. coli* their incubation mixture contained 0·3M malonate at pH 7·0 and 0·5M-arabinose. A method for the autolysis of yeast has been described by Nečas (1956).

The optimum conditions for the autolysis of any given species must be determined empirically, the method is especially useful for organisms that are not attacked by lysozyme. Temperature and length of incubation, pH and molarity of buffer, and the metabolic state of the cells are all parameters which must be explored in order to find the best conditions for the autolytic process to proceed for the efficient production of extracts.

D. Inhibition of cell wall synthesis

Whilst not used for preparing cell extracts to any extent, inhibition of cell wall synthesis could provide a relatively simple and gentle method for disrupting certain species of bacteria. The main objection is that the inhibitor might alter the system under investigation.

The antibiotics based on penicillin are the most specific and most active inhibitors that could be used, since there is no evidence that they interfere with any reactions other than those involved in integrating mucopeptide units into the cell wall (Strominger, 1962).

Penicillins only act on growing bacteria and therefore conditions for lysis will be favourable only in the presence of nutrients at cell densities, temperatures and pH where biosynthesis and reproduction can take place. Otherwise this technique can be used in the same way as breakage with lytic enzymes.

E. Enzymic attack on cell walls

Enzymes which attack specific bonds in the cell wall structure of micro-organisms afford a valuable method for the preparation of cell free extracts. This method probably constitutes the most gentle method known for the liberation of the cell membrane in an intact and undegraded manner, and also provides a means of release of soluble components and organelles with a minimum of mechanical damage. However the use of enzymes for the production of extracts is limited to a fairly small scale mainly for economic reasons.

1. *Egg-white lysozyme*

The term "lysozyme" originally coined by Fleming (1922) has been applied to enzymes from a variety of bacterial, animal and plant sources. Here we confine our remarks to the purified egg-white enzyme which is commercially available. It has been extensively used for the preparation of spheroplasts from certain Gram-positive organisms such as *Bacillus megaterium*, *Micrococcus lysodeikticus*, *Sarcina lutea*, *Streptococcus faecalis* 9790 and also (in the presence of chelating agents) to the partial dissolution of the cell wall of certain Gram-negative bacterial species, *Escherichia coli*, *Proteus*, *Aerobacter*, *Pseudomonas* and *Rhodospirillum rubrum*. The cell wall of many *Streptomyces* spp. are susceptible to lysis by lysozyme but not members of the genus *Nocardia*. The methods described in the literature consist in preparation of spheroplasts in a suitable osmotic buffer, gentle washing and finally lysis, usually by osmotic shock. Heppel (1967) has listed hydrolytic enzymes which are selectively released during protoplast formation, prior to osmotic lysis.

Details of procedure for specific organisms

1. *Bacillus megaterium* (Weibull, 1953) and *Bacillus subtilis* (Wiame, Storek and Wanderwinckel, 1955). Washed cells are resuspended in 0·03M

phosphate buffer, pH 7·0 containing either 7·5% (w/v) polethylene glycol (Carbowax 4000) or 0·2 to 0·6M sucrose. The optimal concentration of stabilizer has to be determined empirically. Densities of bacteria up to 50 mg dry weight/ml have been used and lysozyme concentrations of between 50–500 μg/ml. Activation of the enzyme occurs by Na^+. High salt concentrations (above 0·2M) have sometimes produced inhibition of the enzyme although McQuillen (1956) has used 0·5M phosphate buffer, pH 7·0 as a stabilizing agent. Glucose (0·3M), raffinose, melibiose or trehalose (0·15M) have also been successfully used. The process is complete if conditions are correct in about 30 min. Considerable variation in susceptibility occurs, all of 54 stains of *B. megaterium* tested by Tomcsik and Baumann-Grace (1956) were sensitive whereas 35 strains of *Bacillus cereus* were insensitive.

2. *Micrococcus lysodeikticus and Sarcina lutea*. (Mitchell and Moyle, 1956). Suspensions of cells (10 mg dry weight/ml) in NaCl or sucrose (1·0M), are treated with lysozyme (10 μg/ml for *M. lysodeikticus* and 40 μg/ml for *S. lutea*) and incubated for 30 min at room temperature. The essential difference between this procedure and the preceeding one is the higher osmolality of the medium necessary to prevent lysis during protoplast formation.

3. *Escherichia coli*. (Spizizen, 1957). This method also works with a number of other Gram-negative organisms such as *Proteus, Aerobacter* and *Pseudomonas*. Washed cells are resuspended in 0·01M tris, pH 9, and 0·5M sucrose. Ten-minute incubation at 37°C with 50 μg/ml lysozyme. Another method (Fraser and Mahler, 1957) employs 0·5M sucrose in 0·3M tris, pH 8·0 and 5 μg/ml of egg white lysozyme for 2 min at room temperature with swirling. Then 5 μl 4% EDTA solution/ml of suspension is added with mixing and the process is complete after 10 min. Later modification to this method increased the lysozyme concentration to 100 μg/ml and the EDTA to 200 μg/ml (Fraser and Mahler, 1957).

4. *Azotobacter vinelandii* (P. W. Wilson, unpublished cited by McQuillen, 1960). Cells from a 14 to 16 h culture were washed with distilled water and suspended in tris-buffered sucrose to which K citrate and lysozyme were added. Final concentrations were tris 0·033M, pH 8·0; K citrate 8 mg/ml; lysozyme 13·3 μg/ml; sucrose 0·025–0·05M. After 10 min at room temperature Mg^{2+} equivalent to the citrate was added, and 5 min later the preparation was spun at 400 g. "Protoplasts" were resuspended in buffered media containing 0·05M sucrose.

5. *Rhodospirillum rubrum* (Karunairatnam, Spizizen and Gest, 1958). Suspensions of this organism (0·4 mg dry weight/ml) in 10% (w/v) sucrose

containing phosphate buffer (0·01M, pH 6·8) lysozyme (0·5 mg/ml), and EDTA (1·6 mg/ml, pH 7·0) were incubated at 30°C for 30 min. MgSO$_4$ was added to give 0·05M and after a few minutes, $\frac{1}{2}$ volume of 40% (w/v) sucrose was added.

Eberthella typhi and *Salmonella* para-B become susceptible to attack by lysozyme after heat treatment (Colbert, 1958), while *Aerobacter aerogenes* is not lysed under a variety of conditions of incubation (Gebicki and James, 1958).

2. *Enzymes from Leucocytes*

Enzymes similar to lysozyme have been isolated from leucocytes by Amano *et al.* (1956) and by Hirsch (1956). The former known as leucozyme C was used to produce "protoplasts" of *E. coli* while the latter, "phagocytin" rendered *E. coli* cells suspended in hypertonic media at pH 5·0 sensitive to osmotic shock.

3. Streptomyces *enzymes*

Two distinct bacteriolytic enzymes from *Streptomyces* species have been purified by Ghuysen (1957); these digest walls of staphylococci, *Micrococcus lysodeikticus*, *Sarcina* spp., *Bacillus* spp. and *Clostridium welchii*. A different enzyme has been obtained by McCarty (1952) from *Streptomyces albus* which attacks the walls of streptococci. More recently streptomycete enzymes have been found which lyse yeast cell walls, Elorza *et al.* (1966). Another chitinase preparation from a *Streptomyces* sp. used in conjunction with an enzyme from sporulating *Bacillus circulans* has been used by Horikoshi and Iida (1959) to digest the cell walls of *Aspergillus oryzae*.

4. *Enzymes from the digestive juice of the snail*

The digestive juice from the gut of *Helix pomatia* was shown to contain enzymes that attack yeast cell walls, Giaja (1922), and Eddy and Williamson (1957) showed that these enzymes could be used to prepare protoplasts from *Saccharomyces carlsbergensis* and *Saccharomyces cerevisiae*. The technique described by these workers required long periods of treatment (5 h) and were successful only with dilute suspensions of log phase cells. Other workers (Zajdela, Heyman-Blanchet and Chaix, 1959; Nagasaki, 1960) used modification of the technique to prepare extracts of yeast containing intact mitochondria, and Burger, Bacon and Bacon (1961) discovered that the presence of cysteine accelerated the action of the snail enzymes. A method for the formation of spheroplasts from yeast is fully described by Duell, Inoue and Utter (1964), and we quote their method.

A solution (2·5 ml) containing 0·14M 2-mercaptoethylamine and 0·04M

EDTA was used per g of cells. After incubation for 30 min at 30°C the cells were centrifuged at 1000 g for 10 min at 0°C. After decanting the supernatant, the pellet was resuspended in 1·7 ml of 0·72M sorbitol, 0·02M citrate – KPO$_4$= (pH 5·8), and 33 mg of Glusulase (snail digestive enzyme commercially available from Endo Laboratories Inc., Richmond Hill, New York) per g of cells. The cells were thoroughly mixed in a Vortex mixer and incubated at 30°C from 15 to 30 min with occasional stirring. The formation of spheroplasts can be followed roughly by diluting a sample of the incubation mixture 10-fold with water and observing the remaining cells. Only a part of the cell wall was lost in most cases, and spheroplasts, instead of true protoplasts were formed.

After incubation, the cells and the spheroplasts were centrifuged at 1000 g for 10 min at 0°C and the supernatant decanted. The pellet was washed with 3 ml of 0·9M sorbitol per g of original yeast and recentrifuged. Lysis of the spheroplasts was accomplished with 0·25M sucrose, containing 0·5M K$_2$HPO$_4$/KH$_2$PO$_4$ (pH 6·8) and 1 mM EDTA, by mixing the spheroplasts with a glass rod and then on the Vortex mixer. This method works with log or stationary phase cells and a similar method has been successfully applied to *Candida albicans* (Sugawara, 1966). Various strains of yeast vary greatly in their susceptibility to treatment, variation is also found in resistance to attack by snail juice enzymes when susceptible strains are grown under different nutritional conditions. Another method using partially purified snail juice has been described by Ohnishi, Kawaguchi, and Hagihara (1965). A European source of snail gut enzymes is Industrial Biologique Francaise S.A., 35 a 49 Quai du Moulin de Cage, Gennevilliers (Seine), France.

Emerson and Emerson (1958) have prepared "protoplasts" from strains of *Neurospora crassa*. Treatment of hyphae with snail hepatic enzymes gave rise to spherical forms which lyse in distilled water. Other reports of the preparation of *Neurospora* protoplasts are those by Bachmann and Bonner (1959) Weiss (1963) and by Kinsky (1962). Methods for the preparation of fungal protoplasts have been reviewed by Villanueva (1966).

5. *Proteolytic Enzymes*

Price and Bourke (1966) have described a method for the enzymic preparation of spheroplasts from *Euglena gracilis* using Pan protease, a mixture of trypsin and chymotrypsin available from Worthington Biochemical Corporation, Freehold, New Jersey. Their incubation mixture consisted of 0·8M mannitol, 0·04M phosphate buffer, pH 2 containing 3 mg/ml of Pan protease. After 2 h many of the organisms could be disrupted by osmotic shock.

F. Bacteriophage and other lytic factors

Most groups of bacteria are sensitive to specific bacteriophage. Infection by a bacteriophage consists of adsorption of the phage to the bacterial cell wall. Hydrolysis by an enzyme carried by the phage leads to penetration and passage of phage nucleic acid into the cell. Heavy multiple infections can produce lysis before penetration of phage nucleic acids (lysis from without). Alternatively cell lysis (from within) occurs after phage reproduction: this releases free bacteriophage and the remaining intracellular constituents of the cell. Other lethal factors (colicines from strains of *E. coli*, kappa factor from *Paramecium*) can cause cell lysis. Once more these might form the basis of rather specific and sophisticated breakage methods, as was suggested by Sylvester (Umbreit *et al.*, 1957). Sher and Mallette (1953) obtained the highest yields of any breakage method tested for lysine and arginine decarboxylase from bacteriophage lysates of *E. coli* B.

Although the intracellular constituents of the host cell will be a good deal altered, especially with bacteriophage induced lysis from within, this method seems promising for specific applications, and might be very useful for large scale breakage procedures.

G. Ionizing radiation

Bacteria subjected to ionizing radiation and then incubated often lyse after an initial period in which cells increase in size but do not divide. This lysis has been described as the induction of "radiospheroplasts" since if post irradiation growth is continued in sucrose, spheroplasts are formed which can be lysed subsequently in the usual manner. The number of spheroplasts formed is proportional to the radiation dosage and in the case of *Escherichia coli* ca. 300 Krad in a Cobalt 60 source produced almost 100% lysis. The method has not yet been applied to enzyme preparation (Ali, 1965; Hughes, 1968).

IV. CHOICE OF METHODS

As yet there is no approach other than the empirical one in choosing a method for disintegrating microbes for a specific problem. The approach taken in our own laboratory is to choose the gentlest method which disrupts the particular organisms in question with the least comminution of its parts. Special methods have been described in the foregoing section which can be applied to the more delicate organisms such as colourless algae and protozoa, but for most bacteria, yeasts, algae and moulds, these are not applicable. The choice here is either to produce spheroplasts or their equivalent enzymatically or to disintegrate the cells by the Hughes or

French press. The crude preparation is then treated with DN'ase to facilitate the fractionation into soluble and particulate material by centrifugation at a relatively low speed (8000–10,000 g). The initial fractions generally consist of a cell wall membrane (outer husk) and a supernatant containing fine particles which can be fractionated by centrifugation at increasingly high speeds and by density gradients until a clear supernatant which contains no particles separating below 125,000 g (ca. 50Å diam.) is obtained. Each soluble and particulate fraction is examined separately for enzymic activity.

The crude outer husk preparation is further purified by washing and centrifugation and its purity judged by electron microscopy, chemical and enzymic criteria. The cell wall can then be removed by enzymic methods (e.g. lysozyme) and the membrane generally then fragments and forms microsomal-like vesicles which can be further treated either by solvents such as butanol, enzymes such as lipase (snake venoms) or ultrasonics or vibratory mills to further fragment and to release bound enzymes into solution (Hunt, Rodgers, and Hughes, 1959; Cole and Hughes, 1965).

It is worthwhile to re-emphasize that information about the properties of the soluble enzymes obtained in this way can often be applied to extract larger amounts of cells more easily by drying etc. It must also be borne in mind that fractionation and purification of bound enzymes in particulate fractions is wasteful and large amounts of cells are often used to obtain small amounts of purified fractions. This is of particular importance in studies of enzymes during growth and enzymic induction in continuous culture where small volume culture vessels (500–1000 ml) are generally inadequate to supply sufficient cells for such fractionations.

It is quite apparent that no one method of cell disintegration is in itself sufficient for studies in microbial enzymology especially for studies on the relationship between structure and function. We regard it as essential to have available those which have now become generally accepted such as, French and Hughes press, a vibratory mill (Nossal, Sonomec, Braun or Mickle). Ultrasonic disintegrator, homogenizers of various types, abrasives and beads for hard grinding and a range of enzymes including lysozyme, snail enzymes, lipases etc.

REFERENCES

Ali, D. (1965). Ph.D. Thesis, University of Wales.
Amano, T., Seki, Y., Fujikawa, K., Kashiba, S., Morioka, T., and Ichikawa, S. (1956). *Med. J. Osakallniv*, **7**, 245.
Bachmann, B. J., and Bonner, D. M. (1959). *J. Bacteriol.*, **78**, 550.
Barnard, F. J., and Hewlett, R. T. (1912). *Proc. R. Soc. B*, **84**, 57.
Booth, V. E., and Green, D. E. (1938). *Biochem. J.*, **32**, 855.

Bridgman, P. W. (1912). *Proc. Am. Acad. Arts. Sci.*, **47**, 439.
Bridgman, P. W. (1937). *J. chem. Phys.*, **5**, 964.
Buetow, D. E., and Buchanan, P. J. (1964). *Expl. Cell. Res.*, **36**, 204.
Buchner, E. (1897). *Ber. deut. Chem. Gies*, **30**, 117–124, 1110–1113.
Burger, M., Bacon, E., and Bacon, J. S. D. (1961). *Biochem. J.*, **78**, 504.
Chambers, L. A., and Flosdorf, E. W. (1936). *J. biol. Chem.*, **114**, 75.
Charm, S., and Kurland, G. S. (1962). *Am. J. Physiol.*, **203**, 417.
Colbert, L. (1958). *Annls Inst. Pasteur, Paris*, **95**, 156.
Cole, J. A., and Hughes, D. E. (1965). *J. gen. Microbiol.*, **38**, 65–72.
Cooper, P. D. (1953). *J. gen. Microbiol.*, **9**, 199.
Dark, F. A., and Strange, R. E. (1957). *Nature, Lond.*, **180**, 759.
Davies, R. (1959). *Biochim. biophys. Acta*, **33**, 48.
Dolin, M. I. (1950). Thesis Indiana University.
Duell, E. A., Inoue, S., and Utter, M. F. (1964). *J. Bacteriol.*, **88**, 176?
Edebo, L. (1960). *J. biochem. microbiol. Technol Engng*, **2**, 453.
Edebo, L. (1961). *Acta path. microbiol. scand.*, **52**, 300.
Edebo, L. (1961). *Acta path. microbiol. scand.*, **52**, 361.
Edebo, L. (1967). *Biotech. Bioengng.*, **9**, 267.
Edebo, L., and Heden, C. G. (1960). *J. biochem. microbiol. Technol. Engng.*, **2**, 113.
Eddy, A. A., and Williamson, D. H. (1957). *Nature, London.*, **179**, 1252.
Elder, S. A. (1959). *J. acoust. Soc. Am.*, **37**, 54.
Elorza, M. V., Munoz-Ruiz, E., and Villanueva, J. R. (1966). *Nature, Lond.*, **210**, 442.
El 'Piner, E. E. (1964). "Ultrasound Physical, Chemical and Biological Effects". Consultants Bureau, New York.
Emanuel, C. F., and Chaikoff, I. L. (1957). *Biochim. biophys. Acta*, **24**, 254.
Emerson, S., and Emerson, M. R. (1958). *Proc. natn. Acad. Sci. U.S.A.*, **44**, 668.
Esche, R. (1952). *Acoustica*, **208**, 1.
Fleming, A. (1922). *Proc. R. Soc.*, **93**, 306.
Flynn, H. G. (1964). "Physical Acoustics" (Ed. W. P. Mason), Academic Press, New York, Vol. 1B. Ch. 9.
Foster, J. W., Cowan, R. M., and Maag, T. A. (1962). *J. Bact.*, **83**, 330.
Fraser, D. (1951). *Nature., Lond.*, **167**, 33.
Fraser, D., and Mahler, H. R. (1957). *Archs Biochem. Biophys.*, **69**, 166.
Furness, G. (1952). *J. gen. Microbiol.*, **1**, 335.
Gebicki, J. M., and James, A. M. (1958). *Nature, Lond.*, **182**, 725.
Ghuysen, J. M. (1957). *Arch. intern. physiol. biochim.*, **65**, 173.
Giaja, J. (1922). *C.r. Séanc. Soc. Biol.*, **86**, 708.
Glaser, L., Melo, A., and Paul, R. (1967). *J. biol. Chem.*, **242**, 1944.
Gunsalus, I. C. (1955). In "Methods in Enzymology", (Eds S. P. Colowick and N. O. Kaplan) Vol. 1, p. 51. Academic Press Inc., New York.
Greenawalt, J. W., Hall, D. O., and Wallis, O. C. (1967). *Meth. Enzymol.*, **N**, p. 142.
Harden, A., and Young, W. J. (1906a). *Proc. R. Soc.*, B**77**, 405 (1906b), **78**, 369.
Harvey, E. N., and Loomis, A. L. (1929). *J. Bacteriol.*, **17**, 373–376.
Hawley, S. A., McCleod, R. M., and Dunn, F. (1963). *J. Acoust. Soc. Amer.*, **35**, 1285.
Heden, C. G. (1964). *Bacteriol. Rev.*, **28**, 14.
Heppel, L. A. (1967). *Science*, **156**, 1451.
Hirsch, J. G. (1956). *J. exp. Med.*, **103**, 598.

Hogeboom, G. H., Kuff, E. L., and Schneider, W. C. (1957). *Int. Rev. Cytol.*, **6**, 425.
Hogg, J. F., and Kornberg, H. L. (1963). *Biochem. J.*, **86**, 462.
Holt, S. J., and O'Sullivan, D. G. (1958). *Proc. Roy. Soc.* B, **13** (London), **148**, 465.
Horikoshi, K., and Iida, S. (1959). *Nature, Lond.*, **183**, 186.
Hughes, D. E. (1950). *Biochem. J.*, **46**, 231.
Hughes, D. E. (1951). *Br. J. exp. Path.*, **32**, 97.
Hughes, D. E. (1961). *J. biochem. microbiol. Technol, Engng*, **3**, 405.
Hughes, D. E. (1968). "Energetics and Mechanisms in Radiation Biology". (Ed. G. O. Phillips), Academic Press, London, p. 409.
Hughes, D. E., and Cunningham, V. C. (1962). *Biochem. Soc. Symp.*, 7023, p. 8.
Hughes, D. E., and Galli, E. (1966). *J. gen. Microbiol.*, **39**, 845–853.
Hughes, D. E., and Nyborg, W. L. (1962). *Science*, **138**, 108.
Hughes, D. E., and Williamson, D. H. (1952). *Biochem. J.*, **51**, 45–54.
Hugo, W. B. (1954). *Bact. Rev.*, **18**, 87.
Hunter, M. J., and Commerford, S. L. (1961). *Biochim. biophys. Acta*, **47**, 580.
Hunt, A. L., Rodgers, A., and Hughes, D. E. (1959). *Biochim. biophys. Acta*, **34**, 354.
Ingram, M. (1957). *In* "Microbial Ecology" (Eds R. E. O. Williams and C. C. Spicer), Cambridge Univ. Press, London.
Joffe, S. (1928). "The Physics of Crystals". McGraw Hill, London.
Kalnitsky, G., Utter, M. J., and Werkman, C. H. (1945). *J. Bacteriol.*, **49**, 595.
Karunairatnam, M. C., Spizizen, J., and Gest, H. (1958). *Biochim. biophs. Acta*, **29**, 649.
Kaufmann, W., and Bauer, K. (1958). *J. gen. Microbiol.*, **18**, IX.
Kinsky, S. C. (1962). *J. Bacteriol.*, **83**, 351.
Klucis, E., Lloyd, D., and Roach, G. I. (1968). *Biotech. Bioengng*, **10**, 321.
Korkes, S., del Campillo, A., Gunsalus, I. C., and Ochoa, S. (1951). *J. biol. Chem.*, **193**, 721.
Kronish, D. P., and Mohan, R. R., Epstein, R. L., Pianotti, R. S., and Church, B. D. (1960). *Bact. Proc.*, p. 63.
Layne, E. (1957). *Methods in Enzymol.*, **3**, 443.
Lamanna, C., and Mallette, M. F. (1954). *J. Bacteriol.*, **67**, 503.
Levinthal, C., and Davison, P. F. (1961). *J. molec. Biol.*, **3**, 674.
Lissenden, A. (1965). *Chem. Process Engng*, **46**, No. 4, 203.
Lloyd, D. (1966). *Expl. Cell Res.*, **45**, 120.
Lloyd, D. (1965). *Biochim. biophys. Acta*, **110**, 425.
Lowry, O. H., Rosenbrough, M. I., Farr, A. L., and Randall, R. J. (1951). *J. biol. Chem.*, **193**, 265.
Mahler, H. R., and Fraser, D. (1956). *Biochim. biophys. Acta*, **22**, 197.
Mandell, N. and Roberts, C. F. (1966). *Science*, **152**, 799.
Marr, H. G. (1960). *Ann. Rev. Microbiol.*, **14**, 241.
Marr, A. G., and Cota Robles, E. H. (1957). *J. Bact.*, **74**, 79.
Marshall, V. C. (1966). *Chem. Process Engng*, **47**, No. 4, 177.
Mattoon, J. R., and Balcavage, W. A. (1967). *Meth. Enzym.*, **10**, 135.
McCarty, M. (1952). *J. exp. Med.*, **96**, 555.
McCleod, R. M., and Dunn, F. (1966). *J. acoust. Soc. Am.*, **40**, 1202.
Macfayden, A., and Rowland, S. (1903). *Zentbl. Bakt. ParasitKde, Abt. J. Orig.*, **34**, 767–771.
McIlwain, H. (1948). *J. gen. Microbiol.*, **2**, 288.

McQuillen, K. (1956). *Symposium J. gen. Microbiol.*, **6**, 127.
McQuillen, K. (1960). *In* "The Bacteria", Vol. I p. 294 (Eds I. C. Gunsalus and R. Y. Stanier), Academic Press Inc. New York.
Mickle, H., and Ray, J. (1948). *J. micros. Soc.*, **68**, 10.
Milner, H. W., Lawrence, N. S., and French, G. S. (1950). *Science*, **111**, 633.
Mitchell, P., and Moyle, J. (1957). *J. gen. Microbiol.*, **16**, 184.
Mitchell, P., and Moyle, J. (1956), *Nature, Lond.*, **178**, 993.
Morton, R. K. (1950). *Nature, Lond.*, **166**, 1092–1095.
Mudd, S., Shaw, C. H., Czarnetzky, E. J., and Flosdorf, E. W. (1937). *U.S. Public Health Reports*, **52**, 887.
Muhhammed, A. (1959). D.Phil. Thesis. Oxford University.
Muys, G. T. (1949). *Leeuwen. Loek. ned. Tijdschr.*, **15**, 203.
Nagasaki, K. (1960). *Mem. Fac. Agr. Kochi Univ.*, **8**, 1.
Necas, O. (1956). *Nature, Lond.*, **177**, 898.
Neppiras, E. A., and Hughes, D. E. (1964). *Biotech. Bioengng*, **6**, 247.
Neppiras, E. A. (1960). *Br. J. appl. Phys.*, **11**, 143.
Neu, H. C., and Heppel, L. A. (1964). *J. biol. Chem.*, **239**, 3893. (1965) **240**, 3685.
Noltingk, B. E., and Neppiras, E. A. (1950). *Proc. phys. Soc. Lond.*, **63B**, 674. (1951) **64B**, 1032.
Nomura, M., and Hosoda, J. (1956). *J. Bacteriol.*, **72**, 573.
Nomura, M., and Hosoda, J. (1956). *Nature, Lond.*, **177**, 1037.
Norris, J. R. (1957). *J. gen. Microbiol.*, **16**, 1.
Nossal, P. M. (1953). *Aust. J. exp. Biol. med. Sci.*, **31**, 583.
Nossal, N. C., and Heppel, L. A. (1966). *J. biol. Chem.*, **241**, 3055.
Nyborg, W. L. (1966). *Abstract 5th Congress. Internat. d-Acoustique*, Liege.
Nyborg, W. L., and Hughes, D. E. (1967). *J. acoust. Soc. Am.*, **42**, 891.
Ohnishi, T., Kawaguchi, K., and Hagihara, B. (1965). *J. biol. Chem.*, **241**, 1797.
Potter, V. R. (1955). *In* "Methods in Enzymology" (Eds S. P. Colowick and N. O. Kaplan), Vol. 1, p. 10. Academic Press Inc., New York.
Price, C. A., and Bourke, M. E. (1966). *J. Protozool.*, **13**, 474.
Pritchard, M. J., Peacocke, A. R., and Hughes, D. E. (1966). *Biopolymers*, **4**, 259–273.
Raper, J. R., and Hyatt, E. A. (1963). *J. Bacteriol.*, **85**, 712.
Reith, A., Schuler, B., Vogell, W., and Klingenberg, M. (1967). *Histochemie*, 11, p. 33.
Richmond, M. H. (1957). *J. gen. Microbiol.*, **16**, iv.
Robrish, S. A., and Marr, A. G. (1957). *Bact. Proc. (Soc. Am. Bacteriologists)*.
Rodgers, A., Nyborg, W. L., and Hughes, D. E. (1961). *J. acoust. Soc. Am.*, **33**, 1672.
Rodgers, A., and Nyborg, W. L. (1967). *Biotech. Bioeng*, **10**, 235.
Rodgers, A., and Hughes, D. E. (1960). *J. biochem. microbiol. Tech. Engng.*, **2**, 49.
Runscheidt, F. D., and Mason, S. G. (1961). *J. Colloid Sci.*, **16**, 238.
Sher, I. H., and Mallette, M. F. (1953). *J. biol. Chem.*, **200**, 257–262.
Sohler, A., Romano, A. H., and Nickerson, W. J. (1958). *J. Bact.*, **75**, 283.
Spizizen, J. (1957). *Proc. natn. Acad. Sci. U.S.A.*, **43**, 694.
Spizizen, J. (1962). *In* "Methods in Enzymol". (Eds S. P. Colowick and N. O. Kaplan, Vol. 5, p. 122. Academic Press, New York.
Stephenson, M. (1928). *Biochem. J.*, **22**, 605 (1958).
Strange, R. E., and Dark, F. A. (1957a). *J. gen. Microbiol.*, **16**, 236.
Strange, R. E., and Dark, F. A. (1957b). *J. gen. Microbiol.*, **17**, 525.

Strominger, J. L. (1962). *Fedn. Proc. Fedn. Am. Socs exp. Biol.*, **21**, 134.
Sugawara, S. (1966). *Nature, London.*, **219**, 92.
Sutton, C. R., and King, H. K. (1966). *Biochem. J.*, **101**, 20p.
Suyama, Y. (1966). *Biochemistry*, **5**, 2214.
Taylor, G. I. (1935). *Proc. R. Soc. A*, 151 p. 494–512. (1936). *Proc. R. Soc. A*, 157 p. 546–578.
Travis, P. M. (1929). *Ind. Engng. Chem.*, **21**, 421.
Tomcsik, J., and Baumann-Grace, J. B. (1956). *Verh. naturf. Ges. Basel*, **67**, 218.
Umbreit, W. W., Burris, R. H., and Stauff, J. F. (1957). "Manometric Methods". Burgess, Minn., p. 163.
Villanueva, J. R. (1966). *In* "The Fungi; an Advanced Treatise". (Eds G. C. Ainsworth and A. S. Sussman), Academic Press Inc. New York. Vol. 2 p. 3.
Vitols, E., and Linnane, A. W. (1961). *J. biophys. biochem. Cytol.*, **9**, 701.
Watson, K., and Smith, J. E. (1967). *Biochem. J.*, **104**, 332.
Weibull, C. (1953). *J. Bact.*, **66**, 688.
Weidel, W. (1951). V. *Naturf.*, **6b**, 251.
Weiss, B. (1963). *J. gen. Microbiol.*, **39**, 85.
Wiame, J. M., Storck, R., and Vanderwinckel (1955). *Biochim. biophys. Acta*, **18**, 353.
Yoshida, A., and Heden, C. G. (1962). *J. Immunol.*, **88**, 389.
Zajdela, F., Heyman-Blanchet, T., and Chaix, P. (1959). *C. R. Acad. Sci., Paris*, **235**, 1268.

APPENDIX

Name and address of suppliers of cell disintegration equipment.

Sonicators

100 W generator and magnetostriction transducer	Measuring and Scientific Equipment, 25–28 Buckingham Gate, London, S.W. 1.
350 W generator and ceramic transducer	Dawes Instruments Ltd., Concord Road, Western Avenue, London, W.3.
and	Branston Sonic Power Co., Eagle Road, Danbury, Connecticut 06810

Presses

X Press	BioTec Ltd., LKB House, 232 Addington Road, Croydon
Hughes Press (as Fig. 13)	Microbiology Department, University College, Cardiff
,, ,, (Older type)	Shandon Scientific Co., 65 Pound Lane, London, N.W.10

French Press
  Aminco
                                    American Instrument Co. Ltd.,
                                    8030 Georgia Ave., Silver Springs,
                                    Maryland 20910, U.S.A.
                                    G. E. Moore (Birmingham) Ltd.,
                                    33 Wake Green Road,
                                    Birmingham, 13
  Yeda Press
                                    Yeda, Researchand Development Co. Ltd.,
                                    Yeda Sccientific Instruments at the Weiz-
                                    mann Institute of Science, P.O. Box 95,
                                    Rehovol, Israel.

Shakers
  Mickle shaker
                                    Mickle Ltd.,
                                    Mill Works, Gomshall,
                                    Surrey
  Braun shaker
                                    Shandon Scientific Co.,
                                    65 Pound Lane,
                                    London, N.W.10.
  Centrifuge shaker
                                    International Centrifuge Co.,
                                    Cambridge, Mass.
                                    U.S.A.

Explosive decompression
                                    Artisan Metal Products,
                                    Waltham, Mass.,
                                    U.S.A.
                                    Baskerville and Lindsay Ltd.,
                                    Manchester

Homogenizers
  Hand type
                                    Arthur H. Thomas Company,
                                    Vine Street at Third,
                                    P.O. Box 779,
                                    Philadelphia 5, P.A.
                                    U.S.A.
                                    Jencons,
                                    Hemel Hempstead,
                                    Hertfordshire.
                                    Kontes Glass Co.,
                                    Vineland, New Jersey.
  Manton Goring (15M–8BA)
                                    AVP Co. Ltd.,
                                    Manor Royal, Crawley,
                                    Sussex

Hydraulic Press
                                    Apex Construction Co. Ltd.,
                                    15 Soho Square,
                                    London, W.1
                                    Tel: 01 437 6328
                                    G. E. Moore (Birmingham) Ltd.,
                                    33 Wake Green Road,
                                    Birmingham, 13.

Colloid Mills

Dyno-Mill,
Willy A Bachofer,
4000 Basle 5, Switzerland
MV-6-3 Micro-Mill,
Gifford-Wood Co.,
Hudson, New York.

CHAPTER II

Centrifugal Techniques
for the
Isolation and Characterisation
of Sub-Cellular Components from Bacteria

J. SYKES

Department of Biochemistry, University of Sheffield, Sheffield S10 2TN, England

I.	Introduction	55
II.	Basic Modes of Centrifugation	58
III.	Velocity Sedimentation	60
	A. Moving boundary	61
	B. Moving zone	81
IV.	Equilibrium Sedimentation	102
	A. Introduction	102
	B. Sedimentation equilibrium	103
	C. *Iso*-density sedimentation equilibrium	112
V.	Isolation and Characterization of Sub-cellular Components from Bacteria	128
	A. Introduction	128
	B. Ribosomes and polyribosomes	129
	C. Cytomembranes, including chromatophores and sub-chromatophore units	148
	D. Poly β-hydroxybutyrate and polyphosphate granules	159
	References	197

I. INTRODUCTION

The problems presented to the microbiologist interested in a precise sedimentation analysis of bacterial cells and their sub-structure require the application of the complete range of centrifuge instrumentation and techniques for their solution. The preparative and analytical centrifuge systems employed must be capable of resolving the complete size range from whole cells down to macromolecules. Furthermore the initial objective is usually the isolation of the object of interest in representative yield and reasonable purity. The centrifuge technique adopted at this stage must, therefore, combine capacity with resolution. Once this objective has been attained the subsequent demand may be to realize a maximum of information con-cerning the preparation e.g. its sedimentation constant, diffusion constant, molecular weight, homogeneity, etc., from a minimum of experimental material. Fortunately developments in the theory of sedimentation and the design of centrifuges and rotors have combined to satisfy most of these demands, although certain of the experimental situations likely to be encountered do not have a precise theoretical background and an empirical approach must suffice for the moment.

When a bacterial cell is subjected to controlled disintegration (Hughes, Wimpenny, and Lloyd (this volume, page 1) the range and properties of organelles and macromolecules likely to be encountered may be grouped according to the scheme proposed by Anderson (1966).

(1) The "ideal" particle which is smooth, spherical, uncharged, unhydrated and impermeable. Such a particle will not interact with the medium and the physical properties of the surrounding medium have no influence upon the physical properties of the particle. The sedi-mentation of this type of particle is very adequately described by the classical theory of sedimentation (see later). This idealized situation is rarely, if ever, encountered in cell fractionations, but serves as a useful model to recognize points of departure from the ideal into areas where theory is inadequate and experimental information scant.

(2) Macromolecules that are charged, hydrated and may not be round or smooth. Protein and nucleic acid molecules would be included here and may usually be prepared homogeneous with respect to molecular weight. Macromolecular aggregates with discrete molecular weight ranges could also be included here e.g. ribosomes and viruses.

(3) Macromolecular aggregates occurring in a range of sizes and hence varying degrees of asymmetry e.g. polysaccharides.

(4) Particulate cell elements bounded by a semi-permeable membrane and therefore exhibiting osmotic behaviour. This behaviour usually results in pronounced shape changes and in certain instances, e.g. during sedimentation in density gradients, the speed of osmotic equilibration between particle and environment will be important. The mitochondria and chromatophores may be quoted as examples of this group and with such particles a further complication may arise due to the known physiological variation in these structures.

The purpose of this article is to outline the general considerations which apply to the sedimentation process in the hope that the accent upon the limitations and practical aspects will be helpful in devising suitable procedures for the isolation and physical characterization of cell structures. A rigorous theoretical discussion is not included since a number of excellent reviews deal with this aspect (Creeth and Pain, 1967; Fujita, 1962; Schachman, 1959, Trautman, 1964).

Centrifugal procedures are usually applied very early in many microbiological investigations. Centrifugation offers a rapid, controlled and reproducible method for the concentration of micro-organisms from their growth environment (see Thompson and Foster, this Series, Vol. 2). For small volumes of culture fluid (e.g. from 1 to 2 l) the micro-organisms may be harvested in a single batch process in a variety of swing-out or fixed angle rotors. The majority of manufacturers of centrifuge equipment produce instruments operating at controlled temperatures with rotors of this capacity and capable of developing the necessary gravitational fields to sediment most micro organisms. For larger volumes of culture fluid a continuous flow rotor is to be preferred. Such rotors may be of the cream-separator type (e.g. Sharples supercentrifuge, Camberley, Surrey) or a rotor designed (e.g. Measuring and Scientific Equipment Ltd., London, continuous action rotor) or modified (Szent-Gyorgyi and Blum rotor, I. Sorvall, Inc., Norwalk, Connecticut) for continuous operation in a conventional laboratory centrifuge. In both continuous and batch harvesting, although the fields applied will theoretically sediment all of the organisms from the culture, the yield of cells is sometimes sacrificed to speed the harvest. Continuous action rotors are usually capable of clarifying up to 200 ml of culture fluid/min. In view of the wide size distributions known to occur in populations of organisms in batch culture and the physiological and biochemical variations with size some selection of cells may be occasioned by this procedure (see also Helmstetter, this Series, Vol. 1). The additional factor of the geometry of the sedimenting system (see p. 60, Section IIIA.1.1) must also be taken into account with certain designs of continuous action rotors.

A pre-requisite for the detailed centrifugal analysis of a cellular activity or sub-cellular component is a knowledge of its rate of sedimentation. A useful estimate of this variable may be obtained fairly readily provided methods are available for the identification of the object of interest, e.g. by an activity determination, characteristic absorbance or composition, etc. If a preparation containing the object is then subjected to varying periods of centrifugation in a swinging bucket rotor under controlled conditions of temperature and rotor speed and the boundary position of the object is determined after varying periods of time, then the rate of movement per unit of gravitational field may be estimated i.e. the sedimentation coefficient (see p. 66, Section III.A.1.1). With this preliminary information suitable centrifugal procedures may be devised to isolate the object in reasonable yield and with minimal contamination (see Sections III.A.1.1, III.B.2–4) and IV.C (e). With the additional knowledge of the density of the object there are many ways in which the application of centrifugal force will assist in the isolation and complete physical (and chemical) characterization of the object.

II. BASIC MODES OF CENTRIFUGATION

Developments in the theory of sedimentation and the design of rotors, cells and accessories have permitted the full exploitation of three basic physical parameters of biological objects in order to separate and characterize them by the application of centrifugal force. Separations may be achieved

(i) on the basis of their size and shape, as in the *Velocity* method;
(ii) by their density differences as in the *Iso-density* (iso-pycnic) method or
(iii) by their mass differences as in the *Equilibrium* method.

These three basic methods of velocity, iso-density and equilibrium centrifugation may sometimes be used in combination and the sharp distinction made above may not be possible. Equally the sharp distinction originally drawn between analytical and preparative centrifugation has become blurred with the development of swinging bucket rotors and centrifuges capable of developing extremely high gravitational fields (e.g. 420,000 *g* (max.) in the Spinco SW 65 rotor) under controlled conditions coupled with techniques for the detailed analysis of tube contents at the end of the run (see Section III.B.2–4 and IV.C (e)). In this article a distinction between "preparative" and "analytical" ultra-centrifugation will be made simply on the basis of whether the centrifugal separation is monitored during the course of the run (analytical centrifugation) or at the conclusion of the run (preparative centrifugation).

For analytical ultracentrifugation optical techniques are used to monitor the separation during the course of the run. These techniques are based upon refraction, interference, or absorption methods. The cells used are of special design (see Sections III.A.2, IV.B and Figs. 1 and 2 (d)) and many are designed for specific purposes, e.g. separation cells, synthetic boundary cells, etc. (e.g. see Fig. 1 and Beckman Instruments pamphlet No. SB 200). A common and necessary feature of all cells is an optical path formed by two quartz discs enclosing a solution column(s) which may be of variable length (Fig. 2 (d)). Refraction and interference optics are extremely useful but limited in sensitivity and lacking the discrimination of absorption methods. The recent commercial development of automatic recording, split beam, scanning absorption systems with a mono-chromator has overcome the major limitations of the classical absorption system (see Section III.A.2).

The distinction which has been made between preparative and analytical centrifugation is clearly arbitrary. Provided that the assay techniques are

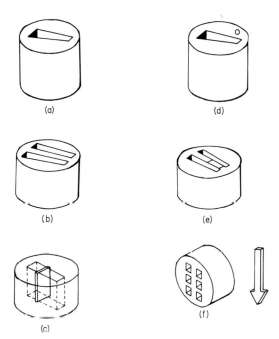

FIG. 1. Analytical ultracentrifuge cell centrepieces (a) Standard 4° sector, 12 mm lightpath centrepiece. (b) Double sector centrepiece, $2\frac{1}{2}$° sectors, 12 mm lightpath. (c) Fixed partition, mechanical separation centrepiece. (d) Type I band forming centrepiece for band sedimentation studies. (e) Double sector, capillary, synthetic boundary centrepiece. (f) Multi-channel, short column equilibrium centrepiece for 3 solvent : solution pairs.

sufficiently sensitive special centrepieces designed for the analytical centrifuge will separate the contents of the cell for assay at the end of the run (see Fig. 1). Further, some of the refinements in preparative technique are capable of yielding information of a predominantly analytical character, e.g. rate-zonal sedimentations may yield precise information on sedimentation coefficients even when the object of interest constitutes only a slight, but detectable portion of the load. Quartz tubes have been produced [Beckman Instruments Ltd.] which may be used at relatively high gravitational fields (170,000 g (max.)) in preparative rotors and the contents scanned in a spectrophotometer at the conclusion of the run.

In all three basic modes of centrifugation the period of centrifugation and the initial state of loading of the cell or tube are of paramount importance. The time of centrifugation is important since both the transient states and the final state of a centrifugation may give valuable information (see later). In the case of the larger cell particulates the periods of acceleration and deceleration of the rotor are important additions to the time of centrifugation. The initial state of loading of the cell [cf. moving boundary (Section III.A.1) and moving zone sedimentation (Section III.B)] may determine the degree of separation achieved. With the above factors in mind, the three basic approaches of velocity iso-density and classical equilibrium centrifugation may be considered in further detail.

III. VELOCITY SEDIMENTATION

The basic premise upon which all velocity sedimentation methods are based is that the gravitational field applied induces a terminal velocity of the object such that movement due to sedimentation (or flotation) is large compared with that due to diffusion. Diffusion is not entirely negligible in these circumstances and its effects may be seen on the boundary shape in the recordings of boundary movement (see later). Two main adaptations of the velocity method are commonly practiced.

  (i) moving boundary method, preparative and analytical
  (ii) moving zone method, preparative and analytical.

A. 1. Moving boundary

1. *Preparative*

This is by far the most commonly adopted method of centrifugation and in its "differential" application was well exploited by Hogeboom, Kuff and Schneider (1957) in their classic studies on the centrifugal fractionation of rat liver homogenates.

In this method the initial state of loading is a uniform distribution of the materials throughout the tube. It is evident that this method of centrifugation, whilst effecting a considerable concentration of a component by repeated cycles of centrifugation (differential method), can only yield one fraction, the slowest sedimenting, free from contamination by other fractions. For all other completely sedimented species the degree of cross contamination by other species is approximately proportional to the ratio of their sedimentation rates.

This method is very useful at the preparative level since it usually combines the necessary capacity to yield a fraction in reasonable yield and purity. The method however suffers from an additional drawback in practice owing to the design of many rotors and containers and their placement in relation to the direction of the centrifugal field, i.e. the geometry of the sedimenting system (see Fig. 2). For preparative work the usual systems encountered are:—

  (i) angle rotors with a fixed angle of inclination of the tubes to the axis of rotation and usually drilled to accept parallel-sided tubes (Fig. 2(a))

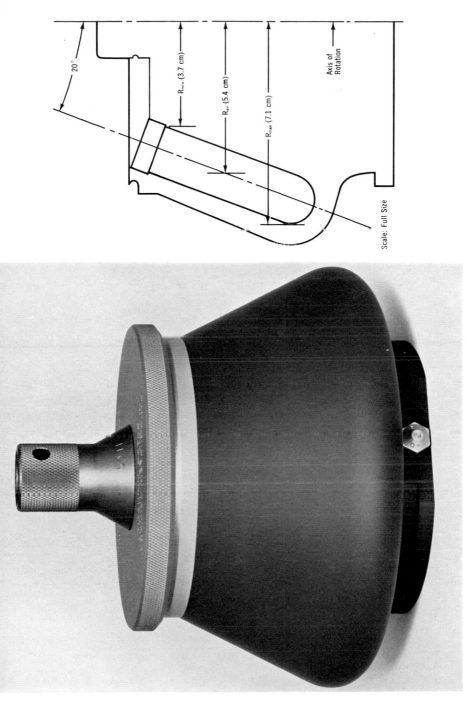

20°

R_{min} (3.7 cm)

R_{av} (5.4 cm)

R_{max} (7.1 cm)

Axis of Rotation

FIG. 2(a)

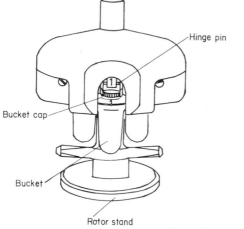

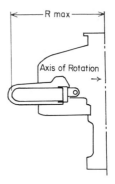

Hinge pin

Bucket cap

Bucket

Rotor stand

R max

Axis of Rotation

FIG.2 (b)

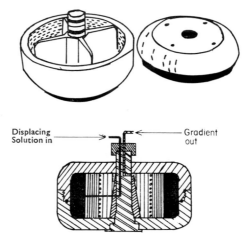

Displacing Solution in

Gradient out

FIG. 2(c)

FIG. 2 (a) Angle rotor and section through the rotor to indicate rotor characteristics. (b) Swinging bucket rotor and section through the rotor to show the position
of the buckets at speed and the rotor characteristics. (c) A zonal rotor. The bowl,
sector divider and lid in an exploded view and a section through the rotor to
indicate the inner and peripheral fluid lines for the input and recovery of the
gradient and particle zones. (d) An analytical rotor and an exploded view of the
components of a typical analytical cell.

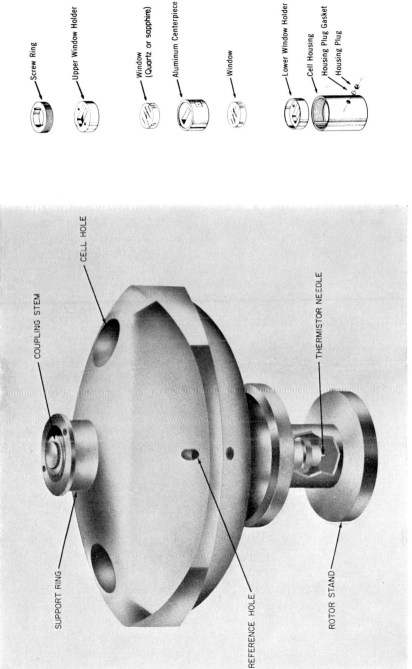

Fig. 2(d)

(ii) swing-out rotors in which parallel-sided tubes are parallel to the axis of rotation when at rest and perpendicular to the axis at speed (Fig. 2 (b)). This reorientation of the tube at the beginning and the end of the run may introduce further complications which will not be considered here.

All the objects sedimenting in a centrifugal field move in a radial direction and the area swept out by the object increases directly with the radial distance from the centre of rotation. This tendency gives rise to the so called "wall-effects" in sedimentation. In the fixed-angle rotor it is evident that the radial mean sedimentation path is not identical with the tube length (Fig. 2 (a)). The radial movement in such an inclined tube results in a clearing of the material from the centripetal wall and a concentration at the opposing wall; a condition is thereby generated which is unstable with respect to density. Convective transport of material ensues, material descends along the outer wall and ascends towards the meniscus on the inner wall as portions of the fluid seek an iso-density position. This phenomenon is frequently exaggerated by the tendency of certain materials (particularly common with sub-cellular particles) to clump and aggregate. The net result of these effects is that the sedimentation is unpredictable. For this reason, *inter alia*, it is advisable to check the distribution of representative materials between supernatant and pellet when this method of centrifugation is used to remove unbroken cells and cell debris from bacterial homogenates.

The use of a swinging bucket rotor reduces the above "wall-effects" although even here if the R_{max} (radial distance from the axis of rotation to the bottom of the tube) is twice that of R_{min} (radial distance from the axis of rotation to the liquid meniscus) approximately 25% of the material may still be expected to collide with the tube wall (Kahler and Lloyd, 1951). This is due to the fact that the tubes are generally parallel sided and not, as in the analytical cells, sectors of an arc. Attempts have been made to design sector shaped vessels for preparative work but unfortunately they appear to be unstable at high gravitational fields (Anderson, 1956). The design of the zonal centrifuge rotor (Anderson, 1966) overcomes these difficulties of rotor geometry on the preparative scale (see Fig. 2 (c) and Section III.B.3).

Differential centrifugation has proved so useful in the past since the separations required have involved materials with large differences in sedimentation rates, e.g. nuclei and ribosomes. It is not a satisfactory method when the objects to be separated have similar sedimentation rates. Thus for a three component mixture with sedimentation rates of S, $0.33\ S$ and $0.1\ S$ then when the fastest component (S) is pelleted it will be contaminated with 40% of 0·33 S and 13% of 0·1 S (de Duve and Berthet, 1964).

Pickels (1943) observed that some amelioration of the convection arising from wall effects could be obtained by including a slight gradient of density (e.g. using sucrose) throughout the material in the tube, increasing with radial distance. The region of increased density at the wall cannot sink down the tube but will then return to the lower density region at the centre i.e. tends to move across the field and will not seriously affect radial movement. Kahler and Lloyd (1951) used polystyrene latex particles as models of a defined sedimenting species and showed that this application of moving boundary sedimentation (usually designated as **stabilized moving boundary centrifugation**) when carried out in a swing-out rotor gave predictable results. The sedimentation rate observed for the latex spheres closely approximated the value obtained in the analytical centrifuge. These observations were later extended to a series of purified proteins having widely different sedimentation rates (Kuff and Hogeboom, 1956) For preparative moving boundary sedimentation swing-out heads are to be preferred to angle heads although the former may be found to have a more limited capacity at high gravitational fields. In both cases the inclusion of a gradient of density within the material in the tube will improve the separation.

It will be evident from the preceding remarks that the commonly adopted form of reporting centrifugal procedures (i.e. gravitational field × time) may prove to be grossly misleading. The rotor characteristics (angle of inclination to the axis of rotation, R_{min}, $R_{average}$ or R_{max}, see Fig. 2) are also required or at least the manufacturer's identification of the rotor so that the variation in gravitational field along the tube may be calculated. With such information the procedures reported for a particular rotor may be transcribed for any other rotor on the basis of their relative *Performance indices*. If we assume that the conditions of sedimentation are reasonably predictable, i.e. convection free, then the rate of migration of the object will be governed by the size, shape and density of the object (dp), by the density and viscosity (ds, η) of the suspending medium and the applied gravitational field. The rate of movement per unit of gravitational field is defined as the sedimentation coefficient (s);

$$ s = \frac{dx}{dt} \cdot \frac{1}{\omega^2 x} \tag{1} $$

ω = angular velocity in radians/sec, 1 revolution equals 2π radians.

x = distance from axis of rotation.

Integrate $\int_0^t$

$$s = \frac{\log_e x_t - \log_e x_0}{\omega^2 (t_t - t_0)} \qquad (2)$$

$(t_t - t_0)$ = time interval during which boundary moves from x_0 to x_t. Then, for most practical purposes, a plot of $\log_{10} x$ (i.e. boundary position measured in cm from centre of rotation) vs. t (in seconds) will give a straight line. The slope of this plot gives s since

$$s = \frac{(\text{slope}) \, (2 \cdot 303)}{\left(2\pi \dfrac{\text{rpm}}{60}\right)^2} \qquad (3)$$

The uncorrected sedimentation coefficient calculated in this manner is usually corrected to the conditions obtaining in a medium corresponding to the density and viscosity of water at 20°C by

$$s_{20,\,w} = s_{\text{uncorr}} \left(\frac{\eta_t}{\eta_{20}}\right) \left(\frac{\eta}{\eta_0}\right) \left(\frac{1 - \bar{v} \, dw_{20}}{1 - \bar{v} \, ds}\right) \qquad (4)$$

The major correction term, (η_t/η_{20}), corrects for the viscosity of water at the temperature of the run relative to that at 20°C and the terms (η/η_0) are the relative viscosities of solvent and water at $t°$. dw_{20} and ds are the densities of water at 20°C and the solvent at $t°C$ and $\bar{v}$ is the partial specific volume (see Section III.A.2).

Sedimentation coefficients generally lie in the range $0 \cdot 25$–500×10^{-13} sec and are commonly expressed in Svedberg (S) units where one Svedberg unit equals 1×10^{-13} sec e.g. 50×10^{-13} sec $\equiv 50S$. The sedimentation coefficient is generally dependent upon concentration, sometimes markedly as in the case of the nucleic acids. When sedimentation coefficients are reported the conditions attaching to their determination should always be included. The $S_{20,\,w}$ values obtained are usually extrapolated (in a plot of $1/S_{w,\,20}$ vs. concentration) to zero concentration to yield the *sedimentation constant* which is a characteristic of the sedimenting material. This constant is most accurately determined in the sector shaped cells of the analytical centrifuge (See Section III.A.2 and Fig. 2 (d)). A study of the variation in $S_{20,\,w}$ with concentration may frequently reveal association/dissociation phenomena with certain macromolecules. If $S_{20,\,w}$ increases and then decreases as concentration falls dissociation may be indicated (Massey et al., 1955).

Furthermore the dependence of sedimentation upon concentration may give rise to additional complications in sedimentation velocity experiments viz: (i) the sedimentation coefficient may not be constant during the experiment owing to radial dilution and inhomogeneity of field along the tube and (ii) the shape of the boundary may be distorted by "self-sharpening". Theories involving the sedimentation coefficient usually require the value of s at infinite dilution (the sedimentation constant) and if the experimental data for the determination of the sedimentation coefficient are treated to give an average value throughout the experiment (as described on p. 66) then this value must be plotted against the average value of the concentration of that solution in the plateau region in plots of s vs. c.

If equation (2) is written in the form

$$t_t - t_0 = \frac{1}{s} \frac{\log_e x_t - \log_e x_0}{\omega^2} \tag{5}$$

then if x_t is made equal to R_{max} and x_0 to R_{min} then the total *precipitation time* (T) for an object becomes

$$T = \frac{1}{s} \frac{\log_e R_{max} - \log_e R_{min}}{\omega^2} \tag{6}$$

T is clearly the product of two factors one, s, encompasses the properties of the object and the medium (see below) and the other the design characteristics and speed of operation of the rotor. The latter term is designated the rotor *performance index* P_i

$$P_i = \frac{(\text{rpm})^2}{\log_e R_{max} - \log_e R_{min}} \tag{7}$$

Values of P_i are readily calculated from eqn. (7) or provided by instrument manuals and are usually based on completely filled tubes. If the tube is only partially filled then an appropriate correction must be made in the R_{min} term. Various rotors may therefore be compared and procedures transcribed on the basis of their performance indices. If assumptions are made with regard to the density and viscosity of the medium and the density of an object for a "typical biological preparation" values for T may be calculated and presented in the form of a nomogram (e.g. Beckman instrument manuals). Whilst such assumptions and data are useful in charting a preliminary fractionation procedure they should not exclude due consideration of the factors influencing s in equation (5) in addition to the rotor geometry and concentration effects already mentioned.

The terminal velocity (or s-rate) of an object is determined by the resultant of the centrifugal force (F_c), the buoyant force (F_b), the frictional force (F_f) and the force of gravity (F_g) acting downwards. The primary determinant is the resultant of F_c, promoting sedimentation, and F_f and F_b which oppose sedimentation. F_g may be neglected for most cases. The frictional force (F_f) is the resistance the object experiences in moving through the solvent and increases with increasing speed of movement and asymmetry of the object. For a constant, terminal velocity

$$F_c = F_f + F_b \tag{8}$$

In the centrifuge

$$F_c = m \cdot \omega^2 x \tag{9}$$

m = mass of the object
and the opposing

$$F_b = \frac{m}{dp} \cdot dm \cdot \omega^2 x \tag{10}$$

dp = density of the object
dm = density of the solution

An object will not move beyond a position of x where dp and dm are equal since here $F_c - F_b = 0$ (isodensity condition—see Section IV.C.)

The frictional force $F_f = f dx/dt$ where f is the frictional coefficient. Substituting these values of F_c, F_b and F_f in equation (8)

$$m\omega^2 x - \frac{m}{dp} \cdot dm \cdot \omega^2 x = f \cdot \frac{dx}{dt} \tag{11}$$

$$m\omega^2 x \left(1 - \frac{dm}{dp}\right) = f \cdot \frac{dx}{dt} \tag{12}$$

Since $s = dx/dt \cdot 1/\omega^2 x$ (equation 1) and $1/dp$ is equal to the partial specific volume $\bar{v}$ we may write

$$s = \frac{m \cdot (1 - \bar{v} \, dm)}{f} \tag{13}$$

or per gram molecule

$$s = \frac{M (1 - \bar{v} \, dm)}{Nf} \tag{14}$$

N = Avogadro number
M = molecular weight

The dependence of s on factors other than those already mentioned may now be seen. Equation (14) embodies terms characteristic of the object $(M, \bar{v}$ and $f)$ and characteristic of the medium $(dm$ and $f)$ and variations in these parameters and their interaction may be expected to influence s.

If we make the usual simplifying assumption that the sedimenting object is spherical, uncharged, impermeable and unhydrated, then according to Stokes' Law the frictional coefficient for a sphere (f_0) is

$$f_0 = 6\pi\eta r \qquad (15)$$

r = radius of the sphere
η = viscosity of the medium.
Substituting for f equation (14) becomes

$$s = M\frac{(1 - \bar{v}\mathrm{d}m)}{6\pi\eta r} \qquad (16)$$

Deviations from the spherical shape may be accommodated in equation (16) by the introduction of a shape factor, θ, the frictional ratio into the denominator

$$\theta = \frac{f}{f_0} \qquad (17)$$

i.e. the ratio of the frictional coefficient of the object of a particular shape to that of a sphere of the same volume. Hydrodynamic shape is rather an indefinite concept with biological macromolecules. They may be rigid with little internal space accessible to solvent, e.g. certain proteins, or highly flexible structures with a high degree of interaction with the environment e.g. ribonucleic acids. In general theory has dealt with a limited number of hydrodynamically equivalent shapes which can be reasonably described in quantitative terms, e.g. ellipsoids of revolution generated by rotating an ellipse about either of its two axes. Rotation of an ellipse about its long axis produces a prolate ellipsoid and about the short axis an oblate ellipsoid. Values of θ for these two mathematically described shapes may be found in Svedberg and Pedersen's (1940) classic monograph. The later section on analytical moving boundary sedimentation deals with methods for the elimination of f from equation (14).

s and T therefore depend upon the characteristics of the object, the medium and their interaction. Increasing asymmetry of the object will tend to decrease s and increase T. T is inversely proportional to the difference in density between the object and the medium and an increase in this difference will decrease T.

An increase in the effective viscosity of the medium may be achieved by altering the properties of the medium or increasing the concentration of solute(s). Increasing the effective viscosity decreases s and increases T. Many bacterial extracts are extremely viscous due to the high concentrations of nucleic acids, particularly DNA. If the investigator is not interested in the nucleic acids or structures likely to embody them it is frequently useful to lower the viscosity of the preparation by suitable enzyme treatment prior to centrifugation. The need to extrapolate sedimentation coefficient data to zero concentration (see above) can now be seen and the simple reporting of a sedimentation coefficient without a precise definition of the conditions can clearly lead to serious ambiguities. The concentration dependence of sedimentation rate for many biological materials e.g. nucleic acids and proteins is well documented, but the corresponding behaviour of cell organelles is not. The sedimentation coefficients of bacterial ribosomes and chromatophores are known to be concentration dependent as are rat liver mitochondria. The dependence of sedimentation upon concentration, even for a homogeneous solute, is a complex phenomenon which has not been fully evaluated in terms of the relative contributions of viscosity, density, shape of object and backward flow of solvent in a closed system. In multi-component systems the influence of one component upon the sedimentation of another is also important. Some excellent examples of this latter phenomenon are given by Schachman (1959) together with a full discussion of this subject.

In the case of sub-cellular organelles the further possibilities of interactions between the structure and the medium must not be ignored (see list of particle types in the introduction to this article). An organelle bounded by a semi-permeable membrane, e.g. mitochondria, lysosomes, may behave as an osmotic system and the volume, radius, density and structural integrity of such a unit may be influenced by the composition of the medium. In moving boundary sedimentations, apart from differences arising from the choice of the suspending solvent, this effect should not change during the course of the centrifugation since a feature of this type of sedimentation is that the particle remains in a reasonably constant environment. In zonal centrifugations in density gradients (i.e. rate zonal or iso-density techniques, see Sections III.B and IV.C) the osmotic behaviour of the organelle will vary as it moves into environments of increasing concentration of solute. Rat liver mitochondria have been extensively studied in this connection (de Duve et al., 1959; Tedeschi and Harris, 1955). The considerations to be applied to this type of situation are complex and depend inter alia upon the initial and final environment of the organelle and the speed of osmotic equilibration between organelle and environment. Rat liver mitochondria appear to equilibrate almost instantaneously and on this basis de Duve et al. (1959) have devised simple equations which describe the observed behaviour of these organelles. Such equations are useful in planning and interpreting iso-density centrifugations of the organelle in media of different composition. Once certain physical properties of the organelle are defined for one environment (e.g. volume, mass and density) the properties of solutions in which the organelle will have the same density and be iso-pycnic with the surrounding medium may be predicted. For organelles behaving as osmometers the extrapolation of s values to the standard conditions defined above may be inappropriate. De Duve et al. (1959) have proposed the use of reference conditions compatible with the osmotic stability of the organelle.

A further aspect of the interaction between the sedimenting object and the medium is the possible influence of variations in the degree of hydration of the object in various media. Most macromolecules of biological interest are hydrated in solution. Ribosomes are extensively hydrated (1·0 gm water/gm dry weight) and many proteins are hydrated to the extent of 0·2 gm water/gm protein. Solvation may be expected to influence two of the terms in equation (14) viz: the effective mass $[M(1 - \bar{v} \, dm)]$ and the frictional ratio. The increase in mass due to solvation is exactly compensated by the additional displacement of the medium (i.e. the buoyancy term) and the overall effect of solvation on the effective mass term leaves the sedimentation coefficient unaffected. Since the frictional coefficient is a function of both the volume and shape of the kinetic unit an alteration in these due to hydration (or the reverse) will influence the value of s. Hydration will increase the molecular volume by the factor $(1 + dp.h)$ where h equals the degree of hydration (gm/gm protein of density dp.). The radius of the equivalent sphere, and hence the frictional coefficient, is thereby enlarged by the factor $(1 + dp.h)^{\frac{1}{3}}$. The combined effects of molecular asymmetry and solvation on the frictional ratio are given by

$$\frac{f}{f_0} = \theta = \left(\frac{f}{f_0}\right)_u . (1 + dp.h)^{\frac{1}{3}} \tag{18}$$

u refers to the unsolvated state.

Although s will therefore be influenced by solvation of the object, it is to be noted that the diffusion coefficient D is also affected. When s, D, and $\bar{v}$ are determined in the same experimental circumstances and combined in the Svedberg velocity equation (eqn. 25) the same value for the anhydrous molecular weight of the object will always be given.

Finally, two other factors influencing sedimentation rate must be noted, i.e. charge and temperature. The critical temperature sensitive factor in equation (16) is η the viscosity of the medium, raising the temperature will lower the viscosity and increase s. Temperature gradients within the medium may also give rise to convection. Temperature fluctuations are minimal in most modern ultracentrifuges since the rotors are usually operated in vacuo and facilities are incorporated into the design of the instrument for the precise monitoring and control of rotor temperature to $\pm 0.1°$ during the run. In the analytical ultracentrifuge the temperature control is sufficiently precise to be used to actuate a thermal valve during the course of a run (Fessler and Vinograd, 1965). The wide use of density gradients in preparative centrifugation also minimises convection arising from thermal fluctuations.

The primary charge effect in sedimentation arises from the differential rates of sedimentation of an ionized macromolecule and its counter ions leading to the formation of a potential gradient within the system. This separation of opposing charges serves to accelerate the sedimentation of the counter ion and reduce that of the charged macromolecule. In such a situation the s-rate of a charged macromolecule is much reduced when compared with an unionized molecule of the same physical characteristics. This primary charge effect may be allowed for if the charge on the macro-molecule is known. However, in practice this is rarely done and the potential gradient effect is supressed by the addition of an excess of low molecular weight, neutral electrolyte (usually 0·2M is sufficient). Depending upon the type of electrolyte used a secondary charge effect may also be manifest, arising from a difference in sedimentation rate which may exist between the ions in the electrolyte. The sedimentation constant ($s^\circ_{20, w}$) for bovine serum albumin was found to vary between 4·6 and 4·1 when the cation was changed from lithium to caesium bromide (Pedersen, 1958). When the concentration of added salt is particularly high, e.g. the practice of centrifuging nucleic acids in density gradients formed from high concentrations of Cs or Rb salts, additional interactions may take place between the object and the electrolyte. In equilibrium density gradient systems interactions between gradient materials and sedimenting objects have already been observed (Konrad and Stent, 1964; Lerman *et al.*, 1966) and may give rise to misleading observations. Furthermore the classical theory of sedimentation is based upon the simplifying assumption that the solvent may be considered as a one-component system. Whereas this is reasonable for dilute solutions it is not valid for strong solutions of electrolyte or in systems where a third solvent component may be present in high concentration, e.g. density gradients.

2. *Analytical*

The careful application of the preparative moving boundary technique will usually meet the primary objective of a fractionation procedure, i.e. the isolation of the object of interest in representative yield and a fair degree of purity. The further physical characterization of the object (s-rate, D, M etc.) may be achieved, using a minimum of experimental material by the application of analytical moving boundary techniques.

At the outset of a moving boundary centrifugation a uniform distribution of concentration of material exists throughout the tube. Once sedimentation commences a region of uniform concentration persists, but its radial extent and magnitude are time dependent (Fig. 3). This region of

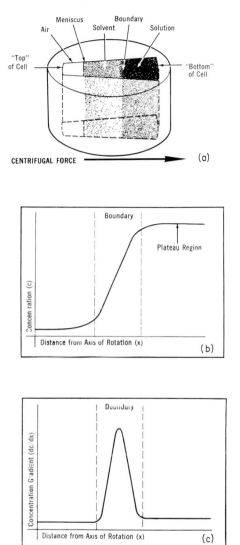

FIG. 3. (a) Diagrammatic representation of boundary velocity sedimentation in the analytical cell and the recording of the boundary position by (b) the variation in concentration *vs.* distance, e.g. absorption optical system and (c) the change in concentration with distance, e.g. the Schlieren optical system. (See Plate 1 for actual photographic records).

uniform concentration is usually referred to as the plateau region and the optical basis of analytical centrifugation rests upon the detection of the boundary region between this plateau and the cleared supernatant fluid

to the centripetal side of the boundary. One is essentially following sedimentation from the meniscus. Initially the boundary region will be sharply
defined, but the progressive influence of diffusion and other factors tends
to spread the boundary shape.

Refractometric methods for following sedimentation rely upon the change
in refractive index between regions of uniform concentration and regions
of changing concentration within the cell. The boundary region is a region
in which concentration varies rapidly with radial distance. The refractive
change is maximal at the mid-point of the boundary and light rays passing
through the boundary will suffer deviations proportional to the refractive
index gradient at any point. The common refractometric method for detecting boundary regions within the cell is the Schlieren system and the
photographic records here are differential plots of refractive index with
radial distance i.e. dn/dx vs. x. The migrating boundary region is therefore
presented as a "peak" whose maximum coincides with the centre of the
boundary (see Fig. 3 and Plate 1a). Measurement of the rate of movement
of the peak maximum down the cell is therefore a measure of s provided
the rotor speed and optical constants of the system are known. The rotor
speed is now electronically controlled and accurately known. The optical
constants may be obtained by photographing a ruled scale or a special
calibration cell in the position of the centrifuge cell. Distances measured
on the photographic records from the known reference points (supplied by
the manufacturer and usually incorporated into the counterpoise) may
then be converted into true cell distances. s may then be calculated as
previously described.

Since refractive index is directly proportional to concentration the area
beneath a Schlieren peak (e.g. Plate 1a) may provide an accurate estimation
of the concentration of the sedimenting material since

$$c_0 = \frac{A \cdot \tan \theta}{a \cdot l \cdot M_c \cdot M_x \cdot \Delta n \cdot E^2} \left(\frac{x_t}{x_m}\right)^2 \qquad (19)$$

A = measured area

a = cell optical path length in cm

l = optical lever arm i.e. the distance between the upper collimating
lens and the plane of the Schlieren diaphragm.

M_c, M_x = magnification factors of the camera and cylindrical lenses,
respectively.

E = magnification of the enlarger if the original record is enlarged.

Δn = specific refractive increment, i.e. the difference in refractive
index between the solvent and the solution at a concentration
of 1%

θ = Schlieren diaphragm angle.

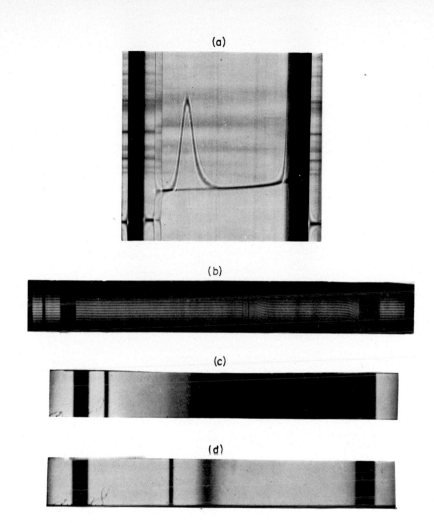

(a)

(b)

(c)

(d)

PLATE 1. Optical patterns from the analytical ultracentrifuge. (a) a Schlieren pattern from a sedimentation velocity study using a double sector cell to provide simultaneous registration of the solvent baseline. The liquid menisci in the cell sectors are seen to the left of the diagram and the outer edges of the broad, vertical dark bands are the images of the counterpoise reference edges. (b) An interference fringe pattern taken at an early stage in a low speed sedimentation equilibrium study. The cell is only part filled. (c) A photograph taken with the ultraviolet absorption optical system during a sedimentation velocity run with a bacterial RNA preparation. Three boundaries are visible and the densitometer tracing of this photograph is shown in Fig. 4. (d) A photograph taken with the ultraviolet absorption optical system during a band sedimentation study with 1 μg of 16S ribosomal RNA layered over a 0·01M acetate-0·1M NaCl buffer, pH 4·6, containing 5% sucrose. The band of ultraviolet absorbing material has migrated down the cell away from the meniscus starting zone.

x_t, x_m = positions of the boundary at time t and time o respectively (see later for radial dilution correction).

The area, A, is usually measured on an enlargement of the Schlieren pattern by weighing, planimetry or trapezoidal integration using a two dimensional micro-comparator. Where the determination of the initial concentration is of prime interest, e.g. in molecular weight determinations by transient or equilibrium methods, it is preferable to use the special double sector layering cells which permit the formation of the boundary in the middle of the cell at low gravitational fields (see Fig. 1). The boundary shape is then not distorted by the restriction imposed upon diffusion by the limits of the cell liquid column i.e. the meniscus and cell bottom. In addition a reference solvent base-line is included (see Plate 1 (a)) and the area under the peak clearly defined.

With the multi-component systems frequently encountered in biological preparations care has to be exercised in determining concentrations of material from the non-specific refractometric plots. In a multi-component moving boundary system, e.g. where a slow component is sedimenting with a faster component not two but three sedimentation coefficients have to be considered i.e. one for the fast species s_f, one for the slow, s_s, and one for the slow component in the presence of the fast, s_x. The movement of the faster species uncovers certain slower components and overtakes others. The uncovered slower species, now in a lower viscosity environment due to the removal of the faster species, speed up their rate of sedimentation and the converse applies to the slow species migrating in the presence of the fast. The combined result of these effects tends to change the concentration of the slower component across the boundary of the faster component. This effect (the Johnston-Ogston effect (1946)) is magnified as the sedimentation coefficients of the two components approach one another, e.g. by increasing the concentration. Area measurements to determine the concentration of species based upon refraction methods will therefore be erroneous unless the measured areas are corrected for this effect or the complication avoided by performing the run at much lower concentrations and a more specific optical method is used (see later discussion of absorption optics). The original correction derived by Johnston and Ogston (1946) involves a knowledge of s_x. However since a boundary corresponding to s_x does not appear in the diagrams a more practical solution to this problem was derived by Trautman (1956)

$$c_s^0 = \frac{\left(\dfrac{x_f}{x_m}\right)^2 - \left(\dfrac{x_s}{x_m}\right)^2 . c_s^{\text{obs}}}{\left[\left(\dfrac{x_f}{x_m}\right)^2\right]^{1-\sigma} - 1} \tag{20}$$

x_f, x_s and x_m are the distances from the centre of rotation of the fast and slow components and the meniscus respectively. $\sigma = s_s^0/s_f^0$ i.e. the ratio of the sedimentation constants of the slow and the fast species for the experimental conditions. c_s^0 is the true initial concentration of the slow component and c_s^{obs}, is the observed concentration. The concentration of the faster species may then be calculated by difference. If the radially corrected area (see p. 77) of the slower species remains constant throughout then the Johnston-Ogston effect is not varying during the run. When this is so Trautman (1956) has shown that a chart may be constructed for the ready estimation of corrections.

For apparently single component systems an analysis of the boundary shape observed in the Schlieren plots of analytical moving boundary sedimentations may give further valuable information concerning the species. The effects of diffusion and polydispersity spread the boundary and the dependence of s upon concentration tends to sharpen the boundary (self-sharpening). If the boundary appears to be a single sedimenting component in dilute solution with little dependence of s upon concentration then an apparent diffusion coefficient (D_{app}) may be calculated from

$$D_{app} = \frac{1}{4\pi t}\left(\frac{A}{H_{max}}\right)^2 (1 - \omega^2 st) \tag{21}$$

t is the time in seconds and H_{max} the height of the maximum ordinate of the gradient curve. The term $(1 - \omega^2 st)$ corrects for the effect of the centrifugal field on different parts of the boundary. In computing the term $(A/H_{max})^2$ the measured area (A) must first be corrected for the effect of radial dilution (p. 77). If the apparently single sedimenting component is really heterogeneous then D_{app} values measured throughout the run will remain constant or increase with time. If it is truly homogeneous D_{app} will tend to decrease with time. In a sedimentation velocity study the measurements of A and H_{max} for this calculation should only be made on diagrams in which a plateau region exists on either side of the peak, i.e. there is no restriction upon diffusion offered by the top (meniscus) or the bottom of the cell. Where detection of heterogeneity and the evaluation of D by this method are the paramount interests a double sector, capillary synthetic boundary cell (see Fig. 1) should be used to form the boundary near the centre of the cell and to introduce a solvent baseline for precision in area measurement. The spreading of the boundary may then be studied at low gravitational fields and a series of different concentrations of solute.

D may also be determined from moving boundary data by taking the second moment, σ^2 of the gradient curve

$$\sigma^2 = \frac{1}{2\pi} \left(\frac{A}{H_{max}} \right)^2 \tag{22}$$

A plot of σ^2 vs. time yields a line with a slope equal to $2D$. A careful analysis of the information from a Schlieren peak may therefore provide information concerning s, D and the homogeneity of the sample.

Equation (21) is only applicable to those systems where s and D may be assumed constant throughout the entire boundary and where s is large and D small to ensure no restriction of diffusion at the meniscus or cell bottom. The equation does not take into account the dependence of s upon concentration, an important determinant of boundary shape in many instances. Whenever s depends perceptibly upon concentration then Fujita's equation should be applied and not equation (21). An excellent practical example of the use of Fujita's equation is given in a paper by Baldwin (1957) in which the value of D determined in this fashion for bovine serum albumin was within 1% of the value obtained by conventional free diffusion methods.

Terms in equations 19, 20, 21 and 22 all require corrections involving the term $(x_t/x_0)^2$. This factor is termed the "radial dilution correction". The basic analytical ultracentrifuge cell centrepiece is a sector shaped vessel designed to form a segment of a circle whose centre is the axis of rotation (Fig. 1). The geometry of such a system thereby approximates the ideal situation since the cell contents are aligned at $90°$ to the axis of rotation (c.f. swinging bucket rotors) and unimpeded radial sedimentation of particles may take place, assuming the walls of the cell are not scratched and the cell is correctly aligned in the rotor. Alignment of the cell within the rotor is made by a keyway or by matching scribed lines on rotor and cell with the aid of a magnifier. The wedge shape of the cell therefore results in a progressive dilution of the contents of the boundary region as it advances down the cell. Svedberg and Rinde (1924) showed that the concentration (c_t) at any point in the cell (x_t) was related to the initial concentration (c_0) by

$$c_0 = c_t \left(\frac{x_t}{x_0} \right)^2 \tag{23}$$

A consequence of this dilution effect, together with the increasing field down the cell, may be a perceptible increase in sedimentation coefficient down the cell. The method of determining s from the slope of the $\log x$ vs. t plot gives an average value and does not usually reveal this phenomenon. We may, therefore, add a further critical test of homogeneity, viz. that the area beneath a single symmetrical sedimenting boundary accounts for all the material in solution throughout the run. A single sedimenting boundary is not in itself a sufficient criterion for homogeneity since without further tests such a situation could be given by (i) a homogeneous solute (ii) a polydisperse solute with a continuous distribution and concentration dependence of s values (iii) systems in rapid reversible equilibrium and (iv) two solutes with very close s values.

The other optical systems in common use avoid many of the pitfalls outlined above for the Schlieren system. The absorption optical system— (see Plate 1 (c)) is considerably more specific and sensitive, particularly with the introduction of a monochromator with wavelength selection between 236 and 440 nm. This system can therefore be used to exploit the characteristic properties of the object, usually at extremely low concentrations, hence avoiding many of the sedimentation anomalies described above which are associated with the relatively high concentrations of materials. Nucleic acids are a good example since amounts as low as 1 μg may be detected in zone experiments (Section III.B.3 and Plate 1 (d)) and 50 μg/ml in moving boundary analyses with absorption optics (Plate 1c). The paper by Schumaker and Schachman (1957) deals with many practical aspects of the earlier UV absorption system and in particular demonstrates that the anticipated instability of sedimenting boundaries in very dilute solutions is not realized in practice. Hitherto the great disadvantage of the absorption system, limiting its application, has been the requirement for "blind" photographic registration of the sedimentation process. The operator cannot "see" the progress of the run and at the end of the run care is required in the development of the photographic record to ensure linearity of response between film blackening and concentration of material. The photographic records showing variation in absorbance along the cell vs. time are then scanned in a densitomter to give an integral trace of variation in concentration with radial distance (Fig. 4). Concentrations are simply determined from the height of the step(s) above a superimposed solvent (baseline) trace. The s-rate may be determined from the rate of movement of a point in the boundary corresponding to 50% of the height. Diffusion coefficients may also be measured from plots of the square of the difference between the 25% and 75% boundary height positions (Schumaker and Schachman, 1957).

The practical drawbacks to this useful technique have now been overcome with the development of a split-beam photo-electric scanning system (Schachman, 1963; Schachman and Edelstein, 1966; Spragg et al., 1965). In this system a photomultiplier, scanner unit and recorder replace the camera unit. With a double-sector analytical cell with solvent in one sector and solution in the other, both sectors are uniformly illuminated by the same monochromatic light source. The photo-multiplier scans an image of the cell magnified through the quartz optical system. The scan speed may be varied together with the scanning slit width and length. As the

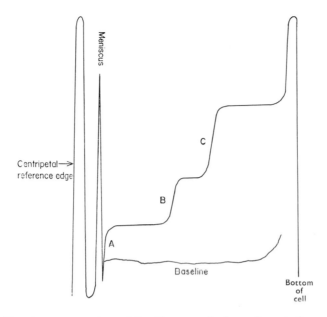

FIG. 4. Densitometer tracing of the film record of a sedimentation velocity run using ultraviolet absorption optics with a total RNA preparation (50 μg/ml) from *Aerobacter aerogenes*. The heights of the "steps" in the trace above the superimposed solvent baseline, when corrected for radial dilution, are proportional to the relative concentrations of soluble (A), 16S (B) and 23S (C) RNA species.

cell sweeps past the optical system two quick bursts of light (separated by only 6 μsec at 60,000 rpm) are followed by a relatively long dark period for each full revolution of the rotor. The scanner may be assumed to be looking at solvent and solution compartments almost simultaneously. However the associated circuitry is designed to store the first light pulse and compare it with the rapidly following conjugate signal from the second sector. Both sectors are therefore scanned under identical conditions of light intensity, wavelength, temperature and field strength. Any minor fluctuation in the experimental environment will be automatically compensated for when one signal is subtracted from the other (c.f. split beam spectrophotometer, the rotor acting as a light chopper). With the immediate read-out provided by the recorder this system achieves an instant concentration curve for the solution cell. Circuitry may be incorporated to transcribe this to the corresponding derivative curve. The latter is a useful facility since the integral curves sometimes fail to show the presence of small contaminants which are revealed in the derivative curves. This photo-electric scanning system has now been elaborated to the point where the sensitivity and accuracy of the recording system are suited to the determination of molecular weights by sedimentation equilibrium (Section IV.B.2) with only 4 μg protein/ml (Schachman and Edelstein, 1966).

The third widely used optical system is based upon the Rayleigh interferometer (Plate 1 (b)). This system, whilst more sensitive than the Schlieren system shares its lack of specificity. This system has found extensive use in sedimentation equilibrium studies since the concentration at any position in a boundary is more readily and accurately ascertained by counting the fringes in the interferograms from a known reference (meniscus) to that point (Richards et al., 1968; van Holde, 1967; Yphantis, 1964). In many analytical ultracentrifuges it is usually a simple matter to change from schlieren to interference optics during the course of a run and Chervenka (1966) has described a simply constructed light source mask which permits the simultaneous recording of the interferometric and Schlieren patterns. Combined optical registrations are a useful facility in sedimentation equibilibrium studies for molecular weight determinations since the interference fringe pattern gives values for the changes in concentration across the cell and the Schlieren pattern provides the concentration gradients (see Section IV.A and B). The interferometric optical system has also proved useful in studies aimed at detecting small differences in sedimentation coefficients, e.g. those arising from minor modifications due to chemical or limited enzyme treatments. The treated preparation is placed in one sector and the untreated preparation in the other sector of a double sector cell appropriately masked for interference work. If, throughout sedimentation, there is no difference in the two specimens then the fringe pattern will remain undeviated. If the treatment has produced a minor change in s then deviation of the fringe pattern will occur in the usual fashion (Richards and Schachman, 1959). It is interesting to note that the development of the split-beam absorption scanning system (p. 78) has made this approach possible using absorption optics (Schachman, 1963).

The determination of s and D together with the detection of heterogeneity and relative concentrations of multiple component systems are valuable features of analytical moving boundary sedimentations. However for homogeneous systems a combination of the sedimentation and diffusion data may be used to eliminate the frictional coefficient from equation (14) and permit an estimate of the anhydrous molecular weight. Einstein and Sutherland showed that

$$D = \frac{RT}{Nf} \tag{24}$$

where R is the gas constant and T the absolute temperature. If we assume that the values of D and f are identical for free and induced translation of the object then substituting in equation (14).

$$M = \frac{RTs}{D\,(1 - \bar{v}\,dm)} \tag{25}$$

Equation (25) is the well known Svedberg velocity equation and strictly applies to the situation at infinite dilution, i.e. $s^0{}_{20,\,w}$ and $D^0{}_{20,\,w}$. This equation, with data provided by sedimentation experiments for s and D, or by combining sedimentation data with free diffusion determinations of D has been widely used, and sometimes abused, for the determination of the anhydrous molecular weight of an object. The calculation of M requires the additional knowledge of $\bar{v}$, the partial specific volume. M is exceedingly sensitive to errors in $\bar{v}$ since equations (14) and (25) include the term $(1 - \bar{v}\,dm)$. Since dm is usually close to unity and for proteins $\bar{v}$ lies in the range 0·70–0·75 a small error in $\bar{v}$ is magnified threefold in M. Sensitive methods for the determination of densities are therefore required to give the necessary precision in $\bar{v}$. It is regrettable, but understandable, that values for $\bar{v}$ are frequently assumed. s and D may be determined with precision using milligramme amounts of material but classical density determinations, e.g. pycnometry (Markham, 1967; Schachman, 1957), require many hundreds of milligrammes. Attempts have therefore been made to circumvent this problem by devising methods which require less material, e.g. the density gradient columns used by Linderstrøm–Lang and Lantz (1938) and the magnetic densitometer used for the determination of $\bar{v}$ for ribonuclease (Ulrich et al., 1964). Using a refinement of the scanning absorption system described earlier Edelstein and Schachman (1967) have proposed and tested a useful solution to this problem. It is based upon the fact that when the density of the solution is increased by changing the solvent from H_2O to D_2O the sedimentation equilibrium distribution (Section IV) changes within the cell, i.e. it is a differential sedimentation equilibrium technique. Since sedimentation equilibrium directly measures the quantity $M\,(1 - \bar{v}\,dm)$ measurements at a number of different solvent densities allow the determination of both $\bar{v}$ and M by solving the two simultaneous equations containing the two unknowns or by plotting $M\,(1 - \bar{v}\,dm)$ vs. dm. Corrections have to be applied to allow for any increase in molecular weight likely to arise from possible deuterium exchange. This method, whilst not quite achieving the accuracy of standard methods gives a determination of $\bar{v}$ with microgram quantities of material.

B. Moving zone

1. Introduction

A major disadvantage of the conventional preparative and analytical moving boundary technique described in the previous Section is that all but the slowest sedimenting components in a mixture sediment in the presence of other components. On the preparative scale this gives rise to lack of resolution and on the analytical scale to difficulties of interpretation of the sedimenting process.

These difficulties may be overcome by sedimenting a thin band of solution through the solvent, although it may be noted that the objects will then, in contrast to the moving boundary situation, move into progressively different environments. On theoretical grounds one might anticipate that a system of this type would be intrinsically unstable due to convective disturbance arising from density inversion; the leading edge of a zone of solute constitutes an effective density inversion. However, it is now well established that this situation does not arise in practice provided that the solute concentration in the zone is small and a sufficient gradient of density is present within the solvent.

A zone of the material to be fractionated is layered over a solvent of continuously increasing density along the tube. Layering may be performed with the tube or cell at rest, e.g. via a hypodermic syringe or as a further check against density inversion, in a small gradient-forming device (p. 85) by incorporating the material into a small reversed gradient (p. 87) or whilst moving, e.g. by the use of special layering caps, by pumping into zonal rotors (p. 89) or by the use of layering cells in the analytical centrifuge (p. 90).

Because of the inclusion of the density gradient into the tube/cell this technique has been commonly referred to as "density gradient centrifugation". This is really a misnomer since the density gradient is incidental and present as a zone stabilizer. The separation process depends primarily upon the different s-rates of the objects within the starting zone and the range of density offered by the solvent should not encompass the density of any component within the zone. From a multi-component starting zone the objects move down the tube in a series of zones according to their sedimentation coefficients. Centrifugation is halted before the most rapidly moving zone reaches the bottom of the container; this may be predicted from a knowledge of the s-rate, the characteristics of the rotor and the gravitational field (see p. 67). The materials in the original zone are therefore characterized by their transport properties and not their densities (c.f. *iso*-density centrifugation, Section IV.C) and this method should strictly be described as "rate-zonal centrifugation".

Rate-zonal centrifugation has found wide application in biochemical and microbiological separations since it is an extremely versatile technique. It is uniquely suited to the problem of separating (analytically or preparatively) and identifying all the components of multicomponent systems, e.g. cell homogenates. The scope of the method is now so great that this Section will be confined to the principles, basic experimental methods and limitations of the technique, together with some examples. Many other articles in these Volumes will deal with specific applications to items of microbiological interest (e.g. see the articles by Work (Vol. 5a) and Eaton (Vol. 6). An excellent source of reference material and the experimental and theoretical background to zonal centrifugation and the design of zonal rotors is to be found in the monograph by Anderson (1966).

Broadly, the rate-zonal methods may be divided into:—

(1) Preparative
    (a) Small scale separations in swing out or angle rotors. These have limited, but sometimes useful, capacity for preparative work. These rotors are useful for analytical studies involving milligramme amounts of material.
    (b) Large scale separations performed in the special zonal rotors designed by Anderson (1966). (e.g. see Fig. 1.iii).
(2) Analytical
    i.e. rate-zonal separations performed in the layering cells (Fig. 1(d)) designed for use in the analytical ultracentrifuge. This method is useful for separating and examining μg amounts of materials, using the absorption optical system to follow the progress of sedimentation (see plate 1 (d)). This method has been described as "band sedimentation" and the gradient is not usually preformed (Vinograd and Bruner, 1966; Vinograd et al., 1963).

2. *Choice of gradient materials and factors influencing resolution in zone sedimentation velocity experiments*

It is simpler to define the ideal characteristics of the material forming the stabilizing gradient than to achieve such standards in practice. The material should be soluble in water/electrolyte solutions to an extent which will produce a reasonably wide range of density coupled with low viscosity. The material should be of high molecular weight and not interact with the objects to be fractionated, be dialysable and have negligible absorption at wavelengths of interest.

The density range of biological materials, particularly those likely to be encountered in an unfractionated cell homogenate, is very large ranging from 0·9 gm/cm³ (e.g. lipid materials) to 2·2 gm/cm³ (e.g. nucleic acids). Clearly there is no universal material for gradients which will satisfy the basic condition that the density range offered by the gradient in rate-zonal separations does not encompass the density of any particle within the initial zone. A compromise has to be made at this point regarding the area of particular interest and where maximum resolution may be required. In general separation will be enhanced by altering the relative buoyancy factors in an appropriate fashion (see equation (14)). A short, steep gradient will therefore be useful in situations where objects have similar s values but widely different density terms. For objects with similar densities, but small differences in s, resolution will be maximal when the density of the gradient is close to that of the objects. Unfortunately for many suitable materials, e.g. sucrose and glycerol, the concentrations required to give an appropriate density range produce solutions of high viscosity. This reduces resolution and compounds the problem of obtaining reliable sedimentation coefficients from rate-zonal experiments (Section III.B.6).

The requirement for a relatively high molecular weight gradient material, preferably non-ionizable is designed to minimize interaction between the object and the stabilizing medium. Zone centrifugations transport the various objects from the initial zone, an environment in which they are presumed stable, to a zone of higher concentration of supporting solutes. The potential hazards of this procedure, expecially with respect to certain types of cell organelle have already been considered (p. 70). In many experiments it may be necessary to remove the gradient material from the fractionated zone at the end of the run and it is therefore useful if the material will dialyse.

Sucrose has been most widely used for rate-zonal sedimentation studies since it meets many of the above requirements and is readily available in a high degree of purity. For certain investigations e.g. involving nucleic acids and ribosomes it is advisable to subject A.R. grade sucrose to charcoal treatment (to remove residual UV absorbing materials) and a nuclease inhibitor, e.g. bentonite. Sucrose is exceedingly soluble, neutral and dialysable. A reasonable density range is encompassed but at the higher concentrations required to give useful density increments two major disadvantages become apparent viz. the high osmotic pressure and viscosity of the solutions. De Duve *et al.* (1959) and Barber (in Anderson (1966)) give useful tables of data for the density and viscosity of various sucrose solutions at 0°C, 5°C and 20°C.

Ficoll (Pharmacia Fine Chemicals, Uppsala, Sweden) offers a number of advantages for rate zonal studies with cell fractions. It is a synthetic, high molecular weight (400,000 $\pm$ 100,000) polymer of sucrose and epichloro-hydrin and is very soluble in aqueous media and stable to autoclaving at neutral pH values. The osmotic pressure of a 5% (w/v) solution is an order of magnitude less than that for 5% sucrose but the viscosity of Ficoll solutions is appreciably higher than that for a solution of sucrose of equivalent density. Ficoll is non-dialysable and does not penetrate cell membranes so that transport of zones of material to higher concentrations of Ficoll does not result in an increase in the density of an organelle.

For zone separation of nucleic acids sulpholane (tetrahydrothiophen-1,

1-dioxide), trimethyl phosphate and urea have all been used (Parish *et al.*, 1966). Sulpholane has some residual absorption at 260 nm but does not interfere with the orcinol and diphenylamine colour reactions for RNA and DNA. Trimethyl phosphate tends to precipitate nucleic acids at concentrations in excess of 40% (v/v).

Glycerol, polyvinylpyrrolidone and even colloidal thorium oxide have all been used to construct density gradients for rate zonal centrifugations.

3. *Gradient forming devices*

With the exception of "band sedimentation" (see later) gradients for rate-zonal investigations are preformed. A suitable gradient may be formed by carefully layering solutions of different concentration in a tube and allowing diffusion to blur the interfaces and produce a gradient which is both smooth and reasonably linear. The process may be speeded by tilting and rotating the tube gently or by carefully stirring with a saw-toothed wire.

Most commonly, gradients are constructed by means of a mixing device. The general principles of mixer design have been outlined by Bock and Ling (1954) in connection with the production of smooth gradients of variable character for the elution of chromatographic columns. However in the latter connection it is not usual to require the mixing of two solutions of widely differing densities as is demanded in the construction of a density gradient. When two such solutions are in communication with each other at the point of mixing (e.g. see Fig. 5) withdrawal of a small volume of liquid from one vessel will produce changes in meniscus heights in the vessels in proportion to their respective densities if the chambers are of identical cross section. The character of the concentration vs. volume relationship of the effluent will not then be predictable as the emptying proceeds. If the menisci are maintained at the same height as the point of mixing then the anomalous mixing pattern arising from the density difference will not arise. Thus to give predictable concentration vs. volume relationships the point of mixing must move at the same rate as the menisci. A mixer for the construction of density gradients using solutions of very unequal density (one equal to the lower and the other the higher limit of the gradient) must therefore have efficient mixing in the mixing chamber. If the chamber dimensions are also varied then gradients of varying character may be produced. Variation in the relative cross sectional areas (A_1 and A_2) of the chambers 1 and 2 (Fig. 5) will result in gradients of varying character such that when $A_2 = 2A_1$, gradient 1 is obtained, when $A_2 = A_1$ a linear gradient 2 is produced and when $2A_2 = A_1$ gradient 3 (Fig. 6).

For the production of linear gradients the device originally suggested by Britten and Roberts (1960) is shown in Fig. 5. This apparatus may be

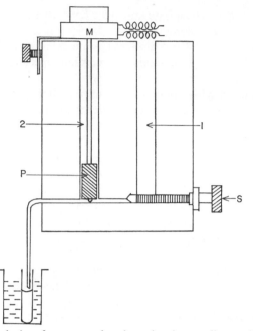

Fig. 5. A basic design for a two chamber, density gradient mixer (see Britten and Roberts, 1960). This mixer is designed to produce linear gradients and the two chambers (1 and 2) are of equal cross sectional area. The mixer may be machined from a block of solid Perspex. The miniature motor (M) drives the stainless steel paddle (P) in the mixing chamber. The stainless steel screw valve (S) terminates in a Teflon tip and is opened to allow the two solutions, of concentrations equal to the limits of the gradient, to be mixed and flow into the centrifuge tube.

constructed from a block of Perspex. The chamber screw valve is made from stainless steel and terminates in a pointed Teflon tip. The stainless steel stirrer may be conveniently driven by a readily demountable miniature motor, preferably mains operated via a transformer and rheostat. The exit tube and chamber screw valve are closed. Chamber 2 is loaded with a suitable volume (i.e. approximately 50% of the total gradient volume) of the highest density solution and chamber 1 is loaded with a slight excess of the lowest density solution. These two solutions define the limits of the gradient. The stirrer is started, the screw valve opened and the exit tube placed against the upper part of the wall of a suitable centrifuge tube. Initially the highest density solution emerges and is followed by the successively lighter mixture which floats on the underlying liquid in the tube with little mixing. It is advisable to make a preliminary check on the character of the gradient, and the operation of the apparatus, by including a dye or some readily detected material in one of the chambers. The

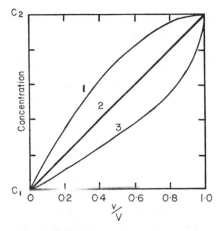

FIG. 6. Theoretical concentration *vs.* volume relationships for two chamber mixing devices of the basic design shown in Fig. 5. c_1 and c_2 are the concentration limits of the gradient and v/V the fraction of the gradient already mixed. Plots 1, 2 and 3 are obtained by making the cross sectional areas in chambers 2 and 1 in the ratios 1 : 2, 1 : 1 and 2 : 1 respectively.

distribution of this material in the tube is then determined by sampling the tube contents as described later. A refractometer may also be used to check the character of the gradient. Once formed density gradients are stable for many hours and require only moderate care in handling. A further advantage of this design of mixer is that its size may be varied and it may be used to load the sample into a further lower gradient of density onto the main gradients. In enlarged versions of this device (e.g. dealing with 30–40 ml total volume) care is required to ensure efficient mixing and the stirrer may be profitably replaced by an Archimedes screw. In all mixers of this type gross liquid disturbance e.g. vortex formation or surface agitation, should be avoided. Excessive agitation of the liquid may also induce pumping/backflow between the chambers and produce gross irregularities in the gradient. A further danger in this design of mixer is that as the junction between the chambers is opened a small amount of light solution will pass through the channel and rise to the top of the heavy solution. Some workers have therefore interchanged the light and heavy solutions having the former in the mixing chamber and then delivering the gradient to the bottom of the centrifuge tube. However when the delivery tube tip is removed from the bottom of the gradient a drop of heavy solution may fall into the gradient and disturb its character. Salo and Kouns (1965) have described the construction and operation of a similar glass mixing device which attempts to overcome these difficulties. Samis (1966) has described a mixing device in which the two chambers are made from

5 ml disposable syringes linked by polythene tubing. This may prove useful when handling pathogenic materials since it can be discarded after use.

Noll (1967) and Henderson (1969) have described the construction and operation of a simple, convex exponential gradient mixer. Gradients of this character ("*iso*-kinetic gradients") are particularly useful when it is desired to keep the sedimentation rate of particles of a given density constant over the full tube length.

A number of workers have described simply arranged mixing devices which employ peristaltic pumps. These devices are particularly useful when the preparation of multiple gradients or large volume gradients (e.g. for zonal rotors) is involved since the multichannel head and high capacity peristaltic pumps now available make this possible. The plan of a simple linear gradient device described by Ayad *et al.* (1967) is shown in Fig. 7 and Birnie and Harvey (1968) have described an arrangement for loading large capacity (400–2000 ml) zonal rotors. Gradient machines for loading zonal rotors are also offered by the rotor manufacturers (e.g. Beckman Instruments, Measuring and Scientific Equipment Ltd.) and Anderson and Rutenberg (1967) have described a mixing apparatus based upon a centrifugal pump.

Density gradient formers based upon two syringes leading into a common line emptying into the centrifuge tube have also been widely used (Anderson

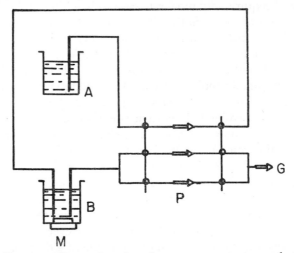

Fig. 7. A linear gradient mixer based upon a constant speed, three channel peristaltic pump (Ayad *et al.*, 1967). A, is a beaker containing the solution of highest density and *B* contains the solution of lowest density mounted over a magnetic stirrer, *M*. *P* represents a peristaltic pump with three channels each fitted with tubing of identical length and diameter. The gradient is collected at *G*.

1955; de Duve *et al.*, 1959; Kuff *et al.*, 1956). The output of the two syringes is controlled by cams which force the solutions into the common, mixing line at a predetermined rate. The character of the gradient may therefore be varied from convex to linear to concave by simply adjusting the individual outputs of the syringes. A version of this device is available commercially.

An elaborate device for the mass production of density gradients in a rotating, multiple swinging bucket rotor has been described by Siakotos and Wirth (1967). Here the outflow from a large capacity gradient machine is fed to a loading head mounted on the rotor with a number of exits equal to the number of swinging buckets on the rotor. The rotor is spun at 1200 rpm and as many as 12 tubes may be simultaneously loaded with a gradient. A similar gradient apportioning device for a multiple place angle rotor has been described by Candler *et al.* (1967).

The development of large scale preparative rate zonal techniques by Anderson and his colleagues (1965, 1966) has necessitated the designing of special cylindrical or disc shaped rotors for use in conventional preparative ultracentrifuges (Fig. 2 (c)). These zonal rotors are designed so that (i) sedimentation takes place in sector shaped compartments; a vaned core is fitted into the hollow rotor bowl to achieve this (ii) the rotor capacity and sample load capacity are increased by at least an order of magnitude compared with conventional rotors (iii) gravitational fields up to 102,000 *g* (max.) are realized. The density gradient is pumped into the rotor whilst it is running at low speed (< 5000 rpm). Input/output fluid lines run to both the centre and the periphery of the rotor and the light end of the gradient is pumped into the rotor via the line to the periphery (see Fig. 2 (c). As the denser solutions follow they spread evenly about the periphery of the rotor and successively displace the lighter end of the gradient towards the centre. When the rotor has been almost filled with the gradient in this fashion the sample is pumped in via the centre fluid line and the rotor then accelerated to the required speed. The centrifugal migration of the components in the sample generates a series of concentric zones of individual components and at the conclusion of the separation the rotor is decelerated to 5000 rpm. The rotor contents are then pumped out via the centre liquid line by pumping in denser liquid via the peripheral line. The effluent may be monitored by a flow analyser as described below (Section III.B.5) for conventional separations and subsequently collected in drop fractions. This preparative rate zonal technique is proving particularly useful for the large scale separation of cell organelles from cell homogenates including polyribosomes and ribosome sub-units (Klucis and Gould, 1966). The separation of viruses (Anderson and Cline, 1967), phages and ribosomal RNA species on a preparative scale has also been reported.

For *band sedimentation* studies, (Vinograd and Bruner, 1966; Vinograd *et al.*, 1963) i.e. rate-zonal centrifugation performed in the analytical ultracentrifuge, a thin layer of sample solution is layered onto a much larger volume of denser, miscible solvent in a rotating cell fitted with a special layering centrepiece (Fig. 1 (d)). The density gradients required to stabilize this type of rate-zonal system are therefore generated during the run. The main contributions to these self-generating gradients are the diffusion gradients arising from the difference in chemical potential between the small molecules in the sample and the bulk solvent and the field gradients arising from the redistribution of these molecules in the field. The size of the diffusion gradient is governed by the volume of the sample and the density difference between the sample and the solvent. For the common 12 mm centrepiece a density difference of $0.02-0.04$ g cm^{-3} has been found adequate to support the very small amounts of material (e.g. 1 μg of DNA/RNA) required for this technique. The choice of solvent i.e. pH, ionic strength etc. will be dictated by these considerations and the nature of the experiment. The theory of self-generating, zone-stabilizing density gradients has been developed by Vinograd and Bruner (1966). The progress of a band sedimentation experiment may be readily followed during the run by means of the absorption optical system (Plate 1 (d)). The sensitivity of the technique and the low shear conditions of transfer make it particularly suitable for the study of DNA and RNA preparations e.g. viruses.

4. *Sample layering, band capacity and density inversions*

Rate-zonal centrifugations are usually performed in the wide variety of recently developed high speed swinging bucket rotors (Fig. 2 (b)). These rotors commonly accommodate three or six buckets with an individual capacity ranging from 3 ml to 60 mls and some may be operated at speeds which develop 420,000 *g* (max.).

With the preformed gradient in the bucket the sample is usually layered carefully onto the top of the gradient by means of a syringe or via a miniature gradient mixer (see Section III.B.3)). The sample initially occupies a thin zone at the top of the gradient. For band sedimentation this thin lamella of sample is introduced from a separate channel in a special layering centrepiece as the rotor accelerates and the zone layering is thereby additionally stabilized under the influence of the gravitational field. This form of layering appears to offer advantages in zone definition and resolution when compared with manual loading. Accordingly band forming caps have been developed for swinging bucket rotors (Beckman Instruments Ltd.). These caps are similar in design to certain analytical layering cells i.e. the sample is placed in a cup mounted within the cap which communicates with the centrifuge tube via a compressible rubber gasket. As the rotor accelerates the gasket is compressed and the sample flows from the cup onto the gradient (Gropper and Griffith, 1966).

An essential condition which must be maintained throughout zone stabilized centrifugations is that the density at every level in the tube must always be less than the level immediately underlying it. Density inversions arising from thermal gradients down the tube are not usually a source of trouble if care is taken over the thermal equilibration of tubes, solutions and rotors and the centrifuge is equipped with a reliable temperature control system to control the temperature during the run. However density inversion may arise at the time of loading if the starting zone is not lighter than the top of the gradient or during the run if the density of any migrating zone exceeds the density of the gradient level immediately preceding it. Density inversions of this type give rise to droplet formation and "streaming" of material through the gradient. For these reasons the loading capacity of the majority of systems is limited and despite the widespread use of this technique very few theoretical and practical explorations of this problem have been reported. A partial practical solution to this problem is to use convex gradients, i.e. where the rate of change of concentration of gradient material is initially great but diminishes with increasing radial distance down the tube. Such a gradient contributes to zone stability in two ways (i) the greatest particle carrying capacity in the gradient is required just below the loading point and the steeper gradient at this point provides the required sharp increment in density between the initial non-diffused concentrated zone of particles and the gradient (ii) as sedimentation proceeds radial dilution, diffusion and zone separation effectively diminish the initial concentration of material and a diminishing steepness of gradient is required to support the zone. Since the migration and separation of zones of material brings about a change in concentration during the run this may also produce a change in the distribution of density as the run proceeds and the total density (i.e. solute plus solvent) in a given zone may actually decrease with radial distance creating a density inversion. Thus an initially stable band may remain stable throughout the separation or become unstable as soon as it begins to migrate or at some critical time during sedimentation. Berman (in Anderson (1966)) has given a theoretical treatment of the problem of density inversion arising during centrifugation in convex gradients and the requirements to achieve stability.

The theory of zone capacity was first considered by Svensson et al (1957) for electrophoretic separations in liquid density gradients. Brakke (1964) made an experimental test of their equations when applied to zone stabilized sedimentation and found that the capacity of the zones was only a small percentage (1–2%) of the predicted values. The amount of material carried by a zone was shown to be a function of the zone width. Berman (see

Anderson (1966)) has re-examined the theoretical aspects of this problem
in connection with the operation of the zonal rotors and derived the following equation relating zone capacity, zone width, particle and gradient density and rotor characteristics:

$$M_{\max} = \frac{\pi L \, \mathrm{d}p \left[\dfrac{\mathrm{d} \cdot \mathrm{d}m}{\mathrm{d}r}\right] \bar{r} \, (\Delta r)^2}{[\mathrm{d}p - \mathrm{d}m]} \tag{26}$$

$M_{\max}$ is the maximum amount of material held in a zone consistent with
stability, i.e. the zone will migrate as an undistorted Gaussian peak.
$\bar{r}$ is the mean radial dimension of the zone and Δr is the zone width.
$\mathrm{d}p$ and $\mathrm{d}m$ are the particle and gradient densities in the zone and L is the
rotor height (for zonal rotors). This equation, which is equivalent to the
one devised by Svensson et al. (1957) is applicable to situations where the
effects of diffusion upon zone stability may be neglected, e.g. large particles
with low diffusion coefficients. Spragg and Rankin (1967) therefore used
purified T3 phage particles, a zonal rotor and sucrose gradients to experimentally test the equation. With this system 85% of the loading capacity,
the zone widths and positions were all found as predicted by the Svensson-
Berman equation. The low figures reported by Brakke (1964) were attributed in part to the construction of gradients by a layering technique since
minor discontinuities in the gradient and overloading were shown to have
profound effects upon the zone shape. In addition, Berman has shown that
band capacity can fall to a few percent of the predicted values as a result
of diffusion of the solute molecules. The stability of the migrating band
then becomes time dependent. Deductions concerning the characteristics
of particles observed in zone sedimentations must therefore be made with
caution (see Section III.B.6).

5. *The location of zones and the sampling of gradients*

(a) *Band sedimentation studies.* Since these rate zonal separations are carried
out in the analytical cell the progress of the separation may be followed
throughout by means of the absorption optical system (Plate 1 (d)).

(b) *Preparative rate-zonal centrifugations.* For these separations the positions
of the zones within the gradient are determined at the conclusion of the
run. The fractionation of the tube contents is made at the same temperature
as the run to minimize the possibility of density inversion and the tube
contents may be sampled by (i) sectioning (ii) pipetting/siphoning (iii)
extrusion and (iv) controlled outflow via the bottom of the tube.

For sampling by sectioning the tube is mounted in a special vice and cut at a pre-determined level with a sharp, flat, triangular blade (Beckman Instruments Ltd.). The tube contents above the blade level are thereby isolated and may be removed for analysis. This procedure may be repeated down the length of the tube. This method is not particularly sensitive if one wishes to construct an accurate profile of tube contents vs. depth (see Fig. 8).

The contents of a tube may also be removed layer by layer by means of a pipette, usually mounted on a lowering device and with its tip maintained below the liquid meniscus to avoid the entry of air bubbles.

Air or a liquid denser than the tube contents may be used to extrude the

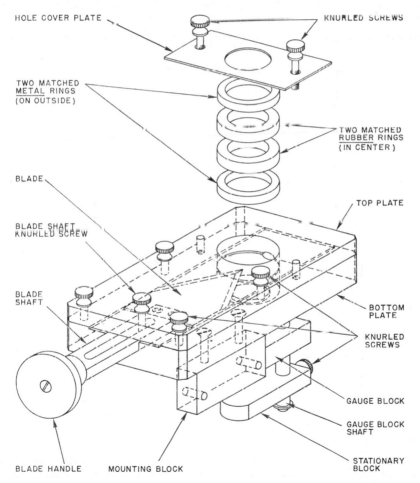

FIG. 8. A device for slicing centrifuge tubes containing density gradients (manufactured by Beckman Instruments Inc.).

tube contents for analysis. The Isco gradient analyser (Shandon Scientific Co., Ltd., London) is based upon the extrusion of the tube contents by the injection of denser solution at the base of the tube. The tube is fitted with a specially designed adaptor cap which may include an optical flow through cell in order to monitor the absorption of the exudate at a predetermined wavelength. The drops emerging from the cap may then be collected in an automatic fraction collector and subjected to further analyses. A syringe is used to inject the denser liquid into the side of the tube near its base in order to minimize a potential fountain effect and to avoid contamination by material pelleted at the bottom of the tube. A liquid seal prevents leakage about the entry point of the hypodermic needle (see Fig. 9). Brakke (1963) has shown that this method of sampling gives minimal mixing of contents, particularly at low extrusion speeds, and excellent recoveries of material.

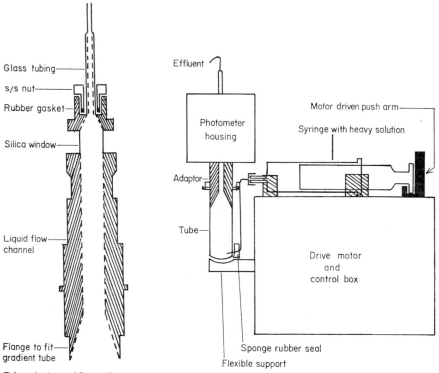

FIG. 9. A diagrammatic representation of an apparatus for extruding density gradients from centrifuge tubes by injecting a solution of higher density (Brakke, 1963) including an enlarged view of the lucite tube adaptor with optical flow cell. (Manufactured by ISCO Instruments, Shandon Scientific Co., London).

The flow cell has a volume of 0·35 ml and therefore represents a scan of a disc of liquid of 0·7 mm depth in a 2·5 cm diam. centrifuge tube; an accurate profile of distribution of material vs. depth is therefore possible with suitable correction for volume dead time.

Oumi and Osawa (1966) have described a simple method of extruding the centrifuge tube contents by air pressure. After centrifugation the tube is sealed with an airtight stopper carrying a controlled air inlet and a fine needle reaching to the bottom of the tube. Air pressure applied to the inlet forces the contents out from the bottom through the needle to an appropriate sampling device. The rate of expulsion of the contents may be controlled by varying the air pressure. This method avoids the use of another liquid and preserves the tube for further use; it suffers from the disadvantage that any pelleted material may be seriously disturbed (see Fig. 10).

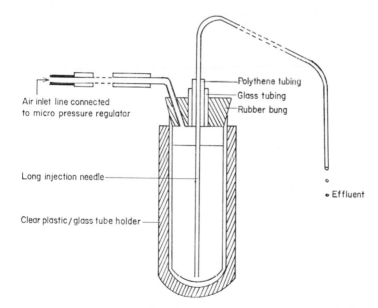

FIG. 10. A density gradient fractionator based upon controlled air pressure extrusion of the tube contents (Oumi and Osawa, 1966).

The most widely used method for sampling density gradients is to pierce the bottom of the tube with a hypodermic needle and collect the drops as they form on the tube. This may be simply accomplished by sealing the top of the tube with a bung carrying a controllable air inlet and placing a small piece of "Parafilm" around the bottom of the tube. The tube is pierced through the "Parafilm" with a fine hypodermic needle and the effluent drops collected in a suitable series of tubes. A number of sampling devices have been described which are elaborations of this basic approach. A complete analyser system based upon such a device and the L.K.B. Uvicord flow monitor (L.K.B. Instruments, Stockholm, Sweden) is shown in Fig. 11. This system is based upon a simply machined Perspex sampling device and common laboratory apparatus which may be readily assembled

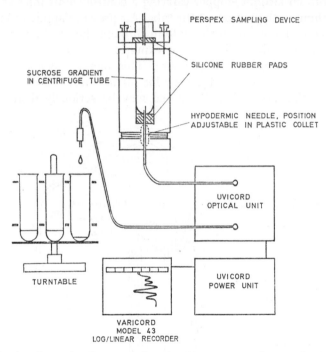

FIG. 11. A schematic diagram of a simply constructed, complete, automatic density gradient analyser and recording system (Coleman and Sykes, 1966).

into the arrangement shown. Figure 12 shows that the resolution obtained using the 0·1 ml, 3 mm path flow cell compares favourably with an analytical centrifuge run and is to be preferred for detecting polyribosomal components (Coleman and Sykes, 1966). The flow rate is held reasonably constant by fixing the distance between the average liquid height in the tube and the drop point. An accurate, constant flow rate throughout sampling could be achieved by including a finger-type peristaltic pump on the flow line. The effluent from the UV scanner may be collected for subsequent analyses by adding an automatic fraction collector equipped with a drop counting head. If a radioactivity vs. depth profile is to be superimposed upon the absorbancy profile samples may be collected directly

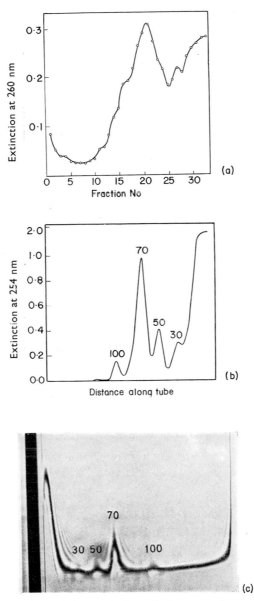

FIG. 12. A comparison of the resolution of ribosomes from cell-free extracts of *Bacillus subtilis* in rate-zonal density gradients fractionated (a) manually and (b) automatically in the analyser shown in Fig. 11. The corresponding boundary sedimentation velocity Schlieren diagram from the analytical ultracentrifuge is shown in (c). In (a) and (b) sedimentation is from right to left and in (c) it is from left to right. The figures above the peaks are their sedimentation constants in Svedberg units.

on to planchettes or into vials. Many modern spectrophotometers are equipped for flow-cell operation and could be used to replace the "Uvicord" and thereby extend the scanning wavelength range. The flow cell volumes may be chosen so that they represent a scan of a very narrow disc of liquid within the original tube.

6. *The evaluation of rate-zonal sedimentation data*

(a) *From preparative rotors.* The technique of rate-zonal centrifugation has found extremely wide application and has catalysed research in Microbiology. It is ideally suited to the speed, precision and small scale operation so frequently required to characterize the sub-cellular organization of bacteria and its response to growth conditions, etc.

The majority of applications have been qualitative in character and despite its widespread use there have been relatively few attempts to integrate theory and practice and thereby realize the full information content of a rate-zonal centrifugation.

For reproducible and quantitative assessment of rate-zonal data the time, temperature and speed of centrifugation must all be carefully controlled. For the purpose of most theoretical treatments and practical applications it is assumed that serious redistribution of the gradient material does not take place under the influence of the applied field. McEwen (1967a) has shown that this assumption is justified in practice except for relatively prolonged runs (24 h) at high gravitational fields (420,000 g). However preformed sucrose gradients are more shallow than equilibrium gradients (see Section IV.C) and at gravitational fields above 70,000 g (max.) sedimentation of sucrose will occur. It is also assumed that the solute does not interact with the gradient or with other solutes since in the latter case incomplete separations may occur if the solutes are in rapid chemical equilibrium and this situation necessarily complicates the interpretation of data.

Analysis of preparative rate-zonal data. In addition to the considerations mentioned above, three further specific points arise in connection with the analysis of data from rate-zonal centrifugations in the conventional swing-out rotors:

(i) The geometry of the sedimenting system. This would not appear to be a serious complicating factor in practice, e.g. see earlier discussion (Section IIIA) and the results of Martin and Ames (1961) discussed below.

(ii) the increasing gravitational field with increasing distance from the centre of rotation down the tube. This factor could be a source of anomalous boundary (zone) shape with materials having a high dependence of s

upon c due to variation in the field across the zone. In practice the narrowness of the zone and the low concentrations of material within a zone make this unlikely (with the possible exception of DNA preparations) and the commonest source of boundary anomalies are discontinuities in the gradient (Spragg and Rankin, 1967). However, the increasing field down the tube will accelerate the movement of a zone but this will be counteracted by

(iii) the increase in viscosity with radial distance as a consequence of the increasing gradient of density. This factor will tend to retard the rate of sedimentation and "*iso*-kinetic gradients" are frequently constructed to avoid this complication (Henderson, 1969; Noll, 1967). In the situation where (ii) and (iii) above exactly match the rate of movement of a zone is linear with time, i.e. log x vs. time is linear and will give a true measure of the sedimentation coefficient as previously described (Section III.A.1).

Martin and Ames (1961) were able to show that the distances migrated by certain enzymes in 5–20% linear sucrose gradients were linear with respect to time and in these circumstances, with marker proteins of known $S^0_{20, w}$ value and assuming the same partial specific volumes for like materials,

$$R = \frac{\text{distance moved from meniscus by unknown}}{\text{distance moved from meniscus by standard}} \tag{27}$$

$$= \frac{S^0_{20, w} \text{ unknown}}{S^0_{20, w} \text{ standard}} \tag{28}$$

Furthermore, if a spherical shape is assumed for the materials then

$$R = \left[\frac{MW_1}{MW_2}\right]^{\frac{2}{3}} \tag{29}$$

i.e. a *rough* estimation of the molecular weight may be made. In a subsequent paper Winzor (1967) has pointed out that where the conditions are as described by Martin and Ames (1961) (constant s and presumably constant D) then a value of s and D may be obtained and combined in the Svedberg equation (eqn. 25) to yield a molecular weight directly. Values of D calculated by Winzor from appropriate items of Martin and Ames' data and shown to be consistent with the literature as were the molecular weights. The preferred conditions for the application of Winzor's treatment, i.e. s and D are constants throughout, may be realized in shallow pre-formed density gradients and in the self generating density gradients used for band centrifugation. However, even where s varies then an average sedimentation coefficient $(\bar{s})$, defined as the distance migrated in time of a solute in a medium with the viscosity and density applying midway between the initial and final zone positions, gives a very close fit of data. Thus s, D and molecular weight may be obtained from a single rate-zonal experiment, starting with a crude mixture of materials. The close agreement in the s values obtained from rate-zonal data and conventional analytical ultracentrifugation also implies that there is no net convective transport of materials in a radial direction in rate-zonal centrifugations. Based upon this observation, i.e. that sedimentation and diffusion (and not convection) are responsible for the mass transport, Schumaker and Rosenbloom (1965) have developed fundamental mass transport equations for rate-zonal sedimentation in centrifuge cells of radial shape (i.e. band centrifugation) and of uniform cross sectional area (i.e. preparative rate-zonal tubes). For the latter situation they were able to show that

$$\frac{d \ln \langle r \rangle}{dt} = s\omega^2 \tag{30}$$

i.e. the s value may be obtained from the rate of movement of the centre of gravity of the zone $\langle r \rangle$. Provided that the migrating zones are symmetrical in shape the usual practice (e.g. Martin and Ames) of using the peak values to measure $s_{20, w}$ gives an identical result since the peak position has the same x co-ordinate as the centre of gravity. When the material in the zone is not homogeneous, the movement of the centre of the zone yields a weight average sedimentation coefficient. The concentration dependence of the diffusion coefficient was shown to have no effect upon the velocity of the centre of gravity. In experiments where the gradient is not of an iso-kinetic character the analysis of the effect of the super-imposed gradient upon boundary location would require a knowledge of the boundary shape and the viscosities and the densities of the solvent as a function of time and position within the tube. Although this is not a practical proposition for a rate-zonal preparative run the viscosity and density of the solvent may be determined as a function of tube depth. If it is assumed that (i) these quantities do not vary during the run (ii) the movement of the centre of gravity of the boundary suffices as an accurate measure of s and (iii) the correction term to be considered for the dependence of the diffusion coefficient (D) upon the changing viscosity may be neglected, then Schumaker and Rosenbloom (1965) showed that:

$$\frac{d \ln \langle r \rangle}{dt} = s_{20, w} \cdot \omega^2 \frac{(1 - \bar{v} \, d_c)}{(1 - \bar{v} \, d_{20, w})} \frac{\eta_{20, w}}{\eta_c} \tag{31}$$

When all the terms have their previously assigned meanings and the subscript c refers to values of d and η at the centre of gravity of the migrating zone. Between the limits of the starting position $\langle r \rangle_0$ and the final position $\langle r \rangle_x$ of the zone equation (31) integrates:

$$\frac{1}{\omega t_{eff}} \int_{\langle r \rangle_0}^{\langle r \rangle_x} \left[\frac{(1 - \bar{v} \, \mathrm{d}_{20,\, w})}{(1 - \bar{v} \, \mathrm{d}_c)} \frac{\eta_c}{\eta_{20,\, w}} \right] \mathrm{d} \ln \langle r \rangle = s_{20,\, w} \tag{32}$$

t_{eff} = effective time of centrifugation.

The integral in equation (32) may be evaluated by plotting

$$\frac{(1 - \bar{v} \, \mathrm{d}_{20,\, w})}{(1 - \bar{v} \, \mathrm{d}_c)} \frac{\eta_c}{\eta_{20,\, w}} \quad \text{vs. } \ln \langle r \rangle$$

and determining the area under the resultant curve. Knowing t_{eff} then $s_{20,\, w}$ may be calculated. McEwen (1967b) has constructed a set of useful tables for estimating sedimentation of inert, non-diffusing particles through linear concentration gradients of sucrose solution. The retarding effect of the sucrose solution is taken into account and the tables may be used for both the design (e.g. predicting the position of a zone of material of known characteristics) and interpretation (e.g. evaluation of s from position in a given gradient) of experiments involving linear sucrose gradients.

(b) *Analysis of analytical rate-zonal data* (band centrifugation). This problem has been discussed by a number of authors (Rubin and Katchalsky, 1966; Schumaker and Rosenbloom, 1965; Vinograd and Bruner, 1966; Vinograd et al., 1963) although the number of published, practical applications of the technique is limited. All of the potentially major complications to the interpretation of data from the preparative technique (see p. 98) are avoided by the basic design of band centrifugation experiments. In this case rate-zonal centrifugation takes place in sector shaped, small path length analytical cells in shallow, field generated density gradients; hence with increasing radial distance the increments in the viscosity and density of the gradient material are negligible. The effect of the inhomogeneous field down the cell is to increase the width of the migrating band by approximately 15% during a run (Vinograd et al., 1963). This broadening of the band is generally negligible compared with the effect of diffusion. The practical observations of Vinograd et al (1963) are in agreement with their accompanying theoretical treatment and the subsequent derivations of Schumaker and Rosenbloom (1965). The rate of movement of the centre of the band is linear with time and a measure of this for a Gaussian peak gives the sedimentation coefficient when used in equation (30). Concentration dependence of s yields a band which is sharp on the trailing edge and forward spreading on the leading edge. In these circumstances the concentration dependence of $s_{20,\, w}$ (and hence $s^0_{20,\, w}$) may be obtained from the data generated in a single run. If the usual type of concentration dependence of s is assumed, i.e.

$$s_{20, w} = s^0_{20, w} (1 - kc) \tag{33}$$

where k is a constant and c the concentration, and the band is fully resolved from the meniscus and base of the cell then

$$x_{max} - x^0_{max} = s^0_{20, w} \omega^2 \bar{x} (t - t_0) (1 - 2kc_{max}) \tag{34}$$

Here x_{max} and x^0_{max} refer to the positions of the maximum ordinate of the band at two timed intervals (t and t^0) and $\bar{x}$ is the distance of the centre of the band from the centre of rotation. A plot of $(x_{max} - x^0_{max})/(t - t^0)$ vs. c_{max} (suitably corrected for radial dilution using equation (23)) will give both $s^0_{20, w}$ and k. Vinograd et al (1963) found that the experimental data from band sedimentation studies when used in equation (34) gave $s^0_{20, w}$ values which were in excellent agreement with the corresponding values from boundary sedimentation, even for materials with a pronounced dependence of sedimentation upon concentration, e.g. T7DNA.

Although the zone spreading due to the inhomogeneous field is greater than that experienced in boundary sedimentation (see above) Vinograd et al. (1963) have shown that it should be possible to calculate the diffusion coefficient from a band sedimentation run using the equation:

$$2D(t - t^0) = \sigma^2 \left(\frac{x^0_{max}}{x_{max}}\right) - (\sigma^0)^2 \left(\frac{x_{max}}{x^0_{max}}\right) \tag{35}$$

in which σ is the standard deviation of the curve of c vs. x. A mathematical treatment for the evaluation of D from band centrifugation data has also been presented by Rubin and Katchalsky (1966).

IV. EQUILIBRIUM SEDIMENTATION

A. Introduction

The sedimentation velocity methods described in the preceding Sections employ high gravitational fields to impart a terminal velocity to a particle such that transport due to sedimentation is large compared with that due to diffusion. On the other hand if a solution is centrifuged for a sufficient period of time at a relatively low speed then the sedimentation of materials gives rise to concentration gradients which in turn lead to back diffusion; ultimately a sedimentation-diffusion equilibrium is established. The equations describing the final concentration distribution at equilibrium may be derived from a consideration of the transport of material (i.e. by sedimentation and diffusion) or thermodynamics. Despite the rigor, lack of assumptions and relative simplicity of the measurements required in these equations the practical applications of the method were few until recent improvements in the design of instruments, theory and technique overcame the problems arising from the long centrifugations necessary to attain equilibrium.

The basic condition of sedimentation equilibrium has now been exploited not only to give information of a "static" character, e.g. molecular weights, heterogeneity, buoyant density (and hence composition e.g. for DNA), but also of a dynamic character, e.g. the study of DNA and RNA replication employing isotopic or other density labels. This centrifugal technique is now of major importance in microbiological research to the extent that the DNA composition data derived from buoyant density determinations (i.e. *iso*-density sedimentation equilibrium) has been used as a basis for bacterial classifications (Hill, 1966; Marmur *et al.*, 1963).

At the present time the two main practical applications of the method are

(i) sedimentation equilibrium, i.e. the straightforward equilibrium state between transport of solute by sedimentation and diffusion as described above.

(ii) *iso*-density sedimentation equilibrium, i.e. the equilibrium attained in the presence of an added, strong density gradient. This technique has been used for both analytical and preparative purposes.

B. Sedimentation equilibrium

1. *Theoretical and practical aspects*

Excellent practical and theoretical treatments of this technique have been published by van Holde (1967), Mazzone (1967) and Richards *et al.* (1968). This technique is the most accurate and hence the preferred method for the estimation of molecular weights. The immense, controlled speed ranges of modern ultracentrifuges enables equilibrium conditions to be achieved with molecules with molecular weights ranging from several hundred to several million daltons, e.g. from simple proteins to virus or ribosomal particles. In addition an analysis of the data will also indicate the homogeneity or otherwise of a sample suspected of being pure.

The fundamental equation describing the state of sedimentation equilibrium may be rigorously derived from thermodynamic treatments (e.g. Fujita (1962) and Williams *et al* (1958)) or by a kinetic approach from a consideration of the transport of the solute as follows:

The transport of the solute by sedimentation $\vec{J}$, (i.e. no. of grams of solute crossing unit area in unit time) is the product of their velocity (V) and the concentration c_x

$$\vec{J} = Vc_x = s\omega^2\bar{x}c_x \tag{36}$$

where $\bar{x}$ is the distance from the centre of rotation. The transport by diffusion, $\overleftarrow{J}$, is given by Ficks 1st law:

$$\overleftarrow{J} = -\frac{D.dc_x}{d\bar{x}} \tag{37}$$

$$\text{At equilibrium } \overrightarrow{J} + \overleftarrow{J} = 0 = s\omega^2 x c_x - \frac{D\, dc_x}{d\bar{x}} \tag{38}$$

Since $s = M(1 - \bar{v}dm)/Nf$ (see equation 14) and $D = RT/Nf$ (equation 24) then substituting these values in the above expression (equation 38):—

$$0 = \frac{M(1 - \bar{v}\, dm)}{Nf}\omega^2 \bar{x} c_x - \frac{RT}{Nf}\frac{dc_x}{d\bar{x}} \tag{39}$$

i.e.

$$\frac{1}{c_x}\frac{dc_x}{d\bar{x}} = \frac{M(1 - \bar{v}\, dm)\omega^2 \bar{x}}{RT} \tag{40}$$

Since this equation (40) describes the concentration gradient existing at a point in the cell integration between the limits set by the meniscus (x_m) and this point, $\bar{x}$, gives the actual concentration:

$$\ln\frac{c_x}{c_m} = \frac{M(1 - \bar{v}\, dm)\omega^2}{RT} \cdot \frac{(\bar{x}^2 - x_m{}^2)}{2} \tag{41}$$

The equation above is the basic equation for sedimentation equilibrium work and a plot of $\ln c_x$ vs. $(\bar{x})^2$ will yield M provided $\bar{v}$ and dm are known. Here lies one of the great advantages of the equilibrium method compared with the velocity method, i.e. it does not require the knowledge of the diffusion coefficient. However a knowledge of c_x (or some quantity directly proportional to it) is required at each point in the cell at equilibrium. This will be given directly by using absorption optics (see p. 98 *et seq.*) to follow the run since the densitometer trace of the U.V. film and the direct photoelectric scanner plot both give an optical density vs. x profile. Since optical density is commonly proportional to concentration this optical system not only confers the advantages of sensitivity and selectivity (by scanning at a preselected wavelength) to identify the component of interest, but also from the trace directly gives all the data required for the calculation of its molecular weight. The development of the automatic scanning system, working in conjunction with a monochromator, will probably lead to a wider future use of this optical system for equilibrium studies (Schachman and Edelstein, 1966).

The Schlieren optical system records the concentration gradient, dc/dx vs. x and this requires integration for use in equation (41). This optical system is less sensitive and relatively high concentrations of material

(>5 mg/ml) are required for precision in the measurements. At these concentrations departures from ideal behaviour become pronounced. For these reasons this system has fallen into disuse for sedimentation equilibrium studies (but note the approach to equilibrium technique described on p. 110). The combined use of Schlieren optics (to measure the concentration gradient) and interference optics (to give the concentration directly) has been suggested for equilibrium studies. The design, or minor modifications (Chervenka, 1966), of most analytical ultracentrifuge optical systems makes this a practical possibility.

The Rayleigh interferometric optical system has the required sensitivity for equilibrium work but records the difference in refractive index between the solution and a reference column of solvent (Plate 1 (b)) by a displacement of interference fringes formed by placing narrow slits behind the liquid columns (Richards et al., 1968; Schachman, 1959). The fringe number, N, between certain levels, which may not be an integral number, is related to the change in refractive index Δn as follows:

$$N = \frac{a\Delta n}{\lambda} \tag{42}$$

where a is the cell optical path length in cm and λ is the wavelength of light (usually 546 nm). The accuracy of this method may be appreciated by considering the common analytical cell of path length 1·2 cm and a value of $N = 1$ at 546 nm. This gives a value for $\Delta n = 4·5 \times 10^{-5}$. Since for most proteins the refractive index increment (Δn) for a 1% solution is $1·86 \times 10^{-3}$ a displacement of one fringe therefore corresponds to a concentration difference of approximately 0·25 mg/ml. In practice it is possible to measure displacements to 0·02 fringes and valuable practical details for the accurate measurement of Rayleigh interferograms using a two-dimensional micro-comparator are to be found in the articles by van Holde (1967) and Richards et al (1968). The interferograms give the concentration at a point relative to that at some other point and the assignment of an absolute concentration value to each of the fringes depends upon determining which fringe corresponds to c_0, the initial concentration, or a knowledge of the concentration corresponding to any fringe in the pattern. In practice these absolute concentrations are not deduced readily and this problem is commonly solved by adopting one of the two main sedimentation equilibrium techniques:

  (i) high speed method
  (ii) low speed method.

In the high speed method, suggested by Yphantis (1964), a low concentration solution is centrifuged at a sufficiently high speed to reduce the concentration of material at the meniscus to zero. Alternatively, to achieve the same effect, the speed may be increased after equilibrium has been reached (Le Bar, 1965). A zero concentration reference at the meniscus is then provided and the increments then measure the actual concentrations at the various points (since the Rayleigh interferograms give the difference in concentration between any cell position and the meniscus). The information required to evaluate equation (41), i.e. for a plot of ln c_x vs. $(\bar{x})^2$ is then directly obtained, hence M at any point in the cell or the molecular weight average in selected regions. This provides a critical test of homogeneity (Stellwagen and Schachman, 1962).

For the alternative, i.e. low-speed method, equation (40) is rearranged and integrated between the limits of the meniscus (x_m) and cell bottom (x_b):

$$\int_{x_m}^{x_b} \frac{dc_x}{d\bar{x}} \cdot d\bar{x} = M \cdot \frac{(1 - \bar{v}dm)}{RT} \omega^2 \int_{x_m}^{x_b} \bar{x} c_x d\bar{x} \tag{43}$$

The right hand side integral is satisfied by $c_0 (x_b{}^2 - x_m{}^2)/2$ and that on the left directly yields $[c_b - c_m]$, and so equation (43) becomes

$$\frac{c_b - c_m}{c_0} = \frac{M (1 - \bar{v}dm)\omega^2}{2RT} (x_b{}^2 - x_m{}^2) \tag{44}$$

The above equation is the basic equation of the classical low speed method in which a finite meniscus concentration is always maintained throughout the run.

The ratio $c_b - c_m/c_0$ does not require a knowledge of absolute concentrations and the necessary values for the numerator are determined from the equilibrium run and those for the denominator from a separate synthetic boundary cell run. The total number of fringes crossed in traversing the Rayleigh interferograms of the cell from meniscus to the bottom yields a figure which is proportional to the quantity $(c_b - c_m)$ at the point of equilibrium c_0 is determined by making a similar total fringe count across a boundary formed in a layering cell (i.e. valve type or double sector, capillary synthetic boundary cell) with a sample of the same material. Great care should be exercised here to ensure that zero gradient regions exist above and below the boundary, that no premature mixing occurs and that there are no leaks from the cell during the run. x_m and x_b are simply measured from the photographs of the known reference points which are built into the counterpoise and appear on the interferograms (Plate 1 (b)). The molecular weight may then be calculated by substitution in equation (44). This straightforward treatment of the data avoids the problems associated with the determination of absolute concentrations and consequently does not realize the total information potential of the equilibrium run (Lansing and Kraemer, 1935). Furthermore extrapolations of the plots of fringe count vs. cell distance are required at the solution column ends in order that the fractional fringe, e.g. between the first observable fringe and the meniscus, may be estimated. For a full analysis of the equilibrium run plots of $\ln c$ vs. $(\bar{x})^2$ are to be preferred as described for the high speed method using equation (41). A straight line plot then indicates homogeneity, an upward curvature heterogeneity and a downward curvature a departure from ideal behaviour. Since the latter may be encountered at the relatively higher concentrations and non-ideal behaviour could combine with heterogeneity to give a straight line plot of $\ln c$ vs. $(\bar{x})^2$ equilibrium runs are best performed at a number of different concentrations to detect heterogeneity. In the case of heterogeneity of the monomer/dimer type where the larger species will concentrate in the lower half of the cell then integration over the whole cell will simply give an M_w average. However plots of $\ln c$ vs. $(\bar{x})^2$ can be used to give the slope at different positions in the cell to yield the particular value of M at certain points.

For such a full evaluation of the data from the low speed method two additional procedures have been adopted to translate the fringe counts into data expressing concentration as a function of distance; one involves determining c_m and the other the fringe corresponding to c_0. The knowledge of a concentration existing at any one point then suffices to give the concentration at all other points in the cell and equation (41) may then be used. For the determination of c_m Richards and Schachman (1959) adopted the conservation of mass approach to show that the value of the concentration at the meniscus was given by

$$c_m = c_0 - \frac{x_b^2(c_b - c_m) - \int_{x_m}^{x_b} x^2 dc}{x_b^2 - x_m^2} \tag{45}$$

Archibald (1947) was the first to note that at one position in the cell during an equilibrium run (the "hinge point"), the concentration was independent of time. For the practical determination of this point, i.e. the fringe in the equilibrium interferogram corresponding to c_0, Richards and Schachman (1959) have suggested photographing the "white light fringe". This is the fringe pattern formed with the monochromatic (546 nm) light filter removed from the optical system and is photographed when the rotor has just reached speed and again when the cell contents have reached equilibrium. The intersection of this achromic fringe (actually one prominent fringe with a less well defined fringe on either side is usually obtained) on the equilibrium pattern with the corresponding fringe on the pattern at speed locates the fringe corresponding to c_0. Since the relative location of the achromic fringes depends upon the relative refractive indices of the solvent and solute columns the fringe shift may be considerable in certain experiments (e.g. 40 fringes with a 1% solution of protein in the solution chamber). In such cases it is common practice to add a material to the solvent chamber which will raise the refractive index to equal that of the solution but will not redistribute in the field. 1,3-butanediol has been widely used for this purpose. When c_0 is determined by this technique the usual synthetic boundary run for c_0 is not required.

2. *Some additional observations relating to practical aspects of equilibrium centrifugation*

The optical components of the ultracentrifuge should be clean, individually adjustable and free from oil smear. The latter point is particularly important and is a problem with the components of the system within the rotor chamber during the long runs required to attain equilibrium.

With the interference optics a source of monochromatic light is required (the 546 nm line in the high pressure mercury arc source is usually chosen) and obtained by placing a suitable filter over the source, e.g. Kodak Wratten 77A, Balzers B40 or Baird Atomic B9. With the absorption optics the addition of the monochromator unit allows the operator a wide selection of wavelengths which may be used to selectively follow the object of interest in an impure system.

The aperture slits behind the double sector cell in interference optics should preferably be offset with respect to the cell sectors. This ensures better fringe registration but does not allow a comparison of conjugate levels which is possible with symmetrical slit apertures. Fine slits also improve fringe registration but, with the filters, reduce the available light and considerably extend the exposure times required to record the fringes. Sapphire cell windows are less susceptible to distortion and therefore reduce the amount of fringe distortion to be corrected in subsequent calculations. Quartz windows may be used for low speed runs but sapphire windows are essential for high speed runs.

Considerable care is required in filling the cell chambers (e.g. see van Holde, 1967) and reproducible filling is assisted by using Hamilton syringes with Chayney adaptors. In the filled cell the bottom of the cell and the meniscus must be visible and the solution column must be just overlapped at each end by a reference column of solvent. A little inert oil [e.g. F.C. 43] is usually placed at the bottom of each chamber to assist in the definition of the bottom of the cell and the presence of aggregates. Evaporation of the solution may be a problem during cell filling operations, particularly with the multi-channel, short column cells designed by Yphantis (see Fig. 1 (f)). These cells are useful since as many as four different solution : solvent pairs may be studied at the same time. The precautions to be observed in filling these cells are described by Yphantis (1964) and van Holde (1967).

Equilibrium runs are preferably only performed on dialysed, suspected pure preparations although the use of the monochromator and absorption optics will allow the detection of a component in a mixture. The high speed technique may also be employed to advantage with preparations containing heavier contaminants.

The choice between a high speed or low speed equilibrium run will be made on the basis of (i) the amount of material available, (ii) the expected molecular weight range and (iii) the stability of the preparation to prolonged centrifugation at defined temperature. The low speed technique requires slightly more material (total approx. 0·5 mg) than the higher speed run, (since a separate run is required to determine c_0), is limited to a molecular weight maximum of 5×10^6 and takes longer. The high speed method requires less material ($\simeq 0·05$ mg), is quicker but not suited to molecular weights below 10,000. The low speed method provides interferograms which may be analysed to exploit the full accuracy of the Rayleigh optical system whereas in the high speed method the relatively high gradients at the bottom of the cell may give rise to unresolved fringes and a corresponding sacrifice in accuracy and information. The shallow concentration gradients may also give rise to convection, although sucrose may be added to produce a stabilizing gradient. With high speed runs a photograph should be taken at speed to provide a "blank" (correction) photograph showing the fringe pattern before the establishment of any concentration gradients. This may then be used to correct for fringe deviations arising from cell window or centrepiece distortion occurring at high speeds.

The actual speed chosen for a run will be dictated by the anticipated molecular weight of the sample; an estimate of $\bar{v}$ must also be made. For a low speed experiment the equilibrium ratio of c_m/c_b is usually 3–4 and so, depending upon the liquid column height in the cell (see below), equation (41) may then be solved for ω^2. For the high speed technique Yphantis (1964) suggests a value of 5 for the quantity $\omega^2 M(1-\bar{v}\mathrm{d}m)/RT$. The speed selected here is critical since not only must meniscus depletion be ensured but also a fringe count of $< 200/\mathrm{cm}$ is desirable at the bottom of the cell. Figure (13) taken from van Holde's article (1967) gives a useful relationship between anticipated molecular weight and rotor speed for high speed and low speed runs.

Heavy rotors are usually selected for low speed runs to reduce precession and since ω^2 features in the equations for molecular weight a precise control and knowledge of the operating speed is essential. The electronic speed control units with modern analytical ultracentrifuges are well suited to this purpose.

The solution column height clearly has a profound effect upon (i) the time to attain equilibrium and (ii) the detection of heterogeneity. Since the

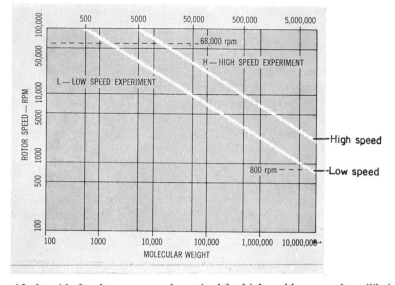

Fig. 13. A guide for the rotor speed required for high and low-speed equilibrium experiments with materials of molecular weights in the range 10^2–10^7 (from Van Holde, 1967).

time to approximate equilibrium is directly proportional to the square of the column depth $(x_b - x_m)^2$, the use of shorter liquid columns will considerably shorten the time to reach equilibrium. Short columns have the additional advantages that less sample is required and labile materials are not subjected to prolonged centrifugation. The disadvantages of these columns are loss of solvent by evaporation, adsorption of material on the cell surfaces and fewer opportunities for the detection of polydispersity. Other experimental techniques which have been adopted to reduce the time taken to attain equilibrium include (i) the specially designed layering cell described by Griffith (1967) and (ii) the use of a short, high-speed run followed by a low speed run (Hexner et al., 1961; Richards et al., 1968). In the former a conventional double sector, capillary synthetic boundary cell has three small feeder reservoirs in line and communicating by capillaries with the solution sector. These reservoirs hold small volumes of increasing concentrations of the sample (from top to bottom of the centrepiece) and on centrifugation these solutions migrate from the reservoirs into the solution chamber to create a steep concentration distribution early in the run. In this manner the time to reach equilibrium may be reduced by a factor of 8.

The Archibald method (Archibald, 1947; Klainer and Kegeles, 1955) was introduced as a means of reducing the time involved in classical equilibrium experiments and is frequently, but perhaps misleadingly, referred to as an "approach to equilibrium" technique. The method is based upon transient states and the appreciation that the cell is a closed system and no transport of material will take place across the meniscus and bottom of the cell. Since the gradient of total potential at these points must be zero an equilibrium type relationship exists at these locations at all times although this does not imply that depletion of material from the meniscus and accumulation at the cell bottom does not take place; the concentration of material at these two surface changes continually. According to Archibald (1947) the situation at these two points is described at all times by the same formal equation as that which obtains elsewhere in the cell at equilibrium, i.e. at the meniscus

$$M = \frac{RT[dc/dx]x_m}{(1 - \bar{v}dm)\omega^2 x_m c_m} \tag{46}$$

and a corresponding equation (with x_b and c_b replacing x_m and c_m) applies at the cell bottom. For the usual refractive index gradient data (i.e. photographs of Schlieren runs) c_b and c_m are given by the equations (Klainer and Kegeles, 1955)

$$c_m = c_0 - \frac{1}{x_m^2} \int_{x_m}^{X} x^2 \cdot \left(\frac{dc}{dx}\right) . dx \tag{47}$$

and

$$c_b = c_0 + \frac{1}{x_b^2} \int_{X}^{x_b} x^2 \left(\frac{dc}{dx}\right) dx \tag{48}$$

Where X is a radius of rotation in the cell at which the concentration is not a function of the radius ($dc/dx = 0$) i.e. a plateau region must exist in the cell at all times for the application of these equations. The integrals in equations (47) and (48) are evaluated by measuring the area bounded by the Schlieren trace, the solvent base-line and the meniscus, or cell bottom in the double sector run. With Schlieren optics the dc/dx term in equation (47) and (48) is given directly. Step by step details of the experimental procedure and the subsequent calculations have been described by Klainer and Kegeles (1955), Schachman (1957) and Mazzone (1967) and will not be repeated here except to note that the use of a double sector cell for the Archibald run avoids one of the runs described in these methods. c_0 is again determined by a separate run in a synthetic boundary cell and is calculated as described on p. 74.

The Archibald method requires accurate focussing of the optical system (see Trautman, 1958) and allied to this problems emerge in the analysis of the Schlieren photographs since both the gradient and the absolute concentration must be measured. On the usual enlargements of the photo-graphs extrapolation of the Schlieren trace to the limits of the liquid column is therefore required. This may be somewhat subjective since the magnified image facilitates area measurements but also magnifies the meniscus image and the necessary, precise identification of the true meniscus position is frequently difficult for this and other optical reasons. These difficulties severely limit the precision of the technique and have been a source of the recent decline in its use. The mid-point of the meniscus is usually taken and a linear extrapolation of the gradient is made. In theory it is possible to calculate M from data obtained at both the meniscus and bottom of the cell (see equations 46, 47 and 48 above). Identical values of M are then taken as evidence of homogeneity, but the converse may not be true. In practice it should be noted that the gradient at the bottom of the cell is extremely sensitive to aggregated material and identical values of M may not then be obtained with a material which is essentially pure. It is safer to determine M from the meniscus data at various times during the run and extrapolate these to zero time. This is a sensitive test of homogeneity. Despite the disadvantages associated with the extrapolation of the Schlieren trace the Archibald method has the particular advantages that molecular weights may be determined after very short periods of centrifugation with small amounts of material using simple calculations and without resort to separate diffusion experiments. However, as in all equilibrium determina-tions of M a knowledge of $\bar{v}$ is required and the difficulties associated with this have been described earlier (p. 81).

C. *Iso*-density sedimentation equilibrium

1. *Introduction*

This technique is often loosely referred to as "density gradient sedimen-tation" but should not be confused with the kinetic density gradient methods of rate-zonal sedimentation (Section III.B) and band velocity sedimentation (Section III.B). *Iso*-density sedimentation equilibrium depends upon the separation of the solvated macromolecules according to their buoyant densities in a field-generated density gradient with a strong salt solution. Thus if a mixture of a suitable concentration of a salt (e.g. a solution of CsCl adjusted to be ultimately denser at the bottom and less dense at the top of the vessel than the macromolecular species) and macro-molecules are centrifuged together then the salt redistributes in the field to form a density gradient. The macromolecular species will also redis-tribute and come to rest in the gradient as bands with centres about the point where $(1 - \bar{v}dm)$ is zero i.e. their buoyant, *iso*-density point with

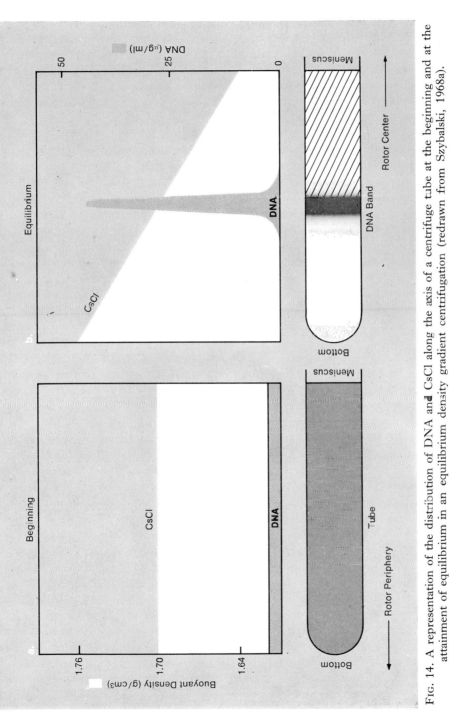

Fig. 14. A representation of the distribution of DNA and CsCl along the axis of a centrifuge tube at the beginning and at the attainment of equilibrium in an equilibrium density gradient centrifugation (redrawn from Szybalski, 1968a).

the salt gradient. A quasi- or true equilibrium is then established in which effective centrifugal forces vanish. Figure (14) reproduced from Szybalski's (1968a) useful, practical treatment of this technique clearly illustrates this phenomenon. The width of the band is determined by diffusion and larger molecules will produce a sharper band. The shape of the band and the number of bands found within a gradient depend upon the number of macromolecular species of different buoyant density present in the sample introduced with the salt solution. A single, skewed (i.e. non-Gaussian) band indicates density heterogeneity within that band.

This technique is an extremely sensitive detector of changes in buoyant density (hence chemical composition) e.g. for DNA in CsCl gradients density differences of $0 \cdot 001$ gm/cm^3 may be detected. It is therefore not only possible to distinguish DNA species of different base composition but also DNAs from various deletion mutants (Weigle et al., 1959) or containing heavy and light isotopes (Meselson and Stahl, 1958) or unusual bases etc. (Vinograd and Hearst, 1962). The iso-density equilibrium technique therefore has the great advantage of being a non-destructive method of extremely high resolving power for the determination of chemical composition which can be applied at the analytical or preparative level. Furthermore, the width of the band at equilibrium may be used to estimate the molecular weight (Meselson et al., 1957). The theoretical background to this technique is complicated by the presence of the third component (i.e. the salt) at high concentrations since this gives rise to non-ideal behaviour and, being an electrolyte, to charge effects. The high speeds employed in this technique also generate pressure effects which are important and complicate the theoretical interpretation.

The recent applications of this technique in microbiological research have given results which have had a profound influence upon, and extended our understanding of, the structure and function of bacterial and viral nucleic acids. Data for the composition of bacterial DNAs determined in this manner are in excellent agreement with the chemical analyses, are more readily obtained and require much less material. These results have provided a new basis for systematic and taxonomic microbiology (Hill, 1966; Marmur et al., 1963). Denatured (i.e. complementary) DNA strands may be separated in iso-density equilibrium gradients and this has been exploited to show the asymmetric synthesis of RNA (Tocchini-Valentini et al., 1963) and other properties of the codogenic DNA strand in relation to its role in RNA synthesis (Syzbalski et al., 1966).

The detection and study of DNA and RNA hybrids (homologous and heterologous) was initiated and facilitated by the resolution offered by this technique. (Spiegelman et al., 1961). Care is required in practice, however, for in addition to the potential depolymerization of the ribonucleic acids by trace heavy metal contaminants in the salt, the strong salt solutions may also bring about the formation of spurious aggregates. Konrad and Stent (1964) have studied this aggregation problem and incline to the view that hybrids involving RNA are best detected in rate-zonal sucrose gradients or by the other methods now available for the detection of hybrids (see De Ley (this Series, Volume 5a) and Midgley (this Series, Volume 5a).

Meselson and Stahl's (1958) experiments utilizing this technique not only demonstrated the semi-conservative mechanism for the replication of DNA in bacteria but also gave the method an important kinetic approach, i.e. the so-called "density transfer experiment", using stable isotopes. This important variation opens for study the general biological problem of the transfer of material between parent and progeny, or simply the behaviour of cellular components following a labelling period. Their well known study of the replication of DNA in *E. coli* is a classic example of this approach and has been extended to the replication of DNA in higher forms. This technique has also been used to show that ribosomal RNA maintains its integrity for at least three generations since fully labelled r-RNA molecules persisted for this time whilst further unlabelled RNA molecules were synthesized. No intermediate density (i.e. partially labelled) RNA was found indicating that new ribosomal RNA does not derive from pre-existing strands (Davern and Meselson, 1960). A study on the behaviour of bacterial ribosomes during protein synthesis using this technique has suggested that ribosomes (70S) must dissociate into their sub-units between successive rounds of translation (Kaempfer *et al.*, 1968).

The iso-density equilibrium technique appears to be particularly useful for the preparation and study of viruses. These large particles give sharp bands and their infectivity is usually maintained and manifest when the salt is dialysed away, e.g. see the *in vitro* synthesis of infective $\Phi \times 174$ DNA by Goulian *et al.* (1967).

The iso-density equilibrium centrifugation is therefore of immense value in providing information of both a static (i.e. molecular weight, composition, heterogeneity) and a dynamic character. It has the previously noted advantage that it may be employed at the analytical level (with micro-gram amounts of material) or at the preparative level (in swing-out, angle or zonal rotors). In all cases it should be borne in mind that the strong salt solutions are corrosive to the common aluminium alloy rotors and parts. Centrepieces machined from synthetic materials (e.g. Kel F, or filled Epon) are preferable for analytical work and titanium rotors for preparative work. The high densities employed also require a reduction in the maximum operating speeds of the rotors. In large scale preparative work (e.g. with zonal rotors) the recovery of the caesium salt is economically worthwhile.

2. Practical aspects of the iso-density equilibrium technique

(a) *The calculation of molecular weights*. Although the theoretical background of the *iso*-density equilibrium method is complex (see above) the basic application of the technique to the determination of buoyant density is relatively simple and the most generally used in biological research (see Section IV.C.(d)). The calculation of molecular weights from the bandwidth requires a knowledge of the partial specific volume, polymer-solvent interaction coefficients and their detailed dependence upon solvent activity and pressure at various points within the centrifuge cell. Meselson *et al.* (1957) have derived the expression

$$\sigma^2 = \frac{RTd_0}{M_0[dd/dx]_{eff}\omega^2 x_0} \tag{49}$$

where the suffix, 0, refers to the position (x_0) and the buoyant density d_0 of the solvated species at the band centre, M_0 is the molecular weight of the solvated macromolecule at the band centre. σ is the standard deviation of the Gaussian concentration distribution of the macromolecule. $(dd/dx)_{eff}$ is the effective density gradient which governs the distribution of the macromolecules (i.e. the density gradient at the centre of the band). The effective density gradient is the summation of (i) the composition density gradient given by the distribution of the salt (ii) the effective compression density gradient due to the difference in compressibilities of the salt and solvated polymer and (iii) the solvated polymer density gradient. Casassa and Eisenberg (1960, 1961) and Eisenberg (1967) have derived alternative equations for the determination of M but these are hard to apply in practice since their evaluation depends upon a very high degree of accuracy in the data. Heterogeneity in molecular weight will result in the different Gaussian distributions centering about the same radial distance to give a distribution which is symmetrical but non-Gaussian. A plot of the logarithm of the concentration in a band against the square of the distance from the maximum will then provide a test for homogeneity. A straight line indicates homogeneity with respect to both M and density, a concave plot heterogeneity with respect to M and a convex plot heterogeneity with respect to density. In rare instances it is conceivable that a straight line (indicating homogeneity) could be obtained when heterogeneities in molecular weight and density are combined. Large errors in the value of M may arise with this method when density heterogeneity is small or continuous since a density distribution with a standard deviation of 0.003 g cm^{-3} will give rise to an underestimate of M by a factor of 2. A further disadvantage of the *iso*-density equilibrium estimation of M is that the value obtained is for the differentially solvated species and not the anhydrous, *iso*-ionic molecular weight. The concentration of the macromolecule and optical effects (e.g. the narrowing of the band width with increasing wavelength of light) may also contribute to the anomalous values of M frequently observed (Cummings, 1963).

The remainder of this Section will therefore be concerned with the practical aspects of the determination of buoyant densities and the separation and concentration of macromolecules and viruses for preparative work with this technique.

(b) *General.* The choice of the buoyant medium is dictated by the prime requirement that the macromolecule should be generally soluble in the solution. The salt chosen must also redistribute in the applied field and be sufficiently soluble to give solutions with a range of densities which will encompass the densities of the macromolecules. For the latter reason the salts which have found major use in biological work have been Cs_2SO_4 and CsCl. Tables I in the articles by Szybalski (1968a) and Vinograd and Hearst (1962) and the paper by Ludlum and Warner (1965) give useful data for the effective density ranges at 25°C and the corresponding refractive indices for solutions of these and other salts. In preparations known to contain materials of different buoyant densities the question of resolving power is important in the choice of salt. Ifft *et al.* (1961) have defined the resolution λ as

$$\lambda = \frac{\Delta x}{\sigma_1 + \sigma_2} \tag{50}$$

Where Δx is the distance between the maximum ordinates of the bands and σ_1 and σ_2 their standard deviations. λ was shown to depend upon the solvated molecular weights of the species present, the solution density and a parameter β where

$$\Delta x = \Delta p . \frac{\beta}{\omega^2 x} \tag{51}$$

and Δp is the difference in the buoyant densities of the macromolecules. β largely depends upon the properties of the salt solution and a suitable choice of salt will therefore improve the resolution. Solvents with the largest β values will offer the greatest resolution at a given density e.g. Table 2 in the article by Vinograd and Hearst shows that in the lower ranges of density KBr and RbCl solutions will resolve much better than CsCl despite their smaller gradients. Some useful β values are recorded in this table and other values may be calculated by following the procedure described by Ifft *et al* (1961). Since both Δx and σ are proportional to $(\omega^2 x)^{-1}$ then in a given salt gradient the resolution is independent of the angular acceleration. As the bands become narrower they approach each other. It may be shown that at values of $\lambda = 3$ total separation is achieved.

At similar rotor speeds (Cs_2SO_4 forms a gradient which is twice as steep as that with CsCl and it is therefore potentially useful for RNA preparations and DNA species with widely divergent buoyant densities. Unfortunately Cs_2SO_4 solutions precipitate single-stranded RNA or DNA but not double stranded RNA or DNA. By working at higher temperatures (40°–60°C) it is possible to solubilize CsCl to give solutions of suitable density for work with RNA. Unfortunately at these temperatures the degradation of RNA may be enhanced and practical problems, such as oil fogging of the lenses in the analytical system, also occur. Mixed salt systems (Cs_2SO_4 + CsCl) have been used to band single stranded RNA. CsCl is preferred for investigations of the % guanine plus cytosine content of DNA since, for DNAs without substituted/unusual bases, there is a linear relationship between the % G + C and the buoyant density (e.g., see Szybalski (1968a, b). Cs_2SO_4 may be used at lower concentrations than CsCl with DNA since DNA is more heavily hydrated in the former salt (i.e., density 1·4 gm/cm^3 c.f. 1·7 gm/cm^3). The formate and acetate salts of caesium also provide useful density ranges for biological materials (particularly RNA) but their strong solutions are viscous.

The density of the solutions used may be determined both before and after the experiment by conventional methods, e.g., using a refractometer or by pycnometry. The salts used for analytical work should be purified and free of any UV absorbing material. Optical grade caesium salts may be purchased from the usual chemical suppliers and procedures for purifying technical grade salts have been described (Szybalski, 1968b).

(c) *Analytical* iso-*density equilibrium analyses.* The absorption optical system is most commonly used and the attainment of equilibrium is recorded by photo-electric scanner or photographic registration. In the latter case the films have to be scanned in a recording densitometer, e.g., Joyce-Loebl double beam micro-densitometer or Beckman Instruments Analytrol with film scanning attachment. Accurate focussing and complete cleanliness of all components of the optical system is required since specks of dust, etc., on the optical components may give rise to spurious "bands" on the densitometer trace. To avoid errors arising from this source it is advisable to photograph the formation of the bands as well as the final equilibrium position.

Kel F or charcoal-filled Epon centrepieces are used. Centrepieces with 2° sectors require less material (0·4–0·5 ml) than the conventional 4° sector (0·8–1·0 ml) but the former require additional care in alignment in the rotor. Since a single run may take 20–40 h to attain equilibrium special rotors, cells and optical components have been devised for multi-cell operation to achieve efficient and economic use of the centrifuge. Procedures etc., for multi-cell operation are described in detail by Szybalski (1968b) and in the instrument maker's technical manual (Beckman Instruments Ltd., Glenrothes, Scotland).

In certain single cell operations a negative (1°) upper wedge window may be required to replace the standard cell window to compensate for the refraction of light out of the optical system by the strong salt solutions. The filling of the cell requires care to avoid depolymerization of the nucleic acids by shearing during this operation. It is preferable to assemble and tighten the cell in the usual manner and secure the centrepiece with the screw plug over the filling hole. The tightening ring and upper window and holder are then removed and the necessary amount of material to fill the cell is allowed to flow carefully into the exposed centrepiece chamber via a wide syringe needle. The upper window and holder are then replaced and the cell retightened with the screw ring.

For analytical runs with UV optics the amount of nucleic acid required is small (0.5–1.0 μg/band). This is usually added, in a suitable buffer, to a small volume of saturated solution of the gradient salt. The density of the final solution is then adjusted to the required density by the addition of salt solution or buffer as required. This step is most conveniently followed using a refractometer operating at controlled temperature. For DNA in CsCl gradients the density is adjusted to be in the range 1.69–1.75 gm/cm^3 and in Cs$_2$SO$_4$ gradients to 1.41–1.45 gm/cm^3. In all runs it is preferable to arrange the density to permit banding of the material at equilibrium close to the centre of the cell (see Section IV.C (d) re position of band and calibration of gradients). In general, in order to add the correct amount of macromolecular material to achieve a desired, detectable band concentration with absorption optics, Vinograd and Hearst (1962) have shown that

$$c_x = 0.40 \frac{L}{\sigma} c_0 \tag{52}$$

where c_x is the concentration at the band centre, c_0 its initial concentration in the cell of liquid column L. σ is the standard deviation and for work with absorption optics a value of 0.1–0.5 mm is convenient. For the well filled, analytical cell centrepiece L is usually about 1.2 cm, and so to achieve an optical density of 1.0 at the band centre an initial optical density of 0.04 would be required. Once again increasing the length of the liquid column increases the time required to attain equilibrium and will also influence the buoyant density because of pressure. Usually the cell is filled to 90% of its capacity.,

Since the time taken to reach equilibrium is inversely proportional to the angular velocity it is preferable to run at the maximum speed possible, bearing in mind the reductions which have to be made in the maximum permissible rotor speeds owing to the increased density of the cell/tube contents. The equation derived by Meselson

$$t = \frac{\sigma^2}{D} \left(\ln \frac{L}{\sigma} + 1 \cdot 26 \right) \tag{53}$$

gives a useful prediction of the time, t, required to attain equilibrium in a liquid column of L cm with a macromolecule of diffusion coefficient D. A comparison of the photographic or scanner records will show when equilibrium has been reached, the position of the band will not change between two intervals of time. The centrifugations are usually carried out at 25°C because most of the required data for the salt solutions have been recorded at this temperature.

At the conclusion of the run all cell parts and rotors should be carefully rinsed to ensure effective removal of all traces of salts since the strong salt solutions are extremely corrosive to the aluminium.

(d) *The analysis of the film densitometer or scanner tracings to determine the buoyant density.* It will be assumed throughout the following Section that if the photo-electric scanning system has been used to record the *iso*-density equilibrium position that a derivative scan has been obtained and the presentation of the experimental data will therefore be in the same form as the densitometer tracing of the film.

In order to determine the buoyant density a knowledge of the magnitude of the effective density gradient throughout the cell is required. For this a point of known density is needed. At present all data are recorded as buoyant densities on a composition density scale since the theoretical and practical treatments have generally neglected the effects of pressure. In the usual analysis of the experimental data the buoyant density is determined on the basis of either (i) the position of the band in the cell and the average density of the gradient or (ii) the relative buoyant density of the species compared with a density marker included within the same gradient (e.g., for DNA determinations DNA extracted from *E. coli* B or K12 has been widely used as a reference density equal to $1 \cdot 704$ gm/cm^3). According to Szybalski (1968b) the essential steps in the measurement of the tracings (see Fig. 15) for a semi-empirical method for calculating buoyant densities are as follows:

(i) determination of the magnification factor (M_f) involved in the preparation of the trace to be measured and the extent of distortion of rotor and/or the cell centre-piece due to stretching at speed. The total magnification factor arises from the magnification due to the optical system of the centrifuge and the enlargement due to the

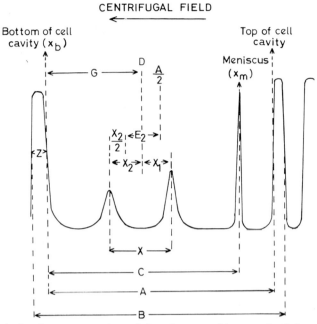

FIG. 15. A densitometer tracing of the photographic record of the equilibrium position in an *iso*-density sedimentation equilibrium centrifugation banding two DNA species in Cs_2SO_4 solution in the analytical cell. A = depth of the cell chamber; B = distance between the counterpoise reference edges; C = height of the liquid column ($= r_b - r_m$); X = distance between the two DNA bands; Z = distance between the bottom of the cell cavity (x_b) and the inner edge of the bottom reference hole. For the explanation of the other symbols see Section N.C(d).

scanning/densitometer tracing. These combined effects are determined by measuring the distance between the counterpoise reference edges (B) which appear on the trace. This value is divided by the real distance between these two points which is supplied in the manufacturer's specification of the rotor. The errors arising from rotor stretching are small (values of 0·02–0·04 cm stretching have been noted) but may be determined from the movement of the reference holes on changing speed by means of a double exposure at high and low speeds. The positions of the meniscus (x_m) and the cell bottom (x_b) from the axis of rotation at the speed of the run may then be corrected.

(ii) the measured position of the bands in the cell have then to be corrected for the combined distorting effects of their distance from the centre of rotation and the corresponding influence of hydrostatic pressure. Since the density gradient is a function of distance from the centre of rotation and steepens towards the bottom of the cell two species differing in buoyant density will be closer together if their *iso*-density points are nearer to the bottom of the cell than if these points were closer to the top. Furthermore macro-molecules banded at different depths are subjected to different hydrostatic pressures and the distance between the bands throughout the cell is not necessarily inversely proportional to the distance from the axis of rotation. The precise nature of this relationship between band separation and the position of the bands in a cell is determined experimentally. This may be done by minor adjustments in the density gradient so that two samples are caused to band at four or more different positions. From the densitometer/scanner records the distance E, (which may be positive or negative) between the mid-point of the cell chamber $(A/2)$ and the mid-point between the two bands $(X/2)$ may be determined for each of the runs. A correction factor, k, may then be determined where

$$k = \frac{X_0 - X}{X} \tag{54}$$

and X is the measured distance between the two bands and X_0 this distance when $E = 0$ (i.e. when the two bands are spaced symmetrically about the centre of the cell). These results for E vs. k (expressed in $(cm)/X(cm)$) may then be plotted for the gradient salt and speed in question. All future measured distances, X, in a similar system, may then be corrected to the true distance X_0, by measuring E, interpolating for k on the graph and substituting this value in equation (54) above.

A knowledge of the corrections and measurements outlined in (i) and (ii) above then permits the calculation of the absolute buoyant density or the relative buoyant density (to a marker) as follows:

(a) for the calculation of the actual density gradient and absolute buoyant density of a macromolecule two runs are performed, each at different starting densities, so that the molecule bands at two different positions in the cell. The starting densities of the mixed cell contents ,p_1 and p_2, are determined by pycnometry at 25°C. A suitable micro method for this using a piece of a serological pipette as a capillary pycnometer is described by Szybalksi (1968a).

A vertical line is drawn on each densitometer tracing (D in Fig. 15) in a position equal to the geometric mean between the distances from the axis of rotation to the bottom of the cell (x_b) and the cell liquid-air meniscus

(x_m) i.e.

$$D = \sqrt{\frac{x_m^2 + x_b^2}{2}} \qquad (55)$$

The line D is assumed to correspond to the "*iso*-concentration position" where the density in the cell corresponds to the measured density of the salt solution. The distance G, i.e. from D to the cell bottom, is measured, or may be calculated from the measurement C by use of the equation given by Szybalski:

$$G = x_b . M_f - Z - \sqrt{\frac{(x_b . M_f - Z)^2 + (x_b . M_f - Z - C)^2}{2}} \qquad (56)$$

For convenience a graph/table of G vs. C may be made, i.e., relating cell liquid column height to *iso*-concentration position. Then for the differently positioned bands from the two runs measure the distances corresponding to X_1 and X_2 on the traces, i.e., the distance from the centre of the band to the *iso*-concentration line D on each trace. The distances X_1 and X_2 are corrected by measuring the distance $X_2/2 + E_2$ (and correspondingly $X_1/2 + E_1$, both of which are equivalent to the E values described in Section (ii) on p. 121 i.e., the extent to which the band centre is removed from the cell chamber centre), determining the k factor, and then the corrected values X_1^0 and X_2^0. Since the same macromolecule of buoyant density, dp, was banded in two different positions in solutions of slightly different initial densities p_1 and p_2, then at equilibrium the actual density gradient, y, for the magnification M_f is given for p_1 by

$$dp - p_1 = yX_1^0 \qquad (57)$$

and p_2 by

$$dp - p_2 = yX_2^0 \qquad (58)$$

Hence

$$y - \frac{p_2 - p_1}{X_1^0 - X_2^0} \qquad (59)$$

Once y, the actual density gradient, is known then since X_1^0, p_1 and X_2^0 and p_2 are also known equations (57) and (58) may be solved for dp, the absolute buoyant density of the macromolecule.

(b) The calculation of the buoyant density of a macromolecule with reference to a known density marker in the same gradient follows a similar procedure to that described above.

The distance, X, between the band centres is measured, also the dimension A (Fig. 15). Then $(A/2 - X/2)$ gives the E value (see p. 121) and interpolation on the graph of E vs. k for the salt and speed in question gives the correction factor k. The measurement X may then be corrected to X_0 using equation (54), and knowing M_f, X_0/M_f will convert this value to the corrected value in actual cell distances. The increment in density, y, must be known for the salt, speed and magnification used in the experiment or determined as described above and corrected by the appropriate magnification factors involved to density increments in true cell distances. The marker buoyant density, d_k, is known and the unknown buoyant density, d_p, is given by

$$d_p = d_k - yX_0 \tag{60}$$

The sign of X_0 will depend upon the position of the unknown density band in relation to the standard.

In the particular case of the determination of the base composition of DNA *via* buoyant density determination in CsCl gradients including a reference density marker DNA, Sueoka (1961) has shown that the following relationship holds

$$dp = d_k + 4 \cdot 2\omega^2(x_p^2 - x_k^2) \cdot 10^{-10} \tag{61}$$

where x_p and x_k are radial distances from the axis of rotation of the band centres corresponding to unknown and marker DNA respectively. For DNA specimens with unsubstituted bases the mole fraction of guanine and cytosine (GC content) is then given by the linear equation describing the observations of Schildkraut *et al* (1962)

$$dp = 1 \cdot 66 + (0 \cdot 098 \text{ GC}) \tag{62}$$

Since polysaccharide materials are frequent contaminants of DNA and RNA preparations and will band in CsCl gradients in the region of DNA care should be taken in the assignment of bands. Additional identification, e.g., by specific enzyme digestion should be made whenever possible. Teichoic acid has also been reported (Young and Jackson, 1966) to band in CsCl gradients and, although the band is diffuse, in preparative experiments with a P^{32} marker for the identification of nucleic acids this polymer could give misleading results.

(e) Preparative iso-*density equilibrium centrifugation*

(i) *General.* A particular advantage of *iso*-density equilibrium centrifugation is that only the materials with densities encompassed by the gradient will be banded in the gradient. This is useful at the analytical level since impure preparations may be examined e.g. cell lysates for DNA buoyant density determinations. On the semi-analytical or preparative scale it enables the component of interest to be resolved from other materials and thereby purified and concentrated within a band. The preparative application of the technique has the additional advantage for biological work that the bands may be localized at the conclusion of the run by their biological activity or other suitable marker. Furthermore the object of interest is not packed as a pellet in the centrifuge tube, a procedure which frequently results in an alteration in biological or physical properties.

The choice of gradient material is governed by similar considerations to those outlined previously. Particular problems and effects may arise with the use of high salt concentrations, e.g., the inactivation and precipitation of certain viruses has been noted and also the removal of protein sub-units from unfixed ribosomes (Lerman et al., 1966).

The sample may be introduced into the centrifuge tube thoroughly mixed with the gradient material or layered onto the top of a solution of the salt. It is useful to set up a rotor with some tubes filled by the former and others by the latter method since at equilibrium the zones should occupy the same position in all tubes. On analysis an observation to this effect confirms that equilibrium has been reached. In certain instances the contents of the tube may be prepared by adding the sample in a suitable quantity of buffer to a weighed amount of the solid salt in the tube.

For straightforward preparative purposes i.e. to simply separate species on the basis of density a satisfactory and quicker separation may be obtained by preforming the gradient. Alternatively a mixture of particles may be layered over a solution which is iso-pycnic with the particle of interest and only contaminants iso-pycnic or of lower density than this particle will fail to be resolved (Bachrach et al., 1964). In either case the centrifugation need only be continued until the objective is achieved and this will be shorter than the time to attain equilibrium.

Semi-analytical and preparative iso-density equilibrium centrifugations may be performed in all types of preparative rotors i.e.

(a) swinging bucket rotors
(b) angle-head rotors
(c) titanium zonal rotors; and
(d) specially designed continuous action rotors.

(a) and (b) *Swinging bucket and angle head rotors.* The centrifugations are usually performed in cellulose nitrate ("lusteroid") tubes for ease of sampling at the conclusion of the run. These tubes are also transparent and allow the location of certain banded materials by light scattering, e.g., viruses. A thin layer of inert oil (e.g. F.C. 43 or Dow Corning No. 555 Silicone oil, Beckman Instruments, Glenrothes, Scotland) may be placed at the bottom of the tube to ensure cylindrical symmetry. For the higher density ranges, e.g., when caesium salts are used, the tubes are only slightly over half-filled and the remaining volume in the tube is filled with paraffin oil to prevent collapse of the tube during centrifugation at high speed. Equilibrium is also reached faster with shorter columns and the use of long liquid columns is only indicated when a wider density range is desired, e.g., in preliminary experiments where there is greater uncertainty in the estimate of the density of the particle.

In connection with loading capacity and resolution the studies of Fisher *et al.*, (1964) and Flamm *et al.*, (1966) have shown that a substantial difference exists between swinging bucket rotors and angle head rotors for *iso*-density centrifugations. In the first instance there is the obvious point that the average angle-head rotor will accommodate many more tubes (2–4×) than the swinging bucket rotors and therefore more material may be prepared/analysed during the course of a single, prolonged centrifugation. In addition to this Flamm *et al.*, (1966) were able to show that the upper limit of loading for a 3·0 ml gradient in a 5 ml swinging bucket tube to give an acceptable resolution of two DNA bands was less than 45 μg DNA. In a 10·0 ml tube for an angle head rotor with a similar internal diamater (1·6 cm vs. 1·3 cm) and a 4·5 ml gradient giving a vertical liquid height of 2·6 cm (cf. 2·8 cm in the swinging bucket tube) superior resolution was obtained at 10× this loading (315 μg total DNA/tube). The reasons for this profound difference in resolution and loading emerge from a consideration of the geometry of the two rotor types and the fluid translations accompanying centrifugation, deceleration and positioning for sampling. With the exception of the brief period of translation from the horizontal (at speed) to the vertical (at rest and sampling) position the liquid

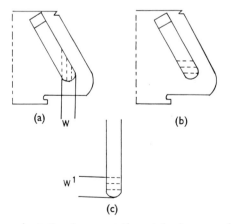

FIG. 16. Liquid re-orientation in a centrifuge tube in an angle rotor (a) at speed (b) at rest and (c) removed for sampling.

contents of a tube in a swinging bucket rotor do not experience profound fluid translation. On the other hand in an angle head rotor the tube contents are inclined as shown in Fig. 16 (a) when at speed, i.e. with the bands parallel to the axis of rotation. A smooth translation takes place on deceleration to reorient the liquid bands (Fig. 16 (b)) finally to the positions shown in Fig. 16 (c) for sampling. In the particular case of a 4·5 ml gradient in a Spinco 40, angle head rotor Flamm *et al.*, (1966) showed that the gradient effectively expands from a 1·6 cm width (W) at speed to a 2·6 cm height (W^1) at sampling. This reorientation of the liquid column has important practical consequences. In the angle head rotor density is a non-linear function of tube volume (owing to a non-linear volume vs. radius relationship) whereas it is linear for swing-out rotors. Both types of rotor generate density gradients which are linear with radial distance. Furthermore the expansion of the density gradient on resuming the vertical position coupled with the non-linear density vs. volume profile effectively produces considerably shallower (by a factor of 10) density gradients in angle-head rotors and this materially influences the degree of resolution. Extending this observation the resolution should be further improved by choosing rotors in which the angle of inclination of the tubes to the axis of rotation is as shallow as possible. Therefore for *iso*-density equilibrium studies angle head rotors with narrow angles of tube inclination to the rotor axis are to be preferred for their vastly increased capacity and resolution. Convective transport of material also takes place during centrifugation with angle rotors and this will enhance the rate of attainment of the equilibrium.

The sampling of the tubes at the conclusion of the run is most readily done using a device which involves pricking the bottom of the tube and direct scanning of the effluent in a flow-through cell. Equal drop fractions may then be taken for further analysis (see Section III.B.5). The density of each fraction may be determined using a refractometer to give the character of the gradient and the position (i.e., buoyant density) of the macromolecules. For additional analyses, e.g., on pooled fractions from the gradient it will probably be necessary to remove the gradient material by dialysis or to lower its concentration by extensive dilution. Both of these procedures will bring about profound changes in the osmolarity and this in turn may affect the properties of the banded material.

(c) *Titanium zonal rotors.* The mode of operation and sampling from these rotors has been described previously (p. 89). The operating range (5000–60,000 rpm) of the B-XIV and B-XV series of zonal rotors coupled with their great capacity make these rotors particularly suited to the isopycnic banding of small cell particulates and molecular components. It is preferable whenever using these larger rotors to preform the gradient to a large measure by pumping in the gradient salt and sample together. Whenever strong salt solutions are employed the rotors and parts should be machined from titanium.

(d) *Continuous flow rotors with* iso-*pycnic banding*. Anderson (1966) and his colleagues have designed a series of rotors (designated B–VIII, B–IX, B–XVL, K–I and K–II) which combine *iso*-pycnic banding with continuous flow of sample. In this way pelleting of sedimentable material (as is usual in bulk, continuous flow clarification) is avoided, a point which may be particularly important with certain materials, e.g., bulk preparation of viruses or bacteriophage. These rotors are basically zonal rotors in which the gradient is held in a vertical position in a tall cylindrical rotor by the centrifugal force. A tapered core, with septae to ensure radial flow, defines a zone in which the sample fluid flows through the rotor over the density gradient. The particles then sediment out of the stream and band in the gradient. Certain of these rotors are available commercially but for further details the reader should consult the monograph by Anderson (1966).

V. ISOLATION AND CHARACTERIZATION OF SUB-CELLULAR COMPONENTS FROM BACTERIA

A. Introduction

The techniques for fractionation, identification and analysis by centrifugal means which are described in Sections I–IV above may be illustrated by the methods for the particular cell components of microbiological origin described in this Section.

With microbiological material, homogeneity of cell type can usually be assured and the procedures adopted for the isolation of a particular component from a cell homogenate should conform to the general principles established for tissue fractionation during the studies of mammalian tissue homogenates (Hogeboom *et al.*, 1957). Thus balanced frequency distribution curves are required to ensure a truly representative yield of a fraction identified by a biochemical marker and to detect "latency" of certain activities sequestered within organelles. Since our direct knowledge of bacterial anatomy is limited, the best approach to the centrifugal fractionation of bacterial homogenates is a purely analytical one, i.e., without reference to preconceived ideas of the cytological composition but simply based upon the identification of biochemical properties as functions of a physical parameter such as size or density. This is ideal since it fosters the wish to find out rather than the will to believe which is an ever present hazard when fractionating cell homogenates in the light of our present knowledge of the cytological structure of larger cells. Cytological identification may follow when a fraction of defined physical characteristics has been isolated and the ultimate goal of the correlation of a biochemical character with a sub-cellular component may then be achieved. The artefacts likely to arise at the biochemical and cytological level during cell homogenization are well recognized (e.g., see Hughes *et al.*, this Volume p. 1; De Duve (1964) and will not be discussed here. However genuine differences in structure and function may also arise from the previous growth history of the popula- tion for bacteria are undoubtedly "the most plastic of living material" (Herbert, 1961; Stephenson, 1949). Environments permitting rapid growth, yield cells with elevated RNA contents and high, functional ribosome popu- lations. Control over the growth conditions and the disruption procedure is therefore essential in order to give reproducible results.

For an anlytical approach to microbial cell sub-structure, following cell harvesting and controlled disruption (see Hughes *et al.*, this volume p. 1 rate-zonal density gradient centrifugations will provide the maximum of information. The analysis of the gradients at the conclusion of the run will allow the construction of graphs recording the distribution of a variety of chosen biochemical parameters vs. the distance moved (roughly proportional to *s*-rate) of the various sub-cellular components. A comparison of the distri- bution of an unlocalized constituent against the profile of a known consti- tuent under a variety of experimental conditions (i.e., disruption methods, suspending media, etc.) will permit some prediction of the cellular locale. This information, together with the estimate of sedimentation coefficient, will provide a basis for charting future procedures for the isolation of crude, concentrated preparations of the constituent on a preparative scale, e.g., by differential or large scale rate-zonal centrifugations. A knowledge of the approximate composition of the fraction then opens the possibility of using either *iso*-density or rate-zonal gradient techniques for the further purification of the concentrated, crude preparations. Finally the purified fraction may be checked for homogeneity by analytical centrifugation techniques (velocity, equilibrium or *iso*-density equilibrium) and accurate information regarding *s*-rate, molecular weight, diffusion coefficient, etc., may be generated from a minimum of the pure material. The often used designation of a cell fraction as a "2 h $\times$ 100,000 g pellet" etc., is clearly meaningless in investigations of this type. The precise physical and bio- chemical characteristics of a fraction should be defined and accompanied whenever appropriate by a morphological description.

B. Ribosomes and polyribosomes

Fortunately the endoplasmic reticulum which complicates the isolation of ribosomes from mammalian cells does not appear to be highly developed in the majority of micro-organisms. Bacterial ribosomes are predominantly free cytoplasmic elements and are therefore released directly into the suspending fluid when the cell wall is ruptured. The majority of the ribosomes remain in suspension when the unbroken cells, broken cell walls and protoplast membranes are removed by a brief period of low speed centrifugation. A small proportion of the ribosomes remain with the cell membranes and claims have been made that these ribosomes are particularly active in protein synthesis (Hendler, 1965; Schlessinger, 1963). Subsequent observations do not lend support to these claims. For example, in a strain of *Bacillus amyloliquefaciens* capable of extensive exo-cellular enzyme production 95% of the ribosomes were found to be free and only 5% remained firmly bound to the membranes after repeated washing with a suitable suspending medium (Coleman, 1969). The concentration of K^+ ions in the extracting fluid appears to influence the extent of binding but at concentrations which are optimal for subsequent amino-acid incorporation studies not more than 8% of the total ribosomal material is membrane bound (Coleman, 1969). No difference was found between the free or membrane bound ribosome/polyribosome rate-zonal density gradient profiles in the same suspending medium. In cases where bacteria have an extensive intra-cytoplasmic membrane structure (e.g., certain photosynthetic bacteria, see Section V.C.4) the bulk of the ribosomes are still free within the cytoplasm and have been used as an indicator for the release of cell contents (Sykes *et al.*, 1965). Thus, in general the isolation of bacterial ribosomes is not complicated by the procedures which are required to remove the residual membrane fragments from mammalian microsomes, e.g., lipase digestion or treatment with surfactants. Their sedimentation behaviour is correspondingly more predictable since variable size clusters of ribosomes bound to membrane fragments are not found. For the purposes of the remainder of this Section the small ribosome population which remains with the cell membrane fraction will be ignored.

The original observations of Schachman, Pardee and Stanier (1952) indicated a certain uniformity in the ribosomal species present in cell-free extracts from bacteria, i.e., the optically clear supernatant after the removal of the unbroken cells, broken cell walls and membranes by a brief low speed centrifugation. These workers observed that there was a basic pattern which was common to the total macromolecular organization of cell-free

FIG. 17. Schlieren diagrams from analytical ultracentrifuge runs with cell-free extracts from (a) *Rhodopseudomonas spheroides* grown in the dark and (b) *E. coli B*. Sedimentation is from left to right and the sedimentation coefficients of the boundaries are (a) 5·9, 23·1, 27·7 and 43·1S and (b) 5·7, 11·0, 15·9, 25·7 and 42·5S. The material sedimenting as a "spike" at 11·0S in diagram (b) is DNA and is not present in (a) since the *Rh. spheroides* extract was treated with desoxyribonuclease.

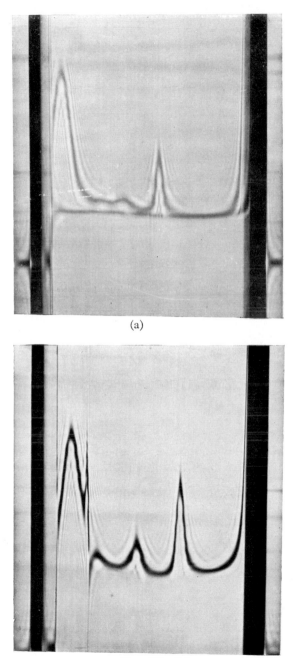

(a)

FIG. 17   (b)

extracts from all bacteria (compare Fig. 17(a) and (b)). In the analytical ultracentrifuge sedimentation velocity profiles the fastest sedimenting species (at 43·1 S, 27·7 S and 23·1 S in Fig. 17 (a) and 42·5, 25·7 S in Fig. 17 (b)) are the ribosomal species and the large, slowly sedimenting boundary comprises the bulk of the soluble cellular protein, the t-RNA etc. The cellular DNA is usually depolymerized in the preparation of the extract and appears in the profiles as a sharp "spike" sedimenting before the soluble protein boundary. Rate-zonal density gradient profiles show a similar distribution of species. The subsequent extensive literature on ribosomes is replete with a variety of ribosome species frequently identified on the slender basis of a sedimentation coefficient determined in undefined conditions (see Petermann (1964) for an excellent summary of this data). It is now reasonable to conclude (e.g., see De Ley, 1964) that all normally growing bacteria and actinomycetes will contain two or more ribosomal species from the following range with *sedimentation constants* ($S^0{}_{20, w}$) of $29·5 \pm 2$, $37·5 \pm 2·8$, $56·3 \pm 1·9$ and $76·7 \pm 2·5$. These are sometimes referred to as the ribosome sub-units of 20S, 30S, 50S and 70S. They are considered to be sub-units since 30 and 50S will combine in the presence of Mg^{2+} to give 70S and 70S units will aggregate in the presence of m-RNA to form polysomes (see later). This common basic pattern of sedimentation coefficients for bacterial ribosomes clearly precludes the use of ribosomal sedimentation coefficients for species classification. However the sub-unit sedimentation constants for bacterial ribosomes (30S, 50S and 70S) distinguish them from the corresponding species in yeasts, higher animals and plants which have values of 40, 60 and 80S. Whenever the growth conditions are restricted or sharply altered various ribosomal precursors or unusual ribonucleoprotein particles may be observed which are not normally present or detectable in cell-free extracts (see later in this Section).

The ribosome content and activity of ribosomes in protein synthesis are functions of growth rate and the latter is a well-known function of the composition of the growth medium (Sykes, 1966, 1968). A precise control over the growth conditions and the point of harvest is therefore desirable. In this connection a continuous culture device such as the chemostat (See Evans *et al.* and Tempest, this Series, Volume 2) has particular advantages since the growth rate and conditions may be accurately controlled and indefinitely maintained by a predetermined growth-limiting concentration of an essential nutrient. The growth rate of the culture may then be varied at will and the influence of this variable and that of the growth limiting nutrient upon the structure, function and amounts of bacterial ribosomes may be determined (Sykes and Tempest, 1965; Sykes and Young, 1968; Tempest and Hunter, 1965; Tempest *et al.*, 1965; Young and Sykes. 1968).

In a bacterial cell-free extract the profile of normal ribosomal sub-units as described above is also dependent upon the method of cell disruption and the composition of the suspending fluid. The former may give rise to both qualitative and quantitive differences in the yield of ribosomes e.g. certain ribosomes, and particularly their precursors, are sensitive to sonic disintegration of the cells but survive cell disruption by Hughes press crushing (see De Ley, 1964 and Dagley and Sykes, 1960). Polysomes are also particularly sensitive to the mechanical methods for breaking cells since many of these involve high shearing forces. The preferred method of disintegration for the preservation of polyribosomes is via lysozyme digestion of the cell wall for Gram-positive bacteria and lysozyme and ethylene diamine tetra-acetic acid (EDTA) treatment for Gram-negative bacteria. Osmotic stabilization of the protoplast (or spheroplast) is not essential and lysis may then take place directly in the buffer of choice. The requirement for the addition of EDTA to ensure spheroplast formation with Gram-negative strains is unfortunate since EDTA is a strong chelating agent and will bring about the complete dissociation of ribosomes. Care should be exercised when using this reagent to keep the concentration as low as possible and to remove residual traces before lysing the spheroplasts. An alternative method of gently breaking cells to preserve polyribosome structures is to grow "fragile cells" e.g., by growing suitably sensitive cells in the presence of penicillin and/or $0.5M$ Na_2SO_4. The cells, presumably with defective cell walls, may then be harvested and suspended in a medium containing 0.5% w/v desoxycholate. Using this method Mangiarotti and Schlessinger (1966) found that 65% of the ribosomal material sedimented as heterodisperse polyribosome species. Free 70S species were not found in this study. Polyribosomes may be transients (see later) and speed is therefore required in the harvesting and gentle disruption of the cells to maximize their yield. Coleman (1969) rapidly centrifuges the cells (*B. amyloliquefaciens*) from the growth medium and immediately freezes the pellets in liquid nitrogen. The cell pellets are then thawed in a suitable buffer saturated with nitrogen and lysed with lysozyme under anaerobic conditions. For Gram-negative bacteria, [*E. coli* strains were used], Godson and Sinsheimer (1967) chilled the growing cells rapidly to 2–4°C by swirling the culture in a flask in an alcohol–CO_2 bath. The cells were rapidly centrifuged from the culture and their cell walls digested by lysozyme–EDTA treatment at pH 8.1 in the presence of 25% (w/v) sucrose. The stabilised spheroplasts were then lysed by diluting with a lysing medium to produce a final concentration of 10% sucrose, 10 mM $MgSO_4$, 0.2% desoxycholate and 0.5% Brij-58 (a neutral detergent, Honeywell and Stein Ltd, Mill Lane, Carshalton, Surrey). Speed and low temperature also minimize endogenous ribonuclease action. This is desirable since both polyribosomes and ribosomes are degraded by this enzyme. With polyribosomes the action of the enzyme is dramatic for as the m-RNA thread binding the 70S units is degraded the rapidly sedimenting polysome peaks in rate-zonal gradients are lost and 70S particles accumulate. This is one of the few diagnostic tests for polysomes which distinguishes them from spurious convection effects or non-specific, rapidly sedimenting aggregates with which they can be readily confused. Failure to control ribonuclease activity in the cell homogenate will yield a cell-free extract without polysomes. The primary effect of ribonuclease upon ribosomes is not so readily detected. The ribosomal RNA component in ribosomes is frequently degraded but this does not give an immediate alteration in their characteristic sedimentation properties (Shakulov *et al.*, 1962). Furthermore their cell-free biological activity does not appear to be impaired since, for example, the ribosomes prepared for studies on the coding problem by the method of Nirenberg and Matthaei (1961) have a degraded RNA component. However future investigations on *in vitro* protein synthesis may reveal that this degradation of the RNA may be one of the factors responsible for the extremely low level of *in vitro* activity compared with the *in vivo* level. Extensive ribonuclease digestion of ribosomes leads to the precipitation of the ribosomal protein since it is very insoluble and aggregates in the absence of the RNA. Ribonuclease inhibitors such as macaloid (Stanley and Bock, 1965), bentonite (Petermann and Pavolec, 1963) or polyvinyl sulphate (General Biochemicals, Chagrin Falls, Ohio) should therefore be added to homogenates during the preparation of ribosomes, although some ribosomes may be precipitated by the alumino-silicate earths. Alternatively the ribonuclease-free strains of bacteria, e.g., *Escherichia coli* MRE 600 (Cammack and Wade, 1965) are useful starting points. Since desoxyribonuclease is frequently added to bacterial homogenates to lower their viscosity for centrifugation, care should be taken to ensure that the enzyme used is ribonuclease-free.

The composition of the suspending fluid has a profound effect upon the quantitative and qualitative distribution of ribosomal species in crude cell-free extracts. The early investigations upon bacterial ribosomes quickly established the importance of Mg^{2+} ions for the stability of ribosomes (Bowen *et al.*, 1959; Roberts, 1964; Tissieres *et al.*, 1959) and the dissociation produced by solutions of high ionic strength or chelating agents. In low ionic strength buffers (e.g., 0·005M tris) increasing the Mg^{2+} ion concentration was shown to bring about the reversible association of the smaller ribosome units in a 1 : 1 ratio i.e.

$$1 \times 30S + 1 \times 50S \rightleftharpoons 1 \times 70S$$

It has frequently been assumed that the 30 and 50S ribosomes invariably occur in bacterial cells in this combining ratio but this is not supported by the majority of the experimental evidence (Sykes and Young, 1968; Young and Sykes, 1968). If the Mg^{2+} ion concentration is raised above the level to produce 70S units then further aggregates are produced, e.g., 100S, which appears to be a dimer of 70S, and 85S which may be a dimer of 50S. These aggregates do not appear to have biological significance and therefore the level of Mg^{2+} producing 70S units would seem to be the *maximum* concentration for this ion in any conditions. Subsequent observations have shown that the level of Mg^{2+} ions in the suspending medium is not the only factor determining the stability and distribution of the ribosome units. Studies involving the banding of ribosomes in *iso*-pycnic CsCl gradients (Meselson *et al.*, 1964) and the salt gradient elution of ribosomes from DEAE-cellulose columns (Furano, 1966) revealed that high mono-valent cation concentrations brought about the removal of protein from ribosomes. The *in vitro* studies of Cammack and Wade (1965) using ribosomes from a ribonuclease-free strain of *E. coli* clearly demonstrated the important contribution of the monovalent : divalent cation ratio to ribosome stability. At a constant ionic strength (I = 0·16) and pH the percentage distributions of the ribosome sub-units in $NaCl$–$MgCl_2$ solutions are shown in Table I. Felsenfeld (1962) has expressed the view that the

TABLE I

The proportions of the different classes of *E. coli* ribosomes present with varying ratios of Na^+ : Mg^{2+} in a suspending medium of constant pH and Ionic strength[*]

Na^+/Mg^{2+} ratio	"30S"	"50S"	"70–100S"
637	Extensive degradation		
13	26%	74%	Trace
3·4	8%	30%	62%
0·2	3%	10%	87%

[*] Abbreviated from Cammack and Wade (1965).

stabilization of ordered polynucleotide structures requires that much of the phosphate backbone charge be neutralized or screened and that the interaction of divalent ions with polynucleotides is a strong interaction characteristic of site binding whereas that with monovalent ions is a weak, non-specific interaction. Goldberg (1966) found that *E. coli* ribosomes preferentially bound Mg^{2+} (to their RNA component) and that increasing the K^+ ion concentration decreased the Mg^{2+} binding by competing for the anionic binding sites. Most of the phosphate groups in the RNA within ribosomes appear to be available for Mg^{2+} binding since removal of the protein only increases the number of sites by 12%. These observations are pertinent to a further important phenomenon which was initially observed by Cammack and Wade (1965) at a Na^+/Mg^{2+} ratio of 29 i.e. a small but perceptible lowering of the sedimentation coefficients of the 30 and 50S species. The subsequent studies of Gesteland (1966) have illuminated the primary events in this transition and those of Lerman *et al* (1966) and Gavrilova *et al*. (1966) the later steps. Gesteland (1966) confirmed the earlier observation of Spirin *et al*. (1963) that the effective lowering of the Mg^{2+} ion concentration in a suspension of *E. coli* 50 and 30S ribosomes converts them into more asymmetric particles, i.e., they have the same overall composition and intact r-RNA but lowered sedimentation coefficients. This phenomenon was interpreted as an unfolding of the ribosome units and was found to be reversible with respect to sedimentation characteristics and biological activity when the Mg^{2+} was slowly dialysed back. Further careful lowering of the Mg^{2+} level (e.g., by raising the monovalent : divalent ion ratio) produces a progressive loss of discrete protein sub-units from these ribosomes. The 50 and 30S particles each give rise to a family of protein-deficient ribosomes (referred to as "core-particles") and ultimately to their respective, protein-free r-RNA units (23S and 16S). For example, for *E. coli* ribosomes Spirin (1968) has described the following disassembly sequences:

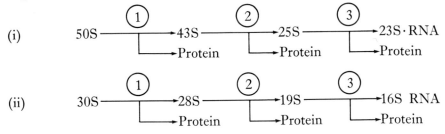

(i)   50S ⎯①⎯→ 43S ⎯②⎯→ 25S ⎯③⎯→ 23S·RNA
                                →Protein   →Protein   →Protein

(ii)   30S ⎯①⎯→ 28S ⎯②⎯→ 19S ⎯③⎯→ 16S RNA
                                →Protein   →Protein   →Protein

At stages 1, 2 and 3 approximately 20%, 30% and 50% respectively of the original ribosomal protein is removed in co-operative groups, i.e., the process is an all or none one in which, for example, a 30S particle will abruptly change to a 28S unit. Once again (cf. unfolding) this process appears to be reversible with respect to both sedimentation coefficient and biological activity. These observations on the dissociation of sub-units and the unfolding and disassembly of the ribosomes are of immense practical importance. In the first place this knowledge provides an experimental tool for the investigation of the detailed role of the ribosome in the activity of the cell. Stage 1 in the above transitions results in the loss of certain characteristic ribosome functions (Nomura and Traub, 1967; Raskas and Staehelin, 1967). The disassembly process also provides an *in vitro* model

for the study of the biogenesis of ribosomes and the specificity of interactions between ribosomal RNA species and the variety of ribosomal
proteins. (Nomura et al., 1968; Traub and Nomura, 1968; Traub et al.,
1968). These observations may also provide a partial explanation for the
wide range of sedimentation constants which have been reported. The degree
of unfolding of the ribosome may also determine the sensitivity of the
ribosome to ribonuclease action. For the precise definition of a ribosome
it is clearly necessary to specify both its sedimentation coefficient and its
chemical composition in the given medium. The s-rate alone is not a
sensitive measure of the minor chemical differences and the convenient
assumption that a ribosome population is homogeneous and static is
dubious in the light of the known flux in the protein components with the
ionic environment. Although Na^+ (and Cs^+) ions have been employed to
bring about the in vitro effects described above, in vivo K^+ ions are probably
the main functional monovalent cations involved in maintaining ribosome
and polyribosome structure and activity. Tempest et al (1966) have shown
that a precise cellular balance of Mg^{2+} : K^+ : PO_4^{3-} of 1 : 4 : 8 is maintained at all growth rates by Aerobacter aerogenes growing in defined, continuous culture. The bulk of the magnesium in the cell is associated with the
ribosomes (Tempest and Strange, 1966). At the reported optimal ionic
conditions for sustaining in vitro protein synthesis with E. coli ribosomes
(10^{-2}M Mg^{2+} and 10^{-1}M K^+) Goldberg (1966) found that the ratio of
bound magnesium to phosphate was 0·5. This is above the ratio required
(0·4) to stabilize 70S ribosomes. Coleman (1969) has made a similar study
in respect of polysome preparations from B. amyloliquefaciens and here
maximum in vitro incorporation of ^{14}C-amino-acids again occurred at a
similar ratio of K^+ : Mg^{2+}, i.e., 10 : 1.

It will be apparent from the preceding discussion of the effects of the
ionic environment upon ribosomes that no universal prescription is possible
for a medium in which to isolate bacterial ribosomes. An appropriate
balance of Mg^{2+} and K^+ ions to stabilize the structure and activity of
ribosomes from one strain of a bacterium will not necessarily apply for
another strain and almost certainly not for another species of bacterium.
For example, the 70S ribosome from many E. coli strains is stable in
0·01M Mg^{2+} and 0·1M KCl whereas the 70S particles from Halobacterium
cutirubrum (an extreme halophile) require 4M KCl and 0·4M Mg^{2+} and
dissociate when the KCl is lowered to 1–2M or the Mg^{2+} to 0·01M (Bayley,
1966). Whilst these may be extreme cases they serve to illustrate the practical importance of preliminary studies to determine the ionic conditions for
the stability of the ribosomes of the bacterial species in question. This
problem will be inter-related with the requirement to maintain the biological activity of the particles. The latter criterion may be a little suspect since

at the present time the rate (and extent) of *in vitro* amino-acid incorporation into protein catalysed by ribosomes is less than 1% of the *in vivo* rate of protein synthesis. It is not clear whether the low activity of the *in vitro* system is due to the ribosomes and their preparation, the other components of this complex system or the accumulation of products in the system. In general a dilute tris-HCl buffer ($0.5-1 \times 10^{-3}$M tris, pH 7-8) forms the basis of the medium for many bacterial ribosome isolations reported in the literature and the main contributions to the final ionic strength are the concentrations of magnesium and potassium salts which are required to stabilize the ribosomes. Finally, in the present context, it should be noted that other cations have been implicated in the stabilization of microbial ribosomes. Chao (1957) has described the involvement of Ca^{2+} ions in maintaining the 80S ribosome structure from *Saccharomyces cerevisiae*. In the writer's experience Ca^{2+} and other divalent metal cations of similar ionic radius to magnesium will not replace Mg^{2+} in the *in vivo* reconstitution of dissociated 50S ribosomes from *E. coli* starved for Mg^{2+}. The polyamines putrescine, spermine and spermidine will bring about the association of ribosomal sub-units in the absence of Mg^{2+} or in the presence of Mg^{2+} levels otherwise insufficient for the stabilization of some 70S species (Cohen and Lichtenstein, 1960; Moller and Kim, 1965; Norton *et al.*, 1968). In the case of *E. coli* 15% of the total cellular polyamine content is associated with the ribosomes and certain strains of Pseudomonads are rich in these amines. It has been suggested that these amines may partially replace the role of Mg^{2+} since, as observed above, the $Mg^{2+} : PO_4^{3-}$ ratio is only 0.5. However the ribosomal associations produced with putrescine added to *Pseudomonas* sp. cell extracts are reported to be less effective in promoting amino-acid incorporation than the corresponding associations stabilized by Mg^{2+} (Moller and Kim, 1965). Furthermore although these polyamines compete with Mg^{2+} and K^+, their amino groups (and those of the ribosome protein) may be reacted with dinitrofluorobenzene without affecting their role in 70S formation. The precise role of these amines is therefore in doubt and in general they are not added to media for the isolation of ribosomes.

Before embarking upon the isolation of ribosomes from a bacterial cell-free extract it is useful to make an estimate of the s-rates of the ribosomal species which are present, e.g., by a sedimentation velocity run in the analytical ultracentrifuge (see Fig. 17) or by a series of rate-zonal density gradient runs for different times. With this information, and a knowledge of the preparative rotor characteristics, procedures may be devised for the preparative scale centrifugal isolation of the ribosomes by repeated cycles of differential centrifugation or rate-zonal density gradient centrifugations in zonal or swing-out rotors. The s-rates for the bacterial ribosomes (e.g., 30, 50 and 70S) are sufficiently different to justify the extensive use made by investigators of repeated cycles of differential centrifugation for their bulk isolation. Here the gravitational fields and run times employed are usually calculated to bring about the complete sedimentation of the fastest sedimenting species. This pellet of material then requires gentle homogenization with fresh suspending medium to redisperse it and the centrifugation may then be repeated with the concentrated suspension. The various supernatant fluids will then serve as a source of the progressively slower sedimenting species which may then be similarly isolated by increasing the gravitational field applied and/or the period of centrifugation. Ribosome pellets are usually difficult to disperse and material which does not resuspend on homogenization or thorough irrigation with the medium is usually removed by a low-speed centrifugation. A differential centrifugation procedure of this type (e.g., see Tissières et al (1959) for a widely adopted method) will not yield a completely uncontaminated preparation of one species from another and when this is required it is advisable to subject the final preparation to rate-zonal sucrose density gradient centrifugation. For the gradient the sucrose should be dissolved in the medium required to stabilize the ribosomal species in question. The use of rate-zonal density gradient techniques throughout the isolation of ribosomes is clearly preferable and the zonal rotors have the loading and gravitational capacity for the majority of the initial bulk isolations which are undertaken. The smaller capacity swing-out rotors may then be used to re-run small samples of the product to ensure uncontaminated ribosomal species for further investigations. The ribosomes are recovered from the pooled gradient fractions by dialysis against a suitable buffer (to remove the sucrose) and, if required, concentrated by resedimentation from the buffer. In principle iso-pycnic equilibrium methods could be used to isolate ribosomes. In practice their use is extremely limited since (i) the overall compositions of the normal ribosomal sub-units are very close (Tissières et al., 1959) and (ii) the high concentrations of Cs or Rb salts required to give suitable density ranges for ribosome work distort the monovalent : divalent cation ratio and give rise to disassembly of the ribosome (Meselson et al., 1964). Iso-density gradient centrifugations with high salt concentrations are therefore most commonly used to prepare "core particles" and their "split" proteins from ribosomes, (Traub and Nomura, 1968), although differential centrifugation in high salt solutions will also suffice for this purpose (Itoh et al., 1968). If the Mg^{2+} ion concentration in the high salt gradients is raised then the ribosomes may be partially protected against this disassembly but two bands were still found by Brenner et al. (1961); the denser band containing the protein deficient "cores" of 30 and 50S ribosomes and the lighter band the remaining intact ribosomes (Meselson et al., 1964). Fixation of the ribosomes with formaldehyde will prevent this disassembly in high salt gradients (Guthrie and Nomura, 1968).

Centrifugally packed pellets of bacterial ribosomes are frequently glassy in appearance, difficult to disperse and brownish in colour. The colour may be due to "impurities" e.g., proteins which have been adventitiously bound. A clear, readily dispersed ribosome pellet may be obtained by resuspending and recentrifuging the ribosomes from buffers of higher pH and electrolyte concentration or buffers containing up to 0·6M NH4 Cl. The latter procedure has been shown to give ribosome preparations which are free from "contamination" by G.T.P. (9-β-D-ribofuranosyl guanine 5'triphosphate), t-RNA and cell supernatant proteins. Such preparations of ribosomes will still catalyze the specific binding of t-RNA species in the presence of a synthetic m-RNA (Kurland, 1966) but certain binding properties are frequently retained by ribosomes which can be shown to be deficient in certain protein components (Nomura and Traub, 1967).

For many investigations, e.g., the construction of *in vitro* amino-acid incorporation systems, mixed, concentrated preparations of ribosomes or polysomes will often suffice. In other investigations crude cell-free extracts may be examined directly. The studies on the biosynthesis of ribosomes by specific pulse-labelling techniques (Roberts, 1964) their behaviour in step-up and step-down transitions, the action of drugs, e.g., chloramphenicol, puromycin (Dagley *et al.*, 1962; Hosokawa and Nomura 1965; Nomura and Watson, 1959; and Sells, 1964) and the influence of the growth environment (Dagley and Sykes, 1958; Sykes and Tempest, 1965; Sykes and Young, 1968, and Young and Sykes, 1968) have all been studied, in the first instance, by analytical or rate-zonal sedimentation analysis of the crude cell-free extract. For more detailed studies, including the study of the ribosome itself "pure" preparations of the individual species are required. Ribosomes are macromolecular aggregates and therefore may never be considered as pure in the strict chemical sense. Furthermore, although this Section has been based upon the assumption that ribosomes may be prepared homogeneous with respect to size it must be noted that this assumption could still disguise a considerable degree of heterogeneity with respect to the nature of the RNA and protein units contributing to the make-up of the individual ribosomes. It is possible to discuss a basic pattern in the RNA make-up of ribosomes (e.g., 23S and 5S RNA from 50S ribosomes and 16S RNA from 30S ribosomes) but even this may mask a microheterogeneity with respect to its overall base composition and sequence (Aronson, 1962; McIlreavy and Midgley, 1967; Midgley and McIlreavy, 1967; Santer *et al.*, 1961). Furthermore studies on the reconstruction of ribosomes from the RNA cores and the split proteins have indicated a high degree of specificity of interaction between these two components (Nomura *et al.*, 1968). Also, in the case of the protein components, it is difficult to distinguish between the basic structural proteins of the ribosome and the functional proteins since the latter may or may not be identical with the former and could also be transiently associated with the basic structure (see Elson (1967), for a discussion of this problem in relation to the enzymes of protein synthesis etc.). The ribosomes are not bounded by a membrane and all the components of the cytoplasm may have unrestricted access. This, coupled with the high negative charge on the RNA, leads to considerable non-specific binding of general cell proteins during the isolation of the ribosomes. The problem in "purifying" ribosomes is to detach these extraneous proteins without disturbing the basic structure. The basic ribosome structure may be viewed as a polynucleotide chain to which are attached a defined number of molecules of structural proteins. Under suitable conditions this structure then assumes a configuration which provides a suitable surface of oriented groups to facilitate the associations necessary for protein synthesis, i.e., ribosomes (70S) are the "inert work-benches of protein synthesis". This view does not match much of the recent evidence concerning ribosome structure and function. The 50 and 30S ribosomes have unique roles in binding certain of the components required for protein synthesis, in particular 30S specifically binds N-formyl methionine t-RNA and m-RNA to form an initiation complex. This complex then combines with 50S to form the initial 70S ribosome on the m-RNA. 70S ribosomes dissociate after reading a strand of m-RNA and on this view may be considered as transients. The normal cell ribosomal elements are therefore 30S, 50S and polysomal structures (Mangiarotti and Schlessinger, 1966). The evidence discussed earlier shows that the 50 and 30S units are not inert structures but may be readily and reversibly changed to provide, at the functional level, an immediate specific (Ozaki *et al.*, 1969) or non-specific point of control over protein synthesis, i.e., a population of ribosomes may be homogeneous with respect to size and overall composition but not function (Guthrie and Nomura, 1968; Sykes and Young, 1968; Young and Sykes, 1968). The ideal approach to a definition of the "purity" of a ribosome preparation must therefore be from all three aspects, i.e., physical, chemical and functional.

Centrifugal isolation of ribosomes provides one physical criterion, i.e., size homogeneity. This may be checked by an analytical ultracentrifuge run and if the total concentration of material is known from independent analyses then the area beneath the single sedimenting boundary should give an estimate of the concentration (see p. 74) within a few percent of this figure. Electron-microscopy and moving boundary electrophoresis are also useful techniques for checking the physical homogeneity of ribosome preparations.

The overall chemical composition and the total concentration of ribosomal material in a preparation may be conveniently estimated by their absorbance in the ultraviolet. The 260/280 nm absorbance ratio may be used to estimate the relative concentrations of protein and nucleic acid present (Warburg and Christian, 1942) or alternatively the absorbance at 260 nm compared with that for a solution of ribosomes of known concentration may be used, e.g., $E^{1\%}_{1\,cm}$ at 260 nm for 50S and 30S ribosome preparations from $E.\,coli$ are 182 and 165 respectively (Tissières $et\,al.$, 1959). It is preferable to check these estimations by direct chemical analyses for RNA and protein. The RNA concentration may be measured on direct, hot acid extracts of the preparation (5% trichloroacetic acid at 100°C or 0·5N perchloric acid at 70°C for 20 min.) by means of the orcinol test (Schneider, 1957). It is essential that all sucrose is removed from the preparations before this test is applied. This test, and its various modifications, measures purine bound ribose and the most suitable standard is, therefore, a purified sample of ribosomal RNA extracted directly from the same organism grown under identical conditions. Commercial preparations of purified "highly polymerised" RNA from other organisms [e.g., yeast or $E.\,coli$] may be used but the assumption made, i.e., that the purine bound ribose content of the standard RNA is the same as the test, is usually not justified (see Petermann, 1964 for a list of base analyses). D-ribose may be used as a standard but, once again, a true estimate of the RNA concentration requires a knowledge of the base composition of the test RNA. Desoxyribose interferes slightly with the orcinol test and, since the diphenylamine reaction (Burton, 1956) for desoxyribose (and hence the DNA contamination) in the preparation may be performed upon the same hot acid extract as used for the orcinol test, it may be conveniently estimated and the orcinol results corrected. The complete characterization of the RNA components of the ribosome requires their isolation from the ribosome. This may be achieved by the direct phenol and detergent treatment of the ribosome preparation to denature the proteins (Midgley, 1965a; Sykes and Young, 1968). The RNA species are found in the aqueous phase of the phenol : buffer system. The high molecular weight species are precipitated by 1–2M NaCl and may be redissolved in buffer and separated by rate-zonal gradient centrifugation. Alternatively, the total RNA species in the aqueous phase may be fractionated on methylated albumin—Kieselguhr columns (Mandell and Hershey, 1960). This chromatographic separation is most suitable when the presence of low molecular weight RNA species is suspected, e.g., 5S ribosomal RNA (Rosset and Monier, 1963). The complete characterization of the individual ribosomal-RNA components then requires a determination of their sedimentation constants and molecular weight, their overall base composition and, ideally, a base sequence determination. The sedimentation constant for ribosomal RNA species may be determined by the sedimentation velocity techniques described in Section III. In view of their strong dependence of sedimentation upon concentration it is essential to make a number of runs at different RNA concentrations and extrapolate the plot of $1/S_{w, 20}$ vs. concentration to zero concentration to obtain the sedimentation constant ($S^0_{w, 20}$). Alternatively a single run may be made with the ultraviolet optical system (260 nm) at concentrations as low as 50 μg/ml and the $S_{w, 20}$ value obtained may be reasonably assumed to equal the sedimentation constant. The molecular weight of ribosomal RNA species has been determined by (i) sedimentation equilibrium (e.g., for the 23S RNA component of *E. coli* 50S ribosomes (Stanley and Bock, 1965) and the 17S and 26S RNA species from yeast 83S ribosomes (Bruening and Bock, 1967), and (ii) by a combination of sedimentation constant and viscosity determinations using the Scheraga-Mandelkern (1953) equation:

$$M^{\frac{2}{3}} = \frac{s^0{}_{20, \text{w}}[\eta]^{\frac{1}{3}} \, [\eta_0]N}{\beta(1 - \bar{v}\text{d}m)} \tag{63}$$

η is the intrinsic viscosity of the preparation at 20°C and η_0 is the viscosity of the solvent at 20°C, all other terms have their previously assigned meanings. Kurland (1960) obtained a value of 0·57 ml/g for $\bar{v}$ with a mixed population of *E. coli* ribosomal RNA species and values in the range 2·21–2·32 × 10⁶ for β (the shape-volume coefficient) for use in the above equation. The molecular weights of 23S and 16S RNA were then calculated to be 1·1 × 10⁶ and 5·6 × 10⁵ respectively. On the other hand, Stanley and Bock (1965) assumed a $\bar{v}$ value of 0·53 ml/g and a value for β of 2·16 × 10⁶ in their study with purified 23S RNA and this gave an identical molecular weight of 1·07 × 10⁶. The assumption of values for $\bar{v}$ and particularly β, introduces a considerable degree of uncertainty into molecular weights determined by this combination of data. A β value of 2·1 × 10⁶ applies to spheres whereas a value in the range up to 2·6 × 10⁶ might be anticipated for a randomly coiled polymer (Creeth and Knight, 1965; Scheraga, 1961) Furthermore errors in the determination of s produce 50% greater errors in the estimation of M. Despite these sources of error the published molecular weights for *E. coli* 23S ribosomal RNA obtained via sedimentation equilibrium (Stanley and Bock, 1965) and sedimentation-viscosity measurements (Kurland, 1960) are in substantial agreement. The molecular weight of the RNA from 30S bacterial ribosomes has only been determined by sedimentation and viscosity measurements for one organism, *E. coli* (Kurland, 1960). Empirical equations relating molecular weight to viscosity (by assuming constant $\bar{v}$ and β terms in equation 63) and other hydrodynamic parameters, e.g., sedimentation coefficient have been employed by certain workers (Spirin, 1964). These equations are based upon measurements made with one type of RNA in a single solvent. However, since the conformation of RNA varies markedly with the solvent, the use of such equations should be strictly limited to the RNA and solvent system for which they were derived. The direct translation of sedimentation constants into estimates of molecular weights is always a doubtful procedure and impossible for RNA species unless the solvent used completely disrupts secondary structure.

A chemical method for the determination of the molecular weights of RNA species has recently been described (Midgley, 1965b; 1965c). The method is based upon the fact that RNA chains are single, unbranched, covalently linked polynucleotide chains formed from ribonucleoside 5′-phosphate units linked via 3′–5′ phosphodiester bonds. Hence, at one terminus there is a ribose unit with vicinal, unsubstituted hydroxyl groups on carbons C–2′ and C–3′. There is only one such unit per RNA chain and only this residue will be susceptible to periodate oxidation to yield the corresponding dialdehyde. One mole of this dialdehyde (equivalent to one mole of oxidized RNA) will then react with one mole of iso-nicotinic acid hydrazide (INH) and give an RNA-iso-nicotinic acid hydrazone. This hydrazone is acid stable and is quantitatively precipitated by cold 10% w/v trichloroacetic acid. The RNA, in acetate buffer pH 4·6 is reacted in the dark with 0·1M $NaIO_4$ for 1 h at 20°C and excess ethylene glycol is then added to remove unconsumed periodate. The solution is dialysed against buffer and incubated for 20 h at 20°C with a 250 fold excess (based on detectable dialdehyde) of [carbonyl-^{14}C] or [^{3}H]INH of known specific activity. The RNA-iso-nicotinic acid hydrazone is then precipitated with cold 10% trichloroacetic acid, filtered onto membrane filters and well washed with acid. The filters are gummed to planchettes for counting. The number of INH-reactive nucleotides per total nucleotides in the specimen may then be determined (assuming an average molecular weight of a nucleotide) and hence a minimum molecular weight. Polysaccharides, which are frequent contaminants of RNA preparations, are clearly a source of error in this method and extreme care must be taken to check the purity of the RNA specimen (McIlreavy and Midgley, 1967). The molecular weights determined in this way for 16S ribosomal RNA from *E. coli* are in good agreement with the physical estimations, but the revised analysis for 23S RNA from this organism indicates a frequency of 3′-hydroxy terminal groups of 1 per 1500 nucleotides, i.e., giving a molecular weight which is one half of the value determined by physical techniques and equal to that of 16S RNA. This result reopens the problem of the relationship, if any, between the 16S and 23S ribosomal RNA species. The analysis of this problem is aided by the studies on the base composition and sequence of the larger

RNA species. Techniques for tackling this problem have now been devised and are the subjects of other articles in these volumes.

The overall protein content of the ribosomes may be estimated by the biuret method or one of its more sensitive modifications. Crystalline bovine serum albumin is often used as a standard although the actual ribosomal protein is a better standard. The methods for the preparation of ribosomal proteins are given below but their insolubility in buffer solutions restricts their use. Spitnik–Elson (1962) has reported that 1 ml of 1M tris-HCl buffer, pH 7·4, will dissolve up to 17 mg of *E. coli* ribosome proteins. The dry weight of a ribosome preparation is best obtained by lyophilization followed by a brief exposure to 100°C to denature any enzymes and avoid the loss of volatile nitrogen. The preparation is then dried to constant weight at 105°C. After correction for the contribution of buffer salts to the dry weight approximately 95% of the weight will be accounted for by the combined totals of the protein and RNA analyses. The remainder of the dry weight is usually accounted for by the inorganic and organic cations associated with the preparation. The complete analysis of the ribosomal proteins requires that they be separated from the RNA. This may be accomplished by:

(i) treatment with glacial acetic acid (Waller and Harris, 1961). In this method two volumes of 66% acetic acid are added to a 1–3% (w/v) suspension of ribosomes at 2–5°C. The mixture is stirred during the addition and for a further 45 min in the cold. The RNA precipitates and may be removed by a brief, low speed centrifugation. The protein solution is dialysed to remove the acid and at this step will usually precipitate (see below). The preparation may be freeze-dried if required.

(ii) the ribosomes are exposed to digestion by added ribonuclease [E.C. 2.7.7.16] or, if present, their latent ribonuclease is activated by dialysis against 1M NaCl (Spitnik–Elson, 1965) or 4–6M urea (Spahr, 1962). For this procedure the ribosomes are placed in a dialysis sac, the dialysis medium is changed regularly and the treatment continued until no more ultraviolet (260 nm) absorbing materials leave the sac. At this point less than 1% of the original RNA remains within the sac with the ribosomal protein.

(iii) the ribosomes are treated with 2–4M LiCl solutions or LiCl and urea. 2M LiCl alone will precipitate the RNA from yeast ribosomes (Chao, 1961) and leave the protein in solution. However *E. coli* ribosomes are resistant to 5M LiCl and in such cases complete precipitation of the RNA may be accomplished by a final concentration of 3M LiCl and 4M urea (Spitnik–Elson, 1965). The supernatant contains all the ribosomal protein. The ribosomal proteins are soluble in neutral and other solutions of high ionic strength ($I > 0.75$) and in concentrated urea solutions. Unless precipitation of the proteins is desired the dialysis to remove the urea/high salt following certain of the above procedures should be against a suitable buffer solution, e.g., 0·75M tris-HCl, pH 7·4 (Spitnik–Elson, 1962). The final preparations may be treated with ribonucleases to ensure the removal of RNA.

Ribosomal proteins are now known to be an extremely heterogeneous group of proteins. However, in the present context two important conclusions have emerged from recent studies, viz: (i) A ribosomal protein is present at the frequency of one copy per particle of 50 or 30S, each protein having a different primary structure. On this basis each 30S ribosome probably has 23 protein sub-units and each 50S particle 43 proteins (Moore et al., 1968), some of which are common to 30S (Traut, 1966); (ii) The loss or modification of certain protein components gives rise to alterations in the functional properties of the ribosome (Nomura and Traub, 1967; Ozaki et al., 1969) or its sensitivity to drugs, e.g., streptomycin, and the association between the ribosomal RNA or ribosomal "cores" and ribosomal proteins is a specific, spontaneous process (Nomura et al., 1968). The examination of the ribosome for the "purity" of its protein complement may therefore be coupled with the analysis of the functional properties of the ribosome. What is required therefore is a ready means of revealing the "profile" of ribosomal protein components and separating the individual proteins. This may be achieved by disc polyacrylamide gel electrophoresis or column chromatography of the ribosomal proteins. Disc gel electrophoresis is usually performed in the presence of buffers containing high concentrations of urea (approximately 6–8M) at a pH of 3–5 (Traub and Nomura, 1968; Traut, 1966), following the procedures described by Reisfeld et al (1962) and using the split gel technique suggested by Le Boy et al (1964). Under these conditions all the proteins migrate towards the cathode. The detailed techniques for polymerization, developing and staining the gels are given in the instruction manuals with the apparatus [e.g., "Chemical Formulation of Disc Electrophoresis", Canalco, 5635 Fisher Lane, Rockville, Maryland, U.S.A.]. The stained bands within the gels may be scanned with a densitometer or, if the proteins are radioactively labelled during cell growth, autoradiographs may be made with dried slices of the gel [see Traut (1966) for a description and comparison of these techniques]. C-M.—cellulose (Möller and Widdowson, 1967) and phosphocellulose column chromatography (Ozaki, et al., 1969) have also been used to fractionate ribosomal proteins by eluting from the column with increasing salt gradients at pH 8·0 in the presence of 6–8M urea. The disc electrophoretic separation appears to offer better resolution in a shorter time with many samples under identical running conditions. It is therefore ideally suited to studies comparing the protein profiles from normal ribosomes from different sources, ribosome precursors, "core particles", "split" proteins, etc. The individual proteins from E. coli 30S ribosomes have been purified by CM-cellulose chromatography followed by DEAE or Sephadex G-100 chromatography (Moore et al., 1968). Proteins eluting as a single peak and running as a single band during electrophoresis on polyacrylamide gels may then be subjected to detailed analysis to determine their molecular weight, composition and primary sequence. Moore et al. (1968) employed the high-speed sedimentation equilibrium technique (Section IV.B) and interference optics to determine the molecular weights of 13 of the E. coli 30S proteins. 6M urea or guandinium hydrochloride must be added to the buffer for these runs. The problem of $\bar{v}$ for each protein may be overcome by the determination of its amino-acid composition. Once this further item of interest is known then, with the known specific volumes of the amino-acids, the partial specific volume of the protein may be calculated (Cohn and Edsall, 1943; Schachman, 1957). The molecular weight range for the E. coli 30S ribosomal protein sub-units was found to be 3300–46,500. Future investigations with ribosomal protein preparations will undoubtedly employ the full range of centrifugal techniques described in Sections I–IV.

Detailed physical studies upon the intact ribosome units have been relatively neglected. Tissières et al (1959) determined the molecular weights of 30, 50, 70 and 100S ribosomes from E. coli by combining sedimentation and diffusion coefficients [in equation (25)] and sedimentation and viscosity data [in equation (63)]. Sedimentation constants were determined using Schlieren and ultraviolet optical systems and an average value of $\bar{v}$ equal to 0·64 was determined by pycnometry and used in all calculations. A spherical shape was assumed for β in equation (63). This assumption was based upon a companion study of the particles in the electron microscope in which the estimated molecular weights for 30 and 50S were found to be 10% higher (Hall and Slayter, 1959) than those found by Tissières et al. (1959). Hess et al. (1967) have published a detailed study of 50S ribosomes from Streptococcus pyogenes in which they determined the molecular weight of the ribosomes by sedimentation-diffusion, Archibald approach to equilibrium (Section IV.B), sedimentation-viscosity and light scattering techniques. All of these methods gave essentially the same value for the molecular weight of $1·9 \times 10^6$. The partial specific volume was found to be 0·65 dl g^{-1} and the $E^{1\%}_{1cm}$ at 260 nm 138 ± 3. Chao and Schachman (1956) estimated the molecular weight of 80S ribosomes from S. cerevisiae to be $4·1 \times 10^6$. This estimation was based upon their sedimentation, viscosity and partial specific volume determinations but assumed a spherical shape. Tashiro and Yphantis (1965) have applied the high-speed sedimentation-equilibrium technique (Section IV.B) to the determination of the molecular weights of hepatic ribosomes. Since ribosome preparations are normally stable for at least a week at 0–5°C and may be paucidisperse, this equilibrium technique is particularly suited to the determination of their molecular weights.

C. Cytomembianes, including chromatophores and sub-chromatophore units

1. *Introduction*

The cytomembranes of bacteria comprise the plasma membrane, the mesosomal elements, the intra-cytoplasmic membranes and the specialized intracytoplasmic membranes of the photosynthetic bacteria (the "chromatophores") (Cohen-Bazire and Sistrom, 1966; Lascelles, 1968) and the nitrifying bacteria (Murray and Watson, 1965; Remsen *et al.*, 1967). Other chemolithotrophes, e.g., the sulphur (Mahoney and Edwards, 1966) and iron-oxidizing bacteria, (Remsen and Lundgren, 1966) do not appear to possess additional cytomembranes to those found in the majority of the Gram-negative bacteria. Since most of the complex intra-cytoplasmic and mesosomal membrane structures in bacteria exhibit some point of continuity with the plasma membrane it seems very likely that the former represent extensions, and possibly differentiation of the plasma membrane. A consequence of this is that the enzymic and chemical make-up of the plasma and intra-cytoplasmic membranes may differ. Cytochemical studies have indicated that this may be the case in *Bacillus subtilis* (Vanderwinkel and Murray, 1962; Van Iterson and Leene, 1964) and careful centrifugal fractionation of membrane preparations from *E. coli* (Miura and Mizushima, 1968) and *B. megaterium* (Mizushima *et al.*, 1967) have confirmed that it is possible to isolate membrane fractions which differ in composition and activity. The cytomembranes of the photosynthetic bacteria also differ in their pigment content (see later Section C.4). Nonetheless a striking common feature of all bacterial cytomembranes is their high content of phospholipids and protein.

The existence of a permeability barrier at the periphery of the bacterial cytoplasm was inferred from the observations that bacteria could concentrate nutrients from the environment, undergo plasmolysis and exhibit selective permeability. The lipid nature of this barrier was indicated by the sensitivity of bacterial cells to the action of organic solvents. The electron microscope studies (e.g., by Ryter and Kellenberger (1958) and subsequently many others) demonstrated the existence of an osmiophilic layer between the cell wall and the bulk of the cytoplasm conforming in its structure and dimensions to the generalized type of osmotically active membrane described for other cell types. The actual presence of the plasma membrane was elegantly confirmed by Weibull's (1953) demonstration that the cell wall of the rod-shaped *B. megaterium* could be digested with lysozyme to give a spherical, osmotically sensitive protoplast. These protoplasts were shown to exhibit the majority of the activities of the intact cell but, unlike the cell, were uniquely sensitive to changes in the osmolarity of the suspending medium (see McQuillen, 1960 for a review). The protoplasts rapidly burst when the stabilizing osmotic pressure is lowered releasing their contents into suspension and leaving ruptured plasma membranes ("ghosts"). The latter may be removed from the mixture by a brief low-speed centrifugation. Weibull obtained one "ghost" per bacterium using this procedure. In a detailed study of the distribution of enzymes and other cell constituents in fractions obtained by differential centrifugation of these lysates Weibull *et al.* (1959) demonstrated that the bulk of the succinic dehydrogenase, $NADH_2$ oxidase and cytochrome pigments was associated with the membrane fragments. Subsequent observations have confirmed that the cytomembranes are the site of much of the oxidative activity of bacteria. The high oxidative activity of many "particulate" preparations obtained from bacteria may be attributed to the presence of membrane fragments.

Extension of the plasma membrane into the cytoplasm has been reported for many non-photosynthetic bacteria. Such extensions may have the characteristic appearance of whorls of membrane (the "mesosomes", which are commonly found in Gram-positive bacteria (Fitz-James, 1960) but have also been reported in Gram-negative species), others may simply resemble a primitive, smooth endoplasmic reticulum and comprise a membrane bounded vesicular element ramifying through the cytoplasm (Pangborn, *et al.*, 1962). The latter may sometimes only be revealed in electron-micrographs of thin sections of cells by previously starving the cells to deplete the electron-dense ribosome content which otherwise obscures this detail (Smith, 1960). The regular mesosomal elements have not been isolated as intact structures free from the plasma membrane since they are readily extruded from the protoplast or via mechanical breaks in the cell wall and are difficult to resolve from the membrane (Fitz-James, 1968; Salton and Chapman, 1962). There have been speculations that these membranes effectively increase the surface area (and activity) of the plasma membrane (cf. cristae mitochondriales). The observations of Pangborn *et al.* (1962) certainly revealed the association of 60% of the $NADH_2$ oxidase activity with the intracytoplasmic membrane of *Azotobacter agilis*. However Fitz-James (1968) has evidence that the mesosomal elements in *B. megaterium* may be sites of synthesis of new plasma membrane.

Mesosomal elements have also been observed in certain photo-synthetic bacteria (e.g. *Chlorobium thiosulphatophilum* (Cohen-Bazire, 1963)) in addition to the variety of membrane or vesicular structures which carry the photo-apparatus. The term "chromatophore" has frequently been applied to the membrane structures bearing the entire complement of photo-synthetic pigments in photosynthetic bacteria. As originally defined by Schachman, Pardee and Stainer (1952) this term was applied to the rela-tively homogeneous particles 600Å in diam., sedimenting at 160S in schlieren analytical ultracentrifuge diagrams of cell-free extracts from *Rhodospirillum rubrum* prepared by a variety of methods. These particles were shown to carry the bulk of the cellular photopigments. A number of subsequent investigations have revealed the variety of membrane structures associated with photosynthesis in the bacteria and blue-green algae (Cohen-Bazire and Sistrom, 1966; Lascelles, 1968). The photosynthetic apparatus may consist of stacked lamellae continuous with the plasma membrane (e.g., *Rhodospirillum molischianum*, Hickman and Frenkel, 1965a, and the blue-green algae Cohen-Bazire and Sistrom, 1966), peripheral or cytoplas-mic vesicles with few obvious connections with the membrane (e.g. *C. thio-sulphatophilum*, Cohen-Bazire, 1963; Cohen-Bazire *et al.*, 1964; Sykes *et al.*, 1965; Sykes and Gibbon, 1967), or inter-connected vesicular elements either unconnected (e.g., *C. ethylicum*, Holt *et al.*, 1966) or connected (e.g., *R. rubrum*, Holt and Marr, 1965a) with the plasma membrane. The dimensions of the unit membranes surrounding all of these structures are the same as those of the plasma membrane. The original definition of the term "chrom-atophore" clearly does not generally apply to the bacterial photosynthetic apparatus. Equally many so called "chromatophore" preparations may have arisen by comminution during the isolation of these complex intra-cyto-plasmic elements, although their function, e.g., photophosphorylation may survive a certain loss of structure. A surprising feature of many of these preparations (e.g., from *R. rubrum*, comparing the results of Schachman *et al.* (1952) and Holt and Marr (1965b)) remains the uniformity of the membrane fragments. This unexplained feature has doubtless fostered the retention of the term "chromatophore" for the bacterial photo-apparatus.

It is essential to control the growth conditions and point of harvest to ensure reproducibility in studies with bacterial membranes. For *St. faecalis* the composition of the plasma membrane fraction was shown to be a function of the growth phase and the growth limiting nutrient (Schockman *et al.*, 1963). In a number of Gram-positive bacteria the plasma membrane fraction was variable and accounted for 10–25% of the total cell dry weight (Salton and Freer, 1965). In the case of the photosynthetic bacteria the growth conditions exert a profound effect upon the composition and extent of the cytomembrane fraction bearing the photosynthetic pigments

(Hickman and Frenkel, 1965b; Holt et al., 1966; Holt and Marr, 1965c; Lascelles and Szilagyi, 1965; Sistrom, 1965; Trentini and Starr, 1967). For many photosynthetic organisms, e.g., R. rubrum, Rhodopseudomonas palustris, the specific bacteriochlorophyll content of the cell falls with increasing light intensity and growth rate. The total cell phospholipid per unit of DNA also falls but in some cases, e.g., Rh. palustris, the bacterio-chlorophyll : lipid ratio is maintained constant whereas in others, e.g., R. rubrum the change from low to high light intensity results in an increase in this ratio. A photosynthetic organism may therefore respond to a lowered light intensity by increasing the amount of photo-pigmented membrane, keeping the pigment : membrane ratio constant, or by increasing the amount of membrane and raising the pigment : membrane ratio (Cohen-Bazire and Sistrom, 1966). The intensity and colour of the incident light also influences the pigmentation of the blue-green algae.

The isolation of the cytomembranes will therefore be considered in the following groups: cytomembranes from Gram-positive bacteria, cyto-membranes from Gram-negative bacteria, cytomembranes from the photo-synthetic bacteria, the nitrifying bacteria and the blue-green algae and finally the preparation of sub-chromatophore particles from the photo-synthetic bacteria.

2. Cytomembranes from Gram-positive bacteria

The majority of the methods for breaking bacteria, and particularly the mechanical methods, usually comminute the cell membranes. When such a preparation is differentially centrifuged (e.g., for 30 min at 20,000 g (av.)) the membranes are deposited either attached to the broken cell wall fraction or as a discernible layer ("particulate") above the cell walls. The "particu-late" layer may be obtained via further cycles of centrifugation and is essentially a comminuted membrane preparation which frequently main-tains many of the functional characteristics of the intact membrane (Marr, 1960). However, since the cell wall of Gram-positive bacteria is completely removed by the action of lysozyme (Weibull, 1968) and intact membranes are a desirable starting point for study, the use of this enzyme is the preferred method of preparing cytomembranes from Gram-positive bacteria. The detailed procedures for the use of this and other muralytic enzymes against these bacteria are described in the articles by Hughes et al. (This Volume, p. 1) and McQuillen (1960) and will not be repeated here. Additional details will be found in the specific references cited below. Certain L-forms of bacteria (frequently described as "stable" L-forms) are usually without the cell wall rigid component. Procedures designed to produce L-forms, e.g., the addition of penicillin to the growth medium (Marston, 1968) or treatment of cells with muralytic enzymes (Gooder,

1968), yield cultures which may be propagated as stable L-forms and these represent useful starting points for plasma membrane isolations.

In Weibull's (1953) original description the membrane fraction was differentially centrifuged from lysed protoplast preparations of *B. megaterium* with gravitational fields in the range 590–14,800 *g* for 15 min. The yellowish pellet of membrane material was resuspended in 0·03M phosphate, pH 7·0 and resedimented at 105,000 *g*. In subsequent experiments, designed to demonstrate the equivalence between "ghosts" and cells, Weibull (1956) found that the presence of 5–10 mM Mg^{2+} was required to maintain the integrity of the membranes. Without Mg^{2+} the "ghost" : cell ratio rose to 1·35. The role of Mg^{2+} ions in stabilizing membrane structure is unknown but Sykes and Tempest (1965) also found that Mg^{2+}-limited, continuously grown cells of *P. putida* were exceedingly susceptible to lysis when compared with the corresponding carbon-limited cells. Weibull and Bergstrom (1958) have used gravitational fields of 20–28,000 *g* for 20 min to differentially centrifuge *B. megaterium* membrane fractions from lysed protoplast preparations. The viscosity of lysed protoplast preparations is usually high since lysozyme digestion is usually carried out with thick cell suspensions and high concentrations of enzyme. The gentle nature of this procedure also preserves the structure of the DNA within the protoplast and this adds considerably to the viscosity and may complicate the centrifugal isolation of the membranes. Salton and Freer (1965) and Shockman *et al.* (1963) have therefore used desoxyribonuclease [E.C. 3.1.4.5] at concentrations in the range 2–40 μg/ml of lysate to depolymerize the DNA and lower the viscosity. Salton and Freer (1965) and Salton and Ehtisham-ud-Din (1965) lysed suspensions of *B. licheniformis, B. stearothermophilus, Sarcina lutea* and *Micrococcus lysodeikticus* in 0·1M tris-HCl, pH 7·5 and removed the membranes by centrifugation at 0°C for 30 min at 20,000 *g*. The membrane pellets were then washed three times by resuspension and sedimentation at 45,000 *g* from 0·05M tris-HCl, pH 7·5. The preparations were finally dialysed against distilled water. The basic method adopted by Shockman *et al.* (1963) was to differentially centrifuge the membranes from a lysate of *St. faecalis* protoplasts at 25,000 *g* for 1 h and then wash the pellet six times by sedimentation from 0·1M phosphate, pH 7·4 at 25,000 *g* for 30 min. The membranes were finally washed in 0·05M NaCl and distilled water. For their preparation of membranes from *M. lysodeikticus* Gilby *et al.* (1958) lysozyme digested the cell walls in the presence of 0·1M NaCl and deposited the membranes by differential centrifugation at 20,000 *g* for 30 min. The membranes were finally washed three times in 0·05M NaCl and centrifuged for 10 min at 20,000 *g*. Freimer (1968) has used a group C phage-associated lysin, combined with desoxyribonuclease and ribonuclease treatments to prepare integrated protoplast membranes from Group A streptococci. Log-phase cells were incubated for 1 h at 37°C with the lysin in the presence of 10^{-2}M phosphate, pH 6·5 and 7·5% NaCl. The protoplasts were washed in 7·5% NaCl and were shown by serological and chemical tests to be without cell wall carbohydrates. The protoplasts were lysed in 10^{-2}M phosphate, pH 7·3 in the presence of desoxyribonuclease (0·2 mg/ml). The suspension was incubated for 30 min at 37°C and the membranes collected by centrifugation at 8000 g. The membranes were washed in phosphate-saline and again incubated at 37°C for 30 min in phosphate pH 7·2 containing 0·02 mg/ml of both ribonuclease and desoxyribonuclease. The phage-associated lysin treatment was now repeated with the membrane preparation in 0·067M phosphate-saline, pH6 followed by a retreatment with the nucleases. Finally the membranes were washed three times by centrifugation from phosphate saline and then washed once with distilled water. The electron micrographs of this preparation showed intact protoplast membranes without the electron dense particles which are associated with untreated membrane preparations.

In one of the rare isolation procedures described for mesosomal and plasma membranes Salton and Chapman (1962) prepared protoplasts of *M. lysodeikticus* in 0·067M phosphate buffer, pH 7·0 and 1·5M sucrose as stabilizer. Lysozyme was added (50 μg/ml) and when protoplast formation was judged complete the protoplasts were lysed by diluting the suspension with two volumes of distilled water. The membrane-mesosome complex was sedimented by 20 min centrifugation of the lysate at 9000 g. The pellet was resuspended in 0·067M phosphate pH 7·0 containing 0·5M sucrose and treated with desoxyribonuclease (20 μg/ml) at room temperature for 30 min. The membrane complex was resedimented at 36,000 g for 30 min and washed twice with the phosphate-sucrose solution. The same preparation was also carried out in the complete absence of sucrose and gave a membrane complex which was indistinguishable in the electron micrographs from preparations made in sucrose. The characteristic morphology of the mesosomal elements was preserved but the resolution of mesosomes from the plasma membrane was not attempted. Fitz-James (1968) has also described the preparation of mesosomes extruded from *B. megaterium KM* protoplasts. Intact protoplasts of this organism were centrifuged in a combined rate-zonal and *iso*-pycnic gradient. The rate-zonal density gradient was a 2–20% Ficoll gradient containing 0·25–0·3M sucrose as a stabilizer. This gradient was constructed over a cushion of 40% Ficoll and on centrifugation the extruded mesosomal elements were concentrated at the 20–40% Ficoll interface and the remaining material sedimented to the bottom of the tube. Alternatively lysed protoplasts were used and layered onto a 5–20% sucrose rate zonal gradient overlying a cushion of 60% sucrose. The sucrose solutions were prepared in 10^{-2}M tris-HCl, pH 7·4 containing 10^{-2}M KCl and $1·5 \times 10^{-3}$M MgCl$_2$.

Since the mesosomal and plasma membranes tend (inevitably?) to separate together the interpretation of data on the function and composition of the total complex will thereby be complicated until the complex is resolved. The difference (if any) between these two cytomembranes of Gram-positive bacteria may then be determined. In this connection the preliminary results and experimental approach adopted to membrane separation by Mizushima *et al* (1967) are of interest. A crude membrane fraction was obtained from lysed protoplasts of *B. megaterium* by centrifugation at 20,000 *g* for 15 min, resuspending the pellet in 5 mM MgCl$_2$ containing desoxyribonuclease and recentrifuging at 20,000 *g* for 15 min. The preparation was dialysed against distilled water and then adjusted to pH 8·5 with 0·1N NaOH. This preparation was then subjected to two cycles of differential centrifugation. The pellet from the first centrifugation at 7000 rpm (*g* values unspecified) for 15 min was discarded and the supernatant was recentrifuged for 1 h at 30,000 rpm. The sediment from this step was shown by electron microscopy to be mainly unit membrane material and by chemical analysis to be rich in phospholipid and protein. Succinic dehydrogenase and NADH$_2$ oxidase were also localized in this fraction (Mizushima *et al.*, 1966). This sediment was then subjected to rate-zonal centrifugation in a 0–30% linear sucrose gradient. A 2 h centrifugation at 39,000 rpm in a swinging-bucket rotor [RPS 40, Hitachi] completely sedimented the membrane material. The sediment was diluted so that the concentration of sucrose was less than 30%, adjusted to pH 8·5 with NaOH and layered onto a 30–50% linear sucrose gradient and centrifuged for 4 h at 39,000 rpm. This time three distinct bands, A, B and C, were separated in the gradient. The materials in bands A and B were separately recovered, adjusted to pH 8·5 and material A re-run on a 25–40% linear sucrose gradient and material B on a 30–45% sucrose gradient. Material A resolved into two further bands, A$_1$ and A$_2$, and material B into B$_1$ and B$_2$. The separation of fractions A and B in the 30–50% sucrose gradient was on the basis of density (i.e., an isopycnic gradient) since A was richer in lipid than B. Furthermore, if we assume an average density for phospholipid of 0·97 and for proteins 1·35, an exclusively phospholipoprotein membrane with a 50–75% protein content will have a density in the range 1·14–1·24. The density range encompassed by 5–50% (w/w) sucrose solutions at 0°C is 1·02–1·24 and gradients of sucrose will therefore encompass the range of densities likely to be encountered with phospholipoproteins. *Iso*-pycnic equilibrium centrifugation in preformed sucrose gradients would seem to offer a useful method of fractionating cytomembrane preparations according to their composition. This approach has already been adopted with mitochondrial and erythrocyte membrane fractions. Although Mizushima *et al.* (1966) found differences in chemical composition and morphology between fractions A, B and C they were not able to establish significant differences in the distribution of succinic dehydrogenase, $NADH_2$ dehydrogenase, cytochrome *a* and cytochrome *b* between sub-fractions A_1, A_2, B_1, B_2 and C. Furthermore the precise relationship between these membrane fractions and the original plasma-mesosome membrane complex remains obscure. More investigations of this type, designed to resolve plasma-intracytoplasmic membrane complexes, e.g., by a preliminary mechanical or chemical step followed by *iso*-pycnic or rate zonal centrifugations, are required to characterize biochemically the individual membranes of the complex.

The biochemical properties, e.g., the enzymic make-up and the overall chemical composition of plasma membranes (plus mesosomes?) have been widely reported (Mizushima *et al.*, 1966; Salton, 1967) . Salton and Freer (1965) found the membranes of four species of Gram-positive bacteria to be predominantly protein (53–75%) and lipid (20–30%). The lipid content of *M. lysodeikticus* membranes was found to vary between 14·6% and 29·5% of the dry weight depending upon the phase of growth at harvest and the nature of the growth medium. The lipid contents were determined by pooling several 95% methanol extracts and two chloroform-methanol (2 : 1 v/v) extracts and evaporating to dryness. The residue was taken up in chloroform-methanol (2 : 1) washed, (Folch *et al.*, 1957), evaporated to dryness and weighed. Refluxing with 95% methanol is usually sufficient to extract lipids from lipo-protein complexes although preliminary hydrolysis with 6N HCl, followed by ether extraction, has been used for *B. megaterium* (Weibull and Bergström, 1958) and *B. subtilis* membranes (Salton, 1953). Preliminary hydrolysis released some "firmly bound lipid". The phosphorous content of the membrane lipids may also be determined following the acid digestion. The membrane protein is usually determined by the biuret reaction with ultra-sonically dispersed preparations made alkaline with NaOH and heated to 100°C for 5 min before addition of the copper reagent.

In common with membrane preparations from most micro-organisms the membranes from Gram-positive bacteria almost invariably contain carbohydrate and a variable amount of orcinol reacting material. The latter is usually reported as RNA and may vary between 1·5% (*M. lysodeikticus*) and 7·5% (*Sarcina lutea*) of the total membrane dry weight. This "RNA" may simply represent material occluded (or sedimented) in the membranes since the final preparations have rarely been treated with ribonuclease or prepared via density gradient centrifugations. For example, in the case of the photosynthetic organism *R. rubrum*, it has been shown that the residual RNA content of the membrane preparations is due to ribosomes and may be removed by ribonuclease digestion or *iso*-pycnic centrifugation in RbCl gradients (Worden and Sistrom, 1964). The detailed composition of the phospholipids from bacterial membranes have not yet revealed a definite pattern. Phosphatidic acid, phosphatidyl ethanolamine, phosphatidyl choline and phosphatidyl glycerol have all been found. One type of phospholipid usually tends to predominate.

There is a dearth of information regarding the proteins and the physical properties of Gram-positive cell membranes. Salton and Netschey (1965) were able to disaggregate the plasma-mesosomal membrane complexes from *Sa. lutea* and *M. lysodeikticus* by treatment with 1% (w/w) Nonidet 40 (a non-ionic detergent). The product in each case was material of low sedimentation coefficient ($S^{1\%}_{20, w}$ = 3·1 for *M. lysodeikticus* and 3·3 for *Sa. lutea*). The membranes from *M. lysodeikticus* could be disaggregated by 3–4 min exposure to ultrasound to a product sedimenting at 4·2S. Unfortunately the biochemical properties of these membrane fragments were not reported, although they were presumably detergent stabilized. The procedures adopted for the resolution of proteins from the lipo-protein structures in higher organisms would seem to be a useful starting point for the study of the proteins associated with bacterial mem-branes.

3. *Cytomembranes from Gram-negative bacteria*

The isolation of cytomembranes from Gram-negative bacteria (other than the photosynthetic and nitrifying bacteria) is generally complicated by the increased complexity in the structure of the outer cell layers com-pared with the Gram-positive strains. It is therefore difficult to ensure complete removal of the cell wall components without degrading the plasma membrane. The distinction between cell wall and cell membrane is readily appreciated with Gram-positive bacteria; the lysozyme sensitive wall appears in electron micrographs as an amorphous layer 200–800Å thick without the electron dense, layered appearance of the plasma mem-brane which it overlies. Furthermore, the cell may be plasmolysed and the membrane shown to contract from the cell wall. The outer layers of Gram-negative bacteria show a much more complex structure with a rigid component (a mucopeptide 20–30Å thick) sandwiched between an outer double track layer (possibly lipopolysaccharide) and the inner, true plasma membrane (Murray *et al.*, 1965). Furthermore, electron micrographs of plasmolysed *E. coli* show that the plasma membrane remains attached to the wall at many points (Bayer, 1968). Many of the methods for forming spheroplasts from Gram-negative organisms share the common feature of attack upon the structure of the rigid mucopeptide component, e.g., (i) lysozyme with or without the addition of EDTA (ii) phage enzymes (iii) growth in the presence of penicillin or (iv) the growth of diaminopimelic acid (DAP) requiring mutants with limiting amounts of DAP (see Hughes *et al.* (this Volume, p. 1) and McQuillen (1960) for practical details of these procedures also Work, this Series, Vol. 6). These methods, together with osmotic lysis from suspensions in glycerol and the mechanical methods of cell breakage, share the common disadvantage that the plasma membrane is either comminuted or contaminated with other wall components. It is therefore questionable whether a procedure has yet been described for the isolation of intact cytomembranes from Gram-negative bacteria uncontaminated by the wall components. Possible exceptions to this generalization are the isolation of a membrane structure from *Mycoplasma* species (Razin *et al.*, 1963) and the procedure described for *E. coli* by Birdsell and Cota-Robles (1967) (see below). For the remainder of this Section the term "envelope" will be used to distinguish preparations of plasma membrane contaminated with cell wall from the true membrane preparations.

Mesosomal and other intracytoplasmic elements are not common in Gram-negative bacteria grown under their usual cultural conditions. However Schaitman and Greenawalt (1966) have observed a mesosomal like structure in *E. coli* and Smith (1960) has observed a rudimentary intracytoplasmic membrane system in methionine-starved *E. coli* and unusual membrane elements were observed by Morgan *et al.* (1966) in magnesium starved cells. To the writer's knowledge there are no reports detailing the separate isolation of these particular membranes.

Salton and Ehtisham-ud-Din (1965) prepared cell envelopes from *E. coli* by disintegration of the cells with glass beads in a Mickle shaker (10×1 min shaking periods at 4°C). The beads were removed by filtration, the extract treated with desoxyribonuclease (20–40 μg/ml) and the envelopes sedimented by differential centrifugation at 20,000 g for 30 min. The packed pellet of cell envelopes was washed by resuspension and sedimentation in 0·05M tris-HCl, pH 7·5. Apart from noting that this fraction contained a higher proportion of cytochromes and NADH$_2$ oxidase activity than the cytoplasm no further characterization was attempted. Birdsell and Cota-Robles (1967) first plasmolysed late-lag/early-log phase cells of *E. coli* (5×10^8 cells/ml) in 0·5M sucrose and 0·01M tris-HCl, pH 7–9. These cells were converted to plasmolysed, osmotically sensitive *rods* ("osmoplasts") by treatment with 20 μg/ml lysozyme. The osmoplasts could be converted to spheroplasts by diluting 1 : 1 with tris buffer. The spheroplasts were then lysed by further dilution (1 : 5) with distilled water to give "ghosts". Direct lysis of the osmoplast, by dilution, would also give "ghosts". The electron micrographs of the spheroplasts revealed coiled cell wall elements almost detached from the cytoplasmic membrane which was clearly visible as a unit membrane surrounding the cytoplasm (i.e., a true protoplast). This technique would appear to offer a method to resolve the plasma membrane and cell wall components in this organism. A brief mechanical or ultrasound treatment of the "ghosts" from the lysates may detach the coiled wall elements from the plasma membrane. The plasma membranes may then be recovered in a reasonably intact form by gradient centrifugation.

Norton et al (1963) and Gray and Thurman (1967) have described the preparation of envelopes from P. aeruginosa. Norton et al. (1963) directly lysed the cells by suspending 10·0 gm (wet weight) in 300 ml of 0·1M tris-HCl, pH 8·0 and adding 48 mg of lysozyme in 32 ml of 4 mg/ml EDTA. 8 mg of desoxyribonuclease were added and after 30 min at room temperature the mixture was not viscous. The envelopes were recovered by differential centrifugation at 37,000 g for 15 min and the pellet washed 3–4× by resuspension and sedimentation at 37,000 g for 15 min from 0·1M tris-HCl, pH 8·0, containing 0·01M MgSO$_4$. The final suspension was centrifuged at 4300 g to remove large contaminants. The total dry weight of this preparation was accounted for by protein (63%), lipid (35%), hexose (1·7%) and RNA (0·5%). Electron micrographs of the preparation showed spherical "ghosts" about 1–2 μ in diam. although the proportion of the preparation in this form was not recorded. Gray and Thurman (1967) used ballotini in a Mickle disintegrator to break the cells and consequently did not claim an intact membrane preparation although their detailed supporting analyses strongly indicate that their procedure gave a comminuted, true membrane fraction. The cells of P. aeruginosa were suspended in ion-depleted water and shaken with glass beads for 10 min. The ballotini were removed by passing the mixture through a glass sinter and the remaining whole cells removed by a centrifugation at 5000 g for 15 min. A subsequent differential centrifugation at 17,000 g for 30 min was employed to sediment the ruptured cell walls and the comminuted membrane fragments were deposited by a further centrifugation at 40,000 g for 30 min. The membrane fragments were washed first by resuspension in ion-depleted water then twice in 1M NaCl and finally by a further four resuspensions in ion-depleted water. After each resuspension the membranes were sedimented by a 30 min centrifugation at 40,000 g. This procedure gave a very reproducible preparation of membranes judged by the accompanying chemical analyses. The detailed analyses revealed that diaminopimelic acid and muramic acid were absent. Protein accounted for 52% of the dry weight, lipid 38%, firmly bound lipid 7% and hexose 2·3%. Orcinol reacting material was absent. A complete amino acid analysis for the membrane protein was also reported. Burrous and Wood (1962) used lysozyme-EDTA treatment at pH 7·6 in the presence of 0·32M sucrose to obtain "spheroplasts" of P. fluorescens, approximately 2·8 spheroplasts being produced per rod. The spheroplasts were recovered by centrifugation at 1000 g, resuspended and lysed by the addition of two volumes of 0·005M sodium citrate. From this lysate by centrifugation first at 17–25,000g for 30 min and then at 40,000 g for 30 min they obtained cell envelope preparations. These sediments of "envelopes" were combined, resuspended in 0·55M sucrose and centrifuged at 80–100,000 g for 30–40 min. This final step was repeated to yield a hard pink pellet of material which contained 22% protein, 13% RNA and 4% DNA. This preparation would appear to give a comminuted cell envelope preparation which is heavily contaminated with nucleic acids.

In the course of their demonstration of the intracytoplasmic membrane of *Azotobacter agile* Pangborn *et al.* (1962) described a procedure for osmotically lysing this organism to preserve the intracytoplasmic membrane structure attached to cell envelopes. A pellet of cells was dispersed in 3M glycerol and ejected via a syringe into 10 volumes of vigorously stirred 0·05M tris-HCl, pH 7·5 containing 0·001M $MgSO_4$. 96% of the cells were lysed. Desoxyribonuclease was added and after a brief period the preparation was centrifuged at 10,000 g for 15 min to sediment the cell envelopes with their attached intracytoplasmic membranes. This preparation was washed once with the buffer and resedimented at 10,000 g. The cell wall material was clearly not removed in this preparation but it was noted that if sonic or glass bead disintegration were adopted they tended to destroy the intracytoplasmic elements and leave the cell envelopes intact.

The *Mycoplasma* and *Halobacteria* are organisms which give a Gram-negative staining reaction but the structure and organization of their outer cell layers differs from the majority of Gram-negative bacteria. *Mycoplasma* species are organisms without cell walls and do not contain diaminopimelic acid or muramic acid. The cytoplasm of these organisms is simply bounded by a unit membrane. *Mycoplasma laidlawii* and *My. bovigenitalium* are both sensitive to osmotic lysis when log-phase cells are suspended in solutions of low tonicity (Razin, 1963). Razin *et al.* (1963) were therefore able to prepare a true plasma membrane fraction from these organisms. Log-phase cells were harvested and washed with 0·25M NaCl. Approximately 1–2 g wet weight of cells were then suspended in 300 ml of deionized water to lyse the cells. The viscous suspension was centrifuged at 20,000 g for 5 min. The sediment was suspended in deionized water, centrifuged at 20,000 g for 5 min and the small sediment discarded. The supernatants from the two centrifugations were pooled and centrifuged at 30,000 g for 40 min. The sediment was washed four times with deionized water and the final resuspension was the membrane preparation. This membrane fraction accounted for 32–36% of the dry weight of the organisms. The membrane dry weight was made up of 47·3–59·2% protein, 36–37·3%

lipid, 2–3% RNA and 1% DNA. The membrane lipid was predominantly acetone insoluble lipid and represented almost all the cell lipid of the *Mycoplasma*.

Halobacterium salinarium, cutirubrum and *halobium* and a marine pseudomonad NCMB 845 have all received intensive study in view of the unusual behaviour of their outer cell layers. In thin sections of permanganate fixed *H. halobium* the cell envelope appears to be a single unit membrane (90–120Å thick) with approximately equal contributions by the two electron opaque layers and the transparent layer. No outer cell wall layers are visible. However in a limited number of preparations a 5 layered membrane structure was observed quite distinct in appearance from the 5 layered structure regularly observed in *H. salinarium* (Brown and Shorey, 1963). In the latter organism the cell envelope appeared to consist of a triple layered, thin cell wall (100–120Å) overlying a typical bilayer plasma membrane. Cell envelopes were prepared from both of these organisms as follows. Following growth in a high salt medium (containing 25% w/v NaCl) the cells were suspended in a half strength salts medium and disintegrated by shaking with glass beads for 20 min in a Mickle disintegrator at 4°C. The beads were removed and the broken cell envelopes isolated by four cycles of alternate high (12,200 g for 30 min) and low-speed (1000 g for 10 min) centrifugation in the half strength salts. Electron micrographs of the final envelope preparations from both organisms revealed a well-defined simple unit membrane structure (however see below). These preparations did not contain diaminopimelic acid. Proteins amounted to 65% of the envelope dry weight in *H. halobium* and 75% in *H. salinarium*. The cell envelopes from *Pseudomonas* NCMB 845, obtained by an identical preparation, autolysed proteolytically (Brown, 1961). The envelopes from *H. halobium* retained their unit membrane appearance in 2·2M NaCl but dissolved rapidly if the salt solution was diluted. In this case the dissolution was shown to be non-enzymatic and probably due to an excess negative charge on the envelope proteins. Macromolecular sub-units were produced which, in 0·1M NaCl at 18·8°C, sedimented as a single, sharp boundary with an $S_{20,w} = 3·9$ (Brown, 1963).

Mohr and Larsen (1963) also found that with *H. salinarium* the rod shaped organisms in 4·3M NaCl were converted to spheres and then lysed as the salt concentration was lowered. Isolated cell envelope fractions showed a similar effect. The cell envelope fractions for this study were prepared by shearing the cells with glass beads in an homogenizer for 5 min in the presence of a salts solution containing (% w/v) 25 NaCl, 0·5 $MgSO_4.7H_2O$, 0·02 $CaCl_2$ at pH 7·0. The beads were removed by a sintered glass filter and the envelope fragments sedimented by centrifugation at 7–15,000 g for 30 min. The sediment was washed by resuspension in the salts medium followed by centrifugation at 30,000 g for 40 min.

The cells and the isolated envelopes of *H. cutirubrum* also dissociate non-enzymically in the absence of salt to form smaller units (Kushner, 1964). Since the organisms described as *H. salinarium* and *H. cutirubrum* may be the same species this result probably confirms the observation of Mohr and Larsen (1963). Kushner *et al.* (1964) have also adopted a similar method to Mohr and Larsen (1963) for the preparation of cell envelopes from *H. cutirubrum*. The organisms were suspended in 0·01M tris-HCl, pH 7·2 containing 4·5M NaCl and 0·1M MgCl$_2$ to a density of 20 mg dry weight of cells/ml. Ballotini were added and the mixture shaken at 250 cycles per min for 1 h at 0°C. About 20% of the cells were ruptured and the remaining whole cells were removed by 2 or 3 centrifugations at 6000 g for 10 min. The cell envelopes were then sedimented from the supernatant by a 30 min centrifugation at 15,000 g. The pellet was washed four times in the buffered salts solution and the envelopes finally recovered by centrifugation at 15,000 g. If this preparation of cell envelopes was exposed to a medium of low ionic strength (e.g., <0·1M NaCl) the envelopes dissociated and the products were only incompletely sedimented by two hours centrifugation at 105,000 g. The addition of NaCl to a dissociated preparation of envelopes gives polydisperse aggregates with a mean sedimentation coefficient of 5·2S (Onishi and Kushner, 1966). Kushner *et al.* (1964) found that the majority of the cellular phosphatides were associated with the envelope fraction but could not detect diaminopimelic acid or muramic acid. Hexose and hexosamines were detected also small amounts of nucleic acids and the RNA : DNA ratio was higher than that in the intact cells.

The precise definition of the envelope preparations from the *Halobacteria* is difficult. The electron microscope observations, the chemical analyses and the non-enzymic disaggregation of the outer layers strongly suggest that these organisms lack the typical rigid mucopeptide component found in other bacteria. The preparations which have been described above may therefore consist of the plasma membrane plus an outer wall and may not be the homogeneous plasma membrane system originally suggested by Brown and Shorey's (1963) electron micrographs.

4. *Cytomembranes from the photosynthetic bacteria, the nitrifying bacteria and the blue green algae*

The wide range of cultural conditions in which the photosynthetic bacteria may be grown (i.e., ranging from strictly anaerobic, photoautotrophic through photo-organotrophic to highly aerobic heterotrophic growth) and the consequent profound effect upon the composition and extent of the intracytoplasmic membrane structure make it difficult to specify a cytomembrane preparation from these organisms. The facultative phototrophy of certain organisms, e.g., *Rh. spheroides* provides a useful experimental system for the study of the development of the photo-apparatus; the transition from aerobic, heterotrophic growth to semi-anaerobic, photosynthetic activity is accompanied by the development of an extensive intracytoplasmic membrane system infiltrated with the photopigments synthesized in response to light and lowered oxygen tension. The conditions of culture are therefore of paramount importance in defining the cytomembranes of the photosynthetic bacteria.

The photosynthetic bacteria are Gram-negative organisms and their outer layers generally exhibit the basic multilayered cell wall structure over-lying a typical plasma membrane common to other Gram-negative bacteria. Although there are few reports specifically concerned with the nature of the cell wall and its relationship with the plasma membrane in the photo-synthetic bacteria it may be assumed that similar problems will be encoun-tered with these organisms to those described for the Gram-negative heterotrophs (Section V.C.3). Thus very few of the procedures des-cribed for the isolation of membranes have deliberately attempted to ensure removal of cell wall material and it is not surprising that in some early studies Newton (1958) found that antisera prepared against intact cells of *Chromatium* cross reacted with membrane preparations bearing the photo-pigments from this organism. Furthermore, since the great majority of the pigmented intracytoplasmic membranes (and vesicles) exhibit some point of continuity with the plasma membrane, a mixed membrane preparation is frequently obtained and has rarely been resolved. The membranes/vesicles with which the photopigments are associated are not the only intracytoplasmic membrane structures which have been reported in photo-synthetic bacteria. In the *Chlorobacteriaceae* mesosomal elements (unpig-mented?) are frequently present.

Many of the procedures described for the isolation of cytomembranes from photosynthetic bacteria have been dominated by the desire to prepare a photochemically active unit bearing the bulk of the photopigments. The relationship of such preparations to the intracytoplasmic membrane skeleton has therefore been obscure. The preponderance of regular, small vesicles in photochemically active preparations from many sources has lead to the retention of the term "chromatophore". In view of the variety of intracytoplasmic membrane arrangements found in the photosynthetic bacteria (see Section V.C.4 (i–iii), also Lascelles (1968) and Cohen-Bazire and Sistrom, 1966) it is evident that for many organisms the preparative procedure which has been adopted has comminuted the complex network of membranes to give rise to the uniform preparation of vesicles. It is of course possible that these regular "chromatophore" units are the minimal active sub-units derived from the membranes but the observations of Holt and Marr (1965a) leave little doubt that "chromatophores" arise by comminution of the membrane from *R. rubrum*. The rationale behind their experiments, i.e., that under controlled conditions of cell breakage all independent structures within the cytoplasm should be released at the same rate as cells are disrupted, provided the breach in the cell envelope is adequate, could be profitably applied to other photosynthetic bacteria (e.g. see Sykes *et al.*, 1965). This would then indicate whether the photo-pigments are contained within an extensive or independent cytoplasmic structure. Ribosomes would appear to provide a universal "marker" to monitor the release of an independent cytoplasmic structure. Kurunairat-nam *et al.* (1958) and Tuttle and Gest (1959) have also shown that *R. rubrum* may be induced to form spheroplasts by treatment with lysozyme and EDTA without release of pigment. When the spheroplasts are osmotically lysed the pigment is released as part of a complex which is sedimented during a low speed centrifugation (14,500 g for 20 min). This centrifuga-tion may be compared with that used by Schachman *et al* (1952) (20,000 rpm for 1 h) to give "chromatophore" preparations from this organism. The isolation of an intact, intracytoplasmic membrane array from *Rhodo-pseudomonas viridis* has been reported by Giesbrecht and Drews (1966). The method of cell disintegration must therefore be carefully applied when membrane preparations are to be isolated from the photosynthetic bacteria. The controlled disintegration of the cells may indicate the organization of the photosynthetic apparatus, more protracted application of the disintegra-tion method may comminute a complex organization but give functional units and excessive disintegration may lead to loss of function as well as structure. Once again the use of muralytic enzymes to give osmotically fragile spheroplasts would seem to offer the safest method of cell rupture if preservation of the intracytoplasmic membrane/vesicle structure is required.

Pfennig (1967) has attempted to group the photosynthetic bacteria accor-ding to their intracytoplasmic membrane structures. However for the purpose of describing procedures for the isolation of these membranes, the bacteria will be grouped conventionally, i.e.,

(*i*) **Chlorobacteriaceae**
(*ii*) **Athiorhodaceae**
(*iii*) **Thiorhodaceae**

whilst recognizing that within groups (ii) and (iii) a wide variety of intra-cytoplasmic membrane arrangements are encountered.

(*i*) **Chlorobacteriaceae** The fine structure of *Chlorobium thiosulphato-philum* and *Chlorobium limicola* has been studied by Cohen-Bazire *et al.*

(1964) and *Chloropseudomonas ethylicum* by Holt *et al.* (1966). These bacteria all share a common basic pattern of membrane structure, i.e., a plasma membrane below the cell wall and within this a series of peripheral vesicles each 300–400Å wide and 1000–1500Å long. The number of these vesicles fluctuates with the light intensity during growth and accordingly with the amount of *Chlorobium* chlorophyll. When isolated these vesicles are found to carry the bulk of the cellular photopigments and in the case of *C. thiosulphatophilum* have been shown to engage in cyclic photophosphorylation (Sykes and Gibbon, 1967). The vesicles may be isolated intact and free from plasma membrane elements (cf. *Athiorhodaceae*) and a clear distinction in structure and function may be drawn between these membrane fractions. The rates of release of the vesicles during cell breakage have been studied with *C. thiosulphatophilum* (Sykes *et al.*, 1965) and *C. ethylicum* (Holt *et al.*, 1966). It would seem that the vesicles are free cytoplasmic elements in both organisms although in *C. ethylicum* some small connection may exist between vesicles and/or the plasma membrane. The mesosomal elements observed in *C. thiosulphatophilum* by Cohen-Bazire *et al* (1964) have not been isolated as independent membrane units.

The photosynthetic vesicles from *Ch. ethylicum* were prepared by Holt *et al.* (1966) as follows. The cells (approximately 5 mg dry weight/ml) were suspended in 0.02M phosphate buffer pH 7.0 containing 10^{-3}M $MgSO_4$ and broken by passage through a French pressure cell at 20,000 p.s.i. The cell homogenate was incubated with 0.5 μg/ml desoxyribonuclease for 30 min and then centrifuged at 18,000 g for 30 min to sediment whole cells and broken cell envelopes. The majority of the pigment was in the supernatant which was centrifuged at 104,000 g for 1 h. The pellet was resuspended in the phosphate-magnesium buffer and then subjected to rate-zonal centrifugation or electrophoresis in a sucrose gradient (Holt and Marr, 1965b). For rate zonal centrifugation the gradient was constructed by layering into a centrifuge tube (6.5×51 mm) 2.0 ml each of 2.0, 1.75, 1.50, 1.25, 1.0, 0.75, 0.50 and 0.25M sucrose solutions in 0.02M phosphate buffer, pH 7.0 containing 10^{-2}M $MgSO_4$. A linear gradient was established by leaving the tube to stand for 16 h and a 3 mm deep sample band of the resuspended pellet fraction was layered on the top of the gradient. The tubes were centrifuged in a swinging bucket rotor for 90 min at 30,000 rpm. Two pigmented bands were obtained, (cf *R. rubrum* later), the band nearest the origin (i.e., centre of rotation) contained the vesicles with some contamination by ribosomes whilst the other band consisted of membrane and cell wall fragments. The vesicle fraction purified by density gradient electrophoresis showed a significant reduction in contamination by cell membrane and wall fragments although some contamination by ribosomes still persisted. For this technique Holt and Marr (1965b) overlaid the bottom of an electrophoresis cell with a solution of 3M sucrose in 0·02M tris-HCl, pH 7·0 and added 0·1M L-cysteine to prevent oxidation of the chlorophyll. Each limb of the cell was then filled with a gradient of 0·25 to 2M sucrose in tris-HCl-cysteine buffer and a 2 mm band of the resuspended pellet fraction layered over this. A potential of 250 volts at 25 m.a. was then applied to platinum-stainless steel electrodes linked to each limb of the cell via an agar-KCl bridge and the circuit was completed across the bottom of the cell by a 3M sucrose solution saturated with KCl. The temperature of the cell was held at 4°C. It is of interest to note that alternative methods of cell disintegration, e.g., shaking with glass beads and sonication also gave preparations of intact vesicles from this organism.

Sykes *et al.* (1965) also found that intact vesicle preparations could be obtained in good yield using a variety of disintegration methods against *C. thiosulphatophilum.* Thus, crushing the cell paste in the Hughes press, 90 sec of sonic treatment and lysozyme digestion and lysis of the cell wall all gave cell extracts from which active, intact vesicles with identical sedimentation characteristics could be obtained. These vesicles corresponded in dimensions to the vesicles observed in electron micrographs of thin sections of this organism. In all cases the vesicles were isolated from the cell homogenate by first centrifuging at 20,000 g (max) for 30 min to remove unbroken cells and cell debris. The supernatant which contained the bulk of the *Chlorobium* chlorophyll was then centrifuged for 120 min at 68,475 g (max) in a swinging bucket rotor. The pellet retained the pigment and was resuspended in buffer and the centrifugation at 68,475 g (max) repeated for 90 min. Once again the pigmented pellet was carefully dispersed in buffer and given a 10 min centrifugation in a bench centrifuge (3600 rpm) to remove aggregated material. The supernatant from this step contained the bulk of the cellular chlorophyll and consisted of vesicular elements with some contaminating ribosomes. Final purification was effected by density gradient centrifugation (86,300 g (max.), 90 min, 3°C) through linear 10–40% sucrose gradients. The buffer used throughout this preparation was 0·01M tris-0·001M magnesium acetate pH 7·5. At the later stages of the preparation the vesicles showed a strong tendency to aggregate. The vesicles had $S^0_{20, w}$ values of 116 and their sedimentation was unaffected by prior ribonuclease treatment but was strongly dependent upon concentration. The preparations were found to contain 4·3% of orcinol reacting material and this may have been due to contaminating ribosomes. The gross chemical composition of the vesicles was very similar to that described by Bull and Lascelles (1963) for *Rh. spheroides* chromatophores. This *Chlorobium* vesicle preparation was subsequently shown to catalyse cyclic photophosphorylation and also contained an active inorganic polyphosphatase producing ATP in the dark from endogenous polyphosphate

(Sykes and Gibbon, 1967). The photophosphorylation activity decayed as the preparation was stored in foil wrapped tubes in an atmosphere of nitrogen in the refrigerator at 4°C. *Chlorobium* vesicle preparations obtained *via* lysozyme digestion of the cells gave an additional pigmented band when layered over 10–60% linear sucrose gradients and centrifuged for 60 min at 72,300 *g* (max.). This band accounted for 17% of the total pigment in the final preparation and sedimented more rapidly than the bulk of the pigment. A similar observation was made during their preparation of *C. thiosulphatophilum* vesicles by Cohen-Bazire *et al* (1964) and the material was tentatively identified as cell envelope fragments. The vesicles were prepared by these workers starting with a cell homogenate prepared in the French pressure cell. Unbroken cells were removed by a brief low speed centrifugation and the cell extract was then centrifuged for 2 h at 100,000 *g*. The pellet contained the bulk of the chlorophyll and was resuspended in the buffer (0·01M tris-HCl pH 7·5 containing 10^{-3}M $MgSO_4$) and a sample layered on a linear sucrose gradient (0·5 to 2M). 87% of the chlorophyll was recovered from the gradient in a single band after centrifugation for 2 h at 25,000 rpm. The material in this band was heterogeneous but predominantly consisted of *Chlorobium* vesicles.

(*ii*) **Athiorhodaceae.** The diversity and complexity in the structure of the intracytoplasmic photopigmented membranes of the *Athiorhodaceae* rules out the possibility of a unified method for their preparation. Furthermore the isolation of an intact membrane complex from any one species is difficult since these highly structured complexes are readily comminuted, quite often to regular, active sub-units (the "chromatophores"). This subsection will therefore group the organisms according to their membrane structures, as suggested by Pfennig (1967), and will survey the methods of preparation and results which have been obtained with individual species within each group.

(a) *Rhodopseudomonas viridis, Rhodopseudomonas palustris and Rhodomicrobium vanniellii.* These species share the common characteristics of multiplying by budding and intracytoplasmic, photopigmented membranes organized as a series of peripheral lamellae arranged parallel to the long axis of the cell. The lamellae may extend around the periphery and finally close to form a series of concentric rings (Boatman and Douglas, 1961; Giesbrecht and Drews, 1962; Trentini and Starr, 1967).

The isolation of this membrane structure in an intact form from *Rh. viridis* has been reported by Giesbrecht and Drews (1966). Here the bacteria were suspended in 5×10^{-3}M phosphate buffer, pH 7·1 containing 10^{-3}M $MgSO_4$ and 10^{-5}M mercaptoethanol and homogenized with 0·1 mm glass beads for 22 min at 0°C. The glass beads and cell debris were removed by a preliminary filtration through a glass sinter followed by a 10 min centrifugation at 4300 g. The crude cell free extract was then centrifuged at 90,000 g for 45 min and the pellet dispersed in distilled water and layered over a linear 15–60% sucrose gradient in a swinging bucket rotor. After 60 min centrifugation at 60,000 g the pigmented membrane fraction was confined to a 5 mm wide zone in the upper third of the tube. Surprisingly, this method of preparation gave a better yield and a cleaner preparation of the intact membrane complex when compared with an alternative preparation in which the bacteria were broken in a French pressure cell in the presence of 0·015M tris-HCl buffer, pH 7·5 containing 0·2% KCl and 0·05% $MgSO_4$. In this case the unbroken cells and debris were removed by a 5 min centrifugation at 9000 g and the supernatant (1 vol.) mixed with 1·5 volumes of a solution containing 58% (w/v) RbCl, 0·15M tris, and 0·077M $MgSO_4$, pH 7·8. Nine ml of this mixture were placed in a centrifuge tube and overlaid with 3 ml of buffer having the same composition as the mixture but without RbCl and centrifuged at 8000 g [see Gibbs et al. (1965) and Section below for R. molischianum]. The pigmented membrane complex was again located as a band in the upper part of the tube. Both methods of preparation, although differing in yield, preserved the intact membrane arrangement and fine structure observed in the whole cell. Unfortunately the biochemical properties of these preparations were not recorded.

Garcia et al. (1968a) prepared a pigmented membrane fraction from Rh. palustris by first suspending the cells in 0·01M tris-HCl buffer (pH 7·5) and exposing the suspension to ultrasound for two minutes. The material which sedimented from this preparation between 15 and 115,000 g (time unspecified) was resuspended in the buffer and applied to the top of a 60% (w/v) CsCl solution in a centrifuge tube which was then centrifuged for 60 min at 100,000 g. The pigmented material remained on the top of the CsCl and was recovered. Electron micrographs of negatively stained preparations showed many large flattened vesicles 0·2 μ or more in diameter together with smaller contaminants.

(b) *Rhodospirillum molischianum, Rhodospirillum fulvum and Rhodospirillum photometricum.* These organisms multiply by binary fission, are strictly anaerobic and their intracytoplasmic, photopigmented membranes consist of infolded plasma membrane elements giving rise to short, discrete, stacked lamellar systems about the periphery of the cell. Initially these tubular intrusions occur independently but then become flattened discs and further intrusions give rise to stacks of discs (2–10 in number) in older cells (Gibbs et al., 1965; Giesbrecht and Drews, 1962; Hickman and Frenkel, 1965a). Gibbs et al. (1965) have described a method for the preparation of a pigmented membrane fraction from *R. molischianum* as follows. The bacteria were suspended in 0·01M tris-HCl, pH 7·8, containing 0·01M $MgSO_4$ and 0·08M NaCl and then ruptured by passage through a French press at 20,000 p.s.i. Cell debris was removed by a 10 min centrifugation at 20,000 *g* and the supernatant contained 95% of the original bacteriochlorophyll. One volume of this supernatant was mixed with 1·5 volumes of the RbCl-tris-$MgSO_4$ mixture and placed in centrifuge tubes as described in the above Section for *Rh. viridis*. The mixture was centrifuged for 4 h at 80,000 *g* when the ribosomes were sedimented and the pigmented, membrane fraction was retained in a band in the upper part of the tube. This material was removed, diluted 5× with the tris-$MgSO_4$-NaCl buffer used for the original cell suspension and centrifuged at 78,000 *g* for 2 h. The pelleted material was resuspended in the buffer, recentrifuged at 78,000 *g* and finally resuspended in the buffer. Whilst this preparation was shown to quantitatively account for the total cellular bacteriochlorophyll its biochemical properties were not recorded and electron micrographs of the preparation were not presented. The extent to which this preparation preserves the structure and function of the original cell membranes is therefore unknown. For instance, an almost identical method of preparation was adopted by Worden and Sistrom (1964) for *Rh. spheroides* (see later Section) and when the final preparation was subjected to zone centrifugation on a linear sucrose gradient two pigmented bands of material were observed. The material in one band was monodisperse having an S value of 153 whereas the other material was heterogeneous with a density varying between 1·15 and 1·18 gm/ml.

Hickman *et al* (1963) have used the same procedure described by Frenkel and Hickman (1959) for *R. rubrum* to isolate a photochemically active preparation from sonically disintegrated *R. molischianum*. The cells, suspended in 0·1M potassium glycylglycine pH 7·5, were disintegrated by 1·5–2·5 min exposure to a 10 Kc Raytheon magnetostriction oscillator. Cell debris was removed by a 30 min centrifugation at 25,000 *g* and the supernatant centrifuged at 54,000 *g* for 50 min to sediment the bulk of the pigmented material. The sediment was dispersed in the buffer and resedimented at 46,000 *g* for 50 min. This sedimented, pigmented material was further purified (i.e., judged by an increase in the 880/280 nm ratio) by layering over and centrifuging through discontinuous sucrose gradients. The first gradient used varied from 0·88M sucrose through 0·59 to 0·29M and finally to buffer. After 90 min centrifugation in a swinging bucket rotor at 24,000 *g* three layers were observed, a red layer in the 0·29M sucrose, a pink layer in the 0·59M sucrose and a non-pigmented band in the 0·88M sucrose. The material in the red layer was then layered over another discontinuous gradient varying from 0·29, 0·59, 0·88 to 1·17M sucrose. A further period of centrifugation at 24,000 g gave a red-pigmented band in the 0·59M sucrose. The material in this band was collected, diluted 1 : 10 with the glycylglycine buffer and sedimented by centrifugation at 50,000 g for 30 min. This final preparation catalysed the light dependent phosphorylation of ADP (150–200 μmoles orthophosphate esterified as ATP/hr/μmole bacteriochlorophyll) and the electron micrographs of negatively stained preparations showed the predominance of regular (0·1 μ diam.), disc shaped units. It was concluded that these discs were individual elements derived from the arrays of flattened discs seen in sections of the whole cells and would, therefore, correspond to "chromatophores".

(c) *Rhodospirillum rubrum and Rhodopseudomonas spheroides.* These two species multiply by binary fission and have intra-cytoplasmic membranes which originate as vesicular or tubular invaginations at many points about the plasma membrane. As these membraneous intrusions into the cytoplasm become more extensive they become bulged at intervals along their length to form a system of linked vesicles. Thin sections of cells therefore reveal invaginations at many points about the plasma membrane and the cytoplasm is packed with regular vesicles when examined in the electron microscope. (Cohen-Bazire and Kunisawa, 1963; Gibson, 1965d; Hickman and Frenkel, 1965b; Holt and Marr, 1965a; Vatter and Wolfe, 1958).

The preparation of photochemically active, pigmented membrane preparations from these organisms has been described by many investigators However the majority of the preparations, whilst yielding an active preparation give rise to disruption of the complex intra-cytoplasmic membranes into regular "chromatophore" units, e.g., the initial, simple preparation of Schachman *et al* (1952) and the later, elaborate preparation by Frenkel and Hickman (1959) also used, as described above, for *R. molischianum.* The appearance of densely packed vesicles throughout the cytoplasm in thin sections of these organisms has no doubt given rise to the anticipation of obtaining a photochemically active vesicle preparation. However since these vesicles are now known to form part of an inter-connected system alternative procedures are required if the intact system of vesicles is to be isolated. In the case of *R. rubrum* the method of preparation described by Karunairatnam *et al.* (1958) and Tuttle and Gest (1959) would appear to isolate the intact membrane complex. For this procedure the cells were washed in 0·1M phosphate buffer (pH 6·8) and resuspended to a final density of 0·4 mg dry weight/ml in the same buffer containing 10% sucrose. Lysozyme (final concentration 0·05%) and EDTA (pH 7·0, final concentration 0·16%) were then added whilst the suspension was stirred. Incubation of the mixture at 30°C for 30 min gave almost total conversion of the spiral shaped cells to non-motile spheroplasts although variations in the cell density, sucrose, lysozyme and EDTA concentrations influenced the rate of this conversion. Magnesium sulphate (final concentration 0·05M) and 0·5 vol. 40% sucrose solution were then added, the former to minimize the tendency of the spheroplasts to clump and the sucrose to ensure stability of the spheroplasts during a brief, recovery centrifugation. The sedimented spheroplasts were lysed by resuspending in water. The viscosity of the lysed preparations varied but in general the "ghosts" could be sedimented by centrifugation at 14,500 g for 20 min (cf. centrifugations for other preparations of "chromatophores" from this organism) and the red, pelleted material contained the majority of the photopigments and would catalyse active photophosphorylation. The sedimentation of this material in a relatively low gravitational field indicates a large structure but unfortunately electron micrographs of this preparation have not been presented.

The preparation of the intracytoplasmic, pigmented membranes from *R. rubrum* described by Holt and Marr (1965b) does undoubtedly preserve some of the interconnections between the vesicles. Their method closely follows their procedure which has been described earlier (p. 164) for vesicles from *Ch. ethylicum*. The centrifugation at 18,000 g to remove cell debris was omitted from this procedure and the *R. rubrum* cells were ruptured by osmotic shock from 6M glycerol suspensions (Robrish and Marr, 1962). The cleanest preparation of membranes, i.e., with negligible contamination by ribosomes was obtained via their technique of electrophoresis in a sucrose gradient (p. 164). Preparations of membranes obtained from French pressure cell extracts of the organism gave two pigmented bands in density gradient centrifugations (cf. preparations described below by Cohen-Bazire and Kunisawa (1960) and Worden and Sistrom, 1964). The material banding at the lower density consisted of pigmented, disc shaped membranes, 700–900Å in diam. occasionally connected together like the similar vesicles seen in thin sections of whole cells. Extensive comminution of the intracytoplasmic membranes (to "chromatophores") had therefore occurred. This preparation was also contaminated by ribosomes (cf. the corresponding preparation obtained via electrophoresis). The pigmented material banding at higher density contained intact cell envelopes with partially extruded intracytoplasmic membrane systems. When this material was given a brief exposure to ultrasound all the pigmented material then came to equilibrium at the lower density position on centrifugation and now consisted of disc shaped membrane units with the same dimensions as the chromatophores. These observations strongly indicate that the chromatophores originate from the comminution of the intracytoplasmic membrane complex and this was confirmed by studying the kinetics of their release (Holt and Marr, 1965a).

Cohen-Bazire and Kunisawa (1960) also found two pigmented bands of differing density in their preparations of "chromatophores" from *R. rubrum*. Here the cells were resuspended in 0·2M phosphate (pH 7·0) containing 0·01M MgSO$_4$ and ruptured in a French pressure cell at 20,000 p.s.i. The unbroken cells and cell debris were removed by centrifugation at 18,000 *g* for 30 min and the supernatant recentrifuged at 104,000 *g* for 60 min. The crude "chromatophore" pellet from this centrifugation was dispersed in tris-HCl (0·33M, pH 7·6) containing 0·01M MgSO$_4$ and layered over a linear density gradient of sucrose (0·5 to 2M). After 90 min centrifugation at 25,000 *g* in a swinging bucket rotor two pigmented bands were observed, one at the density level of 1·77–1·8M sucrose and the other at 1·08M sucrose. The pigmented material banding at 1·08M sucrose was collected, diluted 3–5 × in tris-HCl and magnesium sulphate and repelleted at 104,000 *g*. In the electron microscope this preparation comprised cup/hemisphere shaped membranes flattened by drying and having dimensions approximating those described by Holt and Marr (1965a). This preparation would catalyse active photophosphorylation of added ADP. Other crude preparations of photochemically active "chromatophores" from *R. rubrum* have been obtained by grinding the cells with alumina (Horio *et al.*, 1965) or by sonic disintegration of the cells followed by simple differential centrifugation at 100,000 *g* for 60–90 min (Garcia *et al.*, 1966a). Finally, in connection with *R. rubrum* two papers by Oelze *et al.* (1969) and Oelze and Drews (1969) describe the resolution, in Ficoll gradients, of the cytoplasmic membrane from the chromatophore fraction. The bacteria were washed in tris-HCl buffer and suspended in this buffer for disintegration in the French pressure cell. The unbroken cells and cell debris were removed by a centrifugation at 15,900 *g* for 20 min and the cell-free extract then centrifuged for 1 h at 54,491 *g* or 313,934 *g* .The pellets from these centrifugations were dispersed in buffer to form the crude chromatophore preparations for further investigation. 0·3 ml samples of this preparation (obtained by a 54,491 *g* spin) when layered over 1·02–1·06 g/cc Ficoll gradients and centrifuged for 120 min at 82,860–159,203 *g* gave two pigmented bands. The lower (denser) band contained "chromatophores" in a purified form. The particles were 730Å in diam. and sedimented as a single peak in the analytical ultracentrifuge. The material in the upper band consisted of particles of average diameter 360Å and this material was also obtained from aerobic, dark grown cells and carotenoidless mutants of *R. rubrum*. Thus whilst this method does not offer an improved way of preparing intact cytoplasmic membranes from *R. rubrum* the resolution of a membrane fraction, identified in this and a subsequent paper (Oelze *et al.*, 1969) as plasma membrane, from the conventional chromatophore material opens up the practical possibility of studying the interrrelationships of these membranes. Newton (1960) had earlier attempted to tackle this problem by the preparation of antisera against *R. rubrum* chromatophores. These antisera were shown to contain antibodies which precipitated with colourless antigens present in dark grown cells as well as those specific to the pigmented material of light grown cells. Oelze *et al.* (1969) were able to show that, following treatment with a phenol-formic acid-water mixture (2 : 1 : 1 v/v), the chromatophores gave 5 bands on polyacrylamide gel electrophoresis whereas the material in the other fraction gave only one of these bands, i.e., that characteristic of cytoplasmic membrane material. Direct electrophoresis of the cytoplasmic membrane fraction in 0·5% agar gels gave two zones, the fastest migrating contained protein only and the other protein, pigment, phospholipid and succinic dehydrogenase.

Bull and Lascelles (1963) employed the same method for preparing "chromatophores" from *Rh. spheroides* and *Rh. capsulata*. Whole cells suspended in 40 mM tris-HCl (pH 7·5) were ruptured by 4 min exposure to ultrasound and any unbroken cells and cell debris removed by a 10 min centrifugation at 25,000 *g*. The supernatant was then centrifuged at 105,000 *g* for 60 min and the material which sedimented was resuspended in the tris buffer and layered over a discontinuous sucrose gradient formed by layering 1·5 ml portions of 0·29, 0·59 and 1·17M sucrose solutions made in tris-HCl buffer. After 60 min centrifugation at 125,000 *g* in a swinging bucket rotor a band, containing the bulk of the applied pigment, was found at the 0·59–1·17M sucrose interface. The ribosomes in the layered aliquot were concentrated in the 0·29–0·59M sucrose layers. The pigmented band consisted of "chromatophores", homogeneous by electron microscopy and, with the exception of the pigments, having a chemical composition similar to non-pigmented bacterial membranes [58% protein, 24·6% lipid and approximately 1% of the dry weight as nucleic acid]. The nucleic acid, usually RNA, may represent contamination by ribosomes and may be removed by ribonuclease digestion without affecting the other properties of the chromatophores. Ribosomes are a frequent contaminant in chromatophore and vesicle preparations since the sedimentation coefficients of the latter are a strong function of concentration (e.g. see Sykes *et al.* (1965) and Worden and Sistrom, 1964) and therefore in crude extracts their sedimentation coefficients may sufficiently approach those of ribosomes to make resolution difficult. In their preparation of *Rh. spheroides* chromatophores Worden and Sistrom (1964) used a RbCl gradient *iso*-pycnic with the pigmented membranes to resolve them from ribosomes. The organisms were suspended in 0·02M tris-HCl (pH 7·8) containing 0·08M NaCl and 0·01M MgSO$_4$ and ruptured in a French pressure cell. Centrifugation at 20,000 *g* for 15 min gave a crude cell free extract and this was mixed with a volume of tris-HCl (0·1M, pH 7·8) containing sufficient RbCl and

$MgSO_4$ to give a final concentration of 27% RbCl and 0·05M $MgSO_4$. The mixture was centrifuged for a maximum of 2 h at 40,000 rpm when the pigmented material was concentrated at the top of the tube and the ribosomes were packed at the bottom. The pigmented material was removed, diluted with the tris-chloride-NaCl-$MgSO_4$ buffer and resedimented at 40,000 rpm. The sedimented material was again dispersed in buffer and layered on a 35–47% linear sucrose gradient prepared in the buffer After 7–12 h centrifugation at 23,000 rpm in a swinging bucket rotor (SW 25) the pigmented material resolved into two bands, one at a density of 1·14 g/ml and the other at 1·15–1·18 g/ml (cf. *R. rubrum* above). The material in the lower density band was monodisperse in the ultracentrifuge with a sedimentation constant of 153S (cf. Gibson's value below). This material (and the heavier fraction) contained a negligible amount of RNA and muramic acid and amino sugars were not detectable. The amount of the light fraction depended upon the pigment content of the cells and was absent in cells devoid of bacteriochlorophyll. This was the chromatophore fraction. The heavier fraction was presumed to be less dense than the corresponding fraction observed in *R. rubrum* by Cohen-Bazire and Kunisawa (1960) since it contained less RNA. This fraction was grossly heterogeneous, membraneous material with a lower bacteriochlorophyll : carotenoid ratio than the lighter fraction. If this material was again passed through the French pressure cell all the pigmented material then banded at the lower density (1·14 g/ml), was monodisperse and had a sedimentation constant of 153S. However, it differed in its bacteriochlorophyll spectrum from the light fraction. The photophosphorylating activity of these preparations was not recorded.

Gibson (1965a; 1965b) has used *iso*-pycnic centrifugation in linear CsCl gradients to obtain good yields of highly purified "chromatophores" from *Rh. spheroides*. For this the organisms were washed with water, suspended in tris-chloride buffer (0·002M–0·01M pH 7·5) and ruptured in the French pressure cell. The homogenate was centrifuged at 40,000 rpm for 2 h to sediment unbroken cells, cell envelopes and ribosomes together with all the pigmented structures. The pelleted material was resuspended in tris-chloride buffer (one twentieth the volume of the original extract) and a maximum of 2·0 ml layered over linear CsCl gradients constructed from equal volumes of 0·6M and 1·66M CsCl in a Spinco SW 25·1 tube. The gradients were centrifuged at 25,000 rpm for 2·5–3 h. The 0·5 ml fractions from these tubes which contained the pigmented material were pooled, diluted with an equal volume of tris-chloride buffer and the material resedimented by centrifugation at 40,000 rpm for 2 h. The pelleted material was dispersed in buffer and stored at 5°C. The nucleic acid content of this preparation was twice as high as that recorded by Worden and Sistrom (1964), the bulk of which was DNA. When examined in the electron microscope negatively stained preparations of this material were found to consist entirely of particles of similar size, circular in profile and resembling discs on edge. The dimensions of these pigmented particles exactly co-incided with the dimensions of the membrane bounded vesicles observed by Gibson (1965d) in thin sections of the whole cell (average diameter 570Å). Gibson found that the sedimentation constant for this preparation was 168S. Considerable concentration dependence of sedimentation was again found for these chromatophores and they exhibited a strong tendency to aggregate in the presence of divalent metal ions. The chromatophore preparation accounted for 85% of the cellular bacteriochlorophyll and would actively photophosphorylate.

Although this preparation corresponds in most major respects with the "chromatophore" preparation of Worden and Sistrom (1964) from this organism, and that of Cohen-Bazire and Kunisawa (1960) from *R. rubrum*, Gibson's (1965c) interpretation of the structure and origin of the "chromatophore" differs from that presented in this Section. Gibson (1951c) found that the sedimentation coefficient of the purified chromatophores dropped from 160 to 120S following treatment with cholate or deoxycholate. This transition was accompanied by the loss of 70% of the extractable lipid, 20% of the protein and carbohydrate but no loss of pigment from the preparation. The 120S particles retained the ability to catalyse photophosphorylation for a brief period. On the basis of these observations Gibson has suggested that the "chromatophores" are not vesicular fragments of a complex membrane structure but are definite cellular units consisting of a pigmented core surrounded by a shell of non-pigmented lipid. These units may exist singly or, more commonly, as aggregates within the cell.

(*iii*) **Thiorhodaceae.** The intracytoplasmic membranes of the *Thiorhodaceae* are as varied in their form as those from the *Athiorhodaceae*. Some have vesicular, intruded membrane systems of the *R. rubrum* type (e.g., *Chromatium okenii*, *Chromatium weissei* and *Thiospirillum jenense*) whereas others have a mixture of vesicular and lamellar membrane systems (e.g., *Thiocapsa floridiana*) and others distinctive arrays of tubular membrane systems (e.g. *Thiococcus* sp.). Within this group the genus *Chromatium* has received the most intensive study with respect to the isolation of these photosynthetic membrane structures.

The number of detailed chemical analyses of chromatophore preparations is limited (see Lascelles, 1968 for a recent summary) and some of our most detailed information relates to the chromatophores isolated from *Chromatium* strain D. Newton and Newton (1957) presented an early chemical analysis of "large" and "small particle" fractions from this organism and recently Cusanovich and Kamen (1968) have given a detailed physico-chemical analysis of "light" and "heavy" pigmented fractions from *Chromatium* strain D. For their study Newton and Newton (1957) prepared the chromatophores by first suspending 20–30 g of wet cell paste in 20–30 ml of cold, 0·1M tris-chloride, pH 7·4 and rupturing the cells by 20 min exposure to ultrasound at 0°C. The homogenate was centrifuged at 25,000 *g* for 5 min, to remove cell debris and particles of sulphur, and the resultant supernatant was then centrifuged at 25,000 *g* for 1 h to sediment the chromatophores. Three quarters of the bacteriochlorophyll and carotenoid pigments still remained in the supernatant at the conclusion and were completely sedimented by a further centrifugation at 100,000 *g* for 90 min. The pigmented material sedimented by this step was described as the "small particle fraction". The chromatophore fraction was further purified by twice resuspending in buffer and sedimenting at 25,000 *g* for 1 h. The material was then resuspended in buffer and centrifuged for 3 min at 25,000 *g* to remove aggregated material and residual sulphur particles. Following a final sedimentation at 25,000 *g* for 1 h the chromatophores were dispersed in buffer with the aid of an homogenizer.

Although both the chromatophore and small particle fractions catalysed the photosynthetic phosphorylation of ADP and contained bacteriochlorophyll and carotenoids they differed in a number of their chemical and physical properties. The "chromatophores" were 0·1 μ in diameter and contained significant amounts of polysaccharide material in addition to lipid and protein. The small particle fraction was relatively low in polysaccharide and rich in lipid. Furthermore the yield of chromatophores from whole cells increased with increasing periods of sonic treatment of the cells up to 6 min and then slowly decreased whereas the amount of the small particle fraction steadily increased with increasing periods of ultrasound up to 20 min. Sonic treatment of the chromatophores converted a significant proportion to small particles. Coupled with the observed slow release of the chromatophores from broken cells these observations are consistent with an arrangement of the intracytoplasmic, photopigmented membranes in *Chlorobium* similar to that observed in *R. rubrum*. The electron micrographs of thin sections of *Chromatium* sp. show membrane bounded vesicles similar to those observed in *R. rubrum* sections (Cohen-Bazire, 1963). Cohen-Bazire has also briefly reported that when the preparative method adopted by herself and Kunisawa for *R. rubrum* (p. 171) is used with *Chromatium* strain Tassajara it gives an indistinguishable preparation to that from *R. rubrum*. It may, therefore be concluded that the "small particle" fraction described by Newton and Newton (1957) represents fragments of the vesicularly intruded membrane complex (i.e., is the true chromatophore preparation) and their chromatophore fraction is a preparation of fragmented cell envelopes with attached intracytoplasmic membranes.

Cusanovich and Kamen (1968) have also described the resolution of a crude chromatophore preparation from *Chromatium* strain D into "light" and "heavy" fractions. Their procedure was modelled after the procedure described by Worden and Sistrom (1964) using RbCl gradients for the resolution of chromatophores and ribosomes and *iso*-pycnic sucrose gradients for the resolution of the chromatophores. A cell-free extract of *Chromatium* strain D was prepared by passing a suspension of 1 vol. of packed, wet cells in 4 vol. of 100 mM potassium phosphate—10% sucrose, pH 7·5 through a Ribi cell fractionator at 20,000 lbs per sq in. The cell debris was removed by centrifugation at 30,000 g for 1 h and the supernatant combined with a supernatant obtained by washing the debris pellet with one quarter of the starting volume of buffer. The combined supernatants were centrifuged at 144,000 g for 2 h and the pelleted material (described as "classical chromatophores") was suspended in 4 vols. of buffer and the suspension made 35% by weight in RbCl. This mixture was centrifuged for 2 h at 144,000 g to pellet the ribosomes present and leave the pigmented chromatophore material at the top of the tube. The chromatophores were recovered, diluted with 10 vols of buffer and sedimented at 144,000 g for 2 h. The pellet was dispersed in buffer, aliquots layered over linear sucrose gradients (10–50% sucrose) and centrifuged for 90 min at 25,000 rpm in a Spinco SW 25·1 rotor. Two main pigmented bands were obtained, one centering in 26% sucrose ("light") and the other in 48% sucrose ("heavy"). Centrifugation for 20–30 h did not change the position of the heavy band (*iso*-pycnic with 48% sucrose, density 1·22) whereas the material in the light band came to equilibrium at 37% sucrose (density 1·17). The light and heavy fractions were collected, diluted 3 × and centrifuged at 144,000 g for 2 h. The pellets were resuspended in 2 vols of buffer and again layered on 10–50% sucrose gradients, centrifuged for 90 min recovered, diluted, sedimented at 144,000 g and finally resuspended in buffer and given a 10 min centrifugation at 30,000 g to remove aggregated material.

The detailed properties of the purified light and heavy fractions which are reported in this paper (Cusanovitch and Kamen, 1968) show that the fractions differ profoundly and suggest that the light fraction is again regular fragments of the comminuted intracytoplasmic membrane system (i.e., chromatophores) and that the heavy fraction corresponds to slightly more intact fragments associated with plasma membrane. Compared with Newton and Newton's (1957) large fraction considerably less cell envelope material is associated with the heavy fraction. The light fraction was deep red in colour, optically clear, homogeneous in the ultra-centrifuge with a sedimentation constant of 145S and showed no tendency to aggregate. In the electron microscope this fraction consisted of spheres of uniform size with an average hydrated diameter estimated to be 327Å. On the other hand the heavy fraction was seen to be grossly heterogeneous both in the ultracentrifuge and the electron microscope. The fraction was reddish-pink, turbid and showed a strong tendency to aggregate. 1% of the dry weight of the preparation was carbohydrate material, 22–30% was soluble in organic solvents and 62–68% of the dry weight was protein. The light fraction contained only 0·3% carbohydrate, 26–28% protein and 62–72% of the dry weight was soluble in organic solvents. In these respects the composition of the light fraction differs from the chromatophore fractions obtained from the *Athiorhodaceae* and the *Chlorobacteriaceae* which are described above. The light fraction contained the bulk of the total cellular photopigments and its succinic dehydrogenase activity (based on bacteriochlorophyll) was only 1/6 that of the heavy fraction. A subsequent paper by Cusanovich and Kamen (1968) established that the light particle preparation would actively catalyse photophosphorylation.

A number of other methods for the preparation of chromatophore material from *Chlorobium* sp. have been reported (e.g., Anderson and Fuller, 1958 and Garcia *et al.*, 1966a). However, these procedures are based upon simple differential centrifugations and since the relationship between the final preparation and the intact cell structure has not been explored, these procedures offer no practical advantages over the two methods described above.

(*iv*) **Nitrobacteriaceae.** These organisms are dependent upon the oxidation of nitrite or ammonia and studies on their fine structure have strengthened the idea that coupled energy yielding reactions in bacteria are associated with membranes. The studies of Murray and Watson (1965) and Remsen *et al.* (1967) have established that members of the genera *Nitrocystis*, *Nitrosomonas* and *Nitrobacter* have elaborate systems of intracytoplasmic membranes although it remains to be shown that these membranes are associated with the nitrifying activities of these organisms.

Remsen *et al.* (1967) and Murray and Watson (1965) are in substantial agreement concerning the fine structure of *Nitrocystis oceanus*. This Gram-negative organism has a layered cell wall about 250Å wide with an outer ribbed layer made up from 100Å diam. sub-units overlying an inner triplet structure. Below these outer layers is a typical plasma unit membrane about 80Å wide covered on its inner and outer surface with 120Å particles. Prominent within the cell is a stack of intra-cytoplasmic membranes concentrated across the centre of the cell and displacing both the cytoplasm and nucleoplasm. This ordered stack of approximately 20 membranes consists of closely apposed unit membranes which part when the cell is ruptured. The arrangement of membranes in the other nitrifying bacteria differs from this and a common pattern is not observed in either the ammonia or the nitrite oxidizers (see Murray and Watson (1965) for details). An unusual feature of the membranes in *N. oceanus* is their apparent permanence since they undergo binary fission when the cell divides.

In their paper Remsen *et al.* (1967) described a preliminary method for the preparation of the membranes from *N. oceanus* as follows. Exponential phase cells were harvested, washed with filtered sea water and ruptured by passage through a French pressure cell at 16,000 p.s.i. A 10 min centrifugation of the homogenate at 3000 *g* gave a pellet (termed Fraction 4) and the supernatant from this step was centrifuged for 10 min at 39,000 *g* to give a further pellet (Fraction 6). These two centrifugations brought down all the membraneous material. Electron micrographs of fixed or frozen-etched preparations of fraction 4 showed the presence of the intracytoplasmic membranes parted into unit membranes and not recovered as a stack as seen in the intact cell. These unit membranes were observed to have a connection with one or other pole of the cell. These membranes also appeared to have stalked particles about 80Å in diam. along their periphery, morphologically similar to those described for mitochondrial membranes. In fraction 6 membrane fragments were again observed which exhibited a crystalline pattern of sub-units with a periodicity of 40Å.

(*v*) **Blue-green algae.** The fine structure of the blue-green algae has been reviewed by Echlin and Morris (1965) and Lang (1968). Morphologically the intracytoplasmic membranes associated with the chlorophyll *a* in these organisms consist of lamellae built from unit membranes; *Merismopedia glauca* appears to be an exception having a series of peripheral vesicles carrying the chlorophyll. Each lamella is composed of two membranes joined at their ends and thereby enclosing a space of variable width to form a closed, flattened sac. The distribution of these lamellae may be peripheral, e.g., *Anacystis nidulans* or they may be arranged parallel to the long axis of the cell, e.g., *Gleocapsa* and may be tightly packed throughout the cytoplasm or widely separated according to the physiological state of the organism. The origin of these lamellae is not known, although their connection to the plasma membrane has been observed in blue-green algae recovering from chlorosis (Pankratz and Bowen, 1963). However there is also evidence to suggest that the membranes arise *de novo* and at cell division are divided between the daughter cells (Echlin and Morris, 1965).

Although the blue-green algae possesses a procaryotic form of cellular organization they have a eucaryotic photosynthetic system insofar as it is based upon chlorophyll *a* and is oxygen evolving. In addition to the carotenoids, phycocyanins and/or phycoerythrins are present as accessory pigments. These chromoproteins are loosely bound and readily released when the cell is ruptured; this fact poses practical problems in retaining the full photochemical activity in cell-free preparations (see later). A number of the blue-green algae are lysozyme sensitive and, after a few hours' digestion, will give osmotically sensitive protoplasts, e.g., *Fremyella diplosiphon*, *Plectonema calothricoides*, *Oscillatoria tenuis*, *Oscillatoria amoena* and *Phormidium luridum* (Biggins, 1967; Crespi *et al.*, 1962; Fuhs, 1958), or spheroplasts (Fulco *et al.*, 1967). Other blue-green algae, e.g., *Synechococcus lividus* and *Cyanidium caldarium* are unaffected by lysozyme. However the majority of the cell-free preparations from blue-green algae have been obtained by mechanical grinding of the cells with abrasives, e.g., alumina, carborundum or glass. Black *et al.* (1963) vacuum dried cells of *Nostoc muscorum*, *Anacystis nidulans*, *Anabaena variabilis* and *Tolypothrix tenuis* and then ground the dessicated material with buffer to prepare cell-free extracts. These extracts catalysed the photoreduction of NADP.

Unfortunately there has been no systematic study of the morphology of the chlorophyll-pigmented fractions obtained from the blue-green algae and their relationship to the intracytoplasmic membrane system seen in the intact cell. Furthermore the observations of Biggins (1967), using *Ph. luridum*, clearly show that mechanical methods for the disintegration of cells are responsible for the loss of certain photochemical activities from the membrane preparations (see below). These activities may be retained if osmotic lysis of the protoplasts is adopted to prepare the cell-free extracts. Futhermore the ease with which the accessory phycobilin pigment complexes and certain cytochromes are leached from the algae makes the isolation of a total, active photochemical membrane complex an impossible task. However the study of partial reactions and the assignment of functions to the readily leached components is facilitated. Certain investigators, e.g., Fredericks and Jagendorf (1964) and Thomas and De Rover (1955) have shown that the addition of a high molecular weight polymer [e.g. 30% carbowax 4000, 40% dextran or Ficoll] to the extracting solvent preserves the activity of lamellar fragments from *Anacystis nidulans* and *Synechococcus cedrorum* to catalyse the Hill reaction. This activity is lost if simple, aqueous extraction media are adopted although this loss is not entirely attributable to the loss of phycocyanin (Biggins, 1967; Susor and Krogman, 1964).

Schachman, Pardee and Stanier (1952) were the first to show that a crude cell-free extract, obtained by alumina grinding an unidentified blue-green algae and removing the debris by a 10 min centrifugation at 7000 rpm, contained a component sedimenting at 300S which carried the bulk of the chlorophyll. The phycocyanin was observed to sediment much more slowly. Calvin and Lynch (1952) made a similar observation with *Synechococcus cedrorum* and showed that centrifugation of the cell-free extract for 30 min at 36,000 g would not sediment the phycocyanin but sedimented the chlorophyll and carotenoids as a green pellet.

Shatkin (1960) has described the isolation, and also briefly described the morphology, of a cell fraction from *Anabaena variabilis* which retains all the chlorophyll, carotenoids and photophosphorylating activity of the crude cell-free extract. The cell-free extract was obtained by grinding the cell paste with $3 \times$ its weight of alumina in the presence of 5 volumes of 0·1M tris-chloride, pH 8·0, containing 40% ethylene glycol. Two cycles of centrifugation for 10 min at 2000 rpm were used to remove the abrasive, unbroken cells and cell debris. The cell-free extract was then centrifuged at 105,000 g for 1 h. The pellet from this centrifugation contained all the chlorophyll, carotenoids and photophosphorylating activity of the crude extract. The phycocyanin remained with the supernatant. Electron micrographs of thin sections of the pelleted material showed that it consisted of membranes organized into vesicles. Similar vesicles were observed in sections of damaged whole cells and Shatkin concluded that the vesicles in the pelleted cell fraction arose by fragmentation of the lamellar intracytoplasmic membrane system of the intact cell. Petrack and Lipmann (1961) have shown that this preparation will catalyse anaerobic photophosphorylation at pH 7·7 with added ADP, Mg^{2+} and an electron carrier (e.g., phenazine methosulphate) at a rate of 200–400 μmoles inorganic phosphate esterified/hr/mg chlorophyll. The addition of phycocyanin to this preparation, or 40% dextran to the isolation medium, permitted photophosphorylation in air. This observation suggests that the pigment functions to protect the preparation against photo-oxidation.

Fujita and Myers (1965) and Fredericks and Jagendorf (1964) used 0·02M tris-chloride, pH 7·2, containing 0·001M EDTA and 30% carbowax 4000 as the extracting and suspending medium for their preparations from *Anabaena cylindrica* and *Anacystis nidulans*. The cells were broken by grinding with powdered glass and the crude cell-free extracts were obtained following a centrifugation at 7710 g for 10 min. In the case of *An. nidulans* centrifugation of the extract at 105,000 g for 2 h sedimented all the chlorophyll and, when resuspended, the pelleted fraction catalysed the Hill reaction provided the carbowax concentration in the final assay mixture was lowered to 15%. Fujita and Myers (1965) obtained two particle fractions from *Ana. cylindrica*, one sedimented from the homogenate after 20 min at 8000 g and the other by centrifuging the supernatant for a further 3 h at 21,000 g .These two fractions accounted for all of the original chlorophyll, carotenoids and phycocyanin of the cell-free extract. The hydrogenase activity, the ability to catalyse the Hill reaction and the capacity to photo-reduce NADP were also localized in these two fractions.

Susor and Krogmann (1964; 1966) have described two procedures for isolating chlorophyll-containing particles from *Anabaena variabilis*. In the first preparation the cells were washed with water and 10 g of the wet cell paste were homogenized with glass beads for 15 min in a Virtis homogenizer. If phycocyanin was to be retained in the final particle preparations the cells were homogenized in a solution containing 17% sucrose and 20% Ficoll. The homogenate was then centrifuged at 30,000 g for 15 min to remove the glass beads and the supernatant centrifuged for 2 h at 105,000 g. The precipitate was suspended in 0·05M tris-chloride pH 7·3 containing 0·01M NaCl, 17% sucrose and 20% Ficoll to give the final particle preparation. For the isolation of particles containing little phyco-cyanin the medium for the homogenization of the cells and the final suspen-sion was 0·05M tris containing 0·01M NaCl and 0·4M sucrose. The corresponding reduction in the viscosity of the medium then enabled the centrifugations to be reduced to 30,000 g for 10 min and 105,000 g for 1 h respectively. Both types of particle preparation catalyzed the Hill reaction and were stimulated by added Mg^{2+} up to 0·02M. Duane *et al.* (1965), later demonstrated cyclic photophosphorylation with the phyco-cyanin depleted preparations but were unable to show non-cyclic photo-phosphorylation. The preparation of phycocyanin depleted particles was improved in the later communication by changing the washing and resus-pending medium to 0·4M sucrose–0·01M NaCl and using 5 min exposure to ultrasound to break the cells. The two differential centrifugations were retained but 1 ml of the final particle suspension (equivalent to a chlorophyll concentration of 2·5 mg/ml) was layered onto a 15–50% linear sucrose gradient in a Spinco SW 39 rotor. The rotor was then run at 15,000 rpm for 30 min followed by 40,000 rpm for 3 h to achieve density equilibrium. The phycocyanin banded as a distinct blue layer in the upper part of the tube and the chlorophyll pigmented particles formed a band in the lower part of the tube. The material in the latter band was recovered, diluted to bring the sucrose concentration to less than 15% and recentrifuged over a 15–30% sucrose gradient. Phycocyanin was not detectable in the chloro-phyll containing particles obtained from this step, although the enzyme, ferredoxin–NADP reductase, was dislodged from the particles and sedi-mented in the region of the phycocyanin in the first density gradient. These particle preparations have not been further characterized.

The studies of Biggins (1967) with *Ph. luridum* have clearly shown that the mechanical methods for breaking cells are responsible for the persistent failure to retain non-cyclic photophosphorylation activity in cell-free preparations from the blue-green algae. Biggins formed protoplasts from

Ph. luridum by following the lysozyme treatment described by Crespi *et al.* (1962) and although the intact intracytoplasmic membrane complex was not isolated the procedure adopted would undoubtedly serve to isolate such a photochemically active complex from lysozyme sensitive strains of blue-green algae. The cells were harvested, washed and resuspended in 0·3M potassium phosphate, pH 6·8, containing 0·5M mannitol. Lysozyme was added to the suspension to give a final concentration of 0·05% (w/v) and the suspension incubated at 35°C for 2·5 h. Under these conditions about 70% of the filamentous cells were converted to protoplasts; longer periods of digestion gave an increased level of protoplast lysis when judged by phycocyanin release. The protoplasts were separated from unaffected filaments by passing the mixture through loosely packed glass wool and were then collected by a 4 min centrifugation at 500 *g*. Excess lysozyme was removed by gently resuspending and resedimenting the protoplasts in fresh medium. The protoplasts were finally resuspended in tris-maleate, pH 7·2 or tris-chloride, pH 7·5 buffers containing 0·5M mannitol (Biggins, 1967). The protoplasts were lysed directly into the enzyme assay mixtures by ensuring an 8× dilution of the stabilizing mannitol. These lysed preparations were extremely active in catalysing cyclic photophosphorylation and the photoreduction of NADP or ferricyanide. The NADP reduction was sensitive to inhibition by 3-(3-4 dichlorophenyl)-1, 1-dimethyl urea, the inhibition being relieved, as in higher plants, by an added electron donor. Biggins also obtained a "light" and "heavy" particle preparation from *Ph. luridum* by grinding one volume of cell paste and sand and 10 volumes of 0·05M tris-chloride pH 7·5 containing 40% ethylene glycol. The sand and cell debris were removed by a brief centrifugation at 5000 *g*. Centrifugation of the cell-free extract for 15 min at 27,000 *g* gave a heavy particle fraction and the supernatant from this step, when centrifuged for 30 min at 48,000 *g* gave a light particle fraction. The light particles were almost twice as active as the heavier particles in catalysing cyclic photophosphorylation. This activity was retained by the light particles if washed in the tris buffer containing ethylene glycol but if washed in tris alone the particles lost 90% of their original activity. One essential component removed by washing was identified as a protein. The ability to catalyse non-cyclic photophosphorylation was not retained by either the light or heavy preparations obtained from the cells ground with sand.

On the basis of the foregoing studies it would appear that the isolation of an intact, photochemically active intracytoplasmic membrane complex from the blue-green algae could be achieved by selecting a lysozyme sensitive strain, lysing the protoplasts onto an isopycnic sucrose or Ficoll gradient and banding the membrane complex by centrifuging to equilibrium. If sucrose gradients are used the buffer should contain 40% ethylene glycol. The isolation and morphological and biochemical characterization of such a complex from the blue-green algae has not been reported.

(*vi*) **Sub-chromatophore particles from the photosynthetic bacteria.** In an endeavour to understand the detailed structure and function of the individual components contributing to the total "chromatophore" many investigators have attempted to fragment/solubilize this complex array of bacteriochlorophyll, carotenoids, quinones, lipids and proteins. To this end the methods used have included ultrasonic treatment, treatment with reagents designed to rupture disulphide bonds, attack by proteolytic and lipolytic enzymes and treatments with a wide variety of detergents. Furthermore the majority of these studies have been made with relatively crude chromatophore preparations and the products have rarely been characterized beyond indicating changes in size, spectral characteristics of the pigments or activity compared with the starting material.

Attempts to obtain and study subchromatophore particles have been confined mainly to organisms from the *Athio-* and *Thiorhodaceae* and will therefore be considered under these groupings.

(a) *Athiorhodaceae.* During the course of their study on the chromatophores of *R. rubrum* Cohen-Bazire and Kunisawa (1960) (p. 171) observed that treatment of the chromatophore preparations with crude pancreatic lipase had no effect upon their migration in a rate-zonal sucrose density gradient. The bacteriochlorophyll : protein ratio was increased but the succinic dehydrogenase activity and the ability to catalyse cyclic photophosphorylation was lost. On the other hand, the enzyme treatment converted much of the "heavy" pigmented fraction to material sedimenting in the region of the chromatophores.

Frenkel and Hickman (1959) found that their preparation of chromatophores from *R. rubrum*, with an average diameter of 90 nm, could be reduced to material of 25 nm diam. by treatment with ultrasound. The preparation retained its activity. Newton (1962) has used two disulphide cleaving reagents (0·3M sulphite plus 0·1M CuSO$_4$, pH 10 incubating 1 h at 25°C, and 0·5M mercaptoethanol in 8M urea, pH 8·5 for 3 h at 25°C) to convert preparations of serologically divalent, crude *R. rubrum* chromatophores to univalent derivatives. These treatments, although leaving the absorption spectra of the preparations unaltered, converted the turbid suspension of crude chromatophores to an optically clear solution. The pigment still sedimented as rapidly as in the untreated preparation but a protein component, equivalent to 37% of the original chromatophore nitrogen, was released and found to sediment at 7·1S.

In a series of detailed studies concerned with sub-chromatophore fragments from a variety of photosynthetic bacteria Garcia *et al.* (1966b) and

Vernon and Garcia (1967) have employed both detergent treatment [Triton X-100] and enzyme digestion to produce sub-chromatophore units from *R. rubrum* chromatophores. Crude chromatophores were prepared by washing and resuspending the cells in 0·01M tris-chloride pH 8·1, breaking the cells by 3 min ultrasound treatment and collecting the material pelleted by centrifuging the homogenate for 1 h between 20,000–110,000 *g*. After washing and resuspending this material in the buffer Triton X-100, a non-ionic detergent, was added in the proportion of 70 mg detergent/mg bacteriophlorophyll and the mixture gassed with argon and left in the dark for 3 h at 0–5°C. The suspension was diluted to contain 1·5 mg protein/ml and an aliquot applied to a discontinuous sucrose gradient formed by layering 57, 24 and 14% sucrose solutions dissolved in 0·01M tris-chloride pH 8·1 in a Spinco 30 centrifuge tube. Equilibrium was attained after 15 h centrifugation at 110,000 *g* and at this point five different zones were observed in the gradient. The bulk of the pigmented material was localized in two sharp red-purple bands, the light and heavy (density) bands. At the top of the gradient there was a brown band containing pheophytin and at the bottom a precipitate containing 50% of the original chromatophore protein. The material in both the light and heavy bands showed the same absorption spectrum as the intact chromatophores, i.e., there was no separation of the pigments into one or other of the fractions (cf. Bril's (1958) observation below reporting a separation of the pigments using this detergent against *Rh. spheroïdes* chromatophores). The bacterio-chlorophyll and carotenoids were concentrated in the heavy fraction (80%) and the light fraction was relatively richer in protein. Both fractions catalysed cytochrome *c* and N, N, N^1, N^1-tetramethyl *p*-phenylenediamine photo-oxidation coupled to oxygen and the photo-oxidation of reduced phenazine methosulphate coupled to ubiquinone. The light fraction would catalyse the photoreduction of NAD at one quarter the rate observed with the original chromatophores and also retained some succinic dehydrogenase activity. The electron micrographs of these fractions support the suggestion that the heavy fraction consists of the chromatophore membranes and the light fraction the matrix.

Adopting the same experimental methods employed by Garcia *et al.* (1966b), above, Vernon and Garcia (1967) have shown that the combined action of Triton X-100 and pancreatin/α-chymotrypsin will produce three well defined pigment-protein complexes from *R. rubrum* chromatophores. In this case the enzyme (4 mg/ml incubation mixture) was added to the chromatophore preparation containing 0·5% Triton and the incubation continued for 24 h at 36°C before layering samples onto the sucrose gradients. If the incubation was made under anaerobic conditions a centrifuga-tion for 2 h at 110,000 *g* gave three main bands in the gradient, one coloured green, one brown and one blue in order of increasing density. Incubation under aerobic conditions gave only the green and brown bands even though when the material in the blue band was formed it was stable in air. The appearance of the bands was time dependent; after 1 h digestion only the brown band appeared and after 3 h the green band and a little of the blue band. After 24 h none of the original chromatophore material remained. The appearance of the material in the blue band was dependent upon the detergent concentration. In the presence of 4% Triton a 24 h anaerobic incubation did not produce a blue band although the green band was more pronounced. Protein was present in all three pigmented bands and the enzymes would not degrade the sub-chromatophore particles obtained by direct Triton treatment. The simultaneous action of detergent and enzyme is required to produce the pigment-protein complexes. The absorption spectrum of the material in the brown band suspended in acetone was consistent with it being a carotenoid-protein complex and furthermore this material was not obtained when the procedure was used against chromato-phores from blue-green mutants of the organism. Similarly the spectra for the acetone-methanol extracts of the green and blue bands show that the former contains bacteriochlorophyll and the latter bacteriochlorophyll and carotenoids. The distribution of the chlorophyll between these two bands was in the ratio of 3 : 1 (blue : green) but again, there was no resolution of the forms of bacteriochlorophyll between the two bands. The material in the green band was very active in catalyzing the photo-oxidation of reduced phenazine methosulphate coupled with the reduction of ubiqui-none whereas the material in the blue band was inactive. Since the main requirement for this reaction is that the bacteriochlorophyll be readily available to the reagents in the aqueous environment, i.e., "reaction centre" chlorophyll, or be solubilized by detergents, this observation points to a fundamental difference in these two fractions. The material in the green band would also catalyse other simple, photochemical electron transfer reactions like the intact chromatophores, e.g. photoreduction of 5, 5' dithiobis-(2-nitrobenzoic acid) in the presence of ascorbate and 2,6-dich-lorophenol-indophenol as an electron donor system.

In addition to the above methods for producing sub-chromatophore particles from R. rubrum the treatment of the chromatophores with phenol-formic acid-water mixtures (2 : 1 : 1 v/v/v) was found by Oelze et al (1969) to give a number of products which were resolved by gel electro-phoresis.

The preparation of sub-chromatophore units from Rh. spheroides has been reported by Gibson (1965c) and Bril (1958). Bril (1958) used a crude chromatophore preparation obtained via sonic rupture of the cells followed by cycles of differential centrifugation. The chromatophores were then homogenized in 0·1M tris-chloride, pH 8·5 and 0·5% Triton X-100. The suspension was then centrifuged at 144,000 g for 2 h, a procedure which would normally sediment the chromatophores. However, a large amount of the pigment now failed to sediment and the pigment absorbing at 880 nm in the intact chromatophores was not present in the supernatant pigment but was concentrated in the pellet. Unfortunately, although this procedure appears to resolve the forms of bacteriochlorophyll the preparations are unstable.

Gibson (1965c) prepared "core particles" from highly purified *Rh. spheroides* chromatophores (see p. 173) by suspending the equivalent of 10–30 mg/ml chromatophore protein in 0·01M tris-chloride, pH 7·5, and adding sodium cholate and desoxycholate solutions to give a final concentration of 1–2 mg of each per mg chromatophore protein. The suspension was held at 5°C for 10 min, diluted 5 × with tris buffer and centrifuged in a Spinco SW 39 rotor for 2·5 h at 39,000 rpm. The pigment free supernatant was discarded and the pellet of "core particles" resuspended in tris buffer. These particles would catalyse cyclic photophosphorylation as actively as the intact chromatophores. The further properties of these particles and Gibson's interpretation of their relationship to the intact chromatophores have been described on p. 174.

Garcia *et al.* (1968a) have applied their Triton X-100 method to the production of sub-chromatophore fragments from chromatophores of two other members of the *Athiorhodaceae*, i.e. *Rh. palustris* (1968a) and *Rhodopseudomonas* sp N.H.T.C. 133 (1968b). In the case of *Rhodopseudomonas* sp. N.H.T.C. 133 the details of the method, detergent treatment and density gradient centrifugation, etc., are as described above and used by Garcia *et al.* (1966b) for *R. rubrum*. Using a concentration of 4% Triton two main bacteriochlorophyll bands (light and heavy) were obtained in the sucrose gradient. However, as in the case of *R. rubrum* (but see *R. palustris* and *Chromatium* below) no separation of the forms of chlorophyll was obtained, i.e., the detergent separates the chromatophores of this organism into two fragments of differing density but, in each, the environments and relative relationships of the forms of chlorophyll appear to remain the same *Rhodopseudomonas* sp. N.H.T.C. 133 resembles *Rh. viridis* (p. 166) in the arrangements of its intracytoplasmic membranes and in agreement with Giesbrecht and Drews (1966) the electron micrographs of the initial chromatophore preparations revealed a series of stacked, flattened discs with surfaces composed of ordered sub-units about 130Å in diam. The heavy fraction derived from these chromatophores appeared to consist of variously sized fragments of these membranes together with non-specific aggregates of the 130Å diam. sub-units.

A slightly modified procedure was adopted by Garcia *et al.* (1968a) to prepare sub-chromatophore particles from *Rh. palustris*. The crude chromatophore fraction, which sedimented from the cell homogenate between 15,000 and 115,000 *g* was suspended in 0·01M tris-chloride, pH 7·5 and applied to the top of 15 ml of 60% (w/v) CsCl in a centrifuge tube and the tube centrifuged for 60 min at 100,000 *g*. The pigmented material remained at the top of the tube and this material was then suspended in buffer and treated with 4% Triton X-100 for 1 h at 0°C. Samples of the mixture were then applied to a discontinuous sucrose gradient (q.v.) and centrifuged for 12 h at 100,000 *g* in a Spinco 30 rotor. At equilibrium three pigmented bands were observed in the tube together with a small amount of unchanged chromatophores near the bottom of the tube. In order, these bands were at the top of the tube, a red-yellow band of carotenoids followed by two main chlorophyll bands occupying different density positions (light and heavy) further down the tube. The absorption spectrum of the chlorophyll bands, compared with the intact chromatophores, showed a separation of the longer wavelength form of bacteriochlorophyll (880 nm) into the heavy fraction whilst the shorter wavelength forms (857 and 802 nm) remained with the light fraction. The bacteriochlorophyll present in the heavy fraction accounted for 50–60% of the original chromatophore chlorophyll and both fractions retained some carotenoid pigment. A comparison of the structural properties of the fractions revealed that the heavy fraction was particulate and consisted of linear arrays of particles 60–80Å thick. These particles were not observed in the membranes of the intact chromatophores. The light fraction was structureless.

(b) *Thiorhodaceae*. The chromatophores of *Chromatium* strain D have been comminuted by a wide variety of methods although in many instances the experimental details are scant and the products have not been fully characterized.

In a preliminary report Bergeron (1958) noted that HCN reduced *Chromatium* chromatophores to smaller, less readily sedimented units. Unfortunately no experimental details were given. This indication of the possible involvement of disulphide bonds between the chromatophore subunits has not been fully explored. Newton and Levine (1959) using the chromatophore preparation described by Newton and Newton (1957) (p. 175), showed that trypsin and chymotrypsin both digested the chromatophores to smaller units which could only be sedimented after prolonged ultracentrifugation. These sub-chromatophore particles had a reduced reactivity, compared with the chromatophores, with antisera prepared against the intact chromatophores. During the course of the chymotrypsin digestion the characteristic spectral absorption maxima of the intact chromatophores at 890, 850 and 800 nm slowly disappeared, but this could be reversed by the addition of cysteine. Unfortunately the experimental conditions for the proteolysis of the chromatophores were not recorded.

Bril (1960) and Clayton (1962) have used sodium desoxycholate to fragment *Chromatium* chromatophores. Bril (1960) used a chromatophore preparation obtained following carborundum grinding of the cells and a final differential centrifugation for 1 h at 144,000 g to sediment the chromatophores. The addition of the desoxycholate clarified the suspension of chromatophores and, depending upon the ionic strength of the suspending medium, an increasing amount of pigment remained in solution after centrifugation at 144,000 g. In the presence of 0·25M KCl a concentration of 0·25% desoxycholate brought about complete solubilization of the chromatophore pigment. The pigment component absorbing at 850 nm was photo-bleached during this treatment. Clayton (1962) used a cruder preparation of chromatophores (i.e., the lyophilized material pelleted by a 90 min centrifugation at 100,000 g from a 3 min sonically disrupted cell homogenate) but, like Bril, suspended in 0·05M tris-chloride, pH 8·0 containing 0·25M KCl and 0·25% sodium desoxycholate. If this mixture was then centrifuged at 100,000 g for 90 min a pigmented supernatant and pellet were obtained. The 850 nm bacteriochlorophyll was concentrated in the supernatant and the 870 and 890 nm forms in the pellet. This desoxycholate treatment clearly resolves the bacteriochlorophyll components.

Garcia *et al.* (1966a) have used Triton X-100 to produce photochemically active sub-chromatophore fragments from *Chromatium* strain D. The methods of preparation of the chromatophores, the Triton X-100 treatment and the separation of the products in sucrose gradients was the same as that described earlier for *R. rubrum* (p. 184) with the exception that the incubation period with the detergent was reduced to 2 h. With chromatophores from cells grown in high light intensities this treatment gave three pigmented bands in the final sucrose gradient; a yellow band of carotenoids at the top of the tube, next a brown band containing the 800 and 850 nm bacteriochlorophyll, but not the 890 nm form, and finally a red band enriched in the 890 nm form. The 850 nm bacteriochlorophyll was sensitized to photo-bleaching by the Triton treatment. The fact that the 890 nm chlorophyll was concentrated together with the cytochromes in the red band and this material actively catalysed the photo-oxidation of reduced phenazine methosulphate and N, N, N^1, N^1 tetramethyl-*p*-phenylenediamine strongly suggests that the photochemical reaction centre is concentrated in this sub-unit. Electron micrographs of negatively stained preparations of the red coloured fraction were structureless, whereas the photochemically inactive brown fraction appeared to consist of ruptured chromatophore membranes.

(c) *Chlorobacteriaceae*. Olson *et al* (1963) have described and partially characterized a protein-chlorophyll complex from *C. ethylicum* strain 2K in which the chlorophyll component is distinct from the *Chlorobium* chlorophyll having a λ_{max} at 770 nm in ether. The bacteria were harvested, washed and resuspended in 0·2M Na_2CO_3 and broken in a French pressure cell or by repeated freezing and thawing. The cell homogenate was centrifuged at 144,000 *g* for 90 min to deposit the *Chlorobium* chlorophyll. The supernatant was made 20% (w/v) with ammonium sulphate and the precipitate obtained was recovered and dissolved in 0·2M phosphate pH 7·8 and dialysed. After dialysis the material was adsorbed onto DEAE cellulose and the protein-chlorophyll-770 complex eluted with 0·25M NaCl. Further purification of the complex was obtained by reabsorbing on DEAE-cellulose and eluting with a gradient of NaCl in phosphate buffer. The sedimentation coefficient of the complex was 7S and its molecular weight $1·37 \times 10^5$. This complex functions as an acceptor of electronic excitation energy from *Chlorobium* chlorophyll but its structural relationship, if any, to the *Chlorobium* vesicles is unknown. Direct attempts to obtain sub-vesicle fractions from the *Chlorobacteriaceae* have not been reported.

D. Poly β-hydroxy butyrate and polyphosphate granules

1. *Poly β-hydroxybutyrate granules*

Poly β-hydroxybutyrate (PHB), a linear polymeric ester of D(-)-β-hydroxybutyrate with molecular weights up to 256,000, has been demonstrated in a wide variety of bacteria and the blue-green algae. PHB has been shown by cytological or chemical means to be present in certain members of the *Bacillaceae* (e.g., *B. megaterium*) *Pseudomonadaceae* (e.g., *P. saccharophila* and *P. solanacearum*), *Athiorhodaceae* (e.g., *R. rubrum*), *Nitrobacteriaceae* (e.g., *Nitrobacter sp.*, Tobback and Landelout (1965) and *Nitrobacter agilis*, Pope *et al.* (1969)), *Rhizobiaceae* (e.g., *Chromobacterium violaceum* and *Rhizobium sp.*, Forsyth *et al.* (1958)) and in nitrogen starved *Hydrogenomonas* sp. (Schlegel *et al.*, 1961) and heterotrophically grown *Ferrobacillus ferro-oxidans* (Wang and Lundgren, 1969). In these organisms the "lipid granules" stained with Sudan black or appearing as refractile bodies when viewed by phase contrast microscopy have been identified as major deposits of PHB. However the classical "lipid granules" are not always deposits of PHB and attempts to demonstrate PHB in strains of *E. coli*, *A. aerogenes* and *Serratia marcescens* have failed (Forsyth *et al.*, 1958).

The formation and stability of these granules are profoundly influenced by environmental factors and a strict control over the growth of the organism and the preparative procedures adopted is necessary to maximize the yield of granules. In stationary phase, batch cultures the accumulation of PHB is generally favoured by growth cessation due to exhaustion of the nitrogen, sulphur or phosphorus source in the medium and may amount to 40% of the dry weight of the organism (Williamson and Wilkinson, 1958). In organisms which may synthesize either glycogen or PHB as a storage material the nature of the carbon and energy source in the medium will influence their proportions, i.e., substances metabolically related to acetyl CoA favour PHB production (Stanier et al., 1959). Furthermore, since PHB is undoubtedly a reserve source of both carbon and energy for the cell it is readily degraded in the absence of an external substrate; the factors known to influence endogenous metabolism, e.g., the composition of the suspending medium for the cells, pH, temperature, etc., may therefore be expected to influence the PHB content of the cells (Dawes and Ribbons, 1962). A further complication in preparing PHB granules is the close association with the granules from certain organisms of the enzymes responsible for their depolymerization, e.g., R. rubrum. The granules may therefore be degraded in cell-free extracts or in the buffers used to resuspend the granules during their preparation.

The PHB content of a cell may be raised to unusually high levels under certain growth conditions e.g., to 70% of the cell dry weight in Azotobacter beijerinckii under conditions of nitrogen limitation. Although the storage role of PHB has been questioned when formed in these circumstances these growth conditions clearly provide a rich source of starting material for PHB granules. Kominek and Halvorson (1965) have studied the conditions favouring PHB accumulation in stationary phase cultures of B. cereus and Schlegel et al. (1961), have observed that strains of Hydrogenomonas will accumulate large amounts of PHB when their growth is limited by nitrogen exhaustion. Ferrobacillus ferro-oxidans (Wang and Lundgren, 1969) may also be induced to form PHB if grown heterotrophically with glucose as carbon source. Clearly a wide variety of growth conditions may be manipulated to induce PHB production in many organisms and to the writer's knowledge there has been no systematic study of this aspect of PHB synthesis. In an attempt to establish PHB production under "normal conditions" Wilkinson and Munro (1967) have studied PHB formation in an asporogenous variant of B. megaterium growing in continuous culture on a simple synthetic medium with glucose as carbon and energy source. In these conditions PHB accumulated when the simple carbon, nitrogen, sulphur or potassium sources were used to limit the rate of growth. The maximum levels of PHB accumulation were obtained about a dilution rate of 0.4 h^{-1} and this, and the similar levels reached in all cases (approx. 15% of the dry weight), indicates that growth rate may be an important factor in determining the level of accumulation.

Since the processes of synthesis and breakdown of PHB are essentially of an equilibrium character and the polymer is polydisperse in nature, the isolation of an undegraded PHB polymer may be an unrealistic objective, However, a number of procedures have been recorded for its isolation from a variety of organisms and these are described below. The chemical extractions, and to a lesser degree those employing solvents, are known to degrade the polymer to some extent. The procedures based upon gentle cell lysis and limited centrifugations isolate PHB apparently surrounded by a membrane, a form in which PHB has been observed in electron micrographs of thin sections of bacteria gorged with the polymer. A true unit membrane surrounding the deposits of the polymer has not been demonstrated unequivocally in any bacterium. The protein and solvent soluble lipids associated with the purified granules may constitute a unique coat about the granules which is intimately involved with their degradation and biosynthesis [see below].

The physico-chemical properties of chemically purified PHB extracted from *B. cereus* and *Rhizobium* sp. have been described by Alper *et al.* (1963). Lundgren *et al.* (1965) have compared the crystal structure, infrared absorption spectra, intrinsic viscosity and electron microscopy of many PBH preparations obtained from a total of eleven genera.

The widely used chemical isolation procedure for the extraction of PBH from bacteria, based upon the alkaline hypochlorite digestion of whole cells is fully described by Sutherland and Wilkinson (this Volume, p. 345).

In the course of a series of papers Merrick and co-workers have elaborated a gentle method of releasing PHB from *B. megaterium* in the form of "PHB granules" (Griebel *et al.*, 1968; Merrick and Doudoroff, 1961; Merrick and Doudoroff, 1964). Their most recent method of preparation of these granules is as follows; *B. megaterium* is grown to the late logarithmic phase in the 0·3% glucose, 0·05M sodium acetate medium of Macrae and Wilkinson (1958). 20 gm wet weight of cells are then suspended in 60 ml 0·05M tris-chloride pH 8·0 containing $1·67 \times 10^{-2}$M $MgCl_2$ 66 mg lysozyme and 0·7 mg desoxyribonuclease are added to the suspension which is then incubated for 30 min at room temperature, chilled to 0°C and given a 3 min treatment with ultrasound to dislodge the granules from the cell membranes. The suspension is then layered over glycerol and centrifuged in a swinging bucket rotor for 15 min at 1600 *g* at 4°C. The PHB granules collect at the interface and are removed and resuspended in 60 ml 0·05M tris-chloride pH 8·0. The properties of this stable, crude preparation of PHB granules have been investigated in detail by Merrick *et al.* (1965) and Ellar *et al.* (1968). The granules have a smooth appearance and tend to coalesce when prepared for electron microscopy. Unlike chemically purified PHB the PHB in the granules is susceptible to hydrolysis (75–85%) by soluble extracts from PHB-depleted *R. rubrum* cells which contain an active depolymerase. The residual undigested PHB "core" could be increased by freezing and thawing, repeated centrifugation or prolonged storage of the granules. Trypsin pre-treatment of the granules accelerated their digestion. The granules contained 5–10% of their dry weight as protein and this has been attributed to a membrane surrounding the PHB. This membrane appears to be unique to PHB deposits and may incorporate the enzymes for the synthesis and degradation of PHB. The treatments described above and alkaline hypochlorite all damage this membrane. Light scattering studies give a value of 3.57×10^9 for the weight average particle weight and the granule diameters range from 0·5–0·7 μ.

The crude PHB granules may be further purified by partition in a polymer two phase system followed by centrifugation in discontinuous glycerol gradients. The crude granules are subjected to three cycles of differential centrifugation over glycerol, first at 1000 g, then 650 g and finally at 450 g for 45 min. The granules are recovered at the interface each time and finally resuspended in 10 ml 0·05M tris-chloride buffer pH 8·0. The suspension is dialysed for 2 h against 200 vols of 0·02M tris-chloride pH 8·0 containing 5×10^{-4}M EDTA and then for 2 h against 200 vols of 0·02M tris-chloride pH 8·0. The suspension is diluted with this buffer so as to contain 300–500 mg PHB/ml and dispersed by a 1 min exposure to ultrasound before being introduced into a polymer two phase system of 5% (w/w) Dextran 500 and 3·5% (w/w) polyethylene glycol buffered with 0·02M tris-chloride pH 8·0. 1 ml of this system is used for each 10 mg of PHB and the total system is thoroughly mixed by gentle inversion and allowed to stand 30–60 min at 4°C. The PHB granules partition to the lower dextran-rich phase. The upper phase is removed and replaced by fresh upper phase, the system mixed and separated. This washing procedure is performed once more and the lower phase is then removed and diluted 4× with the buffer. The PHB granules settle under gravity and are washed by decantation with the buffer. 15 ml of the final granule suspension in buffer (equivalent to approx. 100 mg PHB) are then layered over a discontinuous glycerol gradient. The gradient is constructed by layering 5 ml portions of each of 10·5, 10·0, 9·5 and 9·0M glycerol in a centrifuge tube and allowing the tube to stand 48 h at room temperature to smooth the interfaces. The loaded gradients are centrifuged for 75 min at 90,000 g and 0°C during which time the PHB granules band between the 9·0 and 9·5M glycerol layers. The granules are collected, dispersed in 0·02M tris-chloride pH 8·0 and dialysed against this buffer. The granules are allowed to settle, washed by decantation with buffer and stored at 4°C in a dialysis sac immersed in buffer. The final highly purified preparation is stable for 2 weeks. PHB accounted for 97–98% of the dry weight the remainder being protein

(1·87%) and solvent extractable lipid (0·46%). PHB synthetase activity is still associated with the granules.

Ritchie and Dawes (1969) have used the above method for preparing *crude* PHB granules from cell homogenates of *Az. beijerinckii*. Since these organisms were crushed in a French pressure cell the lysozyme lysis step in the above procedure may not be an essential requirement for the recovery of intact PHB granules. The PHB granules from *Az. beijerinckii* had a diameter of 0·2–0·7 μ and a protein coat 60–80 Å thick which accounted for 2% of the dry weight of the granules.

The isolation of membrane-bound PHB granules from other organisms has not been reported but the practical difficulties which may be encountered in certain instances are illustrated by *R. rubrum*. In this organism the enzymes for the synthesis and degradation of PHB are active in the crude granule preparations. Accordingly in cell-free extracts of the organism or in buffer suspensions of crude granule preparations appreciable digestion of the granules takes place, particularly if the previous treatment of the whole cells has depleted the carbon source for growth and thereby activated the depolymerizing enzymes (Merrick and Doudoroff, 1964).

PHB is usually estimated by quantitative conversion to crotonic acid and the determination of the absorption of the acid at 235 nm. 5–50 μg of PHB in chloroform are transferred to a tube and the solvent is evaporated. 10 ml concentrated H_2SO_4 are added and the mixture heated for 10 min at 100°), cooled and the absorption at 235 nm recorded against a standard (Law and Slepecky, 1961).

2. *Polyphosphate granules*

Polyphosphate has been detected by chemical and staining techniques in the majority of micro-organisms. Cytologists have used the term "volutin granules", "Babes-Ernst granules" and "metachromatic granules" to signify cellular deposits of this polymer. In thin sections of bacteria examined in the electron microscope polyphosphate deposits appear as highly electron opaque areas, relatively large in relation to the cell, clear at their margins and tending to evaporate in the electron beam to leave a clear vacuole. However, polyphosphate may be detected in the cell extract without the formation of an observable granule in the intact cell (Kornberg, 1957; Sykes and Gibbon, 1967).

Polyphosphate may be detected by staining, e.g., the spectral shift ("metachromasy") accompanying its reaction with certain dyes, like Toluidine blue. It may be quantitatively estimated by chemical or enzymic methods. Structurally the natural polyphosphates are mixtures of linear, unbranched polymers of orthophosphate of variable chain length and with phosphoanhydride links estimated to be thermodynamically equivalent to the terminal phosphate of ATP (Yoshida, 1955). The average chain length is usually estimated by titration of the secondary acid function associated with each end group (pK 6·9).

Polyphosphates are usually extracted from bacteria via two acid extractions as described by Sutherland and Wilkinson (this Volume, p. 345), Ames (1966) and Cole and Hughes (1965). Alternatively the enzyme inorganic pyrophosphatase (polyphosphate kinase E.C.3. 6.1.1) catalyzing the reaction.

$$ADP + (P_i)_{n+1} \rightleftharpoons ATP + (P_i)_n$$

may be used to quantitatively convert the polyphosphate to ATP (Kornberg, 1957). Sykes and Gibbon (1967) have used a hexokinase trap in conjunction with this enzyme to ensure quantitative conversion of polyphosphate to glucose-6-phosphate. The glucose-6-phosphate was then estimated using glucose-6-phosphate dehydrogenase. In this way a specific, accurate and quantitative estimation of natural and synthetic polyphosphates may be obtained.

The volutin granules are not exclusively deposits of polyphosphate. Widra (1959) summarized the then current opinion regarding the composition and origin of these granules. On the basis of cytological observations made with the light microscope, using staining reactions and enzyme digestions, he proposed that the granules in *A. aerogenes* and *Corynebacterium xerose* contained RNA, lipoproteins and polyphosphate. Wiame's (1947) studies confirmed the presence of polyphosphate and the recent isolation of a "polyphosphate granule" from *M. lysodeikticus* has confirmed the presence of lipid, protein and RNA in these granules.

It is well known that the cultural conditions and point of harvest influence the polyphosphate content of bacteria. The polyphosphate content of rapidly growing *A. aerogenes* cells is low and increases as nutritional imbalances develop, e.g., particularly starvation for sulphur and low pH. If inorganic phosphate is added to cultures which have been starved for phosphate then polyphosphate is rapidly accumulated provided that Mg^{2+}, K^+ and a source of energy are also provided. An antagonistic relationship exists between ribonucleic acid synthesis and polyphosphate accumulation since if growth and nucleic acid synthesis halt due to nutritional imbalance then inorganic phosphate is still taken up from the medium and polyphosphate accumulates in the cells. However if growth is halted and nucleic acid synthesis continues (e.g., during chloramphenicol treatment) the accumulation of polyphosphate does not occur. In untreated cells the resumption of nucleic acid synthesis not only halts polyphosphate accumulation, but also stimulates its breakdown. The polyphosphate content of a cell is determined by an interplay of controlled processes involving the factors determining the level of the enzymes involved in its synthesis and degradation (e.g., polyphosphate kinase, polyphosphate-monophosphate phosphotransferase, polyphosphate glucokinase, polyphosphate fructokinase and polyphosphatases), the level of ATP production and nucleic acid synthesis. Harold (1966) has summarized his pioneering studies of these processes by suggesting that the observations regarding polyphosphate accumulation in *A. aerogenes* may be explained by assuming that in growing cells the synthesis of nucleic acids inhibits synthesis and stimulates degradation of polyphosphate. Thus, when growth and nucleic acid synthesis cease competition for ATP is relieved and polyphosphate accumulates at a rate depending upon the activity of the polyphosphate kinase. Similarly accumulation of polyphosphate takes place rapidly in growing cells previously starved for inorganic phosphate due to the high level of the kinase induced by the period of starvation.

Wilkinson and Munro (1967) have studied the accumulation of polyphosphate in *B. megaterium* grown in defined media in a chemostat. Appreciable accumulation of polyphosphate only occurred under sulphur limitation at low dilution rates.

Despite the extensive literature on bacterial polyphosphates [see Harold (1966) for a recent review] there are few recorded attempts to isolate the intact polyphosphate granules seen in whole cells. A number of investigators have noted the association of polyphosphate with a fraction which sedimented from homogenates of bacteria in a low gravitational field. Others have found a more widespread distribution of polyphosphate within the various cell fractions obtained by centrifugation (Cole and Hughes, 1965; Sykes and Gibbon, 1967). The latter type of observation may be attributed to the disruption of the native granules by the preparative conditions employed (see Friedberg *et al.* (1968), p. 196) and/or the pronounced tendency of polyphosphate to bind to proteins and to be precipitated by environments of high ionic strength.

On the basis of its size and assumed composition (polyphosphate and RNA) Martinez (1963) estimated that the majority of the polyphosphate granules from *Spirillum itersonii* would be sedimented from crude cell lysates by a 60 min centrifugation at 105,000 g. However, when the cells were lysed in the presence of 0·02M tris-chloride, pH 8·1, lysozyme, EDTA and desoxyribonuclease the polyphosphate granules seen in the intact cells were not observed in the lysates. The bulk of the polyphosphate was found in the supernatant following the centrifugation at 105,000 g. Harold (1963) also failed to recover polyphosphate granules from *A. aerogenes*. In this preparation advantage was taken of the observation of Williamson and Wilkinson (1958) that the volutin granules, with the PHB, survived the hypochlorite digestion procedure. The cells of *A. aerogenes* were fully grown in a phosphate limited medium and the culture then refreshed with inorganic phosphate and glucose to give a rapid accumulation of polyphosphate. The cell paste from 4 litres of culture was then digested in alkaline hypochlorite (see Sutherland and Wilkinson, this Volume, p. 345) for 45 min at 25°C. The suspension was centrifuged for 10 min at 20,000 g and the pellet washed twice with cold, 1·5M NaCl containing 10^{-3}M EDTA. The pellet was then extracted with a total of 12 ml of water and solid NaCl added to the pooled aqueous extracts to bring the concentration to 1·5M before adding 3 ml of ethanol. The precipitate which formed was redissolved in water, the material reprecipitated as before and again redissolved in water. The material finally dissolving in water was non-dialysable and was freed of nucleic acids by Norit treatment and proteins by shaking with chloroform: octanol (9 : 1). The final polyphosphate preparation was estimated to have a maximum chain length of 600 residues.

The reasons for the failure by Martinez (1963) and Harold (1963) to isolate the polyphosphate granules may be explained by the properties of the granules which have been noted in the recent successful isolation of intact granules by Friedberg and Avigad (1968). Friedberg and Avigad (1968) used *M. lysodeikticus* which was found to accumulate polyphosphate in the exponential phase of growth but lose the deposits in the stationary phase. The polyphosphate granules were also lost if the cells were suspended in water or repeatedly frozen and thawed. The cells were therefore harvested directly from the exponential phase culture and washed 2× with 1 mM tris-chloride pH 7·2 containing 1 mM $MgSO_4$. The $MgSO_4$ was found to be essential to preserve the structure of the granule. The washed cell paste was then dispersed in one tenth of the original culture volume of the tris-chloride-magnesium buffer and 100 μg lysozyme/ml added to the suspension. After 15 min incubation at 37°C to lyse the cells 0·1 μg/ml desoxyribonuclease was added and the lysate incubated for a further 5 min. The lysed, non-viscous suspension was then centrifuged for 1 h at 2000 g and 0°C. The sediment from this centrifugation was the polyphosphate granule preparation and was suspended in the tris-chloride-magnesium buffer. In electron micrographs of thin sections of whole cells the polyphosphate granules (40–80 nm diam.) were seen to be organized about a granular centre approximately 300 nm in diam. and certain of the final granule preparations contained these units. However the majority of the final preparations consisted of exceedingly regular, electron dense polyphosphate granules with surprisingly little contamination by other cell components. The density of the granules was estimated to be 1·23 and their composition was 24% protein, 30% lipid and 27% polyphosphate. Carbohydrate, metal ions and RNA were also present in the granules. The RNA and metals, although minor components appeared to be essential for the structure of the granule since ribonuclease or EDTA treatments degraded the granule. The granules were stable to trypsin, chymotrypsin, desoxyribonuclease, lysozyme and phospholipase C digestion but were destroyed by all types of detergent (e.g., sodium dodecyl sulphate, Triton X-100 and cetyl trimethylammonium bromide). The successful isolation of intact, native polyphosphate granules would therefore appear to depend upon the observance of certain conditions during the growth and harvesting of the cells and the subsequent isolation procedures. The growth conditions and point of harvest must be regulated to ensure maximum accumulation, i.e., the activity of the degradative enzymes is minimal. The medium chosen for washing and resuspending the cells and the cell-free preparations must be consistent with maintaining the structure of the granules. Finally, the use of chelating agents and detergents is to be avoided at all stages in the preparation and ribonuclease activity (e.g., arising as a contaminant in desoxyribonuclease or endogenously in the cell free extract) should be excluded from the reagents or minimized in the extract by the use of inhibitors.

ACKNOWLEDGMENTS

The author is indebted to Mrs. B. Tonks for typing the manuscript and to Mrs. E. Metcalf for preparing the figures and for considerable assistance in checking the typescript and references.
Figures 1, 2, 3, 8, 13 and 14 are reproduced with the assistance and permission of Beckman Instruments Ltd., Glenrothes, Fife, Scotland. Figures 7, 11 and 12 are reproduced with the permission of LKB-Produkter AB, Bromma, Stockholm, Sweden.

REFERENCES

Alper, R., Lundgren, D. G., Marchessault, R., and Cote, W. A. (1963). *Biopolymers*, 1, 545–556.
Ames, B. N. (1966). *In* "Methods in Enzymology" (Eds E. F. Neufeld and V. Ginsburg), Vol. 8, pp. 115–118. Academic Press, New York.
Anderson, I. C., and Fuller, R. C. (1958). *Arch. Biochem. Biophys.*, 76, 168–179.
Anderson, N. G. (1955). *Rev. Sci Instruments*, 26, 891–892.
Anderson, N. G. (1956). *In* "Physical techniques in Biological Research" (Eds G. Oster and A. W. Pollister), Vol. 3, pp. 299–352. Academic Press, New York.
Anderson, N. G. (1965). *In* "Fractions". No. 1, pp. 2–8. Beckman Instruments, Inc., Fullerton, California, U.S.A.
Anderson, N. G. (1966). *In* "Development of Zonal Centrifuges and Ancillary Systems for Tissue Fractionation and Analysis" (Ed. N. G. Anderson), National Cancer Institute Monograph 21, pp. 9–32.
Anderson, N. G., and Cline, G. B. (1967). *In* "Methods in Virology" (Eds K. Maramorosch and H. Koprowski), Vol. 2, pp. 137–178.
Anderson, N. G., and Ruttenberg, E. (1967). *Anal. Biochem.*, 21, 259–265.

Archibald, W. J. (1947). *J. Phys. Chem.*, **51**, 1204–1214.
Aronson, A. I. (1962). *J. Mol. Biol.*, **5**, 453–455.
Ayad, S. R., Bonsall, R. W., and Hunt, S. (1967). *Science Tools*, **14**, 40.
Bachrach, H. L., Trautman, R., and Breese, S. (1964). *Amer. J. Vet. Sci.*, **25**, 333.
Baldwin, R. L. (1957). *Biochem. J.*, **65**, 503–512.
Bayer, M. E. (1968). *J. gen. Microbiol.*, **53**, 395–404.
Bayley, S. T. (1966), *J. Mol. Biol.*, **18**, 330–338.
Bergeron, J. A. (1958). Brookhaven Symp. Biol., **11**, 118–131.
Biggins, J. (1967). *Plant Physiol.*, **42**, 1442–1446.
Biggins, J. (1967). *Plant Physiol.*, **42**, 1447–1456.
Birdsell, D. C., and Cota-Robles, E. H. (1967). *J. Bact.*, **93**, 427–437.
Birnie, G. D., and Harvey, D. R. (1968). *Anal. Biochem.*, **22**, 171–174.
Black, C. C., Fewson, C. A., and Gibbs, M. (1963). *Nature, Lond.*, **198**, 88.
Boatman, E. S., and Douglas, H. C. (1961). *J. Biophys. Biochem. Cytol.*, **11**, 469–483.
Bock, R. M., and Ling, Nan-Sing (1954). *Anal. Chem.* **26**, 1543–1546.
Bowen, T. J., Dagley, S., and Sykes, J. (1959). *Biochem. J.*, **72**, 419–425.
Brakke, M. K. (1963). *Anal. Biochem.*, **5**, 271–283.
Brakke, M. K. (1964). *Arch. Biochem. Biophys.*, **107**, 388–403.
Brenner, S., Jacob, F., and Meselson, M. (1961), *Nature. Lond.*, **190**, 576–581.
Bril, C. (1958). *Biochim. biophys. Acta.*, **29**, 458.
Bril, C. (1960). *Biochim. biophys. Acta.*, **39**, 296–303.
Britten, R. J., and Roberts, R. B. (1960). *Science*, **131**, 32–33.
Brown, A. D. (1961). *Biochim. biophys. Acta*, **48**, 352–361.
Brown, A. D. (1963). *Biochim. biophys. Acta*, **75**, 425–435.
Brown, A. D., and Shorey, C. D. (1963). *J. Cell. Biol.*, **18**, 681–689.
Bruening, G., and Bock, R. M. (1967). *Biochim. biophys. Acta*, **149**, 377–386.
Bull, M. J., and Lascelles, J. (1963). *Biochem. J.*, **87**, 15–28.
Burrous, S. E., and Wood, W. A. (1962). *J. Bact.*, **84**, 364–369.
Burton, K. (1956). *Biochem. J.*, **62**, 315–323.
Calvin, M., and Lynch, V. (1952). *Nature, Lond.*, **169**, 455–456.
Cammack, K. A., and Wade, H. E. (1965). *Biochem. J.*, **96**, 671–680.
Candler, E. L., Nunley, C. E., and Anderson, N. G. (1967). *Anal. Biochem.*, **21**, 253–258.
Casassa, E. F., and Eisenberg, H. (1960). *J. Phys. Chem.*, **64**, 753–756.
Casassa, E. F., and Eisenberg, H. (1961). *J. Phys. Chem.*, **65**, 427–433.
Chao, F-C. (1957). *Arch. Biochem. biophys.*, **70**, 426–431.
Chao, F-C. (1961). *Biochim. biophys. Acta*, **53**, 64–69.
Chao, F-C., and Schachman, H. K. (1956). *Arch. Biochem. Biophys.*, **61**, 220–230.
Chervenka, C. H. (1966). *Anal. Chem.*, **38**, 356–358.
Clayton, R. K. (1962). *Photochem. Photobiol.*, **1**, 201–210.
Cohen-Bazire, G. (1963). *In* "Bacterial Photosynthesis" (Eds. H. Gest, A. San Pietro and L. P. Vernon), pp. 89–110. Antioch Press Ohio.
Cohen-Bazire, G., and Kunisawa, R. (1960). *Proc. Natl. Acad. Sci. U.S.*, **46**, 1543–1553.
Cohen-Bazire, G., and Kunisawa, R. (1963). *J. Cell Biol.*, **16**, 401–419.
Cohen-Bazire, G., Pfennig, N., and Kunisawa, R. (1964). *J. Cell. Biol.*, **22**, 207–225.
Cohen-Bazire, G., and Sistrom, W. R. (1966). *In* "The Chlorophylls" (Eds. L. P. Vernon and G. R. Seely), pp. 313–341. Academic Press, New York.

Cohen, S. S., and Lichtenstein, J. (1960). *J. Biol. Chem.*, **235**, 2112–2116.
Cohn, E. J., and Edsall, J. T. (1943). *In* "Proteins, Amino-acids and Peptides". Reinhold Publishing Corpn. New York.
Cole, J. A., and Hughes, D. E. (1965). *J. gen. Microbiol.*, **38**, 65–72.
Coleman, G. (1969). *Biochim. biophys. Acta*, **174**, 395–397.
Coleman, G. (1969). *Biochem. J.*, **112**, 533–539.
Coleman, G. (1969). *Biochim. biophys. Acta*, **182**, 180–192.
Coleman, G., and Sykes, J. (1966). *Science Tools*, **13**, 43–45.
Creeth, J. M., and Knight, C. G. (1965). *Biochim. biophys. Acta*, **102**, 549–558.
Creeth, J. M., and Pain, R. H. (1967). Progress in Biophysics and Molecular Biology, **17**, 217–287. Pergamon Press, London.
Crespi, H. L., Mandeville, S. E., and Katz, J. J. (1962). *Biochem. Biophys. Res. Commun.*, **9**, 569–573.
Cummings, D. J. (1963). *Biochim. biophys. Acta*, **72**, 475–482.
Cusanovitch, M. A., and Kamen, M. (1968). *Biochem. biophys. Acta*, **153**, 376–396.
Cusanovitch, M. A., and Kamen, M. (1968). *Biochim. biophys. Acta*, **153**, 418–426.
Dagley, S., and Sykes, J. (1958). *In* "Microsomal Particles and Protein Synthesis" (Ed. R. B. Roberts), pp. 62–69. Pergamon Press, London.
Dagley, S., and Sykes, J. (1960). *Biochem. J.*, **74**, 11P.
Dagley, S., White, A. E., Wild, D. G., and Sykes, J. (1962). *Nature, Lond.*, **194**, 25–27.
Davern, C. I., and Meselson, M. (1960). *J. Mol. Biol.*, **2**, 153–160.
Dawes, E. A., and Ribbons, D. W. (1962). *Ann. Rev. Microbiol.*, **16**, 241–264.
de Duve, C. (1964). *J. Theor. Biol.*, **6**, 33–59.
de Duve, C., and Berthet, J. (1954). *In* "International Review of Cytology" (Eds. G. H. Bourne and J. F. Danielli), Vol. 3, pp. 225–275. Academic Press, New York.
de Duve, C., Berthet, J., and Beaufay, H. (1959). Progress in Biophysics, **9**, 326–369. Pergamon Press, London.
De Ley, J. (1964). *J. gen. Microbiol.*, **34**, 219–227.
Duane, W. C., Hohl, M. C., and Krogman, D. W. (1965). *Biochim. biophys. Acta*, **109**, 108–116.
Echlin, P., and Morris, I. (1965). *Biol. Rev.*, **40**, 143–187.
Edelstein, S. J., and Schachman, H. K. (1967). *J. Biol. Chem.*, **242**, 306–311.
Eisenberg, H. (1967). *Biopolymers*, **5**, 681–683.
Ellar, D., Lundgren, D. G., Okamura, K., and Marchessault, R., (1968). *J. Mol. Biol.*, **35**, 489–502.
Elson, D. (1967). *In* "Enzyme Cytology" (Ed. D. B. Roodyn), pp. 407–473. Academic Press, New York.
Felsenfeld, G. (1962). *In* "The Molecular Basis of Neoplasia", p. 104. University of Texas Press, Austin, Texas.
Fessler, J. H., and Vinograd, J. (1965). *Biochim. biophys. Acta*, **103**, 160–173.
Fisher, W. D., Cline, G. B., and Anderson, N. G. (1964). *Anal. Biochem.*, **9**, 477–482.
Fitz-James, P. C. (1960). *J. Biophys. Biochem. Cytol.*, **8**, 507–528.
Fitz-James, P. C. (1968). *In* "Microbial Protoplasts, Spheroplasts and L-forms" (Ed. L. B. Guze), pp. 124–143. Williams and Wilkins, Baltimore.
Flamm, W. G., Bond, H. E., and Burr, H. E. (1966). *Biochim. Biophys. Acta*, **129**, 310–319.
Folch, J., Lees, M., and Sloan-Stanley, G. H. (1957). *J. Biol. Chem.*, **226**, 497–509.

Forsyth, W. G., Hayward, A. C., and Roberts, J. B. (1958). *Nature, Lond.*, **182**, 800–801.
Fredericks, W. W., and Jagendorf, A. T. (1964). *Arch. Biochem. Biophys.*, **104**, 39–49.
Freimer, E. H. (1968). *In* "Microbial Protoplasts, Spheroplasts and L-forms" (Ed. L. Guze), pp. 279–292. Williams and Wilkins, Baltimore.
Frenkel, A. W., and Hickman, D. D. (1959). *J. Biophys. Biochem. Cytol.*, **6**, 285–289.
Friedberg, I., and Avigad, G. (1968). *J. Bact.*, **96**, 544–553.
Fuhs, G. W. (1958). *Arch. Mikrobiol.*, **29**, 51–52.
Fujita, H. (1962). "Mathematical Theory of Sedimentation Analysis". Academic Press, New York.
Fujita, Y., and Myers, J. (1965). *Arch. Biochem. Biophys.*, **111**, 619–625.
Fulco, L., Karfunkel, P., and Aaronson, S. (1967). *J. Phycol.*, **3**, 51–52.
Furano, A. V. (1966). *J. Biol. Chem.*, **241**, 2237–2244.
Garcia, A., Vernon, L. P., and Mollenhauer, H. (1966a). *Biochemistry*, **5**, 2399–2407.
Garcia, A., Vernon, L. P., and Mollenhauer, H. (1966b). *Biochemistry*, **5**, 2408–2416.
Garcia, A., Vernon, L. P., Ke, B., and Mollenhauer, H. (1968a). *Biochemistry*, **7**, 319–325.
Garcia, A., Vernon, L. P., Ke, B., and Mollenhauer, H., (1968b). *Biochemistry*, **7**, 326–332.
Gavrilova, L. P., Ivanov, D. A., and Spirin, A. S. (1966). *J. Mol. Biol.*, **16**, 473–489.
Gesteland, R. F. (1966). *J. Mol. Biol.*, **18**, 356–371.
Gibbs, S. P., Sistrom, W. R., and Worden, P. B. (1965). *J. Cell. Biol.*, **26**, 395–412.
Gibson, K. D. (1965a). *Biochemistry*, **4**, 2027–2041.
Gibson, K. D. (1965b). *Biochemistry*, **4**, 2042–2051.
Gibson, K. D. (1965c). *Biochemistry*, **4**, 2052–2059.
Gibson, K. D. (1965d). *J. Bact.*, **90**, 1059–1072.
Giesbrecht, P., and Drews, G. (1962). *Arch. Mikrobiol.*, **43**, 152–161.
Giesbrecht, P. and Drews, G. (1966). *Arch. Mikrobiol.*, **54**, 297–330.
Gilby, A. R., Few, A. V., and McQuillen, K. (1958). *Biochim. biophys. Acta*, **29**, 21–29.
Godson, G. N., and Sinsheimer, R. L. (1967). *Biochim. biophys. Acta*, **149**, 489–495.
Goldberg, A. (1966). *J. Mol. Biol.*, **15**, 663–673.
Gooder, H. (1968). *In* "Microbial Protoplasts, Spheroplasts and L-forms" (Ed. L. B. Guze), pp. 40–51, Williams and Wilkins, Baltimore.
Goulian, M., Kornberg, A., and Sinsheimer, R. L. (1967). *Proc. Natl. Acad. Sci. U.S.*, **58**, 2321–2328.
Gray, G. W., and Thurman, P. F. (1967). *Biochim. biophys. Acta*, **135**, 947–958.
Griebel, R., Smith, Z., and Merrick, J. M. (1968). *Biochemistry*, **7**, 3676–3681.
Griffith, O. M. (1967). *Anal. Biochem.*, **19**, 243–248.
Gropper, L., and Griffith, O. (1966). *Anal. Biochem.*, **16**, 171–176.
Guthrie, C., and Nomura, M. (1968). *Nature, Lond.*, **219**, 232–235.
Hall, C. E., and Slayter, H. S. (1959). *J. Mol. Biol.*, **1**, 329–332.
Harold, F. M. (1963). *J. Bact.*, **86**, 885–887.
Harold, F. M. (1966). *Bact. Revs.*, **30**, 772–794.
Henderson, A. R. (1969). *Anal. Biochem.*, **27**, 315–318.
Hendler, R. W. (1965). *Nature, Lond.*, **207**, 1053–1054/1071.

Herbert, D. (1961). *In* "Microbial Reaction to Environment" (Eds. G. G. Meynell and H. Gooder). *Symp. Soc. Gen. Microbiol.*, pp. 391–415. Cambridge University Press, Cambridge.
Hess, E. L., Chun, P. W. L., Utsunomiya, T., and Horn, R. (1967). *Biochemistry*, 6, 861–868.
Hexner, P. E., Radford, L. E., and Beams, J. W. (1961). *Proc. Natl. Acad. Sci. U.S.*, 47, 1848–1852.
Hickman, D. D., and Frenkel, A. W. (1965a). *J. Cell. Biol.*, 25, 261–278.
Hickman, D. D., and Frenkel, A. W. (1965b). *J. Cell. Biol.*, 25, 279–291.
Hickman, D. D., Frenkel, A. W., and Cost, K. (1963). *In* "Bacterial Photosynthesis" (Eds. H. Gest, A. SanPietro and L. P. Vernon), pp. 111–114.
Hill, L. R. (1966). *J. gen. Microbiol.*, 44, 419–437.
Hogeboom, G. H., Kuff, E. L., and Schneider, W. C. (1957). *In* "International Review of Cytology" (Eds. G. H. Bourne and J. F. Danielli), Vol. 6, pp. 425–467. Academic Press, New York.
Holt, S. C., Conti, S. F., and Fuller, R. C. (1966). *J. Bact.*, 91, 311–323.
Holt, S. C., and Marr, A. G. (1965a). *J. Bact.*, 89, 1402–1412.
Holt, S. C., and Marr, A. G. (1965b). *J. Bact.*, 89, 1413–1420.
Holt, S. C., and Marr, A. G. (1965c). *J. Bact.*, 89, 1421–1429.
Horio, T., von Stedingk, L. V., and Baltscheffsky, H. (1965). *Acta Chem. Scand.*, 20, 1–10.
Hosokawa, K., and Nomura, M. (1965). *J. Mol. Biol.*, 12, 225–241.
Ift, J. B., Voet, D. H., and Vinograd, J. (1961). *J. Phys. Chem.*, 65, 1138–1145.
Itoh, T., Otaka, E., and Osawa, S. (1968). *J. Mol. Biol.*, 33, 109–122.
Johnston, J. P., and Ogston, A. G. (1946). *Trans. Faraday Soc.*, 42, 789–799.
Kaempfer, R. O. R., Meselson, M., and Raskas, H. J. (1968). *J. Mol. Biol.*, 31, 277–289.
Kahler, H., and Lloyd, B. J. (1951). *J. Phys. Colloid Chem.*, 55, 1344–1350.
Karunairatnam, M. C., Spizizen, J., and Gest, H. (1958). *Biochim. biophys. Acta*, 29, 649–650.
Klainer, S. M., and Kegeles, G. (1955). *J. Phys. Chem.*, 59, 952–955.
Klucis, E. S., and Gould, H. J. (1966). *Science*, 152, 378.
Kominek, L. A., and Halvorson, H. O. (1965). *J. Bact.*, 90, 1251–1259.
Konrad, M. W., and Stent, G. (1964). *Proc. Natl. Acad. Sci. U.S.*, 51, 647–653.
Kornberg, S. R. (1957). *Biochim. biophys. Acta*, 26, 294–300.
Kuff, E. L., and Hogeboom, G. H., (1956). *In* "Enzymes-Units of Biological Structure and Function" (Ed. O. Gaebler), pp. 235–251. Academic Press, New York.
Kuff, E. L., Hogeboom, G. H., and Dalton, A. J. (1956). *J. Biophys. Biochem. Cytol.*, 2, 33–54.
Kurland, C. G. (1960). *J. Mol. Biol.*, 2, 83–91.
Kurland, G. C. (1966). *J. Mol. Biol.*, 18, 90–108.
Kushner, D. J. (1964). *J. Bact.*, 87, 1147–1156.
Kushner, D. J., Bayley, S. T., Boring, J., and Gibbons, N. E. (1964). *Can. J. Microbiol.*, 10, 483–497.
Lansing, W. D., and Kraemer, E. O. (1935). *J. Amer. Chem. Soc.*, 57, 1369–1377.
Lang, N. J. (1968). *Ann. Rev. Microbiol.*, 22, 15–46.
Lascelles, J. (1968). *Adv. in Microbiol. Physiol.*, 2, 1–42.
Lascelles, J., and Szilagyi, J. F. (1965). *J. gen. Microbiol.*, 38, 55–64.
Law, J. H., and Slepecky, R. A. (1961). *J. Bact.*, 82, 33–36.
Le Bar, F. E. (1965). *Proc. Natl. Acad. Sci. U.S.*, 54, 31–36.

Le Boy, P. S., Cox, E. C., and Flaks, J. G. (1964). *Proc. Natl. Acad. Sci. U.S.,* **52**, 1367–1374.

Lerman, M. I., Spirin, A. S., Gavrilova, L. P., and Golov, V. F. (1966). *J. Mol. Biol.,* **15**, 268–281.

Linderstrøm-Lang, K., and Lanz, H., (1938). *Compt. Rend. Trav. Lab. Carlsberg, Ser. Chim.,* **21**, 315.

Lundgren, D. G., Alper, R., Schnaitman, C., Marchessault, R. (1965). *J. Bact.,* **89**, 245–251.

Ludlum, D. B., and Warner, R. C. (1965). *J. Biol. Chem.,* **240**, 2961–2965.

McEwen, C. R. (1967a). *Anal. Biochem.,* **19**, 23–39.

McEwen, C. R. (1967b). *Anal. Biochem.,* **20**, 114–149.

McIlreavy, D. J., and Midgley, J. E. M. (1967). *Biochim. biophys. Acta,* **142**, 47–64.

McQuillen, K. (1960). *In* "The Bacteria" (Eds. I. C. Gunsalus and R. Y. Stanier), Vol. 1, pp. 249–359. Academic Press, New York.

Macrae, R. M., and Wilkinson, J. F. (1958). *J. gen. Microbiol.,* **19**, 210–222.

Mahoney, R. P., and Edwards, M. R. (1966). *J. Bact.,* **92**, 487–495.

Mandell, J. D., and Hershey, A. D. (1960). *Anal. Biochem.,* **1**, 66–77.

Mangiarotti, G., and Schlessinger, D. (1966). *J. Mol. Biol.,* **20**, 123–143.

Markham, R. (1967). *In* "Methods in Virology" (Eds. K. Maramorosch and H. Koprowski) Vol. 2, pp. 275–302. Academic Press, New York.

Marmur, J., Falkow, S., and Mandel, M. (1963). *Ann. Rev. Microbiol.,* **17**, 329–372.

Marr, A. G. (1960). *In* "The Bacteria" (Eds. I. C. Gunsalus and R. Y. Stanier). Vol. 1, pp. 443–468. Academic Press, New York.

Marston, J. H. (1968). *In* "Microbial Protoplasts, Spheroplasts and L-forms" (Ed. L. B. Guze), pp. 212–220. Williams and Wilkins, Baltimore.

Martin, R. G., and Ames, B. N. (1961). *J. Biol. Chem.,* **236**, 1372–1379.

Martinez, R. J. (1963). *Arch. Mikrobiol.,* **44**, 334–343.

Massey, V., Harrington, W. F., and Hartley, B. S. (1955). Discussions Faraday Soc. No. 20, p. 24–32.

Mazzone, H. M. (1967). *In* "Methods in Virology" (Eds K. Maramorosch and H. Koprowski), Vol. **2**, pp. 41–91.

Merrick, J. M., and Doudoroff, M. (1961). *Nature, Lond.,* **189**, 890–892.

Merrick, J. M., and Doudoroff, M. (1964). *J. Bact.,* **88**, 60–71.

Merrick, J. M., Lundgren, D. G., and Pfister, R. M. (1965). *J. Bact.,* **89**, 234–239.

Meselson, M., Nomura, M., Brenner, S., Davern, C., and Schlessinger, D. (1964). *J. Mol. Biol.,* **9**, 696–711.

Meselson, M., and Stahl, F. W. (1958). *Proc. Natl. Acad. Sci. U.S.,* **44**, 671–682.

Meselson, M., Stahl, F. W., and Vinograd, J. (1957). *Proc. Natl. Acad. Sci.U.S.,* **43**, 581–588.

Midgley, J. E. M. (1965a). *Biochim. biophys. Acta,* **95**, 232–243.

Midgley, J. E. M. (1965b). *Biochim. biophys. Acta,* **108**, 340–347.

Midgley, J. E. M. (1965c). *Biochim. biophys. Acta,* **108**, 348–354.

Midgley, J. E. M., and McIlreavy, D. J. (1967). *Biochim. biophys. Acta,* **142**, 345–354.

Miura, T., and Mizushima, S. (1968). *Biochim. biophys. Acta,* **150**, 159–161.

Mizushima, S., Ishida, M., and Kitahara, K. (1966). *J. Biochem.* (Tokyo), **59**, 374–381.

Mizushima, S., Ishida, M., and Miura, T. (1966). *J. Biochem.* (Tokyo). **60**, 256–261.

Mizushima, S., Miura, T., and Ishida, M. (1967). *J. Biochem.* (Tokyo), **61**, 146–148.

Mohr, V., and Larsen, H. (1963). *J. gen. Microbiol.*, **31**, 267–280.
Moller, M. L., and Kim, K. H. (1965). *Biochem. Biophys. Res. Commun.*, **20**, 46–52.
Moller, W., and Widdowson, J. (1967). *J. Mol. Biol.*, **24**, 367–378.
Moore, P. B., Traut, R. R., Noller, H., Pearson, P., and Delius, H. (1968). *J. Mol. Biol.*, **31**, 441–461.
Morgan, C., Rosenkranz, H. S., Chan.B., and Rose, H. M. (1966). *J. Bact.*, **91**, 891–895.
Murray, R. G. E., Steed, P., and Elson, H. E. (1965). *Can. J. Microbiol.*, **11**, 547–560.
Murray, R. G. E., and Watson, S. W., (1965). *J. Bact.*, **89**, 1594–1609.
Newton, J. W. (1962). *Biochim. biophys. Acta*, **58**, 474–485.
Newton, J. W. (1958). *Brookhaven Symp. Biol.*, **11**, 289–295.
Newton, J. W. (1960). *Biochim. biophys. Acta*, **42**, 34–43.
Newton, J. W., and Levine, L. (1959). *Arch. Biochem. Biophys.*, **83**, 456–471.
Newton, J. W., and Newton, G. A. (1957). *Arch. Biochem. Biophys.*, **71**, 250–265.
Nirenberg, M. W., and Matthaei, J. H. (1961). *Proc. Natl. Acad. Sci. U.S.*, **47**, 1588–1602.
Noll, H. (1967). *Nature, Lond.*, **215**, 360–363.
Nomura, M., and Traub, P. (1967). *In* "Organisational Biosynthesis" (Eds. H. J. Vogel, J. O. Lampen and V. Bryson), pp. 459–476. Academic Press, New York.
Nomura, M., Traub, P., Bechmann, H. (1968). *Nature, Lond.*, **219**, 793–799.
Nomura, M., and Watson, J. D. (1959). *J. Mol. Biol.*, **1**, 204–217.
Norton, J. E., Bulmer, G. S., and Sokatch, J. R. (1963). *Biochim. biophys. Acta*, **78**, 136–147.
Norton, J. W., Erdman, V. A., and Herbst, E. J. (1968). *Biochim. biophys. Acta*, **155**, 293–295.
Oelze, J., Biederman, M., and Drews, G. (1969). *Biochim. biophys. Acta*, **173**, 137 177.
Oelze, J., and Drews, G. (1969). *Biochim. biophys. Acta*, **173**, 448–455.
Olson, J. M., Filmer, D. Radloff, R., Romano, C. A., and Sybesma, C. (1963). *In* "Bacterial Photosynthesis" (Eds. H. Gest, A. San Pietro and L. P. Vernon, pp. 423–431. Antioch Press, Ohio.
Onishi, H., and Kushner, D. J. (1966). *J. Bact.*, **91**, 646–652.
Oumi, T., and Osawa, S. (1966). *Anal. Biochem.*, **15**, 539–541.
Ozaki, M., Mizushima, S., and Nomura, M. (1969). *Nature, Lond.*, **222**, 333 339.
Pangborn, J., Marr, A. G., and Robrish, S. A. (1962). *J. Bact.*, **84**, 669–678.
Pankratz, H. S., and Bowen, C. C. (1963). *Amer. J. Bot.*, **50**, 387 399.
Parish, J. M., Hastings, J. R. B., and Kirby, K. S. (1966). *Biochem. J.*, **99**, 19P.
Pedersen, K. O. (1958). *J. Phys. Chem.*, **62**, 1282–1290.
Petermann, M. L. (1964). "The Physical and Chemical Properties of Ribosomes". Elsevier Publishing Co., Amsterdam.
Petermann, M. L., and Pavolec, A. (1963). *J. Biol. Chem.*, **238**, 318–323.
Petrack, B., and Lipmann, F. (1961). *In* "Light and Life" (Eds. W. D. McElroy and B. Glass), pp. 621–630. Johns Hopkins Press. Baltimore.
Pfennig, N. (1967). *Ann. Rev. Microbiol.*, **21**, 285–324.
Pickels, E. G. (1943). *J. gen. Physiol.*, **26**, 341–360.
Pope, L. M., Hoare, D. S., and Smith, A. J. (1969). *J. Bact.*, **97**, 936–939.
Raskas, H. S., and Staehelin, T. (1967). *J. Mol. Biol.*, **23**, 89–97.
Razin, S. (1963). *J. gen. Microbiol.*, **33**, 471–475.

Razin, S., Argaman, M., and Avigan. J. (1963). *J. gen. Microbiol.*, **33**, 477–487.
Reisfeld, R. A., Lewis, U. J., and Williams, D. E. (1962). *Nature, Lond.*, **195**, 281–283.
Remsen, C., and Lundgren, D. G. (1966). *J. Bact.*, **92**, 1765–1771.
Remsen, C., Valois, F., and Watson, S. W. (1967). *J. Bact.*, **94**, 422–433.
Richards, E. G., and Schachman, H. K. (1959). *J. Phys. Chem.*, **63**, 1578–1591.
Richards, E. G., Teller, D. C., and Schachman, H. K. (1968). *Biochemistry*, **7**, 1054–1076.
Ritchie, G. A. F., and Dawes, E. A. (1969). *Biochem. J.*, **112**, 803–805.
Roberts, R. B. (1964). "Macromolecular Biosynthesis". Carnegie Institution of Washington No. 624, Washington, D.C.
Robrish, S. A., and Marr, A. G. (1962). *J. Bact.*, **83**, 158–168.
Rosset, R., and Monier, R. (1963). *Biochim. biophys. Acta*, **68**, 653–656.
Rubin, M. M., and Katchalsky, A. (1966). *Biopolymers*, **4**, 579–593.
Ryter, A., and Kellenberger, E. (1958). *Z. Naturforsch.*, **13b**, 597–605.
Salo, T., and Kouns, D. M. (1965). *Anal. Biochem.*, **13**, 74–79.
Salton, M. R. J. (1953). *Biochim. biophys. Acta*, **10**, 512–523.
Salton, M. R. J. (1967). *Ann. Rev. Microbiol.*, **21**, 417–442.
Salton, M. R. J., and Chapman, J. A. (1962). *J. Ult. Res.*, **6**, 489–498.
Salton, M. R. J., and Ehtisham-ud-Din, A. F. M. (1965). *Aust. J. Exptl. Biol. Med. Sci.*, **43**, 255–264.
Salton, M. R. J., and Freer, J. H. (1965). *Biochim. biophys. Acta*, **107**, 531–538.
Salton, M. R. J., and Netschey, A. (1965). *Biochim. biophys. Acta*, **107**, 539–545.
Samis, H. V. (1966). *Anal. Biochem.*, **15**, 355–357.
Santer, M., Teller, D. C., and Skilna, L. (1961). *Proc. Natl. Acad. Sci. U.S.*, **47**, 1384–1392.
Schachman, H. K. (1957). *In* "Methods in Enzymology" (Eds. S. P. Colowick and N. O. Kaplan), Vol. 4, pp. 32–103. Academic Press, New York.
Schachman, H. K. (1959). "Ultracentrifugation in Biochemistry". Academic Press, New York.
Schachman, H. K. (1963). *Biochemistry*, **2**, 887–905.
Schachman, H. K., and Edelstein, S. J. (1966). *Biochemistry*, **5**, 2681–2705.
Schachman, H. K., Pardee, A. B., and Stanier, R. Y. (1952). *Arch. Biochem. Biophys.*, **38**, 245–260.
Schaitman, C., and Greenawalt, J. W. (1966). *J. Bact.*, **92**, 780–783.
Scheraga, H. A., and Mandelkern, L. (1953). *J. Amer. Chem. Soc.*, **75**, 179–184.
Scheraga, H. A. (1961). "Protein Structure". Academic Press, New York.
Schildkraut, C. L., Marmur, J., and Doty, P. (1962). *J. Mol. Biol.*, **4**, 430–443.
Schlegel, H. G., Gottschalk, G. and von Bartha, R. (1961). *Nature, Lond.*, **191**, 463–465.
Schlessinger, D. (1963). *J. Mol. Biol.*, **7**, 569–582.
Schneider, W. C. (1957). *In* "Methods in Enzymology". (Eds. S. P. Colowick and N. O. Kaplan), Vol. 3, pp. 680–684. Academic Press, New York.
Schockman, G. D., Kolb, J. J., Bakay, B., Conover, M. S., and Toennis, G. (1963). *J. Bact.*, **85**, 168–176.
Schumaker, V., and Rosenbloom, J. (1965). *Biochemistry*, **4**, 1005–1011.
Schumaker, V. N., and Schachman, H. K. (1957). *Biochim. biophys. Acta*, **23**, 628–639.
Sells, B. H. (1964). *Biochim. biophys. Acta*, **80**, 230–241.

Shakulov, R. S., Aitkhozhin, M. A., and Spirin, A. S. (1962). *Biokhimiya*, **27**, 631–636.

Shatkin, A. J. (1960). *J. Biophs. Biochem. Cytol.*, **7**, 583–584.

Siakotos, A. N., and Wirth, M. E. (1967). *Anal. Biochem.*, **19**, 201–210.

Sistrom, W. R. (1965). *J. Bact.*, **89**, 403–408.

Smith, K. R. (1960). *J. Ult. Res.*, **4**, 213–221.

Spahr, P. F. (1962). *J. Mol. Biol.*, **4**, 395–406.

Spiegelman, S., Hall, B. D., and Storck, R. (1961). *Proc. Natl. Acad. Sci. U.S.*, **47**, 1135–1141.

Spirin, A. S. (1964). "Macromolecular Structure of Ribonucleic Acids". Reinhold, New York.

Spirin, A. S. (1968). *In* "Biochemistry of Ribosomes and messenger RNA". (Eds. R. Lindigkeit, P. Langen and J. Richter), pp. 73–93. Akademie-Verlag, Berlin.

Spirin, A. S., Kisselev, N. A., Shakulor, R. S., and Bogdanov, A. A. (1963). *Biokhimiya*, **28**, 765–774.

Spitnik-Elson, P. (1962). *Biochim. biophys. Acta*, **55**, 741–747.

Spitnik-Elson, P. (1965). *Biochem. Biophys. Res. Commun.*, **18**, 557–562.

Spragg, S. P., and Rankin, C. T. (1967). *Biochim. Biophys. Acta*, **141**, 164–173.

Spragg, S. P., Travers, S., and Saxton, T. (1965). *Anal. Biochem.*, **12**, 259–270.

Stanier, R. Y., Doudoroff, M., Kunisawa, R., and Contopoulou, R. (1959). *Proc. Natl. Acad. Sci. U.S.*, **45**, 1246–1260.

Stanley, W. M., and Bock, R. M. (1965). *Biochemistry*, **4**, 1302–1311.

Stellwagen, E., and Schachman, H. K. (1962), *Biochemistry*, **1**, 1056–1069.

Stephenson, M. (1949). "Bacterial Metabolism". Longmans, Green & Co., London.

Sueoka, N. (1961). *J. Mol. Biol.*, **3**, 31–40.

Susor, W. A., and Krogman, D. W. (1964). *Biochim. biophys. Acta*, **88**, 11–19.

Susor, W. A., and Krogman, D. W. (1966). *Biochim. biophys. Acta*, **120**, 65–72.

Svedberg, T., and Pedersen, K. O. (1940). "The Ultracentrifuge". Oxford University Press, London. Johnson Reprint Corporation, New York.

Svedberg, T., and Rinde, H. (1924). *J. Amer. Chem. Soc.*, **46**, 2677–2693.

Svensson, H., Hagdahl, L., and Lerner, D. K. (1957). *Science Tools*, **4**, 1–11.

Sykes, J. (1966). *J. Theor. Biol.*, **12**, 373–384.

Sykes, J. (1968). *In* "Biochemistry of Ribosomes and messenger RNA" (Eds. R. Lindigkeit, P. Langen and J. Richter), pp. 143–150. Akademie-Verlag, Berlin.

Sykes, J., Gibbon, J. A., and Hoare, D. S. (1965). *Biochim. biophys. Acta*, **109**, 409–423.

Sykes, J., and Gibbon, J. A. (1967). *Biochim. biophys. Acta*, **143**, 173–186.

Sykes, J., and Tempest, D. W. (1965). *Biochim. biophys. Acta*, **103**, 93–108.

Sykes, J., and Young, T. W. (1968). *Biochim. biophys. Acta*, **169**, 103–116.

Szybalski, W. (1968a). *In* "Fractions". No. 1, pp. 1–15. Beckman Instruments Inc., Fullerton, California, U.S.A.

Szybalski, W. (1968b). *In* "Methods in Enzymology". (Eds. L. Grossman and K. Moldave. Vol. 12, 330–360. Academic Press, New York.

Szybalski, W., Kubinski, H., and Sheldrick, P. (1966). Cold Spring Harb. Symp. Quant. Biol., **31**, 123–127.

Tashiro, Y., and Yphantis, D. A. (1965). *J. Mol. Biol.*, **11**, 174–186.

Tedeschi, H., and Harris, D. L. (1955). *Arch. Biochem. Biophys.*, **58**, 52–67.

Tempest, D. W., Dicks, J. W., and Hunter, J. R. (1966). *J. gen. Microbiol.*, **45**, 135–146.

Tempest, D. W., and Hunter, J. R. (1965). *J. gen. Microbiol.*, **41**, 267–273.
Tempest, D. W., Hunter, J. R. ,and Sykes, J. (1965). *J. gen. Microbiol.*, **39**, 355–366.
Tempest, D. W., and Strange, R. E. (1966). *J. gen. Microbiol.*, **44**, 273–279.
Thomas, J. B., and De Rover, W. (1955). *Biochim. biophys. Acta*, **16**, 391–395.
Tissières, A., Watson, J. D., Schlessinger, D., and Hollingsworth, B. R. (1959). *J. Mol. Biol.*, **1**, 221–233.
Tobback, P., and Landelout, H. (1965). *Biochim. biophys. Acta*, **97**, 589–590.
Tocchini-Valentini, G. P., Stodolsky, M., Aurisicchio, A., Sarnat, M., Graziosi, F., Weiss, S. B., and Geiduschek, P. (1963). *Proc. Natl. Acad. Sci. U.S.*, **50**, 935–942.
Traub, P., and Nomura, M. (1968). *J. Mol. Biol.*, **34**, 575–593.
Traub, P., Soll, D., and Nomura, M. (1968). *J. Mol. Biol.*, **34**, 595–608.
Traut, R. R. (1966). *J. Mol. Biol.*, **21**, 571–576.
Trautman, R. (1956). *J. Phys. Chem.*, **60**, 1211–1217.
Trautman, R. (1958). *Biochim. biophys. Acta*, **28**, 417–431.
Trautman, R .(1964). *In* "Instrumental Methods of Experimental Biology". (Ed. D. W. Newman), pp. 211–297. The Macmillan Company, New York.
Trentini, W. C., and Starr, M. P. (1967). *J. Bact.*, **93**, 1699–1704.
Tuttle, A. L., and Gest, H. (1959). *Proc. Natl. Acad. Sci. U.S.*, **45**, 1261–1269.
Ulrich, D., Kupke, D. W., and Beams, J. W. (1964). *Proc. Natl. Acad. Sci. U.S.*, **52**, 349–356.
Vanderwinkel, E., and Murray, R. G. E. (1962). *J. Ult. Res.*, **7**, 185–199.
van Holde, K. E. (1967). *In* "Fractions" No. 1. pp. 1–10. Beckman Instruments Ltd., Fullerton, California, U.S.A.
Van Iterson, W., and Leene, W. (1964). *J. Cell. Biol.*, **20**, 361–375.
Vatter, A. E., and Wolfe, R. S. (1958). *J. Bact.*, **75**, 480–488.
Vernon, L. P., and Garcia, A. F. (1967). *Biochim. biophys. Acta*, **143**, 144–153.
Vinograd, J., and Bruner, R. (1966). *In* "Fractions" No. 1. pp. 2–9. Beckman Instruments Inc., Fullerton, California, U.S.A.
Vinograd, J., and Bruner, R. (1966). *Biopolymers*, **4**, 131–156.
Vinograd, J., Bruner, R., Kent, R., and Weigle, J. (1963). *Proc. Natl. Acad. Sci. U.S.*, **49**, 902–910.
Vinograd, J., and Hearst, J. E. (1962). *Progr. Chem. Org. Nat. Prod.*, **20**, 372–422.
Waller, J. P., and Harris, J. I. (1961). *Proc. Natl. Acad. Sci. U.S.*, **47**, 18–23.
Wang, W. S., and Lundgren, D. G. (1969). *J. Bact.*, **97**, 947–950.
Warburg, O., and Christian, W. (1942). *Biochem. Z.*, **310**, 384–421.
Weibull, C. (1953). *J. Bact.*, **66**, 688–695.
Weibull, C. (1953). *J. Bact.*, **66**, 696–702.
Weibull, C. (1956). *In* "Bacterial Anatomy". (Eds. E. T. C. Spooner and B.A.D. Stocker, pp. 111–126. Cambridge University Press.
Weibull, C. (1968). *In* "Microbial Protoplasts, Spheroplasts and L-forms" (Ed. L. B. Guze), pp. 62–73, Williams and Wilkins, Baltimore.
Weibull, C., Beckman, H., and Bergström, J. (1959). *J. gen. Microbiol.*, **20**, 519–531.
Weibull, C., and Bergström, L. (1958). *Biochim. biophys. Acta*, **30**, 340–351.
Weigle, J., Meselson, M., and Paigen, K. (1959). *J. Mol. Biol.*, **1**, 379–386.
Wiame, J. M. (1947). *J. Amer. Chem. Soc.*, **69**, 3146–3147.
Widra, A. (1959). *J. Bact.*, **78**, 664–670.
Wilkinson, J. F., and Munro, A. L. (1967). *In* "Microbial Physiology and Continuous Culture (Eds. E. O. Powell *et al.*), pp. 173–184. H.M.S.O., London.

Williams, J. W., van Holde, K., Baldwin, R. L., and Fujita, H. (1958). *Chem. Revs.*, **58**, 715–806.
Williamson, D. H., and Wilkinson, J. F. (1958). *J. gen. Microbiol.*, **19**, 198–209.
Winzor, D. J. (1967). *Biochim. biophys. Acta*, **133**, 171–173.
Worden, P. B., and Sistrom, W. R. (1964). *J. Cell. Biol.*, **23**, 135–150.
Yoshida, A. (1955). *J. Biochem.* (*Tokyo*), **42**, 163–168.
Young, F. E., and Jackson, A. P. (1966). *Biochem. Biophys. Res. Commun.*, **23**, 490–495.
Young, T. W., and Sykes, J. (1968). *Biochim. biophys. Acta*, **169**, 117–128.
Yphantis, D. A. (1964). *Biochemistry*, **3**, 297–317.

CHAPTER III

Chemical Analysis of Microbial Cells

D. Herbert, P. J. Phipps and R. E. Strange

Microbiological Research Establishment, Porton, Nr. Salisbury, Wilts., England

I. Introduction 210
   A. Scope of Chapter 210
   B. Automatic analysis 211

II. Preparation of Bacteria for Analysis 213
   A. Washing and storage of samples; freeze drying . . . 213
   B. Wet and dry weight of bacteria; cell number . . . 214

III. Elemental Analysis 216
   A. Carbon and hydrogen 216
   B. Nitrogen 217
   C. Phosphorus 224
   D. Sulphur 229
   E. Sodium 231
   F. Potassium 233
   G. Calcium 233
   H. Magnesium 235
   I. Iron 239

IV. Determination of Protein 242
   A. Total nitrogen 243
   B. The biuret method 244
   C. The Folin–Ciocalteau reagent 249
   D. Amino-acids and α-amino nitrogen after acid hydrolysis . 253
   E. Ultraviolet absorption 262
   F. Separation of protein followed by analysis . . . 264

V. Carbohydrate Analysis 265
   A. Determination of "total carbohydrate" 265
   B. Determination of individual polysaccharides . . . 278
   C. Determination of individual sugars 282

VI. Lipid Analysis 302
   A. Extraction methods 303
   B. Total lipid and polyhydroxybutyrate 304
   C. Identification and estimation of lipid constituents . . 307

VII. Determination of Nucleic Acids 308
   A. General principles 309
   B. Standard methods for RNA and DNA determination . 310

        C. Sugar analysis   316
        D. Ultraviolet absorption   320
        E. Determination of nucleic acids in micro-organisms   .   .   322
        F. Quantitative determination of different nucleic acid fractions   328

VIII.   Teichoic Acids   329

  IX.   Polyphosphate   330

   X.   Bacterial "Pool" Constituents   331
        A. Extraction   332
        B. Identification   333
        C. Estimation of amino-acids   334
        D. Estimation of "nucleic acid pool"   335

References   336

LIST OF ABBREVIATIONS

DNA	Desoxyribonucleic acid	G	Guanine
RNA	Ribonucleic acid	C	Cytosine
s-RNA	Soluble ribonucleic acid	T	Thymine
t-RNA	Transfer ribonucleic acid	U	Uracil
r-RNA	Ribosomal ribonucleic acid	ATP	Adenosine triphosphate
A	Adenine	TCA	Trichloracetic acid

I. INTRODUCTION

A. Scope of chapter

This chapter is primarily concerned with the analysis of microbial cells, though many of the methods given are also suitable for the analysis of growth medium constituents. Since a comprehensive treatment would be impossibly lengthy, consideration has been restricted to the major cell components and the most common minor components. The preparation of material for analysis is also considered, because changes in the chemical composition of cells may occur as a result of the washing and storage conditions used. Wet and dry weight determinations of bacterial samples and assay of total cell numbers are described because analytical results must refer to one or other of these values.

Selection of an analytical procedure is usually a subjective process because the number of apparently suitable methods is often large and each will have different merits and defects. Primary considerations are sensitivity, specificity, reproducibility and absolute accuracy. Alternative methods based on different chemical principles are seldom alike in all these respects, and the method of choice will depend on the particular problem and will require exercise of the operator's judgement; relatively non-specific methods may be useful in some cases and give totally misleading results in others. Reported improvements or "modifications" of a method initially based on a good chemical principle need to be sifted and this can be a difficult task.

Two points may be stressed: (a) many of the commonly used biochemical methods do not provide results with the accuracy of those obtained with classical chemical methods; (b) the results obtained are often relative and not absolute. For example, bacterial protein or RNA are usually determined with reference to standards consisting of purified material from animals or yeasts, and the quantitative reaction of the standard substance is not necessarily equivalent to that of the bacterial constituent.

Having selected a method, the reproducibility of the results obtained with it should be established by doing a series of replicate assays with solutions of different concentrations of the standard substance, and computing the coefficient of variation; it must be realized, however, that this will provide only a minimum value of the potential errors. On first experience with a new method or new analytical material it is advisable to carry out "recovery" experiments by adding known amounts of standard substance to some of the analytical samples, as a check on the presence of interfering substances. It is also advisable, particularly with colorimetric methods, to run a set of standards with every batch of unknowns, rather than referring to a once-determined "standard curve", since such curves are often less "standard" than is commonly supposed.

For each microbial constituent, a number of methods are referred to but only a few are described in detail. These are methods of which we have experience and were selected after testing a number of procedures; some are modifications of existing methods. Most of them have been applied to the analysis of a wide variety of bacteria and a smaller number of yeasts and moulds. In cases where we had insufficient experience of determining a given constituent, we have obtained advice from colleagues in the Microbiological Research Establishment who have actually used them.

B. Automatic analysis

Automatic methods for performing biochemical analyses, already widely accepted in hospitals and in industry, are now beginning to make their way into the research laboratory. It seems probable that this trend will continue, though it is not easy to forecast at the time of writing which of the two current types of automatic analyser will be found most acceptable. These are described in detail by Marten (this Series, Vol. 6), and will be discussed here only in so far as their use may affect the choice of analytical methods, or require their modification.

All automatic analysers so far developed may be classified as either "continuous-flow" or "discrete" types; these differ widely on points of detail but all use colorimetric methods exclusively and contain some form of automatic colorimeter for final read-out. The first and best-known is the Technicon "AutoAnalyzer", which is a continuous-flow type. This uses a multi-channel peristaltic pump to force successive samples along a length of narrow-bore plastic tubing and to add reagents at appropriate stages, the samples being passed, after colour development, through a recording colorimeter. Incubation is effected by passing the samples through a coil of tubing immersed in a water-bath, while separation processes such as centrifugation or distillation are replaced by a continuous dialysis process—a unique and valuable feature of this equipment.

In the U.K. development seems to be favouring the "discrete" type of autoanalyser, e.g. the Joyce–Loebl "Meccolab" and the EEL–Griffin "Bioanalyst" (autoanalysers by Quickfit and Quartz and Vickers have also been announced while this Section was being written). These discrete autoanalysers vary considerably in detail but have the following common features: (a) One or more "analytical" units which handle *batches* of samples, performing such operations as automatic withdrawal of aliquots, successive additions of reagents, incubation for timed periods, etc. (b) A colorimeter unit which automatically takes in one sample after another, reads them and prints out the results; these usually contain scale adjustment controls which enable the print-out to be expressed directly as quantity of substance being determined, and may also incorporate a recorder which plots absorbances.

While both types of autoanalyser are still at a relatively early stage of development, it would seem at present that the continuous type is best when large numbers of samples are to be analysed by one or two methods, but the discrete types, which are definitely more flexible, are preferable when a variety of analyses have to be performed on a smaller number of samples. In either case automation imposes a further decision on the analyst, who now has to consider alternative methods in the light of their suitability for the various types of autoanalyser. This of course restricts the choice to colorimetric methods; even when suitable, these usually need a number of minor modifications to adapt them to the continuous auto-analyser, while with discrete analysers the reagent volumes and concentrations may usually be made the same as for manual methods. Methods involving centrifugation (e.g., deproteinization or extraction) or distillation, however, are more easily handled by the continuous autoanalyser where they are replaced by dialysis; with discrete autoanalysers such operations usually cannot be performed entirely automatically. Whichever type of autoanalyser is used, the importance of choosing a thoroughly reliable method and running repeated checks cannot be over-emphasized. It is believed that the majority of the colorimetric methods described in this Chapter are suitable for use in either type of autoanalyser.

II. PREPARATION OF BACTERIA FOR ANALYSIS

A. Washing and storage of samples, freeze drying

The method of preparing microbial cells for analysis depends upon the constituent being determined and whether interfering substances are present in the suspending medium. Harvesting of cultures is dealt with by Thomson and Foster (this Series, Vol. 2). The cells are separated by centrifugation, sedimentation or filtration and washed repeatedly to remove residual medium constituents and products of growth. In the case of bacterial spores, special treatment is usually necessary to free the spores from residual vegetative cell debris (see Gould this Series, Vol. 6). Cell composition may change during harvesting and washing due to breakdown of macromolecules and/or leakage of endocellular pool constituents. An interesting example of a dramatic change in cellular composition that occurs as a result of merely centrifuging a cell suspension is the fall in the ATP content of aerobic bacteria. In the case of *Aerobacter aerogenes*, the ATP content fell to about 20% of the initial value when the organisms were deposited by centrifugation out of aerobic culture medium (Strange *et al.*, 1963); this fall was due to the anaerobic condition of the packed cell pellet.

Processing cell crops in the cold decreases metabolic activity but sudden chilling of certain rapidly growing bacterial populations ("cold shock") may lead to considerable leakage of soluble endocellular pool constituents including amino-acids, nucleic acid precursors and ATP (Strange and Dark, 1962). On the other hand, at higher temperatures, endogenous metabolism occurs and lengthy processing may lead to some degradation of macromolecules, particularly of RNA and bacterial glycogen (Strange *et al.*, 1961).

The concentration of certain endocellular constituents depends on the concentration of these substances in the environment; thus, when certain bacteria are washed in a solution of different composition to that of the growth medium, changes in their contents of potassium, sodium and magnesium may occur (Epstein and Schultz, 1965; Tempest and Strange, 1966). Distilled water is a convenient washing liquid but the fact that certain bacterial populations (e.g. *Pseudomonas* species) lose viability during treatment indicates that changes have occurred and that gross chemical composition may be affected. A solution of sodium chloride (or another salt) of "physiological" concentration (ionic strength about 0·15) is commonly used for washing and re-suspending microbial cells but it should be realized that such solutions are not necessarily equivalent osmotically to microbial protoplasm although they may be to animal cell protoplasm; the osmotic pressure of microbial cells varies with their growth rate and physiological state, and may be as high as 20 atm in some Gram-positive bacteria (Mitchell and Moyle, 1956). There are many reports concerning the effects of distilled water and saline on the viability of bacterial populations but little regarding their effects on the concentration of endocellular solutes. In terms of viability, many bacterial populations appear to be most stable in solutions of slightly acid pH value (about 6·5); 0·11 M sodium chloride solution buffered with 0·02 M phosphate (pH, 6·5) was found suitable for populations of *A. aerogenes* and *Escherichia coli* (Strange *et al.*, 1961). However, maintenance of viability is not evidence of compositional stability although it does indicate the absence of cell lysis. In general, cellular macromolecular composition will not change significantly during harvesting and washing provided that cell lysis does not occur, endogenous metabolism is inhibited by keeping the temperature low and the processing period is kept as short as possible. However, the concentrations of certain cations and pool constituents may change during processing. In such cases, steps must be taken to maintain their concentrations at the level at which they were at the time of harvesting or these constituents must be determined indirectly (see below).

If immediate analysis is not possible, separated cells can be stored in washed suspension in the cold, in the frozen state or freeze dried. Washed suspensions of certain bacteria remain stable for days at 1–5°C according to viability determinations but, as mentioned above, with other bacteria, leakage of soluble endocellular compounds may occur. Storage of cells in the frozen state prevents endogenous metabolism and growth but lysis may occur during subsequent thawing leading to increased enzymic activity in the lysate; macromolecular degradation may result particularly in the case of RNA. Probably the best method of storing microbial cells is in the freeze-dried state; portions of the resulting material can then be weighed out for analysis and treated directly with the appropriate reagents. If proper freeze drying equipment is not available, the sample can be quick frozen in acetone-dry ice and freeze dried in a vacuum desiccator. The high vacuum that is necessary may be obtained with an oil pump (W. Edwards & Co. (London) Ltd.) which is left in operation for at least 30 min after reaching the minimum pressure; the desiccator cock is then closed and freeze drying allowed to proceed overnight. A suitable desiccant is concentrated sulphuric acid with a tray of sodium hydroxide pellets to absorb acid fumes.

B. Wet and dry weight of bacteria; cell number

The concentration of a given constituent in bacteria is determined either with a weighed amount of dried material or with a measured volume of a (washed) suspension of organisms. In the latter case, the equivalent dry or wet weight or the total number of organisms in a unit volume of cell suspension must be determined.

In general, results for the concentration of cellular constituents are less accurate when referred to the wet weight than to the dry weight of microorganisms (see Mallette, this Series, Vol. 1). When cells are packed by centrifugation, some suspending fluid remains in the interstitial spaces, the volume depending on the shape of the cells, the time of centrifugation and the centrifugal force. Allowance can be made for the weight of the extracellular fluid present (e.g., by incorporating a known concentration of a non-penetrating large molecular weight solute into the suspension and determining the amount of this substance present in the packed cell mass) but this value must be determined for different organisms in different phases of growth and physiological states.

A general technique for determining dry weight and potential sources of error is described by Mallette (this Series, Vol. 1). The following points are emphasized here: (a) This gravimetric determination must be made with due regard to the principles of analytical chemistry. (b) Removal of medium constituents by washing is essential and the final washing should preferably be with distilled water to remove salts; if the cells are osmotically fragile, washing must be done throughout with an appropriate salt solution and allowance made for the salt present in the dried cell mass (e.g., estimate Cl$^-$ if NaCl solution used). (c) Special precautions must be taken with living pathogenic organisms (see Darlow, this Series, Vol. 1); a solution of 0·15 M NaCl solution containing neutralized commercial formalin (10% v/v; about 3·5% HCHO, w/v) which kills most vegetative bacteria can be used for washing but such treatment will cause leakage of endocellular solutes from the cells (D. W. Tempest, private communication). (d) The temperature at which the washed cell mass is dried to constant weight must not cause decomposition. For many bacteria and yeasts, 100°–105°C for 16–20 h are suitable conditions and we have found that the results obtained are in fair agreement with those after freeze drying similar amounts of washed bacteria followed by secondary drying over P_2O_5 to constant weight.

Routine dry weight determinations on 5–6 ml volumes of suspensions containing 2–5 mg dry weight bacteria/ml are conveniently done in small hard glass tubes (10 cm long; 1 cm dia.). Duplicate samples of the suspension are measured into clean, dry, weighed tubes and centrifuged (e.g., 3000 rpm, 15 min, MSE bench angle centrifuge No. 3–61); the sedimented bacteria are washed consecutively by centrifugation and decantation with 5 ml volumes of 0·15 M NaCl (containing HCHO if necessary) and distilled water, and dried to constant weight in a hot air oven at 100°–105°C. The tubes are cooled in a desiccator and weighed on an analytical balance sensitive to 0·1 mg; weights of the duplicate tubes should be within 0·2 mg. With some organisms, it is found that the pellet sedimented after washing with distilled water is loose and a proportion of the organisms resuspend during decantation; this difficulty can usually be overcome by using dilute NaCl solution (0·015 M) for the final washing and the weight of salt retained in the deposit does not significantly affect the determination. If the density of the suspension is less than 2–5 mg bacterial dry weight/ml, larger samples, longer centrifuging and/or a more sensitive balance are required. It is common practice to determine the dry weight of a sparse bacterial suspension indirectly by measuring its extinction value at a selected wavelength in a spectrophotometer and referring to a calibration curve that relates extinction at this wavelength to dry weight values which are pre-determined. This method is less accurate than the direct determination of dry weight because the extinction value of a bacterial suspension depends on the size and shape of the bacteria as well as their dry weight value. In addition, the extinction value of a suspension of bacteria depends on the suspending medium and is higher in salt solutions than in distilled water (Mager et al., 1956).

The average amount of a given constituent per cell can be determined if the total number of cells per unit of dry weight or volume of suspension is known. Methods are given elsewhere in this book for determining total cell number with the Coulter Counter (Kubitschek, this Series, Vol. 1) and with counting chambers (Mallette, this Series, Vol. 1). It is emphasized that coefficients of variation for results of total cell counts determined with counting chambers are high unless special precautions are taken; more accurate results are obtained if the actual depth of the chamber is measured by an optical method in the area where the cells are counted (Norris and Powell, 1961).

III. ELEMENTAL ANALYSIS

A. Carbon and hydrogen

Analyses of microbial cells for carbon and hydrogen are undoubtedly best done with the classical Dumas procedure and preferably by skilled analysts; accurate results are obtained on the micro and sub-micro scale with modern versions of this method. If apparatus and analysts are not available, determinations are performed on a commercial basis by certain laboratories (e.g., Dr. G. Weiler and Dr. F. B. Strauss, Microanalytical Laboratory, 164 Banbury Rd, Oxford).

Carbon alone can be accurately determined with a Van Slyke manometer; the sample is heated with a mixture of chromic and iodic acids in a closed system and the liberated carbon dioxide is transferred to the Van Slyke–

Neill manometer and its volume measured at constant pressure (Van Slyke *et al.*, 1940; Van Slyke *et al.*, 1951). The method is not difficult to use and it is particularly suitable if a large number of samples are to be analysed; accurate results are obtained with samples containing 0·1–0·7 mg of carbon.

B. Nitrogen

This element can be determined by either the Dumas or the Kjeldahl procedure (Jacobs, 1965), the choice of method depending on the type of material to be analysed. The Kjeldahl method is the one usually preferred for biological material and depends on the conversion of bound nitrogen into an ammonium salt when the sample is heated with concentrated sulphuric acid and a catalyst; ammonia is then liberated from the salt with alkali and estimated titrimetrically after distillation or it can be reacted directly in the digest with Nessler's reagent and estimated photometrically. The original method gave inaccurate results with a number of compounds (heterocyclic nitrogenous compounds, hydrazones, dinitrohydrazones, osazones, oximes and nitro, nitroso, azo and some diazo compounds). Later modifications of the original method, for example that of Friedrich (1933), involved a preliminary reduction of the sample with hydriodic acid and red phosphorus, and increased the range of materials that could be analysed. The object of the digestion is to quantitatively convert all of the combined nitrogen in the sample into ammonia without release of volatile nitrogen-containing compounds; to achieve this objective, the temperature and the duration of the digestion must be carefully controlled and the catalyst carefully selected. There are numerous versions of the basic Kjeldahl method differing in the conditions of digestion, type of catalyst, method of measuring the ammonia formed, etc.; all of these modifications cannot be dealt with here, but detailed examples of certain of them are described below. A "perfect Kjeldahl method" certainly does not exist although Beet (1955) introduced a procedure which he said ". . . can fairly be claimed as the most perfect Kjeldahl method so far devised". The method involved treatment of the sample with permanganate during digestion and is claimed to give quantitative yields of ammonia from pyridine carboxylic acids, tryptophan and alkaloids.

Attempts have been made to provide Kjeldahl techniques capable of accurately and reproducibly measuring 1–2 μg of nitrogen in the final digest. Brüel *et al.* (1947) discuss the relevant problems and give details of a thoroughly investigated technique suitable for this purpose.

Undoubtedly, methods involving distillation of the ammonia from Kjeldahl digests are the most accurate and a variety of stills have been designed for this purpose. Probably the best and most convenient to use is the Markham (1942) still.

1. *Comparison of various modifications of the Kjeldahl method*

Catalytic mixtures used to induce breakdown of organic matter and formation of ammonia during digestion with sulphuric acid usually include selenium, selenium oxide or sodium selenate, potassium sulphate (to raise the boiling point of the digestion mixture) and copper sulphate; Chibnall, Rees and Williams (1943) also included mercuric chloride. If selenium metal or compounds are omitted from the catalytic mixture, an oxidizing agent (hydrogen peroxide or potassium permanganate) is added at some stage during digestion to clear charred organic matter. When the ammonia formed is distilled from the digest, it is usually collected into either boric acid (Ma and Zuazaga, 1942) or standard mineral acid.

Samples (2 mg) of dried crystalline bovine serum albumin (used as a protein standard) were analysed in duplicate by digesting them in Kjeldahl flasks (50 ml) with sulphuric acid and various catalysts on a stand of electro-thermal heating mantles; blanks consisting of the appropriate digestion mixtures alone were heated in parallel with the tests. After digestion, the solutions were made alkaline and distilled in the Markham still; the distillates were collected into either boric acid or standard mineral acid and titrated with standardized hydrochloric acid or sodium hydroxide, respectively. The results (Table I) show that, with bovine serum albumin at least, the different conditions had little effect on the nitrogen values obtained; the difference between the highest and lowest values was 1·3%.

Further samples (128·5 µg protein) were analysed by digestion with sulphuric acid containing selenium dioxide followed by direct nesslerization of the diluted digests (King, 1951). The nitrogen value obtained was slightly lower than those by the various Kjeldahl-distillation techniques (Table I).

The conclusion to be drawn from these results is that sulphuric acid containing selenium metal or selenium dioxide without further additions is a satisfactory digestion reagent for protein. It is of interest that the same sample of protein was found to contain only 13·8% nitrogen when it was analysed by the Dumas method.

2. *Detailed example: Kjeldahl-distillation method*

The total nitrogen content of dried crystalline bovine serum albumin was determined.

(a) *Reagents.* The reagents are as follows—

  (i) Catalyst: 10% (w/v) selenium powder dissolved in AR grade concentrated sulphuric acid (Fawcett, 1954).
  (ii) Sodium hydroxide solution, 40% (w/v).
  (iii) Standard acid (0·01 N): mix N hydrochloric or sulphuric acid

TABLE I

Determination of the total nitrogen content of bovine serum albumin by various modifications of the Kjeldahl method[a]

Test	Conc. H$_2$SO$_4$ (ml)	Catalyst and other additions	Time of digestion (h)	NH$_3$ release	Distillation into	Titrant	% Nitrogen
				Distillation–Titrimetric Methods			
1	0·67	K$_2$SO$_4$, CuSO$_4$, H$_2$O$_2$ for clearing	16	40% (w/v) NaOH	2% boric acid	0·01 N HCl	15·90
2	0·67	as (1)	16	as (1)	0·01 N HCl	0·01 N NaOH	15·74
3	0·5	K$_2$SO$_4$, CuSO$_4$, HgCl$_2$, SeO$_2$	16	30% (w/v) NaOH + 5% (w/v) Na$_2$S$_2$O$_3$	0·01 N HCl	0·01 N NaOH	15·95
4	0·5	Se	2·5	40% (w/v) NaOH	0·01 N HCl	0·01 N NaOH	15·83
				Direct Nesslerization			
5	0·1	SeO$_2$	4	..	..	..	15·61

[a] 2·0 mg of crystalline bovine serum albumin taken for all tests except 5 (128·5 μg protein).

(10 ml), Tashiro's indicator (0·08% Methyl Red and 0·02% Methylene Blue in ethanol) (10 ml), ethanol (200 ml) and dilute to 1000 ml with water in a volumetric flask.

(iv) Standard alkali (near 0·01 N): the reagent is prepared from stock 40% sodium hydroxide solution that has stood for several days in a closed container to allow precipitated sodium carbonate to sediment; an appropriate volume of the clear supernatant diluted with an equal volume of boiled distilled water is delivered under the surface of the required volume of freshly boiled water to provide an approx. 0·01 N solution and this is stored in the reservoir of an automatic burette, protected by a soda-lime tube in the inlet. The solution is accurately standardized by titration against potassium acid phthalate with Tashiro's indicator.

(b) *Procedure.* A solution containing 230·7 mg protein/100 ml water was prepared and samples (2 ml) were measured into two Kjeldahl flasks (50 ml). Catalyst solution (0·5 ml) was added to these and to two other flasks containing distilled water (2 ml). The flasks were transferred to the electric heater and heated at low temperature until water was removed; the temperature was then raised so that the acid mixture boiled evenly. After digestion for 2·5 h, the flasks were allowed to cool and their contents were diluted with water (5 ml). Flask contents were then transferred into the Markham still with washings, 40% sodium hydroxide (5 ml) added and the liberated ammonia distilled over with steam into 10 ml of 0·01N hydrochloric acid; about 10 ml of distillate was collected. The residual acid in the receiver flasks was determined by titration with standard sodium hydroxide solution (approx. 0·01N).

(c) *Calculation.* 1 ml of 0·01 N NaOH is equivalent to 0·14 mg of nitrogen and the value is adjusted according to the exact titre of the solution. A 10 ml burette calibrated to read to 0·02 ml should be used and readings are taken to the nearest 0·01 ml. If B ml is the reading for the blank and T ml the reading for the test sample then $(B-T) \times 0·14$ (exactly 0·01 N NaOH used, for example) is the mg nitrogen in the sample; this value $\times 100$/mg of sample gives the nitrogen value (%).

(d) *Results.* Samples contained 4·614 mg of protein and 1 ml of the standard alkali used was equivalent to 0·1458 mg of nitrogen. Titrations for the blanks were 9·88 and 9·87 ml, and for the tests 4·86 and 4·86 ml; therefore, $B-T$ is 5·01 ml, the nitrogen in the sample is 0·7305 mg and the % nitrogen is 15·83.

With the above method, true nitrogen values are often not obtained with a number of substances including nucleic acids, purines, pyrimidines and pyridine dicarboxylic acids although the values depend to a large extent on the presence of other substances. For example, if carbohydrate is present, true nitrogen values may be obtained. Bacteria contain a variable, and sometimes large amount (25%) of nucleic acids; bacterial spores contain a considerable amount of dipicolinic acid.

3. Detailed example: Friedrich Kjeldahl-titration method

The Friedrich (1933) digestion procedure gives quantitative yields of ammonia with dipicolinic acid and bovine serum albumin. We suggest that the nitrogen contents found for micro-organisms and mixtures of biological materials with the usual Kjeldahl techniques should be checked with the Friedrich technique.

Samples of washed freeze-dried *A. aerogenes* were analysed for total nitrogen content by the Friedrich method.

(a) *Reagents*. The reagents are as follows—

  (i)   MAR grade hydriodic acid, 66% (from British Drug Houses).
  (ii) Red phosphorus.
  (iii) Mercuric acetate.
  (iv) Solution of 30% (w/v) sodium hydroxide containing 5% (w/v) sodium thiosulphate.
  (v)   Standard acid (0·01 N) containing ethanol and Tashiro's indicator as described for (iii) in detailed example (Section IIIB.2) above.
  (vi) Standard alkali (near 0·01 N) as described for (iv) in detailed example (Section IIIB.2) above.

(b) *Procedure*. Duplicate samples (5 mg) of freeze dried washed *A. aerogenes* cells are weighed into Kjeldahl digestion flasks (50 ml) and hydriodic acid (1 ml) and a few grains of red phosphorus are added. The flask is heated over a small flame to boil the hydriodic acid for 30 min. Then the neck of the flask is washed down with distilled water until the bulb is half filled. MAR grade sulphuric acid (2 ml) is added and the mixture shaken and heated until it boils vigorously. After about 30 min, the hydriodic acid and water will have evaporated off and the solution becomes clear. When the neck of the flask is free of sublimed iodine, a knife tip of mercuric acetate and two to three knife tips of potassium sulphate are added. The flask contents are again boiled for 30 min and then cooled. The digest is diluted with 2–3 ml of water and transferred with washings to the Markham still. Ammonia is released by adding sodium hydroxide–sodium thiosulphate solution (15 ml) and steam distilled into standard acid containing Tashiro's indicator in the receiver flask. Duplicate blanks without sample are run through the whole procedure in parallel with the tests and the residual acid in the receiver flasks is titrated with standard (near 0·01 N) alkali.

(c) *Results.* The calculation is the same as that for the detailed example (Section IIIB.2) above. The total nitrogen of *A. aerogenes* cells depends on the growth medium and conditions. Stationary phase cells grown in Tryptone–glucose medium contained reserve polysaccharide (15–20% total carbohydrate as glucose) and significantly less nitrogen than cells grown in a defined mannitol salts medium or tryptic meat broth. The data in Table II indicate that the total nitrogen of these organisms depends on their protein and nucleic acid contents.

4. *Detailed example: Kjeldahl-nesslerization method*

The Kjeldahl-titration methods described above require 2–5 mg of cells or protein per test but smaller samples can be analysed if microburettes (e.g. Conway, 1947) are used to titrate the distillates. Alkali or acid weaker than 0·01 N should not be used for the titrations because the change in colour of the indicator at the neutralization point is slow. The sensitivity of the Kjeldahl method is increased significantly if Nessler's reagent is

TABLE II

Determination of the total nitrogen content, in g/100 g, by the Friedrich–Kjeldahl method of *Aerobacter aerogenes* grown to the stationary phase in three different media (data from Strange et. al. 1961)

Growth medium	Nitrogen	Total protein (biuret)	Total carbohydrate (excluding pentose)	RNA	DNA
Defined carbon-limiting	13·6	66	5·8	20	3·1
Tryptone glucose	11·6	60	20·0	11·2	3·8
Tryptic meat broth	13·9	72	3·8	12·8	4·9

used to determine the ammonia formed during digestion and 5–10 µg ammonia nitrogen can be estimated. However, some loss of precision and reproducibility must be accepted if this method is used. The method given below is essentially as described by King (1951).

A sample of crystalline bovine serum albumin was analysed for total nitrogen content with a Kjeldahl-nesslerization method.

(a) *Reagents.* The following reagents are used—

   (i) Acid digestion mixture: dissolve selenium oxide (SeO_2; 1 g) in distilled water (50 ml) in a 100 ml flask and add MAR grade concentrated sulphuric acid slowly to 100 ml.

(ii) Nessler's solution: iodine (11·3 g) is dissolved in water (10 ml) containing 15 g potassium iodide. To mercury (15 g) in a glass bottle is added most of the iodine solution, the bottle is stoppered and the mixture shaken with cooling in a water bath until the supernatant liquid looses its yellow colour. The supernatant is then decanted into a flask (100 ml) and a drop is tested for the presence of free iodine with 1 % (w/v) starch solution. If no colour is obtained more of the iodine solution is added until a drop of the mixture gives a faint blue colour with starch. The solution is then diluted to 100 ml and poured into 10% (w/v) sodium hydroxide solution (485 ml). The turbid solution is filtered or the precipitate allowed to settle before use. Note that the alkaline solution should be stored in a hard glass bottle closed with a rubber stopper.

(iii) Standard nitrogen solution: dry AnalaR ammonium chloride (153 mg) is dissolved in 100 ml of distilled water in a volumetric flask; 25 ml of this solution and 10 ml N sulphuric acid are diluted to 1000 ml to provide a solution of which 1 ml is equivalent to 10 μg nitrogen.

(b) *Procedure.* Samples (0·5 ml) of a solution of bovine serum albumin (e.g., about 250 μg protein/ml) and blanks of distilled water are measured into Pyrex test tubes (6 in. × $\frac{5}{8}$ in.) and acid catalyst solution (0·2 ml) added to each. The tubes are heated gently until water evaporates off and then more strongly so that the acid boils. After 4 h the tubes are removed from the heater and allowed to cool; to each is added distilled water (7 ml) and Nessler's reagent (3 ml) and after standing for exactly 5 min at room temperature the extinction values of the solutions are measured at 465 nm in a spectrophotometer set to give 100% transmission with the diluted blank digests. It is advisable to measure the extinction values of the diluted blank digests against distilled water to check that the values are reproducible. Duplicate volumes of the standard ammonium chloride solution equivalent to 5, 10 and 20 μg of nitrogen are diluted to 7 ml with water, treated with Nessler's reagent (3 ml) and the extinction values of the series measured after 5 min against blanks of distilled water treated in a similar manner. The Beer–Lambert law is obeyed over the range 0–20 μg nitrogen in the sample.

(c) *Calculation.* If T is the absorbance value of the test and S that of the standard, then T/S μg of N in standard gives μg N in the test sample; this value × 100/μg sample in test gives the nitrogen value (%).

(d) *Results.* Extinction values for five test samples at 465 nm were 0·318, 0·320, 0·325, 0·315 and 0·315; average, 0·319. Extinction values for three standards each containing 20 μg of nitrogen were 0·320, 0·320 and 0·319; average, 0·320. Therefore, the test sample contained 20·06 μg of nitrogen in 128·5 μg protein and the nitrogen value is 15·61% (Table I).

If sufficient sample is available, it is our experience that results are more precise and reproducible if 5–10 times the amount of sample required to give a suitable extinction value with Nessler's reagent is digested, the digest is diluted with water to 10 ml in a volumetric flask and a volume (1–2 ml) of the diluted digest is taken for the final colour forming step. Besides increasing the accuracy of the method, this procedure eliminates the necessity of digesting a large number of replicate samples of the material; it is sufficient to measure the extinction of one treated sample from each of two diluted digests to check that reproducible results are being obtained.

C. Phosphorus

This element is present in all micro-organisms and is an essential constituent of growth media. It is present mainly in nucleic acids, nucleotides, phosphoproteins, phospholipids, teichoic acids, polymetaphosphate and as inorganic phosphate, and in the case of bacteria, may account for up to 3·0% of the bacterial dry weight. For example, nucleic acids contain roughly 10% phosphorus and fast-growing bacteria may contain 25% of their dry weight as RNA and DNA.

Although other methods have been described, practically all procedures currently used to determine phosphorus in biological material on the semi-micro or micro scale are photometric and based on the same principle. This principle involves release and oxidation of organically combined phosphorus to produce phosphate (or phosphoric acid), formation of phosphomolybdate by reaction with ammonium molybdate and reduction of the phosphomolybdate under controlled conditions to form a blue compound, the concentration of which is measured spectrophotometrically. The numerous reported variations in this basic procedure concern the conditions of acid digestion and oxidation of the biological material, the order of addition of the reagents to the digest and, in particular, the nature of the reducing agent used to form the blue colour. Apparently, the principle was first exploited by Bell and Doisy (1920) who used hydroquinone as a reducing agent and their method was improved by Briggs (1922; 1924) who changed the order of the addition of the reagents. A notable improvement was made by Fiske and SubbaRow (1925) who replaced hydroquinone with aminonaphtholsulphonic acid as the reducing agent. Kuttner and Cohen (1927) used stannous chloride as the reducing agent and claimed that this immediately reduced phosphomolybdate at room temperature; however, Bodansky (1932) reported that trichloroacetic acid interfered with the reaction. Other suggested reducing agents are ferrous sulphate (Sumner, 1944) and ascorbic acid (Lowry and Lopez, 1946; King, 1951; Chen et al., 1956).

Soyenkoff (1947; 1952) reported a different approach to the problem of measuring phosphomolybdate; this compound was reacted with the dye Quinaldine Red to form an insoluble dye-phosphomolybdate complex and the colour change that occurred was measured. When 2-p-dimethyl-aminostyrylquinoline ethosulphonate was substituted for Quinaldine Red, Soyenkoff (1952) claimed that the method was fifteen times more sensitive than the Fiske and SubbaRow (1925) procedure, that there was less inter-ference by acids, but that perchloric acid could not be used for digesting organic material because it interfered.

There is no doubt that measurement of the blue reduced phosphomolyb-date complex provides a reproducible assay for inorganic phosphate. The colour obeys the Beer–Lambert law over a satisfactory working range and inexperienced workers produce linear concentration–extinction curves without difficulty. It must be emphasized, however, that silica can be estimated by a method that differs only in that a higher concentration of mineral acid is present during reduction of the silicomolybdate complex and that silica will interfere with phosphorus determination unless condi-tions are properly controlled.

Determination of total phosphorus involves acid digestion of biological materials and this is achieved with boiling perchloric acid or a mixture of nitric and sulphuric acids; heating is continued for a considerable period after clearing of the digest has occurred.

Distinction between inorganic and organic phosphates is not easily achieved since many organic esters (e.g. ATP) are extremely labile and inorganic phosphate is released from them in the acid medium during colour development. Delory (1938) provided a method that partially solves this problem. Thus, he claimed it is possible to completely precipitate inorganic phosphate as $Ca_3(PO_4)_2$ even though only small amounts of ortho phosphate (10 μg) are present if the calcium phosphate precipitate is entrained with an excess of light magnesium carbonate powder. The precipitate is separated by centrifugation, washed, dissolved in perchloric acid and analysed for phosphate content by the usual photometric method. If a further sample is analysed after oxidative digestion, inorganic and organic phosphorus contents can be estimated, provided that polymeta-phosphate (see later) is not present in the material. Special methods must be applied to detect the presence of and determine the amount of this polymer.

Two variations of the reduced phosphomolybdate method for deter-mining phosphorus are described.

1. *Detailed example*

Inorganic phosphate is determined in the absence and presence of protein and after formation from biological material by oxidative digestion; the procedure is essentially that of Allen (1940) which is based on that of King (1932).

(a) *Reagents.* The reagents are as follows—

   (i) Perchloric acid (72%), AR grade.
   (ii) Dilute perchloric acid solution: dilute 72% perchloric acid 1 : 10 (v/v) with distilled water.
   (iii) Ammonium molybdate (5%, w/v) in distilled water; a small amount of ammonia solution (S.G., 0·88) may be added to facilitate solution.
   (iv) H_2O_2 (100 vol), phosphorus-free MAR grade.
   (v) Reducing agent: amidol (*p*-aminophenol–HCl 1·2 g) and sodium metabisulphite (24 g) solution; add a few grammes of metabisulphite to about 75 ml of water, dissolve the amidol in this solution, add the remainder of the metabisulphite and dilute the solution to 100 ml. The reagent should be stored in a brown glass bottle at 2°–5°C and discarded after 10 days.
   (vi) Standard phosphate solution: anhydrous KH_2PO_4, AR grade, is dried at 105°C and cooled in a desiccator; the dry salt (1·0967 g) is dissolved in deionized water and the solution diluted to 250 ml; this solution (1 mg phosphorus/ml) is stored over chloroform at 2°–5°C and will be stable for at least 2 years; dilutions of this stock solution are prepared as required and discarded after use.

(b) *Procedure.* (Method 1) For solutions not requiring deproteinization: measure up to 10 ml of sample (containing 2–100 μg P) into a stoppered graduated tube and add water if necessary to the 10 ml mark; the sample should be roughly neutralized if necessary. Add in the order given, 72% perchloric acid (1 ml), amidol reducing solution (1 ml) and ammonium molybdate solution (1 ml), mixing between the additions and dilute the mixture to 15 ml.

(Method 2) For solutions requiring deproteinization: to the sample (20 ml) add 72% perchloric acid (2·2 ml), mix, stand for a few minutes to allow the protein precipitate to flocculate and centrifuge the mixture. Measure up to 10 ml of the supernatant fluid into a 15 ml stoppered graduated tube and dilute to 10 ml with dilute perchloric acid if necessary. Add amidol reagent (1 ml) and ammonium molybdate (1 ml) with mixing and dilute the mixture to 15 ml.

(Method 3) For samples containing organically bound phosphorus:

measure a volume of sample (not more than 5 ml) or a weighed amount (in each case containing not more than 100 μg of total P) into a micro Kjeldahl digestion flask and add 72% perchloric acid (1 ml). Heat gently to remove water and then more strongly until the digest is colourless; if charred organic matter persists, cautiously add 1 or 2 drops of hydrogen peroxide (100 vol) but ensure that this reagent is destroyed and driven from the digest. Remove the flask from the heater, cool and wash out the contents quantitatively into a 15 ml stoppered graduated tube. Add amidol reagent (1 ml) and ammonium molybdate (1 ml) with mixing and dilute the mixture to 15 ml.

The procedure is then the same for Methods 1, 2 and 3. The blue colour is allowed to develop in each case for at least 5 and not more than 30 min and the absorbance value of the solution is measured at 730 nm. The absorbance–wavelength curve for reduced phosphomolybdate shows a broad peak between 710 and 750 nm and using light of 730 nm is a satisfactory compromise. With each series of test samples, blanks of distilled water are run in parallel and, although standard curves for known amounts of phosphorus are usually reproducible, it is advisable to include at least two standards with each series. With samples containing 2–20 μg, 20–50 μg and 50–100 μg of P, absorbance values are measured in 4, 2 and 1 cm light path cells, respectively, to obtain the most accurate results.

With undigested samples, colour or turbidity may interfere with determinations and in such cases a butanol extraction of the reduced phosphomolybdate blue colour may be employed. After colour development for 5–30 min, 10% (w/v) oxalic acid solution (1 ml) is added with mixing and the mixture is extracted by shaking in the stoppered tube with butan-1-ol (5 ml). Phase separation may occur readily but if not the tube is centrifuged. The butanol upper phase is carefully removed with a pipette and its absorbance value measured. If this procedure is used, blanks and standards must be treated in a similar manner and absorbance values are measured at 735 nm.

(c) *Calculation.* As with all spectrophotometric measurements, the P-content of the test sample is equal to the absorbance value of the test divided by that of the appropriate standard multiplied by the amount of the substance in the standard. To determine the phosphorus content as a percentage, it is necessary to determine the actual amount of sample taken for the colour development step and ensure that this is expressed in the same units as the standard.

(d) *Results.* Standard curves showing the relationship between concentration of phosphorus and extinction value of the reduced phosphomolybdate complex formed are shown in Fig. 1.

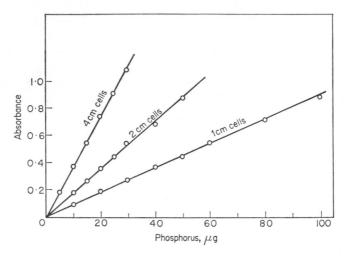

FIG. 1. Determination of phosphate. Values obtained with standard phosphate solutions in the molybdate-amidol method of Allen (1940), as described in the text. Samples of the 15 ml reaction mixture were measured against reagent blanks at 730 nm in the Unicam SP 600 Spectrophotometer, using optical cells of 1, 2 and 4 cm path length.

2. *Detailed example*

This illustrates the use of ascorbic acid for the reduction of phospho-molybdate.

(a) *Reagents.* The reagents are as follows—
  (i)   Perchloric acid (60%).
  (ii) Ammonium molybdate (5%, w/v) in distilled water.
  (iii) Ascorbic acid (10%, w/v) in distilled water; this solution is stored at 2°–5°C and remains effective for many months.
  (iv) Ascorbic acid (0·02%): prepared from the above stock solution and discarded immediately after use.

(b) *Procedure.* For solutions not requiring deproteinization: measure a volume of sample (5–20 μg P) into a tube (6 in. × $\frac{5}{8}$ in.) and dilute to 5 ml if necessary; add 60% perchloric acid (0·4 ml), 0·02% ascorbic acid (0·2 ml) and 5% ammonium molybdate (0·4 ml) with thorough mixing between each addition; allow the colour to develop for 10 min and read the absorbance value of the solution at 730 nm in a spectrophotometer adjusted to give 100% light transmission with a blank solution of distilled water treated in a similar manner to the test.

Samples containing protein are deproteinized as described in Section III C. 1(b), Method 2, above with perchloric acid but a smaller volume of sample is necessary. Similarly, total phosphorus is determined by digesting the sample as described in detailed example Section III C. 1(b), Method 3, above. In each case, the treated sample (5 ml) is allowed to react with ascorbic acid and ammonium molybdate for colour formation. A plot of concentration of phosphorus against absorbance value of the reduced phosphomolybdate complex formed is linear.

D. Sulphur

The amount of sulphur in micro-organisms is relatively small although bacterial spores contain much more than the homologous vegetative cells (Vinter, 1960). The element is present mainly in proteins in the amino-acids methionine and cystine but some is also present in other important compounds, for example α-lipoic acid and glutathione.

The difficulty associated with the determination of total sulphur in biological material is due to the volatility of the element and many of its compounds which may lead to losses during destruction of organic material. Usually sulphur is converted to sulphate ions by fusion with alkaline oxidizing reagents or by wet ashing with oxygenated acids. Alternatively, sulphur can be converted into hydrogen sulphide gas by heating the material in hydrogen (Masters, 1939).

A typical wet ashing procedure for biological material is that described by Masters (1939): the sample (1–5 g) is heated with concentrated nitric acid (5 ml) in a Kjeldahl flask until no more fumes are evolved; more nitric acid (3 ml) and perchloric acid (2 ml) are added, the temperature raised and the digest heated until it clears with further additions of perchloric acid and nitric acid as necessary, the process taking about 16 h; finally, more perchloric acid is added and the digest is heated for another 12 h. The digest is then decreased in volume to 2 ml by evaporation and the sulphate ion concentration is determined. The amount of sample digested depends on the analytical method used; the usual gravimetric procedure in which sulphate is precipitated as the barium salt requires a sample containing about 7 mg sulphur if a balance weighing to 0·1 mg is used but much less if a more sensitive instrument is available. More recently, Blanchar et al. (1965) described a similar but much shorter wet ashing procedure in which the sample was digested with nitric acid for 1 h at 150°C then with perchloric acid for 2 h at 235°C and finally with hydrochloric acid for 20 min at 150°C.

Sulphate ions in the digest can be determined as barium sulphate or as benzidine sulphate. The latter method has high sensitivity since the benzidine moiety can be estimated colorimetrically, for example after diazotization and coupling with thymol (Dodgson and Spencer, 1953; Dodgson et al. 1957; Slack, 1957). Benzidine can also be estimated by titration with alkali (Kolthoff and Stenger, 1947) or with permanganate (Hibbard, 1919). The fact that benzidine and its salts strongly absorb ultraviolet light has been exploited by Anderson (1953) and by Spencer (1960) for the estimation of micro amounts of sulphur. Benzidine sulphate is appreciably soluble in water (98 μg/ml) and precipitation of sulphate ions is done in the presence of ethanol (50% ethanol, 10·1 mg benzidine sulphate/litre; 95% ethanol, 1·16 mg/litre: Anderson, 1953). In fact, the limitation to the sensitivity of the benzidine procedure is not the final colorimetric method but the inability to precipitate small amounts of benzidine sulphate (Spencer, 1960). Irrespective of the merits or otherwise of the benzidine method as an analytical procedure, it must be emphasized that benzidine is a dangerous carcinogen and that its use is discouraged.

Barium chloroanilate has been used as a reagent for determining sulphate ions by Bertolacini and Barney (1957), by Lloyd (1959) and by Spencer (1960). A reaction between sulphate ions and sparingly soluble barium chloroanilate is allowed to occur in aqueous ethanol (50%) when barium sulphate precipitates and the soluble chloroanilic acid which is released is estimated colorimetrically. This method was scaled down by Spencer (1960) for use with 10 μlitre samples containing 0·25–25 μg of sulphate ion.

Atomic absorption spectrometry has been used for the estimation of sulphur or sulphate ions in biological materials (Roe et al., 1966). This method is based on a modified flame spectrometric procedure (Cullum and Thomas, 1959), and determines barium in barium sulphate precipitates separated from digests of oxidized biological materials. Liquid samples are oxidized during digestion with Benedict's (1909) reagent ($Cu(NO_3)_2 . 3H_2O$, 200 g; $KClO_3$, 50 g; water to 1 litre) and solid samples are oxidized in Schöniger (1956) flasks. Sulphate ions in the digests are precipitated with barium chloride and the barium sulphate is separated and dissolved in ammonium ethylenediamine tetra-acetic acid. Barium is then determined in the solution with an atomic absorption spectrometer (e.g. Perkin–Elmer model 303) and sulphate is calculated indirectly. Berglund and Sørbo (1960) describe methods for estimating sulphate ions by precipitation with radioactive barium or by flame photometric procedures.

Sensitive methods depending on precipitation with radioactive barium have been described for the determination of sulphate ions. For example, Picou and Waterlow (1963) treat samples containing 0·5–20 μM sulphate ions with barium chloride in N hydrochloric acid containing 60% (v/v) ethanol and $^{133}BaCl_2$ (0·5 μc/ml) in amounts depending on the sensitivity of the radioactive counter available. The radioactive barium is then separated and counted.

Turbidimetric methods depending on the absorbance values of barium sulphate suspensions have been described. Most methods for the determination of sulphate require a certain minimum amount of sulphate to be present in the sample before the method is quantitative. Dogdson (1961) used gelatin as the barium sulphate cloud stabilizer and found that the amount of sulphate present in certain samples of gelatin fulfilled this requirement; Difco Bacto Gelatin proved to be satisfactory in this respect.

Of the methods referred to for the estimation of the total sulphur in biological material, we have had only limited experience of two of them; these are the classical gravimetric barium sulphate procedure and the benzidine sulphate photometric procedure. Since each of these methods has disadvantages and more attractive methods are available (atomic absorption spectrometric; radioactive barium precipitation), no detailed example of a procedure is given.

E. Sodium

The concentration of this element in micro-organisms varies greatly and in bacteria in general it is not clear whether sodium is essential for growth and survival. However, halophilic bacteria do require sodium for growth and metabolism and it has been established that the activity of certain enzymes in halophiles is sodium ion-dependent. Yeasts, also, have a sodium ion requirement.

Most of the earlier methods for the determination of sodium in biological material involved precipitation of the ion as sodium zinc uranyl acetate which can be estimated gravimetrically (Harrison, 1943) or by measurement of the colour formed after reaction with potassium ferrocyanide to form red uranyl ferrocyanide (King et al., 1942; Noyons, 1939). An alternative colour-forming reaction was described by Stone and Goldzieher (1949) who treated the separated sodium zinc uranyl acetate with hydrogen peroxide in alkaline solution to produce a reddish yellow colour.

The advent of cheap reliable flame photometers has provided a simple, highly sensitive and accurate method of determining sodium and the earlier colorimetric methods are now rarely used. The sample solution is injected as a fine spray in the air supply to the burner of the flame photometer and light produced by the sodium in the flame is passed through suitable filters on to a selenium cell or other photo-sensitive element to produce a current that is measured by a sensitive galvonometer. Any of a number of gases (acetylene, coal gas, propane, butane) may be used to provide the flame but the gas pressure and the air pressure must be carefully regulated so as to maintain a constant steady blue flame with no yellow streaks. A spray of the sample solution is formed by passing compressed air through an atomizer into which the liquid being tested is drawn by suction or gravity.

The "EEL" flame photometer is a suitable instrument for determining sodium (and potassium, see below) and full working instructions are supplied by the manufacturers (Evans Electroselenium Limited, Halstead, Essex). With the instrument set at maximum sensitivity, the range 0–5 μg sodium/ml solution can be estimated. A number of other substances interfere but in many cases means of overcoming such interference are available (Farrow and Hill, 1961). A commonly used method to offset interference is to add a proportional amount of the interfering substance to the series of standards used in preparing calibration curves; this method is appropriate when estimating sodium in the presence of potassium (and vice versa).

An alternative rapid and sensitive method for determining sodium ions in solution is by means of a sodium electrode (Electronic Instruments Limited, Richmond, Surrey). It is claimed that these electrodes will give results as accurate as those obtained with a flame photometer but with rather less sensitivity at extremely low concentrations of sodium. The method is obviously worth considering when continuous monitoring of sodium ion concentration is necessary and can be easily adapted for continuous recording. One would expect about the same order of accuracy as is obtained with pH measurements and this depends entirely on the quality of the particular instrument.

To determine sodium in micro-organisms, the sample is digested with perchloric acid to destroy organic matter and, after dilution, the digest solution containing weak perchloric acid may be directly analysed in the flame photometer. In many cases, sodium is quantitatively released from bacteria by boiling acidified suspensions (1–5 mg dry weight/ml) or by treatment for a short period with 0·5 N perchloric acid at 70°C; in either case, the treated suspensions are freed from cells by centrifugation and the supernatant fluids are analysed in the flame photometer.

It must be emphasized that the concentration of sodium in micro-organisms depends on the concentration of sodium and potassium in the environment; for example, *Aerobacter aerogenes* cells contained extremely small amounts after they were washed with distilled water. Analysis of organisms washed with distilled water probably gives no indication of their sodium content during growth and metabolism and if such information is required, indirect methods must be used. Provided that the sodium content of the environment is not excessive, the sodium content of the organisms can be estimated as follows: duplicate samples of the bacterial suspension or culture are taken and one of them is acidified and boiled to release endocellular sodium; the boiled suspension is freed from cells by centrifugation and the supernatant fluid is analysed. The other suspension is centrifuged immediately and the supernatant fluid analysed. The difference between the first and second determinations gives the amount of sodium in the bacteria and if the dry weight of bacteria/ml of suspension is determined, the sodium content of the bacteria as a percentage of dry weight can be estimated.

F. Potassium

As with sodium, the potassium content of micro-organisms varies and in bacteria, varies markedly in the same organisms according to the conditions of growth and the growth rate (Tempest et al., 1966). Whereas potassium appears to be essential for the growth of bacteria, this probably is not the case with yeasts (D. W. Tempest, personal communication). A difficulty exists with regard to estimating the potassium content of micro-organisms for the ion is lost to the environment to some extent during washing with distilled water or sodium chloride solutions. Therefore, to estimate the true potassium content of bacteria in a given environment an indirect method similar to that described for sodium (see above) must be used.

In the past, potassium ions were estimated photometrically on the micro scale by precipitation of potassium with sodium cobaltinitrite followed by estimation of cobalt in the separated washed precipitate by means of the green colour produced on the addition of choline chloride and sodium ferrocyanide (Jacobs and Hoffman, 1931; Abul-Fadl, 1949). Now, the rapid and sensitive flame photometric method is invariably used and the "EEL" flame photometer is a satisfactory instrument for this purpose. Sodium and other ions may interfere with potassium estimations but as mentioned above with sodium estimations, means are available for offsetting such interference.

The potassium content of bacteria and yeasts may vary from about 0·7 to 2·5 % of the microbial dry weight and with the "EEL" flame photometer adjusted to give a full scale deflection with 20 μg/ml of potassium, extracts derived from suspensions containing about 1 mg microbial dry weight/ml are usually satisfactory.

G. Calcium

Variable amounts of calcium are present in micro-organisms and it is of particular interest that bacterial spores usually contain much more calcium than the homologous vegetative organisms. For example, vegetative cells and spores of Bacillus species contained 0·2–0·54 and 1·45–4·65 % of calcium, respectively (Powell and Strange, 1956). The greater heat resistance of bacterial spores may be related to their calcium content (Murrell and Warth, 1965).

The calcium content of micro-organisms can be determined by ashing a washed sample of the cells at 550°C for 12 h, dissolving the ash in dilute hydrochloric acid and precipitating calcium in the solution as oxalate which is then titrated with potassium permanganate (Kramer and Tisdall, 1921).

Flame photometric methods have been used for determining calcium in biological materials but there are difficulties associated with this method; calcium emits much less light than potassium and the wavelengths at which calcium emits (620 and 554 nm) are near the sodium D lines, so that there is significant interference if sodium is present in relatively high concentration. In order to get accurate results, protein which also interferes must be destroyed and calcium should be separated from other cations. Oxalate depresses the emitted light as does hydrochloric acid if excess is used to dissolve the oxalate precipitate. Powell (1953) found that perchloric acid is most effective in removing interference by oxalate and has the smallest effect in depressing the emitted light. Llaurado (1954) precipitated calcium from biological material with oxalate and incinerated the oxalate precipitate at 300°–400°C for 16 h to convert oxalate to carbonate. The residue was dissolved in 0·2 N hydrochloric acid and the calcium content of the solution measured after dilution 1 : 2 with the EEL flame photometer; calcium in the concentration range of 10–40 μg/ml can be determined by this method.

Zettner and Seligson (1964) have used atomic absorption spectrophotometry to determine calcium in serum but a special diluent had to be used to abolish the effect of absorption depression. The technique they devised provided results with whole serum equivalent to those obtained when calcium was first separated as an oxalate precipitate. Difficulties were encountered during limited attempts by colleagues in this laboratory to estimate calcium in perchloric acid digests of various bacteria with the EEL atomic absorption spectrophotometer; relatively large amounts of bacteria had to be used and the relationship between calcium concentration and absorbance was unsatisfactory.

Finally there are well-established sensitive titrimetric methods with EDTA as the titrant for the determination of calcium (Wilkinson, 1957; Eldjarn et al., 1955). With Eriochrome Black T as the indicator, a colour change occurs when the Ca-Eriochrome Black complex dissociates to give uncomplexed dye. Eriochrome Black T forms coloured complexes with magnesium and calcium so that direct titration with this indicator estimates both cations. Ammonium purpurate complexes specifically with calcium but the endpoint is not sharp. It is possible to determine calcium using Eriochrome Black T by determining the total magnesium and calcium in one sample and the magnesium in a second sample after removing calcium by precipitation with ammonium oxalate (Friedman and Radin, 1955). Such methods can be used for determining calcium in neutralized digests or solutions of the ash from micro-organisms.

H. Magnesium

There is an absolute requirement for magnesium during the growth of micro-organisms and in bacteria the endogenous concentration of this element varies with the growth rate. For example, the magnesium content of *A. aerogenes* varied from 0·1% to 0·35% of the bacterial dry weight with growth rates from 0·1 to 0·6 h⁻¹ in a chemostat with various growth-limiting substrates (Tempest and Strange, 1966). To some extent, studies of the effects of magnesium on bacterial growth, metabolism and survival have in the past been hindered by the lack of a simple, precise, sensitive and reproducible assay procedure for the determination of this element. This difficulty has now been overcome by the availability of reliable and reasonably priced atomic absorption spectrophotometers.

For some time, the only procedures available for the determination of magnesium in biological material were those based on the classical gravimetric method which involved precipitation of magnesium as magnesium ammonium phosphate. The sensitivity of this type of method was improved by Denis (1922) who estimated the phosphate in the precipitated magnesium salt by reacting it with acid molybdate and a reducing agent (Fiske and SubbaRow, 1925) to produce a blue colour (see phosphorus determination, Section III C).

Later, following the observation by Kolthoff (1927), methods based on the properties of the acridine sulpho-dye Titan Yellow were introduced; Kolthoff observed that when magnesium is precipitated as the hydroxide in the presence of Titan Yellow, the dye changes colour from yellow to red. Quantitative adaptations of this reaction were then introduced (Becka, 1931; Hirschfelder and Serles, 1934; Garner, 1946). Although there is a divergence of opinion regarding the accuracy of this method particularly when it is applied to biological material, it has provided results for the magnesium content of bacteria (Tempest and Strange, 1966) which agreed with those obtained by atomic absorption spectrophotometry. It must be accepted, however, that there are inherent disadvantages in the method; the ratio of reacted red "laked" dye to unreacted yellow dye increases with the concentration of magnesium in the sample and hence the "blank" is never a true blank. It is necessary, therefore, that the colour of the test sample is compared with the colour produced by a standard containing about the same amount of magnesium. In addition, various substances, particularly calcium, interfere with the reaction. If a significant amount of calcium is present, this ion must first be removed by precipitating it as calcium oxalate before analysis for magnesium; however, in the case of vegetative bacteria, the amount of calcium present is usually insufficient to interfere.

Magnesium ions can also be estimated by complexometric titration with ethylenediaminetetraacetic acid (EDTA) in the presence of the dye

Eriochrome Black T as indicator (Buckley *et al.*, 1951). In alkaline medium, a colour change occurs when the magnesium–Eriochrome Black complex is converted to magnesium ethylenediaminetetraacetate and free dye. The dye forms a coloured complex with both magnesium and calcium ions so that direct titration with EDTA estimates both; however, calcium can be determined separately by titration with EDTA in the presence of murexide as indicator, or removed by precipitation as calcium oxalate (Kramer and Tisdall, 1921). Harvey *et al.* (1953) and Smith (1955) have also described methods for the determination based on the formation of magnesium–Eriochrome Black T complexes.

Another method suitable for the estimation of small concentrations of magnesium ions depends on the formation of a complex with 8-quinolinol extraction of the complex into chloroform and measurement of the absorbance of the chloroform phase at 380 nm (Jankowski and Freiser, 1961).

Before application of any one of the methods referred to above, the cellular magnesium must be separated in soluble form from the micro-organisms. In the case of a number of different bacteria this is simple and only involves extraction with perchloric acid (see below). It was established in these cases that extraction of magnesium was virtually complete by wet ashing the extracted bacteria and estimation of magnesium in the residue. Other evidence showing the efficiency of extraction with perchloric acid was obtained by continuous culture experiments in which growth was limited by the amount of magnesium added to the culture medium; it was found that the amount of magnesium disappearing from the culture medium was exactly accounted for by the magnesium present in perchloric acid extracts of the bacteria (Tempest and Strange, 1966). Wet ashing is achieved by heating the sample of micro-organisms with nitric or perchloric acid in a Kjeldahl flask or hard glass tube until the organic material is destroyed. Perchloric acid digests can be used directly for the Titan Yellow colorimetric method or for assay in the atomic absorption spectrophotometer; nitric acid digests are taken to dryness on a water bath and the residue is treated with hydrochloric acid. The mixture is again taken to dryness and the residue is triturated with N hydrochloric acid. The resulting suspension is centrifuged and the clear supernatant solution used for the analysis.

Examples of the Titan Yellow and the atomic absorption methods for the estimation of magnesium in bacteria are given below.

1. *Detailed example: The Titan Yellow method*

Duplicate samples (10 ml) of culture containing 1–10 mg bacterial dry weight/ml are centrifuged at 3000 *g* for 10 min. The bacteria now have to be washed and it is important to note that salt solutions may remove some loosely bound magnesium from the bacteria (Tempest and Strange, 1966). Therefore, the bacterial pellet is suspended in either 0·85 % (w/v) sodium chloride solution or distilled water and re-centrifuged. The cells are resuspended in ice-cold distilled water and 2·5 ml of ice-cold 2 N perchloric acid is added. After mixing and standing for 15 min at 4°C, the suspensions are centrifuged and the clear extracts decanted into 10 ml graduated tubes. The pellets are again extracted with perchloric acid (4 ml, N) as before and the supernatant fluids combined with the first extracts. The combined extracts are diluted to 10 ml in each case. If foam-suppressing agents are used in cultures of bacteria, their presence may interfere with the Titan Yellow assay procedure. This occurred with polylgycol P-2000 and this agent had to be removed by treating the perchloric acid extracts twice with an equal volume of AR grade light petroleum (60°–80°C fraction) (Tempest and Strange, 1966).

(a) *Reagents.* The reagents are as follows—

  (i) Titan Yellow (British Drug Houses), 50 mg dissolved in 100 ml distilled water.
 (ii) Perchloric acid, 0·5, 1·0 and 2·0 N.
(iii) 3 N sodium hydroxide.
 (iv) Gum Ghatti (British Drug Houses), a 0·2 % (w/v) solution prepared by dilution with distilled water.
  (v) Stock magnesium solution: dissolve 0·829 g powdered anhydrous AR grade magnesium oxide in 41·5 ml of N hydrochloric acid and dilute the solution to 500 ml with distilled water.
 (vi) Working magnesium standard: dilute 5 ml of the stock magnesium solution to 500 ml in a volumetric flask with 0·5 N perchloric acid; this solution contains 10 μg magnesium/ml.

(b) *Procedure.* Perchloric acid extract of bacteria (1 ml) containing 4–16 μg magnesium is treated with water (1 ml), 0·2% Gum Ghatti (0·2 ml) and Titan Yellow solution (0·3 ml). The components are thoroughly mixed and 3 N sodium hydroxide solution (1 ml) is added. After mixing and standing for exactly 5 min, the absorbance of the mixture is measured at 540 nm. With each series of test samples, "blanks" containing perchloric acid and standards of increasing amounts of magnesium (4–16 μg in perchloric acid) are run in parallel. As the concentration of perchloric acid present affects the colour formation, blank, test and standard samples must contain the same amount of acid.

(c) *Calculation.* The magnesium concentration (% bacterial dry weight) is equal to

$$\frac{\text{Absorbance of test}}{\text{Absorbance of standard}} \times \text{Concentration of standard} (\mu g)$$

$$\times \frac{100}{\text{Dry wt. of bacteria in test} (\mu g)}$$

(d) *Results.* Table III shows the results obtained with the Titan Yellow method for the magnesium content of *Bacillus subtilis* and *Torula utilis* compared with the results obtained on the same samples of organisms with the atomic absorption spectrophotometric technique (see below). The results by the two methods are in fair agreement.

TABLE III

Comparison of chemical and atomic absorption methods for magnesium determination in whole cells[a]

Organism	Magnesium ($\mu g/ml$) determined by:	
	Atomic absorption	Titan Yellow method
B. subtilis	1·60	1·55
T. utilis	3·88	4·05

[a]The organisms were grown in a chemostat under magnesium-limiting conditions (Tempest *et al.*, 1966); samples were prepared and extracted with perchloric acid as described in the text.

2. *Detailed example: The atomic absorption spectrophotometric method*

Organisms are prepared and extracted exactly as described above for the Titan Yellow method and the perchloric extracts are fed directly into the atomic absorption spectrophotometer. The instrument used is the EEL model 140, full details of which are given in the Manufacturer's Handbook. Measurement of the concentration of an element by this and similar instruments depends on the capacity of the element when present in an atomic state in a hot flame, to absorb light of characteristic frequency. The amount of light of this frequency that is absorbed is measured and interference is eliminated by sharp line spectral measurement. The method is therefore specific and it is also highly sensitive and accurate. A stock magnesium standard in 0·5 N perchloric acid is prepared as described for the Titan Yellow method and this is diluted with 0·5 N perchloric acid to provide a working standard containing 5 μg magnesium/ml.

(a) *Procedure.* The atomic absorption spectrophotometer is set up as described in the Manufacturer's Handbook; blank, standard and test samples are then assayed.

(b) *Results.* Typical meter readings obtained by D. W. Tempest for 1, 2 and 3 μg of standard magnesium in 0·5 N perchloric acid are 1·35, 2·45 and 3·35, respectively. The results obtained for the mangesium content of *B. subtilis* and *T. utilis* by this method are shown in Table III.

I. Iron

Small quantities of iron are present in microbial cells, in the haem moieties of the cytochromes of aerobic bacteria and in the enzyme catalase which mediates the decomposition of hydrogen peroxide. Many bacteria will grow in media to which iron is not deliberately added, i.e., traces of iron present in the other medium constituents are adequate for normal bacterial growth. The iron content of certain bacterial spores was found to be significantly higher than that of the homologous vegetative cells (Powell and Strange, 1956).

Although volumetric and gravimetric procedures have been adapted for the determination of iron in microbial materials, such methods are rarely used because the amount of sample required is unrealistic; for example, the iron content of vegetative cells of a laboratory strain of *Bacillus cereus* was only 0·005% (Powell and Strange, 1956). Therefore, colorimetric or atomic absorption spectrophotometric methods are usually chosen. The success of colorimetric methods arises from the fact that iron forms a number of highly coloured salts and complexes which under appropriate conditions are sufficiently stable for the purpose of spectrophotometric determination. Ferric iron reacts with thiocyanate ions to give deep red ferri-thiocyanate ions, with Ferron (7-iodo-8-hydroxyquinoline-5-sulphonic acid) in slightly acid solution to give a green or greenish-blue colour and with acetate to give a reddish brown complex. Ferrous iron reacts with α,α'-dipyridyl to give a deep red complex bivalent cation in mineral acid solution and with O-phenanthroline to give a red coloration. To determine total iron in a sample, ferrous iron must be oxidized before reaction with thiocyanate or Ferron, and ferric iron must be reduced before reaction with α,α'-dipyridyl or O-phenanthroline.

The reaction of ferrous iron with α,α'-dipyridyl was discovered by Blau (1898) and adapted for the quantitative determination of iron by Hill (1930), Koenig and Johnson (1942), Ramsay (1953) and Thorp (1941). Various reducing agents including titanous chloride, hydroquinone, hydroxylamine, ascorbic acid, sodium hyposulphite and hydrazine sulphate have been used to convert ferric iron into ferrous iron; Thorp (1941) used sodium sulphite as the reducing agent and his procedure was capable of estimating 0·0002 mg ferrous iron/ml sample when the reaction occurred at pH 4·0. With hydroxylamine as the reducing agent, maximum colour development with the ferrous iron–dipyridyl complex was found to occur at pH 3–4 in the presence of ammonium acetate buffer; phosphate, if present in relatively high concentration, retards colour formation.

The α,α'-dipyridyl method described below is based on a number of the reported procedures and nearly quantitative recovery of iron added to suspensions of yeast cells was achieved with it.

1. *Detailed example: determination of iron in culture medium and microbial cells*

(a) *Reagents.* The reagents are as follows—

  (i) α,α'-dipyridyl solution: 1 g of reagent is dissolved in 100 ml 1% (w/v) hydrochloric acid.
 (ii) Reducing reagent: 3·5 g hydroxyammonium chloride dissolved in 100 ml distilled water.
(iii) Buffer solution: 355 g AR grade ammonium acetate dissolved in distilled water and diluted to 500 ml.
 (iv) Hydrogen peroxide, 100 vol, AR grade.
  (v) 6 M Perchloric acid: 265 ml of 72% AR grade perchloric acid diluted to 500 ml with distilled water.
 (vi) Stock standard iron solution: 0·4318 g AR grade ferric ammonium sulphate dissolved and diluted to 100 ml with 0·1 N hydrochloric acid; this solution contains 0·5 mg iron/ml.
(vii) Working iron standard: dilute 10 ml of the stock iron solution to 250 ml with distilled water; this solution contains 20 μg iron/ml and a fresh solution is prepared daily.

Great care must be taken to ensure that water and all chemicals used are iron-free; all glassware is cleaned in the usual way, then washed three times with iron-free 1% (w/v) hydrochloric acid and three times with iron-free distilled water.

(b) *Procedure* (Method 1) Determination of iron in culture medium. Culture medium or cell-free culture fluid (10 ml, containing not more than 2 μg iron/ml) is measured into a centrifuge tube (15 ml) and mixed with 1 ml of 6 M perchloric acid. The tube is sealed with "Parafilm" and held at 2°–4°C for 30 min. The mixture is centrifuged at 3500 rpm in a bench angle centrifuge for 15 min and 10 ml of the clear supernatant fluid is decanted into a clean stoppered tube (15 ml). The supernatant fluid is treated in the order given with ammonium acetate solution (1 ml), hydrox-ylamine solution (1 ml) and dipyridyl solution (1 ml) with mixing and the mixture is accurately diluted to 15 ml. After standing for 15 min, the absorbance of the solution is measured at 525 nm against a blank solution containing distilled water in place of the sample but all the other reagents. Standards containing 5, 10, 15 and 20 μg of iron are treated in parallel with the test samples and blanks.

(Method 2) Determination of iron in microbial cells. Culture (2 ml containing 3–5 mg dry weight of bacteria or yeast/ml) is centrifuged at 3500 rpm for 15 min and the cells are washed twice by centrifugation with iron-free distilled water. The washed cells are transferred with two 1 ml volumes of distilled water and 1 ml of 72% perchloric acid, into a Kjeldahl flask (30 ml). The mixture is boiled until clearing occurs, one or two drops of hydrogen peroxide (100 vol) being added if necessary. The digest is allowed to cool, washed into a 15 ml stoppered tube and diluted to 10 ml. The diluted digest is treated in the order given with ammonium acetate solution (1·5 ml), hydroxylamine solution (1·0 ml) and dipyridyl solution

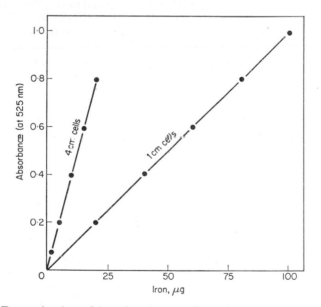

Fig. 2. Determination of iron by the α,α'-dipyridyl method. Values obtained with standard iron solutions by the method described in the text. Samples measured against reagent blanks at 525 nm in the Unicam SP 600 Spectrophotometer, using optical cells of 1 cm and 4 cm path length.

(1 ml) with mixing. The mixture is diluted to 15 ml, allowed to stand for 15 min at room temperature and its absorbance measured at 525 nm against a blank of distilled water run through the procedure.

11–5b

(c) *Calculation.* The iron content of the sample is equal to the extinction value of the test divided by the extinction value of the appropriate standard and multiplied by the amount of iron in the standard. To determine the

TABLE IV

Total iron content of *Torula utilis* and *Saccharomyces cerevisiae*[a]

Sample	Absorbance at 525 nm	Mean	Iron content of sample (μg)	Iron content of yeast (% of dry wt)
Standard (20 μg Fe)	0·207, 0·203	0·205	20·0	..
T. utilis (20 mg)	0·175, 0·171	0·173	16·9	0·087
T. utilis (20 mg) + standard (20 μg Fe)	0·385, 0·393	0·389	38·0	0·090
S. cerevisiae (12·06 mg)	0·017, 0·017	0·017	0·41	0·0034

[a]The yeasts were grown in simple chemically defined media, the *Saccharomyces cerevisiae* under conditions of iron-limitation. Washed cells were digested with perchloric acid and the digests assayed for iron as described in the text.

iron content of the material the weight of material taken for the colour development step is calculated and expressed in the same units as the standard.

(d) *Results.* Depending on the concentration of iron in the sample, a light path of 1, 2, or 4 cm is used for determining the extinction of the sample solutions. Fig. 2 shows representative results for standard amounts of iron when the colour of the dipyridyl complex is measured at 525 nm with the Unicam SP 600 spectrophotometer. Representative results for the iron content of carbon-limited *Torula utilis* with data for the recovery of added iron, and for the iron content of *Saccharomyces cerevisiae* are given in Table IV.

IV. DETERMINATION OF PROTEIN

As with other complex macromolecules, no truly specific methods for the analysis of protein are known; all existing methods are either non-specific or measure a single component of the protein molecule, such as a particular element, grouping or amino-acid. Among those available are the determination of total nitrogen, the biuret reaction, the Folin–Ciocalteu reagent, determination of amino-acids or α-amino nitrogen after acid hydrolysis, and ultraviolet light absorption.

It is unlikely that application of any of these methods to whole microorganisms will provide true protein values in all cases, because other cell constituents will interfere. However, with some of them (e.g., the biuret reaction) interference will often be negligible because of the large proportion of protein in the biomass.

A further analytical possibility is the quantitative separation of protein from all other cell constituents; while satisfactory in principle this is difficult to achieve in practice, particularly on a micro-scale.

In what follows the relative advantages and disadvantages of the above methods will be briefly discussed, and the more suitable ones described in some detail.

A. Determination of total nitrogen

The total nitrogen content of cells is sometimes used as a measure of their protein content; this can give reasonable results with animal tissues but with bacteria the values obtained will be much too high, owing to their much greater content of non-protein nitrogen. The non-protein N in bacteria occurs in amino-acids and other low-molecular-weight "pool" constituents, cell wall mucopeptide, part of the lipid fraction, and nucleic acids; the latter are by far the most important fraction, as the sum of RNA + DNA in fast-growing bacteria may amount to 25% of the dry weight and nucleic acid N is ca. 30% of the total N.

The extent of the error involved in the commonly made assumption that "protein = total N × 6·25" may be illustrated by the following example. A sample of washed *A. aerogenes* cells contained 13·6% of their dry weight as nitrogen determined by the Kjeldahl method (Section III). Multiplying this figure by the conventional factor of 6·25 (i.e., assuming a figure of 16% for the nitrogen content of bacterial protein) gives a value for the protein content of the cells of 85%. The cells were also analysed for protein by the biuret method described below, and for total carbohydrate, RNA and DNA by the methods of Sections V and VII; their lipid content was known to be very low. The values obtained, expressed as percentage of the cellular dry weight, were—

	From total N	Biuret method
Protein	85	66
Carbohydrate	5	5
RNA	20	20
DNA	3	3
Total	113	94

Obviously the value obtained from the total N determination is impossibly high, while that from the biuret method is reasonable.

In spite of the sensitivity and accuracy of the Kjeldahl method itself, determination of the protein content of micro-organisms from their nitrogen content will give grossly inaccurate results unless allowance is made for the non-protein nitrogen, and is not to be recommended. However, if the nucleic acid content is also determined, the approximation—

$$\text{Protein N} = \text{Total N} - \text{Nucleic acid N}$$

may often give reasonably correct results, but in general the direct determination of protein by one of the methods below is recommended.

B. The biuret method

The biuret reaction owes its name to the intense reddish-violet chelation compound formed by cupric ions with biuret ($H_2N.CO.NH.CO.NH_2$). The $-CO.NH-$ linkages of proteins form similar copper chelates in alkaline solution, and this is one of the more satisfactory methods for the colorimetric determination of proteins. It is also reasonably specific; a number of small molecules (amino-acids and peptides, hydroxy-acids, etc.) form similar copper chelates, but these may readily be separated from proteins. Nucleic acids and most other cellular macromolecules do not interfere. The polypeptide component of bacterial cell wall mucopeptides does react, but amounts to only a few percent of the biomass. It is interesting that the capsular polypeptide (poly-γ-glutamyl peptide) of *Bacillus anthracis* and some other Bacilli does *not* give a biuret reaction, presumably because the peptide linkages are γ- and not α-.

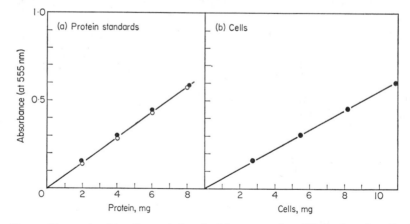

FIG. 3. Determination of protein by the biuret reaction. Method as described in text applied to (a) standard solutions of crystalline ovalbumin (○) and bovine serum albumin (●), and to (b) a washed suspension of *Aerobacter aerogenes* cells of known dry weight. Samples measured against reagent blanks at 555 nm in the Unicam SP 600 Spectrophotometer using 1 cm cuvettes.

Different proteins might be expected to give rather similar colour intensities in the biuret reaction, and this is indeed the case. Minor differences would be expected because (a) the number of peptide linkages in a gramme of protein will vary slightly, being related to the average residue weight, and (b) not all peptide links in the protein molecule react with copper, the average according to Mehl et al. (1949) being about four –CO.NH– linkages to each Cu atom; this ratio too will presumably vary somewhat in different proteins. Such differences are small, however, as shown by the results given in Fig. 3 for crystalline bovine serum albumin and crystalline ovalbumin; further examples are given by Lowry et al. (1951). The fact that all proteins give very similar biuret values is one of the most valuable features of the method.

The biuret reaction has been applied to the determination of proteins in two different ways. One consists of adding to the protein solution a "biuret reagent" containing NaOH, $CuSO_4$ and a complexing agent such as ethylene glycol (Mehl, 1945) or tartrate (Gornal et al., 1949), whose function is to form a Cu complex soluble in NaOH. Added protein displaces Cu from the tartrate or glycol, forming a Cu–protein complex of different colour and greater absorption intensity. A defect of such methods is that the colour intensity developed is not proportional to protein concentration, falling off at high protein values due to competition between protein and complexer for the available Cu. The linear portion of the curve can be extended by increasing the Cu concentration in the reagents, but this results in unduly high blank values.

In the second method, applied to the determination of serum proteins by Robinson and Hogden (1940), NaOH is first added to the protein, followed by aqueous $CuSO_4$ solution; the mixture is centrifuged, when excess Cu is removed as the precipitated hydroxide, leaving the coloured Cu–protein complex in solution. This method allows a large excess of $CuSO_4$ to be added while still giving an almost colourless blank solution, and the biuret colour is strictly proportional to protein added. It is readily adapted to the determination of total protein in whole microbial cells, and forms the basis of the method described below.

1. Application of the biuret reaction to whole microbial cells

The Robinson–Hogden biuret method was devised for soluble proteins, and in applying it to the determination of total protein in whole microorganisms the first step must be the quantitative extraction of their protein in soluble form. Stickland (1951) found, on applying the Robinson–Hogden method to washed suspensions of Gram-negative bacteria, that the first step of this method, namely adding NaOH to a concentration of ca. 1·0 N, appeared to dissolve the cells and solubilize all their protein; on then adding

CuSO$_4$ and centrifuging, the biuret colour appeared in the supernatant and could be determined colorimetrically. With yeasts and Gram-positive cocci, however, the cells were not dissolved and little protein was extracted; on adding CuSO$_4$ and centrifuging the supernatant was almost colourless while the cell deposit showed a strong biuret colour. Stickland found that heating such resistant cells with NaOH at 100°C appeared to extract their proteins and suggested that such a step might make the method generally applicable, though he does not appear to have investigated the matter further. We have therefore re-investigated the extraction of protein from a variety of cells with hot and cold alkali, with the following results.

(i) Cold NaOH (1·0N) extracts most of the protein from Gram-negative organisms and from bacilli, though the extraction is not complete (Fig. 4); the cells appear to dissolve more or less completely, though a small insoluble residue is found on centrifuging, which still contains some protein. Gram-

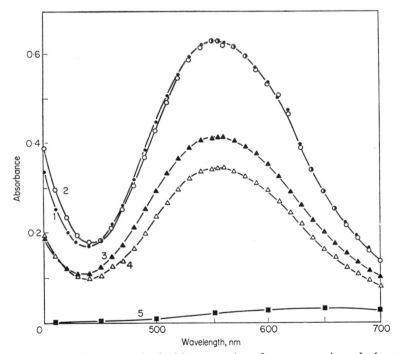

FIG. 4. Absorption spectra in the biuret reaction of a pure protein and a bacterial cell suspension. The reaction as described in the text was carried out with and without the initial heating period of 5 min at 100°C in N NaOH. Samples and reagent blanks were read separately against water in a Unicam SP 600 Spectrophotometer, using 1 cm cuvettes. ●——●, 8·0 mg bovine serum albumin, heated; ○——○, the same, unheated; ▲——▲, 8·9 mg washed *Aerobacter aerogenes* cells, heated; △——△, the same, unheated; ■——■, reagent blank, heated or unheated.

positive bacteria of the *Micrococcus, Sarcina* and *Staphylococcus* groups appeared to retain their cell structure more or less intact and little or no protein was extracted; yeasts and moulds behaved in the same way.

(ii) Hot NaOH (1·0 N, 5 min at 100°C) appeared to effect complete extraction of protein from all bacteria, yeasts and moulds tested. Both Gram-negative and Gram-positive bacteria appeared to dissolve almost completely on this treatment, while with yeasts and moulds a distinct residue remained (presumably insoluble glucan components of the cell wall); this residue contained no protein, however, and was removed with the copper hydroxide precipitate on subsequently adding $CuSO_4$ and centrifuging.

(iii) Heating various pure proteins with NaOH as described above had no effect on the intensity of their biuret reaction on subsequent treatment with $CuSO_4$ (Fig. 4).

This simple modification of the Robinson–Hogden biuret method allows it to be applied to whole microbial cells or to insoluble cell fractions such as cell walls, ribosomes, etc.; it is described in detail below.

(a) *Reagents.* The reagents are as follows—

(i) Aqueous NaOH, 3·0 N.

(ii) Aqueous $CuSO_4.5H_2O$, 2·5% w/v.

(iii) *Protein standard:* bovine serum albumin (Armour Inc.), 5·0 mg/ml (Note 1 below).

(b) *Procedure.* Measure 2·0 ml of washed cell suspension (Note 2) containing 1–5 mg dry weight per ml into a 5 ml centrifuge tube, add 1·0 ml of 3 N NaOH, place in a boiling water bath for 5 min, and cool in cold water. Add 1·0 ml of 2·5% $CuSO_4$, cover the top of the tube with a square of "Parafilm" held down by the thumb and shake thoroughly, allow to stand 5 min and centrifuge. A *reagent blank* containing 2 ml of distilled water instead of cell suspension, and a set of *standard protein solutions* (e.g., 2, 4, 6 and 8 mg protein) are treated in the same way, including the heating stage (Note 3). Read the optical densities of the centrifuged supernatants from the cell suspensions and protein standards against the reagent blank in a Unicam SP 600 (or similar) spectrophotometer, using a wavelength of 555 nm and 1 cm cuvettes (Note 4). Draw a standard curve by plotting optical density against milligrams of standard protein, and use this to read off the amounts of protein in the cell samples.

Figure 3 shows standard curves obtained with crystalline bovine serum albumin and crystalline ovalbumin, which are seen to coincide exactly, and a similar curve obtained with different quantities of a washed suspension of *A. aerogenes* cells; straight-line plots are obtained in each case. Figure 4 shows the absorption spectra obtained from bovine serum albumin and *A. aerogenes* cells treated exactly as described above, and also treated in the same way except for the omission of the heating stage. It will be seen that heating in NaOH is necessary to effect complete extraction of the bacterial protein and also that heating has no effect on the colour given by the protein standard; the shape of the absorption curves is identical in all cases.

Note 1. The ideal protein standard would be a purified preparation of the whole cell protein of the micro-organism being analysed; such preparations are not simple to obtain, however (see Section IV F below), and for most purposes bovine serum albumin or any other readily available pure protein is adequate. Whatever standard is used, it should be checked for total N and moisture content.

Note 2. The cell suspension should be thoroughly washed, particularly when derived from a broth culture, as many culture medium constituents (ammonia, amino-acids, peptides etc.) interfere with the biuret reaction. Preliminary extraction of the cells with trichloracetic acid, to remove free amino-acids, has not been found necessary.

Note 3. As Fig. 4 shows, it is not strictly necessary to heat the protein standards, but it is a good principle to treat samples and standards identically.

Note 4. The cuvettes described (internal dimensions $1 \times 1 \times 4.5$ cm) are a common type, used in Unicam, Beckman and many other spectrophotometers, and hold ca. 3·5 ml. The total volume of sample plus reagents (4 ml) was designed to suit these, but may be scaled up or down for other cuvettes provided the same concentrations of all reagents are maintained.

2. *Interfering substances*

RNA and DNA in large amounts do not interfere, nor do polysaccharides such as starch, glycogen, glucans and mannans. Large amounts of glucose and other reducing sugars (e.g., 10 mg/ml of sample) interfere, owing to their caramelization on heating with alkali with the formation of a strong yellow colour and substances which reduce the $CuSO_4$ to cuprous oxide. Ammonia and free amino-acids interfere relatively weakly; 12 mg NH_3–N or 20 mg acid hydrolysed casein give about the same colour as 1 mg protein. As these figures suggest, the amounts of free "pool" amino-acids present in most cells will cause negligible interference, and we have found that extraction of the cells with trichloracetic acid causes no significant decrease in the value obtained for total protein. More serious interference may be caused by the peptides and peptones found in many bacteriological culture media, and cells grown in such media must be thoroughly washed before analysis.

3. General comments on the method

The biuret method is simple to carry out and gives extremely reproducible results, which seem moreover to be reliable. Values for total protein determined by this method agree well with those obtained by acid hydrolysis followed by determination of α-amino nitrogen (Tables 6, 7). The main defect of the method is a relative lack of sensitivity, 5–10 mg of cells being needed for a determination. When greater sensitivity is needed, the Folin–Ciocalteu method described below is recommended; this requires as little as 50 μg of cells.

C. Determination of protein with the Folin–Ciocalteu reagent

The "phenol reagent" of Folin and Ciocalteu (1927), which is essentially a phosphotungstic–phosphomolybdic acid solution, is reduced by phenols (and by many other substances) to "molybdenum blue" which may be determined colorimetrically. Proteins reduce the phenol reagent, which may therefore be used for their determination; however, the amount of colour produced varies greatly with different proteins, since it is due almost entirely to their content of tyrosine and tryptophan, other amino-acids having little effect.

Herriott (1941) found that pre-treatment of proteins with alkali and a trace of copper salt greatly increased the colour developed with the Folin reagent. Copper salts do not increase the chromogenicity of free tyrosine and tryptophan and the effect is presumably due to Cu complexes of the other amino-acids. This modification of the original Folin method not only increases sensitivity but also reduces somewhat the very unequal chromogenicities of different proteins, and it has been adopted by most subsequent workers. A particularly thorough study of it was made by Lowry et al. (1951) and their method, modified in a few details and adapted for the analysis of whole micro-organisms, is described below.

1. Determination of total protein content of micro-organisms

In the method of Lowry et al. (1951) soluble proteins are treated with a NaOH–Na₂CO₃–CuSO₄–tartrate reagent before adding the Folin–Ciocalteu reagent, while insoluble proteins are first dissolved in NaOH and then treated with a Na₂CO₃–CuSO₄–tartrate reagent. The writers have found that the latter method can be applied to suspensions of whole micro-organisms if these are heated to 100°C in the NaOH, which extracts all the cell proteins in soluble form; this procedure is fully described in the preceding section on the biuret method. Full details of this method are as follows.

(a) *Reagents*. The reagents are as follows—

  (i)   5 % Na_2CO_3.
  (ii) 0·5 % $CuSO_4.5H_2O$ in 1 % sodium potassium tartrate.
  (iii) To 50 ml of reagent (i) add 2 ml of reagent (ii); prepare immediately before use and do not keep.
  (iv) *Diluted Folin–Ciocalteu reagent*. The total acidity of the concentrated reagent prepared according to Folin and Ciocalteu (1927) or purchased from, e.g., British Drug Houses Ltd, is determined by titrating a sample with 1·0 N NaOH using phenolphthalein as indicator. It is then diluted with distilled water to make the total acidity exactly 1·0 N (Note 1).
  (v)   *Protein standard*: bovine serum albumin (Armour), 200 μg/ml (Note 2).

(b) *Procedure*. Measure 0·5 ml of washed cell suspension (ca. 100 μg dry weight of cells) into an ordinary test tube, add 0·5 ml of 1·0 N NaOH, place in a boiling water bath for 5 min, and cool in cold water (Note 3). Add 2·5 ml of reagent (iii), allow to stand 10 min, and *rapidly* add 0·5 ml of reagent (iv) (use a blow-out pipette and mix immediately). A *reagent blank* containing 0·5 ml of distilled water instead of cell suspension, and a set of *standard protein solutions* (50–200 μg protein) are treated in the same way, including the heating stage. After standing 30 min to allow full colour development, measure all optical densities against the reagent blank in a Unicam SP 600 (or similar) spectrophotometer, using a wavelength of 750 nm and 1 cm cuvettes. Draw a standard curve by plotting optical density against micrograms of standard protein, and use this to read off the amounts of protein in the cell samples.

Figure 5 shows a standard curve obtained with bovine serum albumin and a similar curve obtained with a washed suspension of *Pseudomonas fluorescens* cells; the departure from linearity at high optical densities will be noted. Figure 6 shows the absorption spectra of the colours obtained with bovine serum albumin and with a *Ps. fluorescens* cell suspension, which are virtually identical.

Note 1. The total acidity of the Folin–Ciocalteu reagent must be accurately adjusted, as the final colour developed is sensitive to pH.

Note 2. Different proteins give widely different colour values with the Folin reagent (Table V), and if absolute results are required, the relative colour values of bovine serum albumin and the protein being analysed must be determined. This may be done by determining the protein content of one unknown sample by an independent method (e.g., the biuret method); alternatively a purified preparation of the whole cell protein of the micro-

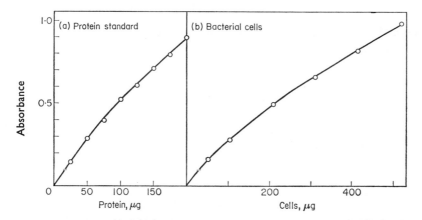

FIG. 5. Determination of protein with the Folin–Ciocalteu reagent. Modified method of Lowry *et al.* (1951) preceded by heating in NaOH for 5 min at 100°C (see text) applied to (a) standard solutions of bovine serum albumin, (b) a washed suspension of *Pseudomonas fluorescens* cells. Samples measured against reagent blanks at 750 nm in a Unicam SP 600 Spectrophotometer, using 1 cm cuvettes.

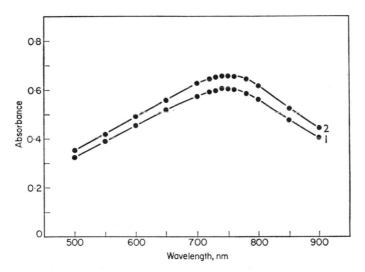

FIG. 6. Absorption spectra in the Folin–Ciocalteu reaction of a pure protein and a bacterial cell suspension. Modified method of Lowry *et al.* (1951) preceded by heating in NaOH for 5 min at 100°C (see text). Curve 1, Bovine serum albumin (125 μg); Curve 2, washed suspension of *Pseudomonas fluorescens* cells (311 μg) dry weight). Samples measured against reagent blanks in a Unicam SP 600 Spectrophotometer using 1 cm cuvettes.

TABLE V

Colour values of different proteins with the Folin–Ciocalteu reagent[a]

| | [b]Values of ε_{750} | | |
Protein	No Cu treatment[c]	Alkaline Cu treatment[d]	Increment with Cu
Trypsin	910	3,600	2,690
Insulin	998	3,000	2,002
Chymotrypsin	425	2,930	2,505
Cytochrome c	738	2,495	1,757
Human serum	365	2,120	1,755
Bovine serum albumin	358	2,050	1,692
Gelatin	78	1,145	1,067
Tyrosine	13,850	15,100	1,250

[a]Values are taken from the paper by Lowry et al. (1951).

[b]$\varepsilon_7^N{}_0$ is defined as the absorbance at 750 nm with 1 g atom of protein-N per litre.

[c]Treated with alkali but no Cu before addition of Folin reagent.

[d]Treated with alkali and Cu before addition of Folin reagent, as described in text.

organism being analysed may be made as described in Section IV F below, and used as a standard instead of the bovine serum albumin.

Note 3. The cell walls of some Gram-positive bacteria and yeasts are not completely dissolved by the hot alkali treatment (Section IV B), but the quantity of cells taken is so small that any resulting turbidity is negligible.

2. *Comments on the method*

The Lowry method is somewhat simpler to carry out than the biuret method, and is easier to adapt to automatic analysis. It is rather less reproducible, however, and much more subject to interference; a wide variety of substances such as glucose, aromatic amines, urea, uric acid, unsaturated aliphatic compounds, sulphides, sulphurous acid, hydrogen peroxide, ferrous ions, etc., all reduce the Folin reagent. It cannot be considered an absolute method since different proteins give widely differing colour values, as shown in Table V. The great advantage of the method is its sensitivity, which is about sixty times greater than the biuret method and allows amounts of protein down to 10 μg to be determined.

D. Determination of amino-acids and α-amino nitrogen after hydrolysis

In this method the amino-acids liberated by hydrolysing proteins are determined as their copper complexes, or by determination of their α-amino or carboxyl groups. These are probably the most accurate and specific methods available for protein determination, though they are less commonly used than the biuret or Lowry methods, probably because they are more time-consuming. Nevertheless they will be described in some detail, as the procedures for amino-acid and amino nitrogen determination have many other useful applications (determination of free amino-acids in metabolic pools, culture media, etc.).

In microbial cells, particularly those of Gram-positive bacteria, amino-acids occur in free form and in the cell wall mucopeptide, as well as in proteins. The free amino-acids are easily extracted from the cells (see below), but cell wall material in the residue may account for 20–30% of the cell dry weight. In different bacteria, the amount of mucopeptide in the cell wall varies and the constituent amino-acids differ qualitatively and quantitatively (see Salton, 1964). However, since mucopeptide also contains a large amount of hexosamine and muramic acid, mucopeptide α-amino nitrogen will rarely account for more than 0·5% (equivalent to 3–4% as protein) of the bacterial dry weight.

Before determination of α-amino nitrogen, proteins must be hydrolysed and, with strong acids, some destruction of the released amino-acids occurs; tryptophan is completely destroyed, serine and threonine are partially destroyed and there is some loss of others including arginine. Factors to correct for losses of amino-acids during acid hydrolysis are not easily determined and the usual procedure is to assay a purified protein (e.g., bovine serum albumin) in parallel; a factor for converting α-amino nitrogen to protein under the hydrolysis conditions used can then be derived. The assumption here is the same as that made with many biochemical assay procedures, i.e., that macromolecules from different sources react similarly under given conditions; in fact, the α-amino nitrogen contents of various purified proteins are different and in the absence of a standard of purified bacterial proteins, some error is inevitable. However, the error is likely to be small and in any case most biochemical analytical data are used for comparative purposes.

Numerous methods of varying specificity have been used to determine amino-acids and α-amino nitrogen.

1. Ninhydrin reaction

A purple colour is produced when solutions containing any of most of the common amino-acids are heated with ninhydrin (triketohydrindene hydrate) (MacFadyen, 1950; MacFadyen and Fowler, 1950) and photometric measurement of the colour provides a basis for determining α-amino nitrogen (Moore and Stein, 1948, 1954; Yemm and Cocking, 1955; Rosen, 1957). At pH 5·0–7·0, the reaction can probably be expressed as follows—

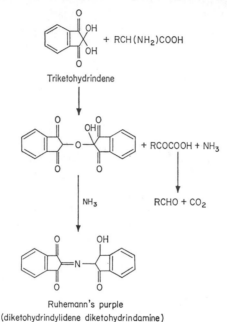

Triketohydrindene

+ RCH(NH₂)COOH

+ RCOCOOH + NH₃

NH₃

RCHO + CO₂

Ruhemann's purple
(diketohydrindylidene diketohydrindamine)

The products of the reaction are an aldehyde with one less carbon atom than the original compound, carbon dioxide, ammonia and an intensely coloured compound (DYDA). Initially, difficulties were encountered during attempts to standardize the reaction in terms of the production of DYDA but these were largely overcome by Moore and Stein (1948; 1952) and Yemm and Cocking (1955). Most α-amino-acids produce a violet colour that absorbs light maximally in the region of 570 nm but α-imino acids (proline and hydroxyproline) give a yellow colour. Most of the common amino-acids also give colour yields equivalent to $100 \pm 1\%$ of that of pure DYDA but tryptophan gives 180% and lysine 110%; ammonia reacts but gives a colour yield only 33% of that of pure DYDA (Yemm and Cocking, 1955). Colour yields depend on the time and exact temperature of heating, cleanliness of glassware, quality of the reagents, etc.; test conditions must be rigorously controlled to obtain reproducible results. In the case of protein hydrolysates, the spectrophotometric reading can be converted to protein concentration by means of a factor obtained with a standard protein (e.g., bovine serum albumin), or to α-amino nitrogen concentration with a factor based on the colour yield obtained with a pure amino-acid, assuming the average yield of DYDA from the amino-acids in the protein is 100%. The advantages of the ninhydrin technique are simplicity and high sensitivity (0·05 μg α-amino nitrogen); the major disadvantage is a lack of specificity since amines including hexosamine produce colours with ninhydrin. The use of this method for determining microbial protein is not recommended unless the small amount of sample available for analysis precludes the use of other methods. Use of the method for the determination of the free amino-acids in bacteria is described in Section X.

2. *Reaction with β-naphthoquinone sulphonate*

The method depends on photometric measurement of the orange-red colour produced when amino compounds react with β-naphthoquinone sulphonate in alkaline solution at 100°C (Folin, 1922; Frame *et al.*, 1943; Russell, 1944). Samples containing 4–40 μg amino nitrogen can be determined with a claimed precision of 1–2% and the results were found to agree with those obtained with the Van Slyke nitrous acid manometric method (see below). This method is not as specific as the ninhydrin method and it is not recommended for determining protein on the basis of amino-acids released by acid hydrolysis.

3. *Reaction with nitrous acid*

The method depends on measurement of the nitrogen liberated from amino groups in amino-acids by nitrous acid in the Van Slyke–Neill manometric apparatus according to the equation

$$RNH_2 + HNO_2 = ROH + H_2O + N_2$$

The ninhydrin–CO_2 procedure (Van Slyke *et al.*, 1941) (see below) is more specific for the determination of amino-acids, is simpler to carry out and in general has supplanted the nitrous acid method. For a discussion of the use and accuracy of the nitrous acid method, see Greenstein and Winitz (1961).

4. *Copper complex method*

The method depends on the formation of a soluble copper complex when amino-acids are reacted with an insoluble copper salt—

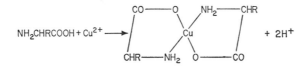

The soluble copper complex, which contains a 1 : 2 ratio of metal to amino-acid, is separated by centrifugation or filtration and determined by iodometric titration after reacting with potassium iodide in acetic acid (Pope and Stevens, 1939; Schroeder et al., 1950), photometric measurement of the yellow colour produced after treatment of the solution with dithio-diethylcarbamate (Woiwod, 1949), or by direct measurement of the absorbance of the clear centrifugate at 230 nm (Spies, 1952). The minimum amounts of α-amino nitrogen that can be determined with the iodometric (Pope and Stevens, 1939), diethyldithiocarbamate and direct absorbance methods are 250, 55 and 22 μg, respectively; if a microburette is used (Conway, 1947) 25 μg α-amino nitrogen can be determined by iodometric titration. It should be noted that the copper salts of phenyalanine, methionine, leucine and tryptophan are relatively insoluble and that copper reacts with peptides (1 : 1 ratio of metal to peptide); there is no interference with ammonia in this reaction. An illustration of the use of the copper complex method to determine bacterial protein is given below.

5. Manometric ninhydrin–CO_2 method

Of the various products formed when ninhydrin reacts with amino-acids (see above), carbon dioxide is uniquely specific because evolution of this gas depends on the presence of both free amino and carboxyl groups (or imino and carboxyl groups as in proline and hydroxyproline). Chloramine T was used earlier as a reagent for quantitatively decarboxylating amino-acids but this has been supplanted by ninhydrin which gives more exact results. When solutions of amino-acids are heated at 100°C with ninhydrin in buffered solution (pH 2·5), no coloured compound is formed but 1 mole carbon dioxide is produced per mole α-amino nitrogen with all the common amino-acids except for aspartic acid, which yields 2 moles. Proteins, peptones, peptides (other than glutathione) and substances other than amino-acids do not react significantly (Van Slyke et al., 1941). The carbon dioxide produced is transferred to the Van Slyke manometer and there determined. The use of this method for determining bacterial protein is illustrated below.

At pH 2·5 and 100°C in the presence of ninhydrin, ammonia is also produced in nearly quantitative yields from most of the common amino-acids and attempts have been made to determine α-amino nitrogen by measuring this product (Saidel, 1957; Sobel et al., 1945). Proline and hydroxyproline evolve no ammonia and the yields from tryptophan and cystine are less than theoretical.

6. Detailed example: The copper complex method (Pope and Stevens, 1939)

The protein content of samples of washed A. aerogenes cells was measured after acid hydrolysis. A volume (3·0 ml) of the suspension (3–4 mg bacterial dry weight/ml) was dried at 105°C and the dry material transferred to a small glass tube (75 mm long; 8 mm diameter). 6 N HCl (0·4 ml) was added and the tube sealed in a flame and heated at 105°C for 18 h. The hydrolysate was neutralized with NaOH solution, diluted to 6 ml and analysed.

(a) *Reagents*. The reagents are as follows—

  (i) Copper chloride, 2·73 g in 100 ml H_2O (0·16 M).

 (ii) Trisodium phosphate: disodium hydrogen phosphate (6·45 g) dissolved in CO_2-free water (50 ml), add NaOH (0·72 g) and when dissolved, dilute the solution to 100 ml.

(iii) Borate buffer: sodium borate (5·721 g) dissolved in water (150 ml) add N HCl (10 ml) and dilute the solution to 200 ml.

(iv) Copper phosphate suspension: mix trisodium phosphate (2 vol) copper chloride (1 vol) and borate buffer (2 vol) solutions. The suspension is stable for several days.

 (v) Thymolphthalein, 0·25 g in 100 ml 50% (v/v) ethanol.

(vi) Sodium thiosulphate: a stock solution is prepared by dissolving $Na_2S_2O_3$ (4·96 g) in CO_2-free water (20 ml) and diluting the solution to 200 ml (approx. 0·1 M). Borate (0·1%) is added to keep the stock solution stable; working solutions (0·01 or 0·005 N) are made from the stock as required.

(vii) Standard iodate: AR quality $NaIO_3$ (0·3568 g) dried at 110°C for 1 h is dissolved in water in a volumetric flask and diluted accurately to 1 litre; this solution is used to standardize the thiosulphate solution.

(viii) Indicator: 2% (w/v) soluble starch solution or starch thioglycollate solution.

(b) *Procedure*. Neutralized hydrolysate of bacteria (6 ml) is treated with copper phosphate suspension (4 ml) and after thorough mixing, the insoluble material is separated by centrifugation, or by filtration through Whatman No. 5 filter paper. If filter paper is used, the smallest possible paper is selected to avoid loss of solution by absorption. A measured volume (8 ml) of the clear supernatant or filtrate is treated with AR quality KI (0·1 g) and glacial acetic acid (0·1 ml) and the liberated iodine is titrated with standardized sodium thiosulphate solution from a 5 ml burette calibrated to read 0·02 ml. The estimation is carried out in duplicate along with a blank of distilled water and weighed amounts of dried pure glycine (5–6 mg).

(c) *Calculation*. 1 ml of 0·1 N thiosulphate is equivalent to 0·28 mg α-amino nitrogen. To convert α-amino nitrogen values to protein, it is assumed that bacterial protein contains the same amount of α-amino nitrogen as bovine serum albumin determined by this method; the factor used was 8·34.

(d) *Results*. Table VI indicates the effect of starvation on the protein content of an aerated suspension of *A. aerogenes* during a period of 43 h at 37°C. Protein was determined at intervals by the biuret method and by the copper complex method after acid hydrolysis of the bacteria. That the results are not in complete agreement is to be expected since with each of the methods the assumption is made that bacterial protein is similar in composition to bovine serum albumin. Nevertheless, in each case the results agree to within less than 5% (with the biuret method, results are consis-

TABLE VI

Comparison of the protein content of *Aerobacter aerogenes* cells by the biuret method and by the copper complex method after acid hydrolysis of the cells[a]

Time (h)	Copper complex method (after acid hydrolysis)		Biuret method	
	Suspension (mg/ml)	Bacteria (% dry wt)	Suspension (mg/ml)	Bacteria (% dry wt)
0	3·19	76·5	3·33	80·0
18	2·28	70·0	2·38	73·4
25	2·26	69·2	2·36	72·2
43	2·12	73·0	2·20	75·9
Total loss (mg/ml)	1·07		1·13	

[a]A washed suspension of stationary phase *Aerobacter aerogenes* cells (grown in tryptic meat broth medium) in saline phosphate buffer, pH 6·5, was aerated at 37°C. Initially the suspension contained 4·16 mg bacterial dry wt/ml ($2·1 \times 10^{10}$ cells/ml) and during 43 h the dry wt decreased to 2·9 mg/ml. At intervals, samples were removed and analysed for protein as described in the text. All analyses were done in duplicate and average results are given.

tently higher than those with the copper complex method) and the estimated changes in protein content of the suspension during the whole period of starvation are very close by the two methods. The bacteria used for these analyses were grown to the stationary phase in tryptic meat broth and such organisms have been shown previously to have a high protein content (Strange *et al.*, 1961).

7. Detailed example: the ninhydrin–CO_2 method

The α-amino nitrogen content of acid-hydrolysed *A. aerogenes* cells was determined by the Van Slyke *et al.* (1941) manometric technique and the value obtained was converted to protein content. Freeze-dried bacteria (20 mg) were suspended in 6 N HCl (0·5 ml) and hydrolysed in a small sealed tube at 105°C for 18 h. The hydrolysate was neutralized with N NaOH, diluted with water to 8 ml and analysed.

(a) *Reagents*. The reagents are as follows—

(i) Citrate buffer: sodium citrate (2·06 g) and citric acid (19·15 g) are pulverized and mixed together by grinding; the mixture is dried in a vacuum desiccator before use. When dissolved in water, the mixture has a pH value of 2·5.

(ii) Ninhydrin powder (British Drug Houses).

(iii) Strong sodium hydroxide solution: NaOH is dissolved in an equal weight of water and the solution is allowed to stand for 2–5 days. After the sodium carbonate has precipitated, a portion of the perfectly clear supernatant solution is standardized by titration with pure dry potassium acid phthalate.

(iv) 0·5 N NaOH solution: a graduated cylinder (100 ml) is filled to the 80 ml mark with CO_2-free saturated sodium chloride solution (boil 250 g NaCl in 750 ml distilled water, allow to cool under reduced pressure and store in a container protected with a soda lime tube). Add the correct amount of (iii) diluted 1 : 1 with distilled water under the surface of the salt solution and add salt solution to the 100 ml mark.

(v) 5 N NaOH: 1 volume of (iii) is diluted with 3 volumes of water.

(vi) 2 N lactic acid: 1 volume of AR quality lactic acid is diluted with 4 volumes of saturated sodium chloride solution.

(b) *Procedure*. A volume of the sample (2 ml) is measured into the reaction tube, 100 mg of dry citrate buffer is added and the mixture is heated to boiling over a micro-burner and boiled for 1 min. A rubber stopper is inserted and after cooling in running water, the tube is chilled in an ice bath. The stopper is removed, 100 mg of ninhydrin is added and a rubber adapter tube with a screw clamp is applied immediately. Air is evacuated from the reaction tube with a vacuum pump, effervescence being minimized by shaking the tube, and the clamp is then tightly screwed up. The tube is immersed in a vigorously boiling water bath and heated for at least 8 min to complete the reaction. The tube is cooled to room temperature. A blank consisting of distilled water is run in parallel with the sample.

The carbon dioxide evolved is measured in the Van Slyke–Neill manometer. Full details of this apparatus are available in a number of textbooks

(Peters and Van Slyke, 1932; Hawke et al., 1947) and are not repeated here. In the original apparatus, the reaction chamber was shaken by means of an electric motor but now a static reaction chamber in which mixing is achieved by magnetic stirring is available (supplied by A. Gallenkamp & Co. Ltd., London). A summary of the procedure for estimating carbon dioxide is as follows: (a) Introduce exactly 2 ml of 0·5 N NaOH into the reaction chamber and lower the mercury until it half fills the reaction chamber; (b) warm the reaction tube to 60°–70°C in a water bath and then attach it by means of the rubber connector to the side arm capillary of the mano-meter chamber; (c) open the tap at the top of the manometer and the screw clamp of the reaction tube to connect the tube and the chamber and absorb the carbon dioxide by ten traverses of the mercury up and down the reac-tion chamber; (d) close the manometer tap, remove the reaction tube, fill the side capillary of the manometer with mercury and eject unabsorbed gases; (e) introduce 1 ml of 2 N lactic acid solution into the reaction chamber, lower the mercury meniscus to the 50 ml mark and stir the solu-tion (or shake the reaction chamber) for 3·5 min, keeping the mercury at the 50 ml mark; (f) stop the mixing or shaking and allow the mercury to rise until the solution meniscus exactly reaches the 2 ml mark etched on the chamber, read the pressure (p_1); (g) introduce 0·5 ml of 0·5 N NaOH into the reaction chamber, mix the solutions to absorb the CO_2 completely and again allow the solution meniscus to rise to the 2 ml mark, read the pressure (p_2); (h) read the temperature of the water jacket surrounding the reaction chamber.

(c) *Calculation.* For each sample, calculate $(p_1 - p_2) - C$, where $C = (p_1 - p_2)$ for the blank solution; $(p_1 - p_2) - C = p_{CO_2}$. The p_{CO_2} value is multiplied by a factor to give either the α-amino nitrogen or the carboxyl carbon in the sample. A complete table of factors which depend on the reaction temperature and the constant volume in the reaction chamber at which the pressure is measured is given by Van Slyke et al. (1941). α-Amino nitrogen is then converted to protein by multiplying with a factor based on the α-amino nitrogen content of bovine serum albumin.

(d) *Results.* Values obtained for the protein content of *A. aerogenes* cells grown in a defined mannitol–salts medium and tryptic meat broth to the stationary phase are given in Table VII. These values are similar to those obtained directly with the biuret method for cells grown in the two media (Strange et al., 1961).

8. Comments on the above methods

Determinations of the protein content of *A. aerogenes* by the copper-

TABLE VII

Protein contents of stationary phase *Aerobacter aerogenes* cells based on α-amino nitrogen determination after hydrolysis[a]

Sample	Weight (mg)	cm Hg				Temp (°C)	α-amino N (%)	Protein (%)	Protein (biuret) (%)
		p_1	p_2	(p_1-p_2)	pCO_2				
Water (blank)	..	15·00	13·79	1·21	—	21·0	..	..	..
Bovine serum albumin	5·83	57·40	13·90	43·50	41·50	21·0	11·90	100	100
		58·85	13·80	45·05	43·06	21·5			
A. aerogenes (tryptic meat broth)	5·30	42·56	13·77	28·79	27·58	20·5	8·60	72·2[b]	72·0
		43·19	13·69	29·50	28·29	22·0			
A. aerogenes (defined medium)	5·35	38·50	13·84	24·66	23·45	22·0	7·18	60·3[b]	66·0
		38·70	13·80	24·90	23·69	22·0			

[a]The bacteria were grown in tryptic meat broth and a defined mannitol–ammonia-salts medium at 37°C in shaken flasks for 20 h, separated and washed with distilled water, and freeze-dried. The dried cells, and a sample of crystalline bovine serum albumin were hydrolysed in sealed tubes (6 N HCl, 20 h, 105°C) and the α-amino nitrogen determined by the van Slyke ninhydrin–CO_2 method (see text). Results with similar bacteria and the biuret method (Strange *et al.*, 1961) are given for comparison.
[b]Assuming % protein = % α-amino N × 8·4, as for bovine serum albumin.

complex and ninhydrin–CO_2 procedures give results close to those obtained by the biuret method. This agreement suggests that either method may be used with confidence; the biuret method will often be preferred on grounds of simplicity, but the determination of amino-N is considerably more sensitive. The Lowry method is more sensitive still, but the results obtained must usually be regarded as relative rather than absolute. Determination of α-amino-N after acid hydrolysis may be made even more sensitive, though rather less specific, by using the colorimetric ninhydrin method of Yemm and Cocking (1955) described in Section X C. With this method as little as 10 μg protein may be determined, and results obtained for the protein content of E. coli have given good agreement with those found by the biuret method.

E. Determination of proteins by ultraviolet absorption

Most unconjugated proteins absorb strongly in the ultraviolet at wavelengths below 300 nm, having an absorption maximum at about 280 nm, a minimum at about 250 nm, and rising end-absorption at shorter wavelengths. Measurement of the absorption at 280 nm is often used to determine the protein content of solutions, and the method has the advantages of speed and sensitivity (most proteins give an easily measurable absorption in a 1 cm cell at a concentration of 200 μg/ml). It also has serious limitations and may give grossly inaccurate results if the underlying principles are not understood; these are well discussed in the comprehensive review of Beaven and Holiday (1952).

The absorption of proteins in the region of 250–300 nm is due almost entirely to the tyrosine and tryptophan they contain. The absorption spectra of these amino-acids differ appreciably and are differently affected by pH. Tryptophan has two absorption peaks at 278 and 288 nm, and its spectrum is little affected by pH. Tyrosine has a single peak at 275 nm at pH values below 8; over the pH range 8–12 this peak shifts to 294 nm and the molecular extinction coefficient increases from 1340 to 2330, these changes being due to ionization of the phenolic –OH group. These facts apply equally whether the amino-acids are in the free form or combined in peptides or proteins.

It is therefore important to control the pH when measuring the absorption spectra of proteins, and the pH value should either be below 8 (when the absorption peak will be about 280 nm), or above 12 (when the peak will be nearer 290 nm and considerably stronger). At any pH, the *shape* of the absorption curve and the exact position of the absorption peak will depend on the tryptophan/tyrosine *ratio*, while the peak absorption intensity will depend on the absolute contents of both amino-acids. These facts may sometimes be turned to useful account, since measurement of the absorption at two different wavelengths (e.g., 280 and 294 nm) allows the tyrosine and tryptophan content of a protein to be determined with fair accuracy (Beaven and Holiday, 1952). But evidently a single determination of the absorption at 280 nm will give only a very approximate value of the protein concentration in solutions containing mixtures of proteins of unknown tryptophan and tyrosine contents; in this respect ultraviolet absorption measurement has the same disadvantages as the Folin–Ciocalteu method.

The presence of even small amounts of nucleic acids in protein solutions can produce large errors in ultraviolet absorption measurements, since they absorb 20–60 times more strongly than most proteins. Warburg and Christian (1941) suggested a method of correcting for nucleic acids by taking absorption measurements at 260 and 280 nm, and this is still widely used. As the method of correction given in the original paper is somewhat involved, it is here explained in a simplified way.

Let c_p and c_n be the concentrations of protein and nucleic acid respectively, in mg/ml. Let ϵ_p^{260} and ϵ_p^{280} be the specific extinction coefficients of the protein at 260 and 280 nm, and let ϵ_n^{260} and ϵ_n^{280} be the corresponding values for nucleic acid. (The specific extinction coefficient is defined as—

$$\epsilon = \frac{1}{c.d} \log \frac{I_0}{I}$$

where c is the concentration in mg/ml and d the solution depth in cm; i.e., ϵ is the optical density of a 1 cm layer of a solution containing 1 mg/ml). Finally let E^{260} and E^{280} be the observed optical densities at 260 and 280 nm of a 1 cm layer of a solution containing both nucleic acid and protein. From the above definitions it follows that—

$$E^{260} = c_p.\epsilon_p^{260} + c_n.\epsilon_n^{260} \tag{1}$$

and
$$E^{280} = c_p.\epsilon_p^{280} + c_n.\epsilon_n^{280} \tag{2}$$

If all the ϵ values are known, these equations can be solved to give both c_p and c_n from the observed values of E^{260} and E^{280}; thus

$$c_p = E^{280}.\frac{\epsilon_n^{260}}{A} - E^{260}.\frac{\epsilon_n^{280}}{A} \tag{3}$$

and
$$c_n = E^{260}.\frac{\epsilon_p^{280}}{A} - E^{280}.\frac{\epsilon_p^{260}}{A} \tag{4}$$

where
$$A = \epsilon_p^{280}.\epsilon_n^{260} - \epsilon_p^{260}.\epsilon_n^{280} \tag{5}$$

Warburg and Christian (1941) gave the following ϵ values†—

Crystalline yeast enolase $\epsilon_p^{260} = 0.513$, $\epsilon_p^{280} = 0.895$

Yeast ribonucleic acid $\epsilon_n^{260} = 22.07$, $\epsilon_n^{280} = 10.77$

Inserting these ϵ values into equations (3)–(5) gives—

$$A = 14.18$$

$$c_p = 1.56 \, E^{280} - 0.76 \, E^{260} \tag{6}$$

and

$$c_n = 0.063 \, E^{260} - 0.036 \, E^{280} \tag{7}$$

Thus, from the observed optical density measurements E^{260} and E^{280}, the protein concentration c_p can easily be calculated from equation 6; if required, the nucleic acid concentration c_n can also be calculated from equation 7, though this is usually unnecessary.

Equation 6, or the rather more involved tables given by Warburg and Christian (1941) are very widely used, but it must be realized that (a) the correction factors become rather uncertain when the contamination with nucleic acid is large, and (b) the results obtained, in terms of mg protein/ml, will be correct only when the protein concerned happens to have the same tyrosine and tryptophan contents as yeast enolase.

A final practical point is that errors due to light-scattering are many times larger at 280 nm than in the visible region, and it is essential that solutions should be optically clear; filtration or high-speed centrifuging may be necessary. A sensitive indication of haze is absorption in the region 320–360 nm; if this is appreciable, light-scattering or absorbing material other than protein should be suspected. Now that recording spectro-photometers are more commonly available, a complete scan of the ultra-violet region from 200–400 nm is recommended, as this facilitates detection of light-scattering material as well as nucleic acid contamination.

In general, ultraviolet absorption methods are best applied to reasonably pure protein solutions, and will be found most useful in enzyme purifica-tion and similar studies, e.g., in the examination of fractions from column chromatography. They are not recommended for the determination of total cell protein, unless this is separated and purified as described in the following section.

F. Separation of total protein followed by analysis

All of the direct methods described above for the determination of protein in micro-organisms are more or less unspecific, and the results

† Warburg and Christian actually express their results in terms of the absorption coefficient β, related to the specific extinction coefficients used here by the factor: $\beta = 2.303 \, \epsilon$.

should be accepted with some reservation. Such results could be unequivocally checked if a method were available for the quantitative separation of total cellular protein, which could then be determined by any convenient method, or even gravimetrically. Unfortunately, such methods have been little investigated.

In the method devised by Schneider (1945) for nucleic acid determination, the cells are extracted with hot trichloroacetic acid (or nowadays usually perchloric acid). This removes RNA, DNA and soluble polysaccharides, and the residue contains all (or nearly all) the cell protein, together with insoluble cell wall constituents such as mucopeptide (bacteria) or glucans (yeasts and moulds). On extraction with N NaOH, the cell wall constituents remain insoluble while the protein is dissolved and may be reprecipitated with cold trichloroacetic or perchloric acid.

An alternative method is to extract the cells with N NaOH at 100°C, as described above under the biuret method, followed by centrifuging to remove insoluble cell wall components. The NaOH extract contains all the protein, together with polysaccharides, DNA, and RNA hydrolysis products. Acidification with trichloroacetic or perchloric acid precipitates protein and DNA (cf. Schmidt and Thannhauser, 1945), and extraction with hot trichloroacetic or perchloric acid dissolves the DNA, leaving the protein as residue.

These methods are presented tentatively, in default of anything better to be found in the literature. Preliminary experiments have shown them to give reasonably pure protein preparations, free from all but traces of carbohydrate, nucleic acids and cell wall components, though the yields are 5–10% lower than those obtained by direct biuret determinations on whole cells. This may in part be due to the direct biuret determination including cell wall mucopeptide, but the main reason for the difference is probably loss of some protein by solubilization on treatment with hot acid; there are some grounds for believing that perchloric acid is worse than trichloroacetic in this respect. Further research on these and other methods for quantitative protein separation is definitely needed.

V. CARBOHYDRATE ANALYSIS

A. Determination of "total carbohydrate"

Micro-organisms contain a great variety of carbohydrates, which may be divided into the following main groups—

(i) Free sugars (pentoses, hexoses, heptoses etc.), the smaller oligosaccharides, and their derivatives such as amino sugars and phosphoric esters. These are usually present in quite small amounts, though yeasts can contain up to 10% of the disaccharide trehalose.

(ii) Simple polysaccharides, i.e., polymers of a single sugar or amino sugar, e.g., glycogen, glucan, mannan, chitin.

(iii) Complex polysaccharides, containing several different sugars, amino sugars, uronic acids, etc.; examples of these are the capsular polysaccharides and lipopolysaccharides of bacteria.

(iv) Complex macromolecules containing both carbohydrate and non-carbohydrate residues; e.g., nucleic acids, mucopeptides, teichoic and teichuronic acids.

Obviously no single reagent will determine all these substances and the so-called "total carbohydrate" reagents in fact determine only the simple sugars and their polymers; these do, however, make up the greater part of the carbohydrate of most cells.

All the common "total carbohydrate" methods are colorimetric and derive from the well-known Molisch test for carbohydrates; they involve heating the material with strong (20 N or more) sulphuric acid and a "colour developer" which is usually an aromatic amine or phenol. The reactions occurring are (a) hydrolysis of the polysaccharides to monosaccharides (b) dehydration and rearrangement of the monosaccharides to form furfural (from pentoses) or hydroxymethyl-furfural from hexoses (c) reaction of the furfural or hydroxymethyl-furfural with the "colour developer" to form a coloured compound which is measured absorptiometrically. Many such methods have been described (reviewed by Dische, 1955a, 1962), differing in such details as sulphuric acid concentration, heating time and temperature, and nature of the colour-developing reagent; examples of the latter are indole, orcinol, carbazole, cysteine, tryptophan, α-naphthol, anthrone and phenol.

Qualitatively, all these reagents are roughly similar; they all give colour reactions with pentoses, hexoses and heptoses, and some of their phosphorylated derivatives, but do not react with trioses or tetroses (since these cannot form furfural derivatives) or with amino sugars. Quantitatively, however, they differ considerably in the extent of the colour reaction given by different classes of sugars, and two have been found particularly useful: (a) the anthrone reagent, which gives a relatively weak reaction with pentoses and heptoses and a much stronger reaction with hexoses; (b) the phenol reagent, which gives approximately the same reaction with all sugars. The different specificities of these two reagents will often be found useful, and both will be described in detail.

1. The anthrone reagent

Dreywood (1946) discovered that carbohydrates give a characteristic green colour on heating with anthrone (a) in sulphuric acid solution.

Sattler and Zerban (1948, 1950) showed that the colour is due to the condensation with anthrone of furfural derivatives formed from the sugars in hot acid; Zipf and Waldo (1952) suggest that anthronol (**b**), the enol tautomer of anthrone, is the actual reactant. Morris (1948) first adapted the reaction to the quantitative determination of carbohydrates, and his method has been used, usually with some modifications, by many later workers (e.g., Koehler, 1952; Shetlar, 1952; Scott and Melvin, 1953).

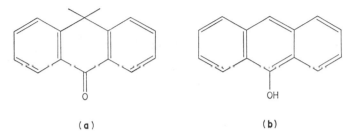

(a)   (b)

Seifter *et al.* (1950) applied the method to the determination of glycogen in animal tissues, and found protein interference to be so slight that the reagent could be applied directly to whole tissues. Fales (1951) and Trevelyan and Harrison (1952, 1956) found that the reagent can be applied directly to whole yeast cells, and the writers have used it successfully with numerous bacteria, and a few yeasts and moulds.

In the method of Morris (1948), two volumes of reagent containing 0·2% anthrone in 95% (v/v) sulphuric acid are added to one volume of sample and allowed to stand 10 min, the heat produced by dilution of the strong sulphuric acid being relied upon to develop the colour. The same "heat of mixing" technique was used by Trevelyan and Harrison (1956), who consider it satisfactory if the addition and mixing technique is sufficiently well standardized, but in our hands and those of others (e.g., Seifter *et al.*, 1950) this method gave occasional erratic results. In the method described below, sample and reagent are cooled to 0°C before mixing, and heated an exact time in a boiling water-bath.

(a) *Reagents.* The reagents are as follows—

   (i) *Standard glucose solution* (1·0 mg/ml in 0·15% benzoic acid). This is stable for long periods if kept at 0°C; working standards containing 100 μg glucose/ml are prepared from it daily by 1/10 dilution.
   (ii) Stock sulphuric acid (75% v/v). Add 750 ml of concentrated sulphuric acid (AnalaR grade) to 250 ml distilled water.
   (iii) *Anthrone reagent.* Pipette 5 ml of absolute ethanol into a 100 ml flask, add 200 mg anthrone and make up to 100 ml with the stock 75% H_2SO_4. Shake until dissolved (it is important to avoid the

presence of floating particles of undissolved anthrone). This reagent is prepared fresh each day, and kept in the refrigerator when not in use. (The incorporation of ethanol in the reagent is due to Fales (1951), who found it stabilized the colour, which otherwise tends to fade on standing.)

(b) *Procedure.* Samples (1·0 ml or less) of washed bacteria or yeast cells, or of standard glucose solution, are pipetted into thin-walled boiling-tubes (6 in. × 1 in.) and made up to 1·0 ml if necessary with distilled water. A reagent blank containing 1·0 ml distilled water only is also prepared.

The tubes are cooled in a rack standing in a large pan of ice-water, in which the anthrone reagent is also cooled. When cooled, each tube in turn is removed from the rack, but kept submerged in the ice-water; 5·0 ml of anthrone reagent are added from a fast-flowing pipette, swirling the tube in the ice-water during this addition. Little or no colour should appear at this stage. When all the tubes have been treated they are allowed to stand a few minutes so that all will cool to 0°C; the whole rack is then transferred to a vigorously boiling water-bath. After exactly 10 min the rack is returned to the ice-bath, and when cool the colours are measured in a Unicam SP. 600 (or similar) spectrophotometer at a wavelength of 625 nm, using 1 cm cells, with the reagent blank in the comparison cell.

Samples are always put up in duplicate, suitable quantities of cells being 0·1–1·0 mg dry weight, according to their carbohydrate content. A set of

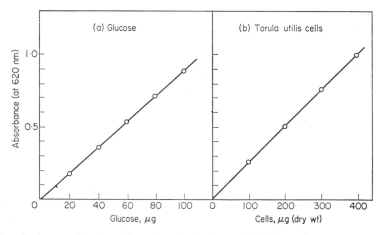

FIG. 7. Determination of total carbohydrate with the anthrone reagent. The method described in the text was applied to (a) a set of pure glucose standards, (b) a set of dilutions of the same washed suspension of *Torula utilis* cells. Samples measured against reagent blank at 625 nm in a Unicam SP 600 Spectrophotometer, using 1 cm cuvettes.

glucose standards (e.g., 20, 50 and 100 μg glucose) is put up with *every* batch of samples, and used to prepare a standard curve from which the quantity of carbohydrate (as glucose) in the unknowns can be read off. Figure 7 shows the results of such measurements with (*a*) a set of glucose standards, and (*b*) different quantities of a suspension of washed *Torula utilis* cells; in both cases Beer's Law is obeyed up to optical densities of at least 1·0. The smallest amount of glucose that can be accurately measured is ca. 10 μg.

No difficulties have been experienced in applying the anthrone reaction directly to whole microbial cells, the suspensions dissolving completely on heating with the reagent to give optically clear green coloured solutions.

(c) *Specificity and interfering substances*. A major advantage of the anthrone method is that free glucose, its disaccharides such as maltose and trehalose, and its polysaccharides such as glycogen, starch, dextrins, glucans and cellulose, all give quantitatively the same colour value per mole of glucose (Morris, 1948; Koehler, 1952); all give identical emerald-green colours, their absorption spectra showing a single main peak at 625 nm (Fig. 8). Galactose and mannose give identical colours and absorption spectra, and that of fructose is almost identical except for a small subsidiary peak at ca. 512 nm; quantitatively, fructose gives a somewhat higher absorption at 625 nm than glucose, while the absorption of galactose and mannose is significantly lower (Table VIII). The 6-desoxyhexoses (e.g., rhamnose, fucose) give green colours which are similar to the eye, though their absorption spectra are different, showing a broad flat-topped band with a centre at approximately 605 nm (Fig. 8).

All other carbohydrates tested give very much less colour than an equal weight of glucose; moreover the colours produced are obviously different to the eye and have very different absorption spectra. Pentoses and ribonucleic acid give a yellow-green colour which is produced more rapidly than that of glucose reaching a maximum in 2–3 min and then decreasing, so that little remains after the standard 10 min heating time. Uronic acids and heptoses give different shades of brownish-green while desoxyribonucleic acid gives a characteristic pink colour. The absorption spectra of all these substances in the anthrone reaction are shown in Figs. 8 and 9, while the absorbance at 625 nm relative to an equal weight of glucose is given in Table VIII.

It will be evident from these results that the anthrone reagent essentially determines total hexose, and other carbohydrates are unlikely to interfere unless present in considerable excess; when this is the case, the difference in colour is evident to the eye and its absorption spectrum will give a good idea of the nature and quantity of the interfering substances. This is

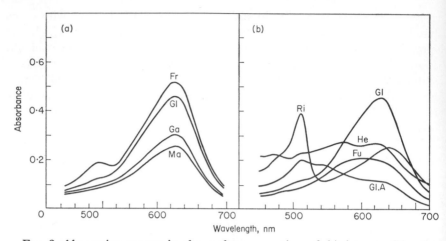

FIG. 8. Absorption spectra in the anthrone reaction of (a) hexoses (b) other carbohydrates. Reaction performed as described in the text, reaction volume 6 ml; samples read against reagent blank in Optica CF4 Recording Spectrophotometer using 1 cm cuvettes. Gl, D-glucose (50 μg); Ga, D-galactose (50 μg); Ma, D-mannose (50 μg); Fr, D-fructose (50 μg); Fu, L-fucose (50 μg); He, D-glycero-D-guloheptose (100 μg); Gl.A., D-glucuronic acid (200 μg); Ri, D-ribose (200 μg). L-rhamnose gives a spectrum virtually identical with that of L-fucose.

substantiated by the writers' experience in applying the anthrone reagent to the determination of total carbohydrate in whole microbial cells; when the carbohydrate content is 10% or more, the anthrone colour produced is indistinguishable from that of glucose, as shown by the absorption spectrum in Fig. 9 (Cells A). However, some Gram-negative bacteria when grown under carbon-limiting conditions may have carbohydrate contents as low as 2–3% and nucleic acid contents as high as 20%, and nucleic acid interference is then appreciable. Such cells give obviously abnormal colours with the anthrone reagent, whose absorption spectra show prominent peaks at 510 and 565 nm due to RNA and DNA respectively; an example of such a spectrum, produced by *A. aerogenes* cells with a low carbohydrate/ nucleic acid ratio, is shown in Fig. 9 (Cells B). When such spectra are observed the nucleic acids are contributing significantly to the measured absorption at 625 nm and apparent carbohydrate values will be spuriously high; however, approximate corrections can be made from the data in Table VIII if the RNA and DNA are separately determined.

Hexosamines give no colour with the anthrone reagent, nor do trioses or sugar alcohols (glycerol, mannitol, ribitol, etc.). Amino-acids also give no colour, with the sole exception of tryptophan, which gives a pale yellow colour with an absorption maximum at ca. 530 nm but negligible absorption at 625 nm. Mixtures of tryptophan and glucose, however, give a blue-

TABLE VIII

Reactions of various carbohydrates and nucleic acids with the anthrone reagent[a]

Substance	Absorption maxima[b] (nm)	Absorbance at 625 nm relative to glucose
D-glucose	625	100
D-galactose	625	66
D-mannose	625	55
D-fructose	625, 512	114
L-rhamnose	605	43
L-fucose	605	45
D-ribose	638, 510	15
D-glucuronic acid	(625), (545), 510	6·5
D-glycero-D-guloheptose	625, 570, (512), (470)	28
D-mannitol		0
D-glucosamine		0
Ribonucleic acid (yeast)	638, 510	6·2
Desoxyribonucleic acid (calf thymus)	560, 525	4·7

[a]Reaction performed as described in text.
[b]Main peaks underlined; figures in parentheses indicate "shoulders" rather than true peaks (see Fig. 8).

violet colour in which the peak at 530 nm is greatly enhanced while the glucose peak at 625 nm is slightly depressed (Fig. 10; cf. also Shetlar, 1952). It is known that tryptophan gives coloured compounds with glucose on heating in strong sulphuric acid, this being the basis of a method of carbohydrate determination (Shetlar et al., 1948). In this reaction, however, the tryptophan-glucose colour has an absorption peak at 460 nm, and it can be shown not to occur under the conditions of the anthrone reaction (62·5% H_2SO_4, 10 min at 100°C) by experiments with anthrone omitted from the reagent. The coloured compound absorbing at 530 nm is apparently produced by a complex reaction in which glucose, tryptophan and anthrone are all concerned.

The tryptophan in proteins reacts in the same way, and large amounts of protein (e.g., 5–10 mg) may cause some interference with the anthrone reaction, producing colours similar to the spectra of Fig. 10. However, (a) the anthrone method described requires only 0·1–1 mg of cells for a determination, and this amount of protein causes negligible interference; (b) even with much larger amounts of protein, the error is not large if the glucose peak at 625 nm is measured with a spectrophotometer of narrow

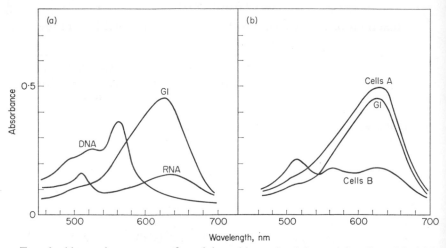

FIG. 9. Absorption spectra of nucleic acids and of bacterial cells with high and low carbohydrate content in the anthrone reaction. Reaction performed as described in text; samples read against reagent blanks in an Optica CF4 Recording Spectrophotometer, using 1 cm cuvettes. (a) RNA (250 μg) and DNA (200 μg), (b) washed suspensions of *Aerobacter aerogenes*; A, 500 μg cells containing 14·4% carbohydrate, 6·1% RNA and 3·4% DNA; B, 1·0 mg cells containing 2·0% carbohydrate, 13·3% RNA and 3·2% DNA. The spectrum given by 50 μg glucose (Gl) is included for comparison.

band-width. (In general the use of a spectrophotometer rather than a colorimeter using filters is recommended for all applications of the anthrone reagent, and greatly increases its specificity.)

Two further interfering substances of some importance are formaldehyde and phenol, since these are often used to sterilize cultures of pathogenic micro-organisms prior to analysis. Formaldehyde gives a strong red-brown colour with the anthrone reagent; phenol alone gives no colour but it depresses the colour produced by glucose, presumably by reacting with some of the hydroxymethylfurfural formed (cf. the following section on phenol method). For this reason the anthrone method should *not* be used for the determination of sugars eluted from paper chromatograms where phenol has been used as a developing solvent.

Hydrazoic acid or azides, often used as metabolic inhibitors in fermentation studies, also reduce the colour given by glucose with the anthrone reagent (Trevelyan and Harrison, 1952).

2. *The phenol method*

This method is comparatively recent, and was so thoroughly studied by its originators (Dubois *et al.*, 1956) that few later modifications have been

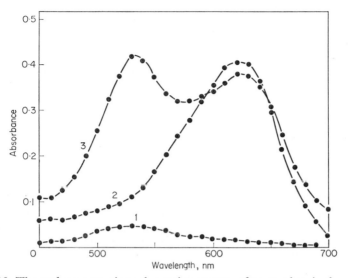

FIG. 10. The anthrone reaction; absorption spectra of tryptophan in the presence and absence of glucose. Reaction performed as described in text; samples read against reagent blank in a Unicam SP 600 Spectrophotometer using 1 cm cuvettes. Curve 1, Tryptophan (50 μg); Curve 2, glucose (50 μg); Curve 3, tryptophan (50 μg) plus glucose (50 μg).

proposed. Although it was originally developed for the analysis of pure sugars eluted from paper chromatograms, the present writers have found it can also be applied to the analysis of whole microbial cells. Both reagents and method have the advantage of great simplicity.

(a) *Reagents.* The following reagents, both of which should be Analytical Reagent grade, are used—

   (i) 5% (w/v) solution of phenol in water.
   (ii) Concentrated sulphuric acid (sp. gr. 1·84).

(b) *Procedure.* Into thick-walled test tubes of 16–20 mm dia. (Note 1) pipette 1·0 ml of sample containing the equivalent of 20–100 μg glucose. A reagent blank containing 1 ml of water, and a set of glucose standards (e.g., 25, 50 and 75 μg glucose, in a volume of 1 ml) are prepared at the same time. To all tubes add 1 ml of 5% phenol (Note 2) and mix; then from a fast-flowing pipette add 5 ml of concentrated sulphuric acid, directing the stream of acid on to the surface of the liquid and shaking the tube simultaneously, to effect fast and complete mixing (Note 3). The tubes are allowed to stand 10 min, shaken, and placed in a water-bath at 25° to 30°C for 10 to 20 min before readings are taken. The colour is stable for several hours and readings may be made later if necessary. The absorbance

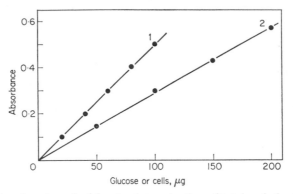

FIG. 11. The phenol method for the determination of total carbohydrate. Method as described in text; reaction volume 7 ml. Samples measured against reagent blanks at 588 nm in a Univam SP 600 Spectrophotometer, using 1 cm cuvettes. Curve 1, Standard glucose solutions; Curve 2, different quantities of the same washed suspension of *Torula utilis* cells.

of the characteristic yellow colour is measured at 488 nm for hexoses and 480 nm for pentoses (Note 4) and follows Beer's Law up to optical densities of at least 1·0 (Fig. 11).

Note 1. With some types of colorimeter employing matched test-tubes the reaction may be carried out in these; their diameter should allow good mixing without dissipating the heat too rapidly.

Note 2. The intensity of the colour developed is a function of phenol concentration. As the amount of phenol is increased, the absorbance rises to a maximum and then usually falls off. The optimum amount of phenol varies considerably with different sugars; the amount used above is a compromise value which gives maximum colour with hexoses and reasonably high values with most other sugars. When the method is applied to known pure sugars (e.g., after elution from paper chromatograms, cf. Section V C) the optimum amount of phenol can be selected from the data of Dubois *et al.* (1956).

Note 3. This method employs the "heat of mixing" technique which does not work too well with the anthrone reagent. However, if the manner of adding and mixing the sulphuric acid are well standardized, results are reproducible; presumably the phenol reagent is less sensitive than the anthrone reagent to small differences in temperature and heating time.

Note 4. The absorption peaks are very sharp (Fig. 12) and great care should be taken over exact wavelength settings if spectrophotometers are used; colorimeters with suitable narrow-band filters may be more convenient for routine analyses.

(c) *Specificity.* Table IX gives the wavelength of maximum absorption of a number of sugars and nucleic acids, and their relative absorbance compared with glucose; a standard curve obtained with glucose is shown in Fig. 11.

Pentoses, methyl-pentoses and uronic acids have an absorption maximum at 480 nm, while that of hexoses, hexose polymers and heptoses is at 486–490 nm. Their absorption curves are similar in shape, showing a single sharp peak (Fig. 12). Further data for less common sugars and their

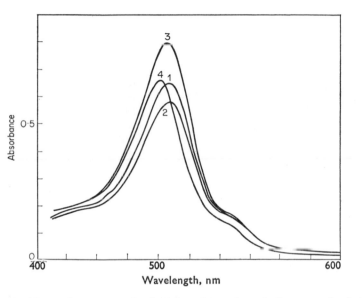

FIG. 12. Absorption spectra in the phenol reaction of glucose and whole cells of yeast and bacteria of high and low carbohydrate content. Method as described in text; samples measured against reagent blank in an Optica CF 4 Recording Spectrophotometer, using 1 cm cuvettes. Curve 1, 60 μg glucose; Curve 2, 100 μg of washed *Torula utilis* cells (52% total carbohydrate); Curve 3, 400 μg of washed *Aerobacter aerogenes* cells containing 18·0% carbohydrate, 6·1% RNA and 3·4% DNA; Curve 4, 1 mg of washed *A. aerogenes* cells containing 6·5% carbohydrate, 13·3% RNA and 3·2% DNA.

derivatives are given by Dubois *et al.* (1956), and it is interesting that partially or completely methylated sugars (e.g., 2, 3, 4, 6-tetra-*O*-methyl glucose) react.

Amino sugars give no colour with the phenol reagent, though they may be deaminated by nitrous acid to 2,6-anhydrohexoses which do react (Lee and Montgomery, 1961).

(d) *Application to total carbohydrate determination in microbial cells.* The

TABLE IX

Reaction of various carbohydrates and nucleic acids with the phenol–sulphuric acid reagent[a]

Substance	λ_{max} (nm)	Absorbance relative to glucose	
		at λ_{max}	at 488 nm
D-glucose	488	100	100
D-galactose	487	71·3	..
D-mannose	487	117·4	..
D-fructose	490	77·4	..
Sucrose	490	96·5	..
Starch	488	113·5	113·5
L-rhamnose	480	90·8	..
L-fucose	480	31·1	..
L-arabinose	480	105·5	..
D-xylose	480	164·5	..
D-ribose[b]	480	197	173
D-glycero-D-guloheptose[b]	486	128	127
D-glucuronic acid[b]	480	42·7	40·6
D-galacturonic acid	480	62·4	..
Ribonucleic acid (yeast)[b]	480	52·9	44·5
Desoxyribonucleic acid[b] (calf thymus)	482	28·1	26·4

[a]Reaction performed as described in text, using 1·0 ml of 5% phenol; for absorbance values at other phenol concentrations see Dubois et al. (1956).

[b]Values determined by present writers; remainder are Dubois et al. (1956).

phenol method is very insensitive to interference by proteins (Montgomery, 1957), and the writers have found no difficulty in applying it to washed suspensions of bacteria and yeasts. The cells dissolve completely to give optically clear solutions with the characteristic yellow colour, the intensity of which is strictly proportional to the weight of cells taken (Fig. 11).

In the writers' experience, the results obtained by the phenol reagent are always higher than those given by the anthrone reagent, presumably because the former reacts strongly with many sugars to which the latter is

TABLE X

Analysis of A. aerogenes with phenol and anthrone reagents

A. aerogenes sample	RNA (%)	DNA (%)	Total carbohydrate %		Phenol method / Anthrone method
			Anthrone method	Phenol method	
A	6·1	3·4	14·4	18·0	1·25
B	13·5	3·2	2·0	6·5	3·25

relatively insensitive. In bacteria these include cell-wall components such as heptoses, methylpentoses and uronic acids, and intracellular components such as the ribose and desoxyribose components of RNA and DNA. The nucleic acids are particularly important, since they react so much more strongly with the phenol reagent than with the anthrone reagent (cf. Tables VIII and IX) and may be present in high concentrations in bacteria. The examples in Table X on two samples of *A. aerogenes* with high and low carbohydrate/nucleic acid ratios are illustrative.

Evidently in sample B the "total carbohydrate" measured by the phenol method is largely due to the nucleic acids present. This sample gave with the anthrone reagent a blue-purple colour obviously different from that of glucose; its spectrum (Fig. 9) clearly indicates the presence of RNA and DNA. With the phenol reagent the colour was not noticeably different from that of glucose to the eye, though its spectrum (Fig. 12) shows a peak at 480 nm indicative of pentoses.

In analyses of yeasts the phenol method also gave results consistently higher than the anthrone method even when the carbohydrate/nucleic acid ratio was very high; e.g., a sample of *T. utilis* containing only 4% total nucleic acids gave total carbohydrate values of 42·6% by the anthrone method and 52·4% by the phenol method, though the absorption spectrum in both cases was identical with that of glucose (Figs. 9 and 12). The reason for the discrepancy in this case is probably the mannan content of the yeast (ca. 10%), since mannose gives only 55% of the colour of glucose with the anthrone reagent, while with the phenol reagent their colours are nearly equal (Tables VIII and IX); here the phenol reagent probably gives a more correct result while the anthrone reagent underestimates.

3. General comment on "total carbohydrate" methods

It will be obvious from the results quoted above that no "fool-proof" method for determining total carbohydrate in whole cells exists, and that any single method can only be relied on if the nature of the main cell carbohydrates and the nucleic acid contents are known. The anthrone method *approximates* to a "total hexose" method, giving identical colours (though not of equal intensity) with all hexoses, while other sugars give different and much weaker colours. The phenol method *approximates* to a "universal reagent" for all types of sugar, including the carbohydrate residues of nucleic acids; this also gives identical colours with all hexoses.

Other methods are available which give different colours with different hexoses such as glucose, mannose and galactose. Examples of these are the orcinol–H_2SO_4 reagent (Tillmans and Philippi, 1929; Sørensen and Haugaard, 1933; Bell and Robinson, 1934) and the carbazole–H_2SO_4 reagent (Dische, 1927; Seibert and Atno, 1946). Dische's cysteine–H_2SO_4

reagent may be applied in several variations (reviewed by Dische, 1962) to react preferentially with different types of sugar. All these methods may be found useful when unknown material is first examined, and will usually give a good idea of the main types of carbohydrate present and a rough estimate of their quantities, if absorption measurements are made at two or more wavelengths. Complete absorption spectra can be very informative with all these reagents, as with the anthrone reagent (cf. Figs. 8–10).

B. Determination of individual polysaccharides

Several hundred microbial polysaccharides are known, but only a few of them can be determined quantitatively and of these some are peculiar to individual micro-organisms. This section is therefore restricted to three polysaccharides which are (a) fairly widely distributed, (b) quantitatively important in that they may amount to 10–25 % of the cell biomass; these are glycogen, glucan and mannan.

Glycogen is very widely distributed and is thought to function as a reserve or "storage" polysaccharide; its concentration in any species may vary from almost zero to as much as 25 %, according to the conditions of growth. Glucan and mannan are cell wall constituents of almost all yeasts, and their concentrations are less variable.

All three polysaccharides are very resistant to hydrolysis by strong alkalis; they are fairly readily hydrolysed by strong acids at 100°C but are relatively unaffected by weak acids at lower temperatures. Glycogen and mannan are water-soluble, while glucan resembles cellulose in its almost complete insolubility. Mannan is unusual in forming an insoluble copper complex. These properties have been used in methods for quantitative separation of the three polysaccharides.

1. *Determination of glycogen in bacteria*

Glycogen can be determined in many bacteria by a method commonly used for animal tissues; namely digestion of the tissue with concentrated KOH at 100°C, followed by precipitation of the glycogen with 60% ethanol. (Commonly known as "Pfluger's method", this is in fact due to Claude Bernard, 1887). The precipitated glycogen may either be hydrolysed with HCl and determined as glucose by a "reducing sugars" method (Good et al., 1933), or determined directly by the anthrone reagent.

To quote an actual example, Palmstierna (1956) isolated glycogen from E. coli by treating the previously washed and lyophilized cells with 30% KOH (1 ml/100 mg cells) for three hours at 100°C, followed by the addition of 3 volumes of water and 8 volumes of ethanol. The precipitated glycogen was centrifuged, washed twice with 60% ethanol, then dried with absolute ethanol followed by ether. Recovery of the glycogen is quantitative; it can be purified if necessary by dissolving in water (100 mg/ml) and precipitating at 0°C with glacial acetic acid to a concentration of 80%. If the cells contain only small amounts of glycogen, the author recommends extracting the lyophilized cells with boiling ether to remove lipids before the KOH treatment.

Such drastic treatment with strong alkali is not recommended as a preparative method as it causes some degradation of the glycogen, but is probably quite satisfactory as an analytical procedure for the majority of Gram-negative bacteria, though not universally applicable. Possible difficulties are: (a) If other alkali-soluble polysaccharides (e.g. mannans) are present, they will be precipitated with the glycogen by 60% ethanol. This can be checked by HCl hydrolysis of the precipitated polysaccharide, followed by paper chromatography. (b) Glucans, which are major cell-wall components of yeasts and moulds, will remain insoluble even after strong alkali treatment, and will need to be separated by centrifuging before precipitation of the alkali-soluble polysaccharide. (c) When *whole* yeast cells are extracted with alkali, the glucan cell wall remains structurally intact and retains some of the glycogen. This can be extracted by dilute acetic or perchloric acid (Trevelyan and Harrison, 1952, 1956); alternatively the cell walls may first be mechanically disrupted (Northcote and Horne, 1952; Northcote, 1953), which renders all the glycogen extractable.

The latter procedure was used by Strange et al. (1961), who isolated glycogen from *A. aerogenes* after disintegrating by shaking with glass "ballotini" in the Mickle (1948) apparatus. After removing the cell wall fraction by centrifugation, protein was precipitated by trichloroacetic acid (2·5%) in the cold and 3 volumes of ethanol added to the neutralized protein-free extract to precipitate the glycogen.

The resistance of the cell walls of Gram-positive bacteria to alkali was mentioned in Section IV B (cf. also Section VII); though they are probably less resistant than yeasts in this respect, it is advisable, if the KOH extraction method is used, to check that all the glycogen is in fact extracted; if not, mechanical disruption of the cells is advisable.

2. *Combined determination of glycogen, mannan and glucan*

Trevelyan and Harrison (1956) evolved a method for the quantitative determination of these three polysaccharides in *Saccharomyces cereviseae* (bakers' yeast), which the writers have found to work well with other yeast species. The method was designed for application to whole yeast cells, which necessitates separate extractions with alkali and acid to solubilize all the glycogen. There is no chemical or physiological difference between these "alkali-soluble" and "acid-soluble" glycogen fractions (Northcote and Horne, 1952; Northcote, 1953), and all the glycogen can be extracted in one step if the cells are first mechanically disrupted; it can reasonably be argued, however, that this is more laborious than an extra extraction step. The determination of the three polysaccharides, as well as trehalose and RNA, can be performed on 25 mg (dry weight) of yeast; an abbreviated description follows.

(i) Washed yeast cells are extracted with 0·5 M trichloroacetic acid for one hour at 0°C with frequent shaking, and centrifuged; the extract contains all the trehalose, which may vary from 1–10% of the yeast dry weight according to growth conditions, and can be determined with the anthrone reagent.

(ii) The cells are now extracted with 0·5 M perchloric acid for 90 min at 38°C with shaking, and centrifuged; the supernatant contains all the RNA which can be determined by the orcinol reagent, but negligible carbohydrate. (This step can be omitted if RNA determination is not required.)

(iii) The cells are extracted with 0·25 M Na_2CO_3 for 45 min at 100°C, and centrifuged. The extract contains all the mannan and the "alkali-soluble" glycogen fraction, and is divided into two parts. One is used for the determination of the sum of both polysaccharides by the anthrone reagent; the other is used for the separate determination of mannan by adding KOH to a final concentration of 1 M, followed by Fehling's solution. After standing overnight, the insoluble mannan–copper complex is separated by centrifuging, dissolved in 1 M H_2SO_4, and determined by the anthrone reagent.

(iv) The cell residue from step (iii) is extracted with 0·5 M $HClO_4$ for 30 min at 100°C, and centrifuged. The extract contains the "acid-soluble glycogen", which is determined with the anthrone reagent and added to the "alkali-soluble glycogen" determined in step (iii) to obtain the total glycogen.

(v) The final residue from step (iv) contains only the cell-wall glucan; this is suspended in 2 M NaOH with shaking to produce an homogeneous dispersion, an aliquot of which is taken for determination of glucan by the anthrone reagent.

The sum of the three polysaccharides and trehalose determined by this method agrees well with the value for "total carbohydrate" determined directly on whole cells by the anthrone method, provided that allowance is made for the different colour values of glucose and mannose in the anthrone reaction.

This fractionation scheme has so far been applied only to yeasts, but would probably be applicable to many moulds; it is scarcely worth applying to bacteria since they do not have glucan cell walls and rarely contain mannans.

3. Physical methods for polysaccharide determination

These methods are as yet insufficiently developed to be classed as routine analytical procedures; nevertheless they will be briefly mentioned since they are likely to become the methods of the future.

(a) *Solvent fractionation on cellulose columns.* Adsorption chromatography of polysaccharides has so far met with little success, owing to the lack of suitable adsorbents. However, columns of an inert porous material have been used as supports for the fractional precipitation of polysaccharides from aqueous solution by organic solvents. The columns are saturated with a solvent mixture which will precipitate all the polysaccharides present, and the polysaccharides added to the top in a solvent mixture which does not precipitate any of them. While passing down the column, the polysaccharides are precipitated on the cellulose; different polysaccharides are then washed out from the precipitate by continuous changes in the composition of the eluting solvent. The method thus combines fractional precipitation with fractional elution, and gives higher resolution than ordinary fractional precipitation methods. The technique has been reviewed by Gardell (1965).

(b) *Ion-exchange chromatography on DEAE cellulose columns.* Anion exchangers such as DEAE(diethylaminoethyl)-cellulose can be used effectively for the fractionation of acid polysaccharides (e.g., those containing free uronic acid –COOH groups). These are readily adsorbed on DEAE-cellulose columns at pH values close to 6 and are usually eluted by acidic solutions of gradually increasing strength. Neutral polysaccharides may also be fractionated through their ability to form negatively charged borate complexes. The usual procedure is first to convert the DEAE-cellulose into its borate form; polysaccharides forming borate complexes are then retained and can be eluted with borate solutions of increasing strength. This method is reviewed by Neukom and Kuendig (1965).

(c) *Zone electrophoresis.* Electrophoretic separation of polysaccharides was at first restricted to acidic polysaccharides; the method was extended to neutral polysaccharides when Northcote (1954) discovered that these could be separated as their anionic borate complexes. This allows rapid and quantitative separations on a micro scale using strip supports of silk or glass paper, or on a preparative scale by the use of glass powder columns The method has been reviewed by Northcote (1965).

C. Determination of individual sugars

Sugars of almost any type may be determined with comparative ease when they occur singly, but the biochemist usually has to cope with mixtures. For example, hydrolysis of the cell wall polysaccharides and lipopolysaccharides of Gram-negative bacteria may yield mixtures of 6 to 12 different sugars, including hexoses, hexosamines, heptoses, 6-deoxyhexoses, 3,6-dideoxyhexoses, hexuronic and amino-hexuronic acids, and 2-keto-3-deoxyoctonate (KDO). Such mixtures present difficult analytical problems. which may be approached in three ways: (*a*) enzymatic methods (unfortunately only a few) whose specificity allows individual sugars to be determined; (*b*) chemical methods which are relatively specific for a particular *class* of sugars; (*c*) physical separation by paper or column chromatography, and more recently by gas–liquid chromatography.

1. *Enzymatic methods*

(a) *Determination of* D-*glucose*. The enzyme glucose oxidase is a flavoprotein present in various moulds (e.g., *Penicillium notatum*, *Aspergillus niger*) which catalyses the reaction—

$$\beta\text{-D-glucose} + O_2 \rightarrow \delta\text{-gluconolactone} + H_2O_2$$

The oxygen consumed may be determined manometrically, but a more convenient method is to measure the H_2O_2 by adding peroxidase and a suitable chromogen which is oxidized to a coloured compound—

$$H_2O_2 + \text{chromogen} \rightarrow H_2O + \text{oxidized chromogen}$$

A suitable chromogen is *o*-dianisidine, which is oxidized to a brown dye that may conveniently be determined colorimetrically (Huggett and Nixon, 1957).

Glucose oxidase preparations of varying purity are available from several firms, and the Worthington Biochemical Corp. markets under the name of "Glucostat" an enzyme kit consisting of phials of buffered enzyme mixture and chromogen with detailed instructions for performing the assay. The method is exceedingly simple and almost completely specific for D-glucose, the only other sugar known to be oxidized (at 12% of the rate of glucose) being 2-deoxy-D-glucose. (Pure glucose oxidase in fact oxdizes only the β-anomer of glucose, but commercial preparations contain a "mutarotase" which catalyses the interconverson of α- and β-glucose). The less pure grades of enzyme contain some maltase, invertase and amylase; these will produce glucose from maltose, sucrose or starch, if these are present, which is oxidized by the glucose oxidase. Purer grades of glucose oxidase which are free from these contaminating enzymes are commercially available. (The use of these purer preparations together with purified maltase, lactase and invertase should allow the specific determination of maltose, lactose and sucrose, though the writers are not aware of this having been done.)

(b) *Determination of D-galactose.* The enzyme galactose oxidase, isolated from the fungus *Dactylium dendroides* (Amaral *et al.*, 1962) oxidizes the $-CH_2OH$ group of D-galactose to $-CHO$, with the production of hydrogen peroxide—

$$\text{D-galactose} + O_2 \rightarrow \text{D-galactohexodialdose} + H_2O_2$$

The H_2O_2 formed can be determined colorimetrically with peroxidase and a chromogen in the same way as described for glucose oxidase, and a similar enzyme kit (using in this case *o*-tolidine as the chromogen) is supplied by the Worthington Corp. under the name of "Galactostat".

This enzyme is by no means as specific as glucose oxidase (Avigad *et al.*, 1962); it also oxidizes D-talose, D-galactosamine and *N*-acetyl-D-galactosamine at rates comparable with galactose. Under suitable conditions, it can also be used for the assay of D-galactosamine, and provides a useful way of differentiating this from D-glucosamine (which is not oxidized).

A D-galactose dehydrogenase produced by *Pseudomonas saccharophilia* (De Ley and Doudoroff, 1957) catalyses the reaction—

$$\text{D-galactose} + \text{NAD} \rightarrow \text{D-galactono-}\gamma\text{-lactone} + \text{NADH}_2$$

This enzyme may be used for the determination of galactose by measuring the absorption peak at 340 nm of the $NADH_2$ formed, using an ultraviolet spectrophotometer. The enzyme is considerably more specific than galactose oxidase, though it does oxidize D-galactosamine very slowly; it is not at present commercially available.

2. *Chemical methods*

(a) *Determination of hexosamines.* The hexosamines (or 2-amino-2-deoxyhexoses) are not determined by any of the carbohydrate methods mentioned in Section V A, since the presence of the 2-amino group prevents the formation of furfural derivatives on treatment with H_2SO_4. In micro-organisms they are not found in significant amounts in the free form, but are common constituents of macromolecules, mostly associated with the cell wall, such as the lipopolysaccharides of Gram-negative bacteria, muco-peptide and teichoic acids of Gram-positive bacteria, and chitin in yeasts and moulds. The commonest are D-glucosamine, D-galactosamine and muramic acid (3-*O*-D-carboxyethyl-D-glucosamine), though a number of others are known (Sharon, 1965).

Before analysis, the hexosamines must be liberated from the parent macromolecules by acid hydrolysis; the correct conditions for complete hydrolysis without destruction should be separately established for each polymer (Spiro, 1966; Davidson, 1966), but 4–8 h in 4 N HCl at 100°C (in sealed tubes to exclude oxygen) is often satisfactory. Separation from other sugars and amino-acids may often be necessary and can be effected on ion-exchange columns (Wheat, 1966).

(1) *Determination of hexosamines by the Elson–Morgan reaction.* Elson and Morgan (1933) described a method for analysis of glucosamine and galactosamine which depends on their reaction with acetylacetone to form a chromogenic material which gives a red colour with p-dimethylaminobenzaldehyde. The method was improved by Rondle and Morgan (1955) whose method is given below, modified only by the use of a reagent containing 5% acetylacetone in a $NaHCO_3$–Na_2CO_3 buffer instead of 0·5 M Na_2CO_3; this gives a pH of 9·6 instead of the original pH 10·1–10·3, which minimizes interference by mixtures of certain amino-acids and reducing sugars (Immers and Vasseur, 1950).

Reagents. The reagents are as follows—

(i) pH 9·6 buffer, prepared by mixing equal volumes of M $NaHCO_3$ and M Na_2CO_3.

(ii) Acetylacetone reagent: Add 2·5 ml of redistilled acetylacetone (B.P. 138°–140°C) to 50 ml of the above buffer (i). The reagent is stable for 2–3 h at 18°C.

(iii) Ethanol, dried 24 h over freshly heated CaO and distilled.

(iv) A pure sample of p-dimethylaminobenzaldehyde (0·8 g) is dissolved in 30 ml of ethanol and 30 ml of conc. HCl is added. The reagent should be clear and pale yellow; it keeps well at $-10°C$.

Procedure. Into 10 ml graduated glass-stoppered test-tubes measure 1·0 ml of sample (5–150 μg hexosamine), 1 drop of phenolphthalein indicator, and add N NaOH to give a faint pink colour, followed by water to make 2·0 ml. Add 1·0 ml of the acetylacetone reagent (ii), mix well, stopper loosely, heat for 20 min in a boiling water-bath, and cool in cold water. Add ca. 5 ml ethanol followed by 1·0 ml of reagent (iv) and ethanol to make to 10·0 ml; mix gently and place the tubes in a water-bath at 65°–70°C for 10 min to accelerate liberation of CO_2 and finally cool to room temperature (water-bath). A reagent blank and standards containing known amounts of glucosamine are treated in the same way. The optical densities of unknowns and standards are read against the reagent blank in a suitable spectrophotometer at a wavelength of 530 nm, using 2 cm cells. The colour obeys Beer's Law over the range 5–150 μg glucosamine.

Comments on the method. Glucosamine and galactosamine give colours of identical intensity and absorption spectrum and cannot be distinguished by this reaction. Muramic acid shows maximum absorption at 510–515 nm (Kent and Strange, 1962) and can be determined in the presence of glucosamine by the absorption ratio at 530 and 510 nm, if the amounts of the two hexosamines are not too unequal.

Moderate amounts of either hexoses or amino-acids cause little interference, but mixtures of glucose and amino-acids (particularly lysine and glycine) interfere considerably (Immers and Vasseur, 1950). Such mixtures of sugars and amino-acids often occur with hexosamines in hydrolysates of glycoproteins and bacterial mucopeptides, and chromatographic separation of the individual hexosamines may be the only way to obtain accurate analyses in some cases (Gardell, 1958).

(2) *Other methods for determination of hexosamines.* Lee and Montgomery (1961) describe a method for the determination of hexosamines based on their deamination with nitrous acid to form 2,5-anhydrohexoses, which are then determined by the phenol reagent described in Section V A. The method cannot be applied in the presence of other sugars, but is useful for the determination of small amounts of hexosamines separated by paper or column chromatography.

Dische and Borenfreund (1950) also describe a method based on deamination of the hexosamines followed by reaction of the anhydrohexose with an indole–HCl reagent; other sugars give similar colours with this indole reagent, but these are not affected by the deamination procedure and the hexosamine can be determined by difference.

The galactose oxidase described above also oxidizes ᴅ-galactosamine, but not glucosamine, and can be used to determine galactosamine in mixtures if galactose is known to be absent.

(b) *The determination of pentoses.* Numerous methods have been described for the determination of pentoses, nearly all based on the fact that pentoses are converted to furfural on treatment with hot acids much more readily than hexoses are converted to hydroxymethyl-furfural; under appropriate conditions the reaction can be made relatively specific for pentoses. The method most commonly used is based on Bial's qualitative colour test for pentoses by heating with orcinol and HCl, which produces a green colour; the reaction is catalysed by ferric or cupric salts. The orcinol method is important since it is widely used for the determination of free pentoses, pentosans, and combined forms of pentose such as pentose phosphates, nucleosides, and nucleotides and in particular ribonucleic acid.

(1) *Determination of pentoses by the orcinol method.* Since Dische and

TABLE XI

The orcinol reaction for pentose determination; conditions used by different authors[a]

Final concentrations				Heating time (min)	Other details	Reference
HCl (N)	Fe³⁺ (mM)	Cu²⁺ (mM)	Orcinol (%)			
5·7	0·88	..	0·48	3	"Reaction 1"—heated at 100°C	Dische and Schwartz (1937)
9·3	1·92	..	0·63	3	"Reaction 2"—heated at 80°C	Dische and Schwartz (1937)
6·0	1·85	..	0·5	20		Mejbaum (1939)
6·0	..	0·2	0·1	10		Barrenscheen and Peham (1941)
7·1	2·1	..	0·15	20	Xylose requires 30 min heating	McRary and Slattery (1945), Brown (1946)
7·1	5·5	..	0·15	5	n-butanol (24 vol) added after heating	Militzer (1946)
6·0	1·85	..	0·5	40	Pigment extracted into isoamyl alcohol	Schlenk and Waldvogel (1947)
5·7	1·76	..	0·48	45	Data on pentose-phosphates, nucleotides	Albaum and Umbreit (1947)
5·7	0·98	..	0·48	45	Hexose interference noted	Drury (1948)
5·7	17·76	..	0·48	45		Beljanski and Macheboeuf (1949)
8·2	0·51	..	0·41	20	For RNA determination in bacteria	Morse and Carter (1949)
4·0	1·48	..	6·0	15–45	For RNA determination in bacteria	Caldwell and Hinshelwood (1950)
6·0	1·85	..	0·5	60		Lusena (1951)
8·6	6·66	..	0·18	8	n-butanol (3·5 vol) added after heating	Markham (1955)
7·7	2·38	..	0·14	3	For RNA determination	Dische (1955b)
6·0	..	0·2	0·1	40	Pigment extracted into iso-amyl alcohol	Ceriotti (1955)
5·9	..	0·2	0·2	20	Ethanol (0·25 vol) added after heating	Trevelyan and Harrison (1956)
9·0	0·56	..	0·015	15	Heated at 90°C	Stuy (1958)
6·0	0·37	..	0·3	30	For RNA determination	Munro et al. (1962)
7·2	2·0	..	0·15	20	n-butanol (2·75 vol) added after heating	This article (cf. Herbert and Phipps, 1971)

[a] Final concentrations are those in the reaction mixture at the time of heating (100°C unless otherwise stated). HCl normalities are calculated on the assumption that "concentrated HCl" (sp. gr. 1·186) is 12·0 N.

Schwartz (1937) first adapted Bial's colour test into a quantitative method, the method has undergone numerous variations, some of which are summarized in Table XI. The wide variation in conditions will be noted; HCl concentration has been varied from 4 N to 9 N, heating time from 3 to 60 min, orcinol concentrations from 0·015% to 6·0%, ferric chloride concentrations from 0·37 mM to 17·8 mM, while others have used cupric salts instead; some have extracted the pigment into a water-immiscible solvent such as amyl alcohol, while others have diluted after heating with miscible solvents such as ethanol or n-butanol. Doubtless most of these methods will give satisfactory results with fairly pure pentose solutions; however, the writers have found that different reaction conditions greatly affect the *specificity* of the reagent, particularly with respect to interference by hexoses.

This aspect of the orcinol reagent has been little studied, though McRary and Slattery (1945) noted that galactose and fructose interfered much more strongly than glucose (mannose also interferes strongly; see Table XII). Hexoses and pentoses give different colours with the reagent (Fig. 13), and Brown (1946) recommended reading at 520 and 670 nm to correct for

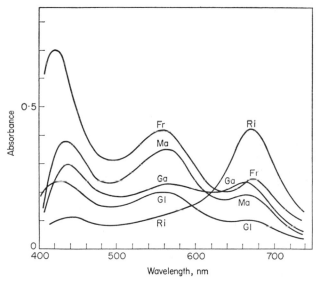

FIG. 13. Absorption spectra of ribose and some hexoses in the orcinol reaction for pentoses. Method of Herbert and Phipps (1971), as described in text, with 20 min heating at 100°C. Samples read against reagent blanks in an Optica CF4 Recording Spectrophotometer, using 1 cm cuvettes. Ri, ribose (50 μg); Gl, glucose (500 μg); Ga, galactose (500 μg), Ma, mannose (500 μg); Fr, fructose (500 μg).

hexose interference. Drury (1948) used a more complex method involving reading at two wavelengths and several heating times, and also noted that large amounts of hexose gave precipitates which made colorimetric readings impossible. This difficulty has been noted by several workers and we have found it particularly troublesome with high-acidity reagents such as that of Morse and Carter (1949). It can be overcome by adding several volumes of n-butanol after heating, as recommended by Militzer (1946) and endorsed by Markham (1955); this dissolves any precipitates to give optically clear solutions and has the further advantages of stabilizing the colour and suppressing the ionization of ferric chloride, thus producing virtually colourless, instead of strongly yellow, reagent blanks. The writers strongly advocate this procedure. While the same result can be achieved by extracting the pigment into an immiscible solvent such as amyl alcohol (Schlenk and Waldvogel, 1947; Ceriotti, 1955), this is more laborious and gives no gain in specificity, since hexose and pentose pigments are both extracted.

TABLE XII

Reactions of various carbohydrates and nucleic acids with the orcinol reagent[a]

Substance	Relative absorbance after heating times of:			
	10 min	20 min	30 min	40 min
D-ribose	78	100	99·5	98·2
D-xylose	70·1	93·7	98·5	100
L-arabinose	60	88·5	95·2	100
D-glucuronic acid	19	35	45	54
D-glucose[b]	0·6	1·5	2·6	3·9
D-mannose[b]	2·8	4·4	—	7·1
D-galactose[b]	3·1	5·2	—	8·7
D-fructose[b]	4·2	5·7	6·6	7·2
Ribonucleic acid (yeast)[c]	18·3	23·8	24·5	24·7
Desoxyribonucleic acid[b] (calf thymus)	2·5	2·7	—	—

[a]Method of Herbert and Phipps (1971), as described in text. Absorbances at 672 nm after different heating times are expressed as percentages of the value given by an equal weight of ribose with a heating time of 20 min. (For actual absorbances obtained with ribose and RNA, see Fig. 14.)

[b]The absorbance of these substances at 672 nm does not follow Beer's Law, and the absorbance relative to ribose depends on the level at which they are tested; this was 500 μg for the hexoses and 1 mg for DNA.

[c]Commercial sample; theoretical purine-ribose content 24·0% from base analysis (performed by Dr. H. E. Wade).

Using a scaled-down version of Militzer's method, Herbert and Phipps (1971) found that high HCl concentrations and long heating times greatly increase the colour produced by hexoses in the orcinol reaction. With pentoses, HCl concentrations above 8 N rapidly produce an intense colour which then fades, while with concentrations below 6·6 N the colour is produced much more slowly and is still increasing after 40 min heating. At the intermediate concentration of 7·2 N HCl, colour development from ribose (and from RNA) is 95% complete in 15 min and almost stable from 20 to 40 min (Fig. 14). Similar experiments with varying $FeCl_3$ and orcinol concentrations led to a reagent containing (final concentrations): 7·2 N HCl, 2·0 mM $FeCl_3$, 0·15% orcinol and a heating time of 20 min followed by the addition of 2·75 volumes of n-butanol. These conditions give the maximum stable colour with pentoses in a reasonably short heating time, combined with low sensitivity to hexoses and a negligible reagent blank; they are also very suitable for the determination of RNA, giving "theoretical" yields of purine-bound ribose (see Section VII).

Reagents. The reagents are as follows—

(i) Dissolve 0·90 g $FeCl_3.6H_2O$ in 1 litre of concentrated hydrochloric acid (sp. gr. 1·186), AnalaR grade; this keeps indefinitely.

(ii) Dissolve 1 g orcinol in 100 ml distilled water; this is stable for a few weeks if kept at 0°C.

(iii) Orcinol reagent: add 1 volume reagent (ii) to 4 volumes of reagent (i). Prepare immediately before use and do not keep.

Procedure

The reaction is conveniently carried out in glass-stoppered test-tubes graduated to 15 ml. Into these measure 1·0 ml of sample (3–120 μg pentose) followed by 3·0 ml of freshly prepared orcinol reagent (iii); a reagent blank and standard ribose solutions are put up simultaneously. Heat in a boiling water-bath for 20 min (standard period, but see below for effect of different heating times), cool in cold tap water, and make up to 15 ml with n-butanol (AnalaR grade). Read in a Unicam SP 600 (or similar) spectrophotometer at a wavelength of 672 nm, using 1 cm cuvettes for the range 20–120 μg pentose, or 4 cm cuvettes (which are just filled by the 15 ml reaction volume) for quantities of pentose below 20 μg; the minimum amount of pentose measurable with accuracy is ca. 3 μg, which gives an optical density of ca. 0·1 in a 4 cm cuvette. The reaction follows Beer's Law up to 120 μg pentose.

Comments on the method. Table XII shows the relative absorbances at

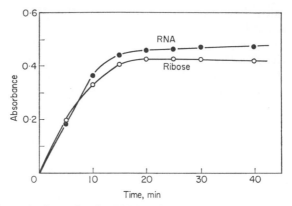

FIG. 14. The orcinol reaction for RNA and ribose; effect of heating time. Method of Herbert and Phipps (1971) as described in text, applied to ribose (50 μg) and RNA (240 μg); the latter was a pure sample of r-RNA from *Escherichia coli* (MRE 600), isolated by Robinson and Wade (1968).

672 nm given with the above reagent by various pentoses, hexoses, glucuronic acid, RNA and DNA, at heating times ranging from 10 to 40 min. Ribose gives a maximum colour after 20 min heating, while xylose and arabinose react more slowly (cf. Albaum and Umbreit, 1947) though the final colour developed is the same. If these pentoses are present, heating times of 30 or 40 min may be necessary, although this increases the interference from hexoses and uronic acids. At the standard heating period of 20 min. the reagent is relatively insensitive to hexoses; in comparative tests the reagents of Ceriotti (1955) and Morse and Carter (1949) gave respectively twice and 3–4 times the colour with hexoses (relative to a given weight of pentose) as the above reagent. The colours given by hexoses are obviously different to the eye and their absorption spectra show distinctive peaks in the regions 420–438 nm and 560–566 nm (Fig. 13), so that interference from this source is easily recognized. If very large amounts of hexoses are present, their interference may be reduced by decreasing the period of heating to 10 min (Table XII).

Hexuronic acids represent a more serious source of interference since on heating with HCl under the conditions of the orcinol reaction they undergo appreciable decarboxylation to form pentoses; as the colour produced is identical with that of pentoses, their presence may go unrecognized, though it can be deduced by the rate of colour development at different heating times (Table XII).

RNA gives a colour with the orcinol reagent at the standard heating time of 20 min which is exactly equivalent to its content of purine-bound ribose (see Section VII and Fig. 14). DNA reacts only slightly, large

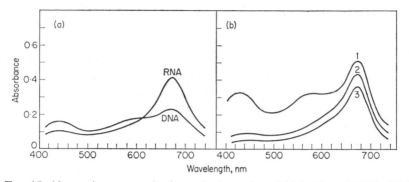

FIG. 15. Absorption spectra in the orcinol reaction of (a) RNA and DNA, (b) hot perchloric acid extracts of bacteria and yeast. Orcinol reaction as in Figs. 13 and 14; cells extracted with 0·5 N HClO₄ at 70°C as described in text. Measurements in Optica CF4 Recording Spectrophotometer using 1 cm cuvettes. (a) Yeast RNA (200 μg); calf thymus DNA (2 mg). (b) HClO₄ extracts from (Curve 1) 8·0 mg *Torula utilis* cells containing 42·6% total carbohydrate and 3·04% RNA, (Curve 2) 2·0 mg *Aerobacter aerogenes* cells containing 12·3% total carbohydrate and 12·8% RNA, (Curve 3) 1·0 mg *A. aerogenes* cells containing 3·8% total carbohydrate and 20·4% RNA. Note the obvious carbohydrate interference in the yeast sample (Curve 1).

amounts giving a colour similar to the eye to that given by pentoses, though its absorption spectrum is appreciably different (Fig. 15).

(ii) *Other methods for the determination of pentoses.* The production of a red colour on heating with phloroglucinol in acid solution, long used as a qualitative colour test for pentoses, has been made into a quantitative method by Dische and Borenfreund (1957). An unusual feature of this method is the inclusion of glucose in the reagent, presumably to stabilize the chromophore and to minimize interfering side reactions.

Reagent. This is prepared immediately before use by mixing 110 ml glacial acetic acid, 2 ml concentrated HCl, 5 ml of a freshly prepared 5% solution of phloroglucinol in ethanol, and 1 ml of a 0·8% solution of glucose.

Procedure. To 0·4 ml of sample containing 10–50 μg of pentose add 5 ml of the above reagent; reagent blanks and standard pentose solutions are put up simultaneously. Heat for 15 min in a boiling water bath, cool in cold tap water, and measure in a spectrophotometer the absorption at 552 nm; the measurements should be made rapidly, as the red colour slowly fades.

The colours produced by the four aldo-pentoses are of almost equal intensity; keto-pentoses (e.g., D-ribulose) produce a green colour with an absorption maximum at 470 nm. Hexuronic acids react as aldo-pentoses, but to a much smaller extent than in the orcinol reaction. The reagent is rather insensitive to hexoses, which in large amounts produce a brown colour; heptoses give a brownish-red. These contribute to the absorption at 552 nm, but their interference can be reduced by dichromatic readings at 552 and 510 nm. The value of $(E_{552} - E_{510})$ is very small for hexoses and heptoses, but is a linear function of the pentose concentration.

This method appears to be little used, but in the writers' limited experience, it appears to work well; we have not found it greatly superior to the orcinol method with respect to interference by hexoses such as mannose, but it is considerably less sensitive to hexuronic acids. Its sensitivity is only half that of the orcinol reagent and the instability of the colour is a disadvantage; nevertheless we consider this method merits further investigation.

(c) *Determination of Hexuronic acids*. Hexuronic acids are decarboxylated to pentoses by boiling HCl (Mann and Tollens, 1896), and the orcinol reagent has been used for their determination; this method is satisfactory only when pentoses are known to be absent. If instead of the pentose, the CO_2 produced by the decarboxylation is measured, the method becomes quite specific and yields of CO_2 under correct conditions are quantitative.

This method was studied by Tracey (1948). Samples are heated in sealed tubes in HCl of final concentration 12% (w/w) for 5 h at 111°C (boiling toluene bath). The sealed tubes are opened in such a way that the generated CO_2 is trapped directly in a van Slyke apparatus and measured manometrically; yields of CO_2 are theoretical. The method determines uronides and uronic acid polymers as well as free uronic acids, but very few other substances give appreciable yields of CO_2 under these conditions; the specificity of the reaction was very thoroughly investigated by Tracey (1948). A slight modification, for those who dislike using the van Slyke apparatus, is the method of Davidson (1966), who traps the CO_2 in a standard base solution and determines it by titration.

Colorimetric methods for hexuronic acid determination are on the whole less satisfactory. The well-known naphthoresorcinol colour test of Tollens is not readily adapted for quantitative work, though methods have been described by Deichmann (1943) and Hanson *et al.* (1944). The carbazole method of Dische (1947*a*) is more satisfactory, and has been improved by Bitter and Muir (1962) by the incorporation of borate; their method is given below.

Reagents. The reagents are as follows—

(i)   Sodium tetraborate, 0·025 M, in concentrated sulphuric acid.
(ii) Carbazole, 0·125%, in ethanol.

Procedure. A 1·0 ml sample containing 0·02–0·2 μmoles of uronic acid is treated with 5 ml of reagent (i), well mixed, heated 10 min in a boiling water bath and cooled to room temperature; 0·2 ml of reagent (ii) is added and the mixture again heated for 15 min in a boiling water bath and cooled. The red colour produced by uronic acids is stable for 16 h and has an absorption maximum at 530 nm; samples are read at this wavelength against a similarly treated reagent blank. With this reagent iduronic acid gives 83% of the colour given by glucuronic acid, as compared with 29% by Dische's method, the sensitivity is approximately doubled, and the colour is more stable.

This reagent was intended to give as nearly as possible equal colours with all uronic acids; more specific reagents can also be useful, and the methods of Dische (1947b) and (1948) are virtually specific for glucuronic and galacturonic acids respectively.

(d) *Differentiation of aldoses and ketoses.* There are no simple colorimetric methods for determination of aldoses in the presence of ketoses; this is because such methods nearly all depend on the formation of furfural derivatives on treatment with hot acid, and ketoses undergo this reaction more readily than aldoses. This fact does allow colorimetric methods for ketoses to be devised, in which aldoses do not react. Aldoses may be determined titrimetrically by reaction with alkaline hypoiodite, under conditions in which ketoses do not react. The aldonic acids thus formed are unreactive in any of the ordinary colour tests for sugars, and determination of total sugar before and after hypoiodite (or better, hypobromite) oxidation will give aldoses by difference. These methods will be briefly discussed.

(1) *Determination of aldoses by hypoiodite oxidation.* In alkaline solution, iodine forms hypoiodite according to equation 8; the hypoiodite oxidizes aldoses to the corresponding aldonic acids according to equation 9, the overall result being equation 10, which shows that 2 gram-atoms of iodine are equivalent to 1 gram-mole of aldose. Under the correct conditions, ketoses are not oxidized.

$$I_2 + 2OH^- \rightarrow I^- + IO^- + H_2O \qquad (8)$$

$$R.CHO + IO^- + OH^- \rightarrow R.COO^- + I^- + H_2O \qquad (9)$$

$$\overline{R.CHO + I_2 + 3OH^- \rightarrow R.COO^- + 2I^- + 2H_2O} \qquad (10)$$

In practice, a known excess of iodine is added, and the amount unused is determined after completion of the reaction by acidification and back-titration with standard thiosulphate. Correct pH is important; if insufficiently alkaline, reaction 9 does not go to completion, while if too alkaline the aldoses are "over-oxidized" and some oxidation of ketoses occurs. The latter was a defect of the original method of Willstätter and Schudel (1918) in which oxidation was effected in dilute NaOH solution; the method of MacLeod and Robison (1929) using $M/60$ Na_2CO_3 and an approximately three-fold excess of iodine gives almost theoretical results. Details are as follows—

Into small glass-stoppered flasks measure 1–2 ml of sample containing 1–2 mg glucose (or equivalent); add 3·0 ml of 0·02 N iodine and water to make 6 ml, followed by 0·2 ml of 5% Na_2CO_3. Stopper the flasks and incubate 30 min at 21°C; a blank containing the same volume of iodine but no glucose is similarly treated. Acidify with 1 ml of 0·5 N H_2SO_4 and titrate the residual iodine with 0·005 N thiosulphate using a starch indicator. Results are almost theoretical: 2·25 ml of 0·005 N $Na_2S_2O_3 \equiv 1$ mg glucose (the theoretical value is 2·22 ml).

Miller and Burton (1959) describe a similar procedure in which the iodine is determined spectrophotometrically, which permits accurate determinations on 0·2 mg glucose. They used stronger Na_2CO_3 solutions (0·15 M) giving a pH of ca. 11·3, and 0·02 N iodine dissolved in 10% KI; at lower KI concentrations the reaction was slower in reaching completion.

Ingles and Israel (1948, 1949) measured the velocities of oxidation of various aldoses at pH values betwen 10·15 and 13·45, and found the optimum pH for all sugars was pH 11·3, the reaction being slower above and below this value. This is the pH at which the concentration of unionized hypoiodous acid (pK ca. 11) is a maximum, and they suggest that the reaction mechanism is—

$$R.CHO + HIO \rightarrow R.COO^- + I^- + 2H^+$$

rather than equation 9 above. They also found that oxidation rates in carbonate buffers in the range pH 10·8–11·4 were slower than in phosphate buffers of the same pH; the rates in borate buffers also were slower than in phosphate. This was ascribed to the formation of sugar-carbonate and borate complexes. This suggests that the most favourable conditions for hypoiodite oxidation of aldoses might be in phosphate buffer at pH 11·3.

Both Miller and Burton (1959) and Ingles and Israel (1948) found that aldohexoses with a *trans* configuration of the C_2 and C_3 hydroxyl groups (e.g., glucose, galactose) were oxidized faster than those with a *cis* configuration (e.g., mannose). Aldopentoses were oxidized at the same rates as the hexoses with the same configuration on C_2, C_3 and C_4; e.g., xylose at the same rate as glucose, arabinose at the same rate as galactose, and lyxose at the same rate as mannose.

(2) *Hypobromite oxidation combined with colorimetric methods.* Horecker *et al.* (1951) applied this method to the analysis of mixtures of ribose and ribulose, both of which react with the orcinol reagent for pentoses. The

sugar mixture (10–100 μg aldopentose) is incubated with an excess of Br_2 (either as saturated Br_2-water or as 0·5 M ethanolic solution) at room temperature for 30 min, solid $BaCO_3$ or phosphate buffer pH 7·5 being added to prevent excess acidity. The unreacted Br_2 is then blown off by an inert gas (not air). Ribose is completely oxidized while ribulose is unaffected; orcinol determinations before and after Br_2 oxidation gave the sum of both sugars (before) and ribulose alone (after).

Slein and Schnell (1953) subsequently showed that this method could be applied to mixtures of aldo- and ketohexoses and also to aldo- and ketoheptoses (colorimetric determination by the cysteine–H_2SO_4 method of Dische and Shettles, 1948).

(3) *Resorcinol method for the determination of ketoses.* The well-known Seliwanoff colour test for ketohexoses (production of a red colour on heating with resorcinol and HCl) has been adapted for their quantitative determination by Roe et al. (1949).

Reagents. The reagents are as follows—

(i) Dissolve 0·1 g of resorcinol and 0·25 g of thiourea in 100 ml of glacial acetic acid; this solution is stable indefinitely in the dark.

(ii) 30% (w/w) HCl, prepared by adding 5 volumes of concentrated HCl to 1 volume of distilled water.

Procedure. To 2 ml of solution containing 20–80 μg fructose (or other ketohexose) add 1 ml of reagent (i) followed by 7 ml of reagent (ii); reagent blanks and standard fructose solutions are put up simultaneously. The tubes are heated 10 min in a water bath at 80°C, cooled, and read within 30 min at a wavelength of 520 nm; the colour follows Beer's Law up to at least 100 μg fructose.

The absorbance at 520 nm is almost the same for fructose, sorbose and tagatose, though they differ significantly in the blue spectral region. Fructose-6-phosphate and fructose-1,6-diphosphate give the same absorbance as an equivalent amount of fructose (Dische, 1951), while that of glucose-6-phosphate is seventy times weaker. Glucose in quantities up to 1 mg gives a negligible colour. Pentoses give a green colour, with an absorption at 520 nm only 1% of that of fructose; ribose-5-phosphate gives a colour similar to that of fructose, but 100 times weaker. Ketopentoses and their phosphates, however, can interfere seriously, as do ketoheptoses. Fructosides and fructosans (e.g., inulin) are hydrolysed under the reaction conditions and react as free fructose.

(4) *Cysteine–H_2SO_4 reaction for ketohexoses.* Dische and Bohrenfreund (1951) used a cysteine–carbazole–H_2SO_4 reagent for the determination of

ketohexoses which is more sensitive than the resorcinol method described above, but which also reacts with ketopentoses. More recently, Dische and Devi (1960) described a cysteine–H_2SO_4 reagent which is virtually specific for ketohexoses.

Reagents. The reagents are as follows—

(i) 20% cysteine hydrochloride monohydrate (freshly prepared).
(ii) Concentrated H_2SO_4 (3 vols) + water (1 vol).

Procedure. To 0·4 ml of a solution containing 50–250 μg of ketohexose add 0·1 ml of reagent (i) followed by 5 ml of reagent (ii); mix by vigorous shaking and stand 3–4 h at room temperature. Readings are made in a spectrophotometer at the two absorption maxima of 380 and 412 nm. Fructose, sorbose and tagatose all produce an intense yellow colour which changes to blue, in the case of sorbose very quickly, but more slowly with the other two. Aldohexoses give less than 1% of the colour of ketohexoses; ketopentoses and ketoheptoses give weak pink colours which are barely visible at 100 μg/ml, and interfere negligibly if the difference in readings at the two wavelengths, $E_{412} - E_{380}$, is measured. The reaction is thus virtually specific for ketohexoses.

(5) *Other sugars (heptoses, 6-desoxyhexoses, 2-desoxypentoses). Heptoses* do not occur in large amounts in micro-organisms, but they are important components of many glycoproteins and lipopolysaccharides (e.g., the "O"-antigens of Gram-negative bacteria). They react to some extent with many sugar reagents, e.g., the anthrone, phenol and orcinol reagents (see above). A reasonably specific method for their determination was evolved by Dische (1953) and is given here in a slightly modified form due to Osborne (1963).

Reagents. The reagents are as follows—

(i) 3% solution of cysteine hydrochloride (freshly prepared).
(ii) Concentrated H_2SO_4 (6 vol) + water (1 vol).

Procedure. Duplicate 0·5 ml samples of heptose solution (0·1–1 μM) are cooled in an ice bath and 2·25 ml of reagent (ii) is slowly added to each and mixed by shaking in the cold. After 3 min, the tubes are transferred to a water bath at 20°C for a further 3 min, and then heated in a vigorously boiling water bath for 10 min. After cooling, 0·05 ml of reagent (i) is added to one of the tubes, the other serving as a blank. Heptoses produce a purple colour with an absorption maximum at 545 nm. Exactly 2 h after adding the cysteine, the sample is read against the blank in a spectro-photometer at wavelengths 545 and 505 nm; the difference is a linear function of the heptose concentration. Under these conditions, $1 \cdot 0 \, \mu\text{M}$ of L-glycero-D-mannoheptose gives a value of $E_{545} - E_{505}$ equal to $1 \cdot 07$; the value is slightly different for other heptoses, and the heptose used as standard should be the same as that in the sample, if this can be identified.

6-*Desoxyhexoses* (or 5-*methylpentoses*) are also important components of many glycoproteins and lipopolysaccharides, and react with both the anthrone and phenol reagents (see above). They may be determined with reasonable specificity by another version of Dische's cysteine–H_2SO_4 reaction (Dische and Shettles, 1948).

Reagents. The reagents are as follows—

(i) Concentrated H_2SO_4 (6 vols) + water (1 vol).

(ii) 3% aqueous solution of cysteine hydrochloride (freshly prepared).

Procedure. 1 ml of solution containing 50 μg or more of 6-desoxyhexose is measured into a wide-mouthed test-tube and cooled in an ice bath. 4·5 ml of reagent (i) are added slowly while shaking in the ice bath. The mixture is then warmed to room temperature for a few minutes, heated in a boiling water bath for 10 min, and immediately cooled in tap water. To the cold solution is added 0·1 ml of reagent (ii) with shaking. 6-Desoxy-hexoses give a greenish yellow colour with a strong absorption maximum at 396 nm; the optical density at this wavelength (E_{396}) is read in a spectro-photometer after standing for 2 h, and is then stable for 24 h.

Hexoses, pentoses and uronic acids give much weaker pink colours with absorption maxima around 415 nm; their absorption is almost the same at 396 nm and 427 nm, while 6-desoxyhexoses have little absorption at the latter wavelength. The difference in optical densities at the two wavelengths, $E_{396} - E_{427}$, is proportional to the amount of 6-desoxyhexose and is virtually zero for other sugars.

The cysteine reaction carried out with 10 min heating as above, desig-nated CyR10 by Dische, is virtually specific for 6-desoxyhexoses. With a heating time of 3 min (reaction CyR3), hexoses give much stronger colours, and it is possible by differential wavelength readings to determine hexoses and 6-desoxyhexoses simultaneously, which may be useful in some cases; the original paper should be consulted for details.

In a later paper, Dische and Shettles (1951) describe a modification of the reaction which, although more complicated, is even more specific for 6-desoxyhexoses; this depends on the rate of breakdown of the 6-desoxy-hexose colour produced in reaction CyR3 after dilution with water, and involves taking differential wavelength readings before, and at various times after, diluting the reaction mixture with water. A method for the determination of 6-desoxyhexoses using a thioglycollic acid–H_2SO_4 reagent has also been described (Gibbons, 1955).

2-Desoxypentoses. Rather specific methods are known for the determination of these sugars; as the only one of natural importance is 2-desoxyribose, these are mainly used for the determination of DNA, and are therefore discussed in Section VII.

(e) *Methods for the determination of reducing sugars.* While these methods may seem somewhat old-fashioned, they still have a place in carbohydrate analysis, particularly when high precision rather than maximum specificity is required. With the better titrimetric methods a reproducibility of better than 1 % is easily achieved in routine analyses, which is more than can be said for most colorimetric methods. The subject has been ably reviewed by Hodge and Hofreiter (1962) and only selected methods will be described here.

(1) *Alkaline ferricyanide methods.* Many such reagents have been described since Hagedorn and Jensen (1923) first introduced one as a reliable method for blood-sugar determination. Sugars are heated in alkaline solution with a known excess of alkaline ferricyanide, which is reduced to ferrocyanide; the ferricyanide remaining is determined by adding potassium iodide after acidification and titrating the liberated iodine with standard thiosulphate. An important feature introduced by Hagedorn and Jensen was the addition of zinc sulphate after acidification to precipitate ferrocyanide as an insoluble zinc salt and prevent its reoxidation by air. Another important improvement was the introduction by Fujita and Iwatake (1931) of reagents buffered to pH 10·6 with K_2HPO_4–K_3PO_4; under these conditions the amount of ferricyanide reduced is strictly proportional to the amount of glucose taken. Earlier methods designed for blood-sugar determinations are no longer used for this purpose and are inconveniently "micro" for many other analyses; the reagent to be described was developed by one of us (D.H.) to deal with up to 2·5 mg glucose, and has been found reliable in use over a number of years.

Reagents. The reagents are as follows—

(i) Dissolve 6·60 g potassium ferricyanide, 70 g K_2HPO_4 and 21 g K_3PO_4 in water previously boiled to expel air, and make to 1 litre.

(ii) Dissolve 50 g of $ZnSO_4.7H_2O$ in ca. 350 ml of previously boiled water, add 70 ml conc. H_2SO_4 and make up to 500 ml.

Procedure. Samples if markedly acid or alkaline should be approximately neutralized (just alkaline to phenolphthalein). If deproteinization is required, the cadmium sulphate–NaOH method of Fujita and Iwatake (1931) can be used.

Measure 5 ml of sample (or a smaller volume made up to 5 ml with water) containing up to 2·5 mg glucose (or equivalent) into thin-walled 6 in. × 1 in. boiling-tubes; add 5·0 ml of ferricyanide reagent (i) (accurately measured, Ostwald pipette recommended), mix, cover the tube with a glass bulb, immerse for exactly 15 min in a boiling water bath, cool in running tap water. With each batch of samples a blank containing 5·0 ml reagent (i) and 5 ml of water is run simultaneously.

Add, while gently shaking the tubes, 2 ml of reagent (ii) followed by 1 ml of 15% KI solution (or a few crystals of solid KI). Titrate the liberated iodine with accurately standardized 0·05 N sodium thiosulphate, using a starch (or better, starch glycollate) indicator. The ferricyanide reduced is found by subtracting the sample titre from the blank titre, and is exactly proportional to the amount of glucose taken over the range 0·25–2·5 mg glucose; 1·0 ml of 0·05 N thiosulphate is equivalent to ca. 1·5 mg glucose. The reagent should be calibrated with standard glucose solutions when first prepared, but needs only infrequent calibration thereafter, as it is very stable.

(2) *Alkaline copper reagents.* Although ferricyanide methods are very simple and accurate, they are less specific than copper methods. Reliable copper reagents are not easily produced and required many years of evolution; the reagent of Somogyi (1952) is generally accepted as the best for titrimetric work, and has the advantage that sugar equivalents are stable over long periods. The earlier reagent of Somogyi (1945) is still useful in some applications, e.g., amyloses are precipitated from the 1952 reagent but are held in solution in the 1945 reagent. Somogyi (1952) also described a special low-alkalinity reagent for colorimetric use with the arseno-molybdate reagent of Nelson (1944); both of these 1952 reagents will be described.

Titrimetric reagent of Somogyi (1952). Rochelle salt (30 g), anhydrous Na_2CO_3 (30 g) and 1·0 N NaOH (40 ml) are dissolved in 200 ml of hot water. A solution of $CuSO_4 . 5H_2O$ (8 g) in 80 ml of water is stirred in, and the solution boiled to expel air. Anhydrous Na_2SO_4 (180 g) is dissolved in 500 ml of hot water, boiled to expel air, and added to the alkaline copper solution. KI (8 g) is dissolved in a small volume of water and added to the combined reagent. 1·0 N KIO_3 solution is now added in a quantity to suit the amount of glucose it is desired to determine: add 5·0 ml for the range 15–500 μg glucose (or equivalent), 12·0 ml for the range 0·5–1·5 mg, or 25 ml to determine up to 3 mg glucose. Finally add boiled distilled water to make 1 litre. The reagent is stored in a 37°C hot room to prevent crystallization of the Na_2SO_4.

Procedure. If the sample is very acid or alkaline is should be neutralized to phenolphthalein. A volume containing an amount of glucose to suit the capacity of the reagent (see above) is measured into 6 in. × 1 in. boiling-tubes and made to 5 ml with distilled water. 5·0 ml of the copper reagent are accurately added and gently mixed; the tubes are capped with glass bulbs, placed in a vigorously boiling water bath for 15 min, and cooled in cold water; the tubes should not be agitated during these procedures. Duplicate reagent blanks should be included with each batch of samples. After cooling, 1·5 ml of 2 N H_2SO_4 is run into each tube with gentle shaking, so that the liberated iodine will oxidize all the cuprous oxide. After standing 5 min, the tubes are again shaken, and the iodine remaining is titrated with standard $Na_2S_2O_3$ (0·005 N to 0·05 N, depending on the amount of iodate added to the reagent), using a starch–glycollate indicator.

Calculations. Reducing sugars react with the alkaline cupric–tartrate complex, reducing it to cuprous oxide. On acidification, the iodate and iodide in the reagent react to form iodine—

$$IO_3^- + 5I^- + 6H^+ = 3I_2 + 3H_2O$$

The iodine then reacts with the cuprous oxide:

$$Cu_2O + I_2 + 2H^+ = 2\ Cu^{2+} + 2I^- + H_2O$$

The difference between the blank titre and the sample titre is proportional to the amount of cuprous oxide formed; i.e., to the amount of glucose. The reagent is initially calibrated with standard glucose solutions at three or more different levels over the working range; this calibration curve (which is in fact a straight line except at the very lowest glucose levels) is used to convert observed thiosulphate titres into mg of glucose. Once calibrated, the glucose equivalents do not change over long periods.

(3) *Colorimetric copper method.* The details of this method are as follows.

Low-alkalinity reagent of Somogyi (1952). Rochelle salt (12 g) and anhydrous Na_2CO_3 (24 g) are dissolved in ca. 250 ml of previously boiled water. $CuSO_4 . 5H_2O$ (4·0 g) dissolved in ca. 50 ml of water is added with stirring, followed by $NaHCO_3$ (16 g). Anhydrous Na_2SO_4 (180 g) in 500 ml of water is boiled to expel air, cooled, and added to the alkaline copper solution; boiled distilled water is added to make 1 litre. The reagent is stored in a 37°C hot room for one week before use, and decanted from any precipitate that forms; it is then stable.

Arsenomolybdate reagent of Nelson (1944). Ammonium molybdate (25 g) is dissolved in 450 ml of water and 96% H_2SO_4 (21 ml) added, followed by a solution of $Na_2HAsO_4 . 7H_2O$ (3·0 g) in 25 ml of water. The mixed solution is incubated 24 h at 37°C and stored in a brown glass bottle.

Procedure. To 1–5 ml of sample containing not more than 0·6 mg of glucose or its equivalent, an equal volume of the copper reagent is added. Samples, blanks and standard glucose solutions are heated for 10 min in a boiling water bath and cooled. 1·0 ml of arsenomolybdate reagent is added to determine less than 0·1 mg of glucose; 2·0 ml are added for 0·1–0·6 mg glucose. When all the cuprous oxide is dissolved after mixing, the solution is diluted to 25 ml (or an appropriate smaller volume) and allowed to stand at least 15 min but not more than 40 min before reading at 500 nm in a spectrophotometer or photoelectric colorimeter with appropriate filter. The glucose equivalent is calculated from a standard curve established with known amounts of glucose.

(4) *Interference by amino-acids in reductimetric methods.* Amino-acids, which are present together with sugars after hydrolysis of glycoproteins, for example, can interfere with reductimetric sugar methods. The effect was studied by Strange *et al.* (1955), who found copper reagents were less affected than ferricyanide reagents. With the Somogyi copper reagent (see above), only cystine, tyrosine and tryptophan gave appreciable reduction when tested alone, but the effect is complex as these amino-acids caused decreased reduction in the presence of glucose. They describe a simple method for removal of amino-acids without loss of sugars by ion-exchange resins.

3. Analysis of sugars by chromatographic procedures

While enzymatic methods and a few colorimetric sugar methods have a high degree of specificity, complete analysis of a complex mixture of sugars requires a combination of these techniques with methods for separation of the individual sugars by paper or column chromatography. It is usually best to get an approximate idea of the types of sugar present by first applying the various colorimetric techniques described above; this will enable the most appropriate chromatographic techniques and solvent mixtures, etc., to be chosen; these will allow the individual sugars to be identified.

Detailed descriptions of chromatographic techniques will not be given as they are dealt with elsewhere in this Volume, but reference to the following review articles dealing specifically with the separation of sugars may be helpful: Whistler and BeMiller, 1961a (paper chromatography); Hough, 1954, and Hough and Jones, 1961 (paper and cellulose-column chromatography); Whistler and BeMiller, 1961b and c), (carbon and cellulose column chromatography); Lemieux, 1961, (celite column chromatography); Gardell, 1958 (paper and ion-exchange column chromatography of hexosamines).

When separation of individual sugars has been effected by chromatographic means, their quantitative determination becomes much easier; either the appropriate specific colorimetric method can be used, or (since the chromatographic technique has usually identified the sugar) a completely non-specific method such as the phenol–H_2SO_4 method of Dubois et al. (1956) may be used. This method (described in detail in Section V A2) is particularly suitable for micro-determination of sugars eluted from paper chromatograms, where phenol is often used as the developing solvent and may remain on the paper in amounts sufficient to interfere with the anthrone and other colorimetric methods. The phenol method is also insensitive to most of the other solvent mixtures used in chromatography; it is slightly affected by borate, which is often used for separation of sugars as their borate complexes, but Lin and Pomeranz (1968) have shown that this effect can be compensated by adding equal amounts of borate to the sugar standards. The phenol method is also useful for structural studies on polysaccharides, as it reacts with fully methylated sugars.

Finally, reference should be made to the review articles by Kircher (1961) and Sweeley et al. (1966) on the determination of sugars by gas–liquid chromatography. This requires a "universal reagent" for converting all sugars to volatile derivatives in quantitative yield; at present the dimethylsilyl derivatives appear most promising. Although still in the developmental stage, this will probably become the method of the future.

VI. LIPID ANALYSIS

Most of the lipid in micro-organisms appears to be associated with cell wall and membrane structures. Lipids are divided into two main classes, "simple" and "complex" lipids. Simple lipids are generally fatty acid esters of glycerol and include monoglycerides, diglycerides and triglycerides, compounds that differ in the number (1–3) of long-chain fatty acid residues present. Complex lipids are polar phosphorylated compounds that include lecithin, phosphatidic acids, phosphatidyl serine, etc.

In the case of bacteria, lipids of Gram-positive organisms are essentially membrane lipids (Weibull, 1957; Salton, 1960; Kolb et al., 1963) whereas Gram-negative organisms contain in addition lipids associated with the cell wall. Many bacterial lipids are simply fatty acid esters of phosphorylated carbohydrates (Lovern, 1957). Ethanolamine is the most commonly found nitrogen-containing component of phospholipids although choline is found in some species (Asselineau and Lederer, 1960; Asselineau, 1962). The complex lipids of bacteria also include lipoamino-acid complexes (Macfarlane, 1962a, 1962b; Hunter and James, 1963; Ikawa, 1963) and special lipid components are found in the mycobacteria in which waxes may account for 10% or more of the dry weight of the organisms (Asselineau, 1962).

Determination of the lipid content of micro-organisms and general methods for the identification and estimation of lipid components are mentioned here but no attempt is made to provide a complete scheme for the analysis of microbial lipids. Some aspects of lipid identification and analysis are discussed by Elizabeth Work (this Series, Volume 5a).

A. Extraction methods

Dehydrating agents (ethanol, acetone, methanol) rupture lipid–protein linkages but since lipids may not be entirely soluble in these solvents, a non-polar solvent such as petroleum ether, chloroform or diethylether is also included in solvent mixtures for extracting lipids from biological material. Suitable mixtures are chloroform–methanol (2 : 1, v/v) and ethanol–diethylether (3 : 1, v/v). Usually, about 10 parts of the solvent mixture are shaken with 1 part of material for several hours at room temperature; in some cases, it may be advisable to homogenize the suspension to achieve efficient extraction. The solvent-extracted residue should be checked for the presence of residual lipid by subjecting it to acid or alkaline hydrolysis at reflux temperatures; the hydrolysate is then examined for the presence of long-chain fatty acids or other lipid-like material.

When lipids are extracted from biological material into organic solvents, non-lipid substances are usually trapped with the lipids. This occurs even if the first extract in ethanol–ether is taken to dryness and the residue is extracted with petroleum ether according to the method of Bloor (1928). To overcome this difficulty, Folch et al. (1951) and Folch et al. (1957) introduced a water-wash treatment of crude lipid extracts in organic solvents. In the later method, the extract is thoroughly mixed with 0·2 vol of water and the mixture is allowed to separate into two phases by standing or centrifuging. The upper phase is removed by siphoning and the interface is rinsed three times with small volumes of pure solvent. The lower phase and the rinsing fluid are mixed and the mixture is treated with methanol to reform one phase. The efficiency of this purification process depends on the presence in crude extracts of sodium, potassium, calcium and magnesium ions; if insufficient of these are present, they are included in the wash fluid.

Direct extraction with organic solvents of certain biological materials removes only a small proportion of the total lipid. For example, Salton (1953) found that less than 40% of the lipid in the isolated cell walls of *E. coli* was removed by repeated extraction with boiling 95% methanol and/or boiling diethylether. Salton therefore introduced the following method for the complete extraction of lipid: the sample (50 mg) with 6N hydrochloric acid (5 ml) is heated in a sealed tube for 2 h in a boiling water bath; the hydrolysate is repeatedly extracted with diethylether in a separating funnel and the extracts are evaporated to dryness *in vacuo* over P_2O_5 and NaOH pellets.

B. Total lipid and polyhydroxybutyrate

To determine the lipid content of micro-organisms, extracts prepared and purified as described above are freed from solvents and dried to constant weight. In practice, the easiest way to do this is to quantitatively transfer the initial residue in the smallest volume of a suitable solvent into a tared weighing bottle, evaporate off the solvent, and dry the residue to constant weight over P_2O_5 *in vacuo*. The dry weight of the lipid is then found.

Polyhydroxybutyric acid. The extraction procedures described above will not remove polyhydroxybutyrate (a lipid?) from microbial cells. This polymer, $(C_4H_6O_2)_n$, was first discovered in *Bacillus megaterium* by Lemoigne (1925, 1927) and later it was found in other *Bacillus* species (Lemoigne *et al.*, 1944; Williamson and Wilkinson, 1958), *Azotobacter chroococcum, Pseudomonas* spp. (Forsyth *et al.*, 1958) and *Micrococcus halodenitrificans* (Smithies *et al.*, 1955). In the context of the gross chemical composition of micro-organisms, the presence of this polymer must be taken into account as it may account for up to 30% of the cell dry weight (Forsyth *et al.*, 1958).

An indication that bacteria contain this polymer is obtained by staining thin films of the organisms with Sudan Black B (G. T. Gurr, Ltd., London) followed by microscopic examination; polyhydroxybutyrate is contained in sudanophilic cytoplasmic granules. No organisms containing this polymer fail to give this reaction although some organisms giving the reaction do not contain the polymer (Forsyth *et al.*, 1958). It is of interest that granules of the polymer isolated from the bacterial cells with alkaline hypochlorite are not sudanophilic suggesting that conventional lipid associated with these inclusions in the cell is responsible for the reaction.

Polyhydroxybutyrate is removed from bacteria with alkaline hypochlorite (see Wilkinson and Sutherland this Volume, p. 345) or hot chloroform

(Forsyth *et al.*, 1958). The crude polymer may be precipitated from chloroform solutions with diethylether or it can be further purified by fractional precipitation with acetone–ether mixture (2 : 1, v/v) (Doudoroff and Stanier, 1959). Granules isolated by digestion of cells with alkaline hypochlorite (Williamson and Wilkinson, 1958) are collected by centrifugation, washed with water, treated with acetone and ether to remove lipid, and extracted with chloroform.

Identification of the extracted material as polyhydroxybutyrate may be based on elemental analysis; $(C_4H_6O_2)_n$ requires C, 55·8; H, 6·8, m.p. 168°C, the formation of crotonic acid on hydrolysis with 2 N sodium hydroxide (crotonic acid is identified by odour, m.p., mixed m.p., and the preparation of the anilide and *p*-bromophenacyl ester derivatives; Forsyth *et al.*, 1958), and infrared spectroscopy (Forsyth *et al.*, 1958).

The cellular content of polyhydroxybutyrate can be estimated approximately by precipitating the polymer in chloroform extracts with diethylether, washing the precipitate with acetone and ether, drying it to constant weight over phosphorus pentoxide and weighing. Or, following the release of cytoplasmic granules with alkaline hypochlorite, the polymer is purified by washing and extraction with chloroform as described above, dried and weighed. In addition, a turbidimetric method can be used after dissolution of the cells with alkaline hypochlorite (Williamson and Wilkinson, 1958).

A more specific method is to convert the polyhydroxybutyrate to crotonic acid, which is determined spectrophotometrically; this will be described in detail.

1. *Determination of polyhydroxybutyric acid by the method of Law and Slepecky* (1961)

This method is based on the finding (Slepecky and Law, 1960) that both β-hydroxybutyric acid and polyhydroxybutyric acid are quantitatively converted to crotonic acid on treatment with concentrated sulphuric acid at 100°C—

$$\left[\begin{array}{c} CH_3 \\ / \\ -O-CH.CH_2.CO- \end{array} \right]_n \xrightarrow{\text{H}_2\text{SO}_4} n \; CH_3.CH : CH.COOH$$

The absorption maximum of crotonic acid in aqueous solution is at 220 nm, which is too low for convenient measurement in some spectrophotometers, but in H_2SO_4 concentrations of 80% or higher, the absorption peak is shifted to 235 nm. In applying the method to whole bacterial cells (Law and Slepecky, 1961), interfering substances are largely removed by digesting

13–5*b*

the cells in sodium hypochlorite, followed by extraction of the poly-hydroxybutyric acid into hot chloroform.

(a) *Procedure.* Freeze-dried bacterial cells (10–20 mg) are weighed directly into small Pyrex centrifuge tubes (12 × 100 mm), treated with 1 ml of commercial sodium hypochlorite solution (British Drug Houses, available Cl content 10–14%) and digested for 1 h at 37 °C. (Alternatively, a suitable volume of bacterial culture may be centrifuged down, washed, and sodium hypochlorite added to the packed cells, the dry weight of cells being determined on a separate aliquot of culture as in Section II B.)

After incubation, 4 ml of water are added, the suspension well mixed, and centrifuged; the supernatant is then decanted from the lipid granules, which adhere to the walls of the centrifuge tube. The lipid granules are then washed with 5 ml of acetone followed by 5 ml of absolute ethanol (this removes water which would interfere with the subsequent extraction of the polymer into chloroform). The washed lipid granules are extracted with 3 ml of chloroform for 1–2 min in a boiling water bath, cooled, centrifuged, and the extract decanted into a 10 ml graduated flask. This extraction is repeated twice and the pooled extracts made up with chloroform to 10 ml.

For the spectrophotometric assay, a suitable volume of the chloroform solution containing 5–50 μg of polymer is pipetted into a dry thin-walled boiling tube which is immersed in a boiling water bath until the chloroform has evaporated. Concentrated sulphuric acid (10 ml) is added, the tube capped with a glass marble, and replaced in the boiling water bath for a further 10 min. After cooling, the sample is transferred to a 1 cm silica cuvette and the absorbance at 235 nm measured against a blank of the same sample of sulphuric acid. According to Law and Slepecky (1961), the molecular extinction coefficient of crotonic acid in conc. H_2SO_4 at 235 nm is 1.56×10^4, and polyhydroxybutyric acid gives an identical value, assuming an average residue weight for the polymer of $C_4H_6O_2 = 86$†. Hence an absorbance of 0.1 at 235 nm corresponds to a polyhydroxybutyric acid concentration of 0.551 μg/ml. Absorbance is proportional to polyhydroxybutyric acid concentration up to at least 3 μg/ml (absorbance 0.544).

(b) *Comments on the method.* Other β-hydroxy acids and some sugars give substances absorbing at 235 nm after treatment with concentrated sulphuric acid (Slepecky and Law, 1960). Such substances will be removed by the

† A misprint occurs on p. 34 of the paper of Law and Slepecky (1961), giving the average residue weight of polyhydroxybutyric acid as 186 instead of 86; however, the data in their Fig. 1 relating absorbance to weight concentration of polyhydroxybutyric acid are identical with the factor given above.

preliminary treatment with hypochlorite followed by chloroform extraction. However, Law and Slepecky (1961) found that with some bacteria which do not contain polyhydroxybutyrate, an insoluble material remains after hypochlorite treatment that contains an unknown interfering substance, though this has an absorption spectrum quite different from that of crotonic acid. When applying the method to unknown material, it is important to check the whole spectrum from 220 to 280 nm. It is also advisable to run spectra on pure samples of polyhydroxybutyric or crotonic acids as a check on stray light in the short ultraviolet, which is appreciable in some spectrophotometers (stray light will give a *low* value for the molecular extinction coefficient). The quality of the sulphuric acid used is also important; some samples have high absorption in the ultraviolet, necessitating unduly large slit width settings of the spectrophotometer.

C. Identification and estimation of lipid constituents

Analytical lipid chemistry is a specialized field and involves the use of various techniques including paper chromatography, thin layer chromatography, column chromatography, gas chromatography, infrared spectroscopy etc.

After the mixed lipid fraction has been separated from biological material as described above, the first step is to separate the simple neutral lipids from the complex phospholipids. Various methods can be used (see Hanahan, 1960), for example, treatment with acetone which under controlled conditions will dissolve the neutral lipids but not phospholipids, or by chromatography on silicic acid columns.

The phospholipids (acetone insoluble fraction) are freed from residual non-lipid components by chromatography on cellulose powder columns (Lea et al., 1955), Sephadex columns (Wells and Dittmer, 1963), silicic acid columns (Borgstrom, 1952) or by paper chromatography and paper electrophoresis (Westley et al., 1957). The mixed phospholipids are separated by chromatography on silicic acid columns (Hanahan et al., 1957) and diethylaminoethyl cellulose columns (Huston et al., 1965). Fractions eluted from the columns are monitored by thin-layer chromatography on which the components after migration are detected with a variety of general and specific reagents. The number of components separated from mixed microbial phospholipids in this way is usually large, for example, Wober et al. (1964) obtained thirty fractions from a silicic acid column to which had been applied the complex lipids of *Brucella abortus* and each of these fractions consisted of several components. Huston et al. (1965) separated thirteen components from the complex lipids of *Sarcina lutea* by column and thin-layer chromatography; Kasai and Yamano (1964) showed that at least eight lipid components were present in the lipid A fractions derived from the endotoxic lipopolysaccharides of several Gram-negative enteric bacteria.

A number of the components can be provisionally identified on the basis of their reactions with ninhydrin and Dragendorff's reagent, and by comparison with authentic marker substances run alone or in admixture on the thin-layer plates. In general, however, the individual components are preparatively isolated from the plates and analysed. As a first step, the substances are confirmed as phospholipids by detecting the presence of phosphorus. The isolated phospholipids are hydrolysed with alkali and the hydrolytic products are examined for the presence of ethanolamine, choline, amino-acids and phosphate esters (e.g., glycerophosphate) by paper or thin-layer chromatography. The results in conjunction with the information obtained with the intact lipids allow some firm identifications to be made (e.g., lecithin, lysolecithin, serine–cephaline, etc.) but other investigations including infrared spectroscopy will have to be made to identify a number of the products.

The fatty acids split off from the lipids are esterified (e.g., by reaction with diazomethane) and the fatty acid methyl esters are identified and estimated by gas chromatography.

VII. DETERMINATION OF NUCLEIC ACIDS

This section will deal mainly with methods for the determination of *total* desoxyribonucleic acid (DNA) and ribonucleic acid (RNA) in micro-organisms. Methods for the determination of the different nucleic acid fractions (e.g., transfer-RNA, the $5S$, $16S$ and $23S$ components of ribosomal-RNA, etc.) are as yet too elaborate to be classed as routine analytical procedures, though they will be briefly mentioned.

The literature on nucleic acid determination is very large; the review of Munro and Fleck (1966) gives 429 references, and more are to be found in earlier reviews by Hutchison and Munro (1961), Webb and Levy (1958), Volkin and Cohn (1954) and the three-volume work on "The Nucleic Acids" (Chargaff and Davidson, 1955, 1960). References will mainly be made to these sources, and only key papers in the original literature will be quoted. Most of this large amount of work has been done on animal tissues, for which fairly satisfactory standard methods have been evolved. Methodological studies on nucleic acid determination in micro-organisms are unfortunately rare, and most workers in this field have used the standard methods devised for animal tissues; it is remarkable how often they have been found unsatisfactory, and references abound to the "special difficulties" encountered in applying them to quite common microbial species.

The reasons for these difficulties will become apparent when the standard methods have been discussed.

A. General principles

1. Composition of nucleic acids and determination of their components

As with most complex macromolecules, there are no specific methods for determination of the nucleic acids as such, and it is possible only to determine their component parts. The methods available are: (a) Determination of phosphorus; this requires quantitative separation of RNA and DNA from each other and from all other phosphorus-containing compounds. (b) Determination of purines and pyrimidines by ultraviolet absorption; this also requires separation of the nucleic acids and removal of other ultraviolet absorbing substances. (c) Determination of ribose and desoxyribose; this does not require separation of RNA and DNA as these sugars can be distinguished by suitable colorimetric methods, but interfering carbohydrates must be removed. (d) Determination of thymine and uracil as measures of DNA and RNA respectively; this method has occasionally been used (Munro and Fleck, 1966) but is more difficult than the other methods and also requires that the base compositions of the DNA and RNA should be known. Variations in base composition will in fact affect all the above methods to some extent, since nucleic acids with different base contents have slightly different phosphorus and sugar contents and ultraviolet absorptions. Variations in the purine/pyrimidine ratio are of particular importance, as most of the available sugar methods determine only the purine-bound ribose or desoxyribose.

The composition of some nucleic acids from microbial and other sources is given in Table XIII; their base compositions can differ considerably from those of the "statistical tetranucleotides" containing equimolar amounts of all four bases. All microbial DNAs appear to be of the double-helix type and conform (within the limits of experimental error) to the "Chargaff rules" which are also demanded by the Crick–Watson structure; namely, adenine = thymine, guanine = cytosine, and total purines = total pyrimidines. The ratio $(A+T)/(G+C)$ varies much more in microbial than in animal DNAs, and in the former can range from 0·38 to 2·3; according to Belozersky and Spirin (1960) this ratio has systematic significance. Table XIII shows, however, that while the nitrogen content of DNA varies appreciably with base composition, the contents of phosphorus and purine-bound desoxyribose are practically constant.

No such simple rules hold for the base composition of RNAs, and in particular the purine/pyrimidine ratio can depart significantly from 1·0. Wade and Robinson (1965) have found that the r-RNAs of a variety of bacteria have very similar base compositions with a purine/pyrimidine ratio of about 1·3, while the s-RNAs also have similar compositions but with a purine/pyrimidine ratio close to 1. Since the ratio of r-RNA to s-RNA can vary with growth conditions, the composition of the total bacterial RNA will therefore vary, though at least within limits. The composition of other microbial RNAs is similar to that of bacteria (Belozersky and Spirin, 1960).

2. *Properties of nucleic acids used in their separation*

Most of the methods for separation of DNA and RNA are based on their different resistances to hydrolysis by alkali and acid. The glycosidic base–sugar linkages of both nucleic acids are resistant to alkaline hydrolysis (which applies to glycosides in general). The sugar–phosphate–sugar linkages of RNA, however, are rather easily hydrolysed at pH values greater than 10, while those of DNA are stable even in strong alkali; this difference is due to the ability of ribose to form 2–3 cyclic phosphates, which is impossible with 2-desoxyribose. RNA therefore is hydrolysed by strong alkalies to nucleotides and eventually to nucleosides, while DNA is resistant. The glycosidic base–sugar linkages of both nucleic acids are acid-labile, but with both RNA and DNA the purine–sugar linkages are much more easily hydrolysed than the pyrimidine–sugar linkages; the sugar–phosphate–sugar linkages of RNA are also more acid-labile than those of DNA.

B. Standard methods for RNA and DNA determination

1. *The method of Schmidt and Thannhauser* (1945)

This method was evolved for the determination of nucleic acids in animal tissues by their phosphorus content; RNA and DNA were separated by treatment with NaOH, the whole procedure being as follows: (i) Extraction of the finely minced tissues with 7 % trichloroacetic acid (TCA) at 0°C; this removes "acid-soluble P compounds" such as orthophosphate, sugar phosphates, nucleotides, ATP, etc. (ii) Removal of lipid by successive extractions with cold and then boiling alcohol–ether (3 : 1) followed by boiling methanol–chloroform (1 : 1); the tissue is then washed with ether and dried. The main purpose of the extractions is to remove phospho-lipids which would interfere with subsequent P analyses. (iii) The tissue is then extracted with N KOH (10 ml/g of tissue) for at least 15 h at 37°C. Under these conditions animal tissues (except bone) are completely dis-solved; RNA is hydrolysed to nucleotides and tissue phosphoproteins to orthophosphate, while DNA remains intact. Total P is determined on an

TABLE XIII

Chemical compositions of nucleic acids from microbial and other sources[a]

Nucleic acid source	Molar ratios							Empirical formula	Average res due weight	N (%)	P (%)	Ribose or desoxyribose	
	A	G	C	T	U	Pu/Py	A+T/C+G					Total (%)	Purine (%)
Desoxyribonucleic acids													
"Statistical tetranucleotide"	0·25	0·25	0·25	0·25	..	1·0	1·0	$C_{9.75}H_{12.25}N_{3.75}O_5P$	308·95	17·00	10·03	43·42	21·71
Micrococcus lysodeikticus	0·144	0·373	0·346	0·137	..	1·07	0·39	$C_{9.61}H_{12.14}N_{3.90}O_{6.03}P$	310·21	17·60	9·98	43·24	22·36
Escherichia coli (K12)	0·260	0·249	0·252	0·239	..	1·04	1·00	$C_{9.72}H_{12.25}N_{3.78}O_{5.98}P$	305·00	17·13	10·02	43·41	22·10
Clostridium perfringens	0·341	0·158	0·151	0·350	..	1·01	2·24	$C_{9.88}H_{12.25}N_{3.65}O_{5.91}P$	308·96	16·54	10·03	43·42	21·66
Calf thymus[b]	0·283	0·219	0·215	0·283	..	1·01	1·30	$C_{9.76}H_{12.28}N_{3.72}O_3P$	305·00	16·87	10·02	43·41	21·79
Ribonucleic acids													
"Statistical tetranucleotide"	0·25	0·25	0·25		0·25	1·0	..	$C_{9.50}H_{11.75}N_{3.75}O_7P$	321·44	16·34	9·64	46·71	23·35
Escherichia coli (MRE 600)[c]													
(a) ribosomal RNA	0·251	0·312	0·222		0·215	1·29	..	$C_{9.48}H_{11.79}N_{3.88}O_{8.96}P$	323·22	16·81	9·58	46·45	26·15
(b) soluble RNA	0·197	0·300	0·304		0·199	0·99	..	$C_{9.40}H_{11.90}N_{3.80}O_{7.0}P$	322·12	16·50	9·62	46·61	23·16
Saccharomyces cerevisiae[d]	0·254	0·246	0·226		0·274	1·0	..	$C_{9.55}H_{11.82}N_{3.73}O_{7.02}P$	321·40	16·24	9·64	46·71	23·56
Saccharomyces cerevisiae[e]	0·258	0·264	0·195		0·283	1·09	..	$C_{9.52}H_{11.81}N_{3.70}O_{7.13}P$	322·23	16·35	9·61	46·59	24·32

[a] A, G, C, T and U designate respectively adenine, guanine, cytosine, thymine and uracil. Pu/Py is the molar ratio of purine to pyrimidine bases.
[b] Average of 17 determinations from Chargaff and Davidson (1955), Vol. 1, p 354.
[c] From Wade and Robinson (1965); ribosomal and soluble RNA from Aerobacter, Proteus and Pseudomonas species gave closely similar values.
[d] Bakers' yeast; determination of total RNA in whole cells (Elson and Chargaff, 1955).
[e] Commercial yeast RNA preparation (Boehringer); base analyses by Dr. H. E. Wade.

aliquot of this solution ($= T_1$). (iv) Another aliquot of the alkaline solution is treated with 0·2 vol. of 6 N HCl and 1 vol. 5% TCA, which precipitates DNA and protein, leaving the RNA split products in solution; after removal of the precipitate by filtration, the filtrate is analysed for total P ($= T_2$) and inorganic P ($= P_i$). $T_1 - T_2$ is taken as a measure of DNA-P, while $T_2 - P_i$ gives the RNA-P; P_i is "phosphoprotein-P".

2. *The method of Schneider* (1945)

In this method, also worked out on animal tissues, RNA and DNA are extracted together with hot TCA and determined colorimetrically, after prior removal of "acid-soluble" compounds and lipids; the steps are as follows. (i) The tissues are homogenized in a Potter–Elvehjem (1938) homogenizer; 1 vol. of 20% homogenate is mixed with 2·5 vol. 10% TCA, centrifuged, and the precipitate re-extracted with 2·5 vol. 10% TCA, all these steps being at 0°C. The TCA extract contains "acid-soluble-P" compounds, but in this case the purpose of the extraction is to remove free sugars, nucleosides, nucleotides and other ribose-containing compounds. (ii) The tissue residue is successively extracted with cold 80% and 95% ethanol to remove TCA, then extracted three times with boiling ethanol-ether (3 : 1) to remove lipids. (iii) The residue is suspended in 5% TCA, heated for 15 min at 90°C and centrifuged. The extract contains all the tissue RNA and DNA, which, although not completely hydrolysed by this treatment, are split into sufficiently small oligonucleotides to remain soluble in acid. DNA is determined as desoxyribose by the diphenylamine reaction of Dische (1930), and RNA is determined as ribose by the orcinol reaction for pentoses.

3. *The Schmidt–Thannhauser–Schneider (STS) method*

The Schmidt–Thannhauser method has the advantage that DNA and RNA are separated, but their determination as phosphorus is unspecific and may be upset by the presence of unknown phosphorus compounds. Schneider (1946) suggested a method combining the Schmidt–Thannhauser separation of RNA and DNA by NaOH treatment, followed by colorimetric determination of ribose and desoxyribose. Later, Schneider *et al.* (1950) introduced the use of perchloric acid as a substitute for TCA; $HClO_4$ is transparent in the ultraviolet (unlike TCA which absorbs strongly) and allows the separated RNA and DNA fractions to be determined by ultraviolet absorption. This method, or some modification of it (many workers have made small changes in detail, see Munro and Fleck, 1966) is currently the procedure most favoured by those working on animal tissues; essential details are as follows. (i) Extraction of "acid-soluble" compounds with cold $HClO_4$, followed by (ii) removal of lipids; these steps are as described above except that $HClO_4$ is substituted for TCA. (iii) The tissue is digested with N KOH for 16–20 h (as in the original Schmidt–Thannhauser method) and then treated with $HClO_4$, which precipitates DNA and protein; after centrifuging, the RNA in the supernatant may be determined by either the orcinol method or by ultraviolet absorption (or both). (iv) The precipitate of DNA plus protein is extracted with hot $HClO_4$ (cf. the original Schneider method) which dissolves the DNA; this is determined either colorimetrically or by ultraviolet absorption.

4. The method of Ogur and Rosen (1950)

Unlike the previous methods, this was devised for use with plant tissues, in which high contents of pentosans and polyuronides often interfere with the determination of RNA by the orcinol reaction. RNA is determined (after removal of lipids and acid-soluble material) by extraction with N $HClO_4$ for 18 h at 4°C, which solubilizes RNA without hydrolysing DNA or extracting large amounts of carbohydrate. The steps are: (i) Removal of lipids by homogenization in 70–95 % ethanol at 0°C, followed by successive extractions with cold 70 % ethanol containing 0·1 % $HClO_4$, and with boiling ethanol–ether (3 : 1); (ii) Removal of acid-soluble material by two brief extractions with cold 0·2 N $HClO_4$; (iii) The residue is suspended in 1 N $HClO_4$, extracted at 4°C for 18 h, centrifuged, and washed twice with cold 1 N $HClO_4$; the combined extracts contain the RNA which may be determined by the orcinol method and/or ultraviolet absorption; (iv) The residue is extracted twice with 0·5 N $HClO_4$ at 70°C (20 min for each extraction); the combined extracts contain all the DNA, which is determined by the diphenylamine reaction.

5. Difficulties in the standard methods

Difficulties may arise at every stage of all the above methods; they have been thoroughly discussed by Munro and Fleck (1966) and Hutchison and Munro (1961), and will be briefly summarized here for the various steps in turn.

(a) *Extraction of acid-soluble materials.* This step is essential, not so much for removal of acid-soluble P (since the original Schmidt–Thannhauser method is now seldom used), but to remove free sugars and other small molecules which either contain sugars or absorb in the ultraviolet (nucleosides, nucleotides, ATP, NAD and NADP come under both categories). The concentration of TCA or $HClO_4$ should be high enough to precipitate proteins, RNA and DNA, but if the acid concentration is too high or the extraction period too long some RNA may be hydrolysed and purine bases may be split from DNA with the formation of apurinic acids; these (unlike native DNA) are alkali-labile and will be hydrolysed by NaOH in the Schmidt–Thannhauser method. With $HClO_4$ the optimum final concentration is 0·2–0·3 M; the temperature should be 0°C and the extraction period brief (not more than 30 min).

(b) *Removal of lipids*. A lipid extraction stage was included in all the earlier methods of nucleic acid determination, but Munro and Fleck (1966) argue that this step is usually superfluous, and indeed may do more harm than good. Removal of phospholipid is unnecessary if nucleic acids are determined by sugar analysis or ultraviolet absorption, while the amount of carbohydrate in glycolipids is seldom enough to interfere. On the other hand, extraction with lipid solvents (particularly at boiling temperature) after the extraction of acid-soluble material with TCA or $HClO_4$ can cause appreciable hydrolysis of RNA and conversion of DNA to apurinic acid. This can be avoided by using ethanol containing sodium acetate as a buffer for the first lipid solvent, but Munro and Fleck (1966) suggest that in most cases it is better to omit the lipid extraction stage altogether; if really necessary (as with tissues of unusually high fat content), then lipid extraction should precede extraction of acid-soluble substances, as in the Ogur–Rosen method.

(c) *Extraction of nucleic acids with hot acids (Schneider procedure)*. This procedure has been criticized on the grounds that extraction conditions are too critical; if acid concentration and temperature are too low the nucleic acids are incompletely extracted, while if they are too high there is appreciable destruction of DNA-desoxyribose. It is generally admitted that the extraction conditions originally used by Schneider are too drastic; the milder conditions used by Ogur and Rosen (1950) and Burton (1956), i.e., several brief extractions with 0·5 N $HClO_4$ at 70°C, appear to effect complete extraction of DNA without loss by destruction. A more valid criticism is that in tissues rich in polysaccharides, sufficient carbohydrate appears in the hot acid extracts to interfere with the determinations of ribose and desoxyribose. This can certainly occur with the orcinol method for RNA determination and the original diphenylamine method of Dische (1930) for DNA; if the much more specific diphenylamine method of Burton (1956) is used, however, we have never observed interference by even large amounts of carbohydrate.

(d) *Alkaline hydrolysis of RNA (Schmidt–Thannhauser procedure)*. The hydrolysis conditions used by Schmidt and Thannhauser (1 N NaOH or KOH for 15–24 h at 37°C) are now considered unnecessarily vigorous; they may

cause deamination of cytidylic acid, considerable extraction of polysaccharides, and some hydrolysis of protein which may interfere with ultraviolet
absorption measurements. If any DNA has been converted to apurinic
acid during the extraction of acid-soluble substances, it will be hydrolysed
under these conditions. Very much milder conditions are advocated by
Munro and Fleck (1966), who point out that it is unnecessary to effect
complete hydrolysis of RNA to free mononucleotides; it is only necessary
to convert it to fragments small enough to be acid-soluble, and this can be
effected (for all animal tissues at least) by hydrolysis for 1 h at 37°C in
0.3 N KOH. After precipitation of DNA and protein from such digests
with $HClO_4$ (0.2 N final concentration of free $HClO_4$ recommended),
determinations of RNA in the acid-soluble fraction by ultraviolet absorption and the orcinol method gave good agreement. The Munro–Fleck
procedure was successfully applied to *E. coli* cells by Dawes and Ribbons
(1965), who give progress curves for the extraction of acid-soluble ribose
and ultraviolet-absorbing substances; extraction was essentially complete
in 60 min.

(e) *Determination of DNA after alkaline hydrolysis (STS procedure)*. In the
original Schmidt–Thannhauser procedure, DNA-P was determined by
difference, a not too reliable procedure. In the method as modified by
Schneider (1946), the precipitate of DNA and protein obtained by acidi-
fying the alkaline digest is extracted with hot acid to obtain the DNA in
solution. This raises exactly the same problems concerning destruction of
DNA–desoxypentose by hot acid that were discussed above under the
Schneider method; the best solution is probably that advocated above,
namely two or three brief extractions with 0.5 N $HClO_4$ at 70°C. An alternative is to dissolve the precipitate in alkali, but the amount of protein
present makes determination of DNA by ultraviolet absorption unreliable
and is often large enough to interfere with the chemical determination of
desoxypentose. Munro and Fleck (1966) suggest that the best method would
be to dissolve the DNA by digestion with DNase; this method has
occasionally been tried, but has not been critically examined.

(f) *Hydrolysis of RNA with cold acid (Ogur–Rosen procedure)*. This method
has been severely criticized (Munro and Fleck, 1966) on the grounds that
the recommended hydrolysis conditions either did not give complete
extraction of RNA, or caused some hydrolysis of DNA. It seems probable
that these criticisms are valid and that a 100% clean separation of RNA and
DNA by cold acid hydrolysis is not possible. Considered solely as a
method for RNA determination, however, the Ogur–Rosen method has, in
the writers' view, much to recommend it, since very much less carbohydrate

is extracted with the RNA than in any other method, which makes it possible to determine RNA by the orcinol method even in polysaccharide-rich material. The method used by Trevelyan and Harrison (1956) for RNA determination in yeast (extraction with 0.5 N $HClO_4$ for 90 min at 37°C) may be regarded as a half-way stage between the Ogur–Rosen and the Schneider extraction procedures.

The reader of this Section may well conclude that every step in every method for nucleic acid determination is beset with difficulties; it is only fair to say that some of these difficulties are less serious in micro-organisms because of their high nucleic acid content. However, micro-organisms present difficulties of their own which are discussed later.

C. Determination of nucleic acids by sugar analysis

1. Determination of RNA by the orcinol method

The orcinol method for the determination of pentoses was described in detail in Section V, and it remains only to discuss its application to RNA determination. Furfural cannot be formed from ribosides until the glycosidic linkage has been split; the purine–ribose links of RNA are easily hydrolysed by hot acid, while the pyrimidine–ribose links are much more resistant, and the orcinol method is commonly supposed to determine only the purine-bound ribose of RNA. This is not precisely true, as shown by the results of Fig. 14, in which the colours produced by our orcinol reagent with ribose and RNA were measured after different heating times. With ribose, maximum colour is produced in 18 min, remains stable up to 30 min, and then slowly decreases. With RNA, the initial rate of colour development is a little slower than that of ribose up to 20 min, after which there is a slow, almost linear increase up to 40 min (and longer); this is almost certainly due to the slow hydrolysis of pyrimidine ribosides.

The apparent purine–ribose content of RNA therefore varies with the heating time, and this is true of all the orcinol reagents we have tested. We have "calibrated" our own orcinol reagent (Section V) with very pure r-RNA preparations of known base composition isolated from *E. coli* by Robinson and Wade (1968), as well as commercial yeast RNA which had been analysed for base composition; with both preparations, theoretical yields of purine–ribose were obtained with a heating period of 20 min. If this heating time is adhered to, ribose may be used as a standard, but with other heating times it is best to use as standard a pure sample of RNA, preferably from the organism being analysed, and having a known purine/pyrimidine ratio. For the determination of RNA in bacterial extracts containing carbohydrates, a heating time of 10 min is recommended as this considerably reduces interference by hexoses (Section V and Table

XII); a RNA standard should then be used. When the reaction is applied to $HClO_4$ extracts of micro-organisms, as in the Schneider method, these should either be neutralized, or an equal amount of $HClO_4$ added to the standards.

2. *Other methods for RNA determination*

The orcinol reagent is rather susceptible to interference by hexoses, and many attempts have been made to find a more specific reagent, so far without success. The furfural formed by heating RNA with HCl can be trapped in xylene and allowed to react with aniline (Davidson and Waymouth, 1944) or with *p*-bromophenylhydrazine (Webb, 1956); these methods are less sensitive and more tedious than the orcinol method and their specificity, though less well studied, would appear to be no better. Dische's cysteine–H_2SO_4 method for pentoses can be applied to RNA determination (Dische, 1955b), but seems to offer no advantages over the orcinol reaction and has seldom been used. Early versions of the phloroglucinol reaction (Euler and Hahn, 1946) were tedious and insensitive, but the reagent of Dische and Bohrenfreund (1957) appears more promising. Originally described as a method for pentoses (see Section V for details), this method seems rarely to have been used for RNA analysis but we have found it gives approximately theoretical yields of purine–pentose from pure RNA preparations. Its sensitivity is only about half that of the orcinol reaction, but it would seem worth investigating on account of its low sensitivity to hexoses. (An alleged improvement of this reagent by Bolognani *et al.* (1961) was found to be much more sensitive to hexoses, and is not recommended.)

All the above methods react only with the purine-bound ribose of RNA. Massart and Hoste (1947) reported that bromination of pyrimidine nucleotides renders them acid-labile and reactive with the orcinol reagent; an attempt to apply this procedure to RNA was unsuccessful, however (Kerr *et al.*, 1949).

3. *Determination of DNA with the diphenylamine reagent*

Dische (1930) found that DNA gives a blue colour on heating at 100°C with a solution of diphenylamine in glacial acetic acid containing 2·75% H_2SO_4; the colour is proportional to the amount of DNA and can be used for its quantitative determination. The reaction is given by free 2-desoxyribose and 2-desoxyxylose; purine desoxyribonucleotides give the reaction with twice the intensity of DNA, while pyrimidine desoxyribonucleotides scarcely react. The reaction mechanism has been investigated by Deriaz *et al.* (1949) who concluded that ω-hydroxylaevulic aldehyde is an essential intermediate, though the results of Burton (1956) do not support this. Hexoses react only weakly with the diphenylamine reagent; pentoses and

RNA react more strongly, but still much less than DNA. Some poly-saccharides containing anhydrosugars are more reactive, and agar in particular gives a strong green colour (Pirie, 1936), which should be borne in mind when agar-grown micro-organisms are analysed. Sialic acids also interfere strongly.

Dische's original method remained virtually unchanged until Burton (1956) introduced a modified diphenylamine reagent containing acetalde-hyde, with which samples are incubated at 30°C for 17 h, instead of heating at 100°C. Burton's method is highly recommended as it is 3·5 times more sensitive and much more specific than the original Dische method; the details are given below.

(a) *Reagents.* The reagents are as follows—

  (i) *DNA standard.* A pure DNA preparation (e.g. from calf thymus) is dissolved in 5 mM NaOH to give 400 μg/ml; this is stable at 4°C for 6 months. Working standards are prepared every 3 weeks by mixing a measured volume of the stock standard with an equal volume of N $HClO_4$ and heating 15 min at 70°C.

  (ii) *Diphenylamine reagent.* This is prepared by dissolving 1·5 g of steam-distilled diphenylamine in 100 ml of redistilled glacial acetic acid and adding 1·5 ml of conc H_2SO_4; it is stored in the dark. On the day of use, 0·1 ml of aqueous acetaldehyde (16 mg/ml) is added for each 20 ml of reagent required.

(b) *Procedure.* In estimating nucleic acids from biological sources, a suitable extract is obtained in 0·5 N $HClO_4$. A measured volume (1 or 2 ml as convenient) of sample is mixed with 2 ml of the diphenylamine reagent containing acetaldehyde; standards containing known amounts of DNA in 0·5 N $HClO_4$, and a reagent blank containing 0·5 N $HClO_4$ but no DNA are also prepared. The tubes are incubated overnight (16–20 h) at 30°C and the optical density at 600 nm is measured against the blank in a suitable spectrophotometer using 1 cm cells, and compared with the values obtained with the standard DNA. Beer's Law is obeyed up to an optical density of at least 1·0 (equivalent to about 100 μg DNA); the limit of accurate measurement is about 5 μg DNA (optical density about 0·05).

(c) *Comments on the method.* The writers have found this method very reliable, provided that really pure reagents are used. While some samples of AnalaR grade diphenylamine and glacial acetic acid are suitable for use without purification, others are not, and it is safest always to purify the diphenylamine by steam distillation and to redistill the acetaldehyde and the acetic acid. When all chemicals are sufficiently pure, the diphenylamine reagent is colourless, and remains so on adding the acetaldehyde; the reagent blank read against water should give an optical density no higher than 0·02–0·03. (A paper by Giles and Myers (1965) describes modifications of the method aimed at reducing the reagent blank, which they found as high as 0·12; as they also state that addition of acetaldehyde to the diphenyl-amine reagent produced a green colour, it is likely that their reagents were impure.)

The specificity of the reagent is considerably higher than that of the original Dische (1930) reagent; among a long list of substances tested by Burton (1956), aldohexoses, 6-desoxyhexoses, hexosamine, ascorbic acid, glutathione, cysteine, tryptophan, bovine serum albumin and adenosine-5'-phosphate gave no detectable colour at a level of 2 mg, while the colour

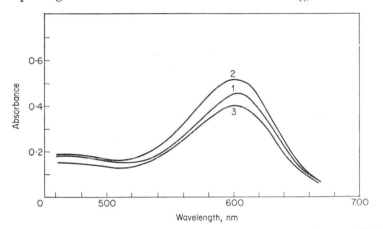

FIG. 16. Absorption spectra in the diphenylamine reaction of pure DNA and hot perchloric acid extracts of bacteria and yeast. Diphenylamine method of Burton (1956); cells extracted with 0·5 N HClO$_4$ at 70°C as described in text. Measurements in Optica CF4 Recording Spectrophotometer using 1 cm cuvettes. Curve 1, calf thymus DNA (50 μg); Curve 2, HClO$_4$ extract from 2·0 mg *Aerobacter aerogenes* cells containing 4·7% total carbohydrate and 2·85% DNA; Curve 3 HClO$_4$ extract from 8·0 mg *Torula utilis* cells containing 42·6% total carbohydrate and 0·53% DNA.

given by 1·5 μg DNA is easily visible. Fructose, pentoses, pyruvic and α-oxoglutaric acids in 2 mg quantities gave a detectable colour which is less than that given by 1·5 μg DNA. RNA (5 mg) gave a colour equivalent to 11 μg DNA; agar (2 mg) gave a colour equivalent to 18 μg DNA.

According to Croft and Lubran (1965), sialic acids interfere somewhat; they give an obviously different colour with an absorption maximum at 540 nm but absorb appreciably at 600 nm, at which wavelength 1·4 mg sialic acid has the same absorbance as 62 μg DNA. These authors found that the reaction of sialic acid could be virtually eliminated by carrying out the reaction at about 10°C for 48 h with a reagent of slightly modified composition. This modification of Burton's procedure is useful for the determination of small amounts of DNA in the presence of large amounts of sialic acids (e.g., in brain tissue), but we have not found it necessary with any micro-organisms we have encountered. Figure 16 shows the absorption spectra given by Burton's diphenylamine reagent with (a) pure DNA, (b) and (c) 0·5 N HClO$_4$ extracts (70°C) of A. aerogenes and T. utilis respectively; it will be seen that there is no indication of any interfering substances.

4. Other methods for DNA determination

Dische (1944) described a reaction of DNA with cysteine and 75 % (v/v) H$_2$SO$_4$, which are mixed at 0°C and then incubated at 40°C; only the purine-bound desoxyribose of DNA reacts. In a modification of this method by Stumpf (1947), carried out at a higher temperature, some, but not all, of the pyrimidine-bound desoxyribose reacts. Cohen (1944) describes a reaction of DNA with tryptophan in perchloric acid at 100°C in which both purine and pyrimidine-bound desoxyribose appear to react. All these reactions, however, are considerably less sensitive and less specific than Burton's diphenylamine reagent. A method due to Ceriotti (1952), in which DNA is heated with indole in 3 N HCl at 100°C, is considerably more sensitive, and formerly was often preferred to the original diphenylamine reagent for the determination of small amounts of DNA; it is no more sensitive than the Burton (1956) reagent, however, and is considerably less specific.

D. Determination of nucleic acids by ultraviolet absorption

All nucleic acids have a strong absorption band in the ultraviolet with a maximum at about 260 nm and a minimum at about 230 nm. Their peak absorption is usually quoted in terms of their ϵ_P value, defined as the optical density in a 1 cm depth of a solution containing 1 g atom of nucleic acid phosphorus per litre.

While optical densities at 260 nm can be measured in modern spectrophotometers with ease and precision, the conversion of such measurements to quantities of nucleic acid is subject to rather large errors; in the writers' view, many workers who use this method greatly over-estimate its accuracy. Errors may arise from the following causes.

1. Variation in base composition

Owing to the considerable differences in absorption spectra of the different purines and pyrimidines and their nucleotides (Beaven et al., 1955), variations in base composition will cause considerable variation in both ϵ_P and λ_{max} (the wavelength of maximum absorption). These are seldom taken into account, and most workers seem to use the literature values for calf thymus DNA and yeast RNA, which may differ appreciably from the nucleic acids they are actually analysing.

2. The hyperchromic effect

All double-stranded DNAs, when isolated by methods avoiding denaturation, exhibit the phenomenon of *hypochromicity*, i.e., their ϵ_P values are considerably *less* than the sum of the ϵ_P values of their constituent nucleotides. On heating, they exhibit the *hyperchromic effect*, i.e., at a fairly well-defined "melting temperature" the absorbance abruptly *increases* by up to 45%; this is attributed to separation of the double helices with resultant breaking of the adenine–thymine and guanine–cytosine hydrogen bonds. The same effect can be produced by other methods of denaturation. Undegraded preparations of single-stranded DNA and RNA exhibit the same effect to a lesser degree and with a broader "melting range", suggesting that the single chain is folded back on itself in a number of places, producing regions of base pairing which have a double-helical structure. Even after denaturation, however, the completely uncoiled chains of DNA or RNA have a "residual hypochromicity", and their ultraviolet absorption will increase still further if completely hydrolysed to their constituent mononucleotides (e.g., by NaOH hydrolysis in the case of RNA). This residual hypochromicity is attributed to "base stacking" effects, and is exhibited even by di- and tri-nucleotides. (These and other changes in nucleic acids occurring on denaturation are reviewed by Felsenfeld and Miles, 1967.)

These effects, which are of considerable magnitude, pose serious problems to those attempting to determine cellular nucleic acids by ultraviolet absorption measurements. The current methods of quantitative RNA and DNA separation cause a rather indeterminate degree of denaturation and partial hydrolysis, with the result that ϵ_P values are not well defined; this is the case, for example, with the Ogur–Rosen method for extraction of RNA by acid hydrolysis at 4°C, which produces rather ill-defined mixtures of oligonucleotides. The original Schmidt–Thannhauser conditions for RNA extraction (N NaOH at 37°C for 18 h) are better in that complete hydrolysis of RNA to mononucleotides is achieved, though with some contamination by ultraviolet absorbing protein degradation products, deamination of cytidylic acid, and in some cases, contamination with DNA fragments. The milder hydrolysis conditions advocated by Munro and Fleck (1966), i.e., 0·3 N KOH for 3 h at 37°C, avoid the latter difficulties, but produce indeterminate mixtures of oligonucleotides.

Extraction with hot TCA or $HClO_4$, as in the Schneider or STS procedures (see above) produces complete denaturation with relatively little splitting of inter-nucleotide linkages in DNA, but appreciable splitting with RNA; it also produces (particularly with DNA) appreciable liberation of free purines, whose ultraviolet absorption is considerably different from that of their nucleotides.

All the above reasons combine to make the determination of nucleic acids by ultraviolet absorption a rather inaccurate procedure. Probably the best method, whatever extraction methods are used, is to compare the observed ultraviolet absorption measurements with those of pure RNA or DNA preparations *isolated from the same tissue or micro-organism*, and subjected to identical manipulative procedures; it is realized that this recommendation is somewhat idealistic. Useful data on the λ_{max} and ϵ_P values of calf thymus DNA and yeast RNA after treatment with NaOH, cold and hot TCA and $HClO_4$, etc., are given by Ceriotti (1955) and Logan et al. (1952).

3. *Contamination with protein degradation products*

Prolonged treatment with alkali at 37°C, as in the Schmidt–Thannhauser method, or brief treatment with TCA or $HClO_4$ at 70°C, as in the Schneider or STS methods, cause some splitting of proteins to acid-soluble peptides which contaminate the nucleic acid fractions, and may have appreciable ultraviolet absorption. Tsanev and Markov (1960) describe a method of correcting for this by taking absorption measurements at 260 and 280 nm, similar in principle to the method used by Warburg and Christian (1941) for determination of protein contamination with nucleic acid (Section IV). Fleck and Begg (1954) use a similar 2-wavelength method, but reading at 260 and 232 nm, the latter being the wavelength of *minimum* nucleic acid absorption; the ratio $E_{260} : E_{232}$ is a sensitive index of protein contamination, as proteins absorb strongly at 232 nm. If a recording spectrophotometer is available, a complete scan of the ultraviolet region from 200–320 nm is recommended whenever a new extraction method or unknown material are investigated.

E. Determination of nucleic acids in micro-organisms

As previously mentioned, nucleic acid analysis of micro-organisms often presents difficulties which are not encountered with animal tissues, and seriously interfere with the standard procedures described above. For example— (i) many micro-organisms contain large amounts of phosphorus-containing macromolecules (polyphosphates, teichoic acids) which seriously interfere with the Schmidt–Thannhauser method, e.g., in yeast (Juni et al., 1948) and mycobacteria (Winder and Denneny, 1956).

(ii) Many micro-organisms contain large amounts of unusual ultraviolet absorbing compounds, e.g., the high dipicolinic acid content of the spores of aerobic bacilli (Fitz-James, 1955), which interfere with nucleic acid determination by ultraviolet absorption. Such substances are less common in vegetative bacteria and yeasts, but frequently occur in moulds.

(iii) Some micro-organisms contain very high contents of polysaccharides which interfere with RNA determination by the orcinol reagent; this is very common in yeasts (Trevelyan and Harrison, 1956).

(iv) The cell walls of many micro-organisms are completely resistant to standard extraction procedures used for nucleic acid analysis of animal tissue. Bacterial spores are an extreme case; thus on applying to *Bacillus cereus* spores the standard methods for extracting "acid-soluble" material and lipids (cold 10% TCA followed by boiling alcohol–ether), nothing was extracted by either solvent and the spores remained 100% viable (Fitz-James, 1955). The resistance to alkali of the cell walls of Gram-positive bacteria, yeasts and moulds has already been mentioned (Sections IV and V), and seriously interferes with the extraction of DNA by the Schmidt–Thannhauser procedure. Thus, in a study of 22 bacterial species (20 genera), Lee *et al.* (1956) found that treatment with N NaOH for 24 h at 37°C extracted only 15–20% of the DNA from most Gram-positive bacteria and none from *Sarcina lutea*; even with Gram-negative bacteria usually only 80–85% of the total DNA was extracted. Considerable undissolved residue remains after alkali extraction, and Sherratt and Thomas (1953), who experienced similar difficulties in extracting DNA from *Streptococcus faecalis*, noted that the cell outlines were still visible and the cell walls apparently intact; this is certainly the case when yeast cells are extracted with alkali, as shown by the electron microscope studies of Northcote and Horne (1952). It seems likely that the cell wall structure acts as a "molecular sieve", hindering the extraction of macromolecules such as DNA; RNA, which is hydrolysed by alkali to mononucleotides, appears to be completely extracted. Confirming this, Lee *et al.* (1956) found that the DNA of Gram-positive bacteria could be completely extracted by alkali if the cells are first disintegrated, either mechanically using the Mickle (1948) apparatus, or enzymically by using lysozyme (not applicable to many species).

Less difficulty is experienced with the Schneider procedure; this may be because extraction with hot TCA or $HClO_4$ causes more extensive disruption of the cell wall, or because the nucleic acids are sufficiently fragmented to become diffusible (this is almost certainly true for RNA). In any case, three 15 minute extractions at 70°C with 0·5 N $HClO_4$ (Burton, 1956) have given complete extraction of RNA and DNA with all micro-organisms we have encountered.

The Ogur–Rosen extraction procedure does not seem to have been applied to micro-organisms, but the method of Trevelyan and Harrison (1956), namely extraction with 0·5 N HClO₄ at 37°C, completely extracts RNA from yeasts in 90–120 min.

1. *Recommended procedure*

Although the Schmidt–Thannhauser or the later STS methods are considered the best for nucleic acid determination in animal tissues, and have the theoretical advantage of separating RNA and DNA, the previous Section shows that they cannot be applied to micro-organisms unless the cell walls are first disintegrated. With a minority of bacterial species this can easily be effected with lysozyme, but the majority require mechanical disintegration, either with the Mickle (1948) apparatus or ultrasonically. This is unacceptably time-consuming when large numbers of samples have to be analysed; moreover, prolonged treatment is necessary to obtain 100% disintegration of Gram-positive cocci, mycobacteria, corynebacteria, etc., with the attendant danger of enzymic breakdown of the nucleic acids. A further disadvantage is that the DNAs of some micro-organisms (as with some animal tissues, cf. McIndoe and Davidson, 1952) are alkali-labile, even with no previous exposure to acid, and partly appear in the RNA fraction (Wade, 1968). For these reasons, the Schmidt–Thannhauser method is *not* recommended for micro-organisms.

The procedure recommended by the present writers is a modified Schneider method, involving three short extractions with 0·5 N HClO₄ at 70°C; DNA is determined by the Burton (1956) method and RNA by the orcinol method described above. For DNA determination, this method has proved successful with all micro-organisms we have encountered. For RNA, the method works well for all bacteria, but fails with yeasts owing to their large content of interfering polysaccharides (the same may apply to some moulds). In such cases, RNA is determined separately on a duplicate sample, which is extracted with 0·5 N HClO₄ at 37°C (Trevelyan and Harrison, 1956). The various steps of this method will be discussed in turn.

(a) *Preliminary treatment of samples.* If samples cannot be analysed immediately, steps must be taken to avoid enzymic breakdown of nucleic acids (particularly RNA) which occurs on storage of cell suspensions even at 0°C. Storage in the frozen state is not recommended, as considerable RNA breakdown may occur on subsequent thawing; we therefore recommend freeze-drying by the following method. Cell suspensions are cooled to 0°C, centrifuged, washed twice with ice-cold 1 mM MgCl₂ and finally resuspended in 1 mM MgCl₂ to give a thick suspension (50–100 mg/ml); a "Whirlimixer" (Fisons Ltd) is convenient for resuspending the centrifuged cells. The suspensions are transferred to one-ounce screw-capped bottles (not more than 6 ml per bottle) and frozen at $-10°C$ or lower with the bottles in the horizontal position to give a thin layer of frozen material. The suspensions can be kept in the frozen state for some time before drying, but must not be allowed to thaw. When convenient, the bottles are rapidly transferred to a desiccator containing a dish of P_2O_5 and connected by wide-bore vacuum tubing to a two-stage vacuum pump of sufficient capacity to reach a pressure of $0·1$ mm Hg or lower within one minute. After 24 h, the samples are dry, and are then kept in a desiccator over P_2O_5 until ready for analysis. The $MgCl_2$ solution (95 $\mu g/ml$) makes no significant contribution to the dry weight.

When samples can be dealt with immediately, freeze-drying may be omitted and the cells extracted directly with cold $0·25$ N $HClO_4$ (stage (c) below); preliminary washing with cold 1 mM $MgCl_2$ may be necessary for cells grown in complex culture media.

(b) *Lipid extraction.* In agreement with Munro and Fleck (1966), we consider this step is best avoided, and it is certainly unnecessary with most micro-organisms. It may prove necessary, however, with mycobacteria and certain "high-fat" yeast and mould strains; if so, it should be performed on freeze-dried cells, *before* extraction of acid-soluble material.

(c) *Extraction of acid-soluble material with cold* $0·25$ N $HClO_4$. Freeze-dried cells (20–60 mg according to nucleic acid content) are weighed directly into small centrifuge tubes (100×10 mm; capacity 6–7 ml), 5 ml of ice-cold $0·25$ N $HClO_4$ are added; the tubes are allowed to stand for 30 min in an ice-water bath, with occasional shaking, centrifuged and drained. (For cells containing unusually large amounts of free sugars, e.g., yeasts, two extractions of 15 min each may be preferable. The $HClO_4$ concentration should not exceed $0·25$ N and the total extraction time should not exceed 30 min, or appreciable hydrolysis of RNA may occur.)

(d) *Extraction of nucleic acids with* $0·5$ N $HClO_4$ *at* $70°C$. The procedure adopted closely follows Burton (1956). The cells from the previous stage are suspended in 4 ml of $0·5$ N $HClO_4$ (use "Whirlimixer"), allowed to stand in a water-bath at $70°C$ for 15 min with occasional shaking, and centrifuged. The extraction is repeated twice with 3 ml volumes of $0·5$ N $HClO_4$, each for 15 min, and the combined extracts made up to 10 ml, if necessary, with $0·5$ N $HClO_4$. The extracts contain all the RNA and $> 98\%$ of the DNA (see below).

DNA is determined by the diphenylamine method of Burton (1956) described above, on 1–2 ml of extract, and RNA by the orcinol method

(Section V) on 0·5–1 ml quantities, according to the DNA and RNA content of the cells. As a check, the total nucleic acid in the extract may be determined by ultraviolet absorption, and should agree with the sum of the chemically determined RNA + DNA.

In agreement with Burton's (1956) observations, this treatment has given virtually complete extraction of DNA in all micro-organisms we have encountered, as illustrated by the following example.

Extraction of DNA from A. aerogenes. Freeze-dried *A. aerogenes* cells (20 mg) were extracted with cold 0·25 N HClO₄, and then extracted three times with 0·5 N HClO₄ at 70°C as described above, except that the extracts were kept separate. The three extracts, and also the cell residue suspended in 0·5 N HClO₄, were separately analysed for DNA by the method of Burton (1956), with the following results—

	DNA (μg)	% of total DNA	
1st extract	342	87·1	
2nd extract	46·2	11·8	= 99·63%
3rd extract	2·9	0·73	
Cell residue	1·5	0·36	
Total	392·6		

It is suggested that this type of experiment should be done when first applying the method to an unfamiliar microbial species.

In agreement with Burton (1956), we have found this extraction procedure causes negligible destruction of DNA; this is in any case compensated for by treating the DNA standard in an identical manner. This can be checked by "recovery" experiments in which known amounts of DNA are added to the cells before extraction.

The specificity of Burton's diphenylamine reagent is such that large amounts of RNA or of polysaccharides (e.g., in yeasts) do not interfere (Fig. 16).

RNA is also completely extracted by this method, > 95 % appearing in the first extract. In all *bacteria* we have examined, RNA could be satisfactorily determined with the orcinol reagent described above, with no significant interference from carbohydrates. This is understandable, since the RNA content of bacteria is seldom less than 7%, while the total carbohydrate is seldom greater than 20–25%; moreover, most of the carbohydrate is glucose, which reacts much less with the orcinol reagent than other hexoses. In *yeasts*, however, the RNA content may be as low as 3% and the total carbohydrate over 50%; a considerable proportion of this is mannose, which reacts more strongly with the orcinol reagent. In such cases, RNA determinations by the orcinol reagent are unreliable, even when the heating time is reduced to 10 min to give increased specificity; fortunately this is immediately obvious by the abnormal colour produced, and an absorption spectrum will often give a good idea of the interfering carbohydrate. This is illustrated by Fig. 15, which shows the absorption spectra given in the orcinol reaction by extracts of *A. aerogenes* and *T. utilis*; the former is identical with that of pure RNA, while the latter shows obvious interference by carbohydrates, mainly mannose (cf. Fig. 13). Figure 15 also shows that no interference from DNA occurs.

When carbohydrates interfere, RNA is determined in a duplicate sample of cells by the following method.

(e) *RNA determination in cells of high carbohydrate content.* Trevelyan and Harrison (1956) showed that the amount of carbohydrate extracted from yeast cells by 0·5 N $HClO_4$ increases markedly with temperature, and is very high at 70°C. Extraction of RNA by NaOH at 37°C (Schmidt–Thannhauser method) also extracts large amounts of carbohydrate. Treatment with 0·5 N $HClO_4$ at 37°C, however, extracts all the RNA in a reasonably short time while extracting very little carbohydrate. DNA is not completely extracted by this method and must be determined separately; the procedure is as follows.

Duplicate cell samples are extracted with cold 0·25 N $HClO_4$ for 30 min, as in stage (c) above. One sample is extracted with 0·5 N $HClO_4$ at 70°C, as in stage (d) above, this is used to determine DNA only. The second sample is extracted with 0·5 N $HClO_4$ at 37°C for 90–120 min, centrifuged, and washed once with 0·5 N $HClO_4$; the combined extracts are used for determination of RNA with the orcinol reagent. The time at 37°C required for complete extraction of RNA should be determined by trial in the first instance; Trevelyan and Harrison (1956) found 90 min sufficient for bakers' yeast, but we found 120 min necessary with *T. utilis*.

This method is probably applicable to carbohydrate-rich micro-organisms other than yeasts. An alternative might be to use the Ogur and Rosen (1950) method for the determination of RNA only, DNA being determined separately as above.

(f) *Nucleic acid determination in bacterial spores.* These present unique problems, and the detailed investigations of Fitz-James (1955) should be studied. With the spores of aerobic bacilli, the resistance of the spore coat is such that treatment with cold 10% TCA does not extract any "acid-soluble" material (and indeed leaves the spores viable). Subsequent treatment with hot (90°C) 5% TCA according to Schneider (1945) does extract RNA and DNA, but they are then accompanied by all the interfering substances normally removed in the "acid-soluble" fraction, as well as dipicolinic acid which interferes with measurements of ultraviolet absorption. Normal extraction procedures can be applied, however, if the spores are first subjected to mechanical disruption, and this appears to be a case where such treatment is unavoidable.

F. Quantitative determination of different nucleic acid fractions

In all bacteria that have been studied the DNA appears to be homogeneous, and it is currently supposed that bacteria contain a single DNA molecule (molecular weight about 5×10^9) in each cell. In more complex micro-organisms that contain mitochondria (e.g., yeasts, *Neurospora*, *Leishmania*, *Tetrahymena*), these have been shown to contain "mitochondrial DNA" which is different from the "nuclear DNA" (Granick and Gibor, 1967). So far, these can be separated only after isolating the mitochondria by differential centrifugation, a procedure which is at best only semi-quantitative.

All microbial RNAs consist of several components; excluding m-RNA which amounts to only a few percent of the whole, the main fractions are t-RNA and r-RNA; the former contains at least twenty and probably nearer forty components all of about the same molecular weight (4 S); the latter consists of at least three components of different molecular weight (5 S, 16 S and 23 S). Quantitative determination of the different components, or at least of their relative proportions, has been achieved by means of (*a*) the analytical ultracentrifuge, (*b*) ultracentrifugation in sucrose gradients, (*c*) chromatography on methylated serum albumin columns (Rosset *et al.*, 1966), and (*d*) polyacrylamide gel electrophoresis (Loening, 1967); the last appears to give the highest resolution and at present appears the most promising for development into a routine analytical procedure.

Before these methods can be applied, the difficult problem arises of obtaining from microbial cells an extract containing all the RNA components in 100% yield, in undegraded form, and free from contaminants such as DNA and protein. Phenol extraction (Kirby, 1964) is the method most commonly used, though it is doubtful if yields of total RNA are often greater than 90%. Moreover, the method is difficult to apply to intact cells; the cell wall appears to remain intact on phenol treatment and act as a molecular sieve, allowing extraction of the smaller s-RNA but retaining much of the r-RNA (e.g., Artmann *et al.*, 1966). A promising approach is that of Robinson and Wade (1968) who found that treatment of *E. coli* cells with 50% acetone makes the cell walls permeable to r-RNA, which can then be extracted by phenol; it remains to be seen whether this method is generally applicable.

VIII. TEICHOIC ACIDS

The characterization of teichoic acid polymers is discussed by Elizabeth Work (this Series, Volume 5a) and here we are concerned with the presence of these polymers in so far as they affect compositional data. They occur in bacteria and in general are composed of residues of glycerol phosphate or ribitol phosphate, D-alanine and a sugar or a 2-acetamido-2-deoxy sugar (Archibald and Baddiley, 1966). Since teichoic acids may account for up to 50% of the dry weight of the cell wall and the cell wall may account for 30% of the total cell dry weight, recognition of their presence is extremely important for the purpose of accounting for the gross composition of bacteria.

As yet, no simple reliable quantitative method for the determination of teichoic acids applicable to whole bacteria has been developed (Hay et al., 1963). Therefore, recognition of the presence of teichoic acids usually depends on analyses of extracts of organisms and quantitative extraction is difficult to achieve. Although quantitative data may be difficult to obtain, extraction and detection of teichoic acids is not difficult. A procedure used by Armstrong et al. (1958) with various bacteria involved treatment of the cells with acetone, ethanol and diethyl ether to remove fat followed by extraction of the fat-free cells with cold 10% trichloroacetic acid in a Waring blender for a short period. The acid extract was clarified by filtration with Supercel and treated with an equal volume of ethanol at 0°C to precipitate crude teichoic acids. A more extensive extraction procedure was necessary to remove teichoic acids from Bacillus subtilis cells (Armstrong et al., 1959b). Critchley et al., (1962) found that during prolonged extraction of fat-free cells of Lactobacillus arabinosus with cold trichloroacetic acid, the first compounds removed and precipitable with ethanol or acetone were ribitol teichoic acid and a rhamnose polysaccharide (both present in the cell walls) whereas polynucleotides and "intracellular" glycerol teichoic acids were extracted later.

When it has been established that a significant amount of material precipitable with ethanol is present in cold trichloroacetic acid extracts of defatted bacteria, identification of teichoic acids in the material depends on the isolation and characterization of the acid and alkaline degradation products of these phosphodiester polymers (Armstrong et al., 1958; 1959a; 1959b).

A suggestion that teichoic acids are present in a given organism may be obtained by determining the total RNA and phosphorus contents (Tempest et al., 1968). Tempest and his colleagues (Tempest et al., 1966; Dicks and Tempest, 1966) have shown that with A. aerogenes (an organism that apparently contains no teichoic acids), the total phosphorus and RNA increased with growth rate in a chemostat with various substances limiting growth but that the molar ratio RNA (as nucleotide) : phosphate remained nearly constant at 5 : 8 irrespective of the growth rate or the growth-limiting substrate. In contrast, this ratio determined for *B. subtilis* organisms growing in a chemostat under magnesium-limiting conditions at various growth rates when teichoic acids were synthesized by the cells was much higher at 5 : 13 (Ellwood and Tempest, 1967; Tempest *et al.*, 1968) although, again, the ratio was nearly independent of growth rate. However, when *B. subtilis* was grown with phosphorus-limitation, the molar ratio RNA (as nucleotide) : phosphate was the same as that for *A. aerogenes* and the significant finding was that the cell walls of these organisms contained very little phosphorus and teichoic acids were not detected in them (although teichuronic acid was present). If these findings are confirmed in other bacteria, RNA (nucleotide): phosphate molar ratios may provide an indication that the bacteria contain teichoic acids although the presence of polyphosphate would have to be excluded.

IX. POLYPHOSPHATE

Polyphosphate (syns. "metaphosphate", "polymetaphosphate") is polymerized inorganic metaphosphate, first observed in *Corynebacterium xerosis* and *Corynebacterium diphtheriae* and later in a variety of bacteria, yeasts, fungi and algae (Mann, 1944; Wiame, 1947, 1948; Juni *et al.*, 1948; Lindegren, 1947; Smith *et al.*, 1954; Wilkinson and Duguid, 1960; Harold, 1963; Harold and Sylvan, 1963; Mallette *et al.*, 1964). It is responsible for the "metachromatic" reaction of micro-organisms stained with Toluidine Blue by Laybourn's modification of Albert's method and is a major constituent of "volutin granules".

The polyphosphate content of bacterial cells is extremely variable and depends on species and on nutritional status; rapidly growing cells usually contain none but nutrient imbalance induces dramatic accumulation of polyphosphate provided that sources of metabolic energy, phosphorus and certain metal ions are available. The presence of this polymer in micro-organisms will be reflected in the amount and distribution of phosphorus-containing compounds and it is necessary to differentiate polyphosphate on this basis from RNA, DNA and teichoic acids.

The extraction of polyphosphate from microbial cells with hypochlorite solution is discussed by Wilkinson and Sutherland, this Series, Vol. 5, but trichloroacetic acid or perchloric acid can also be used. Wiame (1947, 1948) showed that polyphosphate in yeast cells occurs in two forms "soluble" and "insoluble". The soluble polymer is extracted with cold 10% (w/v) trichloroacetic acid (e.g., 250 mg dry weight of yeast extracted three times for 1 h with 4 ml acid at 0°–4°C); metaphosphate is precipitated from the extract by adjusting the pH value to 4–4·5 and adding barium or lead nitrate. To extract insoluble polyphosphate, the cell residue from the cold acid extraction is first defatted (extract with ethanol, 10 ml, then twice with ethanol–ether, 3 : 1, v/v, 10 ml, at 90°C for 3 min) and then extracted once with 10% (w/v) trichloroacetic acid (10 ml at room temperature) and twice with 5% trichloroacetic acid (10 ml at 95°C). The insoluble poly-phosphate in this extract is determined as orthophosphate plus ortho-phosphate liberated after hydrolysis with N hydrochloric acid at 100°C for 7 min ("7 min P") after correcting for the slight hydrolysis of nucleic acid which occurs under these conditions. Both soluble and insoluble forms of polymetaphosphate give the metachromatic reaction with Toluidine Blue.

Harold (1963) extracted polyphosphate from *A. aerogenes* cells with perchloric acid. Dry cells (100 mg) are extracted twice with 0·5 N perchloric acid (5 ml) for 5 min at 20°C with ethanol (5 ml) for 30 min at 20°C with boiling ethanol–ether (3 : 1, v/v; 5 ml), and finally twice with 0·5 N perchloric acid (5 ml) for 15 min at 70°C. The hot acid extract is treated with activated charcoal (50–100 mg Norit A to 6 ml extract) and poly-phosphate is determined as the remaining phosphorus-containing com-pounds that are completely acid-labile (N hydrochloric acid for 15 min at 95°C). Orthophosphate is determined by one or other of the spectrophoto-metric methods described above.

X. BACTERIAL "POOL" CONSTITUENTS

The "pool" constituents of micro-organisms embrace a very large number of soluble, relatively small molecular weight compounds concerned in anabolic and catabolic functions of the cell. For example, the amino-acid pool provides the amino-acids required for protein synthesis and in resting bacteria is apparently in equilibrium with the protein components since protein turnover has been shown to occur involving degradation and synthesis of proteins at similar rates (Mandelstam, 1958; 1960). Or the nucleic acid pool which contains purines, pyrimidines, nucleotides and nucleosides required for nucleic acid synthesis. Detailed analysis of the pool constituents is a complex operation and outside the scope of this Section; for example, thirty three different nucleotides were isolated from the pool constituents of *Penicillium* species and fully identified (Persson, 1963). Here we are mainly concerned with the contribution of pool consti-tuents to the gross composition of micro-organisms but references to methods for the detailed analysis of the individual constituents are given.

A. Extraction

It is generally assumed that provided care is taken to exclude the presence of degradation products from macromolecules, the pool constituents of bacteria and yeasts are those compounds released when the permeability barriers of the cell are destroyed. The extent to which the released compounds include substances that were attached to the outer surfaces of the cells is difficult to determine and usually the only precaution taken to exclude such substances is careful washing of the cells under conditions where endocellular constituents are not released. Macromolecular components themselves are excluded by treating the cells under conditions where such components remain insoluble.

Methods for extracting microbial pool constituents include treatment with cold trichloroacetic acid, or perchloric acid (reagents that precipitate proteins and nucleic acids), hot water and ethanol–water mixtures or butan-1-ol. Possibly the most widely used reagent is 0·6 N perchloric acid at 0°–2°C which has the advantage that the extract can be freed from most of the perchlorate ion by neutralization with potassium hydroxide in the cold when sparingly soluble potassium perchlorate precipitates. Trichloroacetic acid absorbs ultraviolet light significantly and if this reagent is used to extract pool constituents it must be removed by repeated extraction with diethyl ether before UV-measurements on the samples are made. Extraction of cells with boiling water is frequently used but a significant amount of material that is insoluble in cold perchloric or trichloroacetic acid is also extracted from the cells by such treatment.

That the apparent composition of the nucleotide pool of bacteria may vary according to the extraction method used was unequivocally demonstrated by Smith–Kielland (1964). Washed *E. coli* cells were extracted with 3 vol 0·6 N perchloric acid at 0°C for 30 min and the residue was re-extracted with 1 vol of acid; the extracts were combined, neutralized with 5 N potassium hydroxide in an icebath, freed from precipitate by centrifuging and concentrated by freeze-drying. Another sample of the cells was extracted with 3 vol of 50% aqueous ethanol for 6 min at 100°C and the residue re-extracted with 2 vol aqueous ethanol; the combined extracts were freeze-dried. Comparison of the acid and aqueous ethanol extracts showed that there was much more UV-absorbing material, more purines and less nucleotide 5′-phosphates in the latter than in the former. The results could be explained by the acid lability of nucleoside polyphosphates and the author suggests that perhaps pool nucleotides should be extracted with hot aqueous ethanol rather than with perchloric acid. In contrast, Hancock (1958) showed that similar amounts of amino-acids were released from *Staphylococcus aureus* by treatment with water at 100°C for 10 min, 0·2 N perchloric acid at 5°C for 5 min, 5% trichloroacetic acid at

5°C for 15 min and 25% ethanol at 5°C for 10 min. Hot water treatment has been used for extracting the free amino-acid pool from *Saccharomyces cerevisiae* (Halvorson and Spiegelmann, 1953), *Neurospora crassa* (Fuerst and Wagner, 1957) and *E. coli* (Mandelstam, 1958).

B. Identification

1. *Amino-acids*

The individual amino-acids present in the extracts of micro-organisms can be identified by paper, thin-layer or column chromatography or by microbiological methods. The presence of peptides is detected by comparing the number and amounts of amino-acids present before and after hydrolysis with 6 N hydrochloric acid for 16 h at 105°C in a sealed tube and removal of the acid *in vacuo*; or unidentified spots on the chromatograms are eluted, hydrolysed and re-chromatographed. Hot water extracts for chromatographic analysis are freed from cells by centrifugation and/or filtration through bacterial filters (sintered glass, Oxoid or Millipore membrane filters), desalted and concentrated. Removal of salts can be achieved by an electrolytic method but losses of certain amino-acids and the formation of ninhydrin-positive compounds may occur (Stein and Moore, 1951), and treatment with ion-exchange resins is perhaps to be preferred. Mandelstam (1958) passed aqueous extracts of *E. coli* through a column of Zeo-Karb 225 (H⁺ form), washed the column thoroughly with water and eluted the amino-acids with 1·5 N ammonia; the eluates were taken to dryness, the residue dissolved in a small volume of water and chromatographed on Whatman No. 3 paper. For details of conditions and solvents for separating amino-acids by paper or thin layer chromatography, standard works on the subject should be consulted (e.g., Block *et al.*, 1955; Mangold *et al.*, 1964; Randerath, 1963).

Qualitative and quantitative analysis of amino-acids in the pool of *Staphylococcus aureus* is described by Hancock (1958) who chromatographed concentrated extracts on columns of Dowex 50.

2. *Nucleotides, nucleosides and nucleic acid bases*

The identification of these compounds involves their separation usually by column chromatography followed by UV-spectrophotometry, phosphorus and pentose determinations, periodate oxidation studies, enzyme treatment, paper chromatography and paper or membrane electrophoresis of the separated constituents (Markham and Smith, 1952; Bergkvist, 1957, 1958; Hulbert *et al.*, 1954; Persson, 1963; Smith-Kielland, 1964). Neutralized and concentrated perchloric acid extracts can be directly fractionated on Dowex 1 (formate) columns by gradient elution (Hulbert *et al.*, 1954) or preliminary purification of nucleotides is achieved by absorbing them on to charcoal followed by elution with ethanolic ammonia (Persson, 1963).

C. Estimation of amino-acids

The total free amino-acid content of micro-organisms is determined by extraction with cold perchloric acid or hot water followed by analysis of extracts with the ninhydrin method; in our opinion, extraction with cold perchloric acid is preferable to extraction with hot water. The presence of peptides can be detected by analysis of extracts before and after hydrolysis with 6 N hydrochloric acid for 16 h at 105°C in a sealed tube. Ammonia and other amines give colours with ninhydrin and will be estimated as amino-acids but the amounts of these substances in extracts are usually small.

1. *Detailed example*

Determination of the free amino-acids in bacteria by the method of Yemm and Cocking (1955).

(a) *Reagents*. The reagents are as follows—

  (i) 0·2 M citrate buffer (pH 5·0); 21·008 g citric acid monohydrate is dissolved in 400 ml 0·5 N sodium hydroxide and the solution diluted to 500 ml; a crystal of thymol is added and the buffer stored at 2°C.

  (ii) Methyl cellosolve (2-methoxyethanol): Laboratory Reagent grade material is distilled (b.p. 122°–124°C) and about 50% of the starting material is collected, the remainder being discarded to avoid risk of explosion due to peroxides; the distillate must give a clear solution when mixed with an equal volume of water and a faint or negative reaction with 10% (w/v) potassium iodide solution.

  (iii) 0·01 M potassium cyanide: 0·1628 g of potassium cyanide is dissolved in 250 ml water; the solution keeps 3 months at 20°C.

  (iv) KCN–methyl cellosolve: dilute 5 ml of 0·01 M potassium cyanide solution to 250 ml with methyl cellosolve; the solution keeps 1 month at 20°C.

  (v) Ninhydrin–methyl cellosolve: ninhydrin must be pale yellow and odourless and the preparation should be recrystallized if necessary; dissolve 5 g of ninhydrin and dilute to 100 ml with redistilled methyl cellosolve; the *pale yellow* solution will remain stable for 6 months at 20°C.

  (vi) Aqueous ethanol: 60% (v/v) ethanol in distilled water.

(b) *Procedure.* Volumes (equivalent to 5–12 mg bacterial dry wt) of a washed bacterial suspension are centrifuged and the pellets re-suspended in 0·25 N perchloric acid (2·5 ml) at 0°C for 30 min. The suspensions are centrifuged in the cold and the supernatant fluids decanted into 5 ml graduated tubes. The residues are re-extracted with 0·25 N perchloric acid (1·0 ml), the suspensions centrifuged and the supernatant fluids added to the first extracts. The combined extracts are neutralized with N sodium hydroxide to near pH 5·0 and diluted with water to 5 ml. Neutralized extract (0·5–1·0 ml) is mixed in a thin-walled test-tube (6 in. × $\frac{5}{8}$ in.) with citrate buffer (0·5 ml), ninhydrin reagent (0·2 ml) and KCN–methyl cellosolve (1 ml). The tubes are closed with glass marbles, heated in a vigorously boiling water bath for 15 min and cooled in cold water for 5 min. The solutions are diluted with aqueous ethanol (3 ml or any suitable volume) and well mixed. Absorbancies are measured at 570 nm in a spectrophotometer against a blank of distilled water treated as for the tests. The absorbance values of the blanks read against distilled water at 570 nm should be low and consistent (for example, not greater than 0·10 with a 1 cm light path). Erratic blank values probably indicate dirty glassware, and consistently high blanks that the reagents are unsatisfactory. Recrystallized alanine is a suitable standard used in amounts of 2–20 μg.

(c) *Results.* As mentioned above, the majority of the amino-acids give theoretical yields of coloured derivative and to convert the test results in terms of alanine into α-amino nitrogen concentration, the values are multiplied by 14/89; a linear relationship exists between the amount of alanine (0–20 μg) and extinction at 570 nm. The concentration of free amino-acids in bacteria varies with the organism and its physiological state but in general Gram-positive organisms contain more than Gram-negative organisms. For example, the free amino-acids expressed as alanine in *A. aerogenes* and *Staphylococcus epidermidis* account for about 1 and 4%, respectively, of the dry weight of the organisms.

D. Estimation of the "nucleic acid pool" constituents

An approximate estimate of the nucleic acid pool constituents is possible by measuring spectrophotometrically the UV light-absorption of the cell extract (pH about 1·0) against a blank of the extracting fluid in matched quartz cells. To establish that the absorption is due mainly to nucleic acid derivatives, the complete spectrum of the extract in the range 230–290 nm is measured; a sharp peak with a maximum near 260 nm should be evident. In order to convert the peak absorption value of the extract to concentration of UV-absorbing compounds, a mean molar extinction coefficient (ϵ) is used; this coefficient is derived from the ϵ values for the major nucleotides present in nucleic acids which are determined by measuring the UV-absorption of molar solutions of these compounds with a 1 cm light path. An approximate mean value of ϵ is 10,800 and the average nucleotide molecular weight, 340. Thus, if the UV-absorption of the cell extract with a 1 cm light path at the peak wavelength is x, then $(340/10,800) \times x$ gives the weight (mg) of UV-absorbing compounds/ml of extract. By this method, the nucleic acid pool constituents of A. *aerogenes* account for about 1% of the bacterial dry weight.

ACKNOWLEDGMENTS

The writers are grateful to their colleagues Dr. D. C. Ellwood, Dr. A. P. MacLennan, Dr. D. W. Tempest and Dr. H. E. Wade for helpful advice and discussion.

REFERENCES

Abul-Fadl, M. A. M. (1949). *Biochem. J.*, **44**, 282–285.
Albaum, H. G., and Umbreit, W. W. (1947). *J. biol. Chem.*, **167**, 369–376.
Allen, R. J. L. (1940). *Biochem. J.*, **34**, 858–865.
Amaral, G., Bernstein, L., Moore, D., and Horecker, B. L. (1962). *J. biol. Chem.*, **238**, 2281–2284.
Anderson, L. (1953). *Acta Chem. Scand.*, **7**, 689–692.
Archibald, A. R., and Baddiley, J. (1966). *Adv. carbohydrate Chem.*, **21**, 323–375.
Armstrong, J. J., Baddiley, J., Buchanan, J. G., Cass, B., and Greenberg, G. R. (1958). *J. chem. Soc.*, 4344–4354.
Armstrong, J. J., Baddiley, J., Buchanan, J. G., Davidson, A. L., Kelemen, M. V., and Neuhaus, F. C. (1959a). *Nature.*, **184**, 247–248.
Armstrong, J. G., Baddiley, J., and Buchanan, J. G. (1959b). *Nature.*, **184**, 248–249.
Artmann, M., Fry, M., and Engelberg, H. (1966). *Biochem. biophys. res. Comm.*, **25**, 49–53.
Asselineau, J., and Lederer, E. (1960). In "Lipide Metabolism" (Ed. K. Bloch,), pp. 337–406. John Wiley & Sons, New York.
Asselineau, J. (1962). "Les Lipides Bacteriens". Hermann, Paris.
Avigad, G., Amaral, D., Asensio, C., and Horecker, B. L. (1962). *J. biol. Chem.*, **237**, 2736–2743.
Barrenscheen, H. K., and Peham, A. (1941). *Z. physiol. Chem.*, **272**, 81–86.
Becka, J. (1931). *Biochem. Z.*, **233**, 118–128.
Beet, A. E. (1955). *Nature.*, **175**, 513–514.
Beljanski, M., and Macheboeuf, M. (1949). *Compt. rend. Soc. Biol., Paris.*, **143**, 174–182.
Bell, R. D., and Doisy, E. A. (1920). *J. biol. Chem.*, **44**, 55–67.
Bell, J. C., and Robinson, C. G. (1934). *J. chem. Soc.*, 813–819.
Belozersky, A. N., and Spirin, A. S. (1960). In "The Nucleic Acids" (Eds. E. Chargaff and J. N. Davidson,), Vol. 3, pp. 147–185. Academic Press, New York.
Benedict, S. R. (1909). *J. biol. Chem.*, **6**, 363.
Bergkvist, R. (1957). *Acta Chem. Scand.*, **11**, 1465–1472.

Bergkvist, R. (1958). *Acta Chem. Scand.*, **12**, 752–755.
Berglund, F., and Sørbo, B. (1960). *Scand. J. clin. lab. Invest.*, **12**, 147–153.
Bernard, Claude. (1887). "Leçons sur la diabète." Paris.
Bertolacini, R. J., and Barney, J. E. (1957). *Anal. Chem.*, **29**, 281–283.
Beaven, G. H., and Holiday, E. R. (1952). *Adv. protein Chem.*, **7**, 319–386.
Beaven, G. H., Holiday, E. R., and Johnson, E. A. (1955). In "The Nucleic Acids" (Eds. E. Chargaff and J. N. Davidson). Vol. 1, pp. 493–553. Academic Press, New York.
Bitter, T., and Muir, H. M. (1962). *Anal. Biochem.*, **4**, 330–334.
Blanchar, R. W., Rehm, G., and Caldwell, A. C. (1965). *Soil Sci. Soc. Am. Proc.*, **29**, 71.
Blau, F. (1898). *Monatsh.*, **19**, 647–648.
Block, R. J., Durrum, E. L., and Zweig, G. (1955). "A Manual of Paper Chromatography and Paper Electrophoresis". Academic Press, London.
Bloor, W. R. (1928). *J. biol. Chem.*, **77**, 53–73.
Bodansky, A. (1932). *J. biol. Chem.*, **99**, 197–206.
Bolognani, L., Coppi, G., and Zambotti, V. (1961). *Experientia*, **17**, 67–68.
Borgstrom, B. (1952). *Acta Physiol. Scand.*, **25**, 101–110.
Briggs, A. P. (1922). *J. biol. Chem.*, **53**, 13–16.
Briggs, A. P. (1924). *J. biol. Chem.*, **59**, 255–264.
Brown, A. H. (1946). *Arch. Biochem.*, **11**, 269–278.
Brüel, D., Holter, H., Linderstrøm-Lang, K., and Rozits, K. (1947). *Biochim. Biophys. Acta.*, **1**, 101–125.
Buckley, E. S., Gibson, J. G., and Bartolotti, T. R. (1951). *J. lab. clin. Med.*, **38**, 751–761.
Burton, K. (1956). *Biochem. J.*, **62**, 315–322.
Caldwell, P. C., and Hinshelwood, C. (1950). *J. chem. Soc.*, 1415–1418.
Ceriotti, G. (1952). *J. biol. Chem.*, **198**, 297–303.
Ceriotti, G. (1955). *J. biol. Chem.*, **214**, 59–70.
Chargaff, E., and Davidson, J. N. (1955, 1960). "The Nucleic Acids", Vols 1–3. Academic Press, New York.
Chen, P. S., Toribara, T. Y., and Warner, H. (1956). *Anal. Chem.*, **28**, 1756–1758.
Chibnall, A. C., Rees, N. W., and Williams, E. F. (1943). *Biochem. J.*, **37**, 354–359.
Cohen, S. S. (1944). *J. biol. Chem.*, **156**, 691–701.
Conway, E. J. (1947). "Micro Diffusion Analysis and Volumetric Error". Crosby Lockwood & Son, London.
Critchley, P., Archibald, A. R., and Baddiley, J. (1962). *Biochem. J.*, **85**, 420–431.
Croft, D. N., and Lubran, M. (1965). *Biochem. J.*, **95**, 612–701.
Cullum, D. C., and Thomas, D. B. (1959). *Analyst.*, **84**, 113–116.
Dawes, E. A., and Ribbons, D. W. (1965). *Biochem. J.*, **95**, 332–343.
Davidson, E. A. (1966). In "Methods in Enzymology" (Eds. E. F. Neufeld and V. Ginsburg). Vol. 8, pp. 52–60.
Davidson, J. N., and Waymouth, C. (1944). *Biochem. J.*, **38**, 39–49.
Deichmann, W. (1943). *J. lab. clin. Med.*, **28**, 770–775.
DeLey, J., and Doudoroff, M. (1957). *J. biol. Chem.*, **227**, 745–757.
Delory, G. E. (1938). *Biochem. J.*, **32**, 1161–1162.
Denis, W. (1922). *J. biol. Chem.*, **52**, 411–415.
Deriaz, R. E., Stacey, M., Teece, E. G., and Wiggins, L. F. (1949). *J. Chem. Soc.*, 1222–1232.

14–5*b*

338   D. HERBERT, P. J. PHIPPS AND R. E. STRANGE

Dicks, J. W., and Tempest, D. W. (1966). *J. gen. Microbiol.*, **45**, 547–557.
Dische, Z. (1927). *Biochem. Z.*, **189**, 77–80.
Dische, Z. (1930). *Mikrochemie*, **8**, 4–32.
Dische, Z. (1944). *Proc. Soc. exp. Biol. Med.*, **55**, 217–218.
Dische, Z. (1947a). *J. biol. Chem.*, **167**, 189–198.
Dische, Z. (1947b). *J. biol. Chem.*, **171**, 725–730.
Dische, Z. (1948). *Arch. Biochem.*, **16**, 409–414.
Dische, Z. (1951). In "Phosphorus Metabolism" (Eds. W. D. McElroy and B. Glass), Vol. 1, pp. 171–199. Johns Hopkins Press, Baltimore.
Dische, Z. (1953). *J. biol. Chem.*, **204**, 983–997.
Dische, Z. (1955a). *Meth. biochem. Anal.*, **2**, 313–358.
Dische, Z. (1955b). In "The Nucleic Acids" (Eds. E. Chargaff and J. N. Davidson), Vol. 1, pp. 285–305. Academic Press, New York.
Dische, Z. (1962). In "Methods in Carbohydrate Chemistry" (Eds. R. L. Whistler and M. L. Wolfrom), Vol. 1, pp. 477–517, Academic Press, New York.
Dische, Z., and Borenfreund, E. (1950). *J. biol. Chem.*, **184**, 517–522.
Dische, Z., and Borenfreund, E. (1951). *J. biol. Chem.*, **192**, 583–587.
Dische, Z., and Borenfreund, E. (1957). *Biochim. biophys. Acta.*, **23**, 639–642.
Dische, Z., and Devi, A. (1960). *Biochim. biophys. Acta.*, **39**, 140–144.
Dische, Z., and Schwartz, K. (1937). *Mikrochem. Acta.*, **2**, 13–19.
Dische, Z., and Shettles, L. B. (1948). *J. biol. Chem.*, **175**, 595–603.
Dische, Z., and Shettles, L. B. (1951). *J. biol. Chem.*, **192**, 579–582.
Dodgson, K. S. (1961). *Biochem. J.*, **78**, 312–319.
Dodgson, K. S., Lloyd, A. G., and Spencer, B. (1957). *Biochem. J.*, **65**, 131–138.
Dodgson, K. S., and Spencer, B. (1953). *Biochem. J.*, **55**, 436–440.
Doudoroff, M., and Stanier, R. Y. (1959). *Nature*, **183**, 1440–1442.
Dreywood, R. (1946). *Ind. Eng. Chem. (Anal. Ed.)*, **18**, 499–505.
Drury, H. F. (1948). *Arch. Biochem.*, **19**, 455–466.
Dubois, M., Gilles, K. A., Hamilton, J. K., Rebers, P. A., and Smith, F. (1956). *Anal. Chem.*, **28**, 350–356.
Eldjarn, L., Nygaard, O., and Sviensson, S. L. (1955). *Scand. J. Lab. Invest.*, **7**, 92–94.
Ellwood, D. C., and Tempest, D. W. (1967). *Biochem. J.*, **104**, 69P.
Elson, D., and Chargaff, E. (1955). *Biochim. biophys. Acta.*, **17**, 367–376.
Elson, L. A., and Morgan, W. T. J. (1933). *Biochem. J.*, **27**, 1824–1828.
Epstein, W., and Schultz, S. G. (1965). *J. gen. Physiol.*, **49**, 221–234.
Euler, H. v., and Hahn, L. (1946). *Svensk. Kem. Tidskr.*, **58**, 251–264.
Fales, F. W. (1951). *J. biol. Chem.*, **193**, 113–124.
Farrow, R. P. N., and Hill, A. G. (1961). *Talanta*, **8**, 116–128.
Fawcett, J. K. (1954). *J. Inst. med. lab. Technol.*, **12**, 1–22.
Felsenfeld, G., and Miles, T. (1967). *Ann. Rev. Biochem.*, **36**, 407–448.
Fiske, C. H., and SubbaRow, Y. (1925). *J. biol. Chem.*, **66**, 375–400.
Fitz-James, P. C. (1955). *Canad. J. Microbiol.*, **1**, 502–519.
Fleck, A., and Begg, D. (1954). *Biochim. biophys. Acta.*, **108**, 333–339.
Folch, J., Ascoli, I., Lees, M., Meath, J. A., and LeBaron, F. N. (1951). *J. biol. Chem.*, **191**, 833–841.
Folch, J., Lees, M., and Sloane-Stanley, G. H. (1957). *J. biol. Chem.*, **226**, 497–509.
Folin, O. (1922). *J. biol. Chem.*, **51**, 377–391.
Folin, O., and Ciocalteu, V. (1927). *J. biol. Chem.*, **73**, 627–635.

Forsyth, W. G. C., Hayward, A. C., and Roberts, J. B. (1958). *Nature*, **182**, 800–801.
Frame, E. G., Russell, J. A., and Wilhelmi, A. E. (1943). *J. biol. Chem.*, **149**, 255–270.
Friedman, H. S., and Radin, M. A. (1955). *Clin. Chem.*, **1**, 125–133.
Friedrich, A. (1933). *Z. physiol Chem.*, **216**, 68–89.
Fuerst, R., and Wagner, R. P. (1957). *Arch. Biochem. Biophys*, **70**, 311–326.
Fujita, A., and Iwatake, D. (1931). *Biochem. Z.*, **242**, 43–51.
Gardell, S. (1958). *Meth. biochem. Anal.*, **6**, 289–317.
Gardell, S. (1965). *In* "Methods in Carbohydrate Chemistry" (Eds. R. L. Whistler and M. L. Wolfrom), Vol. 5, pp. 9–14. Academic Press, New York.
Garner, R. J. (1946). *Biochem. J.*, **40**, 828–831.
Gibbons, M. N. (1955). *Analyst.*, **80**, 268–273.
Giles, K. W., and Myers, A. (1965). *Nature.*, **206**, 93.
Good, C. A., Kramer, H., and Somogyi, M. (1933). *J. biol. Chem.* **100**, 485–491.
Gornall, A. G., Bardawill, C. J., and David, M. M. (1949). *J. biol. Chem.*, **177**, 751–766.
Granick, S., and Gibor, A. (1967). *In* "Progress in Nucleic Acid Research" (Eds. J. N. Davidson and W. E. Cohn), Vol. 6, pp. 143–186. Academic Press, New York.
Greenstein, J. P., and Winitz, M. (1961). "Chemistry of the Amino Acids". John Wiley & Sons, Inc., New York.
Hagedorn, J. C., and Jensen, B. N. (1923). *Biochem. Z.*, **38**, 274–280.
Halvorson, H. O., and Spiegelman, S. (1953). *J. Bact.*, **65**, 601–608.
Hanahan, D. J. (1960). "Lipid Chemistry", John Wiley and Sons, New York.
Hanahan, D. J., Dittmer, J. C., and Warashina, E. (1957). *J. biol. chem.*, **228**, 685–700.
Hancock, R. (1958). *Biochim. biophys. Acta.*, **28**, 402–412.
Hanson, S. W. F., Mills, G. T., and Williams, R. T. (1944). *Biochem. J.*, **38**, 274–279.
Harrison, G. A. (1943). "Chemical Methods in Clinical Medicine". J. & A. Churchill, London.
Harold, F. M. (1963). *J. Bact.*, **86**, 216–221.
Harold, F. M., and Sylvan, S. (1963). *J. Bact.*, **86**, 222–231.
Harvey, A. E., Komarmy, J. M., and Wyatt, G. M. (1953). *Anal. Chem.*, **25**, 498–500.
Hawke, P. B., Oser, B. L., and Summerson, W. H. (1947). "Practical Physiological Chemistry". J. & A. Churchill, London.
Hay, J. B., Wicken, A. J., and Baddiley, J. (1963). *Biochim. biophys. Acta.*, **71**, 188–190.
Herbert, D., and Phipps, P. J. (1971). In preparation.
Herriott, R. M. (1941). *Proc. Soc. exp. Biol. Med.*, **46**, 642–644.
Hibbard, P. L., (1919). *Soil. Sci.*, **8**, 61–65.
Hill, R. (1930). *Proc. Roy. Soc. B.*, **107**, 205.
Hirschfelder, A. D., and Serles, E. R. (1934). *J. biol. Chem.*, **104**, 635–645.
Hodge, J. E., and Hofreiter, B. T. (1962). *In* "Methods in Carbohydrate Chemistry" (Eds. R. L. Whistler and M. L. Wolfrom), Vol. 1, pp. 380–394. Academic Press, New York.
Horecker, B. L., Smyrniotis, P. Z., and Seegmiller, J. E. (1951). *J. biol. Chem.*, **193**, 383–396.
Hough, L. (1954). *Meth. biochem. Anal.*, **1**, 205–242.

Hough, L., and Jones, J. K. N. (1961). *In* "Methods in Carbohydrate Chemistry" (Eds. R. L. Whistler and M. L. Wolfrom,), Vol. 1, pp. 31–33. Academic Press, New York.

Hugget, A. St-G., and Nixon, D. A. (1957). *Biochem. J.*, **66**, 12P.

Hunter, G. D., and James, A. T. (1963). *Nature*, **198**, 789.

Hurlbert, R. B., Schmitz, H., Brumm, A. F., and Potter, V. R. (1954). *J. biol. Chem.*, **209**, 23–39.

Huston, C. K., Albro, P. W., and Grindley, G. B. (1965). *J. Bact.*, **89**, 768–775.

Hutchison, W. C., and Munro, H. N. (1961). *Analyst.*, **86**, 768–813.

Ikawa, M. (1963). *J. Bact.*, **85**, 772–781.

Immers, J., and Vasseur, E. (1950). *Nature.*, **165**, 898–899.

Ingles, O. G., and Israel, G. C. (1948). *J. chem. Soc.*, 810–814.

Ingles, O. G., and Israel, G. C. (1949). *J. chem. Soc.*, 1213–1216.

Jacobs, H. R. D., and Hoffman, W. S. (1931). *J. biol. Chem.*, **93**, 685–691.

Jacobs, S. (1965). *Meth. biochem. Anal.*, **13**, 241–263.

Jankowski, S. J., and Freiser, H. (1961). *Anal. Chem.*, **33**, 776–787.

Juni, E., Kamen, M. D., Reiner, J. M., and Spiegelman, S. (1948). *Arch. Biochem.*, **18**, 387–408.

Kasai, N., and Yamano, A. (1964). *Japan J. exp. Med.*, **34**, 329–344.

Kent, L. H., and Strange, R. E. (1962). *In* "Methods in Carbohydrate Chemistry" (Eds. R. L. Whistler and M. L. Wolfrom), Vol. 1, pp. 13–21. Academic Press, New York.

Kerr, S. E., Seraidarian, K., and Wargon, M. (1949). *J. biol. Chem.*, **181**, 761–771.

King, E. J. (1932). *Biochem. J.*, **26**, 292–297.

King, E. J. (1951). "Micro Analysis in Medical Biochemistry". J. & A. Churchill, London.

King, E. J., Haslewood, G. A. D., Delory, G. E., and Beal, D. (1942). *Lancet* i, 207–208.

Kirby, K. S. (1964). *In* "Progress in Nucleic Acid Research" (Eds. J. N. Davidson and W. E. Cohn), Vol. 3, pp. 1–31. Academic Press, New York.

Kircher, H. W. (1961). *In* "Methods in Carbohydrate Chemistry" (Eds R. L. Whistler and M. L. Wolfrom), Vol. 1, pp. 13–21. Academic Press, New York.

Koehler, H. L. (1952). *Anal. Chem.*, **24**, 1576–1579.

Koenig, R. A., and Johnson, C. R. (1942). *J. biol. Chem.*, **143**, 159–163.

Kolb, J. J., Weidner, M. A., and Toennies, G. (1963). *Anal. Biochem.*, **5**, 78–82.

Kolthoff, I. M. (1927). *Biochem. Z.*, **185**, 344–348.

Kolthoff, I. M., and Stenger, V. A. (1947). "Volumetric Analysis', Vol. 2, p. 164. Interscience Publishers, New York.

Kramer, B., and Tisdall, F. J. (1921). *J. biol. Chem.*, **48**, 223–232.

Kuttner, T., and Cohen, H. R. (1927). *J. biol. Chem.*, **75**, 517–531.

Law, J. H., and Slepecky, R. A. (1961). *J. Bact.*, **82**, 33–36.

Lea, C. H., Rhodes, D. N., and Stoll, R. D. (1955). *Biochem. J.*, **60**, 353–363.

Lee, K. W., Wahl, R., and Barbu, E. (1956). *Ann. Inst. Pasteur.*, **91**, 212–221.

Lee, Y. C., and Montgomery, R. (1961). *Arch. Biochem. Biophys.*, **93**, 292–296.

Lemieux, R. U. (1961). *In* "Methods in Carbohydrate Chemistry" (Eds. R. L. Whistler and M. L. Wolfrom), Vol. 1, pp. 45–47. Academic Press, New York.

Lemoigne, M. (1925). *Ann. Inst. Pasteur.*, **39**, 144.

Lemoigne, M. (1927). *Ann. Inst. Pasteur.*, **41**, 148.

Lemoigne, M., Delaparte, B., and Croson, M. (1944). *Ann. Inst. Pasteur.*, **70**, 224–233.

Lin, F. M., and Pomeranz, Y. (1968). *Anal. Biochem.*, **24**, 128–131.

Lindegren, C. C. (1947). *Nature.*, **159**, 63–64.

Llaurado, G. J. (1954). *J. clin. Path.*, **7**, 110.

Lloyd, A. G. (1959). *Biochem. J.*, **72**, 133–136.

Loening, U. E. (1967). *Biochem. J.*, **102**, 251–257.

Logan, J. E., Mannell, W. A., and Rossiter, R. J. (1952). *Biochem. J.*, **51**, 470–482.

Lovern, J. A. (1957). *In* "Handbuch der Pflanzenphysiologie" (Ed. W. Ruhland), Vol. 7, pp. 376–392. Springer-Verlag, Berlin.

Lowry, O. H., and Lopez, J. A. (1946). *J. biol. Chem.*, **162**, 421–428.

Lowry, O. H., Rosebrough, N. J., Farr, A. L., and Randall, R. J. (1951). *J. biol. Chem.*, **193**, 265–275.

Lusena, C. V. (1951). *Canad. J. Chem.*, **29**, 107–108.

Ma, T. S., and Zuazaga, G. (1942). *Ind. eng. Chem. (Anal. Ed.)*, **14**, 280–282.

MacFadyen, D. A. (1950). *J. biol. Chem.*, **186**, 1–22.

MacFadyen, D. A., and Fowler, N. (1950). *J. biol. Chem.*, **186**, 13–22.

Macfarlane, M. G. (1962a). *Nature*, **196**, 136–138.

Macfarlane, M. G. (1962b). *Biochem. J.*, **82**, 40P, 41P.

McIndoe, W. M., and Davidson, J. N. (1952). *Brit. J. Cancer.*, **6**, 200–214.

MacLeod, M., and Robison, R. (1929). *Biochem. J.*, **23**, 517–523.

McRary, W. L., and Slattery, M. C. (1945). *Arch. Biochem.*, **6**, 151–156

Mager, J., Kucaynski, M., Schatzberg, G., and Avi-Dor, Y. (1956). *J. gen. Microbiol.*, **14**, 69–75.

Mallettc, M. F., Cowan, C. I., and Campbell, J. J. R. (1964). *J. Bact.*, **87**, 779–785.

Mandelstam, J. (1958). *Biochem. J.*, **69**, 103–110.

Mandelstam, J. (1960). *Bact. Rev.*, **24**, 289–308.

Mangold, H. K., Schmid, H. H. O., and Stahl, E. (1964). *Meth. biochem. Anal.*, **12**, 393–451.

Mann, T. (1944). *Biochem. J.*, **36**, 790–791.

Mann, F., and Tollens, B. (1896). *Ann.* **290**, 155–163.

Markham, R. (1942). *Biochem. J.*, **36**, 790–791.

Markham, R. (1955). *In* "Modern Methods of Plant Analysis" (Eds. K. Paech and M. V. Tracey), Vol. 4, 291. Springer-Verlag, Berlin.

Markham, R., and Smith, J. D. (1952). *Biochem. J.*, **33**, 552–557.

Massart, L., and Hoste, J. (1947). *Biochim. biophys. Acta.*, **1**, 83–86.

Masters, M. (1939). *Biochem. J.*, **33**, 1313–1324.

Mehl, J. W. (1945). *J. biol. Chem.*, **157**, 173–180.

Mehl, J. W., Pacovsk, E., and Winzlcr, R. J. (1949). *J. biol. Chem.*, **177**, 13–21.

Mejbaum, W. (1939). *Z. physiol. Chemie.*, **258**, 117–120.

Mickle, H. (1948). *J. R. micr. Soc.*, **68**, 10–12.

Militzer, W. E. (1946). *Arch. Biochem.*, **9**, 85–90.

Mitchell, P., and Moyle, J. (1956). *In* "Bacterial Anatomy". *Symp. Soc. Gen. Microbiol.*, **6**, 150–180.

Miller, G. L., and Burton, A. L. (1959). *Anal. Chem.*, **31**, 1790–1793.

Moore, S., and Stein, W. H. (1948). *J. biol. Chem.*, **176**, 367–388.

Moore, S., and Stein, W. H. (1954). *J. biol. Chem.*, **211**, 907–913.

Montgomery, R. (1957). *Arch. Biochem. Biophys.*, **67**, 378–386.

Morris, D. L. (1948). *Science*, **107**, 254–255.

Morse, M. L., and Carter, C. E. (1949). *J. Bact.*, **58**, 317–326.

Munro, H. N., and Fleck, A. (1966). *Meth. biochem. Anal.*, **14**, 113–176.

Munro, H. N., Hutchison, W. C., Ramaiah, T. R., and Nielson, F. J. (1962). *Brit. J. Nutr.*, **16**, 387–395.

Murrell, W. G., and Warth, A. D. (1965). *In* "Spores III", pp. 1–24. American Society for Microbiology, Ann Arbor, Michigan.

Nelson, N. (1944). *J. biol. Chem.*, **153**, 375–380.

Neukom, H., and Kuendig, W. (1965). *In* "Methods in Carbohydrate Chemistry" (Eds. R. L. Whistler and M. L. Wolfrom), Vol. 5, pp. 14–17. Academic Press, New York.

Norris, K. P., and Powell, E. O. (1961). *J. R. micr. Soc.*, **80**, 107–119.

Northcote, D. H. (1953). *Biochem. J.*, **53**, 348–352.

Northcote, D. H. (1954). *Biochem. J.*, **58**, 353–358.

Northcote, D. H. (1965). *In* "Methods in Carbohydrate Chemistry" (Eds. R. L. Whistler and M. L. Wolfrom), Vol. 5, pp. 49–53. Academic Press, New York.

Northcote, D. H., and Horne, R. W. (1952). *Biochem. J.*, **51**, 232–236.

Noyons, E. C. (1939). *Pharm. Weekbl.*, **76**, 307–311.

Ogur, M., and Rosen, G. (1950). *Arch. Biochem.*, **25**, 262–276.

Osborne, M. J. (1963). *Proc. Nat. Acad. Sci., N.Y.*, **50**, 499–506.

Palmstierna, H. (1956). *Acta Chem. Scand.*, **10**, 567–577.

Persson, K. O. U. (1963). *Acta Chem. Scand.*, **17**, 2750–2762.

Peters, J. P., and Van Slyke, D. D. (1932). "Quantitative Clinical Chemistry", Vol. 2. Williams and Wilkins, Baltimore.

Picou, D., and Waterlow, J. C. (1963). *Nature.*, **197**, 1103–1104.

Pirie, N. W. (1936). *Brit. J. exp. Path.*, **17**, 269–278.

Pope, C. G., and Stevens, M. F. (1939). *Biochem. J.*, **33**, 1070–1077.

Potter, V. R., and Elvehjem, C. A. (1938). *J. biol. Chem.*, **114**, 495–504.

Powell, F. J. N. (1953). *J. clin. Pathol.*, **6**, 286.

Powell, J. F., and Strange, R. E. (1956). *Biochem. J.*, **63**, 661–668.

Ramsay, W. N. M. (1953). *Biochem. J.*, **53**, 227–231.

Randerath, K. (1963). "Thin Layer Chromatography". Academic Press, London.

Robinson, H. K., and Wade, H. E. (1968). *Biochem. J.*, **106**, 897–903.

Robinson, H. W., and Hogden, C. G. (1940). *J. biol. Chem.*, **135**, 707–725.

Roe, D. A., Miller, P. S., and Lutwak, L. (1966). *Anal. Biochem.*, **15**, 313–322.

Roe, J. H., Epstein, J. H., and Goldstein, N. P. (1949). *J. biol. Chem.*, **178**, 839–845.

Rondle, C. J. M., and Morgan, W. T. J. (1955). *Biochem. J.*, **61**, 586–589.

Rosen, H. (1957). *Arch. Biochem. Biophys.*, **67**, 10–15.

Rosset, R., Julien, J., and Monier, R. (1966). *J. mol. Biol.*, **18**, 308–320.

Russell, J. (1944). *J. biol. Chem.*, **156**, 467–468.

Saidel, L. J. (1957). *J. biol. Chem.*, **224**, 445–451.

Salton, M. R. J. (1953). *Biochim. biophys. Acta.*, **10**, 512–523.

Salton, M. R. J. (1960). "Microbial Cell Walls". John Wiley & Sons, New York.

Salton, M. R. J. (1964). "The Bacterial Cell Wall". Elsevier, Amsterdam.

Sattler, L., and Zerban, F. W. (1948). *Science*, **108**, 207.

Sattler, L., and Zerban, F. W. (1950). *J. Am. chem. Soc.*, **72**, 3184–3188.

Schlenk, F., and Waldvogel, M. J. (1947). *Arch. Biochem.*, **12**, 181–190.

Schmidt, G., and Thannhauser, S. J. (1945). *J. biol. Chem.*, **161**, 83–89.

Schneider, W. C. (1945). *J. biol. Chem.*, **161**, 293–303.

Schneider, W. C. (1946). *J. biol. Chem.*, **164**, 747–751.

Schneider, W. C., Hogeboom, G. N., and Ross, H. E. (1950). *J. nat. cancer Inst.*, **10**, 977–983.

Schöniger, W. (1956). *Mikrochim. Acta.*, 869–876.

Schroeder, W. A., Kay, L. M., and Mills, R. S. (1950). *Anal. Chem.*, **22**, 760–763
Scott, T. A., and Melvin, E. H. (1953). *Anal. Chem.*, **25**, 1656–1661.
Seibert, F. B., and Atno, J. (1946). *J. biol. Chem.*, **163**, 511–522.
Seifter, S., Dayton, S., Novic, B., and Muntwyler, E. (1950). *Arch. Biochem.*, **25**, 191–200.
Sharon, N. (1965). *In* "The Amino Sugars" (Eds. E. A. Balazs and R. W. Jeanloz), Vol. IIA, pp. 2–20. Academic Press, New York.
Sherratt, H. S. A., and Thomas, A. J. (1953). *J. gen. Microbiol.*, **8**, 217–223.
Shetlar, M. R. (1952). *Anal Chem.*, **24**, 1844–1846.
Shetlar, M. R., Foster, J. V., and Everett, M. R. (1948). *Proc. Soc. exp. Biol. Med.*, **67**, 125–130.
Slepecky, R. A., and Law, J. H. (1960). *Anal. Chem.*, **32**, 1697–1699.
Slack, H. G. B. (1957). *Biochem. J.*, **65**, 459–464.
Slein, M. W., and Schnell, G. W. (1953). *Proc. Soc. exp. Biol. Med.*, **82**, 734–738.
Smith, A. J. (1955). *Biochem. J.*, **60**, 522–527.
Smith, I. W., Wilkinson, J. F., and Duguid, J. P. (1954). *J. Bact.*, **68**, 450–463.
Smithies, W. R., Gibbons, N. E., and Bayley, S. T. (1955). *Canad. J. Microbiol.*, **1**, 605–613.
Smith-Kielland, I. (1964). *Acta Chem. Scand.*, **18**, 967–972.
Sobel, A. E., Hirschman, A., and Besman, L. (1945). *J. biol. Chem.*, **161**, 99–103.
Somogyi, M. (1945). *J. biol. Chem.*, **160**, 61–73.
Somogyi, M. (1952). *J. biol. Chem.*, **195**, 19–32.
Sørensen, M., and Haugaard, G. (1933). *Biochem. Z.*, **260**, 247–258.
Soyenkoff, B. C. (1947). *J. biol. Chem.*, **168**, 447–457.
Soyenkoff, B. C. (1952). *J. biol. Chem.*, **198**, 221–227.
Spencer, B. (1960). *Biochem. J.*, **75**, 435–440.
Spies, J. R. (1952). *J. biol. Chem.*, **195**. 65–74.
Spiro, N. (1966). *In* "Methods in Enzymology" (Eds. E. F. Neufeld and V. Ginsburg), Vol. 8, pp. 3–52. Academic Press, New York.
Stein, W. H., and Moore, S. (1951). *J. biol. Chem.*, **190**, 103–106.
Stickland, H. L. (1951). *J. gen. Microbiol.*, **5**, 698–703.
Stone, G. C. H., and Goldzieher, J. W. (1949). *J. biol. Chem.*, **181**, 511–521.
Strange, R. E. and Dark, F. A. (1962). *J. gen. Microbiol.*, **29**, 219–730.
Strange, R. E., Dark, F. A., and Ness, A. G. (1955) *Biochem. J.*, **59**, 172–175.
Strange, R. E., Dark, F. A., and Ness, A. G. (1961). *J. gen. Microbiol.*, **25**, 61–76.
Strange, R. E., Wade, H. E., and Dark, F. A. (1963). *Nature*, **199**, 55–57.
Stumpf, P. (1947). *J. biol. Chem.*, **169**, 367–371.
Stuy, J. H. (1958). *J. Bact.*, **76**, 179–184.
Sumner, J. B. (1944). *Science*, **100**, 413–414.
Sweeley, C. C., Wells, W. W., and Bentley, R. (1966). *In* "Methods in Enzymology" (Eds. E. F. Neufeld and V. Ginsburg), Vol. 8, pp. 95–108. Academic Press, New York.
Tempest, D. W., Dicks, J. W., and Ellwood, D. C. (1968). *Biochem. J.*, **106**, 237–243.
Tempest, D. W., Dicks, J. W., and Hunter, J. R. (1966). *J. gen. Microbiol.*, **45**, 135–146.
Tempest, D. W., and Strange, R. E. (1966). *J. gen. Microbiol.*, **44**, 273–279.
Thorp, R. F. (1941). *Biochem. J.*, **35**, 672–675.
Tillmans, J., and Philippi, K. (1929). *Biochem. Z.*, **215**, 36–48.
Tracey, M. V. (1948). *Biochem. J.*, **43**, 185–189.

Trevelyan, W. E., and Harrison, J. S. (1952). *Biochem. J.*, **50**, 298–303.
Trevelyan, W. E., and Harrison, J. S. (1956). *Biochem. J.*, **63**, 23–33.
Tsanev, R., and Markov, G. G. (1960). *Biochem. biophys. Acta.*, **42**, 442–452.
Van Slyke, D. D., Dillon, R. T., MacFadyen, D. A., and Hamilton, P. (1941). *J. biol. Chem.*, **141**, 627–669.
Van Slyke, D. D., Folch, J., and Plazin, J. (1940). *J. biol. Chem.*, **136**, 509–541.
Van Slyke, D. D., Plazin, J., and Weisiger, J. R. (1951). *J. biol. Chem.*, **191**, 299–304.
Vinter, V. (1960). *Folia Microbiol.*, **5**, 217–230.
Volkin, E., and Cohn, W. E. (1954). *Meth. biochem. Anal.*, **1**, 287–305.
Wade, H. E. (1968). Private communication.
Wade, H. E., and Robinson, H. K. (1965). *Biochem. J.*, **96**, 753–765.
Warburg, O., and Christian, W. (1941). *Biochem. Z.*, **310**, 384–421.
Webb, J. M. (1956). *J. biol. Chem.*, **221**, 635–649.
Webb, J. M., and Levy, H. B. (1958). *Meth. biochem. Anal.*, **6**, 1–30.
Weibull, C. (1957). *Acta Chem. Scand.*, **11**, 881–892.
Wells, M. A., and Dittmer, J. C. (1963). *Biochem.*, **2**, 1259–1263.
Westley, J., Wren, J. J., and Mitchell, H. K. (1957). *J. biol. Chem.*, **229**, 131–138.
Wheat, R. W. (1966). In "Methods in Enzymology" (Eds. E. F. Neufeld and V. Ginsburg), Vol. 8, pp. 60–78. Academic Press, New York.
Whistler, R. L., and BeMiller, J. N. (1961a). In "Methods in Carbohydrate Chemistry" (Eds. R. L. Whistler and M. L. Wolfrom), Vol. 1, pp. 395–399. Academic Press, New York.
Whistler, R. L., and BeMiller, J. N. (1961b). In "Methods in Carbohydrate Chemistry" (Eds. R. L. Whistler and M. L. Wolfrom), Vol. 1, pp. 42–45. Academic Press, New York.
Whistler, R. L., and BeMiller, J. N. (1961c). In "Methods in Carbohydrate Chemistry" (Eds. R. L. Whistler and M. L. Wolfrom), Vol. 1, pp. 45–71. Academic Press, New York.
Wiame, J. M. (1947). *Biochim. biophys. Acta.*, **1**, 234–255.
Wiame, J. M. (1948). *J. biol. Chem.*, **178**, 919–929.
Wilkinson, J. F., and Duguid, J. P. (1960). *Intern. Rev. Cytol.*, **9**, 1–76.
Wilkinson, R. H. (1957). *J. clin. Pathol.*, **10**, 126–135.
Williamson, D. H., and Wilkinson, J. F. (1958). *J. gen. Microbiol.*, **19**, 198–209.
Willstätter, R., and Schudel, G. (1918). *Ber.*, **51**, 780–789.
Winder, F. G., and Denneny, J. (1956). *J. gen. Microbiol.*, **15**, 1–18.
Wober, W., Thiele, O. W., and Urbaschek, B. (1964). *Biochim. biophys. Acta.*, **84**, 376–390.
Woiwod, A. J. (1949). *Biochem. J.*, **45**, 412–417.
Yemm, E. W., and Cocking, E. C. (1955). *Analyst.*, **80**, 209–213.
Zettner, A., and Seligson, D. (1964). *Clin. Chem.*, **10**, 869–890.
Zipf, R. E., and Waldo, A. L. (1952). *J. lab. clin. Med.*, **39**, 497–502.

CHAPTER IV

Chemical Extraction Methods of Microbial Cells

I. W. SUTHERLAND AND J. F. WILKINSON

Department of General Microbiology, University of Edinburgh,
Edinburgh, Scotland

I. Introduction 346

II. Low Molecular Weight Intermediary Metabolites and Coenzymes . 347
    A. Trichloroacetic acid extraction 347
    B. Perchloric acid extraction 348
    C. Ethanol extraction 348
    D. Aqueous extraction 349
    E. Direct extraction of microbial haemins 349
    F. Direct extraction of microbial chlorophylls . . . 350

III. The Extraction of Microbial Lipids 350
    A. Extractable lipid 351
    B. "Bound lipid" or "Total lipid" 353
    C. "Lipid A" 353
    D. Mycobacterial lipids and waxes 354
    E. Lipid-oligosaccharide material 354
    F. Carotenoids 356
    G. Quinonoid electron transfer compounds 357

IV. Poly-β-hydroxybutyrate 358

V. Polysaccharides 359
    A. Aqueous extraction 360
    B. Alkali extraction 362
    C. Phenol extraction 363
    D. Acetic acid extraction 366
    E. Trichloroacetic acid extraction 366
    F. Formamide extraction 367
    G. Citrate extraction of mannan from yeasts 368
    H. Ethylenediamine extraction of yeast glycoproteins . . 368

VI. Inorganic Polyphosphates 368
    A. Acid extraction methods 369
    B. Isolation by alkaline hypochlorite digestion . . . 371

VII.   Nucleic Acids   371
         A. DNA   371
         B. RNA   374
         C. sRNA by phenol extraction   376

VIII.   Proteins   376
         A. Acetate extraction of haemoproteins   376
         B. Citrate extraction   377
         C. Urea extraction   377
         D. Acetone extraction of ferredoxin   377
         E. Acid extraction of basic proteins   378

  IX.   Sequential Methods   378
         A. Extraction Stage I   380
         B. Extraction Stage II   380
         C. Extraction Stage III   380
         D. Extraction Stage IV   380
         E. Extraction Stage V   381

References   381

I. INTRODUCTION

Chemical extraction methods for microbial cells may be used for various purposes. The substances being extracted may be required in their native state and specialized methods are required to maintain this native state. The main purpose of the extraction and subsequent fractionation may be in the estimation of the component substances and the prime consideration is to achieve as quantitative a yield as possible. Alternatively, the purpose may be to fractionate the cell into as many separate components as possible and to study one particular parameter of these fractions—for example— their radioactivity. Obviously the extraction methods used for these different purposes may be the same or different according to the nature of the compound being extracted. We have attempted to give an outline of a variety of methods, choosing those of which we have first-hand experience. As far as possible, we have dealt with extraction methods starting with whole cells without prior separation into different morphological components. We have tried to give an indication of the main groups of compounds in each extract. Our emphasis is mainly on chemical extraction and fractionation of the procaryotic cell, but most of the methods are applicable to cells of eucaryotic micro-organisms.

Since the purpose of extraction is often the isolation of a particular substance in as high a yield as possible and since many of the chemical components of the microbial cell are subject to a wide degree of fluctuation in level, we have given an indication of the kind of cultural conditions suitable for maximum production of a particular material.

In all the methods described, washed cells are used as a starting material and it must be borne in mind that many microbial products will be present in the culture supernatant and washings. (See discussion by Herbert, Phipps and Strange, this Series, Volume 5) Apart from low molecular weight by-products of metabolism and substances produced by any auto-lysis that may have occurred, there may be large amounts of polysaccharide (see pp. 359–360) and smaller amounts of protein and peptide.

II. LOW MOLECULAR WEIGHT INTERMEDIARY METABOLITES AND COENZYMES

In micro-organisms as in other cells, the soluble, intracellular material contains a wide variety of intermediary metabolites and related compounds. These range from free amino-acids, sugar phosphates and sugar nucleo-tides to relatively complex coenzymes such as the folate derivatives. These compounds tend to be present at the highest concentration in cells in the exponential phase of growth. Growth at lowered incubation temperatures may also lead to their accumulation

It is important in the initial harvesting of cells being extracted to avoid changes in pH, ionic strength and temperature since shock of this type often leads to leakage of low molecular weight compounds. In particular, centrifugation and washing should be carried out at a temperature and ionic strength as near as possible identical to that of the growth medium. However, in some cases it may be necessary to lower the temperature in order to prevent loss of or change in the level of intermediary metabolites during endogenous respiration, so a compromise has to be reached.

A. Trichloroacetic acid extraction

The washed microbial cells (about 10 g dry weight or 100 g wet weight) are extracted with 100 ml of 10% (w/v) trichloroacetic acid. The extraction is performed at 0°C for 30 min with stirring, following by centrifugation at 10,000 g for 5 min to remove the cell debris. The clear yellow super-natant is then shaken with cold ether (10 ml/100 ml trichloroacetic acid extract) for 5 min. The ether layer is removed in a separating funnel and the aqueous layer re-extracted with a further aliquot of ether. The process, which removes much of the trichloroacetic acid is repeated until the pH of the aqueous layer is approximately 5·0, the ether solution being dis-carded in each case. Finally neutralization is achieved by the careful addition of 0·01 N KOH. The mixture of nucleotides and other small molecular weight products can be purified and separated further by chromatography and electrophoresis. Such extracts contain nucleosides, nucleotides, nucleotide sugars, nucleotide sugar-peptides involved in cell wall synthesis, together with sugar-phosphates, flavines and other co-enzymes, amino-acids and tricarboxylic acid cycle intermediates.

B. Perchloric acid extraction

The packed wet cells (100 g) are stirred for 10 min in 100 ml of 0·5 N perchloric acid at 4°C. The mixture is centrifuged, then the deposit is extracted with a further 100 ml of cold 0·2 N perchloric acid. After centrifugation, the deposit is extracted with 100 ml of cold 0·2 N perchloric acid and again centrifuged. The three supernatant fluids are pooled and the pH value adjusted to 4·5 by careful addition of 3 N KOH. This causes precipitation of KClO₄ and the precipitate is removed by low speed centrifugation at 0°C. Careful reduction of the volume under reduced pressure at relatively low temperature results in further precipitation of KClO₄ and this is again removed by centrifugation. The final supernatant contains much the same compounds as do TCA or ethanol extracts. The total yield of nucleotides is of the order of 5–15 μmoles/g wet cells, but much variation is experienced depending on the physiological state of the cells and on their species.

C. Ethanol extraction

This method has the advantage that a neutral solvent is used and is therefore particularly suited to the more acid-labile low-molecular-weight intermediates. The heat treatment used is also sufficient to inactivate any de-phosphorylating enzymes.

The microbial cells (c. 100 g wet weight) are suspended in 200 ml of water. The thick suspension is added to 600 ml of 95% ethanol at 70°C and the mixture is heated to 70°C and held at this temperature for 10 min. The cell debris is then recovered by centrifugation and the ethanolic supernant fluid is stored at − 40°C. The ethanol extraction of the cells is repeated and the two extracts are pooled, the cell residue being discarded after the second extraction. The pH value of the extracts is carefully adjusted to 7·0 with 0·01 N HCl. They are then reduced in volume from about 800 ml to 50 ml using a rotary evaporator at 15–20°C. The concentrated solution is deep yellow in colour and somewhat turbid. It contains all the compounds found in TCA extracts, together with protein and peptides and tricarboxylic acid cycle intermediates. The protein can be removed by treating the crude extract with chloroform. To 50 ml to the crude extract is added an equal volume of chloroform. The mixture is shaken vigorously for 5 min and centrifuged at 10,000 g for 10 min. The aqueous layer is separated. The chloroform layer is extracted with an equal volume of water and the aqueous layers pooled. The aqueous solutions are finally extracted with an equal volume of chloroform to give an aqueous solution of intermediates free from peptides and proteins. The yield of nucleotides and similar compounds is again normally of the order of 10 μmoles/g wet cells. For low molecular weight pigments, see Section VIII, pp. 376–378.

D. Aqueous extraction

Although methods such as ethanol, trichloroacetic acid or perchloric acid extraction can be used to obtain the amino-acid pools of micro-organisms, aqueous extraction provides another suitable method. In a comparative study, the total quantity of amino-acids released from *Staphylococcus aureus* was essentially the same for all the methods tested (Hancock, 1958). Aqueous extraction has the advantage that no salts are added to the system, thus minimizing interference if chromatographic separation or identification methods are subsequently applied.

The amino-acid pool of micro-organisms is subject to wide variability and is dependent on the microbial strain used, the composition of the growth medium, the physiological state of the cells when harvested and the physical growth conditions. After growth in the required medium, the cells are harvested by centrifugation and washed twice in distilled water. The wet cell deposit (100 g) is suspended in 1 litre distilled water and heated rapidly at 100°C. The cell suspension is stirred at 100°C for 10 min. The cells are then removed by centrifugation at 10,000 g for 15 min. The supernatant fluid is recovered and lyophilized. As well as amino-acids, it contains small quantities of peptides, inorganic salts, organic phosphates, nucleotides and other co-factors.

E. Direct extraction of microbial haemins

The method used is essentially that described by Barret (1956) in studies on the prosthetic group of cytochrome a_2. The lyophilized cells (10 g) are suspended in a mixture of acetone (140 ml) and water (60 ml) containing 0·6 ml concentrated HCl. The mixture is homogenized by a Waring blendor for 30 sec. The cell debris is deposited by centrifugation at 5000 g for 10 min at 0°C. The supernatant fluid is removed and an equal volume of ether is added. After mixing, the ethereal solution is removed in a separating funnel and washed with 1% (w/v) HCl (i.e. 0·76% v/v) until the washings are free of acetone. The ether solution contains haemins and lipids and the total content of solids is about 1·0–1·5 g/10 g dry cells. It is reduced in volume under vacuum to about 10 ml. The phospholipids present can be precipitated by the careful addition of acetone followed by chromatography of the supernatant fluid to separate and purify the haemins. The yield is dependent on the cytochrome and haemoprotein content of the starting material. This is in turn dependent on the mode of growth of the microbial cells and optimal conditions for cytochrome production are discussed later (Section VIII).

F. Direct extraction of microbial chlorophylls

A wide variety of methods have been employed for the extraction and purification of microbial chlorophylls, depending on the type of cell involved. The following method, based upon that of Jensen et al. (1964), is typical. To avoid any alteration of the chlorophyll, it is advisable to saturate all solvents with gaseous hydrogen sulphide prior to use (for the aqueous solvent, neutralize to pH 7·0 by the addition of 2 N NaOH) and operations should be carried out at low temperatures and in a dim light as far as possible.

Washed cells (equivalent to about 100 g wet weight) are extracted with 80% methanol (50 ml); after centrifugation, the residue is re-extracted with 80% methanol (50 ml portions) until a colourless supernatant is obtained. An equal volume of diethyl ether (peroxide-free) is added to the combined methanolic extracts in a separating funnel and the chlorophylls are removed in the ether layer after the addition of 100 ml of 10% NaCl (w/v) in distilled water. The chlorophyll solution in diethyl ether is washed three times with 10% NaCl, is dried over anhydrous Na_2SO_4 and is concentrated at reduced pressure at low temperatures to about 20 ml. The chlorophyll can then be precipitated by the addition of 80 ml of petroleum ether and collected by filtration or centrifugation.

III. THE EXTRACTION OF MICROBIAL LIPIDS

The yield of lipids extracted from micro-organisms will frequently depend on the cultural conditions employed and on the physiological state of the cells. In order to obtain a high cellular lipid content bacteria are normally grown in a medium in which some factor other than the carbon and energy source (usually the nitrogen source) is the limiting nutrient. The carbon and energy source used for growth is important, replacement of glucose by glycerol resulting in a large decrease in lipid content (Kates, 1964). On the other hand, addition of tricarboxylic acid cycle intermediates to the medium may increase the level of lipid. Variations between the lipid content of cells grown on different utilizable carbohydrates have also been reported. The growth medium may also affect the type of lipids produced, nitrogen limitation yielding cells of Escherichia coli with a relatively high content of saturated fatty acids, while carbohydrate limitation increased the unsaturated acid content (Marr and Ingraham, 1962). It was also found that growth temperature affected the type of fatty acid found. Presumably these environmental factors influence the relative levels of lipid in the cell wall, the cytoplasmic membrane and intracellularly as well as possibly the nature of the lipid at any one site. From these results, together with the relative lack of knowledge of microbial lipids, it is obvious that preliminary experiments with growth media and conditions are essential when optimal yields of lipids and fatty acids from bacteria and other micro-organisms are required.

Several recent review articles give accounts of the chemistry and other features of microbial lipids (e.g. O'Leary, 1962; Kates, 1964; Kates, 1966). From these it can be seen that in micro-organisms, as in other tissue, two preparations of lipid can be obtained, "extractable lipid" and "bound lipid". Into these two rather arbitrary categories fall the material directly extractable with neutral non-polar solvents and that material extractable with similar solvents after acid or alkaline hydrolysis, respectively. In addition certain other groups of lipids are found, peculiar to bacteria. One of those is the "lipid A" found bound in the lipopolysaccharide-protein complex of Gram-negative bacteria, particularly the Enterobacteriaceae. In the genus *Mycobacterium* are to be found glycolipids and waxes. Certain specialized isoprenoid compounds bound to sugar moieties have been cited as intermediates, in the biosynthesis of cell wall mucopeptide in Gram-positive micro-organisms and of the "side chains" of the polysaccharide components of the lipopolysaccharide of Gram-negative bacteria (Higashi *et al.*, 1967; Wright *et al.*, 1967).

A. Extractable lipid

1. *Chloroform/methanol extraction*

Most of the methods used have been developed initially for animal tissue but have proved equally applicable to bacterial material (Bligh and Dyer, 1959; Depinto, 1967). A pellet of frozen bacteria equivalent to 10 g dry weight, 100 ml chloroform and 200 ml methanol is homogenized for 2 min at room temperature in a Waring blendor or similar apparatus. A further 100 ml of chloroform is added and blending continued for 30 sec. Finally 100 ml of water is added to the mixture and another 30 sec blending applied. The contents of the homogenizer are poured through a filter of Whatman No. 12 paper into a separating funnel. The lower layer of chloroform, containing the extracted lipid, is separated off and the solvent removed by evaporation under reduced pressure at 50°C. In a typical experiment using *Rhodopseudomonas palustris* (10 g dry weight) the yield was

0·96 g lipid and was free of chlorophyll and carotenoids. A further small amount of lipid can be extracted by homogenizing the bacterial residue with a further 100 ml of chloroform for 2 min. The blending container and residue are rinsed with 50 ml of chloroform. Although such a re-extraction may be desirable in quantitative work, it was found that with *R. palustris*, almost complete release of carotenoids and chlorophyll from the cells occurred at this stage.

2. *Hot chloroform/methanol extraction*

Higher yields may be obtained using heating along with chloroform/methanol extraction. The heating serves the dual purpose of inactivating any lipases present, and splitting lipid-protein complexes. A comparative study has shown it to be superior to several other mixed solvent extractions (Vorbeck and Marinetta, 1965). Each 100 g wet weight of cells is heated to 65°C for 5 min in 100 ml methanol under reflux. After cooling to room temperature, 200 ml chloroform is added and the mixture stirred at room temperature for 20 min. The cell residue is removed by filtration and re-extracted for 20 min at room temperature with a further 300 ml chloroform/methanol (2 : 1, v/v). The extracts are combined and the solvent evaporated under reduced pressure at 40°C. Both this and the preceding method yield much of the lipid found associated with the cell membrane but not such material as the "lipid A" component of bacterial lipopolysaccharides. The crude extracts are contaminated with amino-acids, nitrogeneous bases, sugars and some inorganic material. Various methods have been used to remove these contaminants such as dialysis, electrophoresis and chromatography. One of the most common has been washing the extract with water, but this can result in considerable losses of phosphatides. Perhaps the most useful method for the separation of these contaminants is by passing through a column of Sephadex LH-20 (Maxwell and Williams, 1967).

3. *Acetone or petroleum extraction*

These methods are particularly applicable when the material required is predominantly triglyceride in nature. Such compounds are soluble in all the fat solvents except cold ethanol. The washed pellet of cells (100 g) is stirred for 16 h at room temperature in 500 ml acetone and the cell debris removed by centrifugation at 5000 *g* for 15 min. The acetone extraction is repeated and the extracts pooled and concentrated under reduced pressure to about 50 ml. The crude extract is clarified by centrifugation. An alternative procedure applicable to dried bacteria (10 g) is to extract with 500 ml petroleum ether (boiling range 60–80°C) using the same conditions as for acetone extraction. The product using this solvent is less likely to be contaminated with non-lipid material than that obtained with acetone.

4. *Soxhlet extraction*

The Soxhlet technique can be applied to microbial material but is probably less useful than the mixed solvent extractions described above. The washed pellet of microbial cells (100 g) is ground in a mortar with 800 g anhydrous Na_2SO_4 and the mixture is extracted for 48 h with 1·0 litre chloroform in the Soxhlet apparatus. The extract is removed by decantation and evaporated to dryness. It is then extracted with 100 ml hot petroleum ether (boiling range 60–80°C) under reflux.

B. "Bound lipid" or "Total lipid"

1. *Acid/acetone extraction*

For these procedures, the starting material can be the dried microbial residue from neutral solvent extraction (see paragraphs A1–4) or alternatively frozen or freeze-dried bacteria. This is suspended in 2 N sulphuric acid (100 ml/10 g dry material) hydrolysed at 120°C for 15 min by autoclaving in an open tube in a sterilizer. After cooling, a filter aid such as "Supercell" (10 g) is added and the whole mixture filtered through a double-layer of acid-resistant paper on a Buchner funnel. The pad is then washed with 2 × 10 ml distilled water. The pad (including the papers) is extracted by refluxing for 2 h with 100 ml boiling acetone. After filtration, the extraction is repeated with two more aliquots of acetone, making three in all, and finally with 100 ml ether. The three acetone extracts are combined, 100 ml water is added and most of the acetone removed under reduced pressure. A turbid suspension results, and this is extracted with 200 ml ether. The ether extract is added to that previously obtained and washed with 5% sodium bicarbonate solution and with water. Finally, it is dried with anhydrous sodium sulphate to yield a crude lipid solution.

2. *Saponification*

An alternative procedure combines alkaline hydrolysis and ethanol extraction in one stage. The dry material (10 g) is suspended in 60 ml water and 100 ml ethanol. Potassium hydroxide pellets (9 g) are added and the mixture refluxed under nitrogen for 5 h. Evaporation under vacuo removes most of the ethanol. The residual liquor is extracted with one volume of ether and the ethereal solution discarded. The aqueous phase is then acidified with HCl to permit isolation of the saponified lipids.

C. "Lipid A"

This unusual lipid containing a relatively large amount of N-β-hydroxy-myristyl-glucosamine is found as a component of the lipopolysaccharide-protein complex present in the cell walls of Gram-negative bacteria. The lipopolysaccharide can be extracted from whole cells or cell walls by the phenol/water method (Westphal, Luderitz, Bister, 1952; see Work, E., this Series, Volume 5a). After purification of the lipopolysaccharide by ultra-centrifugation the freeze dried material is used for preparation of the lipid. The lipopolysaccharide is dissolved in 1% (v/v) acetic acid to give a final concentration of 1% (w/v). Hydrolysis is performed for up to 4 h at 100°C in a sealed vessel, preferably under an atmosphere of nitrogen to minimize destruction of the lipid. After hydrolysis, the lipid A is deposited by low speed centrifugation. It is freed from carbohydrates by thorough washing with distilled water. Finally, it can be dried in a vacuum desiccator or dissolved in chloroform. The yield is of the order of 30% of the lipopoly-saccharide starting material, free from other lipids and non-lipid material.

D. Mycobacterial lipids and waxes

The freeze dried bacterial cells (10 g) are suspended in 100 ml ethanol and an equal volume of ether is added. This mixture is stirred at room temperature for 48 h. The cells are then allowed to sediment and as much as possible of the solvent is carefully withdrawn and retained. A further 200 ml of ethanol/ether (1 : 1, v/v) is added and extraction at room tempera-ture continued for a further 24 h. After sedimentation of the bacteria and removal of the solvent two further aliquots of ethanol/ether are used to perform extraction for 2×4 h periods. The combined extracts are then reduced in volume at 40°C. The aqueous layer is washed with several volumes of ether and all the ethereal solutions pooled. This pool is filtered and evaporated to yield the crude mixture of lipid and waxes. A scheme for extraction and fractionation of these has been devised (Asselineau and Lederer, 1960), and is shown with slight modifications (Fig. 1).

E. Lipid-oligosaccharide material

Lipid-linked oligosaccharides have been cited as intermediates in the biosynthesis of bacterial mucopeptides and lipopolysaccharides. It is probable that they are also involved in the synthesis of other bacterial polysaccharides. They can be extracted either from whole cells or from cells after lysis with tris-EDTA or lysozyme (E.C. 3.2.1.17). The technique is essentially that described by Robbins, Wright, and Bellows (1964). The cells are harvested, washed in ice-cold saline and resuspended in 1 : 20 volume of cold EDTA solution (0·01 M EDTA disodium salt, adjusted to pH 8·0 with TRIS base). The suspension is placed at −20°C and allowed to freeze. For use, it is gradually thawed at room temperature. For cell species susceptible to lysozyme, the micro-organisms are suspended in 0·1 M EDTA; 0·5 M NaCl (pH 8·0) and 40 mg lysozyme added/100 ml

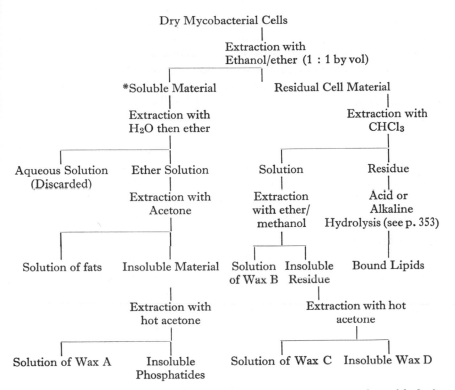

* In all cases, the solutions are taken to dryness prior to extraction with fresh solvent.

FIG. 1. Fractionation of Mycobacterial Lipids and Waxes (after Asselineau and Lederer, 1960).

suspension. After 30 min at 37°C, the particulate material is recovered by high speed centrifugation and washed free of lysozyme.

The cell lysate, before or after incubation with nucleotide-sugars, is extracted with an equal volume of butanol by vigorous shaking at room temperature. Alternatively, the extraction can be made using butanol: 6 M pyridinium acetate (2 : 1, by vol., pH 4·2). The extraction is repeated twice and the extracts combined and washed with a small volume of cold distilled water at 4°C. The butanolic solution is taken to dryness under reduced pressure at 40°C. The dry extract is shaken with a small volume of chloroform and the lipid-oligosaccharides separated from other lipid or sugar-containing entities by chromatography on ion-exchange cellulose (Dankert, M. et al., 1966) DEAE-cellulose in the acetate form is suspended in glacial acetic acid, then washed with methanol containing 1% water.

Columns (30×1 cm) are prepared and the sample dissolved in approximately 600 ml of methanol or butanol is applied with a flow rate of 20 ml/h. After washing the column with 99% methanol a gradient of methanolic ammonium acetate (100 ml 0·4 M ammonium acetate in 99% methanol into an equal volume of 99% methanol) is applied and fractions (3 ml) collected. Chromatography on Sephadex LH-20 is also possible.

The same products can be obtained by direct extraction of unbroken cells (100 g wet weight) with 500 ml butanol at 60°C for 15–30 min. The cell debris is removed by centrifugation at 6000 rpm for 15 min at 0°C. This also serves to separate the aqueous and butanol layers. The butanol layer is removed and taken to dryness. The lipid-oligosaccharides can be dissolved in chloroform or in a small volume of butanol. The yield of these compounds is low, being of the order of 0·002–0·02% cell dry weight.

F. Carotenoids

It may be found necessary to illuminate the culture for maximum carotenoid production since the biosynthetic pathways are sometimes induced by light.

Extraction procedures vary according to the type of cell involved and the polarity of the carotenoids. Some cells with a rigid and relatively impermeable cell wall may need to be exposed to a preliminary mechanical disintegration and although the ideal conditions for extraction would be to avoid heating, in practice it may be required. Further, procedures should avoid exposure to light or air as carotenoids are very liable to isomerization and oxidation once they have been extracted.

Most methods employ extraction by methanol followed by a saponification stage. Methanol has the advantage of disrupting cells and of breaking lipoprotein complexes, it is sufficiently polar to extract even the most polar carotenoids, it is less liable than some solvents to form decomposition products on standing and it allows the succeeding saponification step to be carried out without transference to another solvent.

1. *Method with saponification*

The following method is based on that of Suzue *et al.* (1967).

To about 100 g wet weight of cells is added 200 ml 75% (v/v) methanol, 20 g NaOH and 10 g pyrogallol. The mixture is refluxed for 30 min for extraction and saponification. The carotenoids are extracted, after addition of 200 ml distilled water, by the addition of 100 ml hexane in a separating funnel. This extraction is repeated with two further 100 ml aliquots of hexane and the combined carotenoid solution is washed with five portions of 300 ml distilled water followed by drying over anhydrous Na_2SO_4 and evaporation at reduced pressure to a small volume.

2. Method without saponification

This method, based on that of Eimhjellen and Jensen (1964) is a simple one which involves no saponification step.

The washed microbial cells (about 10 g dry weight or 100 g wet weight) are suspended evenly in water to a volume of 50 ml. To this suspension is added 100 ml acetone and the mixture is left under an atmosphere of nitrogen for about 6 h at 4°C. The cell residue is removed by filtration and is washed with small amounts of acetone until the washings are colourless. The carotenoids are transferred from the acetone solution into petroleum ether (100 ml) in a separating funnel. The hyophase is re-extracted with 50 ml portions of petroleum ether until the epophase is colourless. The combined petroleum ether extract is washed with water and dried over anhydrous Na_2SO_4.

G. Quinonoid electron transfer compounds

The presence of the coenzyme Q and vitamin K type of compound has now been recognized in micro-organisms for some years. Their occurrence and structure in microbial and other cells have recently been reviewed (Morton, 1965). Several of the methods applicable to the extraction of microbial lipids may also be used to obtain quinonoid compounds. However, such methods are not entirely satisfactory on account of the complexity of the extracts, the danger of destroying the quinones and the incomplete extraction of these compounds. Thus, the chloroform/methanol extraction procedures could also yield some quinones if they are present in the microbial cells under examination. Alternative and better extraction methods are available.

1. Acetone extraction

The cells are harvested, washed with water and lyophilized. Acetone (200 ml/10 g dry weight) is added and the suspension heated under reflux for 30 min. The cell debris is removed by centrifugation at 5000 g for 15 min. The supernatant fluid is concentrated under reduced pressure to yield a yellowish-orange oil containing lipid and coenzyme Q.

2. Iso-octane extraction

This procedure is only applicable to lyophilized cells as starting material (Fuller et al., 1961). The cells (10 g) are suspended in 500 ml iso-octane and shaken vigorously for 3 h at room temperature in the dark. The cell debris and other insoluble material is removed by centrifugation and the extraction is repeated twice more. The pooled supernatant fluids contain the quinonoid compounds together with some lipids in iso-octane solution.

This solution is concentrated under vacuum to give an orange-yellow oil which can be extracted with a small volume of iso-octane or of heptane. Separation of the quinones from lipids can be accomplished by chromatography on columns of sodium aluminium silicate.

3. Saponification

An alternative to neutral solvent extraction is saponification (Lester and Crane, 1959). In this the addition of pyrogallol is essential otherwise destruction of the quinones occurs. The method has the advantage that the product is free from non-saponifiable lipids and it is selective for the alkali-stable coenzyme Q type of compound, the vitamin K group being destroyed. The washed cell deposit (100 g) is suspended in 100 ml water and 8 g pyrogallol and 200 ml ethanolic potassium hydroxide (10% w/v) added. The mixture is heated under reflux for 30 min, cooled and centrifuged at 5000 g for 30 min. To every 100 ml of supernatant fluid, 20 ml iso-octane is added and the liquids are shaken vigorously for 5 min. The iso-octane layer is removed in a separating funnel, dried with anhydrous sodium sulphate and evaporated under reduced pressure to give a yellow oil. This is extracted with a small volume of iso-octane following which the coenzyme Q may be purified by chromatography.

IV. POLY-β-HYDROXYBUTYRATE

Poly-β-hydroxybutyrate is produced by a wide range of procaryotic cells and is commonly thought to act as a storage polymer for carbon and energy (Wilkinson, 1963). The amounts formed vary considerably according to the micro-organisms and the environmental conditions. The largest amounts normally occur in the stationary phase in batch culture when cessation of growth is brought on by deficiency of some compound other than the carbon and energy source. In practice, nitrogen limitation is commonly used and the following medium, used for the growth of cells of *Bacillus megaterium* rich in poly-β-hydroxybutyrate, gives an example (Macrae and Wilkinson, 1958).

The organisms are grown at 30°C in a liquid medium containing the following substances in distilled water (g/l):

Na_2HPO_4	6·0
KH_2PO_4	3·0
NaCl	3·0
NH_4Cl	1·0
Na_2SO_4	0·1
$MgCl_2$ $6H_2O$	0·1
Mn Cl_2 $4H_2O$	0·01
Casamino acids	0·01

Glucose is added to a final concentration of 2·0% (w/v). The cells are harvested when they have passed into the stationary phase or when the level of poly-β-hydroxybutyrate, as judged by the number and size of lipid granules staining with Sudan black (Laybourne, 1924) is sufficiently high.

Most methods depend upon two properties of poly-β-hydroxybutyrate— its insolubility in alkaline sodium hypochlorite and its solubility in chloroform from which it can be precipitated by ether.

The alkaline hypochlorite reagent (Williamson and Wilkinson, 1958) is prepared as follows. Fresh bleaching powder (200 g) is thoroughly triturated with a little distilled water and the volume is made up to 1·0 litre. To this is added 1·0 litre of 30% (w/v) Na_2CO_3 with stirring and the mixture is left to stand with occasional shaking for 2–3 h and is filtered through paper. The pH value is adjusted to 9·8 with concentrated HCl, the flocculent precipitate appearing during this adjustment being removed by filtration after warming the solution to 37°C. The resultant clear liquid can be stored in a stoppered bottle in a refrigerator and is stable for several months.

The washed cells (about 10 g dry weight or 100 g wet weight) are suspended in the alkaline hypochlorite reagent and digested for 90 min at 37°C. Alternatively the cells can be suspended first to a concentration of 15% (w/v) in ice cold 0·02 M phosphate buffer (pH 7·2) and disrupted for 10 min in an ultrasonic disintegrator. The suspension is centrifuged and the pellet is then treated with half the amount of alkaline hypochlorite described previously.

The suspension after hypochlorite treatment is diluted with an equal volume of distilled water and dialysed against distilled water for 24 h. The lipid granules are centrifuged and washed twice with acetone and three times with ethyl ether and the granules are dried. Further purification can be effected by dissolving the granules in a small volume of chloroform, removing any undissolved material by centrifugation and then precipitating the poly-β-hydroxybutyrate from chloroform solution by the addition of three volumes of ethyl ether.

V. POLYSACCHARIDES

Several types of polysaccharide may be synthesized by a single microbial strain, some being extracellular—outside the cell wall and cytoplasmic membrane, while others are intracellular. Some, such as lipopolysaccharides and teichoic acids, form part of the cell walls and may be extracted either from the cell walls or from whole cells. The extracellular polysaccharides can themselves be subdivided, on the basis of centrifugation or of India Ink negative staining (Duguid, 1951) into slime and capsule. Some strains produce slime polysaccharide only, while others produce both types of material. Intermediate types are also found. The polysaccharides associated with the surface of Gram-negative bacteria have been studied in considerable detail and this work has been reviewed by Luderitz, Jann, and Wheat (1967). A more general review of microbial polysaccharides has also been published recently (Horecker, 1966).

The amount of polysaccharide produced varies greatly with the cultural conditions used. This is particularly true of the extracellular polysaccharides and of intracellular polysaccharides such as glycogen. In general, polysaccharide production is stimulated by nitrogen deprivation in the presence of excess utilizable carbohydrate, lowered incubation temperature, highly aerobic conditions and the use of solid media. For strains of the family Enterobacteriaceae and similar relatively non-exacting species a suitable medium contains (g/l):

Casamino acids (Difco technical)	1·0
Yeast extract (Oxoid or Difco)	1·0
Na_2HPO_4,	10·0
KH_2PO_4	3·0
$MgSO_4, 7H_2O$	0·2
K_2SO_4	1·0
NaCl	1·0
$CaCl_2$	0·01
$FeSO_4$	0·001
Glucose	20·0

(Sutherland and Wilkinson, 1965)

The components of the medium, less the glucose, are dissolved in 900 ml water and agar (1% w/v final concentration) added. The medium is then sterilized by autoclaving at 120°C for 15 min. The glucose is dissolved in 100 ml water, autoclaved and added aseptically to the molten medium. While Petri dishes may be suitable for small quantities of medium, we have found enamel trays (approximately 40 cm × 30 cm) to be superior for the preparation of large amounts of material. The trays are covered with aluminium or steel covers, and depending on the depth of medium required, contain 0·5–1·0 litre of medium. Where agar contamination is undesirable the medium may be overlaid with sheets of cellophane sterilized by autoclaving. This does, however, reduce the yield of cells and of polysaccharide.

A. Aqueous extraction

1. Cold aqueous extraction of slime polysaccharide

The starting material used may either be the supernatant fluid from liquid cultures or a saline suspension of the whole culture removed from the surface of solid medium. Recovery of the material from solid medium is most readily achieved using an L-shaped glass spreader. Some strains producing very viscous polysaccharide gels can be stripped from the agar surface using forceps and suspended in saline. The addition of formalin (0·5% by volume, final concentration), minimizes the leakage of intracellular components. The suspension is stirred at high speed at room temperature for 30 min or alternatively it may be homogenized in a Waring blendor for 5–10 min. If the latter procedure is chosen, addition of a small quantity of antifoam (e.g. Silicone antifoam MS Emulsion RD, Hopkins-Williams) is advisable. Normally, the cells in the preparations can be deposited by centrifugation at 10,000 g in an angle rotor for 30–60 min. However, certain mutants and strains produce a very viscous slime and require prolonged centrifugation at higher r.c.f. to separate the cells from the polysaccharide solution, e.g. 25,000 g for 2–3 h. Such aqueous extraction is normally sufficient to extract slime polysaccharide from cultures but will not yield significant amounts of material from capsulated micro-organisms. The material obtained can be precipitated by the addition of a suitable non-polar solvent such as acetone or alcohol. To the cold (0–4°C) supernate is added acetone 1·5 volumes at $-40°C$. At this stage the addition of several crystals of sodium acetate may assist precipitation. The mixture is stirred vigorously with a glass rod. The precipitated strands of polysaccharide are thus wound round the rod, and can be dehydrated by washing with several aliquots of pure acetone followed by ether. Finally the material is dried in a freeze dryer. Such preparations will most probably contain contaminating protein along with traces of lipopolysaccharide derived from the cell wall in Gram-negative bacteria, and of intracellular poly-saccharides. There may also be traces of nucleic acids if the cultures have autolysed to any extent during growth and harvesting. Protein can be re-moved from the preparation by the Sevag (1934) deproteinization technique or by digestion with pronase.

The dried polysaccharide is dissolved at a concentration of 5 g/litre in acetate buffer prepared by using equal volumes of 8% (w/v) sodium acetate solution and 4% (v/v) acetic acid. It is essential that the poly-saccharide solution is not more concentrated than 0·5%. The solution in 200 ml aliquots is shaken **vigorously** for 30 min with 40 ml chloroform and 8 ml n-butanol. The organic phase and the interface layer of denatured protein is removed in a separating funnel. The process is repeated until no further precipitate is formed at the interface. The final aqueous solutions are pooled, centrifuged at 5000 g for 30 min and dialysed. The polysac-charide may then be recovered by lyophilization or by precipitation with cold ($-40°C$) acetone. Glycogen and DNA can be hydrolysed by amylase and desoxyribonuclease respectively.

2. Hot aqueous extraction

This can be used for capsular polysaccharides from bacteria and yeasts and to obtain other polysaccharides such as "nigeran" (Barker *et al.*, 1953) from *Aspergillus niger*. The formalized cells from 5·0 litre of medium are suspended in 1·0 litre saline (buffered at pH 7·0) and stirred at 100°C for 15 min in a water bath. The subsequent sedimentation of cells and treatment of the extracted polysaccharide is the same as that used for slime material. However, it may be found that instead of strands of polysaccharide precipitate forming on the addition of acetone, a flocculent precipitate is obtained. Contaminating material, protein and nucleic acid together with other polysaccharides, tends to be present in greater quantities than with preparations of slime polysaccharides.

B. Alkali extraction

Extraction of microbial cells with alkali can be used to obtain (a) capsular material; (b) glycogen or other intracellular polyglucoses; (c) yeast glycans.

1. Alkali extraction of capsular polysaccharides

The cells are suspended for 15 min at room temperature in 1% (w/v) formaldehyde solution. Sodium hydroxide is then added to a final concentration of 1% (w/v). The suspension is then stirred vigorously at room temperature for 30 min, and carefully neutralized with N HCl. The isolation of polysaccharide from the alkali extract is performed in the same manner as for slime polysaccharide. The disadvantages of alkali extraction are the greatly increased contamination of the product with material of intracellular origin, and the possibility of degradation of the polysaccharide. This is most likely to occur in polysaccharides containing acetyl or pyruvyl groups as integral components. A high proportion of these groups may be lost during alkali extraction.

2. Alkali extraction of bacteria for glycogen

The starting material is an acetone powder of the bacteria. This is defatted by refluxing with boiling chloroform/methanol (2 : 1 v/v)for $2\frac{1}{2}$ h followed by refluxing with boiling ether for $2\frac{1}{2}$ h. The solvents are used in the ratio of 120 ml/10 g dry bacteria. After the ether extraction, the cells are dried in air, and extracted with 100 ml 30% (w/v) KOH/10 g cells at 100°C for 1 h. The glycogen is alkali soluble and can be separated from residual microbial material by centrifugation. To 100 ml of the supernatant fluid is added 95% ethanol (80 ml) and the glycogen precipitate removed by centrifugation at 5000 g for 15 min. The crude material is redissolved in water (50 ml/10 g original dry bacteria). This opalescent solution is centrifuged at 5000 g for 15 min. The deposit is washed with a small volume of water and re-centrifuged, the washings being added to the previous supernatant and the sediment discarded. The glycogen is recovered by the addition of ethanol (80 ml/100 ml extract), and centrifugation. It is washed with ethanol and acetone then dried under vacuum, the yield being normally of the order of 5–15% of the original bacterial dry weight. The product may be contaminated with degraded nucleic acid material (see also Herbert, Phipps and Strange, this Volume, page 209).

3. Alkali extraction of yeast cell walls for glucan and mannan

Yeast cell walls are prepared by shaking with Ballotini glass beads (Korn and Northcote, 1960) followed by differential centrifugation. The cell walls are separated from intracellular components by centrifugation at 1500 g for 10 min, washed thoroughly with distilled water and lyophilized. The dried product (1 g) is suspended in 20 ml 3% (w/v) aqueous sodium hydroxide solution and heated at 100°C for 6 h. The alkali-insoluble material is recovered by centrifugation and re-extracted with alkali, then with 20 ml 0·5 N acetic acid at 75 °C for 6 h. The final insoluble residue is washed with ethanol and dried under vacuo, yielding 25–30% of the cell walls as glucan which is free from contaminating material.

The supernatant fluid from the first alkali extraction is rendered acid (pH 5·0–5·5) by the addition of glacial acetic acid. Cold ethanol (4 vols) is added and the precipitated mannan recovered by centrifugation. It is redissolved in water and again precipitated by the addition of 4 vols cold ethanol. The product after drying represents about 30% of the dry cell wall from baker's yeast, but contains some contaminating material. This can be removed if necessary by precipitation of the copper compound.

C. Phenol extraction

1. For capsular material

As well as its more specific application for lipopolysaccharide preparation phenol extraction may be used to extract extracellular polysaccharide, especially when this is of the capsular variety.

The culture from solid medium or cells from liquid medium is suspended in 500 ml physiological (0·9% w/v) saline containing 2% (w/v) phenol. This mixture is stirred overnight at 37°C. To it is then added 1·6 litre of 96% ethanol containing 35 ml of saturated solution of sodium acetate in water/alcohol mixture (7/43, w/v). After addition of the alcohol, the mixture is allowed to sediment by standing a further 20 h at room temperature, to permit separation of the bacterial residue and soluble material.

The clear supernatant fluid is removed by decantation, and the precipitate centrifuged, washed with acetone and finally dried at 37°C.

The original method of phenol-extraction developed by Westphal et al. (1952) can be applied to the extraction of capsular polysaccharides. Although the original procedure involved heating of the mixture at 60°C for 30 min, shorter times (e.g. 5 min) have proved equally effective (Luderitz et al., 1965) and are presumably less destructive towards capsular polysaccharides.

The dried bacteria (10 g) are suspended in 175 ml water and warmed to 65°C. An equal volume of 90% (w/v) aqueous phenol at the same temperature is then added and the mixture stirred vigorously for 5 min. It is then cooled rapidly and centrifuged at 3000 g for 30 min at 0°C to permit phase separation. The upper aqueous layer is carefully removed and dialysed for 48 h against running tap water to remove phenol. The non-diffusible solution contains exopolysaccharide, RNA, glycogen and lipopolysaccharide. The last of these can be removed by ultracentrifugation at 100,000 g for 4 h, when it is found as a clear viscous gel at the foot of the centrifuge tubes. The exopolysaccharide can be freed from RNA and glycogen by treatment with RNAase and amylase respectively. It is then submitted to a repetition of the phenol extraction and dialysis.

2. For the lipopolysaccharides of Gram-negative bacteria

The hot 90% (w/v) phenol extraction is used. The lipopolysaccharide is obtained as a clear viscous gel after ultracentrifugation at 100,000 g for 4 h (Fig. 2). This material is carefully removed with a spatula, suspended in a small volume of water and lyophilized. At this stage contaminating glycogen and nucleic acid are still present. Further purification can be obtained by dissolving the lipopolysaccharide in water to give a 1% (w/v) solution and repeating the ultracentrifugation. The gel is again recovered and lyophilized. One further repetition of the ultracentrifugation procedure, making three in all, yields an almost pure product.

The polysaccharide may be obtained free from lipid by mild acid hydrolysis. The lipopolysaccharide is dissolved in 1% (v/v) acetic acid to give a concentration of 1% (w/v). The solution is gassed with nitrogen and hydrolysed under an atmosphere of nitrogen in a sealed vessel for 4 h at 100°C. The lipid A is deposited by low speed centrifugation. The polysaccharide-containing supernatant solution is dialysed and lyophilized. The product is essentially pure and comprises about 40% of the lipopolysaccharide starting material. The lipopolysaccharide content of bacterial cells is much less variable in quantity than are the other cell components. Yields obtained by the phenol extraction technique are of the order of 0·5–2·0% of the cell dry weight. In terms of pure lipid-free polysaccharide the yield is 0·2–0·8% dry weight.

For certain mutants of species of Enterobacteriaceae, lacking the complete lipopolysaccharide, this material may be precipitated preferentially by the addition of Mg^{++} ions and thus avoiding prolonged ultracentrifugation. After dialysis against distilled water, the aqueous solution containing lipopolysaccharide is mixed with sufficient concentrated MgCl$_2$

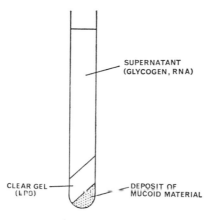

FIG. 2. The purification of lipolysaccharide by ultracentrifugation.

solution to give a final concentration of 0·025 M. Using a 0·5 M MgCl$_2$ solution, 5 ml are required per 100 ml of lipopolysaccharide solution. After 20 min at 4°C, the precipitated lipopolysaccharide is recovered by centrifugation at 30,000 g for 20 min. Ribonucleic acid, glycogen and exopolysaccharide are left in the supernatant fluid.

D. Acetic acid extraction

By using the method of Freeman (1942) whole cells of micro-organisms can be extracted directly with acetic acid to yield a variety of polysaccharides. Gram-negative bacteria yield the cell wall polysaccharide free from lipid A together with other material. A variety of polysaccharides, mainly emanating from the cell wall, are obtained from other micro-organisms. The washed and lyophilized bacteria (10 g) are added to 1·0 litre 0·1 N acetic acid and shaken until a uniform suspension is obtained. It is then refluxed at 100°C for 12 h. The insoluble material is removed by centrifugation at 10,000 g for 30 min leaving a straw-coloured supernatant fluid. This is dialysed against water and concentrated under reduced pressure to about 200 ml. Any material precipitating at this stage is removed by centrifugation at 10,000 g. To each 100 ml of supernatant fluid is added 780 ml cold ($-40°C$) absolute alcohol with stirring. The crude polysaccharide precipitate is recovered by centrifugation at 5000 g for 15 min, washed with absolute alcohol and ether and lyophilized. It is contaminated with protein, glycogen and nucleic acid. Purification can be attempted by digestion with proteases, amylase and ribonuclease or by ethanol fractionation. De-proteinization can also be performed by the Sevag (1932) technique. The yield after purification is about 5% of the dry starting material and is free from lipid and protein.

E. Trichloroacetic acid extraction

1. For the "somatic" antigen of Gram-negative bacteria

For some purposes, the complete lipid-polysaccharide-protein complex found in the cell walls of Gram-negative bacteria is required, rather than the degraded lipopolysaccharide or polysaccharide. Such a product can be obtained by trichloroacetic acid extraction (Boivin et al., 1933; Davies, 1955). The starting material is an acetone powder of bacteria. The cells (10 g) are suspended in 500 ml cold water (2–4°C) and added to an equal volume of N trichloroacetic acid solution at the same temperature. The mixture is stirred at 4°C for 3 h followed by centrifugation at 10,000 g for 30 min to deposit the cell debris. The supernatant fluid is dialysed overnight against running tap water then concentrated under reduced pressure until the solids content is about 1%. The polysaccharide complex can be recovered by collecting the material precipitating between 50% and 85% ethanol concentrations (by volume). The precipitate is almost free from contaminants and contains a mixture of the complete macromolecule (i.e. phospholipid-polysaccharide-protein) together with some polysaccharide free from phospholipid and protein. These two components can be separated by ultracentrifugation at 100,000 g for 4 h. The complete antigen is deposited while the free polysaccharide together with any contaminating polyglucose remains in the supernatant fluid. The yield is 0·5–3% of the dry bacteria.

2. For teichoic acids from Gram-positive bacteria

The starting material is a dried preparation of cell walls (see Work, E., this Series, Volume 5a). The walls (10 g) are suspended in 250 ml water at 2–4°C and 25 g trichloracetic acid is added gradually with stirring. The suspension is stirred for 48 h at 4°C. The wall debris is removed by centrifugation at 100,000 g for 30 min and washed with 50 ml of 10% (w/v) trichloroacetic acid solution. After centrifugation the deposit is discarded unless the degraded mucopeptide is required. The washings are added to the original extract and 235 ml of acetone at 0°C are added. The preparation is held at 0°C for 24 h. The precipitate formed is recovered by centrifugation at 10,000 g for 30 min, washed with acetone and ether and freeze-dried. It is redissolved in a small volume of water (2–3 ml), centrifuged at 2000 g for 15 min and the supernatant fluid freeze-dried. The yield of teichoic acid is variable but can be as high as 5–10% of the cell walls.

A modified procedure has been used to obtain the "teichuronic" acids from the cell walls of *Bacillus licheniformis* (Hughes, 1965). In this the cell walls (10 g) are extracted twice with 500 ml 5% (w/v) trichloroacetic acid for 16 h or longer at 35°C. The debris is removed by centrifugation at 10,000 g for 15 min and the supernatant fluid dialysed against water. The non-diffusible material containing teichoic acid and teichuronic acid is lyophilized and can be fractionated by ethanol precipitation of an aqueous solution. This method gives higher yields than the extraction at 0°C but the product may be more degraded.

F. Formamide extraction

Extraction with hot formamide, although a somewhat drastic procedure, can be used to solubilize polysaccharide-containing material from microbial cells or cell walls resistant to other methods. The cells or cell walls (1 g) are added to 100 ml formamide in an oil bath or isomantle at 150°C. The mixture is maintained at this temperature for 15 min, cooled and 250 ml ethanol containing 5% (by volume) 2 N HCl is added. The mixture is centrifuged at 10,000 g for 15 min and acetone (2·5 litre) at −40°C is added to the supernatant fluid. The polysaccharide precipitate is recovered by centrifugation, washed with acetone and dried under vacuum. When whole cells are used a mixture of polysaccharides of cell wall and intracellular origin is obtained.

G. Citrate extraction of mannan from yeasts

The dried yeast (200 g) is suspended in 500 ml of citrate buffer (0·05 M, pH 7·0) and heated for 2 h in an autoclave at 140°C. Centrifugation at 10,000 g for 15 min yields a supernatant fluid and a semisolid gel. The gel is suspended in 1·0 litre water and autoclaved for a further 2 h at 140°C. After centrifugation as before, the two supernatants are pooled and concentrated under reduced pressure to about 500 ml. Glacial acetic acid (25 ml) is added and the brown sludgy precipitate resulting is removed by centrifugation at 10,000 g for 15 min. The solution is neutralized by the careful addition of 6 N NaOH and 1·0 litre cold ethanol ($-40°C$) added to precipitate the crude mannan. It can be recovered by filtration or by centrifugation and deproteinized by the Sevag technique.

H. Ethylenediamine extraction of yeast glycoproteins

Pure yeast glycans may be prepared by the methods already described but it is also possible to isolate glycoprotein complexes (Korn and Northcote, 1960). Dry yeast cell walls (1 g) are suspended in anhydrous ethylenediamine (250 ml). The suspension is shaken regularly for several hours yielding a gelatinous mixture, then stirred slowly for 72 h at 37°C. The solution is then centrifuged to give a deposit which, after washing with ethylenediamine, methanol and ether, is dried under vacuo. It comprises about 45% of the cell wall of baker's yeast and contains glucose, mannose and protein. The supernatant fluid is concentrated under reduced pressure to 5 ml and 250 ml methanol is added. The resultant precipitate is suspended in water and dialysed for 72 h against distilled water. Any soluble material is removed by centrifugation. The supernatant solution containing mannan and protein is freeze-dried yielding about 35% of the starting material.

VI. INORGANIC POLYPHOSPHATES

Although the ability to accumulate inorganic polyphosphates is widely distributed amongst micro-organisms, there is considerable variation in the range of environmental conditions under which this occurs. Most cells only produce large amounts under a relatively restricted range of growth conditions, and the reader is referred to the review of Harold (1966) for a detailed examination of these effects. In bacteria, largest amounts tend to be produced in stationary phase sulphur-deficient cultures or when inorganic phosphate is added to cells previously subjected to phosphate starvation—"the polyphosphate overplus" reaction. For example, in bacteria growing on a simple synthetic medium, the following methods can be used to produce cells with a maximal accumulation of polyphosphate.

Sulphur-deficient medium

The following components are dissolved in distilled water (g/l):

Glucose	5·0
KH₂PO₄	0·5
NH₄Cl	1·0
NaCl	1·0
MgCl₂	0·01
Na₂SO₄	0·001
Tris-(hydroxymethyl)-methylamine	10·0

The pH is adjusted to 7·5 and the organisms are grown until they are well into the stationary phase.

Phosphorus-deficient medium

The following components are dissolved in distilled water (g/l)

Glucose	5·0
KH$_2$PO$_4$	0·003
NH$_4$Cl	1·0
NaCl	1·0
MgCl$_2$	0·01
Na$_2$SO$_4$	0·1
Tris-(hydroxymethyl)-methylamine,	10·0

The pH is adjusted to 7·5 and the organisms are grown until they are into the stationary phase. At this point, an excess of phosphate (0·5 g KH$_2$PO$_4$/ litre) is added to allow the polyphosphate overplus reaction to occur and the organisms are harvested when the level of polyphosphate is at its maximum as judged by volutin staining (Laybourne, 1924). These growth methods or their equivalent in terms of the level of sulphate or phosphate for micro-organisms with more complex growth requirements, can be used for many bacteria and yeasts. In some organisms such as algae, considerable quantities of polyphosphate may occur under normal growth conditions and the type of medium used is not so critical.

A. Acid extraction methods

The classical methods for the extraction of inorganic polyphosphates from micro-organisms have derived from the general fractionation schemes for phosphorus compounds of Schneider and of Schmidt and Thannhauser. These methods have given two fractions according to their acid solubility.

1. *"Soluble" polyphosphate extractable with cold 5% (w/v) trichloroacetic or perchloric acid*

This polyphosphate, which usually comprises a small fraction of the total polyphosphate of most micro-organisms, is considered to be material of predominantly low chain length.

2. *"Insoluble" polyphosphate which is recovered in the nucleic acid fraction using the general fractionation methods*

This material is of relatively high molecular weight with an average chain length usually in the hundreds, although there may be degradation in the extraction process with hot trichloroacetic or perchloric acids.

15–5*b*

The following method is based on that of Harold (1962, 1963).

The cells are extracted with cold perchloric acid as described previously (p. 348). Although the amount of "soluble" polyphosphate in this fraction is usually low, it can be further purified as follows. Washed Norit A charcoal is added to the perchloric acid extract at the level of about 100 mg per 10 ml. After 10 min, the charcoal is removed by centrifugation. The polyphosphate can be precipitated from the supernatant fluid as the barium salt at pH 4·5.

The "insoluble" polyphosphate is separated as follows.

The residue from cold perchloric acid extraction of 100 g cells is treated as follows:

(a) 100 ml ethanol are added at room temperature for 30 min followed by centrifugation.

(b) The residue is further extracted by 100 ml of an ethanol-ethyl ether mixture (3 : 1) by boiling for 1 min followed by standing at room temperature for 20 min and extraction. The residue is dried by heating in a water bath.

(c) The polyphosphate is extracted from the residue by heating twice with 100 ml portions of 0·5 N perchloric acid at 70°C for 15 min.

(d) Nucleic acids are removed by the addition of 10 g Norit A charcoal and centrifugation. "Insoluble" polyphosphate remains in the supernatant and can be precipitated as the barium salt at pH 4·5. Alternatively, since barium will catalyse some hydrolysis of polyphosphate, precipitation can be carried out by increasing the ionic strength (see p. 351).

As an alternative to this method; polyphosphate is extracted in stage 3 by treatment with 1 N KOH at 35°C according to the Schmidt-Thannhauser procedure. It will again be mixed with nucleic acid which can be removed by charcoal as in stage 2d or by some other means as studied by Ebel and his co-workers. (See Harold, 1966).

B. Isolation by alkaline hypochlorite digestion

This method devised by Harold (1963) depends on the observation that volutin granules are resistant to digestion with alkaline sodium hypochlorite because the high chain length of bacterial polyphosphate renders it virtually insoluble in solutions of high ionic strength.

The washed bacterial cells (about 10 g dry weight or 100 g wet weight) are digested with 500 ml of alkaline hypochlorite reagent (see p. 358) for 45 min at 25°C. The residue is collected by centrifugation at 20,000 g for 10 min in the cold and is washed twice with 100 ml of cold 1·5 M NaCl containing 10^{-3} M ethylenediaminetetraacetic acid. It is extracted twice with 30 ml of distilled water. To the pooled extract solid NaCl is added to give a concentration of 1·5 M; 15 ml of ethanol are then added. Inorganic polyphosphate will be precipitated and can be redissolved and reprecipitated if further purification is needed.

VII. NUCLEIC ACIDS

Nucleic acid extraction from micro-organisms presents several difficulties absent from corresponding preparative methods for cells from higher organisms. One of these problems is associated with the rigid cell wall structures found in microbial cells. Further the lack of a nuclear membrane in the procaryotic cell renders the isolation of intact nuclei impossible. The methods required to disrupt the microbial cell wall are such that the isolation of the nucleic acid in a relatively undenatured form is extremely difficult. Another difference from higher organisms is the presence in most microbial cells of a very much larger quantity of RNA than of DNA, thus rendering the preparation of RNA-free DNA a fairly long and complicated procedure. In general, the level of nucleic acids and, in particular, of RNA is related to the rate of growth. If high yields are required, a medium giving the highest growth rate possible should be used and cells must be harvested in the exponential phase in batch culture or by continuous culture. The preparation and composition of ribonucleoproteins and ribonucleic acids from microbial and other cells has been reviewed in the monograph by Allen (1962). A critical discussion of the quantitative extraction procedures for RNA and DNA and their subsequent analysis is given by Herbert, Phipps and Strange (this Volume, p. 209).

A. DNA

1. *Detergent extraction*

One of the methods most widely used for bacteria has been developed by Marmur (1961) and involves lysis of the cells with detergent with or without prior lysozyme treatment. The cells (approximately 100 g wet weight) are harvested in the logarithmic phase of growth by centrifugation and washed in 2·0 litre EDTA/saline (0·1 M EDTA : 0·15 M NaCl) to inhibit deoxyribonuclease activity. They are again centrifuged and resuspended in 1·0 litre EDTA/saline. Sodium lauryl sulphate (80 ml of 25% w/v solution) is added and the mixture heated at 60°C for 10 min, then cooled. This method has proved suitable with the families Enterobacaceae and Pseudomonadaceae; however other bacterial species may require prior digestion with lysozyme for cell lysis to be complete. Thus certain *Bacillus* species and others are suspended in EDTA/saline and treated with 400 mg lysozyme with stirring for 30–60 min. The sodium lauryl sulphate extraction at 60°C for 10 min is then carried out.

After cooling the extract, sodium perchlorate (5·0 M) is added to give a final concentration of 1·0 M in the cell lysate. The viscous suspension is shaken with an equal volume of chloroform/isoamyl alcohol (24 : 1, v/v) for 30 min at room temperature. An emulsion is formed at this stage but is separated into three layers by centrifugation at 10,000 *g* for 5 min. The nucleic acid is found in the upper aqueous layer and is placed in a narrow tube. Two volumes of absolute ethanol are layered on top of the aqueous solution and stirred with a glass rod. The nucleic acid is precipitated as white strands and wound round the rod. It is transferred to 500 ml of saline/citrate (0·15 M saline : 0·015 M trisodium citrate, pH 7·0) and slowly dissolved. Further de-proteinization is achieved by shaking for 15 min with an equal volume of chloroform/isoamyl alcohol, followed by centrifugation at 5000 *g* for 5 min to separate the phases. The aqueous layer is removed and deproteinization repeated until negligible amounts of protein appear at the interface. Finally the nucleic acid is precipitated with ethanol and redissolved in saline/citrate. The product is a mixture of DNA, RNA and protein.

The RNA is destroyed by the addition of 50 mg RNAase/ml and digestion at 37°C for 30 min. A new series of deproteinization steps is performed again until the interface is effectively free of precipitate. The final supernatant fluid is precipitated with two volumes of ethanol. The nucleic acid is dissolved in 360 ml saline/citrate and the solution is stirred rapidly, with the slow addition of 0·54 volume of isopropyl alcohol. At first a gel forms then a fibrous precipitate is obtained. The precipitate is redissolved in saline/citrate and the precipitation with iso-propanol repeated. The product is washed with increasing concentrations of ethanol (70% to 95%). A yield of 0·1% of the wet weight of cells has been claimed (Marmur, 1961).

2. *Trichloroacetic acid extraction*

The trichloroacetic acid-sodium hydroxide extraction scheme devised by Schmidt and Thannhauser (1945) can be applied to micro-organisms for the preparation of DNA. In its original form it proved satisfactory for bacteria such as *Escherichia coli* but DNA preparations from *Euglena* cells were contaminated with ribonucleotides (Downing and Schweigert, 1956). However, with slight modification it is probably fairly satisfactory.

The washed cell pellet (100 g) is added to 400 ml of 7% (w/v) trichloroacetic acid solution and stirred for 20 min at room temperature. The mixture is filtered using a Buchner funnel with the paper covered with a layer of Supercel or other filter aid. The filter is washed with ice-cold 1% (w/v) trichloroacetic acid solution and water. The residue on the filter is suspended in 800 ml of ethanol/ether (3 : 1, v/v) and boiled under reflux for 5 min. The residual material is recovered on a filter, washed with ether and dried. It is ground to a fine powder in a mortar, and extracted with 800 ml of a mixture of methanol/chloroform (1 : 1) for 30 min under reflux. The insoluble material is recovered using a Buchner funnel, washed with ether and dried under vacuum. It is extracted in a sealed vessel with N KOH (10 ml/g dry material) for 15 h at 37°C. After this extraction, the mixture is centrifuged at 5000 g for 10 min to remove any filter aid present. To each 100 ml of supernatant fluid is added 20 ml 6 N HCl and 100 ml 5% (w/v) trichloroacetic acid solution. The DNA precipitates at this stage, leaving a supernatant fluid containing phosphoprotein and hydrolysis products from RNA. The DNA is dissolved in about 100 ml water and dialysed exhaustively against distilled water. It can then be lyophilized.

3. Perchlorate extraction for DNA and RNA

A disadvantage of the Schmidt-Thannhauser type of procedure is that the RNA is destroyed. An alternative procedure has been proposed in which both DNA and RNA can be isolated (Ogur and Rosen, 1950). This method is applicable to most microbial species but has the slight disadvantage that some types of cell, especially those of some bacterial species, are relatively resistant to perchlorate extraction and the RNA may not be completely extracted. The starting material used is an acetone powder of microbial cells. The cells (10 g) are added to 100 ml of ethanol/ ether (3 : 1, v/v) and boiled under reflux for 5 min. The preparation is allowed to cool and sediment and the solvent removed by decantation. The ethanol/ether extraction is repeated and the cell material suspended in 100 ml N HClO$_4$ at 4°C. The suspension is stirred for 18 h at 4°C. After centrifugation at 5000 g for 15 min, the cell debris is extracted twice more with N HClO$_4$. The three extracts are pooled and dialysed against water until they are free from acid. The RNA can be recovered by precipitation with ethanol or by lyophilization. The residual material, recovered from centrifugation of the perchlorate extracts, is suspended in 100 ml N HClO$_4$ and stirred for 30 min at 80°C. The mixture is centrifuged at 5000 g for 15 min and the deposit is re-extracted with N HClO$_4$ at 80°C. The two extracts are pooled and dialysed against water. The DNA is recovered by ethanol precipitation or by lyophilization. Slight contamination of the products with one another can occur.

B. RNA

1. Detergent extraction

A large beaker, preferably of stainless steel is used. The beaker is covered tightly with aluminium foil and a stirrer is passed through the middle of the foil. The following are added to the beaker: 10 g sodium lauryl sulphate (syn. sodium dodecyl sulphate); 22·5 ml absolute ethanol; 0·975 g NaH_2PO_4. $2H_2O$; 0·887 g Na_2HPO_4 and 500 ml water. This mixture is heated with continuous stirring to 100°C. The dried microbial cells (50 g) are added with stirring and heating continued to bring the temperature to 92–94°C NaH_2PO. $2H_2O$. The beaker is placed on a boiling water bath for 2 min, stirring being continued throughout. It is then cooled rapidly with stirring to 4°C using a solid CO_2/cellosolve or similar coolant mixture. The extract is centrifuged at 5000 g for 15 min. To every 100 ml of the supernatant fluid, 200 ml ethanol is added and the resulting precipitate is recovered by centrifugation at 5000 g for 15 min and washed twice with aliquots of 70% ethanol. To assist flocculation, 5–10 drops of 2 N NaCl solution are added and the precipitate is suspended in 80% ethanol and allowed to stand at 0°C for 16 h. It is dissolved in about 150 ml water and the pH is adjusted to 7·0 with N acetic acid if necessary. The solution is centrifuged at 20,000 g for 30 min at 0°C. Solid NaCl is added to the supernatant fluid to give a 1 N solution. A gel of RNA forms at this stage, and is allowed to stand at 0°C for up to 16 h. It is deposited by centrifugation at 5000 g for 30 min and washed with aliquots of 70% ethanol containing 1% (v/v) 2 M NaCl solution. The precipitate can be dissolved in water, dialysed against water and freeze-dried. The supernatant fluid from NaCl precipitation contains some RNA which can be precipitated by the addition of two volumes of cold (-40°C) ethanol.

2. Sodium chlorate/detergent extraction for RNA

More prolonged heat treatment with detergent may be used (Jones, March, and Rizvi, 1957). The dry cells (10 g) are suspended in 300 ml water. An equal volume of a solution containing 2% (w/v) sodium chlorate in 0·14 M NaCl (pH 7·0–7·5) is heated to 65°C and added. The mixture is stirred for 2 h at 60°C, cooled to 0°C and diluted by the addition of 4 volumes of cold 0·14 M NaCl. The cell debris is removed by centrifugation at 15,000 g for 30 min and re-extracted with several aliquots of 0·14 M NaCl (adjusted to pH 7·0 with $NaHCO_3$). Solid NaCl is added to the combined extracts to give a 0·25 M solution. Centrimide is added until precipitation is complete and the precipitate is recovered by centrifugation at 10,000 g for 15 min. It is suspended in 100 ml M NaCl and centrifuged at 15,000 g for

30 min. The deposit is dissolved in 2% (w/v) sodium lauryl sulphate (100 ml) and an equal volume of 2 M NaCl is added. Any precipitate forming at this stage is removed by centrifugation at 15,000 g for 15 min. The supernatant fluid is deproteinized by the Sevag method (1934). The nucleic acid is precipitated by the addition of 3 volumes of cold ($-40°$C) ethanol.

3. *Phenol extraction of RNA*

The method for extraction of RNA with phenolic solutions from animal cells (Kirby, 1956) can also be applied to some microbial cells or to cell lysates. The washed cells (100 g) are suspended in 500 ml water and mixed thoroughly in a Waring blendor for 2 min. The thick suspension is slowly added with vigorous stirring to 500 ml of 90% (w/v) aqueous phenol solution at room temperature and the resultant mixture is stirred for 1 h at room temperature. It is then allowed to stand for 16 h at the same temperature, followed by centrifugation at 5000 g for 15 min at 0°C. The upper aqueous layer is removed carefully with gentle suction, ensuring that the deposit of insoluble protein at the phase interface is not disturbed. This can be achieved by the use of a glass wash bottle with a fine nozzle to the mouthpiece of which a tube is connected to a water pump at 200 mm Hg pressure. Care must be taken that none of the highly caustic phenolic solutions are spilt on the hands or on working surfaces. After separation of the aqueous solution, potassium acetate is added to give a final concentration of 2% (w/v) and to each 100 ml of such solution, 200 ml cold ethanol is added. The precipitate of RNA is deposited by centrifugation at 5000 g for 15 min, washed with ethanol/water (3 : 1 by volume) and dissolved in 100 ml water. Any residual ethanol is removed using a rotary evaporator. To each 100 ml RNA solution is added 100 ml 2·5 M K_2HPO_4, 5 ml 33·3% (v/v) H_3PO_4 and 100 ml 2-methoxyethanol. The lower layer is separated after standing for 30 min at room temperature. It is washed once with 20 ml of the upper layer from a mixture of 2·5 M K_2HPO_4; 33·3% H_3PO_4; H_2O : 2-methoxyethanol, 20 : 1 : 20 : 20 (by volume). The upper layer from this and the original upper layer are pooled and centrifuged at 10,000 g for 1 h to remove any insoluble material. The resulting supernatant fluid is dialysed exhaustively against water and potassium acetate is added to give a 2% (w/v) concentration. Ethanol (2 vols) is added to precipitate the RNA which is then washed with ethanol/water (3 : 1, v/v) and dried under vacuum. Contamination with microbial polysaccharide can be reduced greatly by using cell-free extracts as the starting material.

For dimension see Herbert, Phipps, and Strange (this Volume, page 209).

C. Phenol extraction of sRNA

The method used is essentially that described by Holley et al. (1961). The washed pellet of micro-organisms (100 g) is mixed in a Waring blendor with 40 ml of 90% (w/v) aqueous phenol solution and 100 ml water. Once the suspension is uniform, the mixture is stirred for 1 h at room temperature and allowed to stand 16 h at room temperature. It is centrifuged at 5000 g for 15 min at 0°C and the aqueous supernatant fluid carefully removed using suction. To this solution, a further 4·5 ml 90% (w/v) phenol is added with stirring and the centrifugation at 5000 g for 15 min repeated. The upper aqueous phase is again separated and to it is added (per 100 ml) 1 ml 20% (w/v) potassium acetate (pH 5·2) and 200 ml cold absolute ethanol (−40°C). After standing for 16 h at 4°C, the sRNA is deposited by centrifugation at 5000 g for 15 min. The deposit is washed with ethanol and may be dried under vacuum or purified further by chromatography on ion-exchange cellulose. The yield is of the order of 20 mg/10 g dry cells.

See also Herbert, Phipps, and Strange (this Volume, page 209).

VIII. PROTEINS

The number of proteins which can be extracted directly from unbroken microbial cells is limited, mainly comprising those which are unaffected by pH values in the range 4·0–6·0. In this group are several of the microbial cytochromes and also some other metalloproteins and simple proteins of less widespread occurrence such as desulphoviridin (Postgate, 1956) and azurin (Sutherland, 1966). The level of metalloproteins within the cells is dependent on the growth conditions employed and on the microbial species. The optimal yield of cytochromes in bacterial species such as *Pseudomonas fluorescens* is found in anaerobically grown cells (Lenhoff et al., 1955), and an adequate iron content in the medium was also required. The methods of extraction following are applicable to whole cells after harvesting by centrifugation, but in most cases higher yields can be obtained if the cells are lyophilized or if an acetone powder is prepared prior to extraction. They can also be used for ultrasonic lysates.

A. Acetate extraction

The washed microbial pellet (100 g) is suspended in 500 ml 0·1 M acetate buffer (pH 5·0). The mixture is homogenized in a Waring blendor for 1–2 min and stirred at 4°C for 16 h. The cell debris is allowed to sediment and the supernatant fluid removed by decantation. The debris is re-extracted with a further 500 ml acetate buffer for 16 h. The two supernatants are pooled and centrifuged at 20,000 g for 30 min the deposit being discarded. The coloured solution is concentrated by placing in a dialysis sac and surrounding this with a 100% (w/v) solution of polyethylene glycol (mol. wt. 6000) for 48 h at 4°C (Kohn, 1959). The contents of the sac are dialysed against several volumes of distilled water or dilute buffer (e.g. 0·02 M phosphate, pH 7·0) for 16 h, centrifuged at 100,000 g for 30 min and stored at −40°C, or lyophilized if no salts are present. The resultant mixture of haemoproteins and similar compounds can be separated by electrophoresis or by ion-exchange chromatography on cellulose or dextran (Sephadex).

B. Citrate extraction

Washed and lyophilized micro-organisms (10 g) are suspended in 800 ml citrate buffer (0·1 M, pH 6·0) and mixed thoroughly for several minutes in a Waring blendor. The pH value of the suspension is checked and adjusted to pH 6·0 with 0·1 M citric acid if necessary. The suspension is stirred for 16 h at 4°C and centrifuged at 20,000 g for 30 min. The deposit is re-extracted in the same way and the pooled supernatant fluids are treated in the same way as acetate extracts.

C. Urea extraction

The washed pellet of microbial cells (100 g) is suspended in 500 ml of 10 M urea solution and frozen at −40°C. It is thawed at 37°C and stirred at room temperature for 2 h. The viscous suspension is ultracentrifuged at 100,000 g for 1 h. The deposit is re-extracted with a further 500 ml of urea solution for 2 h at room temperature and again centrifuged. The two extracts are pooled and dialysed against water or against a dilute buffer solution. The preparation can then be stored in the cold or lyophilized. It contains a mixture of haemoproteins, proteins and nucleic acid which can be separated by protamine sulphate or streptomycin precipitation followed by chromatography.

D. Acetone extraction of ferredoxin

The electron transport protein ferredoxin is soluble in aqueous acetone (1 : 1, v/v) (Mortenson, 1964) a feature which can be utilized for its isolation. The microbial cells are harvested, washed and lyophilized. The cells (10 g) are suspended in 100 ml water and homogenized in a Waring blendor for 30 min. An equal volume of acetone is added to the suspension and the mixture is stirred at 0°C for 15 min. The bulk of the proteins and the cell debris are removed by centrifugation at 10,000 g for 10 min at

$0°C$, leaving the ferredoxin in the amber-coloured supernatant fluid. The electron transport protein can be recovered by passing the solution directly through a column of DEAE-cellulose, when the ferredoxin together with the flavin nucleotides and flavoproteins is adsorbed. The yield of crude ferredoxin is of the order of 12 mg in typical microbial strains such as *Clostridium pasteurianum.*

E. Acid extraction of basic proteins

Basic proteins showing considerable similarity to histones are found in bacteria and other micro-organisms (Leaver and Cruft, 1966) and can be extracted with acid. The cells (100 g wet wt.) are suspended in 500 ml cold ($4°C$) 4% (v/v) acetic acid and centrifuged at 10,000 g for 15 min at $0°C$. The supernatant fluid is discarded. The deposit is resuspended in acetic acid and centrifuged twice more, making three washings in all. The final deposit is washed in ethanol (3 ×) and in ether (3 ×) and dried under vacuum. The dry microbial preparation (10 g) is ground with glass powder in ice cold 0·1 N H_2SO_4 for 5 min. The grinding is repeated several times, while leaving the preparation in the cold for 3 h. A further 30 ml of N H_2SO_4 is added and the extraction allowed to proceed for 3 h in the cold. The cell debris and glass-powder are removed by centrifugation at 25,000 g for 30 min and re-extracted twice more with 30 ml aliquots of acid. The three supernatant fluids are pooled and dialysed against 0·1 N H_2SO_4 (10 g) for 16 h at $4°C$. The protein sulphates can be precipitated from the non-diffusible material by the addition of 8 vols cold ($-40°C$) ethanol. Alter natively, the non-diffusible material is dialysed further against distilled water and against bicarbonate buffer (0·01 M, pH 7·0). Any material precipitating at this stage is removed by centrifugation. Contaminating non-basic protein is removed by chromatography on CM-cellulose. Yields of 30–90 mg/10 g dry microbial cells have been obtained from some species (Leaver and Cruft, 1966).

IX. SEQUENTIAL METHODS

There is a wide variety of sequential methods available and the choice of a suitable one depends upon the purpose of the extraction and the organism being used. Most sequential methods involve the denaturation of some of the components, but for most purposes this is unimportant. Unfortunately, it is impossible to devise a method in which each main class of compound occurs in a particular fraction and which can be applied to any micro-organism. This applies particularly to substances like polysaccharides which may appear in a variety of different fractions and which vary considerably in their properties from one organism to another. The method that we describe below is mainly based on those given previously in this Chapter and can separate most groups of compounds in a typical bacterium. It is outlined in Fig. 3. However, it does not deal with particular groups of compounds such as lipopolysaccharides in Gram-negative organisms and teichoic acids in Gram-positive organisms and will require to be altered according to the organism being used and the needs of a particular experiment by reference to the methods described previously.

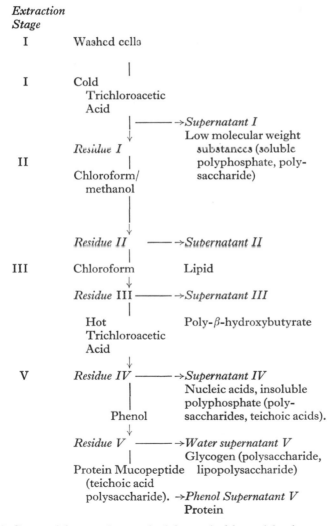

FIG. 3. Sequential extraction method for typical bacterial culture.

The starting material is in the form of washed cells and it must be borne in mind that there may be many soluble extracellular materials present as a result of microbial growth which will occur in the initial washings of the cells, and some intracellular materials may be lost on washing.

A. Extraction Stage I

The washed microbial cells (equivalent to about 100 g wet weight or 10 g dry weight) are extracted with 100 ml of 10% (w/v) trichloroacetic acid. The mixture is stirred at 0°C for 30 min and is centrifuged at 10,000 g for 5 min. The residue is washed with 25 ml cold 10% trichloroacetic acid and the supernatants are combined. Polysaccharide (e.g. some capsular polysaccharide) may be extracted at this stage together with any "soluble" polyphosphate that is present.

As an alternative method, extraction with cold perchloric acid (p. 348) or ethanol (pp. 348–349) can be used. In the latter case, polysaccharide and polyphosphate will remain in the residue.

B. Extraction Stage II

To residue I is added 100 ml methanol and the mixture is refluxed for 5 min. After cooling to room temperature, 200 ml chloroform is added and the mixture is stirred at room temperature for 20 min. The residue is removed by filtration and re-extracted for 20 min at room temperature with a further 300 ml chloroform/methanol (2 : 1, v/v). The supernatants are combined.

This extraction should recover most of the cell lipid in Supernatant II. Alternatively, it may be sufficient for the purpose of the experiment to use cold chloroform/methanol (pp. 351–352) or some further hydrolysis may be required prior to extraction (pp. 353).

C. Extraction Stage III

This stage should only be necessary in cells which store poly-β-hydroxybutyrate which will not be extracted by a chloroform/methanol mixture.

The residue from Stage II is extracted using 100 ml chloroform at room temperature with stirring. After centrifugation, the residue is extracted with a further 50 ml chloroform. The poly-β-hydroxybutyrate in the combined supernatants can be precipitated by the addition of three volumes of ethyl ether.

D. Extraction Stage IV

The residue from Stage III is extracted by treatment with 100 ml of 5% (w/v) trichloroacetic acid at 90°C for 30 min. If necessary, the residue left after centrifugation (5000 g for 15 min) can be extracted with a further 50 ml of 5% (w/v) trichloroacetic acid under similar conditions. The supernatant will contain all the nucleic acid, the insoluble polyphosphate fraction and some polysaccharides including most of any teichoic acids present (a small amount may be extracted in Stage I).

An alternative to the use of hot trichloroacetic acid which will result in destruction of RNA and, to a lesser extent, of DNA, is to use perchlorate extraction. This leads to less breakdown of nucleic acid. The residue from Stage III is suspended in 100 ml N HClO$_4$ and stirred for 18 h at 4°C. After centrifugation at 5000 g for 15 min, the cell debris is extracted twice more with 50 ml N HClO$_4$. The pooled extracts are dialysed and should contain most of the RNA. The residue is suspended in a further 100 ml N HClO$_4$ and stirred for 30 min at 80°C. After centrifugation (5000 g for 15 min), the residue is re-extracted with two further aliquots of 50 ml HClO$_4$ at 80°C. The pooled supernatants will contain most of the DNA.

E. Extraction Stage V

The residue from Stage IV is suspended in 150 ml distilled water and warmed to 65°C. An equal volume of 90% (w/v) aqueous phenol at the same temperature is added and the mixture is stirred vigorously for 5 min. It is then cooled rapidly and centrifuged at 3000 g for 30 min at 0°C to allow phase separation. The upper aqueous layer will contain certain polysaccharides (especially glycogen), the phenol layer certain proteins, while the residue on the interphase will contain protein, mucopeptide and certain other polysaccharides which have not been extracted in previous stages. Slight hydrolysis of glycogen may occur in Stage IV.

Application of the sequential procedure to 2–5 mg dry weight of cell is possible but losses in recovery at some stages would be large and addition of unlabelled carrier, ca. 0·5–1·0 g dry weight, might be advantageous.

ACKNOWLEDGMENT

The authors are grateful for the assistance of Mrs. A. M. Stuart in evaluating several of the methods described.

REFERENCES

Allen, F. W. (1962). "Ribonucleoproteins and Ribonucleic Acids". Elsevier. Amsterdam.

Asselineau, J., and Lederer, E. (1960). *In* "Lipid Metabolism". (Ed. K. Bloch) p. 337. Wiley, New York.

Barker, S. A., Bourne, E. J., and Stacey, M. (1953). *J. Chem. Soc.*, p. 3084.

Barret, J. (1956). *Biochem. J.*, **64**, 626.

Bligh, E. G., and Dyer, W. J. (1959). *Canad. J. Biochem. Physiol.*, **37**, 911.

Boivin, A., Mesrobeanu, I., and Mesrobeanu, L. (1933). *C.r. Séanc. Soc. Biol.*, 114, 307.
Dankert, M., Wright, A., Kelley, W. S., and Robbins, P. W. (1966). *Archs Biochem. Biophys.*, 116, 425.
Davies, D. A. L. (1955). *Biochem. J.*, 59, 696.
Depinto, J. A. (1967). *Biochim. biophys. Acta*, 144, 113.
Duguid, J. P. (1951). *J. Path. Bact.*, 63, 673.
Eimhjellen, K. E., and Jensen, S. L. (1964). *Biochim. biophys. Acta*, 82, 21.
Freeman, G. G. (1942). *Biochem. J.*, 36, 340.
Fuller, R. C., Smillie, R. M., Rigopoulus, N., and Yount, V. (1961). *Archs Biochem. Biophys.*, 95, 197.
Hancock, R. (1958). *Biochim. biophys. Acta*, 28, 402.
Harold, F. M. (1962). *J. Bacteriol.*, 83, 1047.
Harold, F. M. (1963). *J. Bacteriol.*, 86, 885.
Harold, F. M. (1966). *Bact. Rev.*, 30, 772.
Higashi, Y., Strominger, J. L., and Sweeley, C. C. (1967). *Proc. nat. Acad. Sci. U.S.A.*, 57, 1878.
Holley, R. W., Apgar, J., Doctor, B. P., Farrow, J. Marini, M. A., and Merrill, S. H. (1961). *J. biol. Chem.*, 236, 200.
Horecker, B. L. (1966). *Ann Rev. Microbiol.*, 20, 253.
Hughes, R. C. (1965). *Biochem. J.*, 96, 700.
Jensen, S. L., Aasmundrud, O., and Einhjellen, K. E. (1964). *Biochim. biophys. Acta*, 88, 466.
Jones, A. S., Marsh, G. E., and Rizui, S. B. H. (1957) *J. gen. Microbiol.*, 17, 586.
Kates, M. (1964). *Adv. Lipid Research*, 2, 17.
Kates, M. (1966). *Ann Rev. Microbiol.*, 20, 13.
Kirby, K. S. (1956). *Biochem. J.*, 64, 405.
Kohn, J. (1959). *Nature, Lond.*, 183, 1055.
Korn, E. D., and Northcote, D. H. (1960). *Biochem. J.*, 75, 12.
Laybourn, R. L. (1924). *J. Am. med. Ass.*, 82, 121.
Leaver, J. L., and Cruft, H. J. (1966). *Biochem. J.*, 101, 665.
Lenhoff, H. M., Nicholas, D. J. D., and Kaplan, N. O. (1955). *J. biol. Chem.*, 220, 983.
Lester, R. L., and Crane, F. L. (1959). *J. biol. Chem.*, 234, 2169.
Luderitz, O., Jann, K., and Wheat, R. (1967). *In* "Comprehensive Biochemistry". (Eds. Florkin, M. and Stotz, E. H.) Vol. 26A, 105. Elsevier, Amsterdam.
Luderitz, O., Risse, H. J., Schulte-Holthausen, H., Strominger, J. L., Sutherland, I. W., and Westphal, O. (1965). *J. Bact.*, 89, 343.
Macrae, R. M., and Wilkinson, J. F. (1958). *J. gen. Microbiol.*, 19, 210.
Marmur, J. (1961). *J. molec. Biol.*, 3, 208.
Marr, A. G., and Ingraham, J. L. (1962). *J. Bact.*, 84, 1260.
Mortensen, L. E. (1964). *Biochim. biophys. Acta*, 81, 71.
Morton, R. A. (1965). *In* "Biochemistry of Quinones". Academic Press, London.
Ogur, M., and Rosen, G. (1950). *Archs Biochem. Biophys.*, 25, 262.
O'Leary, W. M. (1962). *Bact. Rev.*, 26, 421.
Postgate, J. R. (1956). *J. gen. Microbiol.*, 14, 545.
Robbins, P. W., Wright, A., and Bellows, S. L. (1964). *Proc. natn. Acad. Sci. U.S.A.*, 52, 1302.
Schmidt, G., and Thannhauser, S. J. (1945). *J. biol. Chem.*, 161, 83.
Sevag, M. G. (1934). *Biochem. Z.*, 273, 419.

Sutherland, I. W. (1966). *Archiv. Mikrobiol.*, **54**, 350.
Sutherland, I. W., and Wilkinson, J. F. (1965), *J. gen. Microbiol.*, **39**, 373.
Suzue, G., Toukada, K., and Tanaba, S. (1967). *Biochim. biophys. Acta*, **144**, 186.
Vorbreck, M. L., and Marinetta, G. V. (1965). *J. Lipid. Res.*, **6**, 3.
Westphal, O., Luderitz, O., and Bister, F. (1952). *Z. Naturf.*, **7B**, 148.
Wilkinson, J. F. (1963). *J. gen. Microbiol.*, **32**, 171.
Williamson, D. H., and Wilkinson, J. F. (1958). *J. gen. Microbiol.*, **19**, 198.
Wright, A., Dankert, M. Fennessey, P., and Robbins, P. W. (1967). *Proc. natn. Acad. Sci, U.S.A.*, **57**, 1798.

CHAPTER V

Biphasic Separation of Microbial Particles

PER-ÅKE ALBERTSSON

Department of Biochemistry, University of Umeå, Umeå, Sweden

I.	Introduction	385
II.	Polymer Phase Systems	386
	A. Introduction	386
	B. Phase diagrams	390
	C. Polymers and preparation of polymer solutions	392
	D. Some properties of polymer phase systems	393
III.	Distribution of Macromolecules and Particles	393
	A. Introduction	393
	B. Proteins and nucleic acids	393
	C. Particles	401
IV.	Virus Purification and Concentration	407
	A. Introduction	407
	B. Bacteriophage T2	408
	C. Poliovirus and Echo virus	410
V.	Counter-Current Distribution (CCD)	412
	A. Introduction	412
	B. Thin layer CCD	413
	C. Liquid-interface CCD	414
	D. CCD of bacteria	415
	E. CCD of virus	417
	F. CCD of antigen-antibody complexes	418
VI.	Removal of Polymers	419
VII.	Discussion	420
References		421

I. INTRODUCTION

Separation by liquid-liquid partition is a classical method used in different fields of chemistry. It is attractive because it is both selective and simple. Essentially partition is based on the selective distribution of substances between two immiscible phases. The behaviour of a substance in a given phase pair is determined by the physical and chemical properties of the substance and the phase components in a complicated manner which can usually be interpreted in a qualitative fashion only.

The choice of a suitable phase system is the key step in the development of a partition method. Particularly when dealing with biological particles such as cells or cell organelles special attention has to be paid to the properties of the phase components so that delicate cell structures are not damaged with concomitant loss of biological activity. Conventional phase systems which contain an organic solvent phase are generally not suitable. The risk of denaturation by interaction with the organic solvent and also surface denaturation is great. In addition the distribution usually favours the aqueous phase only which makes utilization of the system for separation purposes impossible.

It is therefore necessary to use phase systems in which both phases are aqueous and thus resemble the natural environment of the material studied. Such systems may be obtained by mixing aqueous solutions of two different polymers. It has been shown (Albertsson, 1960) that in such phase systems biogenic particles and macromolecules distribute in a reproducible manner without loss of biological activity. The distribution is also selective and the polymer phase systems can therefore be used for separation purposes.

II. POLYMER PHASE SYSTEMS

A. Introduction

If solutions of two different polymers are shaken together the mixture often becomes turbid even at low concentrations of the polymer such as a few per cent. After standing for a while such a mixture will separate into two layers. Analysis of the phases at various concentrations of the polymers shows that the system behaves as a three component system (disregarding the polydispersity of the polymers) and that the two layers behave as two phases in equilibrium, conforming to the phase rule. An example of a phase diagram is given in Fig. 3.

Such liquid-liquid phase separation in macromolecular mixtures was first reported long ago by the Dutch microbiologist Beijerinck (1896). He observed that if aqueous solutions of gelatin and agar, for example, were mixed a turbid mixture which separated into two liquid layers was obtained. Dobry and Boyer-Kawenoki (1947) studied the miscibility of a large number of pairs of different polymers and could demonstrate that demixing and phase separation was the rule when two different polymers were put together. This so called incompatability is a very common phenomenon. It has been of a great practical nuisance, for example, when clear polymer mixtures for paints are desired.

When two polymers are incompatible they collect in different phases; the two polymers are only partly miscible. In some cases, however, the two polymers collect together in one of the phases which therefore becomes polymer rich while the other phase consists mainly of solvent. This phenomenon occurs frequently when the two polymers carry opposite charges. This kind of phase separation has been studied in detail by Bungenberg de Jong and co-workers (Bungenberg de Jong, 1949). They called these systems "complex coacervates" in contrast to "simple coacervates", which are the systems of two polymers showing incompatibility. An example of "complex coacervation" is an aqueous mixture of gum arabic which is negatively charged and gelatin at such a pH that this is positively charged.

Some phase systems which have been used in partition studies are listed in Table I. As seen both non ionic polymers and polyelectrolytes may be used. In the systems with polyelectrolytes one obtains liquid ion exchangers. Their properties depend a great deal on the ionic composition of the system.

TABLE I

Pairs of polymers which have been used as phase polymers

Dextran—polyethylene glycol
Dextran—hydroxypropyldextran
Dextran—methylcellulose
Dextran sulphate—polyethylene glycol
Dextran sulphate—methylcellulose

The time of phase separation in polymer phase systems is longer than conventional phase systems. This is caused by the high viscosity of the phases and the low density difference. In a test-tube it usually takes 10–30 min. for the phases to settle but this time may be reduced either by using thin layers of the phases or by a short low speed centrifugation.

Multiphase systems are obtained when several polymers, all mutually incompatible, are mixed above certain concentrations. For example a mixture of dextran sulphate + NaCl, dextran, Ficoll (a polysucrose), hydroxypropyldextran and polyethylene glycol may give a five phase system.

Phase separation can also occur in polymer mixtures where one of the polymers is a protein. In some cases the protein will collect in a liquid, protein rich phase but more often the protein forms a thick gel or solid like phase. This is for example frequently the result when proteins are mixed with high concentrations of polyethylene glycol. Such precipitation of proteins, and also virus, by polyethylene glycol has been reported by many authors. It has also been utilized with success for purification purposes (Polson, 1964; Zeppezauer and Brishammar, 1965). Since proteins are polyelectrolytes their phase separation with polyethylene glycol depends on the ionic composition (Albertsson, 1960) as do all other polyelectrolyte systems; see for example phase diagrams of the dextran-sulphate-polyethylene glycol or carboxymethyldextran-polyethylene glycol systems (Albertsson, 1960).

The theory of phase separation in polymer mixtures has been treated by many authors. Their results are summarized and discussed in the books by Flory (1953), Tompa (1956) and Morawetz (1965).

We may compare the aqueous polymer phases with conventional solvents as is done in Fig. 1. To the left in this figure a number of solvents are listed according to their hydrophobic-hydrophilic nature; at the bottom a salt solution then water, acetone etc., with increasing hydrophobicity, up to heptane. We may consider these as selected solvents from a continuous solvent spectrum with increasing hydrophobic character. If we want to place the polymer solutions in this spectrum it is obvious that, since they all

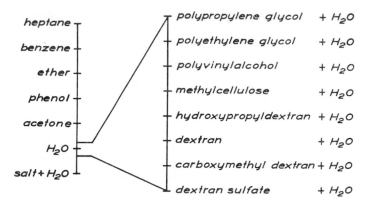

Fig. 1. Hydrophobic ladder. To the left, a number of solvents have been selected from a spectrum of solvents with increasing hydrophobicity. Aqueous solutions of polymers to the right are mutually immiscible but since they all contain mainly water they fall within a narrow part of the solvent spectrum to the left.

contain mainly water, they should fall within a very narrow part of the spectrum. To the right in Fig. 1, a number of polymer solutions are listed according to the hydrophilic-hydrophobic nature of the polymers. The order between the polymers may be somewhat arbitary but the main point is that we have a number of immiscible liquids which all are very close to each other on the solvent spectrum. This means that phase systems formed by these solvents can be expected to be selective in separating substances

TABLE II

Tube No.	20% w/w g Dextran	g 32% PEG 6000	ml 0.4 M NaFB pH 6.8	ml 0.4 mM LiFB pH 6.8	ml 0.4 M KFB pH 6.8	ml 0.4 M NaCl	ml H₂O	ml 0.2 M sucrose	ml Sample in 0.2 M sucrose and 0.004 M NaFB pH 6.8
1	1	0·5	0·1	—	—	—	0·9	0·5	1
2	1	0·5	—	0·1	—	—	0·9	0·5	1
3	1	0·5	—	—	0·1	—	0·9	0·5	1
4	1	0·5	0·1	—	—	0·05	0·85	0·5	1
5	1	0·5	0·1	—	—	0·1	0·8	0·5	1
6	1	0·5	0·1	—	—	0·2	0·7	0·5	1
7	1	0·5	0·1	—	—	0·3	0·6	0·5	1
8	1	0·5	0·1	—	—	0·5	0·4	0·5	1
9	1	0·5	0·1	—	—	0·9	—	0·5	1
10	1	0·5	0·1	—	—	—	1·1	0·5	1
11	1	0·5	0·1	—	—	—	0·5	0·5	1
12	1	0·5	0·1	—	—	—	—	0·5	1

Scheme for investigating the partition behaviour of a substance in the dextran-polyethylene glycol system. In tubes No. 1–3 the effects of Na⁺, Li⁺ and K⁺ ions is compared; in tubes No. 4–9 the effect of addition of NaCl is studied and in 10–12 the effect of polymer composition. In a similar fashion the effect of other salts, pH, or other mol. wt. of polymers may be studied.

which themselves fall within the same aqueous part of the solvent spectrum, such as for example particles and macromolecules of biological origin.

In addition, by varying the properties of the polymers, one should, in theory, be able to vary the selectivity of the systems almost at will. For example polymers with certain conformations and with certain groups—charged, hydrophilic and hydrophobic—should be very effective in selectively extracting from a mixture those components to which the polymers have a specific affinity.

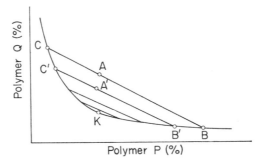

FIG. 2. Phase diagram of a three component system; two components, P and Q, are plotted as abscissa and ordinate.

B. Phase diagrams

The composition of the phases for different polymer concentrations is represented by a phase diagram, Fig. 2. All mixtures represented by points above the curved line (binodial) result in two-phase systems, whereas mixtures represented by points below the curve result in one-phase systems. Thus, a mixture A will split up into two phases represented by points B and C. Similarly, a system with the over-all composition of A^1 has one phase with a composition of B^1 and the other phase with a composition of C^1. Like all other points representing the composition of pure phases, points B, C, B^1, and C^1 lie on the binodial. Pairs of points such as B and C are called nodes and the lines joining them are called tie lines. Point A, representing the overall composition of the phase system, lies on the tie line joining B and C. Any mixture represented by points on the same tie line will give rise to phase systems with the same phase compositions; only the ratio between the phase volumes will vary. The volume ratio B phase/C phase is for example equal to the ratio between the length of line AC and AB if the polymer concentration is expressed as volume per cent (if it is expressed as weight per cent the phase volumes above should be replaced by the phase weights).

The composition of the two phases converge as the overall composition of the system approaches point K, the critical point, Fig. 2. Near this composition the properties of the system are most sensitive to changes in composition or temperature. Phase systems close to a critical point are therefore, if possible, avoided.

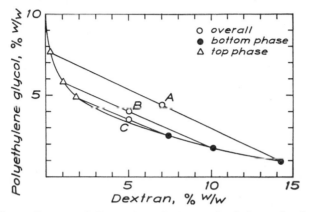

FIG. 3. Phase diagram of the system dextran-polyethylene glycol-water. The compositions represented by A and B have been frequently used for partition studies. They have the following overall composition: A, 7% (w/w) dextran and 4·4% (w/w) polyethylene glycol; B, 5% (w/w) dextran and 4% (w/w) polyethylene glycol. The dextran is "Dextran 500" (Pharmacia, Uppsala). $\overline{M}_w = 460,000$ $\overline{M}_n = 180,000$ and the polyethylene glycol is "Carbowax 6000" (Union Carbide), $\overline{M}_n = 6000$–7500. Temperature 20°C.

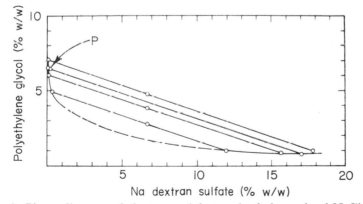

FIG. 4. Phase diagram of dextran sulphate-polyethylene glycol-NaCl-water. The composition represented by point P is 0·2% dextran sulphate and 6·45% Polyethylene glycol. This system has frequently been used for virus concentration. The dextran sulphate is "Dextran sulphate 500" (Pharmacia) and the polyethylene glycol is "Carbowax 6000 (Union Carbide). NaCl concentration is 0·3 M and temperature 4°C.

For a general treatment of different ways of constructing phase diagrams, I refer to the books by Treybal (1951), and Zernike (1955).

Phase diagrams of two systems, which have been widely used for partition studies are given in Figs 3 and 4. Diagram of other phase systems may be found in ref. (Albertsson, 1960).

C. Polymers and the preparation of polymer solutions

Dextran (D). Dextran is branched polyglucose; the glucose units of the main chain are connected by $1 \to 6$ linkages and the branches are connected with this chain through $1 \to 3$ linkages. (The dextran used for the work described here was supplied by Pharmacia, Uppsala, Sweden). Several fractions are available and are characterized by the manufacturer by means of the limiting viscosity number η, number-average molecular weight M_n, and weight-average molecular weight M_w. A fraction which has been used in the studies described here has the commercial name "Dextran 500" and the following data. $\eta = 48$ ml/g; $M_n = 180 \cdot 000$; $M_w = 460 \cdot 000$; These figures may vary a little from batch to batch.

A solution of dextran is prepared in the following way. The dextran is first wetted and mixed to a paste with a small amount of water. The rest of the water is then added and the dextran dissolved by stirring. Commercial dextran usually contain 5–10% moisture. For determination of its concentration, about 10 g of the solution is weighed into a 25 ml measuring flask, which is then filled to the mark with water. The optical rotation of this solution is then determined and the concentration calculated from the (α) D value of dextran, which is $+199°$ at 25°C. For stock solutions, 10–20% (w/w) is a suitable concentration range. Dextran solutions can be sterilized by autoclaving in the usual manner.

Dextran sulphate, sodium salt (NaDS) (supplied by Pharmacia, Uppsala, Sweden). The sample used in the studies described here has the commercial name "Dextran sulphate 500". The sulphur content of NaDS is 17%. It is dried by keeping it over P_2O_5 and dissolved directly in water. Dextran sulphate is not as stable to heat as dextran; it is autoclaved for only 15 min and the pH should first be adjusted to 7.

Polyethylene glycol (PEG). This polymer is marketed by several companies. For the experiments to be described here, Carbowax polyethylene glycol 6000, was used (supplied by Carbon and Carbide Chemicals Co., U.S.A.), and had a molecular weight of 6000–7500. Polyethylene glycol is dissolved directly in water. For a stock solution, 30% (w/w) is a suitable concentration. The Carbowax compounds contain an impurity that has a peak of absorption in UV light at 290 nm. If desirable, this may be removed by reprecipitation in the following way: 300 g PEG 6000 is dissolved by careful warming in 6 litres of acetone. Three litres of ether are then added, with stirring, and the resulting mixture is allowed to stand overnight. Subsequently the precipitate is collected by filtration through a filter paper, washed with an acetone-ether (2 : 1) mixture and dried in air. PEG solutions can be autoclaved in the usual way.

D. Some properties of polymer phase systems

The water content of the two phases is in the range between 85–99%. This high water content makes these phase systems particularly suitable for labile biological material. Due to the presence of polymers the viscosity of the phases is relatively high. The density difference between the phases is small. The time of phase separation is therefore relatively long. The dextran or dextran sulphate-polyethylene glycol systems thus require about 5–30 min for settling of the phases. This time may be reduced to 1 3 min by centrifugation or by employing thin layers (1–2 mm) of the phases (Albertsson, 1965).

The volumes of the phases of the systems containing only non-ionic polymers do not depend on the addition of electrolytes up to about 0·2 M concentration. In contrast, the volumes of the polyelectrolyte systems vary considerably with different electrolytes.

III. DISTRIBUTION OF MACROMOLECULES AND PARTICLES

A. Introduction

When a substance is shaken with a liquid two phase system, and provided it is not precipitated, it may distribute either between the two bulk phases or between the two bulk phases and the interface. The first type of behaviour is usually the case for soluble macromolecules such as proteins and nucleic acids while the second is the case for suspended particles such as cells, cell organelles or large virus particles. It is convenient to describe the two cases separately and in this Section the distribution of macromolecules exemplified by proteins and other molecules, will first be described then the distribution of particles.

B. Proteins and nucleic acids

1. *General behaviour*

A number of partition experiments with proteins (Albertsson and Nyns, 1959, 1961) and nucleic acids (Albertsson, 1965) have been carried out and the data obtained give us an idea about the factors which determine partition. This information is useful for planning fractionation experiments with mixtures of unknown substances.

Size. The particle or molecular size of the partitioned substance has a great influence on partition. Thus, in the dextran-polyethylene glycol systems, low molecular weight substances such as for example salts, amino-acids, nucleotides, distribute evenly between the phases. Larger molecules such as proteins usually have partition coefficients, in the range between 0·1–10. (Partition coefficient = concentration in top phase/concentration in bottom phase). Still larger molecules such as DNA usually give a one-sided distribution, the substance collecting either in the upper or in the lower phase. Which phase the substance prefers depends on factors other than size.

Conformation. Native and denatured DNA display very different parti-tion behaviour (Albertsson, 1962, 1965). This effect of denaturation is probably not only due to the change in conformation alone but also to the exposure of previously hidden groups. The bases in a single stranded DNA are thus more exposed to the surrounding medium than in double stranded DNA. In the case of proteins hydrophobic side chains which are buried in the centre of the native molecule may be more exposed when the protein is denatured and therefore change the partition.

Pettijohn (1967) made the interesting observation that supercoiled DNA molecules differed in partition compared to DNA lacking this topology. In the dextran-PEG system the partition coefficient of supercoiled (form I) polyoma DNA is 6 to 12 fold smaller than those of the open circle and linear forms II and III. This partition difference may arise from the more com-pact conformation of supercoiled DNA relative to the open circle and linear forms or from partially uncoiled helical structures which may exist in the supercoiled form.

Ionic composition and charge. In many phase systems the ionic composi-tion has a strong influence on partition. In the dextran-polyethylene glycol system for example, the partition of proteins (Albertsson and Nyns, 1959, 1961) and nucleic acids (Albertsson, 1965) depends strongly on the salt composition. It is mainly the *ratio* between the different ions which influ-ences the partition coefficient but also for many substances the *overall* concentration of salt. Figure 5 shows the partition coefficient of DNA in the dextran-PEG system with different salts. By changing the ratio between different ions proteins may be more or less transferred from one phase to the other. The same phenomena are observed for RNA and proteins. The transfer of the macromolecules is specific and it can therefore be utilized for separation purposes (see below).

The salt effect on the partition of nucleic acid or protein like that shown in Fig. 5 is characteristic for proteins carrying net negative charges. For positively charged proteins the effect of different ions is reversed. Thus Na^+ and K^+ give higher K values than Li^+ ions for ribonuclease, lysozyme

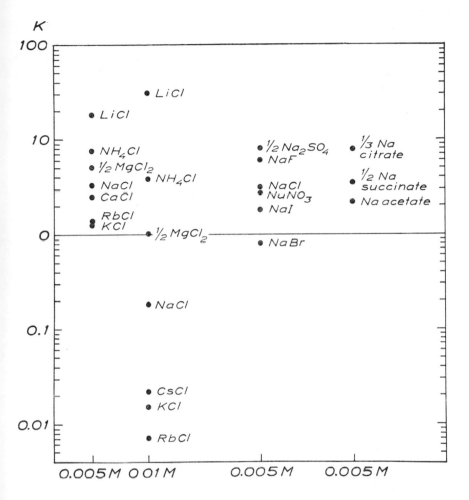

FIG. 5. Partition coefficient, K, of calf thymus DNA with different salts in a dextran-polyethylene glycol phase system. For details see Albertsson (1965b).

or cytochrome C which are positively charged at neutral pH while the opposite is true for egg albumin, serum albumin or phycoerythrin which are negatively charged at the same pH. Close to the isoelectric point the effect of different ions seems to be at a minimum. Figure 6 shows the partition of egg albumin at different pH but in two different salt media. The two curves in Fig. 6 cross close to the isoelectric point of the protein. Studies with other proteins show that this is a general phenomenon and such "cross partition" can therefore be used for the determination of isoelectric points of proteins (Albertsson et al., 1970).

The charged groups of the distributed particle therefore play a role in its distribution. This seems to hold for cell particles too. In the case of red blood cells, for example, Walter and collaborators have shown that there is a correlation between partition and surface charge of the cells (Walter, 1967, 1969).

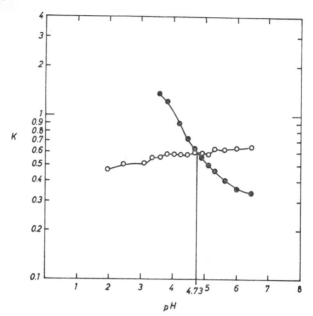

FIG. 6. Partition coefficient of egg albumin as a function of pH in two different salt media, 0·1 M NaCl (filled circles) and 0·05 M Na₂SO₄ (open circles). Glycine and phosphate buffers were used at a concentration of 0·01. The two curves cross each other close to the isoelectric point (Albertsson, 1970).

The reason for the strong influence of ions on the partition is not known. It seems plausible that the ions modify the water layers around the phase polymers and/or that the ions modify the surface layers of the partitioned molecules or particles. For example in LiCl a charged protein molecule would expose a different "surface" to its surroundings than say in NaCl.

Other factors. The type of phase polymer and its molecular weight also influences the partition coefficient. In addition to size and charge other properties of the partitioned particle must determine their partition since particles with the same size and electrophoretic mobility can display rather different partition behaviour. One can expect that the number and location of hydrophobic and hydrophilic groups on the surface of the particle should influence its partition. Our knowledge of the surfaces of particles or large molecules is yet very limited and a comparison between the surface properties and the partition behaviour is therefore difficult to make. However, proteins, whose complete three dimensional structure is known, should offer useful and interesting model substances which would allow such a comparison in order to study the mechanism behind partition.

The characteristic behaviour of macromolecules such as DNA, RNA and proteins of high molecular weight is that the partitioned macromolecules can easily be transferred selectively from one phase to the other by small changes in the phase system composition. Their behaviour can be summarized graphically by Fig. 7 where K values are plotted against change in phase composition (polymer concentration or ionic composition) for different hypothetically macromolecular substances. This behaviour is very favourable for fractionation because only a few extraction steps at different phase compositions are needed to achieve separation.

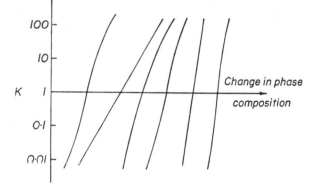

FIG. 7. Typical partition behaviour of large molecules in polymer two-phase systems. A small change in phase composition causes a large change in K value. Low molecular weight substances usually have K values around 1 independent of phase composition.

2. Applications

If the partition coefficients between two substances are sufficiently different a desired separation may be obtained by one single batch operation, for example with a separating funnel. Such a procedure is very easy to scale up. Centrifugation may be used to speed up the settling time.

It is not enough, however, that the partition coefficients are different. The phase volume ratio should also be such that an efficient separation is obtained. Two examples may illustrate this point. In the first the partition coefficients are 100 and 0·01 respectively. An efficient separation will be accomplished if the two phase volumes are equal. In the second example the partition coefficients are 1 and 0·0001 respectively that is the ratio between

the partition coefficients is the same as in the first example. We may therefore say that the phase systems are equally selective two in the cases. However, a poor separation would be obtained in the latter case if the phase
volumes were equal. We have to use an upper phase volume which is 100
times larger than the lower phase in order to get an equally good separation
as in the first case.

Theoretical considerations lead to the conclusion that for one extraction
step and with a given phase system maximum separation between two
substances is obtained when the volumes are arranged such that

$$K_1 . K_2 . \left(\frac{V_t}{V_b}\right)^2 = 1$$

where K_1 and K_2 are the partition coefficients for the two substances and
V_t is the volume of upper phase and V_b the volume of the lower phase.

The experimental technique for batch operations is as follows.

First one has to find out by small partition experiments the behaviour
of the substance under study. This may be done according to the scheme
in Table II. The polymers may be mixed with the other solutions in any
order. They may be added as dry powder or as concentrated solutions. The
main thing is that the *final* concentration of polymer is the one desired.
The contents are then mixed by shaking and allowed to stand for 30 min.
for phase separation. Samples are withdrawn from the two phases by pipette.
A sample from the lower phase can also be obtained by making a small hole
in the bottom of a plastic tube.

Centrifugation may be used to speed up the separation.

Each phase sample is diluted with water or buffer, usually between one
or ten times, before measurement of absorbancy or for enzymatic determination.

Protein is determined by absorbancy at 280 nm or by the method of
Wadell (1956) (Bailey, 1967). The Lowry method (Lowry et al., 1951)
may be used but then the protein must first be free from the polymers.
This is done by precipitation of the proteins by trichloroacetic acid
(5% final concentration). The precipitate is centrifuged down at 15,000 rpm
and washed twice with 5% trichloroacetic acid.

Nucleic acids are determined by absorption at 260 nm or by the Schmidt-
Thannhauser method. Procedures using counting radioactivity in the
presence of polymers are described in reference (Pettijohn, 1967).

For large scale (1–10 l) operations conventional separating funnels may
be used. For smaller scale 10–100 ml it is convenient to use a centrifuge
tube for extraction and to decrease the settling time by centrifugation
(Rudin and Albertsson, 1967).

Here follows some applications on nucleic acids and proteins. The original articles may be consulted for details. The dextran-polyethylene glycol system is suitable for the separation of proteins from nucleic acids. The general partition behaviour of proteins and nucleic acids may be found in Albertsson, 1960, 1965. Okazaki and Kornberg (1964) applied a dextran-polyethylene glycol system for purification of the enzyme DNA polymerase. At high NaCl concentrations the enzyme partitioned into the upper phase while the nucleic acids were found in the lower phase. By selecting a suitable volume ratio a high yield of enzyme was found in the upper phase. This partition step was used in the beginning of the purification procedure and later steps removed the polymers from the enzyme.

Similar procedures were used for purification of an exonuclease induced by phage λ (Little *et al.*, 1967; Radding, 1966) and, for purification of an enzyme hydrolysing N-substitued amino acyl-t-RNA (Cuzin *et al.*, 1967).

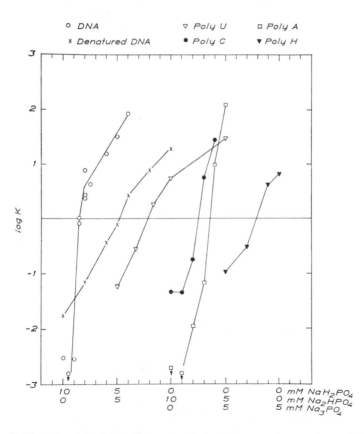

Fig. 8. Partition of high molecular weight nucleic acids in a dextran-polyethylene glycol phase system (Albertsson, 1965b).

Phage Qβ RNA polymerase was also purified using a similar procedure. (Eoyang *et al.*, 1968).

In these applications the partition step was followed with other purification methods such as precipitation and chromatography which removed the polymers from the enzymes in addition to further purifying them from other impurities.

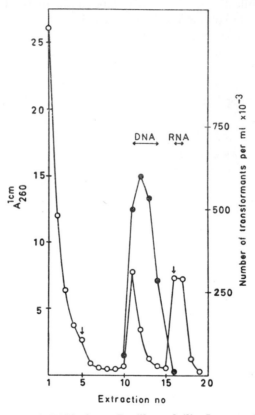

Fig. 9. Extraction of DNA from *Bacillus subtilis*. Lower phase is extracted repeatedly by upper phase. The ionic composition is shifted appropriately to transfer DNA selectively, by one or two extractions into the upper phase. The absorbancy of the upper phase is plotted. Phase mixing and settling takes place in the same centrifuge tube throughout the entire procedure. Open circles: absorbancy at 260 nm; filled circles: transforming activity. (Rudin and Albertsson, 1967).

For other applications in protein purification see Capecchi (1967), Gordon (1969), Loeb (1969) and Hayashi *et al.* (1970).

Babinet (1967) used partition in the dextran-polyethylene glycol system in a procedure for purification of RNA polymerase from a micro-organism.

By using successive partition steps the enzyme was not only separated from nucleic acids but also from other proteins since a substantial increase in the specific activity of the enzyme was obtained. See also Alberts *et al.* (1968).

High molecular weight DNA and RNA can be selectively transferred from one phase to the other by small changes in the ionic composition of the dextran-polyethylene glycol phase system, Fig. 8. This may be used for isolation of for example DNA by a few extraction steps (Rudin and Albertsson, 1967; Favre and Pettijohn, 1967). To speed up the separation the phase systems are centrifuged after each equilibration. Figure 9 shows an example of isolation of DNA from *Escherichia coli* using this technique. All partition steps are carried out with one centrifuge tube and after each centrifugation the upper phase is withdrawn and replaced by a new upper phase the ionic composition of which may be varied appropriately. Favre and Pettijohn (1967) developed a similar procedure. By radioactive labelling they were able to show that the DNA contained less than 0·05% RNA or less than 1 μmole arginine per mmole DNA phosphate. Denatured DNA pieces are also removed from the native DNA by this procedure. Advantages of the method are that the DNA is not precipitated during the procedure and ribonuclease treatment is unnecessary. In addition the method is fast; less than 8 h being required for the complete procedure. A protein-DNA complex could also be isolated by a slightly modified procedure (Pettijohn, 1967). See also Pettijohn (1969).

Single-stranded and double-stranded DNA have very different K values (Albertsson, 1965b) and can be very efficiently separated from each other even by one single partition step. This has been demonstrated by Alberts (1967a, b) who also used the same phase system to separate, in a single step, a small fraction of cross-linked, reversibly denaturable DNA molecules from the single stranded products predominating in normal denatured DNA samples (Alberts and Doty, 1968). The same technique was used by Summers and Szybalski (1967a, b) for detection and quantiation of single-strand breaks in DNA. The addition of solid CsCl to the upper (polyethylene glycol rich) phase selectively salts out the polyethylene glycol. DNA in the upper phase can therefore be examined and characterized by direct equilibrium sedimentation in a CsCl density gradient (Alberts, 1967a, b). A single partition step was also used for the large scale isolation of a satellite DNA (Patterson and Stafford, 1970).

C. Particles

1. *Adsorption at the interface*

Particles such as virus particles or cell particles have a strong tendency to collect at the interface between the two liquid phases. Their adsorption can easily be explained by considering the surface free energies in the system (Albertsson, 1960). The amount of adsorption at the interface is different for different particles and it may therefore be utilized for separation purposes. The capacity of the liquid–liquid interface for adsorption is very large. This is because the interfacial area is very large during shaking and therefore has a high capacity to adsorb particles. After shaking has stopped the phase drops coalesce and the interfacial area diminishes. However, most of the particles already adsorbed seem to stick to the interface and are therefore more and more concentrated into the final interface obtained after the phases have settled. Large amounts of material can be collected in this way as a thick layer between the two bulk phases.

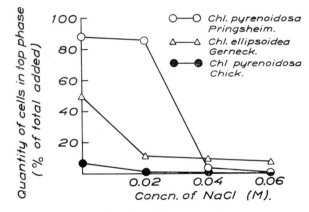

FIG. 10. Distribution of *Chlorella* cells between the upper phase and the interface. Per cent in upper phase is plotted against NaCl concentration of phase system. The rest of the cells are at the interface. For details see Albertsson and Baird (1962).

Distribution of particles in the two phase system is influenced by the ionic composition in much the same way as proteins or other soluble macromolecules. For example, a suspension of cells may be transferred from the upper phase to the interface of the dextran-PEG system by a change from phosphate to chloride ions, see Fig. 10. At intermediate compositions of the two ions the particles distribute between one phase and the interface such that, say 50% of the particles may be free in the upper phase while the rest are attached to the liquid–liquid interface. If now the upper phase is removed and replaced by a fresh upper phase without particles and the system shaken again then the particles at the interface will redistribute themselves so that a fifty-fifty distribution between the upper phase and the interface is again established. Thus, in this respect, the distribution between liquid phase and the interface is analogous to the distribution of soluble substances between two liquid phases.

However, the collection of particles at the interface displays a novel and interesting type of adsorption phenomenon which is quite different from usual adsorption for example on a solid material. Analysis of the distribution of particles at different particle concentrations and with different volume ratios has shown that the ratio between the *amount* of particles in the upper phase and the *amount* of particles at the interface is constant and independent of particle concentration and phase volume ratio (Albertsson, 1958; Albertsson and Baird, 1962). Thus, if C_t is the concentration in the top phase, V_t is the volume of the top phase and a is the amount of particles at the interface then the ratio G_i

$$G_i = \frac{C_t.V_t}{a}$$

is constant, i.e. for a given a the concentration in the upper phase depends on V_t. This is quite different from a linear adsorption isotherm when the ratio between the *concentration*, C in the liquid and the *amount* of adsorbed substance, a_i, is constant.

However, the fact that a substance collects at the interface does not necessarily mean that its particles or molecules are adsorbed by the interface. It may also be the result of a precipitation of the substance with the formation of a third phase which simply separates due to a density difference and collects between the two phases.

A study of the behaviour of a substance in the two individual phases, that is when no interface is present, is a convenient way of checking if the substance really dissolves in the phases or if it is precipitated as a separate phase. If there is no sign of precipitation or aggregation when the substance is suspended in each phase separately but it collects at the interface when the two phases are mixed this is an indication of adsorption at the interface. This is the case of, for example, all bacterial cells studied so far. If, however, precipitation takes place even when the phases are separated and no change in concentration of solute takes place after the two phases are mixed, this is an indication that the material between the two phases consists of a third phase. By studying concentration dependence one may also distinguish between the two extreme cases (Albertsson, 1960).

2. Bacteria

The distribution of bacteria in the dextran-polyethylene glycol system has been studied by Baird *et al.* (1961) and by Albertsson and Baird, 1962). It was found that different bacteria behaved differently in the destranpolyethylene glycol system, Fig. 11. Their distribution could be changed conveniently by varying the ionic composition. Even closely related strains behaved differently.

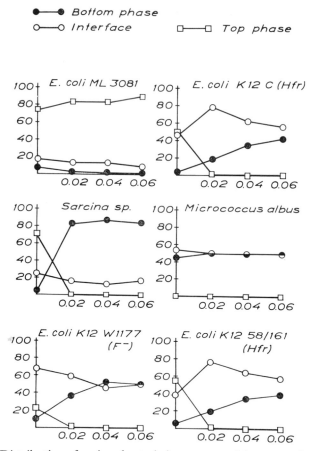

FIG. 11. Distribution of various bacteria in a system of dextran and polyethylene glycol with 0·01 M potassium phosphate buffer pH 6·8 and different amounts of NaCl (Baird *et al.*, 1961).

3. *Algae*

Distribution of unicellular algae was studied by Albertsson and Baird (1962). Figure 10 shows the distribution of different *Chlorella* strains. Counter-current distribution of one strain gave two distinct peaks indicating at least two different classes of cells in a culture. The cells were viable after the treatment.

4. *Viruses*

The distribution of a number of viruses in polymer two-phase systems has been described elsewhere (Albertsson, 1947). In the dextran-polyethylene glycol system partition of virus shows the same strong salt dependence as proteins or nucleic acids. Frequently the virus collects at the interface. This may be due to adsorption at the interface or because the virus is precipitated by polyethylene glycol or both.

Partition of virus in the dextransulphate-polyethylene glycol has been studied extensively by Bengtsson (1968) (Bengtsson *et al.*, 1962; Bengtsson and Philipson, 1963). By changing the polymer composition and salt concentration the virus may be partitioned in favour of the lower phase or the upper phase, Table III. Such systems are useful for concentration and purification purposes; see below. By choosing suitable compositions, K values between 0·2 and 5 may be obtained and such systems are useful for counter-current distribution (Bengtsson *et al.*, 1962; Bengtsson and Philipson, 1963; Bengtsson, 1968), see below. Phase diagrams of the dextransulphate-polyethylene glycol system at different NaCl concentrations are described in Albertsson (1960).

5. *Other cell particles*

Cell walls from unicellular algae and from bacteria have been isolated using the dextran-PEG system or the phosphate-PEG system (Albertsson, 1958; Hofsten and Baird, 1962). The same phase systems were used for isolation of antigens (R or O) attached to granules from cell walls of *Salmonella typhi* (Cherman *et al.*, 1966). Spores and vegetative cells of bacteria behave in strikingly different ways in partition most probably reflecting their different surface properties. This has been utilized by Sacks and Alderton (1961) to isolate spores from an excess of vegetative cells by one single distribution step. The dextran-PEG system was used by Hofsten and Baird (1962) to compare the partition of vegetative cells, cell wall fragments, spores, and spore coates and protoplasts of *Bacillus megaterium*. It was demonstrated that cell walls could be easily separated from the rest of the cell particles by one or a few distribution steps. By a change in salt composition the material could be effectively concentrated at the interface. A similar phase system was used for the purification of poly-β-hydroxybutyrate particles (Griebel *et al.*, 1968). De Ley (1963) used a system of dextran sulphate-methylcellulose to separate bacterial oxidosomes from ribosomes. A similar result might have been obtained by using dextran or dextransulphate-PEG systems which have the advantage of relatively short settling times.

TABLE III

Distribution of polio virus type 2 strain MEF 1 in phase systems with different composition
(Bengtsson et al., 1962)

System No.	Final concentration of polymers in per cent w/w		Final concentration of NaCl in M	Partition coefficient (K)	Per cent recovery of virus	Interface adsorption
	Dextran sulphate (DxS)	Polyethylene glycol (PEG)				
1	7·0	1·0	0·70	>12	19	0
2	7·0	1·2	0·60	0·57	69	0
3	7·0	1·3	0·50	<0·05	60	+
4	7·0	1·3	0·55	0·05	100	+
5	7·0	1·3	0·60	0·2	62	0
6	7·0	1·5	0·50	0·05	110	+
7	7·0	1·5	0·55	0·05	62	+
8	7·0	1·5	0·60	1·25	72	+
9	7·0	1·5	0·70	1·5	52	+

IV. VIRUS PURIFICATION AND CONCENTRATION

A. Introduction

Virus cultures contain very small amount of virus as mg/ml. One therefore has to start with large volumes of a culture in order to prepare pure
virus in amounts sufficient for chemical characterization. The concentration of virus from large volumes into a small volume is therefore a necessary
step in any purification procedure. A two-phase system can accomplish
such a concentration if it can be constructed in such a way that most of the
virus particles are transferred to a phase with a small volume compared
to that of the original virus culture. Since other substances such as proteins
usually distribute in a different manner a purification and concentration is
obtained at the same time.

TABLE IV

**Viruses which have been concentrated and purified by
polymer two-phase systems**

Virus	Reference
Phage T2	Albertsson (1967)
Phage ø X174	Sedat and Sinsheimer (1964)
Phage Qβ	Eoyang and August (1968)
TMV	Venekamp and Mosch (1964)
TYMV	Leberman (1966)
TCV	Leberman (1966)
Polio	Norrby and Albertsson (1960)
ECHO	Philipson et al. (1960)
Japanese encephalitis	Nakai (1965)
Foot-and-mouth	Belgian Patent (1964)
Adeno	Wesslén et al. (1959)
Influenza	Wesslén et al. (1959)
Parotitis	Wesslén et al. (1959)
Newcastle	Wesslén et al. (1959).

This application of polymer two-phase systems for virus concentration
has been described in a number of papers, see Table IV. The general
principle is as follows. Polymers are added to a virus culture to give a
phase system with a phase that is small in volume and contains most of the
virus activity. An efficient concentration is also obtained if the virus collects
at the liquid-liquid interface. An example, where this phenomenon is
utilized, is given below.

One can easily predict the suitable polymer concentrations from the
phase diagram provided the partition behaviour of virus is known. For a

general account on different possibilities, theoretical considerations, and practical details, see Albertsson (1960). Below, two typical examples of virus purification and concentration are given.

B. Bacteriophage T2

The following procedure has been worked out for phage T2 (Albertsson, 1967) but it may also be applied to other bacteriophages and viruses.

Principle. The phages are first concentrated at the interface of the dextran sulphate-polyethylene glycol system. The interfacial material is then suspended and dextran sulphate precipitated by KCl together with other cell particles. The phages stay in solution and have now been concentrated 50–100 times and are also fairly pure. They may be further purified by a centrifugation step.

Materials.

Dextran sulphate 500.

"Carbowax Polyethylene Glycol 6000". The flakes are pulverized by grinding in a mortar.

3 M KCl

NaCl

Separating funnel (3 litres).

Procedure. (for about 2 litres of lysate) Lysate, dextran sulphate, pulverized PEG, and NaCl are mixed in a separating funnel to give a mixture containing 6·5% (w/w) PEG 6000 and 0·2% (w/w) dextran sulphate in 0·3 M NaCl. In a typical experiment, for example, 1778 g T2 lysate was mixed with 126·2 g PEG 6000, 3·87 g dextran sulphate and 33·9 g NaCl.

After mixing, the funnel is allowed to stand in the cold overnight. A heavily turbid bottom layer of about 30 ml is obtained, see Fig. 12. This layer contains both bottom phase and top phase; an emulsion which contains a lot of material forms at the liquid–liquid interface and thus, does not separate by gravity only. The bottom layer is slowly collected into a 50 ml centrifuge tube and centrifuged for 15 min at 2000 rpm.; the turbid material with the phages collects at the interface, Fig. 13. The clear top and bottom phases are removed by pipette. The remaining interface "cake" is suspended in 15–20 ml of a 1% (w/w) dextran sulphate solution. Then, 0·15 ml of a 3 M KCl solution is added per ml suspension. This precipitates the dextran sulphate. Allow to stand for 2 h, then centrifuge 10 min at 2000 rpm. In the pellet the precipitated dextran sulphate is found together with cellular debris and a large amount of other impurities. The phages stay in the supernatant and are now already fairly pure. They may be further purified by a centrifugation at 15,000 rpm for 2 h and resuspension in a small volume.

The concentration obtained is more than a hundred-fold. In a typical experiment the titre of the lysate was $2 \cdot 3 \times 10^{10}$, the titre of the suspension before the last centrifugation (15,000 rpm) was $1 \cdot 3 \times 10^{12}$ and the titre after the same centrifugation and resuspension was $6 \cdot 3 \times 10^{12}$. The yield is more than 90%. The quotient E_{260}^{1cm} corr/10^{12} plaques per ml was $20 \cdot 6$ before and $9 \cdot 4$ after the last centrifugation.

Comments. This method is very simple and rapid. The final product is obtained the day after the lysate is prepared. It is also very reproducible

Fig. 12. Concentration of phage T2 in the small bottom layer of the dextran-sulphate-polyethylene glycol system.

and we have used this for a long time as a routine method. If desirable the procedure may be easily scaled up.

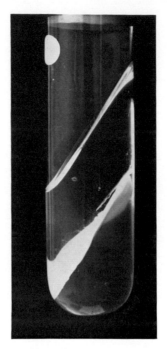

Fig. 13. The bottom layer of Fig. 12 after low speed centrifugation. The phages concentrate at the interface.

C. Polio virus and echo virus

The procedure described below is that of Philipson *et al.* (1960) and Norrby and Albertsson (1960). The same phase system is used as for phage T2 (described above).

Principle. In the dextran sulphate-polyethylene glycol system, most viruses are concentrated in the bottom phase or at the interface at low NaCl concentrations (0·15–0·3 M) (see Table V). In order to obtain a small bottom phase, a system of 0·2% (w/w) NaDS50 and 6% (w/w) PEG 6000 with 0·3 M NaCl (point P in Fig. 5) may be used. The bottom phase volume of this system is less than one hundredth of the total volume. The dextran sulphate of the bottom phase is subsequently removed by precipitation with KCl.

Materials.
Dextran sulphate "Dextran sulphate 500" 20% (w/w)

Polyethylene glycol, PEG 6000, 30% (w/w)
Sodium chloride, 5 M
Potassium chloride, 3 M.

TABLE V

Concentration of Polio Virus (Norrby and Albertsson, 1960) and ECHO (Philipson _et al._, 1960) Virus in a two-phase system of Sodium Dextran Sulphate and Polyethylene Glycol.

Virus	Preparation	Vol., ml.	Infectivity TCD 50/ml log units	Concentration factor
Polio	Orig. virus culture	100	7·5	1
	Bottom phase	1·2	9·4	80
	Top phase	134	5·6	
ECHO	Orig. virus culture	5000	7·3	1
	Bottom phase	50	9·3	100
	Top phase	6400	5·9	

Procedure. The polymers are added to harvested tissue culture fluid in a separating funnel until a final concentration of 0·2% (w/w) of NaDS 50 and 6·45% (w/w) of PEG 6000 is reached. In addition, NaCl is added to give a final concentration of 0·3 M (The 0·15 M NaCl of the culture fluid must be taken into account). In an experiment with 100 g of virus culture, 1·34 g of 20% (w/w) NaDS 70, 29·0 g of 30% (w/w) PEG 6000, and 5·0 g of 5 M NaCl are therefore added. The contents are mixed by inverting the funnel several times. After a 24 h phase separation in the cold (4°C) a bottom phase with a volume of about one hundredth of the original virus culture forms. This (sometimes together with the interface) contains most of the virus activity. The bottom phase is collected and the dextran sulphate, which is at a concentration of about 17% (w/w) (see Fig. 4) is removed by adding either 0·7 ml of 3 M KCl or 0·3 ml of 1 M BaCl$_2$ per gram of bottom phase. A semisolid precipitate of dextran sulphate is formed, while the virus activity remains in suspension. The virus titre and volumes of the phases in experiments with polio virus and ECHO virus are given in Table V.

Comments. The virus may now be collected by centrifugation, or it may be further concentrated by a new distribution step in the same system as that used above. However, in the latter case, KCl should be removed by dialysis before addition of the polymers. It is possible to transfer the virus activity into the top phase of a new dextran sulphate-polyethylene glycol system by the addition of NaCl up to a concentration of 1 M to the bottom phase collected from the first distribution. For details see Albertsson, 1960.

As seen from Table V, most of the virus activity is in the bottom phase and little in the top phase or at the interface. However, frequently a large part of the virus activity is collected at the interface. When the virus activity is collected at the interface, the concentration effect is, however, as good, provided the interface is collected together with the bottom phase.

This procedure may also easily be scaled up to larger volumes. It has found industrial application for large scale concentration of Foot and Mouth disease virus (Belgian Patent 636,492, 1964).

V. COUNTER-CURRENT DISTRIBUTION

A. Introduction

Counter-Current distribution (CCD) is a multistage technique where a number of extraction steps are carried out in order to separate completely substances which are only partially separated by one partition step.

A general treatment of CCD may be found in the books and review articles by Craig and Craig (1956), Craig (1960), Hecker (1955), and Tavel and Signer (1956).

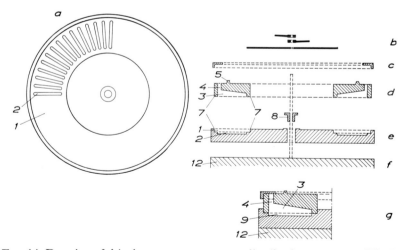

Fig. 14. Drawing of thin-layer counter-current distribution apparatus. The lower phase chamber is indicated by a2 and e2 and the upper phase chamber by d3 and g3. For details see Albertsson, 1965a.

B. Thin-layer CCD

A drawback with polymer phase systems when using conventional CCD machines, for example the Craig glass machine, is the long time needed for phase separation and the high viscosity of the phases. A special counter-current distribution apparatus for use with polymer two-phase systems has therefore been constructed (Albertsson, 1965a) and used with success in many applications of polymer phase systems for fractionation of cell particles and macromolecules. A drawing of the unit is shown in Fig. 14. Its

FIG. 15. Automatic version of the thin-layer counter-current distribution apparatus.

main parts are two cylindrical plates, d and e, which rest upon each other. In each plate there are shallow cavities, 2 and 3, which form the partition cells of lower and upper phase respectively. The phases are mixed by rotary shaking and after settling the upper phases are transferred to adjacent partition cells by turning the upper plate relative to the lower plate. For details see reference. The main point with this apparatus is that the depth of the phase layer is only 1–2 mm. The settling time for polymer phase systems can thereby be decreased to 1–5 min from the 10–30 min needed with conventional types of counter-current distribution apparatus.

An automatic version of the apparatus has recently been constructed, Fig. 15 (This apparatus is marketed by IRD, Box 11074, 161 11 Bromma, Sweden.)

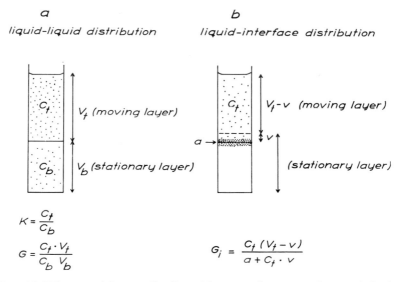

FIG. 16. When particles are distributed between the upper phase and the interface liquid-interface distribution (right) is applied for separation. The interface material is included in the stationary layer. Compare with conventional counter-current distribution (left) when the substance distributes between the bulk phases.

C. Liquid-interface CCD

In the case of particles, such as cells or cell organelles, adsorption at the interface is very frequent. The distribution of particles therefore takes place between one or the other bulk phase and the interface. The adsorption of particles at the interface is both selective and reversible. For separation by a multistage procedure such as CCD the same principle can be applied for particles as for soluble substances. It is only necessary to introduce a practical modification. Thus, when the distribution takes place between the upper phase and the interface most of the upper phase is transferred in the CCD apparatus while the interface material is allowed to remain together with the stationary lower phase. The difference between this so called liquid-interface CCD and conventional CCD is shown diagramatically in Fig. 16. In the practical arrangement and also in the calculation of the theoretical curves the interface is treated in the same way as the lower phase in conventional CCD.

Liquid-interface CCD has been applied for fractionation of bacteria, algae, blood cells and chloroplasts.

D. CCD of bacteria

1. *Introduction*

A detailed analysis of the behaviour of bacteria and also other cells in CCD using the dextran-polyethylene glycol system has been carried out (Albertsson and Baird, 1962). It was found that the cells behaved as if their distribution between top phase and interface was reversible and concentration independent, that is, they demonstrated almost ideal behaviour. Thus each particle species travelled along the distribution train according to what one would expect from its distribution in a single phase pair.

Different strains of bacteria displayed different distribution diagrams. A mixture of different strains could be resolved by CCD, (Fig. 17). Of

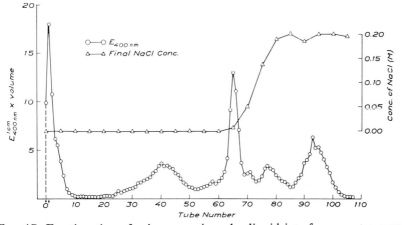

FIG. 17. Fractionation of micro-organisms by liquid-interface counter-current distribution. The peaks represent from left: yeast cells, *Escherichia coli* K12, W1177, *E. coli* K12, 58, *Chlorella* (two small peaks) and *E. coli* ML3081 (Albertson and Baird, 1962).

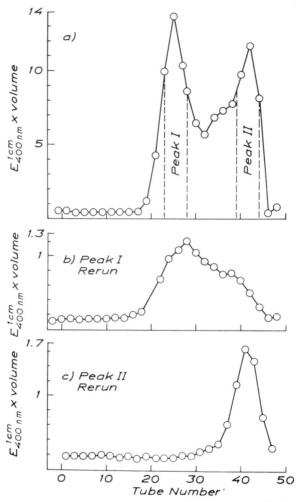

FIG. 18. Counter-current distribution of *E. coli* K12. (b) and (c) the redistribution of peaks I and II respectively under the same conditions as in (a) (Albertsson and Baird, 1962).

particular interest is that if a culture of one strain was subjected to CCD it frequently showed two distinct peaks in the diagram indicating the presence of at least two different classes of bacterial cells (Fig. 18). The difference between the two classes of cells is not known and this phenomenon deserves further study.

2. *Experimental procedure*

Here follows a description of the experimental procedure for a liquid-interface CCD using the thin-layer apparatus shown in Fig. 15 and the phase system dextran-polyethylene glycol.

First the distribution of the cells or cell particles under study between a single phase pair has to be investigated. This is done according to the procedure outlined in Table II. By changing the salt composition (or polymer composition) one can get such a distribution that the amount of material in the upper phase is between about 30–70%. Then, a large volume about 500 ml of the desired phase system is prepared and the two phases are separated. Separately the sample to be fractionated is mixed with polymers and salts to give a phase system with the same final polymer and salt composition as the large phase system. The apparatus is filled in the following way. Suppose it contains 120 partition cells with 0·7 ml capacity of lower phase cavity. The cells are numbered 0, 1, 2, . . . 119. 0·6 ml lower phase is added to each of cell numbers 1–119. (An automatic syringe pipette is suitable). Then 0·6 ml of upper phase is added to cell numbers 1–119. Then, 1·2 ml of the phase system with sample is added to cell no. 0. This sample phase system should consist of 0·6 ml lower phase and 0·6 ml upper phase so that the phase volumes are identical in all partition cells. After this the phases are mixed by shaking, allowed to settle and the CCD run in the usual manner. If more material is needed one can fill more cells with sample phase system for example cells nos. 0–5. If 120 transfers are made one can fill as many as 10 cells without broadening the peaks seriously.

E. CCD of virus

Different polio virus strains have been subjected to CCD using the dextran sulphate-polyethylene glycol system. In this system it appears that the virus is mainly distributed between the two bulk phases and conventional liquid–liquid CCD could therefore be used.

The procedure for selecting a suitable system and experimental details have been published elsewhere (Bengtsson, et al., 1962). Different polio virus strains displayed quite different counter-current distribution diagrams (Fig. 19). The Sabin strain gave a peak in fraction 4 corresponding to a K value of about 0·3 whereas the CHAT, SB and E 206 strains showed peaks in the fractions 15–17 corresponding to partition coefficients of about 4. The 1423/53 strain showed a considerable heterogeneity; the reason for this is probably that this strain was not plaque purified.

In order to correlate the distribution pattern of these and other polio virus type I strains to their genetic characteristics Bengtsson and Philipson (1963) carried out marker tests. It turned out that a clear correlation existed between the partition coefficient and one marker (m); the strains with a partition coefficient of 0·2–0·3 had the character of m^- and strains with a partition coefficient of 4 had the m^+. Further they showed that in a prepar-

ation of Sabin L Sc 2 ab strain which carries the m⁻ character a small
impurity of m⁺ virus (probably as a result of mutation) was present and
that this impurity could be separated from the main fraction by counter-
current distribution.

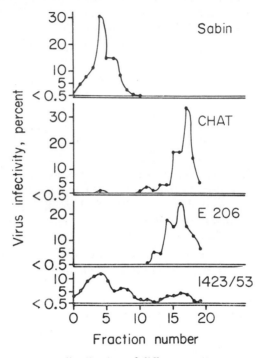

FIG. 19. Counter-current distribution of different polio virus strains. (Bengtsson
and Philipson, 1963).

It is evident from these studies that the polymer two-phase systems are
highly specific and that in combination with counter-current distribution
they can be used as powerful tools for analysis of virus, for their purification,
and for isolation of mutants.

F. CCD of antigen-antibody complexes

If the partition coefficients of antigen, antibody, and antigen-antibody
complexes are different partition may be used for detection of antigen-
antibody reactions (Albertsson and Philipson, 1960) and also to separate
the different particle species present in a antigen-antibody mixture. These
possibilities have been utilized by Philipson *et al.* (1966) to study the
reaction between virus and its antibodies. The redistribution of the virus

as a result of antibody complex formation could be conveniently measured by using P^{32} labelled polio virus. Counter-current distribution (Philipson, 1966) was used to analyse the products obtained when polio virus was inactivated by antibody from human G globulins. Uncombined virus could be separated from the virus-antibody complexes, Fig. 20.

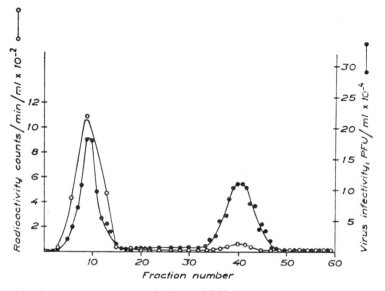

FIG. 20. Counter-current distribution of P^{32} labelled polio virus mixed with antibody. To the right uncombined virus, to the left virus-antibody complex. (Philipson, 1966).

VI. REMOVAL OF POLYMERS

After a fractionation experiment is finished it is often desirable to remove the phase polymers from the substance under study. In the case of large particles such as cells and cell fragments the polymers can be washed away by repeated centrifugations. Also virus particles and DNA can be recovered from the polymers by high speed centrifugations. DNA has been freed from the polymers dextran and polyethylene glycol by chromatography on methylated albumin or hydroxylapatite columns (Bollum, 1963), or by centrifugation in a CsCl gradient (Alberts, 1967). The polyethylene-glycol phase can be layered directly onto a 5–20% w/w sucrose gradient and fast sedimenting proteins and nucleic acids can be sedimented free of the phase polymer. Many proteins can be adsorbed on a column of, for example DEAE-cellulose or hydroxylapatite, while neutral polymers like dextran and polyethylene glycol pass through the column.

The proteins may then be eluted from the columns. However, some proteins may be difficult to elute. It would be desirable, therefore, to have a simple and general method, which does not utilize a solid adsorbing phase, and which could be used routinely for recovering proteins from the phase polymers. Since proteins almost always are charged, electrophoresis should be the method of choice. Recent experiments in our laboratory with a simple electrophoresis apparatus (Albertsson, 1970) also show that this technique is useful for the purpose of removing the polymers.

VII. DISCUSSION

Partition in polymer phase systems is like all partition methods an empirical method. It is difficult to predict in detail the behaviour of different substances. A suitable phase system has therefore to be found by a number of trial and error experiments. However, with the experience now available it is fairly easy to set up a number of different systems, with different ionic composition for example, in order to find the system which gives the desired distribution. Of particular importance is that one polymer pair, can be used for widely different substances. Thus, the dextran-polyethylene glycol combination has been applied on proteins, nucleic acids, nucleic-acid-protein complexes, virus, mitochondria, chloroplasts, bacteria, algae, yeast and blood cells. By using one or two standard compositions of these polymers (System A and B or C in Fig. 3) and by varying the ionic composition one can fairly quickly obtain a suitable phase system. Once a suitable system has been found the procedure for fractionation is simple and highly reliable.

The polymer phase systems appear to be mild towards biological material. Enzymic activities, infectivity of virus nucleic acid, transformation activity of DNA and other biological activities of cell particles are retained after treatment with the polymer two phase system. In many cases the polymers exert a protective effect on the proteins or cell particles.

Of particular importance is that factors other than size, density and charge determine the distribution. The polymer two-phase technique therefore is a useful complement to centrifugation and electrophoresis.

In the case of particles such as cells and virus particles the polymers can be removed by centrifugation. In the case of soluble macromolecules such as proteins and nucleic acids complete removal of the polymers is a problem. By using chromatography, precipitation or electrophoresis this problem can usually be solved. Therefore, for proteins and nucleic acids the two-phase technique is best utilized when it is used as an early step in a purification procedure where different methods are used. Other methods such as chromatography, electrophoresis and centrifugation can then later remove the polymers at the same time as the desired substance is further purified.

REFERENCES

Alberts, B. (1967a). *Biochemistry, N.Y.*, **6**, 2527.

Alberts, B. (1967b). *In* "Methods in Enzymology", (Eds. S. P. Colowick and N. O. Kaplan) Vol. XII, Part A. Pp 566, Academic Press, New York.

Alberts, B., and Doty, P. (1968). *J. molec. Biol.*, **32**, 379.

Alberts, B. M., Amodio, F. J., Jenkins, M., Gutman, E. D., and Ferris, F. L. (1968). *Cold Spring Harb. Symp. quant. Biol.*, **33**, 289.

Albertsson, P.-Å. (1960). "Partition of Cell Particles and Macromolecules". Almquist och Wiksell, Stockholm; Wiley, New York.

Albertsson, P-Å. (1958). *Biochim. biophys. Acta*, **27**, 378.

Albertsson, P-Å., and Nyns, J. (1959). *Nature, Lond.*, **184**, 1465.

Albertsson, P-Å., and Philipson, L. (1960). *Nature, Lond.*, **185**, 38.

Albertsson, P-Å., and Nyns, J. (1961). *Arkiv Kemi*, **17**, 197.

Albertsson, P-Å. (1962). *Archs Biochem. Biophys. Suppl.*, **1**, 264.

Albertsson, P-Å., and Baird, G. D. (1962). *Expl. Cell Res.*, **28**, 296.

Albertsson, P-Å., and Baltscheffsky, H. (1963). *Biophys. Res. Commun.*, **12**, 14.

Albertsson, P-Å. (1965a). *Anal. Biochem.*, **11**, 121.

Albertsson, P-Å. (1965b). *Biochem. biophys. Acta*, **103**, 1.

Albertsson, P-Å. (1967). "Methods in Virology", (Eds K. Maramorosch and H. Koprowski), Vol. 2, Chapter 10.

Albertsson, P A. (1970) In *Adv. Prot. Chem.* Vol. 24, p. 309.

Babinet, C. (1967). *Biochem. biophys. Res. Commun.*, **26**, 639.

Bailey, L. (1967). "Techniques in Protein Chemistry". Elsevier, Amsterdam, 2nd ed.

Baird, G. D., Albertsson, P. Å., and Hofsten, B. (1961). *Nature, Lond.*, **192**, 236.

Beijerinck, M. W. (1896). *Zentbl. Bakt.*, **2**, 627, 698.

Bengtsson, S., Philipson, L., and Albertsson, P. Å. (1962). *Biochem. biophys. Res. Commun.*, **9**, 318.

Bengtsson, S., and Philipson, L. (1963). *Virology*, **20**, 176.

Bengtsson, S., (1968). *Acta Univ. Upsaliens.*, Abstracts of Uppsala Dissertations in Medicine, **50**.

Bollum, F. J., (1963). *Cold Spring Harb. Symp. quant. Biol.*, **28**, 21.

Brønsted, J. N. (1931). *Z. Phys. Chem. Ser. A (Bodenstein-Festband)*. p. 257 Academic Press, New York.

Bungenberg de Jong, H. G. (1949). "Colloid Science". (Ed. H. R. Kruyt). Vol. II. Elsevier Publ. Co., Amsterdam.

Capecchi, M. R. (1967). *Proc. natn. Acad. Science, U.S.A.*, **58**, 1144.

Chermann, J. C., Digeon, M., Raynaud, M., and Giuntini, J. (1966). *Ann. Inst. Pasteur*, **111**, 59.

Craig, L. C., and Craig, D. (1956). *In* "Technique of Organic Chemistry". (Ed. A. Weissberger), Vol. III, Part 1, 2nd ed. Interscience Publishers, New York.

Craig, L. C. (1960). *In* "A Laboratory Manual of Analytical Methods of Protein Chemistry". Academic Press, New York, (Eds P. Alexander and R. J. Block)

Cuzin, F., Kretchmer, N., Greenberg, R. E., Hurwitz, R., and Chapeville, F. (1967). *Proc. natn. Acad. Science, U.S.A.*, **58**, 2079.

De Ley, J. (1963). *Biochem. Soc. Symp. No. 23*, p. 38.

Dobry, A., and Boyer-Kawenoki, F. (1947). *J. Polymer Science*, **2**, 90.
Eoyang, L., and August, J. T. (1968). *In* "Methods in Enzymology". (Eds S. P. Colowick and N. O. Kaplan), Vol. XII, Part B, p. 530, Academic Press, New York.
Falaschi, A., and Kornberg, A. (1966). *J. Biol. Chem.*, **241**, 1478.
Favre, J., and Pettijohn, D. E. (1967). *European J. Biochem.*, **3**, 33.
Flory, P. J. (1953). "Principles of Polymer Chemistry", Cornell University Press, Ithaca, New York.
Gordon, J. (1969). *J. Biol. Chem.*, **244**, 5680.
Griebel, R., Smith, Z., and Merrick, J. M. (1968). *Biochemistry*, **7**, 3676.
Hanzon, V., and Philipson, L. (1960). *J. Ultrastruct. Res.*, **3**, 420.
Hayashi, H., Knowles, J. R., Katze, J. R., Lapointe, J., and Soll, D. (1970). *J. Biol. Chem.*, **275**, 1701.
Hecker, E. (1955). *Verteilungsverfahren in Laboratorium, Monographien zu Angewandte Chemie und Chemie-Ingenieur-Tech.* No. 667. Verlag Chemie, Weinheim/Bergstr., Germany.
Hjertén S. To be published.
Hofsten, B. v., and Baird, G. D. (1962). In "Biotechnology and Bioengineering", IV, 403.
Koningsveld, R. (1968). *Advances Coll. Int. Sci.*, **152**, 2, 2.
Leberman, R. (1966). *Virology*, **30**, 341.
Lehman, J. R. (1960). *J. biol. Chem.*, **235**, 1479.
Little, J. W., Lehman, I. R., and Kaiser, A. D. (1967). *J. biol. Chem.*, **242**, 672.
Loeb, L. A. (1969). *J. Biol. Chem.*, **244**, 1672.
Lowry, O. H., Rosebrough, N. J., Farr, A. L., and Randall, R. J. (1951). *J. biol. Chem.*, **193**, 265.
Mok, C. C., Grant, C. T., and Taborsky, G. (1966). *Biochemistry*, **5**, 2517.
Morawetz, H. (1965). "Macromolecules in Solution; High Polymers", Vol. XXI. Interscience Publishers, New York.
Nakai, H. (1965). *Acta Virol.*, **9**, 89.
Nakayama, T., and Hirata, T. (1968). *Abstr. 9th Int. Leprosy Congress, Lond.*
Norrby, E., and Albertsson, P. Å. (1960). *Nature, Lond.*, **188**, 1047.
Okazaki, T., and Kornberg, A., (1964). *J. biol. Chem.*, **239**, 259.
Patterson, J. B., and Stafford, D. W. (1970). *Biochemistry*, **9**, 1278.
Pettijohn, D. E. (1969). *In* "Progress in Separation and Purification", Vol. II., p. 147.
Pettijohn, D. E. "Progress in Separation and Purification", Vol. II. In press.
Philipson, L., Albertsson, P. Å., and Frick, G. (1960). *Virology*, **11**, 553.
Philipson, L. (1966). *Virology*, **28**, 35.
Polson, A., Potgieter, G. M., Largier, J. F., Mears, G. E. F., and Joubert, F. J. (1964). *Biochim. biophys. Acta.*, **82**, 463.
Radding, C. M. (1966). *J. molec. Biol.*, **18**, 235.
Rudin, L. (1967). *Biochim. biophys. Acta*, **134**, 199.
Rudin, L., and Albertsson, P. Å. (1967). *Biochim. biophys. Acta*, **134**, 37.
Sacks, L. E., and Alderton, G. (1961). *J. Bacteriol.*, **82**, 331.
Sedat, J., and Sinsheimer, R. L. (1964). *J. molec. Biol.*, **9**, 489.
Summers, W. C., and Szybalski, W. (1967a). *J. molec. Biol.*, **26**, 107.
Summers, W. C., and Szybalski, W. (1967b). *J. molec. Biol.*, **26**, 227.
Tavel, P. v., and Signer, R. (1965). *Advances in Protein Chem.*, **11**, 237.
Tompa, H. (1956). "Polymer Solutions". Butterworths Scientific Publications, London.

Treybal, R. E. (1951). "Liquid Extraction". McGraw-Hill Book Co., Inc., New York, Toronto and London.

Waddell, W. J. (1956). *J. Lab. Clin. Med.*, **48**, 311.

Walter, H. (1967). *Protides biol. Fluids*, **15**, 367.

Walter, H. (1969). *In* "Progress in Separation and Purification", Vol. II. p. 121.

Watanabe, M., and August, J. T. (1967). *In* "Methods in Virology", (Eds. K. Maramorosch and H. Kopronski). Vol. 3, p. 99.

Wesslén, T., Albertsson, P. Å., and Philipson, L. (1959). *Arch. Virusforsch.*, **9**, 510.

Vinograd, J., Lebowitz, J., Radloff, R., Watson, R., and Laipis, P. (1965). *Proc. natn. Acad. Sci. U.S.A.*, **53**, 1104.

Zeppezauer, M., and Brishammar, S. (1965). *Biochim. biophys. Acta*, **94**, 581.

Zernike, J. (1955). "Chemical Phase Theory". N.V. Uitgevers-Maatschappij AE. E. Kluwer, Deventer, Antwerp, Djakarta.

Öberg, B., Albertsson, P. Å., and Philipson, L. (1965). *Biochim. biophys. Acta*, **108**, 173.

CHAPTER VI

Separation and Purification of Proteins

Mitsuhiro Nozaki and Osamu Hayaishi

*Department of Medical Chemistry Kyoto University Faculty of Medicine
Kyoto, Japan*

I.	Introduction	425
II.	General Considerations	427
	A. Source materials	427
	B. Stability of enzymes	428
	C. Assay methods	436
	D. Separation of protein from nucleic acid	438
III.	Fractionation of Soluble Enzymes	439
	A. Fractionation by stability	439
	B. Fractionation by solubility	439
	C. Fractional adsorption	440
	D. Fractionation by chromatography	441
	E. Molecular-sieve chromatography	444
	F. Separation by isoelectric focusing	445
IV.	Crystallization	446
V.	Other Methods	448
	A. Desalting.	448
	B. Concentration	449
VI.	Sequence of Fractionation Methods	449
	References	450

I. INTRODUCTION

The structure, function and metabolic activity of an individual cell depend on the presence of specific protein molecules, since proteins play an essential role in all phases of the chemical and physical activities associated with life. In particular, enzymes, one class of proteins, are intimately involved in all types of biological processes; all individual chemical reactions in metabolism, mechanical work and so forth that take place in a cell may each be catalyzed by a specific enzyme.

In order to understand cellular activity in detail, it is desirable to clarify the specific properties of a particular protein involved. Isolation of each enzyme from other cell compounds and obtaining the enzyme as pure as possible facilitates the characterization of individual enzymes.

This article is intended to review the fractionation technique applied to enzymes rather than proteins in general, although common methods may be applied for both. Since much more detailed information for these techniques and specific purification methods for individual enzymes are available in "Methods in Enzymology", we will orient our discussion in a way which will be of specific use to microbiologists rather than to enzymologists as a whole and try to include methods of interest that have been reported recently. Other review articles on the fractionation of proteins are available (Taylor, 1953; Dixon and Webb, 1964; Sober et al., 1965).

Before we go further, we shall discuss the properties of enzyme proteins which will direct the course of enzyme fractionation. Proteins are naturally occurring high molecular weight compounds, hydrolysis of which yields a mixture of amino-acids. Thus, proteins are made up of 20 different amino-acids linked together in a definite sequence by peptide bonds. Some proteins may contain more than one peptide and these are held together by specific cross links, i.e. disulphide bonds. The arrangement or sequence of the amino-acids in the polypeptide chain is called the **primary structure** of the protein.

This polypeptide chain is held in a coiled shape mainly by hydrogen bonds in which the hydrogen atom shares its electrons between a neighbouring nitrogen atom and the double-bonded oxygen of peptide bonds. These helical structures of proteins are called **secondary structure.**

The helix is further strengthened by various types of non-covalent bonds to make up the **tertiary structure** of the protein. The non-covalent bonds that stabilize the three-dimensional protein structure include (a) electrostatic interactions, (b) hydrogen bondings, (c) interactions of non-polar side chains and (d) van der Waals interactions. In some instances enzymes may consist of a number of subunits, which may or may not be identical. The association of subunits to make up biologically active enzymes is called the **quarternary structure** of the protein.

The distinguishing characteristics of individual proteins and their specific biological functions are mainly due to their large size, to unique compositions and sequence of amino-acids and to their secondary, tertiary and quarternary structure that make up their specific three-dimensional structure. Although the peptide linkages of amino-acids (primary structure) are usually hydrolyzed under drastic conditions (i.e. 6 N HCl at 110°C for 2 h), non-covalent bonds which stabilize the three-dimensional structure of the protein are rather weak and may be easily broken down depending upon the environmental conditions. Slight changes in pH, temperature and/or salt concentration may cause conformational changes in protein structure, which are often accompanied by loss of specific biological activity. This phenomenon is called *denaturation* of protein.

The main purpose of the purification of enzymes is to isolate the desired enzyme from a cell with retention of its native configuration and to purify it as far as possible. Therefore, one must be careful to avoid procedures that may cause denaturation of the enzyme protein during purification.

As is to be expected from the structure of proteins described above individual enzyme proteins have their own specific properties in regard to stability, solubility, charge, size, density and so forth. The principle of enzyme purification is, therefore, to separate the desired enzyme from other unwanted components by making use of differences in these properties.

II. GENERAL CONSIDERATIONS

A. Source material

Since the amount of a given enzyme in bacteria may be very different depending on the strain, growth conditions and the stage of growth, it is usually worthwhile spending some time selecting the bacterial strains and searching for optimum growth conditions, unless there are reasons for preparing the enzyme under less favourable conditions. Especially in the case of bacteria, it is often possible to obtain cultures which are greatly enriched in the desired enzyme by making use of induction or mutation phenomena and by growing them under special conditions.

In order to select a desired type of micro-organism, the enrichment culture technique is often applied as follows: A sample of soil, animal faeces, or sometimes sewage water, containing a mixture of various kinds of micro-organisms is incubated with a specific substrate under certain physico-chemical conditions. After several successive transfers in the same medium, single colonies may be isolated by routine methods. Detailed descriptions of enrichment culture techniques have appeared elsewhere (Hayaishi, 1955). (See also Veldkamp, this Series, Vol. 3A).

It has been known for a long time that synthesis of individual proteins may be stimulated or suppressed within a cell under the influence of specific external agents. The best defined examples in which the synthesis of a protein is shown to be controlled by such agents are the specific induction of enzyme electively provoked by a substrate, and the specific repression of enzyme formation brought about by a metabolite. Although the enrichment culture has nothing to do with induction, the organisms obtained by this technique are often of inducible nature and a number of their enzymes are inducible enzymes.

The following are examples of successful enrichments for desired enzymes that may provide ideas for purification experiments.

Protocatechuate 3,4-dioxygenase which catalyzes the conversion of protocatechuic acid to β-carboxymuconic acid was recently purified from

Pseudomonas aeruginosa and obtained in a crystalline form (Fujisawa and Hayaishi, 1968). The bacterium was selected from about 100 strains which were isolated from soil by the use of the enrichment culture technique mentioned above. The culture medium contained the following compounds (per 1 litre of the medium):

p-hydroxybenzoic acid	3·0 g
$(NH_4)_2HPO_4$,	3·0 g
K_2HPO_4	1·2 g
$FeSO_4.7H_2O$	0·1 g
$MgSO_4.7H_2O$	0·2 g

The pH was adjusted to 7·0. Addition of 0·3 g of yeast extract accelerated the growth rate, but it was not essential. The enzyme appears to be inducible since no activity of the enzyme is observed when this organism is grown in a medium containing citrate as a major carbon source.

β-Galactosidase activity in wild type *Escherichia coli*, when the cells are grown in the absence of a galactoside, is about 1 to 10 units per mg dry weight, whereas the activity is increased 1000 to 10,000 fold by the addition of a suitable inducer in the culture medium (Jacob and Monod, 1961; Cohn, 1957).

Formation of alkaline phosphatase in *E. coli* is shown to be repressed specifically by inorganic phosphate and the enzyme is only formed in measurable amounts when inorganic phosphate levels are closely controlled in the medium (Torriani, 1960). This finding suggests that under such conditions the cells might produce the enzyme in high enough concentration to make the purification relatively simple. When *E. coli* K 12 is grown under conditions of phosphate deprivation, which is optimal for the synthesis of alkaline phosphatase, about 6% of the total protein synthesized is alkaline phosphatase (Garen and Levinthal, 1960).

Formation of aspartate transcarbamylase, one of the enzymes responsible for orotic acid synthesis, in *E. coli* is found to be repressed by some product formed from uracil (Yates and Pardee, 1957). Thus, under conditions which limit the amount of uracil in a growth medium, the specific activity of the enzyme increases about 1000-fold over the basal level (Sheperdson and Pardee, 1960).

Tryptophanase in *E. coli* is another example of an inducible enzyme. In the wild type of *E. coli*, this enzyme is synthesized only when tryptophan or tryptophan analogues are present in the medium. However, the spontaneous mutant of this organism, *E. coli* B/lt 7-A, is constitutive for tryptophanases and produces maximal amounts of tryptophanase (Newton and Snell, 1965). The enzyme can then be purified from the mutant cells and obtained in a crystalline form (Newton *et al.*, 1965).

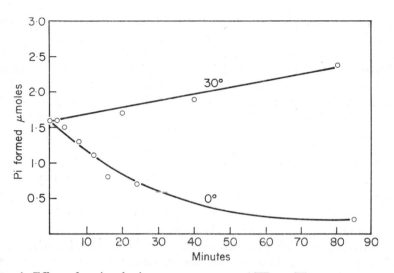

FIG. 1. Effect of preincubation temperature on ATPase. The enzyme was preincubated either at 0°C or 30°C. At the indicated time, aliquots containing 5·2 Hg of protein were removed and the activity was measured at 30°C with the ATP generating system (Reproduced with permission from Pullman *et al.*, 1960).

B. Stability of enzyme

Since, as we discussed above, break down of non-covalent bonds which stabilize the three-dimensional structure often results in the loss of biological activity, it is worthwhile surveying a range of physico-chemical conditions such as pH, temperature, ionic strength, etc., under which a given enzyme is most stable. Needless to say, purification of any one enzyme has to be performed under a specific range of conditions to avoid its denaturation. Care also has to be taken to avoid bacterial contamination during purification. Stability of a given enzyme is sometimes different depending on other components contained in a solution, i.e. stage of purification. Some enzymes are stabilized as the purification proceeds and some unstabilized.

Although enzyme proteins, in general, are more stable at lower temperature, some enzymes are more stable at higher temperature. The first of these enzymes to be reported was glutamic acid dehydrogenase from a mutant cell of *Neurospora crassa* (Fincham, 1957). The enzyme has very little activity at 20°C but becomes active after a few minutes at 35–50°C. This temperature activation is completely reversible; all the extra activity is lost after 2 h at 21°C, but is quantitatively regained on repeated heat treatment. A soluble ATPase from beef heart mitochondria is also cold labile and rapidly loses activity at 0°C whereas it is stable at room temperature (Fig. 1). Attempts to reactivate the cold-inactivated enzyme by incubation at 30°C for varying periods of time in the presence of various cofactors have been unsuccessful (Pullman et al., 1960). Likewise, cold-labile ATPases have also been isolated from yeast (Schatz et al., 1967) and from cytoplasmic membranes of Streptococcus faecalis (Abrams, 1965). Other enzymes which were reported to be cold labile include glutamic acid decarboxylase from E. coli (Shukuya and Schwert, 1960), frog carbamyl phosphate synthetase (Raijman and Grisolia, 1961), $D(-)\beta$-hydroxy-butyric acid dehydrogenase from Rhodospirillum rubrum (Shuster and Doudoroff, 1962), nitrogen-fixing enzyme from Clostridium pasteurianum (Dua and Burris, 1963), pyruvate carboxylase from chicken liver mitochondria (Scrutton and Utter, 1965), glycogen phosphorylase from rabbit muscle (Graves et al., 1965) and acetyl-CoA carboxylase from rat liver (Numa and Ringelman, 1965). Hydrophobic interactions are thought to be involved in stabilizing the three-dimensional configuration of these cold labile enzymes, since these interactions are stronger at higher temperature than at lower temperature (Kauzmann, 1959; Scheraga et al., 1962). Sometimes inactivation is accompanied by the dissociation of enzyme proteins (Scrutton and Utter, 1965; Numa and Ringelman, 1965; Penefsky and Worner, 1965).

Under aerobic conditions, many enzymes are unstable and gradually inactivated. This inactivation is probably due to the oxidation of essential groups in the protein such as sulphydryl groups, and the enzyme is often protected or reactivated by the addition of sulphydryl reagents including cysteine, reduced glutathione, mercaptoethanol, dithiothreitol (DTT) and dithioerythritol (DTE). The latter two compounds, known as Cleland's reagents are the threo and erythro isomers of 2,3-dihydroxy-1,4-dithiol-butane and are found to have desirable properties as protective reagents for sulphydryl groups (Cleland, 1964).

For example, biodegradative threonine deaminase of E. coli is completely inactivated by dialysis at 4°C for 50 h against 25 mM potassium phosphate buffer, pH 7·4. This inactive enzyme can be fully reactivated by incubation with various sulphydryl reagents including DTT (Tokushige et al., 1967). Therefore, purification of the enzyme was carried out with a buffer containing 0·01 M AMP, an allosteric effector of the enzyme, and 2 mM 2-mercaptoethanol, and the enzyme was finally obtained in a crystalline form (Shizuta et al., 1969).

Metapyrocatechase which catalyzes the conversion of catechol to α-hydroxymuconic semialdehyde is reported to be extremely unstable in the presence of oxygen and over 90% of the activity of a crude preparation of the enzyme is lost within 24 h at pH 7·5, even when the preparation is stored at 4°C. This inactivation is, however, partly prevented if the enzyme

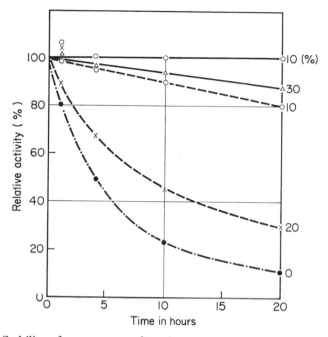

Fig. 2. Stability of metapyrocatechase in an organic solvent. The enzyme was stored at 30°C in 0·05M phosphate buffer, pH 7·5, containing different concentrations (as indicated in the figure in per cent) of an organic solvent. At the indicated time, activity was measured with aliquots by the standard method (Nozaki et al., 1963).

is kept under nitrogen (Kojima et al., 1961). Later, we found that the inactivation of the enzyme by air was completely counteracted by the presence of a low concentration of organic solvent such as 10% acetone or ethanol (Fig. 2). By merely using buffers containing 10% acetone, we have succeeded in purifying the enzyme extensively and finally in obtaining a crystalline preparation (Nozaki et al., 1963).

Nitrogen-fixation systems have been one of the most difficult classes of enzymes to stabilize in cell-free form. Since Carnahan et al. succeeded in preparing one from Cl. pasteurianum (1960), studies of this type of enzyme system have greately expanded. The enzyme system has now been separated into three components, two of which are synthesized by cells only when NH3 is limited and both are very sensitive toward oxygen. Therefore, all treatments throughout the purification are performed under anaerobic conditions (Bulen and LeComte, 1966; Mortenson et al., 1967). Fractionations by column chromatography under anaerobic conditions may be conducted using apparatus and sample collection techniques similar to

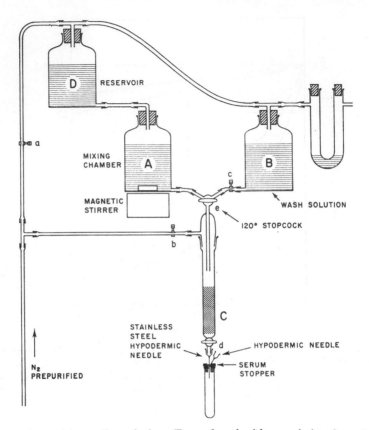

FIG. 3. Anaerobic gradient elution. (Reproduced with permission from Sakami, 1962).

those described previously (Fig. 3) (Sakami, 1962; Munson *et al.*, 1965).

For the fractionation of oxygen-labile enzymes an anaerobic chamber may be employed. Tokyo Air Engineering Company has developed several anaerobic chambers for laboratory use, known as "vacuum dry boxes", which are designed to resist vacuum pressures of the order of 10^{-3} mm Hg, so that the gas phase can be replaced with an inert gas such as nitrogen by repeated evacuation and filling. Manipulations in the chamber can be carried out by the use of airtight gloves when the gas pressure in the chamber is equal to atmospheric.

As another approach to work under anaerobic conditions, an anaerobic room has been set up in Dr. Stadtman's laboratory at National Institutes of Health, Bethesda. The gas phase of the room can be replaced with an inert gas in essentially the same manner as above. One of the authors (O.H.

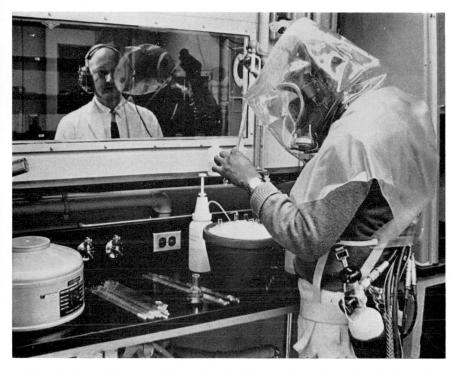

FIG. 4. Anaerobic room. (Reproduced with permission from Dr. Stadtman).

had a chance to visit the laboratory and, through Dr. Stadtman's kind permission, a photograph is reproduced here (Fig. 4).

A major difficulty encountered during the development of satisfactory methods for the purification of enzymes is their instability under conditions commonly employed in enzyme fractionation. Many enzymes lose their activity during purification or storage even if they are kept under optimal conditions. In such cases, it is advisable to attempt the stabilization of a given enzyme with specific compounds. Combination of substrate, substrate analogues, coenzymes or allosteric effectors (see below) with the enzyme results in conformational changes in the protein moiety. Thus, enzymes are sometimes specifically stabilized. There are many instances in which the purification of a given enzyme is carried out in the presence of its specific substrate (Berger *et al.*, 1946; Civen and Knox, 1960; Cho and Pitot, 1967). Besides these specific protectors, an inert protein such as albumin, has often been used as a non-specific protector of proteins (Berger *et al.*, 1946; Taylor *et al.*, 1948). Glycerol has mainly been employed for the protection of biological materials against the deleterious effect of freezing, storage at low temperatures or drying by sublimation *in vacuo* (Greiff *et al.*, 1961; Polge *et al.*, 1949). Recently, dimethyl sulphoxide was been shown to be equal to glycerol as a protective substance and to be superior in certain cases (Greiff and Myers, 1961). The activity of rat liver mitochondria could be maintained by prolonged storage in 10% dimethyl sulphoxide. Hoppel and Cooper pointed out that protein concentration was important for stabilization and the optimum was 30–40 mg of protein per ml of 10% dimethyl sulphoxide. At 100 mg/ml, there was a loss of approximately 50% (Hoppel and Cooper, 1968).

Besides protection against the damaging effects of freezing and thawing, mixed glycerol-water and dimethylsulphoxide-water solvents have been used directly for purifying enzymes. Some cold-labile enzymes including adenosine triphosphatases of beef heart mitochondria and of yeast (Racker *et al.*, 1963), 17β-hydroxysteroid dehydrogenase of human placenta (Jarabak *et al.*, 1966), carbamyl phosphate synthetase (Novoa and Grisolia, 1964) and phosphorylase b (Graves *et al.*, 1965) are reported to be protected from cold inactivation by the presence of suitable concentrations of glycerol or dimethylsulphoxide. Other enzyme purifications which have been performed in a glycerol- or dimethylsulphoxide-water medium are: Glucose-6-phosphate dehydrogenase from mammary gland (Levy, 1963) RNA polymerase from *Micrococcus lysodeikticus* (Nakamoto *et al.*, 1964), 3α-hydroxysteroid dehydrogenase of *Pseudomonas testosteroni* (Boyer *et al.*, 1965) steroid Δ^4-5β-dehydrogenase (Davidson and Talalay, 1956) and glycerokinase (Bublitz and Kennedy, 1954).

Organic solvents such as acetone or ethanol also act as stabilizers for certain enzymes. Since low concentrations of acetone or ethanol were found to protect metapyrocatechase of a *Pseudomonas* sp. from air oxidation (see above) (Nozaki *et al.*, 1963), a number of enzymes have been found to be stabilized by these organic solvents. Benzylalcohol dehydrogenase from *Pseudomonas* sp. has a half-life of a few hours in 0·05 M phosphate buffer, pH 7·5 at 5°C and reducing or chelating agents have no significant retarding effect on the loss of activity. An organic solvent such as acetone or ethanol, however, can cause significant and often complete stabilization of the dehydrogenase. Likewise, yeast alcohol dehydrogenase and bovine liver homogentisicase were also found to be stabilized by these organic solvents (Takemori *et al.*, 1967).

Although the mechanism of the protection by the organic solvent is obscure, 10% acetone protects not only oxygen-labile groups in the protein but also the enzyme from denaturation caused by urea (Nozaki *et al.*, 1968). The advantage of using these organic solvents as a stabilizer is that they can be used without significant interference with chromatographic or other purification procedures.

Specific anion, cation or metal ions are sometimes required for the maximal stabilization of the enzyme. For example, bacterial α-amylase, like other α-amylases, can be markedly protected from denaturation by the presence of calcium ion (Okunuki *et al.*, 1956). Likewise, a bacterial proteinase has been reported to be protected against heat denaturation by calcium ion, which is also essential for full enzymatic activity (Gorini, 1950). Manganase ion is known to stabilize many enzymes, including lysozyme (Gorini and Felix, 1953) and arginase (Mohamed and Greenberg, 1949). Tryptophanase from a mutant of *E. coli* is markedly stabilized against heat treatment by the presence of high concentration of ammonium sulphate (11·4%) (Newton *et al.*, 1965). Likewise, malic dehydrogenase and other enzymes purified from a halophilic bacterium, *Halobacterium salinarium*, requires high concentrations of monovalent ion (approximately 25% NaCl) for stability and activity (Holmes and Halvorson, 1965; Holmes *et al.*, 1965).

Stability is often dependent on the buffer solutions used. Buffers which are widely used for biological research in the pH range 6 to 8 may be inadequate in some cases because of inefficacy, undesired reactivity or toxicity. For example, phosphate buffers poorly above pH 7·5 and precipitates certain useful cations including metal ions. "Tris" is a poor buffer below pH 7·5 and is a potentially reactive primary amine, e.g. "Tris" uncouples oxidative phosphorylation. Other buffers such as borate, glycylglycine, imidazole, veronal (5,5-diethylbarbiturate), maleate, and dimethylglutarate are for one reason or another even less satisfactory.

As a start toward improving the available inventory of buffers in the pH range 6 to 8, Good and his colleagues prepared 12 new or seldom used hydrogen ion buffers with pK values between 6·15 and 8·35 (1966). Ten are zwitterionic amino-acids, either N-substituted taurines or N-substituted glycines and two are cationic primary aliphatic amines. All of the zwitterionic buffers are better than conventional buffers in some biochemical reactions. Two of the zwitterions, N-tris-(hydroxymethyl)-methylaminoethanesulphonic acid and N-2-hydroxyethylpiperazine-N'-2-ethanesulphonic acid, give particularly active and stable mitochondrial preparations. These two also give higher rates of protein synthesis in cell-free bacterial preparations than do tris-(hydroxymethyl)aminomethane (Tris) or phosphate buffers. These buffers are now commercially available from Calbiochem. (3625 Medford St., Los Angeles, California 90063, U.S.A.).

It has been known that enzyme proteins, in general, increase in their susceptibility toward enzymic hydrolysis when they are denatured. Although some enzyme proteins are susceptible to a proteolytic action even in their native state, many globular proteins such as microbial α-amylase, catalase of bovine liver, alcohol dehydrogenase and lactic dehydrogenase are hardly attacked by proteinases unless they are first subjected to some

denaturing treatments (Okunuki, 1961). In fact, susceptibility to a proteinase has been applied as a measure of the "ratio of denaturation"of
some globular proteins. A bacterial proteinase crystallized from the culture
medium of *Bacillus subtilis* (Hagihara *et al.*, 1958a), commercially known
as Nagase, is favourably used for this method (Okunuki *et al.*, 1956)
Although loss of biological activity can sometimes by used as a criterion
for the denaturation of protein, some enzyme proteins may be inactivated
without being denatured under conditions which merely modify a few
essential groups in the protein. For example, as mentioned above, metapyrocatechase is easily inactivated by air or H_2O_2, but this inactivation is
not accompanied by denaturation of the protein moiety as measured by the
above-mentioned method (Nozaki *et al.*, 1968). Therefore, determination
of the susceptibility of a given enzyme to a proteinase may provide some
criteria concerning the rigid structure of the enzyme, and the ease with
which it may be purified. In general, enzyme proteins which are resistant
to proteolytic action are more stable and easier to purify and crystallize.

C. Assay method

In addition to the importance of the separation method itself the detection methods used in conjunction with it may be critical in revealing heterogeneity and in searching for the optimal conditions for a given method.
Since a large part of the time may be spent in testing the activity of the
different fractions during purification, availability of a rapid detection
method speeds up the whole purification process. Speed here is more
important than extreme accuracy and even if the method is semiquantita-
tive, it might be equally helpful in screening the various purification
methods.

When proceeding with the purification of a given enzyme, the enzyme
may lose considerable activity after some particular purification steps.
This does not necessarily mean that the enzyme is inactivated, but may
mean that the enzyme merely loses its cofactor or other factors essential for
the activity, or that it is fractionated into two or more protein fractions, all
of which are essential for catalysis.

The biological activity of many proteins may be controlled by specific
compounds such as metabolites or nucleotides which do not interact
directly with the substrates or products of the reactions. These regulatory
enzymes, known as **allosteric enzymes** often show an absolute requirement for an activator, called an **allosteric effector** for activity to occur.
Glycogen phosphorylase b from rabbit muscle requires AMP for its activity
(Helmreich and Cori, 1964). Biodegradative threonine deaminase from
Clostridium tetanomorphum and *E. coli* are specifically activated by ADP
and AMP, respectively (Hayaishi *et al.*, 1963; Hirata *et al.*, 1965). These

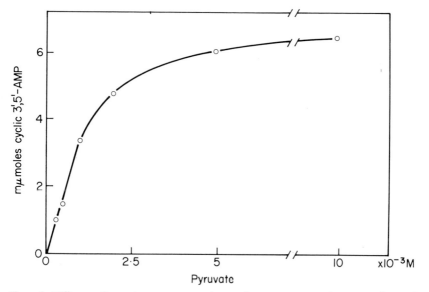

FIG. 5. Effects of varying concentrations of pyruvate on the rate of reaction (Hirata and Hayaishi, 1965).

activations are more pronounced when substrate concentrations are low. Besides nucleotides, some metabolites sometimes act as activators: Citric acid for acetyl CoA carboxylase from rat adipose tissue (Vagelos *et al.*, 1963), acetyl CoA for pyruvate carboxylase from chicken liver (Utter *et al.*, 1964), and acetylglutamate for carbamyl phosphate synthetase (Fahian *et al.*, 1964).

During the course of studies on adenylyl cyclase of *Brevibacterium liquefaciens*, initial attempts at purification of the enzyme by Sephadex column chromatography resulted in almost complete loss of enzyme activity, although the activity was found in the 105,000 *g* supernatant fraction of cell-free extracts. Subsequently, it was discovered that the low recovery of activity was related to the removal of a factor which was identified as pyruvate (Hirata and Hayaishi, 1965) and without this factor the reaction could not proceed (Fig. 5).

Examples in which enzyme activity is split into more than one fraction are as follows: The nitrogen-fixing systems of *Azotobacter* and *Cl. pasteurianum* are fractionated into two major fractions (Bulen and LeComte, 1966; Mortenson *et al.*, 1967). Both fractions are essential for the electron- and ATP-dependent N_2-fixation. Neither of these fractions alone catalyses any part of the overall reaction without the other (Kennedy *et al.*, 1968). The conversion of benzene to catechol is catalysed by an enzyme system from *Pseudomonas putida*. The enzyme system responsible for the conversion can be separated into two protein fractions by ammonium sulphate fractionation, both of which are required for maximal activity (Gibson *et al.*, 1968). The ω-hydroxylation of fatty acid is catalysed by a soluble enzyme system from *Pseudomonas oleovorans* in the presence of molecular oxygen and NADH. The enzyme system may be separated into three protein components: rubredoxin (a red nonheme iron protein containing no labile sulphide), an NADH-rubredoxin reductase and the ω-hydroxylase, all of which are required for the overall reaction (Peterson *et al.*, 1967).

From these facts, it may naturally be concluded that it is important to determine total yield of a given enzyme after each purification step. If the yield is too low, it is advisable to test the activity either by mixing the other inactive fractions, by adding some known coenzymes or a boiled extract of crude preparation. Purity of the various fractions may be quantitatively expressed in terms of specific activity. In most cases specific activity is defined as units of enzyme per mg protein in a certain time under the test conditions. For the determination of protein, detailed descriptions are available (Layne, 1957).

D. Separation of protein from nucleic acid

Extracts of bacterial cells contain much nucleic acid which greatly interferes with the fractionation of proteins. Therefore, it is desirable to separate protein from nucleic acid at an early stage in purification. A number of procedures are available for this purpose. Separation of nucleic acid by $MnCl_2$ from extracts prepared by alumina grinding has been reported (Korkes *et al.*, 1951). The extracts are mixed with 0·05 volume of 1 M $MnCl_2$ and the precipitate removed by centrifugation. Separation of nucleic acids from protein solutions can also be achieved by the addition of either protamine sulphate or streptomycin. Since these compounds complex with nucleic acids causing their precipitation, the supernatant solution can then be fractionated without interference. The amount of protamine sulphate used has to be determined by preliminary tests with solutions of nucleic acid and protamine so as to keep the excess of protamine to the minimum. To the solution, 1 to 5% of protamine solution is added dropwise with stirring until the 280/260 mμ absorption ratio of 1·0 to 1·1 is obtained in a centrifuged aliquot. Usually, 0·1 to 0·3 mg of protamine per mg of protein can be used. Protamine sulphate treatment sometimes removes active components which are occluded with the nucleic acids. This may be prevented by prior treatment with ribonuclease and deoxyribonuclease (Mortenson *et al.*, 1967). Streptomycin treatment sometimes gives better results than protamine sulphate. The precipitation of nucleic acids with streptomycin depends on the salt concentration (Moskowitz, 1963). With an extract of *Lactobacillus plantarum*, 0·5 to 2 mg of streptomycin sulphate per mg of protein gives best separation of nucleic acid from protein (Oxenburgh and Snoswell, 1963). Digestion of DNA or RNA polymer by pretreatment with nucleases may counteract the interference of these polymers during the fractionation.

III. FRACTIONATION OF SOLUBLE ENZYME

A. Fractionation by stability

As we discussed above, stability of proteins is different depending on the individual protein. If the desired protein is more stable than other unwanted proteins in a mixture under certain conditions, the unwanted proteins may be separated by fractional denaturation by heating or by changing the pH. This method is usefully applied as the first step before the main purification begins. By these treatments, it is possible to coagulate much unwanted proteins which can then be centrifuged off and discarded. It is also possible to make use of the fact that the presence of substrate or other reagents frequently has a specific stabilizing effect on the enzyme It may be possible to heat the solution at a higher temperature in the presence of the specific stabilizer. For example, glutamic asparatic transaminase from pig heart is stabilized over a wide pH range in the presence of maleate, a competitive inhibitor of the enzyme. Further addition of α-ketoglutarate, a substrate, gives an appreciable increase in stability and makes it possible to heat the enzyme at 75°C for 10 min. Since many other proteins are denatured under these conditions, heat treatment is one of the most effective isolation methods for the enzyme (Jenkins *et al.*, 1959).

B. Fractionation by solubility

Fractionation by solubility is one of the classical methods but still of great value. It is particularly useful in the early stage of a fractionation because of its technical simplicity and less dependence on the scale of operation.

1. *Fractional precipitation with organic solvents*

Ethanol and acetone, both of which are completely miscible with water, are widely used for fractional precipitations. Since most enzymes are denatured by such organic solvents at room temperature, the operation has to be carried out at low temperature; preferably below 0°C. Protein solubility in the presence of organic solvents usually decreases markedly with decreasing temperature. This marked sensitivity to temperature also necessitates adequate temperature control during fractionation. Provided that the given enzymes are sufficiently stable to the organic solvent, fractionation by organic solvent usually gives better results than that by salts, probably because of the removal of contaminating lipids that interfere with the purification. In general, proteins which contain lipid as a functional unit are particularly sensitive to organic solvents. Salt fractionation is much more mild and facilitates the fractionation of such enzymes.

2. *Fractional precipitation by salts*

At very low ionic strength, the solubility of proteins in general increases with increasing salt concentration ("salting in"). As ionic strength is increased, however, a maximum is reached beyond which solubility decreases continuously ("salting out"). The phenomenon of "salting out" has been widely used for the fractional precipitation of proteins. The salt most employed is ammonium sulphate, because it is highly soluble in water and does not harm most enzymes. Different proteins are precipitated at different fixed salt concentrations provided the pH and temperature are fixed. It is advisable to use good quality ammonium sulphate in order to avoid toxic impurities or free acid. Even pure preparations are slightly acidic and care must be taken to insure proper pH control. There is a time factor in the precipitation of proteins by ammonium sulphate and it is advisable to wait for at least 15 min before centrifuging and longer, if possible; otherwise the fractions will not be clearly separated. The solubility of most proteins go through minima in the vicinity of their isoelectric points. This property can be taken advantage of for more effective salting-out.

Although ammonium sulphate has the most desirable properties and has been commonly used in enzyme fractionation, there are disadvantages in that the pure material is slightly acid, pH control is difficult owing to loss of ammonia and nitrogen estimations cannot be performed without careful removal of the salts. Salts other than ammonium sulphate can be used and, among them, sodium sulphate and potassium phosphate are more effective than ammonium sulphate and give sharp peaks and very good separation. But they are much less soluble and are used only with certain proteins. For a theoretical consideration of "salting out" phenomena and practical use of the method, the readers are referred to appropriate review articles (Dixon and Webb, 1961; Green and Hughes, 1955).

C. Fractional adsorption

Adsorption techniques have been widely employed in the purification of proteins. This procedure involves addition of a selected adsorbent to a solution of protein mixture and then elution of the enzyme from the adsorbent (batch adsorption method). Since, in a general sense, interaction between protein and adsorbent involves any of several types—ionic, hydrophobic, or hydrogen bonds—elution can usually be accomplished by changing pH or salt concentration. The number of adsorbents that can be used in this way is very large and includes a great many that are suitable for column chromatography. The choice has to be made by trial and error over a range of pH values and salt concentrations. Calcium phosphate gel and alumina Cγ have been used for many years as adsorbents. For the preparation of calcium phosphate gel and alumina Cγ, readers are referred to methods described elsewhere (Dixon and Webb, 1964; Colowick, 1955). These gels are mainly used for the batch method and are not suitable for the column method, because their fine particles resist the flow of liquid. One advantage in the use of the batch method is its less dependence on scale of operation; it can be applied for large sample volumes. Hydroxyapatite which is composed of crystals of neutral and basic calcium phosphate gel, has a larger particle size and can be favourably used as an adsorbent for column chromatography. For the preparation of hydroxyapatite readers are referred to the methods previously described (Siegelman et al., 1965; Tiselius et al., 1956; Levin, 1962).

D. Fractionation by chromatography

This technique has recently undergone considerable improvement and now is one of the most important methods for separating protein. Separation may depend on the equilibrium distributions between two phases, solid and liquid. In general, the chromatographic procedure involves the use of a columnar bed of tightly packed particles of adsorbent which is equilibrated with a liquid under specified conditions of pH, salt concentration and temperature. The sample, similarly equilibrated and preferably in a small volume, is applied to the top of the bed and eluted from the column by a selected programme of eluants. The adsorbed proteins may be eluted by either of three different procedures; (1) one step elution, (2) stepwise elution and (3) gradient elution. One step elution is carried out with the same buffer that is in equilibrium with the adsorbent. This method offers the highest resolving power attainable with the material employed but usually gives broad protein peaks. The tendency of protein bands to spread and tail in chromatography can largely be overcome by the use of stepwise or gradient elution. Proteins with widely different affinities for the adsorbent can be eluted separately by discrete changes in the eluant in terms of pH, salt concentration or both (stepwise elution). Elution with a gradient has other important advantages. A gradient can be adjusted over a range of pH and salt within which a given protein is released from the adsorbent and also in shape to suit the protein mixture under study.

Numerous devices are available for the production of gradients for this purpose (Bock and Ling, 1954; Peterson and Sober, 1959).

Thus, the column method is much superior to the batch method in many respects: conditions suitable for a given enzyme in adsorption to and elution from adsorbents can be critically controlled by proper selection of pH, salt concentration and temperature, resulting in much higher resolution. Hydroxyapatite, synthetic ion exchange resins, cellulose ion exchanger and Sephadex ion exchanger have been used with more or less success as adsorbents in column chromatography.

1. *Synthetic ion exchanger*

There are a number of synthetic ion exchange resins available for the chromatography of low molecular compounds but only a few can be applied for the separation of protein. Strong cation or anion exchangers such as Dowex-50 and Dowex-1 often cause the denaturation of protein by local pH effects and the capacity of these resins for macromolecules is small because of their low porosity. Weak cation or anion exchange resins with high porosity may be applied for the separation of protein. Amberlite IRC-50, also known as XE-64 or CG-50, a weak cation exchange resin having carboxyl residues as the active site, has often been used for the separation of basic proteins such as ribonuclease (Crestofield *et al.*, 1963) and cytochrome c (Boardman and Partridge, 1955; Boman, 1955; Prins, 1959; Hagihara *et al.*, 1956; Nozaki *et al.*, 1957; Okunuki, 1959).

2. *Cellulose ion exchanger*

The development of cellulose ion exchangers overcomes the difficulty of the synthetic ion exchange resins (Peterson and Sober, 1956). Their open structure permits ready penetration by large molecules, resulting in very high capacity for the adsorption of protein. Typical examples of the cellulose ion exchanger are diethylaminoethyl (DEAE) cellulose, a weak anion exchanger and carboxymethyl (CM) cellulose, a weak cation exchanger. These celluloses can be applied for the separation of a number of proteins. For the separation of strongly anionic proteins that may be too tightly adsorbed on DEAE-cellulose, ECTEOLA cellulose, an anion exchanger that contains a smaller number of weaker basic groups, can be applied. Phosphorylated cellulose [P cellulose (Peterson and Sober, 1956)] and sulphethyl cellulose [SE cellulose (Porath, 1957)] are cation exchangers with stronger acidic groups that retain a negative charge at very low pH. Therefore these celluloses can be applied for proteins like pepsin which are stable at low pH, since most of the charges on CM cellulose will be suppressed below pH 4 and its capacity for adsorption will be greatly diminished. Likewise, guanidinoethyl cellulose [GE cellulose (Semenza

TABLE I

Properties of cellulose ion exchangers

Description	Types	Ionic groups	Total capacity meq/g
Strong base	GE(Guanidoethyl)	$-O-C_2H_4-NH-C\begin{smallmatrix}\nearrow NH\\\searrow NH_2\end{smallmatrix}$	0·9
	TEAE(Triethylamine)	$-O-C_2H_4-\overset{\oplus}{N}(C_2H_5)_3X^{\ominus}$	0·5–0·8
	DEAE(Diethylamine)	$-O-C_2H_4-N(C_2H_5)_2$	0·8–1·0
Intermediate base	ECTEOLA (Mixed amines)	(undefined) $\oplus X^{\ominus}$	0·2–0·5
	AE(Aminoethyl)	$-O-C_2H_4-NH_2$	0·4–0·8
	PEI(Polyethyleneimine)	$-(CH_2-CH_2-NH)_n$ $-CH_2CH_2-NH_2$	0·2
Weak base	PAB(Paraaminobenzyl)	$-O-CH_2-C_6H_4-NH_2$	0·2–0·3
Weak acid	CM(Carboxymethyl)	$-O-CH_2-COOH$	0·6–0·8
Intermediate acid	P(Phosphoric acid)	$-O-PO_3H_2$	0·7–0·9
	PPM(Phosphomethyl)	$-O-CH_2PO_3H_2$	0·2–0·3
Strong acid	SM(Sulphomethyl)	$-O-CH_2-SO_3H$	0·2–0·3
	SE(Sulphoethyl)	$-O-CH_2-CH_2-SO_3H$	0·2

(1960))] provides a means of chromatographing at very high pH values. Triethylaminoethyl cellulose [TEAE cellulose (Porath 1957)] is another anion exchanger that is capable of retaining its positive charge at higher pH than that of DEAE-cellulose.

Some properties and loading capacities of cellulose ion exchangers are listed in Table I.

3. *Sephadex ion exchanger*

Instead of cellulose, cross-linked dextran (Sephadex) is used for the skeleton of the Sephadex ion exchanger. DEAE, CM, SE and QAE (diethyl-2-hydroxypropylammonium) Sephadex are now available. As discussed below these Sephadexes have the ability to act as molecular sieves. This characteristic may be effective in getting good results, although it may reduce the adsorption capacity. The difficulty of using the Sephadex column is that changes in volume are accompanied by changes in pH and salt concentration; therefore, Sephadex ion exchanger may be most useful for the batch method.

E. Molecular-sieve chromatography

Since the introduction of a cross-linked dextran, commercially known as Sephadex in 1959 (Porath and Flodin, 1959), "Molecular-Sieve" chromatography (also known as "gel-filtration" or "Exclusion" chromatography) has been developed extensively and is now one of the principal techniques in the purification and characterization of proteins (Porath, 1962). Separations by this technique depend on the fact that molecules larger than the largest pores of the swollen Sephadex, i.e. above the exclusion limit, cannot penetrate the gel particles and therefore pass through the bed in the liquid phase outside the particles. They are thus eluted first. Smaller molecules, however, penetrate the gel particles to a varying extent depending on their size and shape. Molecules are therefore eluted from a Sephadex bed in the order of decreasing molecular size.

There are about ten different gels in the Sephadex G-series, ranging from Sephadex G-10 with the lowest molecular weight exclusion limit of approximately 700, up to Sephadex G-200 with an exclusion limit above 200,000 molecular weight. The introduction of polyacrylamide (commercially available as Bio-Gel P) (Hjertèn and Mosbach, 1962; Hjertèn, 1962a) and agarose (Sepharose or Bio-Gel A) has improved the flexibility of the method and extended the exclusion limit up to 300,000 molecular weight with Bio-Gel P-300 and to 150,000,000 with Agarose (Hjertèn, 1962b).

Important publications on the applications of gel filtration and ion exchangers and a comprehensive list of pertinent literature are distributed regularly from Pharmacia Fine Chemicals, Uppsala, Sweden or their local representative. This service is of considerable value to workers in this field.

One of the first applications of molecular sieve chromatography was "desalting". This term refers not only to the removal of salts but also to the removal of other low molecular weight compounds from solutions of macromolecules. Since the introduction of high-porosity granules, molecular-sieve chromatography can be used in both analytical and preparative methods for proteins. Determination of molecular weights is also one of the useful applications of this technique. Over a considerable range, the elution volume is approximately a linear function of the logarithm of the molecular weight (Whitaker, 1963; Andreus, 1964; Leach and O'Shea,

1965). The same principle may be applied to the separation of individual proteins. Resolution may depend on the column size, flow rates and sample volume. For high resolution, long narrow columns and low flow rates should be employed and the sample volume should be small. However, the difficulty in the use of the high-porosity granules is their resistance to the flow of liquid. This difficulty can be minimized by directing the flow upward.

When two or more proteins to be separated are close in molecular weight and the column available is not long enough to accomplish the separation, a recycling method introduced by Porath and Bennich (1962) sometimes gives good results.

F. Separation by isoelectric focusing

Like amino-acids, proteins are ampholytes, i.e. they act as both acids and bases. Since proteins are electrolytes, they migrate in an electric field and the direction of migration will be determined by the net charge of the molecule. The net charge is influenced by pH, and for each protein there is a pH value at which it will not move in an electric field; this pH value is the isoelectric point (pI). The isoelectric point of a given protein is a constant and assists in characterization and separation of proteins.

Provided that the pH gradient is established in an electrolyte system in a way that pH steadily increases from the anode to the cathode and the gradient is sufficiently stable for the duration of the experiment, ampholytes such as proteins and peptides present in the electrolyte system, by imposing a d.c. potential, will collect at the place in the gradient where the pH of the gradient is equal to the isoelectric point of individual ampholytes. A necessary prerequisite for a useful application of isoelectric focusing in practice is to stabilize the electrolyte system against uncontrolled convection and against remixing of focused ampholytes.

The required properties of carrier ampholytes for isolectric fractionation and analysis were defined from a theoretical point of view by Svensson (1961). In 1966, Vesterberg and Svensson developed a method of synthesizing a series of carrier ampholytes whereby they could be used for producing a natural equilibrated pH gradient. Convective disturbance is prevented in a liquid column by using a density gradient consisting of a sucrose solution with gradually decreasing sucrose concentration from the bottom to the top of the column (Kolin, 1954). The synthetic carrier ampholyte consists of a mixture containing very large numbers of aliphatic aminocarboxylic acids probably having an average molecular weight of 300–600 and representing a large number of pI values. A difference of approximately 0·02 units in the pI of two compounds is sufficient for their separation into separate zones.

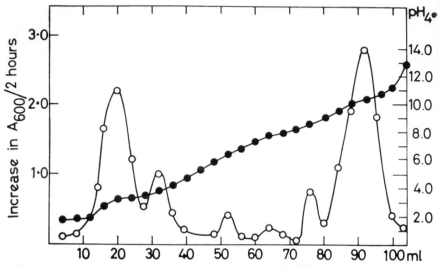

FIG. 6. Deoxyribonuclease activity, ○——○, and pH at 4°, ●——●, of the fractions taken from a 110 ml column after isoelectric focusing. This figure which is more up to date than one which appeared previously (Vesterberg et al. (1967) was kindly provided by Dr. T. Wadström.

The processes of synthesizing and fractioning ampholytes have been improved by LKB-Produkter AB. Stockholm-Bromma, and it is now possible to obtain "Ampholine Carrier Ampolytes" in ready to use mixtures covering the principal pH range (pH 3–10) of interest or small fractions thereof (range less than 2 pH units). Svensson has described an apparatus (1962) for electrofocusing, consisting of a glass column tube with cooling jacket and platinum electrodes. The apparatus and kits of ampholine carrier ampolytes of different pH range are also commercially available from LKB. The technique is described in detail by Vesterberg, (this Volume, p. 595).

In Fig. 6 is shown a heterogeneity pattern obtained by isoelectric focusing of deoxyribonuclease, an extracellular protein produced by *Staphylococcus aureus*. After electrolysis two main peaks and some minor ones could be observed. The two major peaks were isoelectric at pH 2–3 and 10·1 and the minor ones between pH 4 and 9 (Vesterberg et al., 1967; Wadström, 1967).

IV. CRYSTALLIZATION

Crystallization of the protein is one of the final purposes of purification. However, crystallization must not be regarded as evidence that the enzyme is pure. On recrystallization the specific activity of the crystals may increase and it may be necessary to recrystallize several times until the specific activity rises to a constant maximum value. For this reason it is best to regard crystallization as a specific method of fractionation to be used in the final stages of purification.

The most common method of crystallization is from ammonium sulphate solution. This can be done by the addition of finely powdered ammonium sulphate or saturated ammonium sulphate solution to a rather concentrated enzyme solution until a slight turbidity appears. The mixture is then kept for several hours, up to a few days if necessary, until a silky sheen can be seen upon gradual shaking. This is a convenient method for the detection of appearance of crystals. Confirmation is provided by microscopic observation and increase in specific activity of the crystals after repeated crystallization.

Enzymes can also be crystallized by salt extraction. The enzyme is salted out by adding ammonium sulphate and the supernatant is discarded. The residue is sequentially extracted with solutions of decreasing ammonium sulphate concentration. Using this method Kohn et al. have succeeded in crystallizing several enzymes from *Pseudomonas* which are involved in tartaric acid metabolism (Kohn and Jakoby, 1968a; Kohn et al., 1968; Kohn and Jakoby, 1968c)

Crystallization of some enzymes can also be achieved either by the addition of organic solvent such as acetone or ethanol or by simple dialysis against distilled water or a diluted buffer solution. Pepsin (Northrop, 1946) and chymotrypsin (Kunitz, 1948) have been crystallized from alcohol solution, although they may also be crystallized from ammonium sulphate solution. Alpha-amylase from *Bacillus subtilis* (Hagihara, 1951) and from *Aspergillus oryzae* (Akabori et al., 1951) are crystallized from acetone solution. Protocatechuate 3,4-dioxygenase can be crystallized either by the addition of ammonium sulphate or by dialysis against distilled water (Fujisawa and Hayaishi, 1968). In any case, first crystals appear under critical conditions in which a given enzyme is still soluble but almost at the threshold for precipitation. At a given ammonium sulphate concentration solubility of a protein depends on temperature; it increases as temperature decreases. This fact may be used to establish such critical conditions.

Sometimes addition of other components facilitates the case of crystallization: for example, Hg^{++} for phosphopyruvate kinase (Warburg and Christian, 1941) and for lactic dehydrogenase (Kubowitz and Ott, 1943); decanol or related substances for both human and bovine plasma albumins (Cohn et al., 1947); calcium ion for bacterial α-amylase (Hagihara, 1951); magnesium ion for phosphorylase b (Fischer and Krebs, 1958). For the crystallization of protocatechuate 3,4-dioxygenase from ammonium sulphate solution, incubation of the enzyme with mercaptoethanol for

1 week is necessary to obtain rhombic crystals. This compound prevents the polymerization of the enzyme (Fujisawa and Hayaishi, 1968).

It should be noted that the enzyme to be crystallized has to be reasonably pure, not only in the sense of composition but also of tertiary structure of the given protein. For example, the first crystallization of cytochrome c which had been purified to almost homogeneity for a long time was achieved by Bodo in 1955 from Penguin muscle by the addition of hydrosulphite to reduce cytochrome c prior to the addition of ammonium sulphate. By using essentially the same procedure, cytochrome c's from various sources have been crystallized in Okunuki's laboratory (Okunuki, 1959). Later, it was found that reduced cytochrome c has a more rigid structure than the oxidized form in the sense that the former is resistant to proteolytic action whereas the latter is easily susceptible (Nozaki et al., 1957; Nozaki et al., 1958; Yamanaka et al., 1959). In a solution containing both forms of cytochrome c their interaction prevents crystallization. Cytochrome c's can be crystallized from either their reduced or oxidized forms but not from a mixture of the two (Hagihara et al., 1958b).

V. OTHER METHODS

Other important fractionation methods include ultracentrifugal and electrophoretic separations. These techniques are discussed in other Chapters in this Volume. In addition to the methods described in this Chapter and above, removal of small molecular substances from protein and concentration of protein solutions are frequently required prior to some purification procedures.

A. Desalting

Dialysis methods using cellophane tubing have been widely applied for the removal of small molecular substances from protein solutions. The rate of dialysis through the walls of cellophane tubing is relatively slow, but it is possible to increase the average pore size either by treatment with an aqueous solution of zinc chloride or by two-dimensional stretching. For detailed descriptions of these methods readers are referred to a review article (Craig and King, 1962).

The removal of small molecular substances from marcomolecules can also be achieved by use of the gel filtration technique as described above. This method is faster and more efficient than dialysis and may be used for large volumes of sample. Sample volumes up to approximately 30% of the total bed volume may be used (Flodin, 1961).

B. Concentration

Several procedures can be used for the concentration of protein solutions but with each the usual precautions against denaturation must be employed. Protein solution may be simply concentrated by precipitation with salt or organic solvent, followed by resolution and then desalting if required.

Concentration can also be achieved by ultrafiltration by means of selectively permeable membranes under reduced pressure. As for dialysis, cellophane tubing has been widely used for this purpose (Siegelman and Firer, 1962). These authors used a long cylinder of porous polyethylene to support the cellophane bag. Recently Amicon Co. (Lexington, Mass.) has developed synthetic membranes (DIAFLO membranes) which consist of cross-linked dextran. These membranes reject all retained substances at the surface—not in the membrane—so that they are readily swept away be stirring or by fluid flow in the ultrafiltration module. All molecules below the "cut-off" size pass easily through the material. Amicon's proprietary membranes are tailored for a great variety of separation tasks. A wide selection of fluid transport rates and cut-off levels is offered, ranging from a relatively open material capable of rejecting only substances above 100,000 molecular weight to a membrane that will not pass molecules above approximately 500 molecular weight. Therefore, this method can be employed not only for concentration of dilute solutions but also for fractionation of protein, removal of salts or removal of dissolved as well as particulate contaminants from liquids and gases (Blatt et al., 1965; Blatt et al., 1967). Apparatus for ultrafiltration is commercially available from the same company.

Protein can also be concentrated by treatment with dried highly cross-linked gels such as Sephadex G-25 or Bio-Gel P-2. Water and low molecular weight substances are adsorbed by the swelling gels while high molecular weight substances remain in the external solution. After 10 min the gel grains are removed from the mixture by either centrifugation or filtration. In this way an almost three-fold enhancement of high molecular weight solute concentration can be obtained while the ionic strength and the pH are unchanged. The process can be repeated to concentrate the solute even further. This procedure is simple and rapid and is especially useful for very labile proteins (Flodin et al., 1960).

VI. SEQUENCE OF FRACTIONATION METHODS

Since various fractionation methods discussed in this Chapter have some advantages or disadvantages in particular circumstances, the methods which can be employed in any one purification stage will naturally be limited. Obviously crystallization comes last, since it requires purity and a concentrated enzyme solution. Most techniques described above are applied for the fractionation of soluble proteins. Therefore, for the fractionation of particle bound or insoluble enzyme proteins, attempts at solubilization must first be made before the main fractionation can be started.

Since in the early stages of purification rather large volumes of enzyme solution have to be dealt with, techniques which are less dependent on volume should come first. It is advisable to carry out the removal of nucleic acid before the main fractionation is begun. Heat treatment and fractional precipitations are less dependent on volume unless the size of refrigerated centrifuges or filtration capacities are not limiting. The latter techniques are especially useful because they bring about reduction in volume and facilitate the subsequent purification procedures. Although fractional adsorption itself is less dependent on volume, sometimes preliminary dialysis is required to reduce the ionic strength of the solution so as to adsorb a given enzyme on the adsorbent. Dialysis is one of the most time-consuming processes and no convenient method is available for the rapid dialysis of large volumes of liquid. Resort to such processes should be kept to a minimum and restricted as far as possible to the later stages, where volumes are smaller. Nevertheless, adsorption methods facilitate the reduction of volume and are so convenient that they may be worth while in the early stages of a purification.

Other methods including chromatography, molecular-sieve chromatography, isoelectric focusing etc. are all applicable to the crude enzyme preparation, but require preliminary concentration of the enzyme solution prior to application. Therefore, it is advisable to use partially purified and concentrated preparations when employing these techniques as purification steps.

Sometimes the same procedure may be applied repeatedly but one can not expect too good results by merely repeating the same procedure. Separation depends on differences in the properties of individual proteins. It is desirable to combine methods that separate proteins on a basis of their different properties. There is no definite procedural order to follow for individual enzyme purifications; the best procedure must be determined by trial and error in each case.

REFERENCES

Abrams, A. (1965). *J. biol. Chem.*, **240**, 3675–3681.
Akabori, S., Hagihara, B., and Ikenaka, T. (1951). *Proc. Japan Acad.*, **27**, 350–351.
Andreus, P. (1964). *Biochem. J.*, **91**, 222–233.
Berger, L., Slein, M. W., Colowick, S. P., and Cori, C. F. (1946). *J. gen. Physiol.*, **29**, 379–391.
Blatt, W. F., Feinberg, M. P., Hoffenberg, H. B., and Saravis, C. A. (1965). *Science*, **150**, 224–226.

Blatt, W. F., Robinson, S. M., Robbins, F. M., and Saravis, C. A. (1967). *Anal. Biochem.*, **18**, 81–87.

Boardman, N. K., and Partridge, S. M. (1955). *Biochem. J.*, **59**, 543–552.

Bock, R. M., and Ling, N. S. (1954). *Anal. Chem.*, **26**, 1543–1546.

Bodo, G. (1955). *Nature, Lond.*, **176**, 829–830.

Boman, H. G. (1955). *Nature, Lond.*, **175**, 898–899.

Boyer, J., Baron, D. N., and Talalay, P. (1965). *Biochemistry*, **4**, 1825–1833.

Bublitz, C., and Kennedy, E. P. (1954). *J. biol. Chem.*, **211**, 951–961.

Bullen, W. A., and LeComte, J. R. (1966). *Proc. nat. Acad. Sci. U.S.A.*, **56**, 979–986.

Carnahan, J. E., Mortenson, L. E., Mower, H. F., and Castle, J. E. (1960). *Biochim. biophys. Acta*, **44**, 520–535.

Cho, Y. S., and Pitot, H. C. (1967). *J. biol. Chem.*, **242**, 1192–1198.

Civen, M., and Knox, W. E. (1960). *J. biol. Chem.*, **235**, 1716–1718.

Cleland, W. W. (1964). *Biochemistry*, **3**, 480–482.

Cohn, E. J., Hughes, W. L. Jr., and Weare, J. H. (1947). *J. Am. chem. Soc.*, **69**, 1753–1761.

Cohn, M. (1957). *Bact. Rev.*, **21**, 140–168.

Colowick, S. P. (1955). *In* "Methods in Enzymology" (Eds. S. P. Colowick and N. O. Kaplan), Vol. I, pp. 90–98. Academic Press, New York.

Craig, L. C., and King, T. P. (1962). *In* "Methods of Biochemical Analysis" (Ed. D. Glick), Vol. X, pp. 175–199. Interscience Publishers, New York.

Crestofield, A. M., Stein, W. H., and Moore, S. (1963). *J. biol. Chem.*, **238**, 2421–2428.

Davidson, S. J., and Talalay, P. (1956). *J. biol. Chem.*, **241**, 906–915.

Dixon, M., and Webb, E. C. (1961). *Ad. Protein Chem.*, **16**, 197–219.

Dixon, M., and Webb, E. C. (1964). *In* "Enzyme" (Eds. M. Dixon and E. C. Webb), 2nd ed., pp. 27–53. Longmans Green Ltd., London.

Dua, R. D., and Burris, R. H. (1963). *Proc. nat. Acad. Sci. U.S.A.*, **50**, 169–175.

Fahian, L. A., Schooler, J. M., Gehred, G. A., and Cohen, P. P. (1964). *J. biol. Chem.*, **239**, 1935–1941.

Falaschi, A., and Kornberg, A. (1966). *J. biol. Chem.*, **241**, 1478–1482.

Fincham, J. R. S. (1957). *Biochem. J.*, **65**, 721–728.

Fischer, E. H., and Krebs, E. G. (1958). *J. biol. Chem.*, **231**, 65–71.

Flodin, P., Gelotte, B., and Porath, J. (1960). *Nature, Lond.*, **188**, 493–494.

Flodin, P. (1961). *J. Chromat.*, **5**, 103–115.

Fujisawa, H., and Hayaishi, O. (1968). *J. biol. Chem.*, **243**, 2673–2681.

Garen, A., and Levinthal, C. (1960). *Biochim. biophys. Acta*, **38**, 470–483.

Gibson, D. T., Koch, J. R., and Kallio, R. E. (1968). *Biochemistry*, **7**, 2653–2662.

Good, N. E., Winget, G. D., Winter, W., Connolly, T. N., Izawa, S., and Singh, R. M. M. (1966). *Biochemistry*, **5**, 467–477.

Gorini, L. (1950). *Biochim. biophys. Acta*, **6**, 237–255.

Gorini, L., and Felix, F. (1953). *Biochim. biophys. Acta*, **10**, 128–135.

Graves, D. J., Sealock, R. W., and Wang, J. H. (1965). *Biochemistry*, **4**, 290–296.

Green, A. A., and Hughes, W. L. (1955). *In* "Methods in Enzymology" (Eds. S. P. Colowick and N. O. Kaplan), Vol. I, pp. 67–82. Academic Press, New York.

Grieff, D., and Myers, M. (1961). *Nature, Lond.*, **190**, 1202–1204.

Greiff, D., Myers, M., and Privitera, C. A. (1961). *Biochim. biophys. Acta*, **50**, 233–242.

Hagihara, B. (1951). *Proc. Japan Acad.*, **27**, 346–349.

Hagihara, B., Horio, T., Yamashita, J., Nozaki, M., and Okunuki, K. (1956). *Nature, Lond.,* **178**, 629–630.

Hagihara, B., Matsubara, H., Nakai, M., and Oknunki, K. (1958a). *J. Biochem Tokyo,* **45**, 185–194.

Hagihara, B., Sekuzu, I., Tagawa, K., Yoneda, M., and Okunuki, K. (1958b). *Nature, Lond.,* **181**, 1588–1589.

Hayaishi, O. (1955). *In* "Methods in Enzymology" (Eds. S. P. Colowick and N. O. Kaplan), Vol. I. pp. 126–137. Academic Press, New York.

Hayaishi, O., Gefter, M., and Weissbach, H. (1963). *J. biol. Chem.,* **238**, 2040–2044.

Helmreich, E., and Cori, C. F. (1964). *Proc. nat. Acad. Sci. U.S.A.,* **51**, 131–138.

Hirata, M., Tokushige, M., Inagaki, A., and Hayaishi, O. (1965). *J. biol. Chem.,* **240**, 1711–1717.

Hirata, M., and Hayaishi, O. (1965). *Biochem. biophys. Res. Commun.,* **21**, 361–365.

Hjertèn, S. (1962a). *Archs Biochem. Biophys.,* Suppl. **1**, 147–151.

Hjertèn, S. (1962b). *Biochim. biophys. Acta,* **62**, 445–449.

Hjertèn, S., and Mosbach, R. (1962). *Anal. Biochem.,* **3**, 109–118.

Holmes, P. K., and Halvorson, H. O. (1965). *J. Bact.,* **90**, 316–326.

Holmes, P. K., Dundas, I. E. D., and Halvorson, H. O. (1965). *J. Bact.,* **90**, 1159–1160.

Hoppel, C., and Cooper, C. (1968). *Biochem. J.,* **107**, 367–375.

Jacob, F., and Monod, J. (1961). *J. molec. Biol.,* **3**, 318–356.

Jarabak, J., Seeds, A. E. Jr., and Talalay, P. (1966). *Biochemistry,* **5**, 1269–1279.

Jenkins, W. T., Yphantis, D. A., and Sizer, I. W. (1959). *J. biol. Chem.,* **234**, 51–57.

Kauzmann, W. (1959). *Adv. Protein Chem.,* **14**, 1–63.

Kennedy, I. R., Morris, J. A., and Mortenson, L. E. (1968). *Biochim. biophys. Acta,* **153**, 777–786.

Klett, R. P., and Smith, M. (1967). *In* "Methods in Enzymology" (Eds. L. Grossman and K. Moldave), Vol. XII, pp. 566–581. Academic Press, New York.

Kohn, L. D., and Jakoby, W. B. (1968a). *J. biol. Chem.,* **243**, 2472–2478.

Kohn, L. D., Packman, P. M., Allen, R. H., and Jokoby, W. B. (1968). *J. biol. Chem.,* **243**, 2479–2485.

Kohn, L. D., and Jakoby, W. B. (1968b). *J. biol. Chem.,* **243**, 2486–2493.

Kohn, L. D., and Jakoby, W. B. (1968c). *J. biol. Chem.,* **243**, 2494–2499.

Kojima, Y., Itada, N., and Hayaishi, O. (1961). *J. biol. Chem.,* **236**, 2223–2228.

Kolin, A. (1954). *J. Chim. phys.,* **22**, 1628–1629.

Korkes, S., Campillo, A. del, Gunsalus, I. C., and Ochoa, S. (1951). *J. biol. Chem.,* **193**, 721–735.

Kubowitz, F., and Ott, P. (1943). *Biochem. Z.,* **314**, 94–117.

Kunitz, M. (1948). *J. gen. Phys.,* **32**, 265–269.

Layne, E. (1957). *In* "Method in Enzymology" (Eds. S. P. Colowick and N. O. Kaplan), Vol. III, pp. 447–454. Academic Press, New York.

Leach, A. A., and O'Shea, D. C. (1965). *J. Chromat.,* **17**, 245–251.

Levin, Ö. (1962). *In* "Methods in Enzymology" (Eds. S. P. Colowick and N. O. Kaplan), Vol. V, pp. 27–32. Academic Press, New York.

Levy, H. R. (1963). *J. biol. Chem.,* **238**, 775–784.

Mohamed, M. S., and Greenberg, D. M. (1945). *Arch. Biochem.,* **8**, 349–364.

Mortenson, L. E., Morris, J. A., and Jeng, D. Y. (1967). *Biochim. biophys. Acta,* **141**, 516–522.

Moskowitz, M. (1963). *Nature, Lond.,* **200**, 335–337.

Munson, T. O., Dilworth, M. J., and Burris, R. H. (1965). *Biochim. biophys. Acta*, **104**, 278–281.

Nakamoto, T., Fox, C. F., and Weiss, S. B. (1964). *J. biol. Chem.*, **239**, 167–174.

Newton, W. A., and Snell, E. E. (1965). *J. Bact.*, **89**, 355–364.

Newton, W. A., Morino, Y., and Snell, E. E. (1965). *J. biol. Chem.*, **240**, 1211–1218.

Northrop, J. H. (1946). *J. gen. Phys.*, **30**, 177–184.

Novoa, W. B., and Grisolia, S. (1964). *Biochim. biophys. Acta*, **85**, 274–282.

Nozaki, M., Yamanaka, T., Horio, T., and Okunuki, K. (1957). *J. Biochem., Tokyo*, **44**, 453–464.

Nozaki, M., Mizushima, H., Horio, T., and Okunuki, K. (1958). *J. Biochem., Tokyo*, **45**, 815–823.

Nozaki, M., Kagamiyama, H., and Hayaishi, O. (1963). *Biochem. Z.*, **338**, 582–590

Nozaki, M., Ono, K., Nakazawa, T., Kotani, S., and Hayaishi, O. (1968). *J. biol. Chem.*, **243**, 2682–2690.

Numa, S., and Ringelmann, E. (1965). *Biochem. Z.*, **343**, 258–268.

Okazaki, T., and Kornberg, A. (1964). *J. biol. Chem.*, **239**, 259–274.

Okunuki, K., Hagihara, B., Matsubara, H., and Nakayama, T. (1956). *J. Biochem., Tokyo*, **43**, 453–467.

Okunuki, K. (1959). *In* "A Laboratory Manual of Analytical Methods of Protein Chemistry" (Eds. P. Alexander and R. J. Block), pp. 31–64. Pergamon Press, London.

Okunuki, K. (1961). *Adv. Enzymol.*, **23**, 29–82.

Oxenburgh, M. S., and Snoswell, A. M. (1965). *Nature, Lond.*, **207**, 1416–1417.

Penefsky, H. S., and Warner, R. C. (1965). *J. biol. Chem.*, **240**, 4694–4702.

Peterson, E. A., and Sober, H. A. (1956). *J. Am. chem. Soc.*, **78**, 751–755.

Peterson, E. A., and Sober, H. A. (1959). *Anal. Chem.*, **31**, 857–862.

Peterson, J. A., Kusunose, M., Kusunose, E., and Coon, M. J. (1967). *J. biol. Chem.*, **242**, 4334–4340.

Polge, C., Smith, A. U., and Parkes, A. S. (1949). *Nature, Lond.*, **164**, 666–667.

Porath, J. (1957). *Arkiv Kemi*, **11**, 97–106.

Porath, J., and Flodin, F. (1959). *Nature, Lond.*, **183**, 1657–1659.

Porath, J. (1962). *Adv. Protein Chem.*, **17**, 209–226.

Porath, J., and Bennich, H. (1962). *Archs Biochem. Biophys.* Suppl., **1**, 152–156.

Prins, J. (1959). *J. Chromat.*, **2**, 445–486.

Pullman, M. E., Penefsky, H. S., Datta, A., and Racker, E. (1960). *J. biol. Chem.*, **235**, 3322–3329.

Racker, E., Pullman, M. E., Penefsky, H. S., and Silverman, M. (1963). *In* "Proceedings of the Fifth International Congress of Biochemistry", Moscow, 10–16, 1961, Vol. V (Ed. E. C. Slater), pp. 303–312. Pergamon Press, London.

Raijman, L., and Grisolia, S. (1961). *Biochem. biophys. Res. Commun.*, **4**, 262–265.

Sakami, W. (1962). *Anal. Biochem.*, **3**, 358–360.

Schatz, G., Penefsky, H. S., and Racker, E. (1967). *J. biol. Chem.*, **242**, 2552–2560.

Scheraga, H. A., Némethy, G., and Steinberg, I. Z. (1962). *J. biol. Chem.*, **237**, 2506–2508.

Scrutton, M. C., and Utter, M. F. (1965). *J. biol. Chem.*, **240**, 1–9.

Semenza, G. (1960). *Helv. chim. Acta*, **43**, 1057–1068.

Shepherdson, M., and Pardee, A. B. (1960). *J. biol. Chem.*, **235**, 3233–3237.

Shizuta, Y., Nakazawa, A., Tokushige, M., and Hayaishi, O. (1969). *J. biol. Chem.*, **244**, 1883–1889.

Shukuya, R., and Schwert, G. W. (1960). *J. biol. Chem.*, **235**, 1658–1661.

Shuster, C. W., and Doudoroff, M. (1962). *J. biol., Chem.*, **237**, 603–607.

Siegelman, H. W., and Fiver, E. M. (1962). *Anal. Biochem.*, **3**, 435–437.

Siegelman, H. W., Wieczorek, G. A., and Turner, B. C. (1965). *Anal. Biochem.*, **13**, 402–404.

Sober, H. A., Hartley, R. W. Jr., Carroll, W. R., and Peterson, E. A. (1965). *In* "The Protein" (Ed. H. Neurath), 2nd ed. Vol. 3, pp. 1–97. Academic Press, New York.

Svensson, H. (1961). *Acta chem. scand.*, **15**, 325–341.

Svensson, H. (1962). *Archs Biochem. Biophys.* Suppl., **1**, 132–138.

Takemori, S., Furuya, E., Suzuki, H., and Katagiri, M. (1967). *Nature, Lond.*, **215**, 417–419.

Taylor, J. F., Green, A. A., and Cori, G. T. (1948). *J. biol. Chem.*, **173**, 591–604.

Taylor, J. F. (1953). *In* "The Proteins" (Eds. H. Neurath and Bailey, K.), Vol. I, pp. 1–85. Academic Press, New York.

Tiselius, A., Hjertèn, S., and Levin, O. (1956). *Archs Biochem. Biophys.*, **65**, 132–155.

Tokushige, M., Hayaishi, O., and Morita, K. (1967). *Archs Biochem. Biophys.*, **122**, 522–523.

Torriani, A. (1960). *Biochim. biophys. Acta*, **38**, 460–479.

Utter, M. F., Keech, D. B., and Scrutton, M. C. (1964). *In* "Advances in Enzyme Regulation" (Ed. G. Weber), Vol. 2, pp. 49–68. Pergamon Press, Oxford.

Vagelos, P. R., Alberts, A. W., and Martin, D. B. (1963). *J. biol. Chem.*, **238**, 533–540.

Vesterberg, O., and Svensson, H. (1966). *Acta chem. scand.*, **20**, 820–834.

Vesterberg, O., Wadström, T., Vesterberg, K., and Svensson, H. (1967). *Biochim. biophys. Acta*, **133**, 435–445.

Wadström, T. (1967). *Biochim. biophys. Acta*, **147**, 441–452.

Warburg, O., and Christian, W. (1941). *Biochem. Z.*, **310**, 384–421.

Whitaker, J. R. (1963). *Anal. Chem.*, **35**, 1950–1953.

Yamanaka, T., Mizushima, H., Nozaki, M., Horio, T., and Okunuki, K. (1959). *J. Biochem., Tokyo*, **46**, 121–132.

Yates, R. A., and Pardee, A. B. (1957). *J. biol. Chem.*, **227**. 677–692.

CHAPTER VII

Zone Electrophoresis of the Separation of Microbial Cell Components

J. R. SARGENT*

Department of Biochemistry, Marischal College
University of Aberdeen

I.	General Theory	456
	A. Introduction	456
	B. Factors affecting movement of ions	457
	C. Apparatus for Electrophoresis	460
	D. Application of zone electrophoresis	462
II.	Low Voltage Electrophoresis on Paper	463
	A. Serum proteins	463
	B. Other proteins	467
	C. Amino-acids	468
	D. Nucleotides and their derivatives	468
	E. Carbohydrates	472
III.	High Voltage Electrophoresis on Paper	472
	A. Cooled metal plates	473
	B. Liquid-cooled systems	475
	C. Source of current	477
	D. Applications of high voltage electrophoresis . . .	478
IV.	Thin Layer Electrophoresis	481
V.	Electrophoresis on Cellulose Acetate	482
	A. Buffers used	484
	B. Procedure	485
	C. Staining cellulose acetate strips	486
	D. Applications of cellulose acetate electrophoresis . .	489
VI.	Starch Gel Electrophoresis	489
	A. Apparatus used	489
	B. Buffers used	493
	C. Preparation of starch gels	494
	D. Application of the sample and running conditions . .	495
	E. Staining starch gels	496
	F. Applications of starch gel electrophoresis	499

* Present address: N.E.R.C. Fisheries Biochemical Research Unit, University
of Aberdeen [St. Fitticks Road, Torry], Aberdeen, Scotland.

VII.   Agar Gel Electrophoresis   499
       A.   Apparatus used   500
       B.   Buffers used   500
       C.   Preparation of the gel   501
       D.   Application of the sample   501
       E.   Drying and staining the gel   501
       F.   Immunoelectrophoresis using agar gels   502

VIII. Block Electrophoresis   503
       A.   Starch block electrophoresis   503
       B.   Polyvinyl chloride blocks   506
       C.   Glass powder blocks   506
       D.   Sephadex blocks   506
       E.   Applications of block electrophoresis   507

IX.   Choice of Media   508
       Acknowledgment   509
       References   509

I. GENERAL THEORY

A. Introduction

Electrophoresis remains one of the more powerful methods for separating molecules of all sizes. Although the early applications of electrophoresis were concerned with solving chemical and biochemical problems the method has for long been a major tool in many varied fields of which microbiology is one of the more important. The present Chapter is intended to provide an introduction to the uses of electrophoresis in the analysis of microbial cell components and deals with paper, thin layers, cellulose acetate, starch gel and agar gel as media for electrophoresis. Apart from the thin layer media the media dealt with here had all been well documented by the late 1950s. Recent years have seen the introduction of novel and highly efficient media such as polyacrylamide (This Volume Chapter 10) or the development of sophisticated methods such as used in column electrophoresis (This Volume, Chapter 9), free-flow electrophoresis (This Volume, Chapter 8) and ion-focusing electrophoresis (This Volume, Chapter 11). These innovations have complemented rather than displaced the older methods of electrophoresis which at present are finding increasing usage and still possess many advantages, not least of which is their simplicity.

Electrophoresis is the term used to describe the movement of ions in solution under the influence of an electric field. Early electrophoretic separations were carried out for the most part by free movement of ions in solution, in which the moving ions form a boundary which can be detected by measuring changes in refractive index throughout the solution. This type of electrophoresis is known as boundary or moving boundary electrophoresis and has been developed to a high degree of efficiency largely by the work of Tiselius and his colleagues (Tiselius, 1950). Mainly because of the high cost of apparatus for moving boundary electrophoresis and difficulties in isolating the separate zones this is not in widespread use in most laboratories although the method can provide a powerful analytical tool.

Electrophoresis can also be carried out in a system whereby the solution containing the ions to be separated is supported in a more or less inert material such as paper, starch or agar. This type of electrophoresis is known as zone electrophoresis and is based on exactly the same principles as moving boundary electrophoresis. In zone electrophoresis, however, the ions separated remain as discrete areas or zones on the supporting medium and can, therefore, be easily detected by conventional physical, chemical or biochemical means.

B. Factors affecting movement of ions

If a direct electric current is passed through a solution containing two ionic species A^+ and B^- the cationic species A^+ will move towards the cathode and the anionic species B_- will move towards the anode. The rate of movement of both A^+ and B^- will be determined by the motive force to which they are subjected, viz., $Q \times N$, where Q = field strength and N = the net charge on the ion. This force is opposed by the frictional force encountered by the ion as it moves through the solution, this being largely determined by the size and shape of the ion and the viscosity of the solution. The rate of migration of an ion can therefore be said to depend on the applied current, the shape and size of the ion, and the charge carried by the ion. It follows from this that ions which differ from each other in either charge, shape or size can, in principle, be separated by electrophoresis. It should, however, be emphasized that two ions which show the same rates of migration are not necessarily identical.

It is to be noted that separations on the more efficient media in use at present involve separation by molecular sieving as well as separation on the basis of net charge. This is to say that in several instances the molecule separated is retarded by the medium itself even though the molecule is highly charged and has, therefore, a high intrinsic mobility in the buffer used. The medium where molecular sieving plays a major role in separation is, of course, polyacrylamide (see this Volume, Chapter 10). Molecular sieving, however, is of major significance in electrophoretic separations on Sephadexes and on starch gels.

1. *Effects of current and voltage*

Since it is the ions which carry the current, an increase in current will bring about an increased rate of migration of a particular ion. Furthermore, since passage of current through a solution is governed by Ohm's law, viz., $V = R \times I$, where V = volts, R = resistance and I = amps, increased current can be effected in practice by increasing the voltage. Passage of current through a solution, however, also results in generation of heat which in turn results in a decrease in resistance of the solution and an increased rate of evaporation of solvent from the supporting medium. These two effects in general oppose each other and their overall effect is difficult to predict in theory. Heating, however, causes both voltage and current to vary during electrophoresis and these variations are minimized in practice by using an electrical supply which provides either a constant current or a constant voltage. Virtually all optimal conditions of voltage and current in electrophoresis are determined empirically so that considerations of variations of current and voltage are for the most part unnecessary.

It should be noted that if two strips are run in parallel the current required for effective separation will be double that required for one strip alone. Likewise, for three strips three times the current will be required. The voltage will be the same, however, irrespective of the number of strips employed, provided the lengths of the strips are constant.

2. Effects of ionic strength, pH and buffer

Ionic strengths of buffers used in electrophoresis seldom lie outside the range 0·03–0·15 and in fact the majority of ionic strengths used are either 0·05, 0·075 or 0·10. The ionic strength of a solution may be calculated from the equation $\mu = \frac{1}{2} \sigma m c^2$, where μ = the ionic strength, m = the molarity of an ion and c = the charge carried by that ion. In general terms, solutions of high ionic strength result in lower mobilities than solutions of low ionic strength but the former result in sharper and more discrete zones. Values of ionic strength from 0·05–0·10 approximate to the optimum compromise between high mobilities and sharpness of zones. Solutions of high ionic strength also result in more heat being generated during a run.

The pH of the solution plays a major role in determining the mobilities of ions since the net charge carried by most ions is pH-dependent. This is particularly important where zwitterionic species, such as amino-acids or proteins, are concerned. In these cases the molecule can exist as a neutral species, $NH_3^+RCOO^-$ which does not move in an electric field; this species is usually found near neutral pH values. At low pH values the carboxyl group is undissociated so that the molecule bears a net positive charge NH_3^+RCOOH and moves towards the cathode. Alternatively, at high pH values the carboxyl group is dissociated so that the molecule now bears a net negative charge NH_2RCOO^- and moves towards the anode. This property of amino-acids and proteins makes electrophoresis a powerful tool in their separation and is perhaps best exploited when separating two zwitterions at the isoelectric point of one of the two species. In this case the charged molecule moves free from the uncharged molecule so that maximal separation is possible; this situation may be encountered in separating two haemoglobin species at the isoelectric point of one of them.

A great variety of buffers is used in electrophoresis at the present time. These buffers include barbitone, tris, borate, phosphate, acetate, citrate, EDTA, etc., and cover a pH range from about 1·0–11·0. The buffer type in itself can play an important part in the separation processes in electrophoresis, e.g., different patterns are obtained when serum proteins are separated on starch gel in borate buffer or in tris buffer at the same pH. The exact reason for these phenomena is unknown at the present time although it is considered that borate ions probably affect the properties of the starch gel by complexing with hydroxyl groups in the gel. A similar effect of buffer can be seen in such processes as the separation of carbohydrates when uncharged molecules are converted into charged forms by complexing with borate ions. In practice carbohydrates are separated by carrying out electrophoresis in borate buffers.

3. *Electroendosmosis*

So far it has been assumed that the supporting medium is entirely inert as far as the electrical processes involved in electrophoresis are concerned. This is not, however, the case when the supporting medium carries ionizable groups which are in fact ionized at the pH at which electrophoresis is

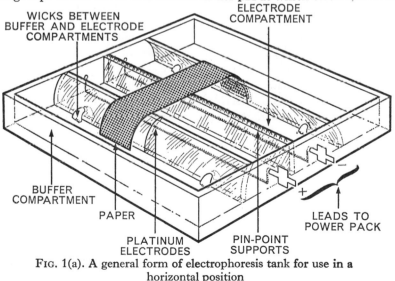

FIG. 1(a). A general form of electrophoresis tank for use in a horizontal position

carried out. Most media used do in fact carry such ionizable groups, e.g., paper has a significant content of carboxyl groups which are negatively charged at neutral and alkaline pH values. In addition to the possible presence of ionizable groups in the medium, most media used in electrophoresis acquire a net negative charge due to zeta potential effects at the junction of the medium and the buffer solution. The natural tendency of these negatively charged groups is to move towards the anode but this of course is impossible since the medium is stationary. To counter this effect a movement of positively charged water molecules (H_3O^+) occurs towards the cathode. This gives the effect of an osmosis of solvent towards the cathode and frequently results in an apparent movement of neutral molecules towards the cathode. This is the reason why the electrically neutral γ-globulin fraction of serum moves towards the cathode during electrophoresis at pH 8·6. It is essential to allow for electroendosmosis when carrying out exact measurements on mobility and this can be effected by measuring the movement of an electrically neutral molecule such as urea, dextran or deoxyribose during the course of electrophoresis.

C. Apparatus for electrophoresis

The two main pieces of equipment for carrying out electrophoresis are a source of electric power and a tank or trough. The power source or power pack should be capable of providing either a constant current or a constant voltage or preferably both. Such power packs capable of providing a voltage range up to 400 volts and a current range up to 80 mA are available commercially (Shandon Scientific Co. Ltd.). These instruments are ideally suited for conventional low voltage separations which include virtually all protein separations.

Electrophoresis tanks are quite varied in practice but all conform to the basic design shown in Figs 1(a) and 1(b). In such an apparatus the supporting medium, impregnated with buffer solution is suspended between the two buffer compartments. Electrophoresis is carried out with the strip in either the vertical or horizontal position. In most cases the choice between horizontal or vertical electrophoresis is a personal one though in certain cases such as starch gel electrophoresis there are specific reasons for choosing between these methods. It is obvious that the vertical arrangement allows longer separations to be carried out but, nevertheless, the horizontal arrangement is most frequently encountered in practice and is almost always used for electrophoresis of proteins and other macromolecules except on starch and polyacrylamide gels. When used in the horizontal position the strip is supported by a thin nylon thread or a series of pointed graphite rods placed midway between the buffer compartments. For vertical electrophoresis the strip is usually supported by means of a thin

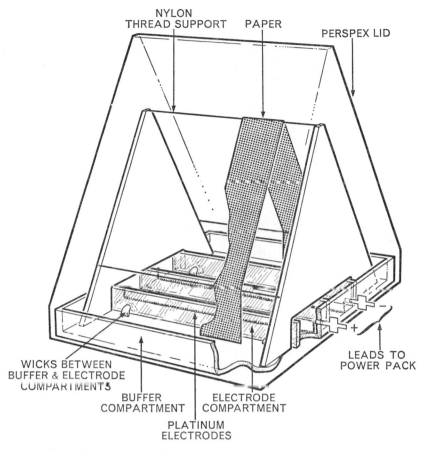

FIG. 1(b). A general design of tank for vertical electrophoresis

nylon thread (a special arrangement is required for gels). The buffer compartments in the tank are separated physically from the electrode compartment, contact between the two being effected by means of paper or cotton wool wicks. This prevents the buffer compartments becoming contaminated with the products of electrolysis formed in the electrode compartments. The electrodes are constructed of either carbon, Ag/AgCl or platinum. Carbon electrodes can be simply constructed from the graphite rods in torch batteries but suffer from the disadvantages of being easily polarized and having to be replaced rather frequently. Ag/AgCl electrodes are excellent for electrophoresis, but possibly the most suitable electrode is the platinum electrode. This has the advantage of being fairly robust, being easily cleaned and giving excellent performance. To prolong the life of buffers it is advisable to reverse the electrodes at the end of each run so that changes due to electrolysis (more marked at the cathode) are evened out.

The entire tank is enclosed by a fairly tight-fitting lid to ensure a reasonable degree of vapour saturation, this being particularly important when dealing with very thin media such as cellulose acetate as these tend to dry out during electrophoresis due to the heat generated. Saturation can be further improved by lining the under surface of the lid with a sheet of Wettex absorbent (obtainable from any hardware store) which has been moistened, but not soaked, with water. It is noteworthy in the present context that the buffer and electrode compartments should be as large as possible in order to expose a large surface area of solution to help maintain vapour saturation.

For further detailed descriptions of designs of electrophoresis tanks the following papers may be consulted: Block *et al* (1955), Lederer (1955), Laurell (1957), and Kohn (1960). It may be mentioned that excellent electrophoresis tanks are available from a variety of commercial firms.

D. Applications of zone electrophoresis

Before considering the methodology and applications of zone electrophoresis in detail it may be well to note that many of the excellent electrophoretic separations reproduced in the literature are achieved with mixtures of compounds that are enormously simplified as compared to the ultra complex intracellular milieu of a micro-organism. For this reason it is essential to consider preliminary fractionation of cellular material before applying electrophoresis. When starting with micro-organisms the disintegration and fractionation of cells are problems in themselves which are dealt with in Chapters 1 and 4 of this Volume. For the present it may be stated that the precise means of disrupting cells and fractionating their contents will place strictures on the different classes of compounds which can be analysed ultimately by electrophoresis.

There is at the present time scarcely any class of charged compound that has not been separated by zone electrophoresis. Likewise a wide variety of supporting media have been used for electrophoresis. Electrophoresis has been used as an analytical tool in clinical and research laboratories and has also found use as a preparative technique, albeit on a relatively small scale. At the present time one of its most wide-spread applications lies in the field of protein separations and it is indeed from this field that most of the major advances in recent years have derived, including starch gel, agar gel, cellulose acetate and polyacrylamide gels. The following Sections will approach the applications of zone electrophoresis from the viewpoint of the media used rather than the classes of compound separated. It is hoped

that this treatment will be of more use from the practical point of view
and at the same time will provide a starting point for those who wish
to consult the original literature. For this reason a fairly large bibliography
has been included with each section.

II. LOW VOLTAGE ELECTROPHORESIS ON PAPER

Paper was first used as a medium for electrophoresis in the late 1930s
but it was not until after the discovery of paper chromatography by Martin
and Synge (1945) that the full importance of paper as a medium for
electrophoresis was realized. The studies of Wieland and Fischer (1948)
on amino-acids and of Durrum (1950, 1951) on proteins led to a rapid
spread in the use of paper electrophoresis whose applications at the present
time are covered by a vast literature. A large part of the literature deals with
the separation of scrum proteins, enzymes and proteins in general but at
the present time it is true to say that paper as a medium for separating
proteins has been largely superseded by other materials. Paper, nevertheless, remains a powerful medium particularly for the separation of compounds with relatively low molecular weights. Some examples of the different classes of compounds which have been separated by paper are described
below.

A. Serum proteins

1. *Buffers used*

Barbitone, pH 8·6. Buffer of $\mu = 0·05$ is prepared by dissolving 1·84 g of
barbitone and 10·30 g of barbitone sodium in 1 litre of distilled water.
Buffer of $\mu = 0·075$ is prepared by dissolving 2·76 g of barbitone and
15·45 g of barbitone sodium in 1 litre of distilled water. For runs of about
6–7 hours buffer of ionic strength 0·05 is used while for overnight runs
of about 16–17 hours buffer of ionic strength 0·075 is used.

Sodium tetraborate, pH 8·6. 8·80 g of sodium tetraborate and 4·65 g of
boric acid are dissolved in 1 litre of distilled water.

Sodium phosphate, pH 7·4. 0·60 g of $NaH_2PO_4.H_2O$ and 2·20 g of
Na_2HPO_4 are dissolved in 1 litre of distilled water.

2. *Procedure*

Both Whatman No. 1 and Whatman No. 3 mm papers have been used,
No. 1 being generally used for analytical purposes and No. 3 mm for
preparative purposes. The paper is cut into strips of suitable length (about
30 cm) and width (about 6 cm). A pencil line is drawn across the paper

about one-third of the distance from one end after which the papers are
dipped in the buffer of choice and blotted lightly between filter paper
strips. The wet papers are then placed in the electrophoresis tank such that
the pencil line is nearest the cathode end. It is preferable though not essential to pass current through the strips for about 15 min in order to allow
equilibrium conditions to be attained.

After this time the serum sample is applied in a straight line along the
pencil mark leaving about 0·5 cm at each end of the paper. Application is
best carried out using a micropipette or capillary, guided by a ruler resting
on the edges of the tank. Approximately 4 μl of undiluted serum is applied
per cm width of paper, i.e., about 20 μl for the standard 6 cm wide strip,
though for detection of minor serum components this volume may be
doubled. The course of electrophoresis is best followed by staining the
serum prior to application by addition of a few crystals of Bromophenol
blue. This dye is fairly tightly bound to serum albumin during the course
of electrophoresis and does not alter the mobility of albumin or other serum
components. Neither does the dye interfere with subsequent staining of
the strip. Albumin is the fastest running component of serum so that the
progress of electrophoresis can be gauged by progress of the blue band.
Excess Bromophenol blue moves ahead of the albumin as a purplish band
and eventually moves off the paper as electrophoresis progresses.

Satisfactory separation of serum proteins is effected by applying a current
of about 1 mA per cm width of paper for 16–17 hours when barbitone
buffer pH 8·6 is used. Alternatively, 2 mA per cm width of paper produces
a satisfactory separation in about 7 hours. Electrophoresis under these
conditions is carried out at room temperature. At the end of a run the
current is switched off and the paper dried in an oven at 100°C for 10 min;
this also serves to denature the proteins prior to staining.

3. *Stains used*

The following stains are used for the detection of protein components
in serum:

Bromophenol blue. Proteins need not be heat-denatured prior to the application of this stain since the mercuric chloride present denatures the
proteins directly. Strips are immersed for 10 min in a solution of Bromophenol blue (1% w/v) in 95% v/v ethanol saturated with mercuric chloride.
The strips are then washed in several rinses of 1% v/v acetic acid until a
relatively clear background is obtained. The paper is finally washed twice
in methanol and air-dried.

Azocarmine B. The dired paper is immersed for 10 min in a solution of
Azocarmine B (0·1% w/v) in 50% v/v methanol containing 10% acetic

acid. Excess dye is removed by rinsing in 10% v/v acetic acid and finally in methanol.

Light green. Strips are soaked in a solution of Light green (0·5% w/v) in 25% v/v ethanol containing 5% v/v acetic acid. Excess dye is removed by washing the strips in 2% v/v acetic acid.

Naphthalene black 12B. Strips are immersed for 10 min in a saturated solution (about 1% w/v) of Naphthalene black in 10% v/v acetic acid. Excess stain is removed by rinsing in 10% v/v acetic acid in methanol until a relatively clear background is obtained. A final wash in methanol is included before the papers are air-dried.

Lipoprotein components may be detected on paper by either of the following methods:

Sudan black B. A saturated solution of the dye in 55% v/v ethanol is prepared by adding excess dye to the warm ethanol. The dye solution is cooled to room temperature and filtered. The dried paper strips are immersed in the saturated dye solution for about 60 min, then washed in 40% v/v ethanol until a relatively clear background is obtained.

Acetylated Sudan black B. Serum proteins are stained with this dye prior to electrophoresis so that the movement of lipoprotein bands can be followed during the course of the run. The stain, however, should be used with only short runs as stained zones tend to lose the dye and become fairly faint after about 4 h.

500 mg of dye is added to 50 ml of warm ethanol and the suspension is stirred thoroughly, cooled to room temperature and filtered. 1 vol of the dye solution is added slowly with stirring to 10 vols of serum. After standing for about 1 h the solution is filtered to remove excess dye particles and any precipitated proteins. The dyed serum is applied to Whatman No. 1 paper using 10 μl per cm width of paper and electrophoresis carried out in the usual manner.

Glycoprotein components in serum are detected by either of the following methods:

Schiff-periodic acid reagent (as described by Block *et al.*, 1955). The following solutions are required:

Periodic acid: 1·2 g of periodic acid are dissolved in 30 ml of water and to this solution are added 15 ml of M/15 sodium acetate and 100 ml of ethanol. This solution is stable for about three days if stored in the dark.

Reducing solution: 5 g of potassium iodide and 5 g of sodium sulphate are dissolved in water then 150 ml absolute ethanol and 2·5 ml of 2N hydrochloric acid are added with stirring; the solution must be made up daily before use.

Schiff's reagent (Fuchsin sulphite): 2 g of Basic fuchsin are dissolved in 400 ml of water and the solution cooled and filtered. After addition of 10 ml of 2N hydrochloric acid and 4 g of potassium metabisulphite the solution is stoppered and left in a cool dark place overnight. 1 g of activated charcoal is then added and the solution stirred and filtered. After this 10 ml, or more, of 2N hydrochloric acid are added until a small test aliquot of the solution does not turn red when dried on a glass slide. The solution is stoppered and stored in a cool, dark place and should be discarded when it turns pink.

Sulphite rinse: 0·4% w/v potassium bisulphite in 1% v/v aqueous hydrochloric acid.

The paper strip is fixed in ethanol and immersed in the periodic acid solution for 5 min. After rinsing with 70% v/v ethanol the strip is immersed for 5–8 min in the reducing solution. The strip is again rinsed in 70% v/v ethanol and soaked for 25–45 min in the Schiff's reagent (Fuchsin sulphite). The strip is then rinsed three times in the sulphite rinse solution, washed finally with absolute ethanol and air-dried. Carbohydrates are revealed as violet-red bands on a pale pink background.

Diphenylamine. The detection of glycoproteins using the diphenylamine reagent has been described by Drevon and Donikan (1955). The papers are dried and washed several times with acetone, then soaked in a boiling solution of diphenylamine (1% w/v) in glacial acetic acid for 5 min. After air-drying the strips are placed in 37% v/v formaldehyde for 2 min then air-dried in the dark. Carbohydrates show up as red zones on a pale background.

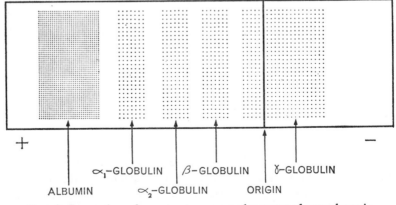

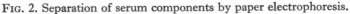

FIG. 2. Separation of serum components by paper electrophoresis.

4. Comments

Normal sera yield 5 protein components on paper electrophoresis, all of which, with the exception of γ-globulins, move towards the anode in alkaline buffers. The γ-globulin fraction appears to move towards the cathode, this being caused by electroendosmosis since the fraction is electrically neutral. Albumin moves furthest towards the anode in alkaline buffers, being followed by α_1-globulins, α_2-globulins, β-globulins and finally the γ-globulin fraction. The approximate positions of these 5 zones are shown diagrammatically in Fig. 2. In general paper electrophoresis of serum does not yield sharp zones and it is particularly difficult to distinguish between the albumin and α_1-globulin zones in certain cases.

B. Other proteins

It is hoped that the foregoing description of paper electrophoresis of serum proteins will serve as an example for the separation of proteins in general. The technique is exactly the same for any particular protein to be separated though, of course, the optimal buffer, pH and running conditions have to be established separately for each protein. It should again be stressed that paper is fast becoming outdated as a medium for the separation of proteins, having largely been superseded by more efficient media such as starch gel, cellulose acetate, polyacrylamide, etc.

Paper, however, can still be a simple and useful medium for the separation of proteins, and will often give valuable preliminary information concerning the composition of cell disintegrates. A wide variety of enzymes has also been separated by paper electrophoresis. It may be mentioned at this point that many of the standard histochemical tests can be used to detect enzymic activity on paper. Enzymic activity can be further detected after eluting the proteins from the paper. This can easily be achieved by macerating the appropriate area of the paper (detected after staining a marker strip run in parallel) in buffer. Paper debris is removed by centrifugation and the supernatant liquid removed and tested for enzymic activity. Quantitative elution of protein from paper is not always obtained, this being a function of the different degrees of binding of different proteins to paper. Among the many enzymes which have been studied by paper electrophoresis may be cited takadiastase components (Wallenfels and Pechmann, 1951), pepsin (Lundquist et al., 1955), trypsin (Nikkila et al., 1952), ribonuclease (Crestfield and Allen, 1952), invertase (Wetter and Corrigal, 1954), lysozyme (Caselli and Schumacher, 1954), enolase (Malmstrom, 1953), adenosine triphosphatase (Toschi, 1954), lipase (Delacourt and Delacourt, 1953), peroxidase (Jermyn and Thomas, 1954), dehydrogenases (Wieland and Pfleiderer, 1957), and cytochrome c (Paleus, 1952).

All the above mentioned proteins are, of course, very soluble under normal physiological conditions. It follows, therefore, that any attempt to survey proteins in tissues by the methods mentioned so far will detect only those proteins remaining in solution after removing particulate fractions of tissues, e.g., by centrifuging. Examination of the soluble phase so obtained may not necessarily give a true estimate, qualitatively or quantitatively, of a particular class of protein within the cell. In an effort to overcome this limitation Bagdasarian et al. (1964) carried out electrophoresis of relatively crude fractions of plant tissue after rendering these soluble in the very powerful solvents, phenol : acetic acid : water (1 : 1 : 1 by volume) and phenol : acetic acid : water (2 : 1 : 1 by volume). In a further attempt to separate peptide intermediates in protein synthesis the same group of workers (Brattstein et al., 1965) modified the above solvents by the addition of either, 1 g of cetylpyridinium bromide or 2 g sodium bromide per 100 ml of solvents. Using these modified solvents it proved possible to separate mixtures of transfer RNA and proteins such as cytochrome c which formed insoluble coacervates in the non-modified solvents. These solvents would appear to hold considerable promise for electrophoretic examination of normally insoluble proteins, e.g., structural proteins of membranes. The solvents have already been used to examine proteins present in ribosomes using polyacrylamide gels. It is important to note that these solvents must be used with an all-glass apparatus since they soften, if not dissolve, Perspex and similar materials.

C. Amino-acids

Although amino-acids, like proteins, are zwitterions, the overall differences in charge between individual amino-acids are generally much less pronounced than with proteins. The subtle charge differences between amino-acids can be fully exploited by high voltage electrophoresis on paper but low voltage separations are of limited value since amino-acids can only be separated under those conditions into the three groups basic, acidic and neutral. Cystine can also be separated from the other amino-acids. Separation into these three groups can, neverless, be of considerable value for routine and clinical purposes. Mobilities of amino-acids at various pH values are quoted by Evered (1959).

D. Nucleotides and their derivatives

Paper electrophoresis is still one of the most valuable techniques for the separation of nucleotides and similar compounds. Some examples of the uses of paper electrophoresis in this field will be described.

1. *Separation of mononucleotides*

This technique is of particular value in the analysis of ribonucleic acid and deoxyribonucleic acid hydrolysates, being of equal value for the identification of both ribose and deoxyribose nucleotides. Original descriptions of the technique will be found in the paper by Davidson and Smellie, (1952). (See also Smith, 1955).

(a) *Buffer used.* 0·02 M citric acid-*tri*-sodium citrate, pH 3·5. This solution is prepared by mixing one volume of M *tri*-sodium citrate and three volumes of M citric acid and diluting the solution 50 times with water.

(b) *Procedure.* The solution to be analysed should be as free of salts as possible and should have its pH adjusted to 3·5 with citric acid. Separation is carried out using the vertical strip method of electrophoresis with a strip of Whatman No. 3 mm paper of dimensions 72×7 cm. The sample (1–5 μl) containing 2–10 μmoles of nucleotides per ml is applied as a band 2·5 cm wide about 10 cm from the cathode end of the strip. The applied sample is dried in a current of warm air and the paper assembled in the electrophoresis tank. The paper is then wetted with buffer applied from a pipette to within 2 cm of either side of the origin. The buffer is allowed to wet the remaining parts of the paper by capillary action, particular care being exercised that the origin is in fact uniformly wetted. A voltage of about 10 volts per cm (700 volts total) is applied across the paper, resulting in a current of not more than 5 mA per cm width of paper. Under these conditions satisfactory separation of AMP, GMP, UMP and CMP is effected in about 18 h. Evaporation during this time may reach considerable proportions, so that care should be exercised in maintaining the strip in as vapour-tight conditions as possible. The electrophoresis tank should be fitted with a tight-fitting cover and electrophoresis should preferably be carried out with the tank enclosed in a fume cupboard. At the end of the run the current is switched off and the paper air-dried.

(c) *Detection of mononucleotides.* Nucleotides and other compounds containing purine or pyrimidine bases can be most easily detected as absorbing areas in ultraviolet light of wavelength 2537 Å. This method has the advantage that nucleotides can then be eluted from the paper by macerating the appropriate area of the paper in citrate buffer, and used for further studies. Note, however, that the eluates will contain *tri*-sodium citrate and citric acid which may interfere with subsequent tests.

An alternative means of detection is the phosphate spray of Hanes and Isherwood (1949). The paper is sprayed with a mixture of 60% w/v perchloric acid: N hydrochloric acid: 4% w/v ammonium molybdate: water (5 : 10 : 25 : 60) and heated at 85°C for 10 min. After this time yellow spots of ammonium phosphomolybdate are apparent. On exposure of the paper to hydrogen sulphide or to general ultraviolet light the yellow spots are converted to much more intense blue spots.

2. *Separation of mono-, di- and trinucleotides*

Markham and Smith (1952) are largely responsible for developing techniques for the separation of the alkaline degradation products of nucleic acids. In addition to separating the mono-, di-, and trinucleotides, the technique is also capable of separating the cyclic nucleotides which are formed during the alkaline degradation of nucleic acids. The voltages used during separations are about 20 volts per cm which is bordering on high voltage conditions. For this reason a feature of the technique is the use of a cooling liquid to help dissipate excess heat generated during the electrophoresis.

(a) *Buffer used*. 0·05 M ammonium acetate or ammonium formate.

(b) *Procedure*. Whatman No. 3 mm paper (56 × 8 cm) is soaked in buffer, blotted lightly and the sample (0·1–0·2 ml containing hydrolysate equivalent to 5–10 mg of RNA) applied as a band 6 cm wide about 12 cm from the cathode end of the strip. The paper is then placed in an apparatus of the type shown in Fig. 3. A feature of the apparatus is the presence of

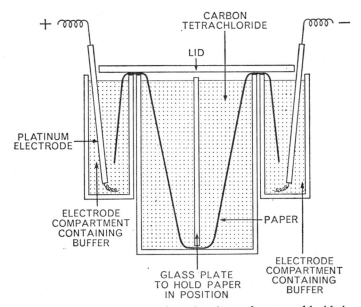

FIG.3. Apparatus for paper electrophoresis using carbon tetrachloride immersion.

carbon tetrachloride (white spirit, chlorobenzene or any similar non-conducting liquid less dense than water is also satisfactory to dissipate heat generated during the run. A voltage of 1000 volts (about 20 volts per cm) is applied for a period of 2 h. At the end of this time the current is switched off and the paper air-dried. Nucleotide products may be detected by viewing at 2537 Å. Permanent records of these strips may be obtained by photography under ultraviolet light as described by Markham and Smith (1949). While the method described above will give a reasonably good separation of relatively simple mixtures of nucleotides, for more complex mixtures it is essential to use much higher voltages.

3. Separation of adenosine mono-, di-, and triphosphates

A simple system for the separation of AMP, ADP and ATP has been described by Wade and Morgan (1954). The system also separates inorganic phosphate and pyrophosphate from these nucleotides.

(a) *Buffer used.* Redistilled *n*-butyric acid (4·6% v/v) and 0·057% w/v sodium hydroxide, pH 3·2.

(b) *Procedure.* The separation is carried out on Whatman No. 3 mm paper of dimensions 24 × 48 cm. The sample (2·5 μl containing about 25 μg of total phosphate) is applied as a spot about 2 cm from the cathode end of the paper. The strip is moistened with buffer using the technique described for mononucleotides and a voltage of 300–400 volts applied for a period of 2 h, resulting in a current of about 0·25 mA per cm width of paper. After this time the paper is air-dried and nucleotides detected by spraying with the phosphate spray or by viewing at 2537 Å. In this case the phosphate spray is more suitable since it is often useful to know the extent to which ATP, ADP and AMP are contaminated with inorganic phosphate and pyrophosphate.

4. Other nucleotide derivatives

A remarkable example of the resolving power of paper electrophoresis may be found in the separation of nicotinamide, riboflavin and pyrodoxin derivatives described by Siliprandi *et al* (1954).

(a) *Buffer used.* 0·05 M sodium acetate, pH 5·1.

(b) *Procedure.* A strip of Whatman No. 1 paper (30 × 6 cm) is moistened with buffer and placed in the standard electrophoresis tank. The sample containing some 15 μg of each of the components to be separated is applied exactly half-way between the electrodes. Separation is effected by application of a current of 3·5 mA per strip for 10 h.

Using this system the following compounds can be separated: FAD, FMN, riboflavin, pyridoxal phosphate, pyridoxamine phosphate, pyridoxal, pyridoxine, pyridoxamine, NAD, NADP and nicotinamide. Flavin compounds can be detected by their marked yellow fluorescence in general ultraviolet light. Derivatives of nicotanamide can be detected by their absorption of light of 2537 Å. Alternatively, nicotinamide compounds can be detected by their intense bluish-white fluorescence after exposure to an atmosphere of ammonia and methyl ethyl ketone for a period of 1 h at room temperature (Kodicek and Reddi, 1951). FMN, FAD, NAD and NADP can also of course be detected by the phosphate spray.

E. Carbohydrates

The separation of carbohydrates by paper electrophoresis illustrates an important aspect of the method, viz., that compounds which do not themselves carry a charge can be separated provided they are first converted to a charged form. Carbohydrates can form a charged complex with borate ions so that separation can be effected by carrying out the electrophoresis in borate buffers. Thus Consden and Stanier (1952) were able to carry out a partial separation of monosaccharides by electrophoresis on Whatman No. 1 paper at 10 volts per cm for 215 min. The buffer used by these authors consisted of 0·2 M boric acid containing 0·05M sodium chloride, the pH of which had been adjusted to 8·6 by adding 0·05M sodium tetraborate. The electrophoretic separation obtained in this way is not very effective, and satisfactory separation of complex sugar mixtures is only obtained by carrying out chromatography at right angles after electrophoresis has been carried out in the first dimension. Better electrophoretic separation of sugars was obtained by Gross (1953, 1954) using high voltage electrophoresis with cooled metal plates (see also Michl, 1952). Reasonably good separation of sugars was obtained on Whatman No. 1 paper in 0·05M sodium tetraborate, pH 9·2 by applying a potential of 40 V cm for $1\frac{1}{2}$ h. Detailed descriptions of high voltage electrophoresis using cooled metal plates are given in Section III. At the present time electrophoresis is seldom used for the separation of sugars since much better resolution can be obtained using paper chromatographic procedures.

III. HIGH VOLTAGE ELECTROPHORESIS ON PAPER

High voltage conditions on paper are usually defined as voltages in excess of 20V/cm. Thus for a paper strip 50 cm long high voltage conditions are normally employed with voltages equal to or greater than 1 kV and in fact voltages up to 10 kV have been used with paper. The main difficulty encountered in applying such voltages to paper is the heat evolved which leads to rapid evaporation of buffer from the paper strips, this in turn leading to marked distortion of the zones and eventually to the applied current dropping to negligible values when migration stops. It is obvious, therefore, that the strip must be cooled during the passage of current and two methods of effecting this are in use at the present time, viz. (i) enclosing the strip between two cooled metal plates or (ii) surrounding the strip with a water-immiscible, non-conducting liquid which is less dense than water and which is cooled by means of a conventional cooling coil. These two methods are described separately below.

A. Cooled metal plates

High voltage electrophoresis using cooled metal plates has been described by a number of authors including Gross (1955, 1956), Efron (1959) and Atfield and Morris (1961).

The apparatus used for this type of high voltage electrophoresis is rather complex and although it can be built in the laboratory it is preferable to use the commercial models of which several excellent types are now available. The essential features of the apparatus are shown in Fig. 4 which is based on the apparatus supplied by The Locarte Co., 24 Emperor's Gate, London, S.W.7. A similar apparatus is supplied by Miles Hivolt Ltd., Shoreham, Sussex. The sample is applied to sheets of Whatman No. 1 or No. 3 mm paper of dimensions slightly less than those of the plates, usually about 40 × 30 cm. The paper is then moistened with buffer using a pipette so that an area approximately an inch from either side of the origin is left dry. A convenient method of carrying out this operation is to fold the paper along the origin line and also about one inch on either side of the origin so that when the paper is laid on a glass plate the origin is off the glass at the top of an inverted "V". In this way all parts of the paper in contact with the glass can be simply wetted by applying buffer from a pipette, and the buffer rises up both sides of the inverted V to meet at the origin. When the origin area is uniformly wet, which may take several minutes, especially if the origin has been heavily loaded with material, excess buffer is removed by pressing between sheets of filter paper and the paper is then placed between the metal plates. This is carried out as follows: a sheet of polythene is placed over the lower plate and the moist paper is placed on top of this sheet. Contact between paper and buffer compartments is effected by means of paper wicks (four thicknesses of Whatman No. 3 mm) moistened with the buffer. The paper is sandwiched between the wick layers with two thicknesses on either side. A further sheet of polythene is then placed over the paper and the top metal plate applied. The metal plates are usually constructed of aluminium alloy and are machined to achieve a uniformity of flatness to within 0·001 in. This

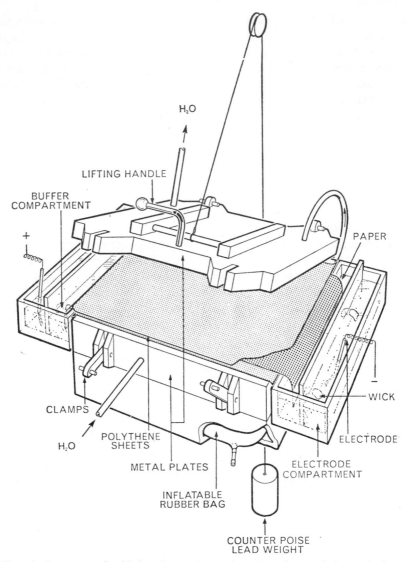

FIG. 4. Apparatus for high voltage electrophoresis using cooled metal plates.

uniformity of flatness is essential if puddles of buffer are not to collect during a run and lead to distortion of zones. A uniform pressure (5 lb/sq. in.) is applied to the plates by means of an inflatable rubber bag. Cooling is effected by circulating water, usually at room temperature though for really high voltage water at 2°C may be required, through channels which have been machined inside the plates. The electrodes are generally Ag/AgCl or, more usually, Pt. It is essential when using this type of high voltage apparatus to employ adequate safety precautions and one way of doing this is to incorporate a safety switch in the lid of the box in which the apparatus is enclosed so that the lid cannot be opened without breaking the circuit. An apparatus of this type is capable of excellent results and is indispensable where studies at really high voltages (up to 10 kV) are contemplated. A minor disadvantage of the apparatus is the relatively high cost and for particular separations it should be considered whether the much simpler liquid-cooled apparatus, which is not capable of accommodating such high voltages as the cooled metal plate system, will do the same job.

It is worth noting that a high voltage apparatus with only a single cooled metal plate is becoming increasingly popular. In this type of apparatus the lower surface, on which the paper rests, is a metal plate very efficiently cooled over its entire surface by means of a labyrinth water cooling system within the plate. The top surface, however, is a flexible foam pad enveloped in polythene such that the pad moulds on to the lower metal surface and so irons out any slight irregularities in the contour of the metal plate. While this apparatus will provide in theory less efficient cooling than is achieved in the apparatus with twin metal plates, the results obtained in practice are highly satisfactory. The major advantage of the single metal plate apparatus is that it is much cheaper than the twin plate version. An excellent version of such an apparatus is supplied by The Shandon Scientific Co. Ltd., Willesden, London.

B. Liquid-cooled systems

While not capable of accommodating such high voltages as the cooled metal plate system, liquid-cooled systems of high voltage electrophoresis have many advantages including ease of construction in any laboratory, cheapness and ease of operation. For these reasons it is recommended that preliminary studies without previous experience of high voltage electrophoresis should be carried out on the liquid-cooled apparatus.

Most of these types of apparatus are based on the apparatus of Michl (1951). A modification of Michl's apparatus as described by Ryle et al (1955) is shown in Fig. 5. The rectangular all-glass chromatography jar contains a trough which is supported near the top of the tank by means of glass plates which fit down the sides of the tank. Buffer is poured into the bottom of the tank to a depth of about half an inch and also into the trough. After application of the sample to the paper in the usual manner the paper is wetted with buffer as described before, excess buffer removed by blotting between filter papers and the uniformly wet paper placed in the tank so that it is supported over a glass rod. The assembled apparatus then conforms essentially to the conventional descending paper chromatography

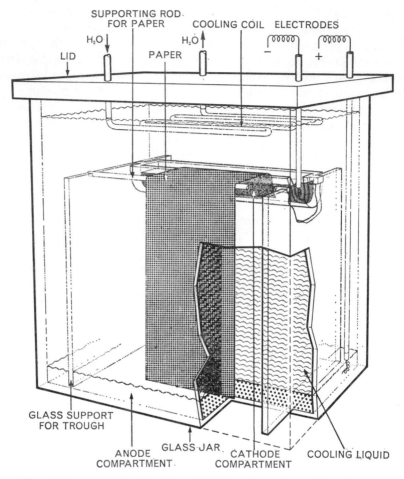

FIG. 5. Apparatus for liquid cooled high voltage electrophoresis.

arrangement. The tank is filled with liquid coolant and for this a variety of liquids, including toluene, carbon tetrachloride, chlorobenzene, heptane or white spirit, have been used. The coolant should be non-conducting, less dense than water and immiscible with water. It should be relatively volatile so that the paper can be satisfactorily dried after electrophoresis, and should be relatively inefficient as a solvent for the usual buffer constituents, e.g., a liquid which readily dissolves pyridine is virtually useless with pyridine acetate buffers. A drawback to such liquids is that they are usually highly inflammable and for the most part toxic. Low viscosity silicone oil suffers none of these disadvantages but has the other disadvantage that it is relatively expensive, an important factor in view of the large volume required. Probably the most widespread coolant in use at the present time is Esso white spirit 100 (Varsol) which has proved to be cheap and efficient and is perfectly safe providing the usual fire precautions are observed. In this context the apparatus should preferably be set up in a fume cupboard in a room in which no naked flames are allowed.

After filling the apparatus with coolant the cooling coil is set in position. The coil is the normal all-glass cooling coil and may be suspended from the underside of the lid of the apparatus so that it is located just over the glass rod which supports the paper. This is the area of the tank which gets warmest during a run because of the warm white spirit rising to the top of the tank. Platinum electrodes are then inserted into the trough and into the bottom of the tank, and the apparatus is ready for use. Once assembled the white spirit can be left in the tank for some time, being replaced as necessary. Contamination of the white spirit, e.g., contamination with pyridine when using pyridine acetate buffers particularly that of pH 6·5, is indicated by a rise in the current during successive runs, this being caused by the coolant acting as a conductor.

It may be noted that several variations of the apparatus described are in use at present. A particularly common modification (see Katz et al. (1959) as modified by Naughton and Hagopian (1962)) is where the paper is set up in the tank essentially as in Fig. 5, but in this case the two buffer compartments are in the bottom of the tank, e.g. as two long rectangular, separate glass containers (Fig. 1(a)). Alternatively the bottom of the tank may be bisected along the short axis by a glass strip about 4 in. high, cemented to the bottom and sides of the tank to give a water-tight seal. A glass frame is constructed in the shape of an inverted V such that the separate arms of the V fit into the separate buffer compartments. The wet paper is laid over the frame and the whole lowered into the tank so that the paper is in contact with buffer in the separate compartments. Electrophoresis is then carried out as above. An advantage of this particular arrangement is that the length of separation during the electrophoretic run may approach twice that obtained with the simple hanging paper arrangement shown in Fig. 5. Moreover, when supported on the glass frame the paper is more easily placed in and removed from the tank.

C. Source of current

Power packs for high voltage electrophoresis capable of delivering up to 10 kV and 500 mA are available commercially. (Savent Instruments Inc., Hicksville, N.Y., U.S.A.; Shandon Scientific Co. Ltd., London, England; Miles Hivolt Ltd., Shoreham, Sussex, England.) It should again be stressed that voltages of the magnitude used in high voltage electrophoresis are lethal and considerable care should be exercised in the use of these power packs. It is advisable that the electrophoresis tank be enclosed in a Perspex box which incorporates a safety switch in the door, which therefore cannot be opened without switching off the current. In this way the tank is completely isolated during a run and accidental touching of electrodes, etc., is avoided. A lethal accident resulting from inadequate safety precautions has been reported (Spencer *et al.*, 1966).

D. Applications of high voltage electrophoresis

By far the most common use of this technique in the past was in the separation of amino-acids and, more particularly, peptides. In recent years, however, the technique has been very successfully applied to the separation of fragments of ribonucleic acids in the form of either mononucleotides or oligonucleotides. Efficient separation of amino-acids in a very short time is possible by high voltage electrophoresis, which yields results at least comparable to the best of those obtained with paper chromatography. The technique is certainly one of the most efficient in use at the present time for the separation of peptides and so has been extensively used for studies of protein sequences. The great advantage of using really high voltages for the separation of amino-acids and peptides, is, of course, that high voltages can exploit the subtle charge differences between these species under conditions where diffusion is minimized. As the majority of proteins are denatured under the conditions used, however, separation of proteins has generally not been possible under high voltage conditions. Such denaturation is almost certainly due to the high local heating generated on the paper. The problems encountered in determining the sequences of ribonucleic acids are essentially the same as those involved in determining the sequences of proteins; that is, methods are required to separate fragments of various sizes arising from partial degradation of the original polymer. Since the constituent mononucleotides contain amino, hydroxyl and phosphate groups with characteristic pK_a values, slight and subtle differences occur between these mononucleotides and also between oligonucleotides. High voltage electrophoresis, therefore, is eminently suitable for solving some of the problems arising during sequence determinations of nucleic acids.

An important aspect of high voltage electrophoresis as applied to the separation of amino-acids, peptides, mononucleotides or oligonucleotides is that the separation obtained can often be markedly improved by carrying out a second separation, using either chromatography or electrophoresis, at right angles to the first. The resulting two-dimensional map or pattern of fragments obtained from a macromolecule is termed a "fingerprint" of the macromolecule.

An extremely useful technique in conjunction with finger-printing studies is that of machine-sewing two pieces of paper together to facilitate subsequent separations. For example, electrophoresis is commonly carried out in the first dimension on narrow paper strips of approximately 60×6 cm. Ascending chromatography or electrophoresis may be conveniently carried out in the second dimension simply by machine-sewing a piece of paper 60×30 cm on to the first strip. The exact procedure, illustrated in Fig. 6(a), is as follows. After carrying out electrophoresis or chromatography in the first dimension on the narrow strip, this is laid on top of a larger sheet so that the two edges of paper are aligned. The pattern of peptides or other material should preferably lie towards the inner edge of the strip. The strip

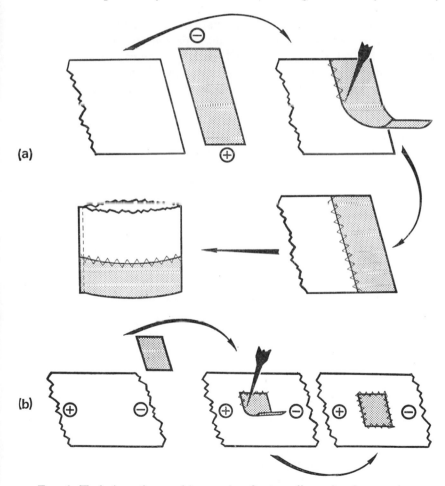

FIG. 6. Technique for machine sewing for two-dimensional separations.

is now machine sewn on to the sheet using a zig-zag stitch with a span of about 0·5 cm. Ordinary cotton thread is satisfactory for this purpose. The ends of the thread are tied off and the area of sheet underlying the strip is cut off as near to the stitching as possible, taking care not to cut or weaken the stitches in any way. If ascending chromatography is to be carried out the stitched paper can be simply folded into a cylinder and sewn up using a straight stitch.

It is obvious that the above method can also be used to sew a further sheet of paper on to the remaining free edge of the strip should descending chromatography or electrophoresis be required in the second dimension. The basic aim of the method is to transfer all the components of a narrow strip on to a larger sheet. Often, however, it is required to transfer only a small zone of the original strip or chromatogram, e.g., a neutral amino-acid zone obtained after electrophoresis. Once this zone has been located, e.g., by using σ-DNP lysine as a coloured marker in the case of neutral amino acids, the area of paper containing the compounds is cut from the strip and placed on top of a strip of slightly larger dimensions than the first. The small area is then sewn on to the larger area, again using a zig-zag stitch, and the paper underlying the small area is cut from the larger area (Fig 6(*b*)). It is apparent that machine-sewing as described enables three or more separations to be carried out under different conditions, both with electrophoresis and chromatography, without the compounds ever being eluted directly from the papers. Solvents and zones will move across stitched papers of the type described with scarcely any distortion of the zones.

A vast literature covers the many applications of high voltage electrophoresis in sequence determinations. Specific examples of such applications are listed by Sargent (1969). An important feature of high voltage electrophoretic separations is the use of volatile buffers which enables separated compounds to be recovered from the paper in the absence of contaminating salts. A list of useful volatile buffers is compiled in Table I.

TABLE I
Volatile buffers for electrophoresis

2·5 formic acid–8·7% acetic acid, pH 1·9
0·5% pyridine –5% acetic acid, pH 3·5
1% pyridine –10% acetic acid, pH 3·5
0·4% pyridine –0·8% acetic acid, pH 4·4
10% pyridine –4% acetic acid, pH 5·3.
10% pyridine –0·4% acetic acid, pH 6·5
1% ammonium bicarbonate, pH 7·9
1% ammonium carbonate, pH 8·9

IV. THIN LAYER ELECTROPHORESIS

Thin layers offer several advantages over paper for conventional chromatography, such advantages including speed, simplicity of manipulations, extremely good resolution, very high sensitivity and also the fact that a wide variety of supporting media can be used. For these reasons thin layer chromatography is a technique in routine use in many laboratories and its applications are covered by a vast literature (see, e.g., Stahl, 1965). It is obvious that electrophoresis on thin layers should possess similar advantages over electrophoresis on paper but, despite general knowledge of this fact, electrophoresis on thin layers has only been developed to a high degree in recent years and is not yet in general use. The major reason for the slow development of thin layer electrophoresis seems to have been the idea that passage of a current through a thin layer would result in generation of sufficient quantities of heat to cause the thin layer to dry out. While heat is indeed generated on thin layers the problem of drying out does not seem to be a very serious limitation in practice, and at the present time several relatively simple methods are available for carrying out electrophoresis on thin layers. (Fig. 7).

The first documented use of thin layer electrophoresis was in the separation of amino-acids and amines (Honegger, 1961). An apparatus which is basically similar to that used by Honegger has been described more recently by Ritschard (1964) for the separation of peptide mixtures derived from proteins. This apparatus is now commercially available from Shandon Scientific Co. Ltd., Willesden, London. Up to 1000 V can be applied

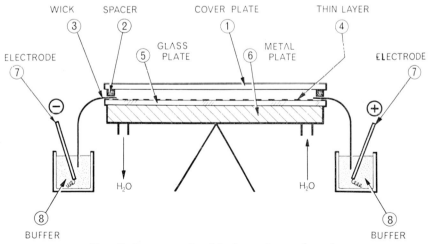

FIG. 7. Apparatus for thin layer electrophoresis.

(30–80 mA) for 30 min using a conventional pyridine-acetate buffer, pH 6·5. Quantities of peptide mixtures ranging from 50–500 μg can be handled. The sample to be separated is applied to a dry layer of silica gel G on 20 × 20 cm glass plates and the layer then sprayed evenly with buffer. A glass covering plate is set in position to rest over paper wicks making contact between the layer and the buffer compartments, and electrophoresis is carried out in the usual way. The plate is dried and electrophoresis at a different pH, or chromatography, can be carried out simply in the second dimension. It is to be emphasized that the method described here is an example of true high voltage electrophoresis applied to thin layers to give essentially "micro-figerprints". The resolving power of the method is underlined by Ritschard (1964) who separated 60 components from a tryptic digest of myosin on a single plate. The method, of course, is ideal for carrying out routine diagonal electrophoresis on an analytical scale.

Variations on the basic technique designed to enable routine finger printing of small amounts of proteins to be carried out and for determination of radioactivity in separated peptides have been described by Sargent and Vadlamudi (1968).

The preceding description has concentrated on the use of thin layers of silica. Apart from the preparation of the thin layer plates the methods described are essentially the same for electrophoresis on a variety of other thin layer media.

V. ELECTROPHORESIS ON CELLULOSE ACETATE

As mentioned in Section II, paper as a medium for electrophoresis of serum and other proteins in general has been almost entirely superseded by other media. Of these, cellulose acetate is one of the more important being of particular value for the electrophoresis of serum proteins. Electrophoresis on cellulose acetate was developed largely through the studies of Kohn (1957). Cellulose acetate offers the advantage of being a homogeneous material containing not more than traces of impurities. Adsorption of proteins on this medium is minimal so that tailing of zones is largely eliminated. This results in sharper bands and also renders more visible the minor components which are frequently obscured on paper by overlapping with larger zones. Further advantages of cellulose acetate are the speed of separation and the small quantities of material which can be handled. Certain sera which are not well separated on paper, e.g., rat and mouse sera, can be successfully separated on cellulose acetate and for all sera the separation of albumin zones from α-globulin zones is much more clear cut on cellulose acetate than on paper. These factors make cellulose acetate the medium of choice for routine electrophoresis of serum proteins.

Cellulose acetate strips are available commercially in a variety of sizes: e.g., $2·4 \times 12$ cm, $2·5 \times 16$ cm, $2·5 \times 10$ cm, $5·0 \times 18$ cm and $5·0 \times 20$ cm. The strips are insoluble in water, alcohol, ether and in most hydrocarbons, but are soluble in phenol and in acetone. The following solvents are of particular value in dissolving cellulose acetate: (a) dichloromethane:

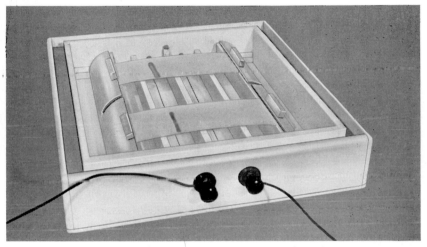

FIG. 8(a). Cellulose acetate electrophoresis.

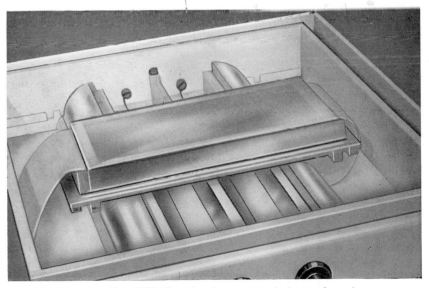

FIG. 8(b). Starch gel or agar gel electrophoresis.

FIG. 8(c). Agar gel electrophoresis.

ethanol (9 : 1), (*b*) chloroform : ethanol (9 : 1), (*c*) dichloromethane: acetone (1 : 1). Strips of cellulose acetate can be rendered translucent prior to optical scanning by immersion in Whitemor oil 120, cotton seed oil, dekalin and liquid paraffin. Of these clearing agents Whitemor oil 120 is the most satisfactory as it provides negligible optical obsorption in the range 250–1000 nm.

The thickness of cellulose acetate strips is about 120 microns; for satisfactory electrophoresis on a medium as thin as this certain precautions must be taken to ensure adequate vapour saturation in the tank, thus avoiding drying out of strips during a run. The type of tank used for cellulose acetate electrophoresis (Kohn, 1960) has electrode and buffer compartments of as large a surface area as possible and is fitted with a fairly tight-fitting lid, covered on the under-surface with a layer of Wettex sponge which is moistened with water prior to carrying out a run. These features, already dealt with generally on pp. 460–462, are particularly important with cellulose acetate. It should be noted of course that a tank of this type is also eminently suitable for most other types of electrophoresis (Fig. 8a, b, c).

A. Buffers used

Any of the following buffer solutions are suitable for use with cellulose acetate:

Barbitone-acetate buffer (Owen, 1956), pH 8·6. Sodium acetate 6·5 g, barbitone sodium 8·87 g and barbitone 1·13 g are dissolved in water and diluted to 1 litre.

Barbitone buffer (Flynn and Mayo, 1951), pH 8·6. Barbitone sodium 10·3 g and barbitone 1·84 g are dissolved in water and diluted to 1 litre.

Tris-EDTA-borate buffer (Aronsson and Gronwall, 1957), pH 9·0. Tris [2-amino-2-(hydroxy-methyl)-propane-1 : 3-diol] 78·65 g, EDTA 7·8 g, and boric acid 5·98 g are dissolved in water and diluted to 1 litre.

B. Procedure

In the past commercially available strips of cellulose acetate (Oxoid) were brittle and some care had to be exercised in their handling. Modern strips of cellulose acetate such as Celagram (supplied by Shandon) do not suffer from this disadvantage and are quite easy to handle. The following procedure should be adhered to closely when soaking strips in buffer. The strip is dropped into a shallow tray containing buffer so that only the lower surface is in contact with the buffer. This is best done by holding the strip at both ends by means of curved forceps bringing the strip as closely over the surface of the buffer as possible, and releasing both ends of the strip at the same time. The strip will then float on the surface of the buffer which will soak in almost immediately. If the buffer is allowed to flow on to the top surface of the strip before the under surface is wetted then opaque areas are apparent on the strip which is now useless for electrophoresis. The strip cannot be simply immersed in the buffer as this results in air bubbles being entrapped in the strip and these may take several hours to disperse. After soaking, the strips are blotted lightly between sheets of filter paper so that no excess buffer is present, this being easily recognized by the sheen on the strips when they are held up to the light. The strips are then immediately placed in position in the tank and the current switched on (Fig. 8a). Drying out during this procedure is apparent by the appearance on the strips of white opaque areas and if these are present the strip should be reimpregnated with buffer. The procedure just described can be simply and rapidly carried out in practice and no difficulties should be encountered in its application.

About 15 min are usually allowed between switching on the current and application of the sample, in order to allow equilibrium conditions to become established in the tank. With the current switched on the sample of serum or other protein is applied to the strip in a straight continuous line about one-third from the cathode end of the strip for alkaline buffers. About 350 μg of protein may be applied to a strip 2·5 cm wide and 700 μg for the 5 cm wide strips. Some 10 μl and 20 μl of fluid can be absorbed by strips 2·5 and 5·0 cm wide respectively. Application of the sample is made with the usual micropipettes and may be facilitated by using a ruler resting along the edges of the tank as a guide line. About 0·5 cm should be left clear at either end of the strip. The most satisfactory runs are obtained when the sample is applied in a continuous line without interruptions. If the application is smeared it is best either to leave the strip and carry on with the run, or to start another strip, as it is impossible to suck up the applied sample in order to start again. As soon as the sample is applied the origin is marked by making two tiny scissors cuts on either side of it. With the usual currents applied this is a quite safe procedure, and with a little care, no electric shocks are experienced. If this procedure does not appeal to the worker the position of the origin can be marked with a ball point pen before soaking the strips. Application of the sera samples is considerably simplified by staining the serum before application with a few crystals of Bromophenol blue. This has the advantage of staining the albumin zone blue so that the progress of separation can be followed visually. With this procedure excess bromophenol blue moves ahead of the albumin zone, eventually moving off the paper as a purple zone.

Satisfactory separation can be achieved by applying a current of 0·4 mA per cm width of paper (6–8 volts per cm) with a buffer of pH 8·6 ($\mu = 0·05$–$0·10$). For strips 20 cm long separation is effected in about 5–7 h, whereas with strips 12 cm long separation is effected in about $2\frac{1}{2}$–3 h. Following electrophoresis Oxoid strips are dried by hanging in an oven at 100°C for 10 min; this also denatures the protein prior to staining. After drying, the strips frequently assume a wrinkled form which, however, disappears after staining. Celagram strips, however, should not be heated excessively, proteins being denatured by brief (2 min) immersion of the moist strip in 10% w/v trichloroacetic acid.

A modified procedure for achieving very fast separations of serum proteins at relatively high voltages has been described by Afonso (1961). For this separation 8·2% v/v glycerol is included in the buffer during a run to minimize evaporation during passage of current.

C. Staining cellulose acetate strips

Exactly the same technique is employed in staining dried Oxoid strips as is used in impregnating strips with buffer. Celagram strips treated with trichloroacetic acid should be blotted lightly and immersed directly in the staining solution. Aqueous stains are preferred to alcoholic stains and the dye concentration is usually less than that used for the conventional stains with paper. After washing, the stained strips are allowed to dry between sheets of filter paper under slight pressure so that the dried strips are perfectly flat and do not assume a wrinkled appearance.

1. *Protein stains*

Ponceau S: The strip is floated on to a solution of Ponceau S (0·20% w/v) in 3% w/v aqueous trichloroacetic acid. Penetration of the dye into the protein areas is apparent within a few seconds. The strip is then immersed in the dye solution when staining is complete within 5–10 min. Complete penetration of the dye into areas of high protein concentration can be helped by moistening the affected parts of the strip with methanol. After staining the strips are washed in 5% v/v acetic acid until the background is perfectly white.

Naphthalene black 12B: Strips are stained with a solution of 0·2% w/v Naphthalene black 12B in 10% v/v acetic acid. Staining is usually complete within 10 min after which the strips are soaked in 10% v/v aqueous acetic acid until a clear background is obtained.

Procion blue: Strips are immersed in a solution of 0·2% w/v Procion blue in 10% v/v acetic acid for 10 min then transferred to fresh 10% v/v acetic acid until the background is white.

Azocarmine B: Strips are immersed in a solution of 0·1% w/v Azocarmine B in 0·45M acetic acid containing 0·45M sodium acetate for 5 min. Excess dye is removed by washing in 10% v/v acetic acid until a clear background is apparent.

Nigrosine: This is the most sensitive of all the protein stains and should not be used for zones containing more than 80 μg of protein (Kohn, 1958). The stain can be used, however, for restaining zones which are not well detected with less sensitive stains. Staining is best carried out by immersing the strips overnight in a solution of Nigrosine (0·001% w/v) in 2% v/v aqueous acetic acid. After this time the strips are rinsed rapidly in tap water (the background remains perfectly clear during the staining process) and air-dried.

Leuco-malachite green (for haemoglobin): Strips are immersed in a solution of Leuco-malachite green (0·1% w/v) in 5% v/v acetic acid to which a few drops 10 vol. hydrogen peroxide are added before use. Haemoglobin zones are apparent within a few minutes. The stained strips are washed briefly in 5% v/v acetic acid and air-dried.

2. *Glycoprotein stains*

The following modification of the Schiff-periodic acid stain has been described by Bodman (1957):

1. The strip is immersed in 95% v/v ethanol for 10 min.

2. The strip is then transferred to the following solution for 10 min:

Periodic acid (50% w/v)	2·5 ml
Sodium acetate (0·2M)	25·0 ml
Water	250·0 ml

3. After this the strips are rinsed in 0·001N hydrochloric acid.

4. The strips are then immersed in the following solution for 2–3 min:

Potassium iodide	40·0 g
Water	400·0 ml
0·1N Hydrochloric acid (added just before use)	10·0 ml

During immersion of the strips the solution assumes a brown colour which disappears after the addition of a few drops of concentrated ammonium thiosulphate solution. The strips are then left in the solution until the colour has just disappeared from the zones.

5. The strips are placed in Schiff's reagent (the fuchsin-sulphite reagent described for staining glycoproteins on paper, p. 465) for about 30 min until the zones become a dark purple.

6. Each strip is then rinsed three times in a solution of sodium dithionite (0·1 w/v in 0·1N hydrochloric acid). The strips are left in this solution until the zones become very faint.

7. The strips are finally transferred to 0·1N hydrochloric acid until the zones regain their deep purple colour, when the strips are blotted between filter paper and air-dried.

3. *Lipoprotein stains*

Until recently lipoproteins to be separated on cellulose acetate were stained prior to electrophoresis. This is done in the case of serum lipoproteins by mixing 5 volumes of serum with 1 volume of Sudan black (1·0 w/v in 10% v/v ethyl acetate and 90% propane-1, 2-diol). After standing for 15 min at room temperature, the solution is centrifuged at 1000 rev./min for 10 min to remove excess dye particles and denatured protein. A sample of the supernatant liquid is used for electrophoresis in the usual way (MacDonald and Kissane, 1960). More recently, however, Charman and Landowne (1967) have described direct staining of lipoprotein components on cellulose acetate. The method will detect general α- and β-lipoprotein classes in serum (see also Scherr (1961), Williamson *et al.*, 1962), and also additional classes of lipoproteins commonly found in clinical conditions. Electrophoresis is carried out in the first instance in a barbital buffer I = 0·05, pH 8·6, which is 1% w/v with respect to serum albumin and 0·001M with respect to EDTA. 5 μl samples of whole serum are applied to strips 6 × 1 in. and separation effected by applying 240 V (2·5 mA per strip) for 90 min at room temperature. A solution of Oil red

O is prepared by adding 0·2 g of solid to 350 ml of methanol and boiling under reflux. 150 ml of water are added with stirring and the suspension stirred until the temperature falls to about 39°C. The suspension is filtered and kept at 39°C. Cellulose acetate strips are transferred directly to the stain after electrophoresis, and kept there for 6–7 h at 37°–39°C. The strips are finally bleached by immersing for 3 min in 5% aqueous sodium hypo-chlorite solution, rinsed several times with water and air-dried.

D. Applications of cellulose acetate electrophoresis

By far the most frequent use of cellulose acetate electrophoresis is for the separation of serum proteins. Kohn (1960) has described the rapid separation of abnormal haemoglobin components on a small scale. Campbell *et al.* (1960) used cellulose acetate electrophoresis as a final step in the purification of albumin from liver microsomes. These authors also describe the elution of albumin from cellulose acetate by chopping the appropriate zone into small pieces in normal saline solution, followed by thorough homogenization of the pieces in a Potter-type homogenizer and removal of the debris by centrifugation at 10,000 g for 10 min. Using this procedure recoveries of at least 70% of the albumin can be achieved.

It may be noted that cellulose acetate has also found use as a medium for immunoelectrophoresis (see Chapter 6, Volume 5a).

VI. STARCH GEL ELECTROPHORESIS

A major development in electrophoresis was the discovery by Smithies (1955) of the remarkable resolving powers of starch gel for serum proteins. The resolving powers of this medium far exceed those of any other medium so far mentioned. At the present time the mechanism for this high resolv-ing power is unknown although it is almost certain that molecular sieving plays a major role. Other advantages in the method lie in the relative ease of manipulation and the sharpness of the zones obtained. Proteins separated by this method can generally be recovered from the gel in reasonable yield so that the method is of considerable value for preparative as well as for analytical purposes. An excellent review of starch gel electrophoresis has been published by Smithies (1959).

A. Apparatus used

Starch gel electrophoresis is conveniently carried out in trays or troughs of the type shown in Fig. 9. These troughs are constructed of Perspex and have internal dimensions of 15×5 cm $\times 6$ mm. The bottom of the trough has a hole cut in it. A false bottom is provided together with three

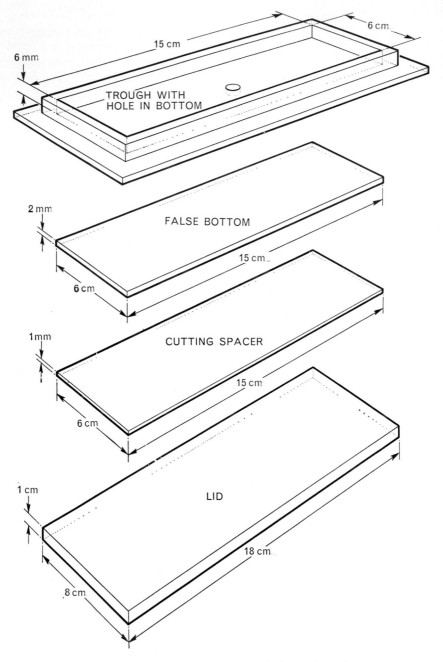

6 cm

15 cm

6 mm

TROUGH WITH
HOLE IN BOTTOM

2 mm

FALSE BOTTOM

15 cm

6 cm

1mm

CUTTING SPACER

15 cm

6 cm

1 cm

LID

18 cm

8 cm

FIG. 9. Apparatus for starch gel electrophoresis.

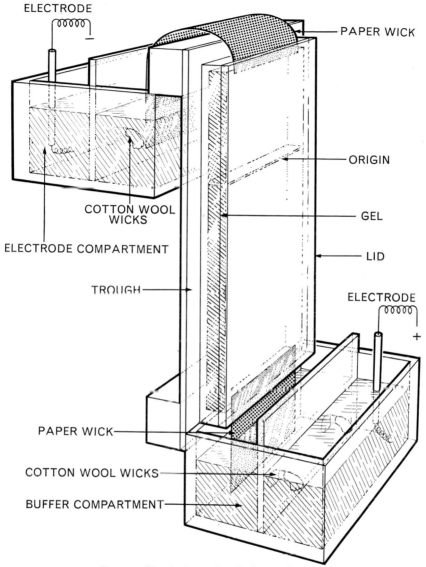

FIG. 10. Vertical starch gel electrophoresis.

or four "cutting spacers". A Perspex lid is also used with the trough. Alternatively simple trays can be constructed from plate glass sheets and 0·25 in × 0·5 in. strips of glass cemented together with high vacuum grease. Electrophoresis on starch gels can be carried out in the vertical or horizontal positions. With the gel in the horizontal position electrophoresis

can be carried out in conventional electrophoresis tanks. A special arrangement is necessary for electrophoresis in the vertical position and this is
shown in Fig. 10. The choice of vertical or horizontal electrophoresis is
determined by the method used to apply the sample. After the gel has been
poured into the trough and allowed to set a slit is cut in it with a razor
blade and the sample applied into the slit. Samples of solution may be
pipetted directly into the slit which is then sealed by means of paraffin
wax or petroleum jelly. Alternatively the sample can be applied into the slit
either by absorption on filter paper or cellulose acetate, or as a slurry in
starch grains (see p. 503). When the sample is applied in liquid form and the
current applied, the proteins migrate to one side of the liquid origin so that
a concentration gradient exists across the origin. This results in protein
sedimenting to the bottom of the trough at the area of highest concentration, i.e., in the area towards which the protein migrates. The overall
result is that at the end of a run the fastest running proteins tend to be
located at the bottom of the gel while the slow running proteins are located
at the top of the gel. This phenomenon, known as electrodecantation, results
in difficulties during the staining of gels, but can be overcome by carrying
out the run with the gel in the vertical position when the protein has no
opportunity to sediment towards the bottom of the trough. When samples
are applied to the gel either as slurries in starch grains or adsorbed on to
paper or other media electrodecantation does not take place so that electrophoresis can then be carried out in the normal position.

A consideration in choosing between the different methods of application
of the sample is that samples applied on paper are sometimes irreversibly
adsorbed on to the paper and consequently do not migrate into the gel.
No such adsorption occurs with starch grains but in this case entry of the
β-lipoprotein fraction of serum into the gel does not always occur satisfactorily. The application of a sample in the liquid form does not suffer
from either of these disadvantages so that for precise work this method of
application together with a vertical arrangement is the method of choice.

A further piece of useful equipment in starch gel electrophoresis is the
former. This is simply a piece of Perspex whose shape corresponds to the
shape of the origin finally required. The former is inserted in the trough
at the position of the origin and the gel is poured round the former. After
the gel has set the former can be removed leaving a ready made slot so that
cutting is unnecessary. The former may also be fixed permanently to the
lid of the gel trough. The formers can be of several shapers depending on
how many samples it is desired to run on a single gel. Where more than one
sample is to be run it is essential to leave spaces of about $\frac{1}{2}$ cm at least
between individual samples, and the former can be designed to meet these
needs.

The final piece of apparatus required for starch gel electrophoresis is a gel cutter. Many variants of this are in use but in essence they consist of a fine wire (e.g., 5 amp fuse wire) which is tensioned either by hand or in a suitable holder. Gels can also be cut very smoothly using a hack-saw blade whose teeth have been removed by grinding and whose cutting edge has been finely honed.

B. Buffers used

Useful buffers for starch gel electrophoresis cover the range pH 2–11, though gels are not often used outside the pH range 4–10. The following buffers are a few of the many which have been used successfully with starch gels.

(a) 0·025M boric acid 0·01M sodium hydroxide, pH 8·48. This is a general purpose buffer and is used with a bridge buffer (i.e., the buffer in which the paper wicks forming contact between the gel and buffer compartments are soaked) containing 0·30M boric acid-0·06M sodium hydroxide, pH 8·48. The latter buffer is also used in the buffer compartments. The bridge and compartment buffer has a higher ionic strength than the gel buffer to prevent shrinkage of the gels during a run.

(b) 0·076M tris-buffer-0·005M citric acid (Poulik, 1957), pH 8·6. This buffer is used in conjunction with a 0·30M boric acid-0·06M sodium hydroxide buffer in the buffer compartments and in the bridges. This, therefore, constitutes a discontinuous buffer system and a brownish boundary moves through the gel during the course of electrophoresis. Tris-citrate buffer, pH 8·6, has also been used in the presence of 7M urea where dissociation of proteins into their sub-units is desired (Bloemendal et al., 1962).

(c) Sodium potassium phosphate buffer (Ashton, 1917), pH 7·8. 42 g of di-sodium hydrogen orthophosphate is dissolved in 1 litre of water and adjusted to pH 7·8 with saturated potassium dihydrogen phosphate.

(d) 0·025M sodium acetate buffer, pH 5·5. This buffer is used with 0·50M sodium acetate buffer, pH 5·5 in the buffer compartments (Hsiao and Putnam, 1961).

(e) 0·05M formic acid-0·01M sodium hydroxide (Poulik, 1960), pH 3·1.

(f) 0·01M hydrochloric acid (Johns et al., 1961), pH 2·3.

The choice of buffer in starch gel electrophoresis is determined by the nature of the protein to be separated. For separation of serum proteins borate and tris-citrate buffers are generally used. The acidic buffers of pH less than about 3·0 are used only for very basic proteins such as histones.

C. Preparation of starch gels

12 g of starch (starch hydrolysed as prepared by the Connaught Laboratories, available from BDH), are added to 100 ml of the appropriate buffer in a 500 ml Buchner flask attached to a water pump. The contents are vigorously stirred, and when a uniform suspension is attained the flask is heated evenly over a Bunsen flame, vigorous stirring being continued. At the gel point of about 65°C the suspension becomes very viscous; a stronger stirring action is required during the subsequent heating stage, this producing a more fluid gel. When the gel becomes more mobile again and the temperature reaches 90°C the heating is discontinued and a vacuum applied to the flask for a few seconds when the contents of the flask boil vigorously with the expulsion of air bubbles. The vacuum is disconnected and the gel immediately poured into the trough with the false bottom (gel support) in position. The inner surfaces of the trough and the false bottom are lightly smeared with liquid paraffin before pouring the gel. The trough is filled until the gel is just above the top, when the Perspex lid, the under surface of which has also been lightly smeared with liquid paraffin, is lowered gently on to the surface of the gel, care being taken to exclude air bubbles during this process. The lid is then pressed down firmly to extrude excess starch and the gel is left for 2–3 h at room temperature to set. Gels can be stored overnight at 2°C before use if necessary. Some workers prefer to cover the molten gel with a sheet of "Melinex" (ICI Polyester Fibre) which is easy to apply, provides a very flat surface and is easy to remove.

Apart from being used with normal buffers, starch gels are often used with buffers containing urea at concentrations of up to 8M. This is to enable normally insoluble proteins to be brought into solution, or to disrupt protein complexes into their sub-units. A problem encountered when preparing starch gels with urea is that in the course of heating the starch suspension in urea, a significant proportion of the urea is converted to ammonium cyanate and ammonium carbonate. At high concentrations of urea this can give rise to sufficient quantities of ammonium ion to cause appreciable changes in the pH of the buffer (particularly important in buffers of low pH) and such changes are not too easily controlled. Recently a simple method has been described to overcome this problem (Melamed, 1967). Using the procedure to be described, 500 ml of gel, 12% with respect to starch and 8M with respect to urea may be prepared. 288 g of urea are stirred with 150 ml of water containing 2 g of Zeo-Karb 225 (H⁺) ion exchange resin (14–52 mesh), and heated at 60°C until solution occurs. 72 g of starch are suspended in 146 ml of water in a separate flask, heated to 60°C and the contents mixed with the urea-ion exchange mixture. The whole is heated in a bath at 80°C with swirling and held there for 10 min

The suspension is then allowed to cool in a water bath at room temperature for 20–30 min. Since the urea-containing solution is slow to gel the Zeo-Karb particles may be removed at this stage by filtering through muslin. To 450 ml of the filtrate is added 50 ml of buffer, e.g., sodium acetate, $I = 0.10$, pH 4.0, and the solution mixed, de-gassed at the water pump and poured into trays in the usual way. The gel is left overnight to set. Using this procedure the pH of the gel can be controlled to within 0.1 of a pH unit.

D. Application of the sample, and running conditions

After the gel has set the lid is removed from the trough and a slit about 1–2 mm thick is cut in the gel at the origin using a razor blade and taking care to leave about $\frac{1}{2}$ cm clear at both sides of the trough. As described previously the slit can be pre-formed in the gel using a former. The sample is then applied to the slit (i) as a solution, (ii) after absorption on paper or (iii) as a slurry in starch grains as described before. Excellent separations can also be effected by applying the sample after absorption on to Sephadex G-25. In this case sufficient dry Sephadex is added to the sample so that the liquid is completely absorbed. The sludge of Sephadex can then be applied to the slit in the gel in the same manner as for starch grains. The slit is then sealed either with soft paraffin or by pouring paraffin wax (coagulating point about 49°C) at just above its congealing point on to the slit and allowing the wax to set. Alternatively the slit may be sealed with a thin strip of glass the under surface of which is smeared with liquid paraffin. Where the sample is applied absorbed on paper the razor slit need not be sealed. An interesting method of covering gels has been described by Bussard (1954). A solution of vinyl chloride (monomer) in 1,2-dichloro-ethane is carefully poured or sprayed on to the gel and the 1,2-dichloro-ethane allowed to evaporate at room temperature. The inert plastic seal which forms over the gel considerably reduces evaporation during a run.

The trough is then transferred to the electrophoresis tank in either the horizontal or vertical position and contact is effected between the gel and buffer compartment by means of a double thickness of Whatman No. 3 mm paper soaked in buffer. Washed medical lint may also be used as wicks. The lid is then replaced and the run is started (see Fig. 8b, 10). Satisfactory separation with a gel 20 cm long is achieved by applying a voltage of 100 V (some 5 mA per tray) for a period of 17–19 h when using pH 8.65 buffer. Voltage of 220 V will result in separations in 7–9 h but higher voltages than this tend to cause distortion of zones. Serum samples can be stained with Bromophenol blue in the usual way prior to application so that the course of separation can be followed visually. For other proteins the optimal conditions of separation should be determined empirically for different buffers.

E. Staining starch gels

After separation is complete the gel is sliced before staining. Slicing is carried out as follows. The false bottom with the gel on top is pushed out from the trough, using the hole in the bottom. A cutting spacer is then inserted in the trough and the false bottom with the gel on top is replaced in the trough. The gel then protrudes from the trough in thickness equal to that of the cutting spacer. The protruding thickness of gel is cut off using the cutter in a continuous horizontal cutting movement, guiding the cutter on the sides of the trough. The slice is carefully removed from the gel; a simple and reliable method of doing this is to float the slice off under buffer. It is usual to discard the top slice and to cut further slices from the gel by inserting further cutting spacers into the trough. In this way a gel of standard thickness (6 mm) will provide at least three slices. Thus a single gel can be stained with several reagents, and samples which are required for elution from the gel can be easily detected by staining one of the slices. Contact staining as described for starch blocks (p. 51) can also be applied to starch gels. This procedure is particularly useful for radio-autographic estimation of radioactivity on starch gels.

A selection of useful stains for starch gels is described below. It may be noted that starch is useless for the application of glycoprotein stains since the starch itself will react with these.

(a) Proteins

Naphthalene black 12B (Amidoschwartz 10B). The slice is immersed in a solution of 1% w/v Naphthalene black 12B in methanol : acetic acid: water (500 : 200 : 500) for some 20 min. The dye solution is decanted and excess dye is removed by washing exhaustively in a solution of methanol: acetic acid : water (500 : 200 : 500) until a relatively clear background is obtained. Note that shrinkage of the gel occurs during the procedure so that care should be exercised when locating zones on unstained slices by comparison with the stained slice.

Nigrosine. The slice is immersed in a solution of 0·2% w/v Nigrosine in 40% v/v aqueous ethanol containing 2·5% w/v trichloroacetic acid. Development of the stain is best carried out by leaving the slice in the dye solution overnight. Excess dye is removed from the gel by washing exhaustively in 2·5% w/v trichloroacetic acid. The stain is much more sensitive than Naphthalene black 12B and has the added advantage that shrinkage of the gel is much less severe during the staining process.

(b) Lipoproteins

Lipids and lipoproteins can be detected by immersing the slice in a saturated solution of either Oil red O or Oil red N in 50% v/v methanol containing 10% w/v trichloroacetic acid. These solutions are prepared by adding an equal volume of 20% w/v trichloroacetic acid to a saturated solution of the dye in methanol, and filtering off the precipitated dye particles. Staining is best carried out by immersing the slice in the dye overnight and excess dye is removed by washing the slice in 50% v/v methanol containing 10% w/v trichloroacetic acid. Serum lipoproteins can also be detected by pre-staining the serum with Sudan black (see p. 11).

(c) Specific enzymes

It is probable that many of the standard histochemical stains can be used on starch gels for the detection of specific enzymes. Among histochemical reagents which have been used so far are 1-naphthyl disodium ortho-phosphate for the detection of phosphatases (Bodman, 1960), Nitro-blue tetrazolium for the detection of lactic dehydrogenases (Allen, 1961), and 2,6-dichlorophenol-indophenol for the detection of nicotinamide adenine dinucleotide-linked dehydrogenases in general (Watts and Donninger, 1962). 2,6-Dichlorophenol-indophenol is particularly useful since bands which have been stained for the enzyme can be counter-stained with any of the protein stains as the histochemical colour is stable for only about 10 min. Hunter and Markert (1957) have used the acetate and butyrate esters of 1-naphthol for the detection of esterases, and 3,4-dihydroxyphenyl-alanine (DOPA) for the detection of tyrosinases.

Of particular importance in microbiology is the application of starch gel electrophoresis for taxonomic classification. Norris (1968) lists applications of starch and polyacrylamide gel electrophoresis of proteins and specific enzymes and discusses their relevance to the classification of several groups of micro-organisms. For example Norris and Burges (1963) have used the supernatant resulting from centrifuging bacteria treated with ultra-sound for direct analysis of esterase activity on starch gels. Extracts were subjected to electrophoresis in tris-citrate/borate buffer and esterase activity detected by flooding the gels with a solution containing tris-maleic acid buffer, pH 6·4 (0·1M), 100 ml 1-naphthyl acetate (1% w/v in 50% v/v acetone) 2 ml and Fast blue B salt, 50 mg. Zones containing esterase activity stained red over a period of 2 h. Such patterns, a typical example of which is reproduced in Fig. 11, can be extremely useful in classifying bacteria.

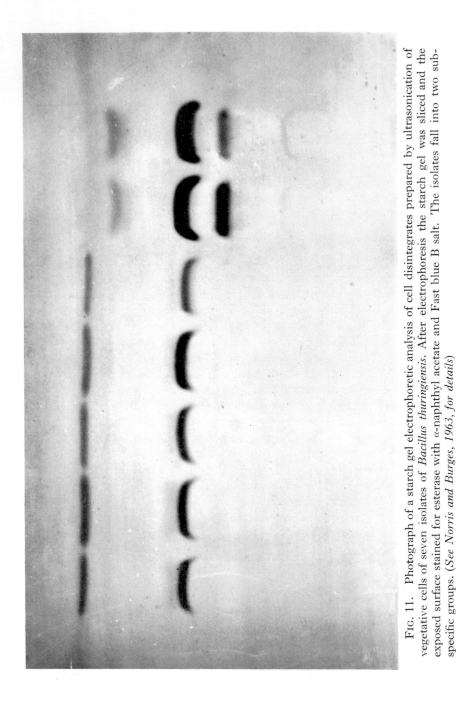

Fig. 11.　Photograph of a starch gel electrophoretic analysis of cell disintegrates prepared by ultrasonication of vegetative cells of seven isolates of *Bacillus thuringiensis*. After electrophoresis the starch gel was sliced and the exposed surface stained for esterase with α-naphthyl acetate and Fast blue B salt. The isolates fall into two sub-specific groups. (*See Norris and Burges, 1963, for details*)

F. Applications of starch gel electrophoresis

A particularly useful form of starch gel electrophoresis when limiting amounts of material are available is the microslide adaptation described by Bloemendal (1963). Two-dimensional separations involving paper and starch (Smithies and Poulik, 1956; Paulik and Smithies, 1958), agar and starch (Ashton, 1957) cellulose acetate and starch (Duke, 1963) and starch itself in two dimensions (Poulik, 1959) have also been described. Probably the most impressive separations on starch to date are those involving a two-dimensional separation on starch followed by immunodiffusion (Poulik, 1959).

It is frequently desirable to recover separated components from starch gels for further study. In the past such recovery presented problems due to some proteins being tightly bound to the gel. Freezing and thawing of the gel followed by centrifuging with precautions to prevent the disrupted gel sedimenting to the bottom of the tube (e.g., by having a mushroom shaped plug in the centrifuge tube) is frequently sufficient to give good recovery. Bocci (1963) states that recovery is improved by the prior incorporation of 1% w/w Pevikon into the gel. At present most efficient recovery is obtained by cross-flow electrophoresis (see Murray, 1962). A more simple but less efficient form of electrophoretic recovery has been described by Gordon (1960).

At the present time a vast literature covers the applications of starch gel electrophoresis and it is true to say that virtually all classes of proteins have been studied by this technique. The very high resolving power of starch gels makes the technique of particular value in determining the purity of protein fractions, e.g., in the purification of enzymes. In fact, the high resolving power of the starch gel has sometimes been treated with caution and it is possible that some of the zones obtained represent artifacts due, for example, to the separation of proteins into sub-units or to the aggregation of proteins into polymer forms. A very useful bibliography of the different classes of proteins separated by starch gel has been compiled by Bloemendal (1963).

VII. AGAR GEL ELECTROPHORESIS

Electrophoresis of colloids in agar gel was described by Field and Teagre (1937) but it was not until 1946 that serious attention was paid to electrophoresis on agar when Consden *et al* (1946) used this medium for the separation of peptides from wool hydrolysates. Proteins were separated in 1% w/v agar gel by Gordon *et al.* (1950), but modern applications of agar gel electrophoresis date from the work of Grabar and Williams (1953) who used the immunodiffusion methods of Ouchterlony (1958) following agar gel electrophoresis to obtain immunoelectrophoretic analysis. The main advantages of agar lie in the relatively low adsorption of proteins to this medium so that sharp zones are obtained. The method furthermore is capable of handling fairly large amounts of protein and is simple to operate. Recovery of proteins from the gel is in general good so that the method is valuable from the preparative point of view, but a minor disadvantage is that little is known about the contaminating materials which may be eluted from the gel along with the protein. A further disadvantage of the method is that, due to the acidic nature of agar, marked electroendosmosis takes place in most alkaline buffers. Agar as a medium for the electrophoresis of proteins is superior to paper but is generally inferior to starch gel. This is largely due to the fact that the agar gel does not possess the molecular sieving properties of starch gels.

A. Apparatus used

Agar gel electrophoresis can be carried out in the same type of troughs as described for starch gel electrophoresis, or on thin glass plates or slides. The standard electrophoresis tanks can accommodate glass plates of various dimensions (Fig. 8b, c).

B. Buffers used

The following buffers represent a useful series for use in agar gel electrophoresis:

Sodium acetate, 0·1M, pH 4·0: 25·6 ml of glacial acetic acid and 13·6 g of sodium acetate are dissolved in water and the solution diluted to 1 litre.

Sodium acetate-EDTA, 0·1M, pH 4·5: 1·0 g of EDTA disodium salt is dissolved in 125 ml of glacial acetic acid and the solution diluted with water to 1 litre.

Sodium-phosphate buffer, 0·1M, pH 7·0: 260 ml of 0·2M disodium hydrogen orthophosphate is mixed with 220 ml 0·2M sodium dihydrogen orthophosphate and diluted to 1 litre with water.

Barbitone-acetate buffer, 0·1M, pH 8·6: 10·0 g of barbitone sodium and 6·5 g of sodium acetate are added to 64·4 ml of 0·1N hydrochloric acid and diluted to 1 litre with water.

Barbitone-calcium lactate, pH 8·6: 12·8 g of barbitone sodium, 1·66 g of barbitone and 0·384 g of calcium lactate are dissolved in water and the solution diluted to 1 litre with water.

Tris-EDTA-borate, pH 9·0: 60·5 g of tris buffer, 6·0 g of EDTA disodium salt and 4·6 g of boric acid are dissolved in water and the solution diluted to 1 litre with water.

Glycine, 0·1M, pH 9·0: 3·0 g of glycine, 0·82 g of sodium hydroxide and 4·68 g of sodium chloride are dissolved in water and diluted to 1 litre with water.

C. Preparation of the gel

Deionized agar is normally used for gel preparation, e.g., Oxoid, Ionagar No. 2. A stock gel containing 2% w/v agar is prepared by adding the agar to distilled water at about 90°C and dissolving the suspension by heating at the same temperature in a water bath. An equal volume of the required buffer of twice the final strength is added while hot to the gel which is then ready for use. The final buffer concentration used in agar gels is usually about 0·5M, buffers of this strength giving good separations of serum proteins, though buffers of 0·10M strength may be used. The warm solution of agar is either poured into the trough or on to the glass plates, which should be just covered. The volume required for this purpose may be determined by pouring a trial plate. After pouring the gel the plates or trough are set aside at room temperature until the gel sets, when the plates or troughs may be transferred to the electrophoresis tank. Contact is made between the gels and buffer compartments with filter paper wicks soaked in buffer. The arrangement for slides is shown in Fig. 8c.

D. Application of the sample

A narrow slit is cut in the surface of the gel with a razor blade, exactly as described for starch gel, taking care to leave a space at both edges of the gel. The sample is applied to the slit either as a solution or after absorption on paper or cellulose acetate. The slit is sealed either with soft paraffin or by placing over it a small strip of glass, the under surface of which is smeared with liquid paraffin. Prior to application serum samples can be stained with Bromophenol blue in the usual way.

Satisfactory separations of serum proteins are obtained on the thin gels by applying a current of about 3 mA per 1 in. width of gel. For a gel 12 in. long separation is complete in about 6–8 h. For gels prepared in troughs of 6 mm depth a current of some 10–15 mA per tray is required.

E. Drying and staining the gel

At the end of a run the entire gel is placed in a bath containing 10% v/v acetic acid in order to denature the proteins. After immersion in the bath for 15 min the gel is removed and covered with a sheet of filter paper of dimensions slightly greater than those of the plate. The gel is then placed in a current of warm air (about 40°C) when the soluble constituents of the gel and buffer pass into the filter paper and the gel eventually dries out. After the paper has completely dried out it is removed from the gel which is now ready for staining. Marked shrinkage of the gel occurs during this process but the gel assumes its original form on staining.

1. *Protein stains*

Any of the conventional protein stains (see p. 42) may be used but probably Nigrosine is one of the most frequently used stains. The gel is immersed in a solution of Nigrosine (0·5% w/v) in 2·5% v/v acetic acid for 24 h or longer until the stain has developed. Excess dye is removed by washing the gel in 5% v/v acetic acid until a relatively clear background is obtained.

2. *Lipoprotein stains*

The following method described by Bodman (1960) offers advantages of speed and simplicity and has the further advantage that the gel can be counter-stained for protein, using nigrosine, after staining for lipoprotein. The plate containing its gel is placed in a saturated solution of Lipid crimson in 60% v/v ethanol and left in an airtight compartment overnight. Washing is carried out in a solution of aqueous 2·5% v/v acetic acid containing 2·5% w/v Teepol. The plates are allowed to air dry and can then be counterstained for protein using Nigrosine as described above.

Gels can also be stained for lipoproteins using a saturated solution of Oil red O in 60% v/v ethanol, the procedure being exactly the same as that described for Lipid crimson.

3. *Glycoprotein stains*

The periodic acid Schiff reagent as described previously can also be applied to agar, but longer times than normal are required to allow the reagents to penetrate the gel. The following method is essentially that described for cellulose acetate (see p. 33).

The dried gel is soaked in 95% v/v ethanol for 30 min when it is transferred to the periodate solution and left for 30 min. After rinsing in water the gel is treated with the iodide-hydrochloric acid solution and the resulting brown colour cleared with concentrated ammonium thiosulphate. The gel is treated with Schiff's reagent (see p. 12) until colour development has occurred, after which the gels are rinsed in the dithionite-hydrochloric acid solution and left in 0·1N hydrochloric acid overnight. After a final rinse in dilute Teepol solution the gels are dried in the usual way.

F. Immunoelectrophoresis using agar gels

By far and away the most common application of agar gels lies in immunoelectrophoresis (Grabar and Williams, 1953; Ouchterlony, 1958).

This technique is described separately by Oakley in Chapter 6 of Volume 5a.

VIII. BLOCK ELECTROPHORESIS

So far we have dealt with media which are capable of handling rather limited amounts of material. While some of these media have, in fact, been used for preparative purposes on a relatively small scale (e.g., starch gel electrophoresis is frequently used for preparative purposes), their main use is in analysis. In the strictest sense electrophoresis on blocks includes separations carried out in the standard gel troughs (Fig. 9), i.e., starch, agar and polyacrylamide electrophoresis. The latter methods are, however, rather more specialized than the general block techniques described in this Chapter and have consequently been dealt with separately in previous Chapters. In the present context, therefore, block media are considered primarily as non-gel media which are used for preparative purposes on a relatively large scale.

A. Starch block electrophoresis

Starch can be used as a medium for electrophoresis in one of two forms, viz., grains which are used for starch blocks, and gels which are used mainly for analytical purposes. Starch gels are formed after starch grains have been disrupted by heating (starch, hydrolysed) and their uses have been described in Section VI. Starch grains are used directly in starch blocks.

Starch block electrophoresis offers the advantages of relative ease of manipulation and the handling of reasonably large amounts of material (up to 1 g of proteins can be separated on large starch blocks). A disadvantage of the method is that proteins which have been eluted from the block are liable to be contaminated with carbohydrate constituents of starch.

1. *Preparation of the starch*

A slurry of potato starch in acetone is poured into a large measuring cylinder and after allowing the larger grains to settle the finer grains and the supernatant fluid are removed by suction. This process is repeated twice more so that eventually all the finer grains are removed and fairly large grains of more or less uniform size are left at the bottom of the container. The acetone is allowed to evaporate at room temperature; to remove soluble constituents the starch is washed several times with water and finally several times more with the buffer to be used. A thick slurry of starch in the buffer to be used is then prepared before pouring into the block.

2. *Apparatus used*

A suitable container, $20 \times 10 \times 1.5$ cm, which can be used for starch and indeed any type of block electrophoresis, may be constructed from glass or Perspex (Fig. 12). This container can also be used with most of the conventional electrophoresis tanks. Troughs of greater capacity than this are not to be recommended since excess heating during electrophoresis then becomes a major problem.

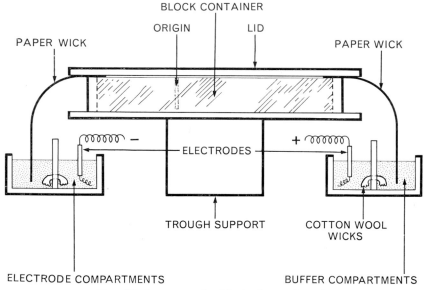

FIG. 12. Apparatus for block electrophoresis.

3. *Preparation of the block*

The slurry of starch grains in buffer (any of the usual buffers used for protein electrophoresis is suitable) is poured into the container and, after allowing some time for the starch to settle, excess buffer is removed by applying several thicknesses of Whatman No. 3 mm paper to both ends of the trough. Buffer is soaked into the paper and so is removed from the trough until the block is of such a consistency that it can be cut easily with a sharp knife or razor blade without losing its shape.

4. *Application of the sample*

A slot about $\frac{1}{2}$ cm wide is cut in the block using a razor blade, taking care to leave about 1 cm uncut at both edges of the block. The starch is then carefully removed from this area using a spatula and the removed starch is mixed with the protein sample to be separated. The mixture should be of as thick a consistency as possible and may contain up to 0·1 g of protein. This slurry is poured back into the slot which is then sealed with paraffin or by a thin strip of glass the under surface of which has been lightly smeared with petroleum jelly. The trough is transferred to the electrophoresis tank and contact is made between the block and the buffer compartments by means of wicks of Whatman No. 3 mm paper soaked in buffer. The trough is then sealed with a Perspex lid the under surface of which may be covered with a layer of Wettex (obtainable from any hardware store) which has been moisted with water.

A current of about 25 mA can be passed through such a block for a period of 16 h (overnight) without undue heating, though it is advisable to carry out the run in a cold room at 2°C.

5. *Elution of proteins from the block*

At the end of the run the current is switched off and the block is cut into transverse sections using a razor blade. These sections can be about 1 cm wide in the first instance though for closely spaced zones they may be cut as small as is conveniently possible. Each section is then transferred separately to small sintered glass funnels and the fluid withdrawn from the section by applying suction to the funnel. The section may be chopped and macerated during this process to aid elution and the starch may be washed with small quantities of buffer to help recovery though this may result in unnecessary dilution of the eluate.

The different fractions eluted in this way can then be assayed for protein in the usual manner, measurement of optical density at 280 nm providing a good index in the first instance though it is advisable to apply a colorimetric estimation or test for enzymatic activity wherever possible. Recoveries of protein by this technique are generally good (in excess of 90%) although yields ultimately depend on the particular protein in question. Elution of proteins which are tightly retained on the starch can frequently be helped by incubating the starch with amylase. It is essential for this purpose to use a pure amylase preparation which is as free of protein contaminants as possible (α-amylase from pig pancreas, an excellent preparation from this point of view, is available from BDH Chemicals Ltd.). It should be noted, however, that the eluate is certain to be contaminated with carbohydrate constituents from the amylase digestion.

Where rapid elution of components is required, sectioning of the entire block may be obviated by contact staining. For this technique a strip of paper or cellulose acetate is placed over the surface of the block, and buffer is allowed to soak into the strip. The strip is then dried at 100°C for 10 min and may be stained in the usual way for protein or other components as described previously. Contact prints obtained in this way can also, of course, be tested histochemically before drying, to detect specific enzymes or else can be subjected to radio-autography on X-ray film to detect radioactive components.

B. Polyvinyl chloride blocks

The main use of these blocks is to avoid contamination of the eluted sample with carbohydrate constituents. Thus polyvinyl chloride is of particular importance in the preparation of glycoprotein components. PVC resins which are available commercially in finely powdered form include Geon 426, 427, 428 and 431 (available from B. F. Goodrich and Co., Cleveland, Ohio, U.S.A.) and Pevikon C-870 (available from Superfosfat Bolaget, Stockholm, Sweden).

The particles are washed extensively before use in N hydrochloric acid then finally in water, and are then suspended in the selected buffer. Preparation of the block, application of the sample and running conditions are then exactly the same as those described for starch block electrophoresis.

Elution from PVC blocks can be carried out as described for starch blocks although it is more convenient to recover the eluate by centrifuging the appropriate sections of the block.

C. Glass powder blocks

Glass powder is conveniently prepared from glass fibre which is cut into small pieces about 1 cm long and finely ground in a mortar. The powder is washed with water in a measuring cylinder, the finer particles which do not sediment after an hour being discarded. After several such washes a preparation of more or less uniform particle size is obtained. The particles are then boiled in nitric acid and washed thoroughly with water until the washings are neutral. A thick slurry of the glass particles is prepared in buffer and the block poured in the usual way. Thereafter the procedure is exactly the same as that described for starch or polyvinyl chloride blocks.

D. Sephadex blocks

Several grades of Sephadex (available from Pharmacia, Uppsala, Sweden) are now available and electrophoresis on this medium has the added advantage that molecular sieving takes place. The block is simply prepared by mixing Sephadex in an appropriate volume of buffer and proceeding in exactly the same way as described for starch blocks. It may be noted that the "G" number of Sephadex is equal to the weight of water regained per 10 g of dry Sephadex. Thus to prepare a block of Sephadex G25 it is sufficient to equilibrate the dry Sephadex with slightly more than 2·5 times its weight of water and to pour off the excess buffer.

E. Applications of block electrophoresis

Most applications of block electrophoresis have used starch as a medium. Among the different classes of compounds separated by this medium may be included peptides and polypeptides, proteins, lipoproteins, nucleoproteins and nucleic acids. Thus the separation of plasma proteins on starch blocks has been reported by Goldsworthy and Volwiler (1958) and Miller and Bale (1954). These authors used barbitone buffers of pH 8·6. Proteolytic enzymes, including trypsin and chymotrypsin, have been studied by Kuzovleva and Chung-Yen (1959), Raacke (1956) and Liener and Viswanata (1956). Among other enzymes which have been successfully separated by starch block electrophoresis are dehydrogenases (Wieland and Pfleiderer, 1957; Grimm and Doherty, 1961), β-galactosidase (Rotman and Spiegelman, 1954), intestinal alkaline phosphatase (Harris and Mehl, 1955), prostatic phosphatase (Lundquist, et al., 1955), aspartase (Seno and Ammo, 1961) and elastase (Dvonch and Alburn, 1959). Tobacco mosaic virus protein was separated on starch blocks by Paigen (1956). Cohen and Lichtenstein (1960) successfully separated bacterial DNA, while bacterial RNA was separated by Pardee et al (1967). Transfer (soluble) RNA was studied by Bloemendal and Bosch (1959) while ribonucleoprotein was studied by Elson (1958, 1959). Among the polypeptide hormones which have been studied on starch blocks are α-corticotrophin (Raacke and Li, 1955), follicle-stimulating hormone (Ellis, 1958), growth hormone (Foons-Bech and Li, 1954) and thyroid-stimulating hormone (Postel, 1956). Serum lipoproteins have been separated by Kunkel and Slater (1952). For a more complete list of the different types of separations carried out on starch blocks and the conditions of separation Bloemendal (1963) should be consulted.

Electrophoresis on polyvinyl chloride has been used to a much more limited extent than electrophoresis on starch, though PVC has found use in separating carbohydrates (Muller-Eberhard and Kunkel, 1956) and serum glycoproteins (Bottiger and Carlson, 1960). Other examples of the uses of PVC blocks may be found in the work of Kunkel and Trautman (1956) on the separation of serum lipoproteins and Weiss et al. (1958) on the separation of polyribonucleotides.

Glass block electrophoresis has found application in the work of Rhodes (1960) on the preparation of ribonucleoproteins, of Allfrey et al. (1954) on the separation of components from the microsome fraction of mouse liver, and of Wieland et al. (1958) on the preparation of lactic dehydrogenase.

Block electrophoresis on Sephadex has not been used to a large extent so far, though in view of the markedly successful use of this medium in recent years for the separation of proteins on columns its more widespread use may be anticipated in block electrophoresis in the near future. So far Kuyama and Pramer (1962) have used Sephadex blocks for isolating a protein with nemin activity while Verwoerd *et al* (1961) used Sephadex block electrophoresis for the separation of the reaction products of hydroxylamine with cytidylic acid and also for separating mono- and oligonucleotides.

IX. CHOICE OF MEDIA

One of the first impressions on undertaking electrophoresis is the large number of techniques available for application to any particular problem of separation. This is particularly true in the field of protein separations and purifications, and indeed it is frequently a problem in itself as to which medium should be used for a particular separation. There is no short answer to this problem since in the last resort it depends on the materials being examined but a few simple comments may perhaps be of some value in this context.

In the first instance when compounds of relatively low molecular weight, e.g., amino-acids, peptides, mono- and oligonucleotides, are being considered the choice is relatively simple. Only two types of media are in use at the present time, viz., conventional paper and thin layer. The choice between these two media is dictated largely by the amounts of material available for analysis and whether it is required to recover separate fractions after separation. Thus paper will be used for large amounts of material, especially if recovery of the component is required whereas thin layer methods will be applied for fast routine analysis of small quantities of material. Thin layers, of course, have the advantage that a variety of well defined supporting media are available. There appears to be no reason why any of the media at present used for thin layer chromatography, e.g, Sephadex, DEAE cellulose, etc., should not be used for thin layer electrophoresis.

As far as proteins are concerned the choice is rather more difficult. At present paper is not to be recommended simply because resolution is poor due to the marked adsorption of proteins and consequent tailing of the zones. The choice of medium is dictated to some extent, of course, by the quantities of material to be separated. Thus for large scale separations block electrophoresis provides a simple method capable of good resolution. Within block media the choice is again dictated largely by the protein to be separated, and in the past starch grains have been the most popular medium. However, it is unwise to use these for the isolation of carbohydrate-containing materials due to the risk of contamination with soluble starch products. At present media such as PVC or Sephadex are superior to starch grains. If on the other hand extremely small quantities of proteins are to be separated then polyacrylamide is undoubtedly the medium of choice, since it provides a fast and very efficient separation on a very small scale. Polyacrylamide, moreover, is inert chemically and well defined. With the availability of preparative polyacrylamide gel electrophoresis it can indeed be stated that this medium will fulfil virtually all requirements for protein separations. As far as RNA separations are concerned polyacrylamide electrophoresis has no real competitor at present. The only other medium which approaches polyacrylamide in its resolving power for proteins is starch gel, still very worthy of consideration especially in view of the very extensive literature available on its application in electrophoresis. Starch gel, however, has the usual limitations when used for the separation of proteins containing carbohydrate groups.

Elution from starch gels, in contrast, is probably easier on a small scale than elution from polyacrylamide. Agar gel is probably still the method of choice when carrying out immunoelectrophoretic studies, and is the most widely used medium for this purpose. Immunoelectrophoresis has been carried out infrequently on starch gels or cellulose acetate, and even less frequently on polyacrylamide. The advantage of using agar for immunoelectrophoresis is that a wealth of experience on its use for this purpose is readily available in the literature. Finally for routine examination of latent fractions, cellulose acetate is still an important medium since it provides the advantages of speed, simplicity and a resolving power probably more than sufficient for most routine scanning of sera.

ACKNOWLEDGMENT

This Chapter is based on "Methods of Zone Electrophoresis" by J. R. Sargent, Second Edition, published by The British Drug Houses, 1969. As in the original publication I again wish to record my thanks to the staff of The British Drug Houses for their interest and encouragement in the present article. I am particularly indebted to The British Drug Houses for their kind permission to reproduce many of the figures used in the original publication.

REFERENCES

Afonso, E. (1961). *Clin. Chim. Acta*, **6**, 883.
Allen, J. M. (1961). *Ann. N.Y. Acad. Sci.*, **94**, 937
Allfrey, V., Daly, M. M., and Mirsky, A. E. (1954). *J. Gen. Physiol.*, **37**, 157.
Aronsson, T., and Gronwall, A., (1957). *Scand. J. Clin. Lab. Invest.*, **9**, 338.
Aschaffenburg, R. (1961). *Nature, Lond.*, **192**, 431.
Ashton, G. C. (1957). *Nature, Lond.*, **180**, 917.
Atfield, G. N., and Morris, C. J. O. R. (1961). *Biochem.* J., **60**, 541.
Bagdasarian, M., *et al.* (1964). *Biochem. J.*, **91**, 91.
Block, R. J., Durrum, E. L., and Zweig, G., (1955) "A Manual of Paper Chromatography and Electrophoresis". Academic Press, New York.

Bloemendal, H. *et al.* (1962). *Nature, Lond.*, **193**, 437.

Bloemendal, H. (1963). "Zone Electrophoresis in Blocks and Columns". Elsevier, Amsterdam, p. 87.

Bloemendal, H., and Bosch, L. (1959). *Biochim. biophys. Acta*, **35**, 244.

Bloemendal, H. (1963). "Zone electrophoresis in blocks and columns" Elsevier Publishing Co., Amsterdam, p. 42.

Bocci, V. (1963). *Nature, Lond.*, **197**, 491.

Bodman, J. (1960). "Chromatographic and Electrophoretic Techniques; Vol. 2, Zone Electrophoresis" (Ed. I. Smith) p. 113. Heinemann, London.

Bodman, J. (1957). *Lab. Pract.*, **13**, 517.

Bodman, J. (1960). "Chromatographic and Electrophoretic Techniques, Vol. 2, Zone Electrophoresis". (Ed. I. Smith) p. 137. Heinemann, London.

Bottiger, L. E., and Carlson, L. A. (1960). *Clin. chim. Acta*, **5**, 812.

Brattstein, I., Synge, R. L. M., and Watt, W. B. (1965). *Biochem. J.*, **97**, 678.

Buzzard, Z. (1954). *Compt. rend.*, **239**, 1702.

Campbell, P. N., Greengard, O., and Kernot, B. A. (1960). *Biochem. J.*, **74**, 107.

Caselli, P., and Schumacher, H. (1954). *Z. ges. exp. Med.*, **124**, 65.

Charman, R. C., and Landowne, R. A. (1967). *Analyt. Biochem.* **19**, 177.

Cohen, S. S., and Lichtenstein, J. (1960). *J. Biol. Chem.*, **235**, PC 55.

Consden, R., Gordon, A. H., and Martin, A. J. P. (1946). *Biochem. J.*, **40**, 33.

Consden, R., and Stainier, W. H. (1952). *Nature, Lond.*, **169**, 783.

Crestfield, A. M., and Allen, F. W. (1954). *J. Biol. Chem.* **211**, 363.

Davidson, J. M., and Smellie, R. H. S. (1952). *Biochem. J.*, **52**, 294.

Delacourt, A., and Delacourt, R., (1953), (*C.r. Séanc. Soc. Biol.*, **147**, 1104.

Drevon, B., and Donikan, R. (1955). *Bull. Soc. Chim. biol.*, **37**, 1321.

Duke, E. J. (1963). *Nature, Lond.*, **197**, 288.

Durrum, E. L. (1950). *J. Am. chem. Soc.*, **72**, 2943.

Durrum, E. L. (1951). *J. Colloid Sci.*, **6**, 274.

Dvonch, W., and Alburn, H. E., (1959). *Archs Biochem. Biophys.*, **79**, 146.

Efron, M. L. (1959). *Biochem. J.*, **72**, 691.

Ellis, S. J. (1958). *Biol. Chem.*, **233**, 53.

Elson, D. (1958). *Biochim. biophys. Acta*, **27**, 207.

Elson, D. (1959). *Biochim. biophys. Acta*, **36**, 362.

Evered, D. F. (1959). *Biochim. biophys. Acta*, **36**, 14.

Field, C. W., and Teagre, O. (1937). *J. exp. Med.*, **9**, 222.

Flynn, F. V., and de Mayo, P. (1951). *Lancet*, ii, 235.

Foons-Bech, P., and Li. C. H. (1954). *J. biol. Chem.*, **207**, 175.

Goldsworthy, P. D., and Volwiler, W. (1958). *J. biol. Chem.* **230**, 817.

Gordon, A. H. (1960). *Biochim. biophys. Acta*, **42**, 23.

Gordon, A. H., *et al.*, (1950). *Coll. Czech. Chem. Comm.*, **15**, 1.

Grabar, P., and Williams, C. A. (1953). *Biochim. biophys. Acta*, **10**, 193.

Grimm, F. L., and Doherty, D. G., (1961). *J. biol. Chem.* **236**, 1980.

Gross, D. (1953). *Nature, Lond.*, **172**, 908.

Gross, D. (1954). *Nature, Lond.*, **173**, 487.

Gross, D. (1955). *Nature, Lond.*, **176**, 72.

Gross, D. (1956). *Nature, Lond.*, **178**, 29.

Hanes, C. S., and Isherwood, F. A. (1949). *Nature, Lond.*, **164**, 1107.

Harris, E., and Mehl, J. W. (1955). *Proc. Soc. exp. Biol. Med.*, **90**, 521.

Honegar, C. G., (1961). *Helv. chim. Acta*, **44**, 173.

Hsias, S. H., and Putnam, F. W., (1961). *J. biol. Chem.* **236**, 122.

Hunter, R. L., and Markert, C. L. (1957). *Science*, **125**, 1294.
Jermyn, M. A., and Thomas, R. (1954). *Biochem. J.*, **56**, 631.
Johns, E. W., *et al.*, (1961). *Biochem. J.*, **80**, 189.
Katz, A. M., Dreyer, W. J., and Anfinsen, C. B. (1959). *J. biol. Chem.*, **234**, 2897.
Kodicek, E., and Reddi, K. K. (1951). *Nature, Lond.*, **168**, 475.
Kohn, J. (1960) "Chromatographic and Electrophoretic Techniques" (Ed. I. Smith). William Heinemann Ltd., London, Vols. 2, p. 58.
Kohn, J. (1957). *Clin. chim. Acta*, **2**, 297.
Kohn, J. (1957). *Biochem. J.*, **65**, 9.
Kohn, J. (1957), *Nature, Lond.*, **180**, 986.
Kohn, J. (1958). *Nature, Lond.*, **181**, 839.
Kohn, J. (1958). *Trans. R. Soc. trop. Med. Hyg.*, **52**, 4.
Kunkel, H. G., and Slater, R. J. (1952). *J. clin. Invest.*, **31**, 677.
Kunkel, H. G., and Slater, R. J. (1952). *Proc. Soc. exp. Biol. Med.*, **80**, 42.
Kunkel, H. G., and Trautman, R. J. (1956). *J. clin. Invest.*, **35**, 641.
Kuyama, S., and Pramer, D. (1962). *Biochim. biophys. Acta*, **56**, 631.
Kuzovleva, O. B., and Chung-Yen, W. (1959). *Biokhimiya*, **24**, 550.
Laurell, A. H. F. (1957). "Methods in Enzymology" (Ed. S. P. Colowick and N. O. Kaplan). Academic Press, New York, Vol. 4, p. 21.
Lederer, M. (1955). "Introduction to Paper Electrophoresis and Related Methods". Elsevier, Publ. Co., Amsterdam.
Liener, I. E., and Viswanata, T. W. (1956). *Biochim. biophys. Acta*, **22**, 299.
Lundquist, F., Thorsteinsson, T., and Buus, O. (1955). *Biochem. J.*, **59**, 69.
MacDonald, H. J., and Kissane, J. Q. (1960). *Analyt. Biochem.*, **1**, 178.
Malmstrom, B. (1953). *Arch. Biochem. Biophys.*, **46**, 345.
Markham, R., and Smith, J. D. (1952). *Biochem. J.*, **52**, 552.
Markham, R., and Smith, J. D. (1949). *Biochem. J.*, **45**, 294.
Martin, A. J. P., and Synge, R. L. M. (1945). *Adv. Protein Chemistry*, **2**, 31.
Melamed, M. B. (1967). *Analyt. Biochem.*, **19**, 187.
Michl, H. (1952). *Mh. Chem.*, **3**, 737.
Michl, H. (1951). *Mh. Chem.*, **82**, 489.
Miller, L. L., and Bale, W. F. (1954). *J. exp. Med.*, **99**, 125.
Muller-Eberhard, H. J., and Kunkel, H. G. (1956). *J. exp. Med.*, **104**, 253.
Murray, K. (1962). *Analyt. Biochem.*, **3**, 415.
Naughton, M. A., and Hagopian, H. (1962). *Analyt. Biochem.*, **3**, 276.
Nikkila, E., Ekholm, K., and Solvok, H. (1952). *Acta chem. Scand.*, **6**, 617.
Norris, J. R., (1968). *In* "Systematics Association Special Volume No. 2. : Chemotaxonomy and Serotaxonomy" (Ed. J. G. Hawkes) p. 49, Academic Press Inc., London.
Norris, J. R., and Burges, H. D. (1963). *J. Insect. Pathol.*, **5**, 460.
Norris, J. R. (1964). *J. appl. Bacteriol.*, **27**, p. 439.
Ouchterlony, O. (1958). *Progress in Allergy*, **5**, 1.
Owen, J. A. (1956). *Analyst*, **81**, 28.
Owen, J. A. Silberman, H. J., and Got, C. (1958). *Nature, Lond.*, **182**, 1373.
Paigen, K. (1966). *Analyt. Chem.*, **28**, 284.
Paleus, S. (1952). *Acta chem. Scand.*, **6**, 969.
Pardee, A. B., Paigen, K., and Prestidge, L. S. (1957). *Biochim. biophys. Acta*, **23**, 163.
Pastuska, G., and Trinks, H. (1962). *Chemiker Ztg.*, **86**, 135.
Postel, S. (1956). *Endocrinology*, **58**, 557.
Poulik, M. D. (1957), *Nature, Lond.*, **180**, 1477.

Poulik, M. D. (1960). *Biochim. biophys. Acta*, **44**, 390.

Poulik, M. D., and Smithies, O. (1958). *Biochem. J.*, **68**, 636.

Poulik, M. D. (1959). *J. Immunol.*, **82**, 502.

Raacke, I. D. (1956). *Arch. Biochem. Biophys.*, **62**, 184.

Raacke, I. D., and Li, C. H., (1955). *J. Biol. Chem.*, **215**, 277.

Rhodes, J. M. (1960). *Biochim. biophys. Acta*, **44**, 209.

Ritschard, W. J. (1964). *J. Chromatog.*, **16**, 327.

Rotman, B., and Spiegelman, S. (1954). *J. Bacteriol.*, **68**, 419.

Ryle, A. P., Sanger, F., Smith, L. F., and Kitai, R. (1955). *Biochem. J.*, **60**, 541.

Sargent, J. R. (1969). "Methods in Zone Electrophoresis". A. B. D. H. publication, 2nd edition.

Sargent, J. R., and Vadlamudi, B. V. (1968). *Biochim. J.*, **107**, 839.

Scherr, G. H. (1961). *Trans. N. Y. Acad. Sci.*, **23**, 519.

Schulte, K. E., and Muller, F. (1955). *Milchwissenschaft*, **10**, 90.

Seno, V., and Anno, K. (1961). *Biochim. biophys. Acta*, **49**, 408.

Siliprandi, N., Siliprandi, D., and Lis, H. (1954). *Biochim. biophys. Acta*, **14**, 2121.

Smith, J. D. (1955). "The Nucleic Acids", Academic Press, New York, Vol. 1., p. 267.

Smithies, O. (1955). *Biochem. J.*, **61**, 629.

Smithies, O. (1959). *Adv. Protein Chemistry*, **14**, 65.

Smithies, O., and Poulik, M. D. (1956). *Nature, Lond.*, **177**, 1033.

Spencer, E. W., Ingram, V. M., and Levinthal, C. (1966). *Science*, **152**, 1722.

Stahl, E. (1965). "Thin Layer Chromatography". Springer Verlag, Berlin.

Tiselius, A. (1950). *Naturwissenschaften*, **37**, 25.

Toschi, G. (1954). *Boll. Soc. Biol. sper.*, **30**, 563.

Verwoerd, D. W., Kohlhage, H., and Zillig, W. (1961). *Nature, Lond.*, **192**, 1038.

Wade, H. E., and Morgan, D. M. (1954). *Biochem. J.*, **56**, 41.

Wallenfels, K., and Pechmann, E. (1951). *Angew. Chem.*, **63**, 44.

Watts, D. C., and Donninger, C. (1962). *Analyt. Biochem.*, **3**, 489.

Weiss, J. B. (1960) "Chromatographic and Electrophoretic Techniques", Vol. 2, Zone Electrophoresis" (Ed. I. Smith), Heinemann, London.

Weiss, S. B., Acs. G., and Lipmann, F. (1958). *Proc. Nat. Acad. Sci. Wash.*, **44**, 189.

Wetter, L. R., and Corrigal, J. J. (1954). *Nature, Lond.*, **174**, 695.

Wieland, T., and Pfleiderer, G. (1957). *Angew. Chem.*, **69**, 199.

Wieland, T., and Fischer, E. (1948). *Naturwissenschaften*, **35**, 29.

Wieland, T., and Pfleiderer, G. (1957). *Agnew. Chem.*, **69**, 194.

Wieland, T., Pfeiderer, G., and Rettig, H. L. (1958). *Angew. Chem.*, **70**, 341.

Williamson, M. B., DiLallo, J., and Haley, H. B. (1962). *Naturwissenschaften*, **49**, 471.

CHAPTER VIII

Free-flow Electrophoresis
(A Technique for Continuous Preparative and Analytical Separation)

K. HANNIG

Max-Planck-Institut für Eiweiss-und Lederforschung, Munich

I.	Introduction	513
II.	Principle and Design of the Apparatus	516
	A. General remarks	516
	B. Description of free-flow electrophoresis apparatuses . .	516
III.	Operational Considerations	521
	A. Instrument and procedure details	521
	B. The separation medium	521
IV.	Applications	532
	A. Soluble biological materials	532
	B. Sub-cellular particles	537
	C. Whole cells	540
	D. Phages	543
	E. Bacterial cells	544
References	. . . .	546

I. INTRODUCTION

The development of a large variety of instruments for electrophoretic separation indicates the importance that this technique has gained amongst the modern physico-chemical research tools in the last two decades. It is one of the few techniques that can be applied to even the most delicate materials, as the only force acting on the particles or molecules under investigation is a mild electrostatic one which causes them to migrate. The "classical" electrophoresis of Tiselius and "paper electrophoresis" are of generally acknowledged value as analytical methods and paved the way for the recent developments in preparative electrophoresis. The application to various chemical, biological and medical problems has led to many new and valuable results as evidenced by the comprehensive literature in this field.

For a full treatment of the theoretical foundations of electrophoresis, i.e. the behaviour of ions, colloids and particles in an electric field, the

21–5b

interested reader may refer to the relevant specialist publications (Ambrose, 1965; Bier, 1967). The laws of basic interest here are easily comprehensible. The migration velocity of a material depends on the size of the electric charge and on the frictional resistance in the medium used. It is unimportant whether the molecule itself is charged or whether it obtains its charge from other ions attached to it. While the charge of simple ions is constant, being equal to the charge of an electron or a multiple of this, depending on the valence state, the charge of colloids, particularly in the case of proteins, amino-acids and other amphoteric substances usually varies over a wide range depending on the pH of the milieu. The migration velocity v of a spherical particle in a defined medium is given by

$$v = \frac{e.H}{6.\pi.\eta.r} \tag{1}$$

where e is the charge of the particle, H the field strength (volt.cm^{-1}) and the expression $6\pi\eta r$ the frictional resistance (Stoke's Law; where η is the viscosity of the medium and r the radius of the particle). The mobility u of a particle migrating in an electric field is defined as the quotient:

$$u = \frac{v}{H} \tag{2}$$

Equations 1 and 2 are strictly valid only for infinite dilution and in the absence of salts. If the solution also contains salts, as is always the case in practice, a layer of ions of opposite change is formed around the particle reducing its effective charge and reducing the migration velocity in the electric field applied.

The effect of the electrolyte ions present is expressed as an additive quantity, the "ionic strength" (μ):

$$\mu = \tfrac{1}{2}\Sigma c_i \times z_i^2 \tag{3}$$

where c_i is the concentration of the ith type of ion and z_i its valence. The dependence of the mobility u of a migrating particle on the ionic strength of the solution is given by the approximate equation

$$u_1 : u_2 \approx \mu_2^{1/2} : \mu_1^{1/2} \tag{4}$$

for small changes in ionic strength. Of practical importance is the calculation of the apparent electrophoretic mobility from the experimental data, namely the current strength i, the conductivity of the solution κ, the distance s travelled in time t and the cross-section of the electrophoresis chamber. Substituting field strength

$$H = \frac{i}{q.\kappa}$$

(Ohm's Law) and $v = s/t$ in equation (2) we obtain:

$$u = \frac{s.q.\kappa}{t.i} \tag{5}$$

or

$$v = \frac{s}{t} = \frac{u.i}{q.\kappa} \tag{6}$$

The equations (5) and (6) are quite generally applicable to mobility measurements. They are valid for investigations in a homogeneous medium under defined and constant conditions. It is mainly material of this kind that we shall be discussing in this contribution. For those electrophoretic arrangements that make use of porous carriers, e.g. paper electrophoresis or the various gel electrophoresis procedures which shall not be discussed further here (cf. Cooksey, this Volume p. 573; Sargeant, this Volume p. 455) additional factors alter the theoretically expected migration velocities. Adsorption or filtering effects, for example, may complicate the separation or even make it impossible if the substances have a high affinity for the carrier material or have a very large particle size.

The classical U-tube procedure of Tiselius mentioned above has proved of great value in the field of protein chemistry. It is, however, a purely analytical method and requires considerable technical effort. The application to the separation of particles (cells, bacteria, etc.) did not prove satisfactory even on an analytical scale.

The so-called analytical cell-electrophoresis (Ambrose, 1965; Rueff, 1964), whose theoretical foundation was laid as early as the turn of the century, is a micromethod for measuring the electrophoretic mobility of individual cells or particles. Comparable migration velocities can be obtained under known experimental conditions. Deviations from the empirical norms permit deductions on changes in the relevant surface structure. The method has been of value in bacteriology (James, 1965) and virology (Forrester, 1965; Straub, 1965) and in recent years in clinical diagnosis (Rueff, 1964).

The difficulties encountered in electrophoresis experiments with microparticles lie in the necessity of working in a carrier-free milieu. Only special precautions will attain the required stabilization of the free fluid film, and avoid disturbances due to thermal convection and gravitational settling of the particles. A number of apparatuses for analytical electrophoretic separation that satisfy the mentioned conditions have been reported. Kolin (1960, 1967) describes a separation apparatus in which the flow of the buffer film is stabilized in the annular gap of a hollow cylinder by a combination of electromagnetic and electric fields. Other arrangements use density gradients (Mel, 1964) or the rotation of a cylindrical electrophoresis tube (Hjertén, 1959) to stabilize the medium.

We have developed a method of separation, based on a different principle and also suitable for preparative problems, called "continuous deflection electrophoresis" (Graßmann and Hannig, 1949; Svensson and Brattsten, 1949). In this electrophoretic procedure the mixture to be separated is injected continuously in a fine stream into the medium which is flowing uniformly at right angles to the electric field. Molecules and particles with differing charge are then deflected from the direction of flow at an angle that depends on the rate of flow of the medium and the electrophoretic mobility of the particle. As it is possible to apply this principle to a buffer film that flows vertically (Hannig, 1964), continuous free-flow electrophoresis can be applied both for analysis and preparation to a whole spectrum of materials from low molecular substances to coarse particles. The method thus permits not only the separation of cells and subcellular particles or organelles, but also the separation and preparative isolation of their enzyme systems without loss of activity.

II. PRINCIPLE AND DESIGN OF THE APPARATUS

A. General remarks

Figure 1 illustrates the principle of free-flow electrophoresis. An electrolyte solution of suitable pH flows across the lines of force of an electric field. The sample mixture to be separated is injected continuously into the streaming medium at a defined point. Components with differing electrophoretic mobility move along different paths and can be collected continuously at different points at the end of the separation chamber. As the streaming velocity of the medium and the field strength can be varied experimentally over a wide range, it is in principle possible to adjust the deflection at will. The deflection angle, designated α in Fig. 1, is obtained from the relation:

$$\tan \alpha = \frac{\text{velocity of electrophoretic migration } v}{\text{streaming velocity of the buffer solution } w}$$

or substituting Equation (6)

$$\tan \alpha = \frac{u.i}{q.\kappa.w} \tag{7}$$

B. Description of free-flow electrophoresis apparatus

1. *Designs for preparative separation*

In the first model of the continuous free flow electrophoresis apparatus the two glass plates forming the separation chamber (50×50 cm) were arranged horizontally or slightly tilted.

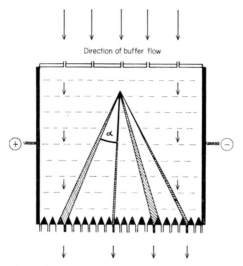

Direction of buffer flow

FIG. 1. *Principle of continuous deflection electrophoresis.*

$$\tan \alpha = \frac{u \cdot i}{q \cdot \kappa \cdot w}$$

u = electrophoretic mobility
i = current strength
q = cross-section of the electrophoresis chamber
κ = specific conductivity of the buffer
w = streaming velocity of the buffer solution.

With this arrangement disturbing thermoconvection within the moving buffer solution is avoided and the separation of low and high molecular weight substances is possible. The separation of particles like cells and cell organelles was not successful, since sedimentation of the particles on the lower glass plate during an experiment made separation impossible.

This problem is solved in the case of the vertical arrangement of the separation chamber as the particles sediment in the direction of the fluid flow. Here, however, the requirement for an uniformly cooled separation chamber is far stronger, as disturbances due to thermal convection can arise very easily. Temperature differences of only a few tenths of a degree give rise to uncontrollable variations in the flow pattern making a separation quite impossible. Suitable constant cooling can be obtained by an electronically regulated arrangement using "Peltier cooling".

Due to the broader spectrum of application possibilities the models with vertically arranged separation chamber are commonly used.

Figure 2 and Fig. 3 show two commercially available instruments. The separation chamber consists of two glass plates held apart 0·5–1·0 mm. The

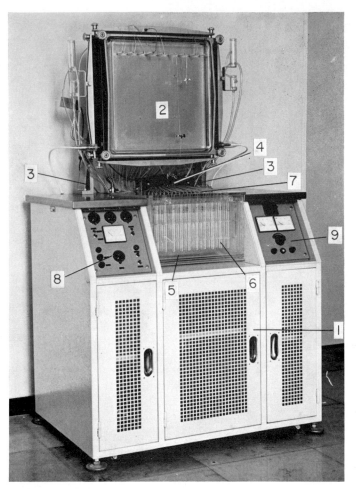

FIG. 2. Free-flow electrophoresis apparatus with vertical separation chamber (Elphor, VaP II)* (Hanning, 1964). 1, Pumps; 2, separation chamber; 3, dosing pump; 4; buffer stock; 5, low-temperature cabinet; 6, rack for test tubes; 7, container for sample mixtures; 8, power supply for Peltier batteries; 9, high-voltage current stabilized power supply.

buffer solution flows between them as an uniform film. Buffer solution also flows continuously through the electrode chambers on either side. They are separated from the separation chamber by membranes which do not break the electrical contact.

In order to achieve reproducible results and sharp separations, it is of

* Bender & Hobein GmbH, 8 Munich, Germany.

FIG. 3. Free-flow electrophoresis apparatus (FF 4).* (Hannig, 1968). 1, separation chamber; 2, framework; 3, instrument panel for Peltier batteries; 4, instrument panel for high voltage; 5, 90-channel peristaltic pump; 6, cooled fraction chamber; 7, dosing pump.

great importance to maintain a constant and laminar fluid flow. This is achieved by three provisions:

(1) The buffer solution is symmetrically fed to the upper rim of the separation chamber with a selectable constant speed through multiple inlets.

(2) The collection of the fractions over the whole breadth of the separation chamber is achieved with the required accuracy by means of a 50- or 90-channel peristaltic pump.

* Desaga GmbH, 69 Heidelberg; Bender u. Hobein GmbH, 8 Munich, Germany; Brinkmann Instruments Inc., Westbury, N.Y., U.S.A.

(3) In order to remove the "Joule Effect" uniformly, the rear glass plate of the electrophoresis chamber is in good thermal contact (contact grease) with a copper plate, cooled by Peltier-batteries attached to the other side. The current supply of these batteries is regulated thermostatically so that a constant temperature is secured over the whole separation area.

The sample mixture is injected continuously into the separation chamber with the aid of a microperistaltic pump at a convenient point. The test tubes containing the collected separated fractions are standing in a cooled chamber.

The model shown in Fig. 2 has an effective separation area of 45 cm height and 50 cm breadth. Alternatively 50 or 90 fractions can be obtained. Field strengths of up to 50 V/cm (a voltage difference of 2800 V) can be applied. The required temperature in the separation chamber can be preset between 2° and 20°C. The cooling arrangement is able to maintain the required temperature for the duration of the experiment against a power dissipation as high as 400 watts. The rate of flow of the buffer film can be kept constant at any value over a continuous range as required. The time required for the buffer front to traverse the chamber runs between 20 and 120 min.

In order to attain still shorter separation times for very delicate cell materials we have developed a separation chamber which is 50 cm long but only 12 cm wide and has 90 outlets (Fig. 3). Using a plate spacing of 1 mm and field strengths of up to 80 V/cm, we are able to reduce the separation time to 1–10 min and obtain a separation as good as that in the apparatus previously described.

The cooling system is able to maintain the temperature close to the freezing point even when 300 watts are applied.

Basically in both types of the continuous free-flow electrophoresis apparatus equipped with vertical separation chambers the separation of dissolved substances as well as of particle suspensions is possible. Which of the two models is preferable depends on the purpose. The model with the broader separation chamber (VAP II) is preferable for the preparative separation of complex low and high molecular weight substances, e.g. proteins and protein derivatives, nucleic aids, nucleotides and other soluble carbohydrate- or nucleic acid-containing biological materials.

Depending on the substance, 20–500 mg (dry weight) of material (0·1–5% sol.) can be separated per hour.

Using comparable separation selectivity, the amount of separable material per hour in the small separation chamber (FF 4) is about 25% less than in the broader chamber. Because of the shorter time used for the separation process, the small chamber is preferable when sensitive materials, e.g. enzymes or cells, are used. The separation capacity is up to

200×10^6 cell/hr, when the suspension should not contain more than 15–30×10^6 cells/ml.

2. Continuous electrophoresis as a rapid analytical tool

It is an obvious step to utilize the advantages of free-flow electrophoresis for the solution of analytical problems. The principle permits a rapid fractionation of complex mixtures and the introduction of different samples in rapid succession without interrupting the continuous performance of the apparatus. This means that the conditions for automation of routine analytical runs are perfectly fulfilled. The separation chamber can be made almost arbitrarily small, the separation performed on a microscale and the results analysed quantitatively by optical or by other physical procedures.

Strickler (1967) has put forward a suggestion to this end. This author separated various bacterial strains and pigments and evaluated the results photographically. A new interesting method to calculate quantitatively the resulting separation (e.g. in the model with the small separation chamber, Fig. 3) is briefly described (Hannig and Wirth, 1968).

A light-beam illuminates a narrow band across the width of the chamber. The transmitted image is enlarged (or reduced as required) and cast onto the screen of a so-called Vidikon (television eye). This scans the picture in lines, the varying light intensity being transformed into successive current pulses. The logarithms of the pulses are formed and integrated by a computer. They can then be observed for example as an extinction distribution curve on an oscillograph.

In this way analytical data on the composition of protein mixtures can be obtained in quick succession at intervals of e.g. 5 min if the evaluation is carried out using UV light of 280 nm. The use of a Vidikon sensitive to the ultraviolet allows bands of proteins or other UV absorbers to be made directly visible by the television process.

III. OPERATIONAL CONSIDERATIONS

A. Instrument and procedure details

An essential prerequisite for attaining well-resolved separations is the absolute constancy of the experimental conditions. The conductivity and pH within the separating chamber are kept constant by means of the following precautions. The electrolysis products produced at the platinum electrodes are completely removed by the rapid flow of buffer solution which circulates through the electrode chambers. The electrode buffer has a higher concentration than the separation buffer. The chambers are separated by membranes which stop the migration of strongly deflected fractions into the electrode chambers. Ion exchange membranes (cation exchange membrane, Nepton CR 61*, at the anode and anion exchange membrane, Nepton AR III at the cathode) are used as they have low resistance and there is only a slight voltage drop between electrode and separation chamber. A value of $\mu \times 0.03–0.05$ has proved to be a suitable ionic strength for the separation buffer.

A homogeneous electric field is essential for the linear flow of the separated components through the chamber and measurements have been made to ensure that this is so (Hannig, 1964). The temperature is very constant in the direction of the lines of force of the electric field, thanks to the electronically controlled cooling. The maximum deviation over the whole separation surface is $0.2°C$ thus keeping thermoconvection sufficiently small.

The sharpness of the separation can be defined as the ratio of the band width to the horizontal migration. While the latter depends on the streaming velocity of the transporting medium and on the field strength, the band width has three main causes (1) the diameter of the sample stream at the injection point; (2) thermal diffusion and (3) the horizontal and vertical velocity profiles. Thermal diffusion is negligible for particles of the size investigated. The diameter of the sample stream depends on the geometry of the sample inlet and on the dose rate, and the velocity profiles depend on the chamber surfaces and to a small extent on the field strength, the streaming velocity and the depth of the chamber. Strickler (1967) arrives at the same conclusions in his theory of a continuous electrophoresis chamber.

A further factor which can affect the sharpness of the separation i.e. the constancy of the migration velocity, is the way in which the electric field is stabilized. If all other parameters are kept constant, namely buffer composition and conductivity, pH and temperature of the buffer solution, the migration velocity of a particular charge carrier (a particular component) will (according to Equation 6) depend only on the magnitude of the current flux. It is in principle irrelevant whether the current or the voltage is stabilized. In practice, however, it has proved advantageous to stabilize the current. It is easily seen that small variations in temperature or conductivity will tend to be compensated (the migration velocity being directly proportional to the current flux). This would not be possible if the voltage were stabilized.

An influence on the buffer flow due to the difference in potential between the surfaces of the glass plates and the surface of the adjacent buffer film cannot be excluded. The resulting so-called electro-osmotic fluid flow can, however, be practically neglected in free-flow electrophoresis because of the very small total surface compared with arrangements using a porous carrier.

* Ionics Incorporated, Cambridge, Mass., U.S.A.

B. The separation medium

1. *Fractionation of dissolved materials*

The calculation of ionic strengths of complicated buffer systems is very laborious. It usually suffices to determine the conductivity of the buffer system suitable for the separation. The range of specific conductivity suitable for the separation of high molecular substances is $\kappa = 0\cdot001 - 0\cdot003 \ \Omega^{-1} \ cm^{-1}$. A lower conductivity causes a loss of sharpness in the separation. In general the ionic strength of the buffer solution should, however, be greater than that of the sample ions. The media described above usually satisfy this condition for the fractionation of e.g. a 1–5% protein or nucleic acid solution.

A higher conductivity of the buffer solution is unfavourable not only due to an increased production of unwanted Joule heat but also as the absolute and relative electrophoretic mobility of the components to be separated drops (cf. equations (6) and (7)). A further important precondition for good separation is that the buffer capacity of the separation milieu, should be as high as the dilution will permit. Buffer salts with low electrophoretic mobility should be used whenever possible. For the same reason the buffer mixture should contain few ballast ions (non-buffering salts).

In enzyme separation additives that may influence the enzyme activity unfavourably must, of course, be avoided. Activating heavy metal ions can, on the other hand, often be added with advantage.

In principle, the runs can be performed at any pH as long as the sample mixture is not sensitive to large differences in pH. For amphoteric substances a higher separation resolution is attained if the pH chosen lies outside their iso-electric range. For proteins, for example, a pH of 8–9 is favourable.

The sample mixture should be dissolved in the same separation buffer. The addition of large amounts of other salts only impairs the separation. If necessary the sample may have to be dialysed against the buffer.

The buffer that flows past the electrodes should have the same buffer salt composition as the separation buffer. In order to minimize changes in pH due to the electrolysis products this continuously circulating buffer should be at least twice or, better, many times as concentrated as the separation buffer and should be renewed from time to time in experiments that run over a number of days.

Table I summarizes buffer combinations that have been found suitable for various separation problems.

2. *Fractionation of particles (cells, sub-cellular particles, bacterial cells).*

The same buffer systems may in general be used for the electrophoretic separation of cells and their substructures as were used for soluble sub-

TABLE I

Buffer systems for the separation of soluble mixtures by free-flow electrophoresis.

pH	No.	Buffers	Examples
		I = Separation chamber II = Electrode vessels	
2–3	1	I = 0·187 M acetic acid, 0·187 M formic acid pH 2·0 (10·7 ml acetic acid + 7·1 ml formic acid (99%)→1 l) κ = 0·002 $\Omega^{-1}.cm^{-1}$ II = (32·1 ml acetic acid + 21·3 ml formic acid (99%)→1 l)	Nucleosides and nucleotides (Matthaei et al, 1966)
	2	I = 0·5 M acetic acid pH 2·4 (28·6 ml acetic acid→1 l) κ = 0·00107 $\Omega^{-1}.cm^{-1}$ II = 1 M acetic acid (57·2 ml acetic acid→1 l)	Mucopolysaccharides and proteinpolysaccharides (Bräumer and Kühn, 1965; Mashburn and Hoffmann, 1966)
	3	I, II = 1 M acetic acid pH 2·2 (57·2 ml acetic acid 1 l) κ = 0·00137 $\Omega^{-1}.cm^{-1}$	Prolamine (Waldschmidt-Leitz and Kling, 1966)
	4	I = Phenol-acetic acid-water (1 : 1 : 1; w/v/v) pH3, 0 κ = 0·000153 $\Omega^{-1}.cm^{-1}$ II = I or acetic acid-water (1 : 1; v/v)	Fractionation of plant material (Brattsten et al., 1964, 1965)
3–5	5	I = Pyridine-acetic acid-water pH 3· 9 (2·58 ml pyridine + 8·58 ml acetic acid→1 l) κ = 0·00216 $\Omega^{-1}.cm^{-1}$ II = (7·73 ml pyridine + 25·75 ml acetic acid→1 l)	Digests of proteins (Guest and Yanofsky, 1965; Hannig, 1961)
	6	I = Pyridine-acetic acid-water pH 4·9 (3·7 ml pyridine + 2·96 ml acetic acid→1 l) κ = 0·0022 $\Omega^{-1}.cm^{-1}$ II = (11·1 ml pyridine + 8·9 ml acetic acid→1 l)	Peptides (Hannig, 1961) Extracts from thymus (Hannig and Comsa, 1963)
	7	I = 0·05 M sodium acetate-acetic acid pH 4·5 (4·53 g sodium acetate + 1·13 ml acetic acid→1 l) κ = 0·0022 $\Omega^{-1}.cm^{-1}$ II = (13·6 g sodium acetate + 3·4 ml acetic acid→1 l)	Peptides (Hannig, 1964) Protein-polysaccharides (Mashburn and Hoffman, 1966)

TABLE I—*continued*

pH	No.	Buffers	Examples
3–5	8a	I = 0·15 M ammonium acetate-acetic acid pH 3·5 (a) 11·56 g ammonium acetate→1 l (b) 8·72 ml acetic acid→1 l (adjust solution (a) with solution (b) to pH 3·5) κ = 0·0011 Ω^{-1}.cm^{-1} II = 0·45 M ammonium acetate-acetic acid pH 3·5 (a) 34·68 g ammonium acetate→1 l (b) 26·16 ml acetic acid→1 l (adjust solution (a) with solution (b) to pH 3·5)	Nucleotides and their derivatives (Sulkowski and Laskowski, 1967)
	8b	I = 0·05 M ammonium acetate-acetic acid pH 5·0 (a) 3·85 g ammonium acetate→1 l (b) 2·90 ml acetic acid→1 l (adjust solution (a) with solution (b) to pH 5·0) κ = 0·0021 Ω^{-1}.cm^{-1} II = 0·15 M ammonium acetate-acetic acid pH 5·0 (a) 11·55 g ammonium acetate→1 l (adjust solution (a) with solution (b) to pH 5·0)	
5–6	9	I = 0·025 M ammonium acetate-acetic acid pH 5·1 (a) 1·925 g ammonium acetate→1 l (b) 1·45 ml acetic acid→1 l (adjust solution (a) with solution (b) to pH 5·1) κ = 0·0019 Ω^{-1}.cm^{-1} II = 0·075 M ammonium acetate-acetic acid pH 5·1 (a) 5·775 g ammonium acetate→1 l (b) 4·53 ml acetic acid→1 l (adjust solution (a) with solution (b) to pH 5·1)	Enzymes (cerebrosidsulfatase) (Mehl and Jatzkewitz, 1964)

TABLE I—*continued*

pH	No.	Buffers	Examples
5–6	10	I = Trimethylacetic acid-NaOH-6 M urea pH 5·25 (1 Vol II + 1 Vol 6 M urea) κ = 0·0008 $\Omega^{-1}.cm^{-1}$ II = 1320 ml water + 34 ml tri-methylacetic acid + 2746 ml 8 M urea, adjust with 6 N NaOH to pH 5·25	Separation of α-chains of collagen (Francois and Glimcher, 1967)
6–8	11	I = 0·01 M phosphate pH 7·3 (a) 1·36 g $KH_2PO_4 \rightarrow$ 1 l (b) 1·78 g $Na_2HPO_4.2H_2O \rightarrow$ 1 l 43·5 Vol (a) + 56·5 Vol (b) κ = 0·00155 $\Omega^{-1}.cm^{-1}$ II = 0·03 M phosphate pH 7·3 (a) 4·08 g $KH_2PO_4 \rightarrow$ 1 l (b) 5·34 g $Na_2HPO_4.2H_2O \rightarrow$ 1 l 43·5 Vol (a) + 56·5 Vol (b)	Growth Factor (Sayre *et al.*, 1967)
	12	I = 0·04 M tris-hydrochloric acid pH 7·3 (adjust 4·85 g tris with 1 N HCl to pH 7·3 → 1 l) κ = 0·0019 $\Omega^{-1}.cm^{-1}$ II = 0·12 M tris-hydrochloric acid pH 7·3 (adjust 14·55 tris with 1 N HCl to pH 7·3 → 1 l)	Enzymes (pyrophosphatase) (Schliselfeld, 1965)
	13	I = Tris-calcium acetate pH 7·6 (1·93 g tris + 1·52 g calcium-acetate → 1 l) κ = 0·0019 $\Omega^{-1}.cm^{-1}$ II = (4·54 g tris + 3·56 g calcium acetate → 1 l)	Enzymes (elastase) (Fujii, 1961)
8–11	14	I = 0·08 M tris-citric acid pH 8·6 (adjust 9·7 g tris with 2 M citric acid to pH 8·6 → 1 l) κ = 0·0016 $\Omega^{-1}.cm^{-1}$ II = 0·24 M tris-citric acid pH 8·6 (adjust 29·1 g tris with 2 M citric acid to pH 8·6 → 1 l)	Serum proteins (Hannig, 1961; Seitz and Eberhagen, 1966; Frölich, 1967)

TABLE I—*continued*

pH	No.	Buffers	Examples
8–11	15	I = 0·08 M tris-EDTA pH 8·7 (adjust 9·7 g tris with 0·12% EDTA to pH 8·7→1 l) κ = 0·0015 Ω^{-1}.cm^{-1} II = (adjust 29·1 g tris with 0·36% EDTA to pH 8·7→1 l)	Brain protein (Rubin and Stenzel, 1965)
	16	I = 0 05 M tris-hydrochloric acid-calcium acetate pH 9·0 adjust 6·06 g tris + 1·58 g calcium acetate with 1 N hydrochloric acid to pH 9·0→1 l κ = 0·0022 II = Add 1 M sodium acetate to solution I until κ = 0·0066 Ω^{-1}.cm^{-1}	Enzymes (collagenase) (Graßmann *et al.*, 1963)
	17	Veronal-acetate pH 8·6 I = (3·27 g veronal (sodium salt) + 2·15 g sodium acetate + 20 ml N hydrochloric acid→1 l) κ = 0·023 Ω^{-1}.cm^{-1} II = (9·8 g veronal (sodium salt) + 6·45 g sodium acetate + 60 ml N hydrochloric acid→1 l)	Proteins from kidney beans (Jaffé and Hannig, 1965)
	18	6 M urea-tris-glycin pH 8·6 I = 7·4 mM tris, 0·575 M glycine II = three times more conc. than I	Myeloperoxydase VIII (Schultz *et al.*, 1967)
	19	I = 0·08 M borate-glycine pH 8·49 (a) 2·47 g H$_3$BO$_3$ + 3, 81 g Na$_2$B$_4$O$_7$.10H$_2$O→1 l (b) 6 g glycine→1 l adjust solution (a) with solution (b) to pH 8·49 κ = 0·0015 Ω^{-1}.cm^{-1} II = (a) 7·4 g H$_3$BO$_3$ + 114·4 g Na$_2$B$_4$O$_7$.10H$_2$O→1 l (b) 18 g glycine→1 l adjust solution (a) with solution (b) to pH 8·49	Serum proteins (Factor VIII) (Bidwell *et al.*, 1966)

TABLE I—*continued*

pH	No.	Buffers	Examples
8–11	20	I = 0·15 M tris, 4·0 M urea adjusted with citric acid to pH 8·5 κ = 0·00316^{-1}.cm^{-1} II = 0·45 M tris adjusted with citric acid to pH 8·5	RNA-protein complex (Schweiger and Hannig, 1968)
	21	I = 0·04 M tris, 4·0 M urea adjusted with hydrochloric acid to pH 8·5 κ = 0·00065^{-1}.cm^{-1} II = 0·45 M tris, adjusted with hydrochloric acid to pH 8·5	Acid proteins (Schweiger and Hannig, 1970)
	22	I = 0·005 M EDTA (Na-Salt), adjusted with NaOH to pH 7·6 or 8·5 κ = 0·002^{-1}.cm^{-1} II = 0·1 M EDTA (Na-Salt) adjusted with NaOH to pH 7·6 or 8·5	
	23	I = 0·03 M tris, adjusted with Hydrochloric acid to pH 8·5 κ = 0·00066Ω^{-1}.cm^{-1} II = 0·45 M tris adjusted with hydrochloric acid to pH 8·5	Proteins from cell-nuclei not associated with RNA (Schweiger and Hannig, 1970)
	24	I = 0·017 M glycine-sodium chloride-NaOH pH 10·5 (adjust 1·27 g glycine with 0·1 N NaOH to pH 10·5→1 l; adjust with M sodium chloride to a proper conductivity) κ = 0·0015Ω^{-1}.cm^{-1} II = (adjust 3·8 g glycine with 0·5 N NaOH to pH 10·5→1 l; adjust with 2 M sodium chloride to a proper conductivity) κ = 0·005 Ω^{-1}.cm^{-1}	Tobacco mosaic virus protein (Sarkar, 1966)

stances. The pH must, of course, be kept within the limits physiologically compatible with the particle. In addition the buffer must constitute an iso-osmotic milieu. As the correct osmolarity cannot be attained with buffer salts alone, since the conductivity would be too high, this must be effected by adding substances which do not themselves contribute to the conductivity, i.e. which do not dissociate. Cane sugar, glucose, ribose, etc., are usually used. It is sometimes of value to add nutrients that promote the metabolic processes in order to maintain the biological activity of the particles.

The electrode buffer, however, contains no additives, being composed only of buffer salts in the same relation as in the separation buffer but at higher concentration. It is expedient to first set up isotonic conditions with the buffer salts. The separation buffer can then be prepared by diluting this electrode buffer with isotonic sugar solutions in a suitable ratio. For separation of cells, care should be taken that the specific density of the cell suspension that is introduced into the separation chamber is never lower than that of the separation buffer. If this is not the case the injected cell suspension will rise in the separation chamber making a separation impossible.

Table II shows some buffer systems suitable for particle separations. The publications and descriptions of experimental results to date show clearly however, that many preliminary experiments are necessary to find a suitable milieu. Recipes of general validity cannot be given, as the most favourable conditions always have to be adjusted to the individual behaviour of the various species investigated.

TABLE II

Buffer systems for separation of particles

pH	No.	Buffers	Examples
		I = Separation chamber; II = Electrode vessels	
5–6	1	I = 0·025 M sodium acetate-acetic acid pH 5·3 1 Vol of solution II + 2 Vol water $\kappa = 0·0015 \ \Omega^{-1}.cm^{-1}$ II = (a) 0·75 M sodium acetate . $3H_2O$ = 10·2 g/l (b) 0·75 N acetic acid = 12·9 ml/l adjust solution (a) with solution (b) to pH 5·3	Phages (Braunitzer *et al.*, 1964)
6–7	2	I = 0·003 M phosphate pH 6·1 1 Vol of solution II + 2 Vol water $\kappa = 0·0021 \ \Omega^{-1}.cm^{-1}$ II = (a) 1·229 g KH_2PO_4/l (b) 1·60 g $Na_2HPO_4 . 2H_2O$/l 845 ml (a) + 155 ml (b)	Bacteria and pigments

TABLE II—*continued*

pH	No.	Buffers	Examples
7–8	3	I = 0·01 M tris-0·01 M magnesium acetate-citric acid pH 7·2 (adjust 1·21 g tris + 2·14 g magnesium-acetate.4H$_2$O with 0·5 M citric acid to pH 7·2→1 l) κ = 0·0008 Ω^{-1}.cm^{-1} II = 0·05 M tris-0·05 M magnesium acetate (adjust 6·05 g tris + 6·6 g magnesium acetate with 0·5 M citric acid to pH 7·2→1 l)	Ribosomes from *Escherichia coli* (Matthaei *et al.*, 1966)
	4	I = 0·04 M tris-0·003 M magnesium acetate-acetic acid pH 8·0 adjust 4·84 g tris + 0·64 g magnesium-acetate.4H$_2$O + 0·005 g polyvinyl-sulfate with 0·5 N acetic acid to pH 8·0→1 l κ = 0·00175 Ω^{-1}.cm^{-1} II = adjust 24·2 g tris with 2 N acetic acid to pH 8·0→1 l	Liver ribosomes (Schweiger and Hannig, 1967)
	5	I = 0·04 M tris-citric acid-0·001 M Na$_2$EDTA-0·33 M sucrose pH 7·4 adjust 4·85 g tris + 112·8 g sucrose + 0·372 g Na$_2$ EDTA with 1 M citric acid to pH 7·4→1 l κ = 0·0014 Ω^{-1}.cm^{-1} II = 0·15 M tris-citric acid adjust 17·17 g tris with 2 M citric acid to pH 7·4→1 l	Organelles from liver cells (Hannig *et al.*, 1969; Stahn, 1969)
	6	I = 10 mM triethanolamine 10 mM acetic acid, 1 mM EDTA, 0·33 M sucrose adjust with 2 N NaOH to pH 7·4 II = 100 mM triethanolamine 100 mM acetic acid adjust with 2 N NaOH to pH 7·4	Lysosomes, mitochondria and mitochondrial membranes (Stahn, 1969; Heidrich *et al.* 1969; Stahn *et al.*, 1969)
	7	I = 26·5 g Na$_2$CO$_3$; 26·35 g 2-amino-2 methyl-1,3 propandiol; 39 g NaH$_2$PO$_4$; 52·5 g citric acid; distilled water to 40 litres II = I distilled water to 4 litres, pH 7·0; adjusted with NaOH or HCl	Spores (Norris, 1969)

TABLE II—*continued*

pH	No.	Buffers	Examples
7–8	8	I = 0·015 M triethanolamine; 0·01 M glucose; 0·004 M potassium acetate; 0·24 M glycine, adjust with acetic acid to pH 7·4 II = 0·075 M triethanolamine; 0·004 M potassium acetate, adjust with acetic acid to pH 7·4	Lymph node cells (Hannig and Zeiller, 1969)
	9	I = 0·024 M tris-citric acid-sucrose-EDTA pH 7·2 adjust 2·92 g tris + 83·3 g sucrose + 0·93 g Na₂EDTA with 0·3 M citric acid to pH 7·2→1 l. Instead of sucrose one can use 44 g glucose κ = 0·0013 $\Omega^{-1}.cm^{-1}$ II = adjust 17·5 g tris with 2 M citric acid to pH 7·2→1 l	Cells (blood cells) (Ganser *et al.*, 1968; Hannig and Krüsmann, 1968)
	10	I = 0·0114 M phosphate-sucrose EDTA pH 7·2 (a) 9·078 g KH₂PO₄→1 l (b) 11·876 g Na₂HPO₄→1 l 12 ml (a) + 131 ml (b) + 0·93 g Na₂EDTA + 857 ml 10% sucrose. Additional 40 ml 5% glucose are recommended. κ = 00·015 $\Omega^{-1}.cm^{-1}$ II = (1680 ml (a) + 4320 ml (b)	
	11	I = 0·0223 M tris-0·32 M Glycine-EDTA pH 7·2 adjust 24 g glycine + 0·93 g Na₂ EDTA + 2·7 g tris with 0·3 M citric acid to pH 7·2→1 l κ = 0.0013 $\Omega^{-1}.cm^{-1}$ II = Adjust 27·0 g tris + 9·3 g Na₂EDTA with 2 M citric acid to pH 7·3→1 l	
	12	I = 0·033 M tris-citric acid pH 7·3 adjust 4 g tris with 0·5 M citric acid to pH 7·3→1 l κ = 0·00169 $\Omega^{-1}.cm^{-1}$ II = Adjust 20 g tris with 2 M citric acid to pH 7·3→1 l	Bacteria (Klofat, 1968)

IV. APPLICATIONS

In this section directions will be given and experimental procedures briefly described for the solution of various separation problems with the aid of free-flow electrophoresis. Applications found in the literature will also be mentioned and the results discussed.

Although little experience has been gained in the application of free-flow electrophoresis in the field of microbiology, the experimental conditions found suitable for other biological materials can serve as a basis for the solution of specific separation problems.

A. Soluble biological materials

1. *Proteins and proteides*

The simplest example of a protein separation is the fractionation of serum. The isolation of homogeneous serum components with the help of free-flow electrophoresis has been variously reported (Hannig, 1961, 1964; Wieck *et al.*, 1964; Bidwell *et al.*, 1966; Seitz and Eberhagen, 1966; Fröhlich, 1967). The serum sample is usually diluted with 2 or 3 times its volume of separation buffer and injected at a rate of 2–3 ml/hr into the separation chamber (50–100 mg protein/hr).

Figures 4 and 5 show how it is possible to isolate or at least enrich serum proteins of similar electrophoretic mobility even if present in small quantities. Figure 4 shows a normal distribution curve of serum proteins. The most suitable buffer is tris-citrate buffer of pH 8·6 (cf. Buffer No. 14, Table 1). The resolution of the α_1-globulin component is at first not very

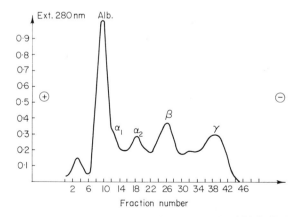

FIG. 4. Separation of serum proteins (from Hannig, 1964) VaP II; tris buffer pH 8·6 (No. 14, Tab. 1) 2300 V; 160 mA; 40 mg serum protein/hr; buffer flow 45 cm/hr; temp +10°C.

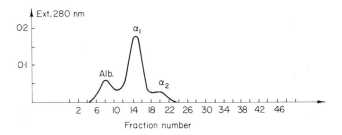

FIG. 5. Isolation of the α_1-fraction. A second fractionation of the fractions 12 and 13 of the experiment illustrated in Fig. 4 (experimental conditions as given in Fig. 4).

successful by free-flow electrophoresis as the adsorption effect, that favours the isolation of this component from albumen in paper electrophoresis, is absent here. A second run was made under the same conditions with fractions 12 and 13 of the preliminary separation, i.e. with the α_1 region. The result, illustrated in Fig. 5, is that the α_1 component has been resolved quite clearly from the accompanying small amounts of albumin and α_2-globulin.

It is not necessary to describe here more examples of separation and isolation of proteins of different origin.

Table I gives references to literature and relating methods applied.

In conclusion it can be said that for the separation of proteins, weak alkaline buffers show the best results. In order to isolate a specific component it may sometimes be necessary to modify the conditions after a first electrophoretic separation and subject the material to a renewed electrophoresis.

2. *Nucleotides, nucleic acids and substances containing nucleic acids.*

The preparative separation of nucleotides and their derivatives by free-flow electrophoresis is described by Sulkowski and Laskowski (1967). Figure 6 shows the separation of a mixture of deoxy-mononucleotides, Fig. 7 the separation of adenosine mono-, di- and triphosphate (AMP, ADP, ATP) using ammonium acetate buffer (Nos. 8a and 8b, Table I). The authors obtained equally good results for the separation of digestion products of thymus DNA (mono- and di-deoxyribonucleotides) digested with micrococcal nuclease (Fig. 8). These and other nucleotides fractionated were of high purity and were suitable for subsequent spetrophotometric and chromatographic investigations. In this way the authors were able to identify and determine quantitatively the terminal groups in short oligonucleotides. Similar results for the separation of nucleotides and nucleosides are reported by Matthaei *et al.* (1966).

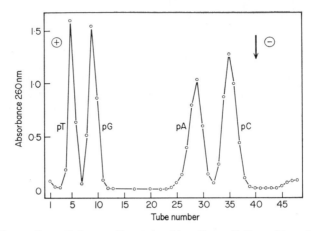

Fig. 6. Separation of desoxyribonucleotides (from Sulkowski and Laskowski, 1967) VaP II; buffer ammonium acetate, pH 3·5 (No. 8a, Tab. 1); 2500 V; 100 mA; buffer flow rate 60 ml/hr. Sample flow rate 0·5 ml/hr (concn. 20 A_{260} units/ml).

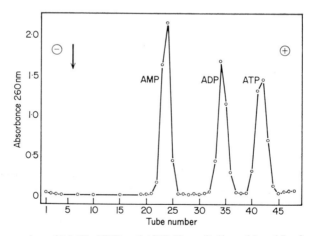

Fig. 7. Separation of AMP, ADP and ATP (from Sulkowski and Laskowski, 1967). VaP I; buffer, ammonium acetate, pH 5·0 (No. 81, Tab. 1) 1400 V; 155 mA; buffer flow rate 50 ml/hr. Sample flow rate 0·5 ml/hr (concn. 35 A_{260} units/ml).

Sarkar (1966) describes a complete preparative separation of RNA and the preparation of pure native tobacco mosaic virus proteins. The virus sample (5–50 mg/ml) is dialysed at 0–4°C for 24 hrs against a 0·05–0·01 M glycine-NaCl-NaOH buffer of pH 10·3–10·6 and subsequently centrifuged at 100,000 g. The profile of separation of A-Protein from RNA in this buffer (No. 24, Table I) is shown as an example in Fig. 9. 20–100 mg of the

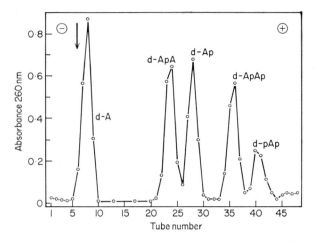

Fig. 8. Separation of d-A, d-Ap, d-ApA, d-ApAp, and d-pAp (from Sulkowski and Laskowski, 1967). VaP II; buffer, ammonium acetate, pH 5·0; (No. 80, Tab. 1) 1600 V; 180 mA; buffer flow rate 50 ml/hr. Sample flow rate 1·5 ml/hr (concn. 15 A_{260} units/ml).

virus sample were introduced into the separation chamber per hr. The protein yield was 90% and it was shown to be in the native state by virtue of its ability to aggregate at pH 5.

Another method of separating RNA and basic protein components is to add large quantities of urea (about 4 M) to the separation buffer. In this way vestiges of RNA can be separated from samples of ribosomal proteins from *E. coli* (Spitzauer *et al.*, unpublished). This procedure was also adopted to identify the protein component with sedimentation rate of 30 S in a mRNA protein complex from the nuclei of rat liver cells (Schweiger and Hannig, 1968) (No. 20, Table I).

Free-flow electrophoresis has also been applied in other cases for the isolation of the constituents of animal tissue. Hannig and Comsa (1963) isolated biologically active components from thymus, and Harbers *et al.* (1967) were able to obtain two fractions from nucleohistones of various tissues (tumours) and are investigating their correlation with eu– and heterochromatin.

3. *Mucopolysaccharides*

Mucopolysaccharides are usually very difficult to separate but they can be well resolved with the aid of free-flow electrophoresis.

Bräumer and Kühn (1965) isolated chondroitin sulphate, hyaluronic acid and heparin from raw extracts of skin tissue in the presence of 0·5 N

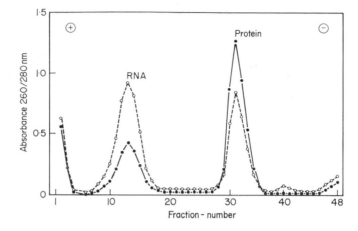

Fɪɢ. 9. Separation of A-protein from RNA (TMV-vulgare) (from Sarkar, 1966). VaP II; 0·017 ᴍ-glycine-NaCl-NaOH buffer, pH 10·5; 1400 V; 100 mA; buffer flow rate 150 ml/hr; sample flow rate 150 mg/hr; temp. 4–6°C.

acetic acid. Mushburn and Hoffman (1966) undertook a systematic investigation of the most favourable conditions for the separation of mucopolysaccharide protein complexes by free-flow electrophoresis. The authors also used 0·5–1 ɴ acetic acid for the separation of mucopolysaccharides and protein polysaccharides.

A 0·05 ᴍ acetate buffer of pH 5·0 (No. 7, Table I) is also suitable for the separation of cartilage extracts and protein polysaccharides. An average of 40–60 mg (dry weight) of these substances could be introduced into the separation chamber per hr. The field strength lay between 30 and 40 V/cm.

4. *Enzymes*

As it does not induce any chemical changes in sensitive biological materials, free-flow electrophoresis is an elegant method for preparing enzymes in pure form and for enriching enzyme solutions. Figure 10 shows the purification of a crude elastase fraction as an example (Fujii, 1967). The sample was an acetate extract of pancreatin that had been prepurified to an activity of 17·5 elastase units by precipitation with ammonium sulphate. The chief contaminating proteins could be removed by electrophoresis at pH 7·6 (tris buffer, No. 13, Table I) in the presence of small amounts of Ca^{++} ions, yielding a specific activity of 220 elastase units. This is a 13-fold purification with a total yield of 96% in a single step. The activity of the lyophilized fraction was more than 25% higher than that of the most active preparations described in the literature to date.

An interesting result was obtained in the purification of a cerebroside

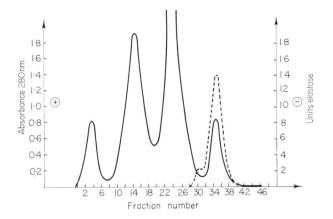

Fig. 10. Purification of an elastase fraction (from Fujii, 1967). VaP I; tris buffer, pH 7·6 (No. 13, Tab. 1); 2000 V; 160 mA; 5°C. Buffer flow rate 140 ml/hr; sample flow rate 250 mg/hr.

—————Absorbance at 280 nm (Protein)

— — — — —activity

sulphatase from pig kidney (Mehl and Jatzkewitz, 1964). After the final purification step, which was free-flow electrophoresis, it was found that the full cerebroside sulphatase activity could only be regained after recombining two high molecular electrophoresis components, one possessing aryl sulphatase activity, the other being enzymatically inactive. An enzyme extract suitable for electrophoresis was prepared from the mitochondrial fraction of the kidney homogenate by precipitation in acetone and centrifugation. The total enrichment of the specific activity was 6000-fold after recombination.

Further enzymes that have been successfully purified with the aid of continuous free-flow electrophoresis are a protenase from *Aspergillus* and a new peptidase from kidney extract reported by Nordwig *et al.* (1967).

Lezius *et al.* (1967) purified a DNA polymerase A from *E. coli* and Schliselfeld *et al.* (1965) a nucleotide pyrophosphatase from rat liver nuclei.

B. Sub-cellular particles

A number of experimental difficulties arise in the electrophoretic separation and isolation of sub-cellular particles. The iso-osmotic buffer systems that necessarily contain electrolytes often lead to the formation of aggregates. It appears that the most favourable experimental conditions always have to be chosen to suit the individual behaviour of the various sample materials. As electrophoresis of a homogenate does not lead to complete separation, simply because of the large variety of particles and soluble constituents, we prefer to begin with a simple and non-destructive pre-fractionation by ultra-centrifugation.

In the so-called microsomal fraction, it had always been difficult to isolate homogeneous ribosomal components free of components of the endoplasmic reticulum or its membrane system. The quantity of protein components present in such fractions was still relatively high. In the purification of ribosomes, in particular, the combination of electrophoresis with centrifugation scored a considerable success (Schweiger and Hannig, 1967).

For electrophoretic purification it was particularly important to exclude all ribonuclease activity and the RNase inhibitor, polyvinyl sulphate (Serva, Heidelberg) was added to the separation buffer (0·04 M tris-acetate, pH 8·0, No. 4, Table II) at a concentration of 5 μg/ml. The ribonucleoprotein particles were first prepared by ultracentrifugation, suspended and then introduced into the electrophoresis chamber. Impurities such as glycogen, ferritin haemoglobin and membrane fragments were successfully removed in the pH region of 7·0–8·5 using a field strength of 40 V/cm. The resulting material was free of adsorbed proteins as indicated by the ratio of absorption at 260 and at 235 nm (1·6–1·65). The diagram obtained in the analytical ultracentrifuge shows that the ribosomes are present mainly in the 80-S form. In addition, small amounts of two heavier aggregates and a subunit are present.

In collaboration with Matthaei et al. (1966) we succeeded in separating ribosomes from E. coli from their subunits (30 S and 50 S particles). Subunits charged with polyuridylic acid also showed different electrophoretic mobilities. To give a complete picture of the capability of the method, we will mention here the successful isolation of lysosomes (Hannig et al., 1969; Stahn, 1969), mitochondria (Stahn et al., 1970) and mitochondrial membranes (Heidrich et al., 1969) from rat liver homogenates.

Bacillus thuringiensis is an aerobic spore-forming bacterium which is pathogenic for insects. Disease in the insect host is caused by the action of a protein toxin synthesized during sporulation in the form of small intracellular crystals of pro-toxin (cf. Norris, 1969). It has proved difficult to separate the spores and crystals which are released by rupture of the vegetative cell at the end of growth of the bacterium on nutrient media such as Lab Lemco broth (Oxoid). The two elements can be separated by free-flow electrophoresis. Spores and crystals are carefully washed by centrifugation in distilled water and finally suspended in cuvette buffer at pH 7·0. The buffer used in the electrode compartment has the following composition—

Na₂CO₃,	26·5 g
2-amino-2-methyl-1,3-propandiol	26·35 g
NaH₂PO₄.2H₂O	39 g
Citric acid	52·5 g
Distilled water	4 litres

pH adjusted to required value with NaOH or HCl.

This is used at a 1 : 10 dilution for the cuvette. Spores and crystals suspended in cuvette buffer to give a concentration of approximately 0·1 mg/ml are introduced into the cuvette at a rate of 2 ml/hr and Figure 11 shows the separation obtained, under a potential difference of 1·5 Kv and 200 mA. Figure 12 shows a phase contrast photomicrograph of the original mixture and Figs. 13 and 14 show similar photomicrographs of samples from the peaks shown in the figure.

The protein crystal can be solubilized by the use of alkaline reducing conditions. When crystals are taken into solution by incubation with a similar buffer to that described above, but adjusted to pH 10 in the presence of 0·1 M mercaptoethanol, a mixture of proteins is produced. Figures 15a and b shows the result of the separation of this mixture in the free-flow

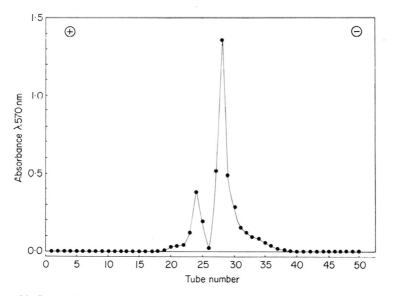

FIG. 11. Separation of spores and protein crystals of *Bacillus thuringiensis* (Norris, 1969; unpublished data). VaP II; buffer, carbonate: 2-amino-2 methyl-1,3-propandiol: citrate, pH 7·0 (No. 7, Table 2); 1500 V; 200 mA; buffer flow rate 150 ml/hr; temperature, +10°C; sample concentration 0·1 mg/ml; sample flow rate 2 ml/hr.

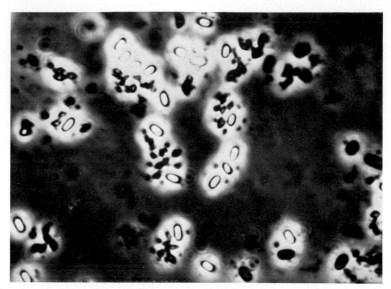

FIG. 12. Phase contrast photomicrograph of the mixture used in the separation illustrated in Fig. 11 (× 2800). Oval refractile bodies are spores, phase-dark bodies are protein crystals.

electrophoresis apparatus using similar buffer at pH 10, but without mercaptoethanol. Figure 15a shows a separation obtained under conditions designed to produce a separation of the three major components and Fig. 15b shows the results of changing the conditions in order to produce a purer preparation of the right-hand peak, shown in Fig. 15a. (Unpublished data from J. R. Norris).

C. Whole cells

General remarks

Very mild and well-defined experimental conditions are necessary for the processing of cells. In general, all operations should be performed rapidly and at temperatures below 5°C. Even brief warming can cause an irreversible inactivation of these complicated systems. Apart from mechanical forces, osmotic influences and pH differences in the suspending medium can damage the sample. The best way of deciding on suitable milieu conditions for the isolation of a particular cell fraction is to observe the morphological behaviour in the optical or electron-microscope. Aggregation is always a sign of damage.

In all events care must be taken during the processing of the cell material not to alter the surface charge of the particles by, for example, adsorption

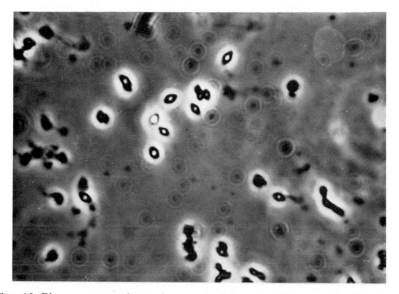

FIG. 13. Phase contrast photomicrograph of the material present in the left hand peak of Fig. 11 (× 2800). Field contains only protein crystals.

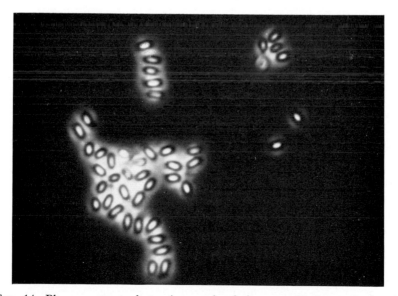

FIG. 14. Phase contrast photomicrograph of the material present in the right hand peak of Fig. 11 (× 2800). Field contains only spores.

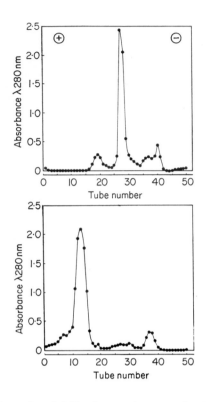

Fig. 15. Separation of solubilized protein crystal of *Bacillus thuringiensis* (Norris, 1969; unpublished data). VaP II; buffer—as Fig. 11 but pH 10·0 (No. 7, Table 2); temperature + 10°C. (a) 1500 V; 200 mA; buffer flow rate 150 ml/hr; sample concentration 20 mg/ml; sample flow rate 2 ml/hr. Introduction point No. 7. (b) 1500 V; 200 mA; buffer flow rate 100 ml/hr; sample concentration 20 mg/ml; sample flow rate 1·4 ml/hr. Introduction point No. 6.

of proteins from the already haemolysed cells or adsorption of complexing electrolyte ions. Washing three times with physiological NaCl solution (for example) has proved an effective pretreatment for electrophoretic separation as it usually removes all adsorbed proteins form the cells.

In reproducible runs we were able to show that the preparative separation of cells from human blood into erythrocytes, lymphocytes, granulocytes and monocytes is quite feasible by free-flow electrophoresis (Hannig and Krüssmann, 1968; Ganser *et al.*, 1968).

Cells of Walker-Ascites tumors could be separated from erythrocytes and unaltered leucocytes added artificially (Hannig and Wrba, 1964).

The cells obtained showed hardly any loss of vitality through fractiona-

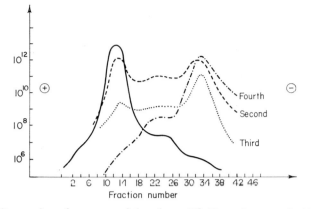

FIG. 16. Separation of mutants of the phage fd³ (Braunitzer *et al.*, 1964). VaP I; abscissa, fraction number; ordinate, N, the number of particle counted in the dilution test. 0·025 M sodium acetate buffer, pH 5·3; 1500 V; 125 mA; 1°C; through put lasted 110 min. 1. primary curve (wild strain). 2–4. mixture of mutants after 1st, 2nd and 3rd culture respectively.

tion. We have recently been able to demonstrate the successful application of the free-flow electrophoresis technique as a new method for the elucidation of immunological events at the cellular level (Hannig and Zeiller, 1969).

D. Phages

The experiment illustrated in Fig. 16 indicates how mutants can be selected with the aid of free-flow electrophoresis. The sample is a DNA phage fd³ (Braunitzer *et al.*, 1964) fractionated in 0·025 M sodium acetate buffer of pH 5·3 (No. 1, Table II). The sample is very uniform but exhibits two side bands in the plaque test (full curve in Fig. 16; primary curve) that differ from the main component by one and two charge units, respectively. It should be noted that the scale of the ordinate is logarithmic. If the fractions of one of the two secondary peaks (e.g. fractions 32–35) are combined and cultured further and the virus then purified by the usual procedures and refractionated by electrophoresis, the new activity distribution differs from the primary curve in that the peak of the wild strain is reduced and that of the mutant is increased. The combination of these enriched fractions of the mutant and its reculturing followed by electrophoretic separation were repeated twice.

Four such steps lead to a decrease of the peak due to the wild strain of 10^7 and an increase of the peak due to the mutant of 10^7 which is equivalent to a relative enrichment of between 10^{13} and 10^{14}.

This new component contains a mixture of mutants of the phage fd³.

The charge of their coat proteins differs from that of the wild strain in the sense of loss of acid or addition of basic amino-acids.

In order to obtain chemically and biologically homogeneous material the sample was cultured from a single plaque after suitable dilution and then submitted to the following processes (a) ammonium sulphate precipitation, (b) frigen treatment (c) incubation with ethylenediaminetetraacetate and a mixture of enzymes, (d) differential centrifugation and (e) electrophoresis at pH 8·2 and 5·3.

Samples prepared by this procedure are extremely pure and, in particular, free of other proteins. The procedure chosen here has proved superior to other methods, in particular to gradient centrifugation.

E. Bacterial cells

The first measurements of electrophoretic mobility of several bacterial strains were reported as early as 1933 by Przyrembel in his doctoral thesis. The experiments were done with an analytical discontinuous cell electrophoresis apparatus. He was, for example, able to show that various strains of the diphtheria bacillus group carry different charges and thus migrate at different speeds in the electric field.

Since preparative separation is not possible using the analytical microcell (one can only obtain data on the electrophoretic mobility of individual cells) further investigations were limited to the problem of how various external influences change the surface charge of bacterial cells (Lerche, 1954; James, 1965).

The alteration of the surface charge by chemical or enzymatic influence on the cell surface combined with the measurement of electrophoretic mobility permits deductions on the nature of the charge carrier. In certain instances it has been possible to correlate changes in mobility on enzymatic treatment with known antigenic structures. Of particular interest in this connection is the investigation of the effects of antibiotics on the cell surfaces. McQuillen (1951), for example, investigated the influence of penicillin on the electrophoretic mobility of *Staphylococcus aureus*. Lerche (1955) succeeded in isolating small amounts of mutants of *Micrococcus pyogenes* and *Staphylococcus aureus* in a modified cell-electrophoresis arrangement. Strickler *et al.* (1966) reported initial results for the separation of artificial mixtures of several bacterial ˙strains in a continuous electrophoresis apparatus designed on a micro scale. Using 0·003 m-phosphate buffer, pH 6·1, the authors separated e.g. *Bacillus globigii* and *Chromobacterium violaceum* and mixtures of bacteria and their pigments. The results were recorded photographically. Although these separations are of only academic interest, they do show in what a well-defined manner certain populations migrate in the electric field. The result is convincing evidence for the constancy of the electric charge of these biological particles.

The association of complex biological processes with spatially defined structures is only possible in homogeneous samples and the preparative isolation of uniform cell populations on a larger scale is an essential pre-requisite. The splitting of charge carriers followed by an electrophoretic information on the nature and mode of action of materials (e.g. receptor substances). We have begun the electrophoretic separation on a preparative scale of bacteria of different species, with different Gram staining reaction, different resistance to antibiotics and of serologically different pathogenic strains of *E. coli* (Dyspepsy coli) (Klofat *et al.* in preparation).

The bacteria were cultured in the usual way in nutrient solution, centri-fuged, suspended in the separation buffer and washed twice. The suspen-sions were diluted to equal optical density in order to keep the number of individuals in the various runs as constant as possible.

Distribution curves of the electrophoretically obtained fractions were determined by measuring the optical density at 660 nm.

Aliquots of the fractions were cultured on agar plates in order to identify the strain and to count the numbers.

1. *Experiments with bacteria of different species*

Suspensions of equal optical density of *Pseudomonas fluorescens* and *Serratia marcescens* in 0·033 M-tris buffer, pH 7·3, were mixed in equal parts and separated at a field strength of 40 V/cm. Three bands were obtained, *P. fluorescens* being split into two bands.

2. *Experiments with bacteria with different Gram-staining reaction*

During the electrophoresis of a mixture of *S. marcescens* (Gram-negative) and *Staphylococcus albus* (Gram-positive) agglutination took place in the separation chamber and this could not be prevented by altering the pH (5·5–8·5) or the field strength. Individual experiments showed that only *S. albus* tends to agglutinate.

3. *Experiments with bacteria and mutants resistent to antibiotics*

Two wild types of *E. coli*, strains of different origin (670 and 737) and their mutants, resistant to nalidixic acid were fractionated in tris-citrate buffer of pH 7·3. While the strains 737 and 737^r proved to be uni-form under these experimental conditions the strains 670 and 670^r were both resolved into two bands.

4. *Further experiments (experiments with* E. coli *double mutants)*

It was not possible to separate *E. coli* 226 (u⁻ his⁺), 227 (u⁺ his⁻) and 228 (u⁻ his⁺) from one another. All three mutants seem to have retained the same surface charge.

Experiments with pathogenic *E. coli* strains (Dyspepsy coli).

While the serologically different strains 04, 018 and 019 agglutinate in the separation chamber under the chosen conditions (tris-citrate buffer, pH 7·3), 095 and 051 do not. It was possible to separate a mixture of the latter two strains electrophoretically into two bands, strain 051 migrating more rapidly to the anode.

A mutant (095ʳ), derived from 095, resistant to kanamycin, was obtained from strain 095. Electrophoresis of a mixture of the two strains produced two main peaks and a number of minor components that migrated more strongly to the cathode.

These experiments are only preliminary but do indicate that the method should prove very useful in the investigation of intact bacterial cells.

REFERENCES

Ambrose, E. J. (1965). "Cell Electrophoresis". J. & A. Churchill Ltd, London.

Barollier, J., Watzke, E., and Gibian, H. (1958). *Z. Naturf.* **13b**, 754–755.

Bidwell, E., Dike, G. W. R., and Denson, K. W. E. (1966). *Br. J. Haemat.,* **12**, 583–594.

Bier, M. (1967). "Electrophoresis (Theory, Methods and Applications)". Academic Press, New York and London.

Bräumer, K., and Kühn, K. (1965). *Lyon Conf. Biochem. et Physiol. on Tissu Conjouctif* pp. 69–73.

Brattsten, I., Synge, R. L. M., and Watt, W. B. (1964, 1965). *Biochem. J.,* **92**, 1–2; **97**, 678–681.

Braunitzer, G., Hobom, G., and Hannig, K. (1964). *Hoppe-Seyler's Z. physiol. Chem.,* **338**, 276–277, 278–280.

Forrester, I. A. (1965). *In* "Cell Electrophoresis" (Ed. E. J. Ambrose), pp. 115–124. J. & A. Churchill Ltd., London.

Francois, C. I., and M. I. Glimcher (1967). *Biochem. biophys. Acta,* **133**, 91; *Biochem. J.,* **102**, 148–152.

Fröhlich, Ch. (1967). *Klin. Wschr.,* **45**, 461–466.

Fujii, T. (1961). *Dissertation Univ. Munich.*

Ganser, M., Hannig, K., Krüsmann, W. F., Pascher, G., and Ruhenstroth-Bauer, G. (1986). *Klin. Wschr.,* **46**, 809–814.

Graßmann, W., and Hannig, K. (1949). BBP 805399.

Graßmann, W., Strauch, L., and Nordwig, A. (1963). *Hoppe-Seyler's Z. physiol. Chem.,* **332**, 325–327.

Guest, J. F., and Yanofsky, C. (1965). *J. Biol. Chem.,* **240**, 679–689.

Hannig, K. (1961). *Z. analyt. Chem.,* **181**, 244–254.

Hannig, K. (1964). *Hoppe-Seyler's Z. physiol. Chem.,* **338**, 211–227; Habilitationsschrift, Universität, Munich.

Hannig, K. (1968). *Jb. Max-Planck-Ges. Förd Wiss.* p. 117–137.

Hannig, K., and Comsa, J. (1963). *C.r. hebd. Séanc. Acad. Sci., Paris*, **256**, 1855–1857.
Hannig, K., and Zeiler, K. (1969). *Hoppe Seyler's Z. physiol. Chem.*, **350**, 467–472; **351**, 435.
Hannig, K., and Wrba, H. (1964). *Z. Naturf.*, **19b**, 303.
Hannig, K. and Krüsmann, W. F. (1968). *Hoppe Seyler's Z. physiol. Chem.*, **349**, 161–170.
Hannig, K., and Wirth, H. (1968). *Z. analyt. Chem.*, **243**, 522–526.
Hannig, K., Stahn, R., and Maier, K. P. (1965). *Hoppe-Seyler's Z. physiol. Chem.*, **350**, 784–786.
Harbers, E., Alan, S., and Vogt, M. (1967). Paper: Meeting Ges. f. biol. Chem., Tübingen, April 4–5th; Alan, S., Alan, R., and Harbers, E. (1970). *Biochem. biophys. Acta*, **209**, 550–558.
Heidrich, H. G., Prokowa, D., and Hannig, K. (1969). *Hoppe Seyler's Z. physiol. Chem.*, **350**, 1430–1436.
Hjerten, H. (1959). *In* "Proteides of the Biological Fluids", Proc. 7th Colloq., pp. 28, Bruges. Elsevier, Amsterdam.
Jaffé, W. G., and Hannig, K. (1965). *Archs. Biochem. Biophys.*, **109**, 80–91.
James, H. M. (1965). *In* "Cell Electrophoresis". (Ed. A. J. Ambrose) pp. 154–170; J. & A. Churchill Ltd., London. *In* "Progress in Biophysics and Biophysical Chemistry" **8**, 95–142.
Klofat, *et al.* (1968) unpublished.
Kolin, A. (1960). *Proc. nat. Acad. Sci. U.S.A.*, **46**, 509–523.
Kolin, A. (1967). *J. Chromat.*, **26**, 164–179.
Lerche, Ch. H. (1954). *Scand. J. clin. Lab. Invest.*, **6**, 250–254.
Lerche, Ch. H. (1955). *Acta path. microbiol. Scand.*, **37**, 99–102.
Lezius, A. G., Hennig, S., Menzel, C., and Metz, E. (1967). Paper: Meeting Ges. biol. Chem., Tübingen April 4–5th.
Mashburn, T. A., Jr. and Hoffman, Ph. (1966). *Analyt. Biochem.*, **16**, 267–276.
McQuillen, K., (1951). *Biochem. Biophys. Acta*, **6**, 534–547; **7**, 54–60.
Matthaei, J. H., Voigt, H. P., Heller, G., Neth, R., Schöch, G., Kübler, H., Amelunxen, F., Sander, G., and Parmeggiani, A. (1966). *Cold Spring Harb. Symp. quant. Biol.*: The Genetic Code, June 2–9th, pp. 25–38.
Mehl, E., and Jatzkewitz, H. (1964). *Hoppe-Seyler's Z. physiol. Chem.*, **339**, 260–275.
Mel, H. C. (1964). *J. theoret. Biol.*, **6**, 159–179, 181–200, 307–324.
Nordwig, H., Dehm, P., and Jahn, W. F. (1967). *In* "Proteides of the Biological Fluids", **15**, 531–534.
Norris, J. R. (1969). unpublished.
Przyrembel, L. R. (1933). Dissertation, Univ. Breslau.
Rueff, F. (1964). "Die Zellelektrophorese in der klinischen Diagnostik". J. F. Lehmann, Munich.
Rubin, A. L., and Stenzel, K. H. (1965). *Proc. nat. Acad. Sci. U.S.A.*, **53**, 963–968.
Sarker, S. (1966). *Z. Naturf.*, **21b**, 1202–1204.
Sayre, F. E., Lee, R. T., Sandman, R. P., and Preez-Mendez, G. (1967). *Arch. Biochem. Biophys.*, **118**, 58–72.
Schliselfeld, L. H., Van Eys, J., and Touster, O. (1965). *J. biol. Chem.*, **240**, 811–818.
Schultz, J., Feldberg, N., and John, S. (1967). *Biochem. biophys. Res. Commun.*, **28**, 543–549.

Schweiger, A., and Hannig, K. (1967). *Hoppe-Seyler's Z. physiol. Chem.*, **348**, 1005–1008.
Schweiger, A., and Hannig, K. (1968). *Hoppe-Seyler's Z. physiol. Chem.*, **349**, 943–944.
Schweiger, A., and Hannig, K. (1970). *Biochem. biophys. Acta*, **204**, 317–324.
Seitz, W., and Eberhagen, D. (1966). *Z. klin. Chem.*, **4**, 22–27, 169–172.
Stahn, R. (1969). Dissertation, Universität München.
Stahn, R., Maier, K. P., and Hannig, K. (1970). *J. cell. Biol.* In press.
Straub, E. (1965). *In* "Cell Electrophoresis" (Ed. E. J. Ambrose) pp. 125–141. J. & A. Churchill Ltd, London.
Strickler, A. (1967). *Separation Science* **2** (3), 335–355.
Strickler, A., Kaplan, A., and Bigh, E. (1966). *Microchem. J.*, **10**, 529–535.
Sulkowski, E., and Laskowski, M., Sr., (1967). *Analyt. Biochem.*, **20**, 94–101.
Svensson, H., and Brattsten, I. (1949). *Ark. Kemi*, **1**, 401–405.
Waldschmidt-Leitz, E., and Kling, H. (1966). *Hoppe-Seyler's Z. physiol. Chem.*, **346**, 17–19.
Wieck, K., von Oldershausen, H. F., and Klaschka, F. (1964). *Klin. Wschr.*, **42**, 357–360. *J. Chromatog.*, **13**, 111–118.

CHAPTER IX

Preparative Zonal Electrophoresis

WILLIAM MANSON

Hannah Dairy Research Institute, Ayr, Scotland

I.	Introduction	549
II.	Zonal Density Gradient Electrophoresis	550
	A. General consideration of the technique	550
	B. Experimental method	555
III.	Zonal Electrophoresis Using Solid Supporting Media . . .	559
	A. Introduction	559
	B. Supporting media	559
	C. Instrument design	563
	D. Experimental method	568
IV.	Conclusion	569
References	. . . .	570

I. INTRODUCTION

Electrophoresis is the movement exhibited by charged particles or molecules suspended or dissolved in a liquid when brought under the influence of an applied electric field. The direction and rate of such migration in a particular medium is dependent on the electric charge of the particle or molecule and also on its size. Since charge is a property associated with the surface of molecules in general and of such biological particles as viruses, bacteria and cells in particular, electrophoretic behaviour can be considered as being primarily determined by the surface structures of such charged objects. Both surface structure and charge may vary considerably with variations in environment and consequently the medium in which the electrophoresis is performed also has an important direct part to play in determining electrophoretic behaviour.

During electrophoresis, a charged particle or molecule is also subject to other than direct electrical effects. Its migration is influenced by gravity, by convection and by diffusion all of which if not controlled may reduce seriously the efficiency of electrophoresis as a process for the separation of charged particles. Gravitational effects are usually controlled by conducting the electrophoresis in chambers designed to permit migration to take place in a vertical direction only. Effects due to convection and to some extent those due to diffusion may also be controlled. It is the object of this Chapter to provide an introduction to two forms of preparative zonal electrophoresis in which the electrophoretic separation is stabilized against the effects of convection. In each of these, electrophoresis is performed in a liquid column. The stabilizing effect is produced either by a density gradient formed within the column or alternatively by the presence in the column of a suitable solid supporting medium. The use of gels prepared from starch and from polyacrylamide as electrophoretic media is discussed elsewhere in this book and is consequently outside the scope of this Chapter.

II. ZONAL DENSITY GRADIENT ELECTROPHORESIS

A. General considerations of the technique

1. *Introduction*

Of all the forms of zonal electrophoresis, density gradient electrophoresis is that which most closely resembles the original analytical method of Tiselius (1937) in that the separation is achieved in a liquid medium without the use of solid supporting material. It differs fundamentally from this method however, since under favourable conditions all the components of a mixture may be completely separated into discrete zones, whereas in the analytical procedure complete resolution is not obtained and only a proportion of those components which migrate fastest and slowest in a mixture may be isolated in an electrophoretically pure state.

In the moving boundary electrophoresis of Tiselius the stability of the migrating zones is produced by small density gradients formed by the migrating charged molecules themselves and in this sense the analytical technique can be considered to be a form of density gradient electrophoresis. The stabilizing effect of density gradients of this kind is, however, small and is usually insufficient to confer gravitational stability to migrating zones except in the specialized conditions to be found in a Tiselius electrophoretic cell. For application on a preparative scale the presence of a density gradient additional to that produced by the migrating molecules themselves is required.

Stabilization by density gradients prepared from electrically neutral solutes was introduced by Philpot (1940) who used glycerol, ethanol and water to form the density gradient. Although this proved to be largely unsuccessful as a preparative technique, later investigators using the same principle proved more successful. Notable examples of such studies were described by Ericson and Nihlen (1953a, b); Brakke (1953; 1955) and Kolin (1958). The last named author also described techniques whereby a pH gradient was superimposed on a sucrose density gradient (1954; 1955) thus producing an early form of the technique of isoelectric focusing. The development of this technique is described in a separate Chapter in this book. The earlier stages of development of density gradient electrophoresis have been fully described by Svensson (1960); Bier (1959) and Bloemendal (1963).

As a preparative method, density gradient electrophoresis has certain advantages to offer in addition to those associated with electrophoretic procedures in general, namely that the conditions employed are mild and the risk of decomposition of the products of the separation are minimal. These are:

(i) It is entirely free from effects of sorption since no solid supporting medium is present.

(ii) The separated zones can be easily detected using relatively simple optical methods and in some instances by visual inspection.

(iii) The isolation of separated zones from the liquid column may be achieved rapidly and easily.

(iv) The recovery of material from the liquid column in general approximates closely to 100% of that applied initially.

On the other hand its main limitation possibly lies in its somewhat restricted capacity. With protein mixtures practical considerations make it impossible in general to perform separations on quantities significantly greater than 500 mg which is considerably below the limit for columns incorporating a solid supporting material. This may not be considered a serious drawback when it is remembered that the separation procedure once started requires virtually no attention thereafter.

2. *Development*

Essentially all equipment designed for density gradient electrophoresis consists of a vertical compartment containing the liquid density gradient connected at one end to a chamber which houses a positive electrode and at the other to a chamber carrying a negative electrode. The final design of the apparatus has often been greatly influenced by the U-tube shape of the standard Tiselius cell used in moving boundary electrophoresis e.g. Brakke (1955) has described an apparatus in which the electrophoresis was performed in simple U-tube cells, a maximum of six of these tubes being connected in parallel to the same pair of electrode vessels. Partridge and Elsden (1961) in a study of the dissociation of a complex of chondroitin sulphate and protein from cartilage used an apparatus which consisted essentially of a single large U-tube. In these two examples density gradients were formed in both limbs of the U-tube cells using variations in sucrose concentration in suitable buffer systems.

It is noteworthy that the development of equipment designed for use in zonal electrophoresis with solid supporting media has also been much influenced by this U-tube shape, and indeed many of the columns designed for that purpose have been satisfactorily adapted for use with the solid supporting medium replaced by a density gradient. A notable example of this was the apparatus described by Svensson (1954; 1960) and marketed by LKB Instruments Ltd. (232 Addington Road, South Croydon, Surrey CR2 8YD). The company has recently replaced this instrument with a more flexible, and more expensive piece of equipment (Bergrahm, 1967; Bergrahm and Harlestam, 1968) the "Uniphor 7900" apparatus. With this all the forms of zone electrophoresis hitherto described in the literature can be performed and a change from one type to another is easily achieved. The versatility of the instrument is probably its greatest asset, since it is expensive and it is limited in capacity to about 50 mg protein per electrophoretic run. It is, however, the only apparatus designed for preparative density gradient electrophoresis at present obtainable commercially. The earlier column electrophoresis apparatus (Svensson, 1954; 1960) which was marketed by that same company, although less versatile than the "Uniphor 7900", has a greater capacity and is capable of producing adequate separations.

In contrast to the marked lack of equipment commercially available a variety of simple pieces of apparatus capable of effecting satisfactory separations has been described. Some have already been mentioned. Many were designed to overcome specific problems associated with particular fractionations and may therefore require adaptation before being generally applied. Nevertheless, wide use has been made of equipment similar to that described by Berg and Beeler (1958); Charlwood and Gordon (1958); Brakke (1955); Partridge and Elsden (1961); Manson (1965) and Choules and Ballentine (1961).

3. *Instrument design*

In constructing a satisfactory apparatus for zonal density gradient electrophoresis attention must be paid to the particular separation which it will be required to perform. Nevertheless the following requirements of design and technique may be regarded as being of a general and fundamental nature.

(a) *Electrodes and electrode vessels.* Consideration of the design and use of electrodes and electrode chambers is necessary since electrode products formed during electrophoresis can seriously affect the quality of the separation obtained if they are permitted to enter the density gradient column (Svensson, 1955). This can to some extent be avoided if the electrode vessels are much larger than the density gradient column, and it has been suggested that to be satisfactory they should have about ten times the capacity of the separatory chamber (Tiselius, 1937). This estimate might be excessive (Gordon and Kay, 1953) but it should be remembered that in prolonged electrophoresis the only certain method of avoiding interference from electrode products is to remove them by renewing completely the buffer in the electrode vessels, whatever their size, at regular intervals throughout the run.

Two types of electrode are commonly used, the reversible silver–silver chloride type and the platinum electrode. Silver–silver chloride electrodes which normally consist of spirals of silver wire coated with silver chloride, if properly formed, are durable and satisfactory in that they generate no harmful electrode products under normal conditions of use. They do, however, have a limited electrical capacity and care must be exercised to ensure that this is not exhausted during the course of the electrophoresis. They may be prepared in the following way.

Silver wire is threaded through a length of thick-walled capillary glass tubing until 2 cm protrude from one end. This serves as a connection for the electrode to the power supply. The wire remaining at the other end is wound into a compact ball or tight spiral and forms the actual electrode itself. Both ends of the glass sheath are sealed to the silver wire using a suitable waterproof resin adhesive, such as Araldite (Ciba (A.R.L.) Ltd., Duxford, Cambridge). The size of the electrode should of course be suited to its particular use. Those used with the apparatus described by Partridge and Elsden (1961) and by Manson (1965) were each prepared from approximately 2 m of 17 S.W.G. silver wire (approx. 1·45 mm diam.) with glass sheaths of 40 cm.

Electrodes are prepared for use in pairs by immersing them in 1 N hydrochloric acid and passing a current of about 30 mA between them. A brown deposit of silver chloride is initially built up on the anode and hydrogen is liberated at the cathode, but after some time chlorine begins to be formed at the anode. The connections on the electrodes should then be exchanged and current passed in the opposite direction until chlorine begins to be liberated at the new anode. The connections are then again reversed and current passed. With each reversal of polarity more silver chloride is formed before liberation of chlorine than before. In this way deposits are built up of porous silver on the cathode and of silver chloride on the anode. These constitute the electrical capacity of the electrodes and the formation process should be continued until a suitable capacity is attained.

The second type of electrode, those made from platinum, requires no special preparation. However, during electrophoresis in which platinum electrodes are employed acid is generated at the anode and alkali at the cathode. Care must be taken to prevent these electrode products from entering the electrophoretic column either by the methods suggested above or by introducing into the apparatus acid and alkali "locks" between the electrode vessels and the density gradient column (Svensson, 1955). The use of platinum electrodes tends therefore to complicate the design of an electrophoresis apparatus.

(b) *Density gradient columns.* A number of substances are commonly used in the construction of density gradients for zonal electrophoresis but before a substance can be regarded as suitable for this purpose it should satisfy the following requirements. It should have

  (i) high solubility in the electrophoretic buffer,
 (ii) a density markedly different from that of the buffer,
(iii) electrical neutrality,
 (iv) no chemical affinity for the substances to be subjected to electrophoresis,
  (v) properties differing sufficiently from these substances for complete separation to be easily attainable.

In addition to these requirements it is advantageous if the substance selected also possesses a viscosity similar to that of water in order that increasing density will not be accompanied by increasing viscosity, since a viscosity gradient tends to retard the faster-migrating components of a mixture to a relatively greater extent than those migrating more slowly. Substances of this type which have been used successfully are methanol and ethanol (Charlwood and Gordon, 1958). Despite viscosity considerations, however, the material of choice has usually been sucrose although glycerol, ethylene glycol and glycine have also been employed satisfactorily. It should be noted that if ultraviolet spectroscopy is used to detect the separated zones in the density gradient column and this is usual with protein mixtures, sucrose will require purification before being incorporated in the column since even the A.R. grade normally contains ultraviolet light-absorbing material. The purification is conveniently done by preparing all the sucrose solutions required by dilution from a stock 65% w/v solution in water which has first been stirred with charcoal (3 g charcoal per 100 ml sucrose solution) at 60°C for 20 min and filtered. This solution exhibits virtually no light absorption at 278 nm.

The preparation of liquid density gradient columns may be accomplished using a gradient mixing device similar to that used in gradient elution chromatography (Bock and Ling, 1954). The characteristics of the density gradient may of course be varied by varying the size and number of mixing devices supplying the gradient column. This has been fully discussed by Svensson (1958; 1960).

(c) *Buffer systems*. It is not possible to describe suitable buffer systems in any but the most general terms. Clearly the most suitable system is that in which the best separation is obtained and consequently its composition will to some extent depend on the properties of the materials to be separated. There are nevertheless certain general considerations which are worthy of note. A satisfactory buffer system should have high buffering capacity and should have a pH value at which all the components of the mixture to be resolved have the same sign of charge. In addition since the electrophoretic current is carried by all the ions present it is desirable that the proportion carried by the buffer ions should be low in order that the rate of migration of the other ions, i.e. those to be separated, may be reasonably high. For this reason the required high buffering capacity should be coupled with a low ionic strength, preferably of not greater than 0·10.

Buffer systems which are suitable for electrophoresis over almost the entire pH range are readily available and are too numerous to be described in detail. The following systems which have been used successfully in conjunction with sucrose gradients for separations of milk proteins (Manson, 1965) are given only as typical examples of the type employed:

(i) Phosphate buffer, pH 7·1, I = 0·06 having the composition per litre, 1·20 g NaH_2PO_4. $2H_2O$; 2·02 g Na_2HPO_4; 0·58 g NaCl.

(ii) Chloride buffer, pH 2·35, I = 0·05 having the composition per litre, 3·20g KCl; 6·5 ml N HCl

In addition 0·05 M Tris buffer pH 8·6 (Berg and Beeler, 1958) and 0·025 M borate buffer, pH 9·2 (Svensson, Hagdahl, and Lerner, 1957) have been used in successful fractionations of serum proteins.

B. Experimental method

1. *The basic procedure*

The basic technique may perhaps be best described by reference to a relatively simple system such as that of Partridge and Elsden (1961) which possesses in addition to its simplicity the merits of efficiency and cheapness. The following adaptation of that method is that described by Manson (1965) for use with the modification of their apparatus illustrated diagrammatically in Fig. 1. The separation takes place in the right limb of the U-tube which is surrounded by a water-jacket and contains the liquid density gradient. This is formed over a solution of 50% sucrose in buffer solution which fills the bend of the U-tube, using a gradient mixing device similar to that employed in gradient elution chroma-

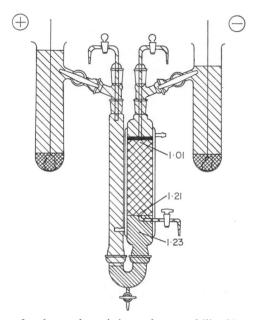

Fig. 1. Apparatus for electrophoresis in a column stabilized by a density gradient. The numbers are density values in g/ml and the black zone at the top of the density gradient represents the initial position of the mixture to be fractionated.

tography (Bock and Ling, 1954) coupled to the side-arm of the right limb of the U-tube. The remainder of the apparatus contains electrophoretic buffer solution. A solution of 58% sucrose in buffer solution is pumped into a closed vessel containing 3% sucrose in the same buffer solution, and the expressed liquid is introduced into the right limb of the U-tube through the side-arm. This produces a diffuse boundary just above that already present between the 50% sucrose solution in the bend of the U-tube and the electrophoretic buffer. As the process continues the concentration of sucrose in the solution emerging from the mixing chamber increases and the less dense buffer is displaced upwards in the column. When the diffuse boundary reaches the lower end of the outlet tube at the top of the column, pumping is temporarily stopped. By this method, a sucrose density gradient is formed on a base of 50% sucrose in a buffer solution whose density is 1·23 g/ml. Its density varies from 1·01 to 1·21 g/ml which for this 26 cm column corresponds to an average increase in density of 0·008 g/ml per cm vertical height. The uniformity of the gradient is dependent on the physical characteristics of the mixing device and the column. For this column of 810 ml capacity, a single mixing chamber of 575 ml capacity and a pumping rate of 200 ml/h proved satisfactory.

The method of application of the sample to the top of the column varies to some extent with the type of sample to be analysed. If it is assumed that the sample is protein and that it is available as a dry solid the procedure is as follows. Pumping of liquid from the gradient mixing chamber through the side-arm of the column is recommenced, and about 40 ml of sucrose solution is displaced from the top of the column through the vertical outlet tube thus producing a "density shelf" visible as a sharp boundary at the head of the column. Sufficient of the solid sample to yield a 1·0–1·2% solution is dissolved in about 20 ml taken from a middle fraction of this expressed solution and deposited gently on the "density shelf" by allowing it to be sucked slowly down through the outlet tube while liquid drips slowly through the tap at the bottom of the bend of the U-tube. As it reaches the "density shelf" the sample solution spreads out over it forming a well-defined cylindrical zone. When all of the sample has been added a small amount of buffer solution is similarly introduced on top of the protein layer to wash out the sample delivery tube.

If the sample is available in solution its composition is adjusted to approximate to that of the buffer solution, usually by dialysis, at the same time taking care to ensure that the final concentration of protein does not exceed 1·2%. A few drops of 50% sucrose in buffer solution are added to bring the density of the sample solution to a value between those existing immediately above and below the "density shelf". It is then applied to the column in the manner already described. A sample of 25 ml yields a zone about 6 mm thick containing about 300 mg protein.

If prepared in this way the initial zone should be stable. If, however, the sample solution does not approximate fairly closely in composition to that of the density gradient in those regions immediately below and above it, instability of the sample zone may occur due to differences in the rates of diffusion of sucrose into the zone from below and out of the zone to the region above (Brakke, 1955). This is characterized by the formation of droplets which sediment slowly in the column with a rapid disintegration of the initial sample zone. With the use of the "density shelf" technique introduced by Svensson, Hagdahl and Lerner (1957) droplet formation is largely excluded.

For the electrophoresis itself reversible silver–silver chloride electrodes are used with this apparatus. They are placed in the outer electrode vessels and in use are covered with saturated sodium chloride solution. Current from a DC power unit is passed through the system, care being taken to ensure that the polarities of the electrodes are such that electrophoretic migration takes place downwards from the sample zone. The duration of the electrophoresis varies with the properties of the sample but runs of 65 h are common with an applied current of 60 mA at 650 V. With this apparatus it has not proved satisfactory to use electrical loadings of more than 50 W since above this level convection effects are produced within the column.

At the conclusion of the separation and after disconnection of the current, the contents of the right limb of the U-tube are run out through its side-arm at a rate of about 6 ml/min. With substances exhibiting selective light absorption the effluent is either monitored at a suitable wavelength directly, using a flow cell in a spectrophotometer or is first collected in fractions of 5–10 ml and the light absorbing properties of the fractions determined. The positions of the separated components in the column having been determined only their isolation remains. The actual procedure selected for this depends however on the physical properties of the separated substances themselves.

An alternative method for detecting separated zones within the column has been described by Kolin (1958). In this, a grid of oblique parallel lines is placed behind the density gradient column and viewed through the column. The positions of migrating zones appear as distortions on the otherwise uniform grid. More recently a procedure for the direct photography of zone electrophoresis columns using ultraviolet light has been described (Luner, 1968). It is intended only for application to the detection of zones having selective light absorbing properties in the ultraviolet region.

2. *Applications*

Density gradient electrophoresis has been used in the separation and purification of a wide range of substances although its main application has been to the separation of proteins. While pure homogeneous preparations may be obtained in some circumstances by this method, more often the product is not homogeneous as judged by such techniques as starch gel or polyacrylamide gel electrophoresis. The resolving power of this technique is nevertheless considerable and large scale density gradient electrophoresis is a valuable tool when used in conjunction with a highly resolving system such as zone electrophoresis on polyacrylamide gel.

In addition to being successfully applied to the separation of certain components of totally soluble protein systems such as blood serum (Svensson, Hagdahl, and Lerner, 1957; Berg and Beeler, 1958) and phosphoproteins from bovine milk (Manson, 1965), density gradient electrophoresis has also been applied to the separation of particulate substances such as cell components and micro-organisms (Kobozev and Azhits'kii, 1968). Thus Norris and Manson (1964) were able to separate a crystalline metabolite of *Bacillus thuringiensis* from cell debris and in an early experiment Kolin (1955) demonstrated the separation into three distinct fractions of a mixture of the algae *Ankistrodesmus* and *Chlorella* with human blood cells. This technique has also been used for the separation of virus particles

(Brakke, 1953; Matheka and Geiss, 1965) and for the purification of bacterial enzymes. Bangham and Dawson (1962) described the separation of the α-toxin from *Clostridium perfringens* (phospholipase C) in a glycerol gradient and Dierickx and Ghuysen (1962) have reported the purification of N-acetylhexosaminidase from *Streptomyces albus G.* in a sucrose gradient.

Choules and Ballentine (1961) using an apparatus of their own design have obtained satisfactory separation of cellular subfractions including a mixture of aldolase, glyceraldehyde phosphate dehydrogenase and cytochrome C, and Bernheimer (1962) has investigated the purification of bacterial extra-cellular proteins and bacterial enzymes.

An interesting development in preparative zonal electrophoresis has recently been described by Kolin (1967). In this stabilization of the separated zones is achieved neither by a density gradient nor by a solid support medium but by electromagnetic rotation.

III. ZONAL ELECTROPHORESIS USING SOLID SUPPORTING MATERIAL

A. Introduction

In this type of zonal electrophoresis a wide range of solid materials has been used in place of a liquid density gradient to help to stabilize migrating zones against the effect of gravity and that of convection produced by hydrostatic differences set up in the column by the migrating zones themselves. Compared with density gradient electrophoresis this technique suffers from several disadvantages. Removal of the separated zones from the column is usually slower and may be accompanied by their broadening and distortion so that resolution is impaired. In addition absorption of migrating substances on the support material in the column can lead to recoveries of much less than 100%. Solid support electrophoresis has, however, one considerable advantage over the density gradient technique in that it inherently possesses the greater capacity. Consequently in those circumstances where a supporting medium can be found which does not interact with any of the components of the mixture to be fractionated and large amounts of material are available for fractionation the solid support technique may be preferred. As with density gradient electrophoresis this technique has found wide application in preliminary stages of purification where the final stage involves zonal electrophoresis on a gel support having a higher resolving power but a lower capacity.

B. Supporting media

The general conditions outlined above for density gradient electrophoresis and those relating in particular to electrodes and buffer composition are equally applicable to solid support electrophoresis. Not surprisingly the main differences in technique and in instrument design apply to the column on which the separation takes place. The most obvious of these is concerned with removal from the column of the discrete zones produced by the electrophoresis. The procedure used in density gradient electrophoresis is clearly not applicable, and an extra step is required whereby the zones are washed from the column with a suitable eluting liquid and the eluate itself fractionated. This is most easily carried out after conclusion of the electrophoresis although as an alternative elution may be started during the run. Whichever procedure is applied this type of zone electrophoresis may be properly regarded as a combination of electrophoresis and column chromatography, and one of the main problems associated with the efficient use of the technique lies in ensuring that the degree of resolution achieved in the electrophoresis is not significantly reduced during the chromatographic elution.

The use of the correct supporting medium is of fundamental importance and while this varies with the substances to be analysed a satisfactory medium should possess the following general characteristics:

  (i) It should be completely insoluble and stable in the electrophoretic buffer system.
 (ii) It should be obtainable in a pure and homogeneous state.
(iii) It should contain a minimum of ionizable groups.
 (iv) Its physical form should permit uniform packing in a column.
  (v) It should not interact with the substances to be separated.

In applying these criteria it should be remembered that they are of a general nature and all may not require to be satisfied in every circumstance. This is particularly so with (v) as is illustrated by the widespread use of gel filtration media such as Sephadex as the supporting medium and also by the recent development of electrochromatography (Nerenberg and Pogojeff, 1969).

A number of materials which in general fulfil these requirements is discussed below. The efficiency of the separation obtained however does not depend only on the types of material selected but also to a considerable extent on the preparation of an evenly packed homogeneous column. Consequently care should be taken to ensure that the column remains vertical throughout the entire procedure from filling to elution and that irregular packing is avoided.

1. Starch

Potato starch granules have been used successfully as a supporting medium particularly for separations of proteins. Tightly packed homogeneous columns are obtainable since starch granules swell markedly in the presence of water. The use of this property however has to be regulated to ensure that the liquid space between the granules is not reduced to the point where the column's capacity is impaired. Satisfactory procedures for the preparation of starch columns have been reported by Flodin and Kupke (1956) and by Carlson (1954) who modified the method described by Stein and Moore (1948). A suspension of starch is made in n-butanol containing water equivalent to 30% of the dry weight of the starch, by gently grinding the starch and the butanol together in a mortar. This slurry is poured into the electrophoretic column and allowed to stand under an applied pressure of approximately 5 cm Hg. After 1 h the butanol remaining above the settled starch in the column is carefully sucked off and replaced by water. This is allowed to percolate slowly through the column under an applied pressure of 5–8 cm Hg for about 48 h. During this time the starch granules absorb water and swell, thereby creating an internal pressure evenly distributed within the column which contributes to its uniformity. Before electrophoresis the column is equilibrated with the electrophoretic buffer.

The size of the intergranular spaces in the column may be varied by varying the water content of the n-butanol with which the starch is initially ground.

Although starch behaves satisfactorily as a supporting medium for protein separations except those involving strongly basic proteins it does suffer from a number of disadvantages. It has been reported (Flodin and Kupke, 1956) that strong electro-osmotic flow results from its use with buffers containing polyvalent ions such as citrate and phosphate, and that substances of low molecular weight spread more on starch than on other supporting materials. Moreover it is difficult to prevent contamination of starch columns by micro-organisms so that their useful life is likely to be severely limited.

2. Cellulose

Columns composed of powdered cellulose have been found to provide a satisfactory medium for the electrophoretic resolution of protein mixtures, again excepting basic proteins. Improved results are obtained generally if ethanolysed cellulose, i.e. partially alkylated cellulose prepared by treating cellulose with dry ethanolic hydrogen chloride (Flodin and Kupke, 1956) is substituted for pure cellulose. Whichever form of cellulose is selected the method of packing the column is virtually that employed in the preparation of a standard chromatographic column. The column is first filled with the electrophoretic buffer solution and cleared of air bubbles. A slurry of cellulose in the same buffer solution (approximately 10 volumes of buffer to 1 volume of cellulose) which had previously been placed under reduced pressure in order to remove occluded air is run slowly into the column while buffer is allowed to drip from the bottom of the column. The cellulose settles slowly and when sedimentation is seen to be complete the column is ready for use.

With columns of cellulose or ethanolysed cellulose (Sorof et al., 1964) certain disadvantages encountered with starch columns are absent. These columns are not readily contaminated by micro-organisms and their re-use is not limited for this reason. It is nevertheless advisable to incorporate sodium azide, to the extent of 0.02%, as a bacteriostatic agent in buffers used with these columns. They have an increased capacity for protein separations over starch columns of similar size and no undue spreading of zones of substances of low molecular weight has been observed. The use of phosphate buffers produces no marked electro-osmotic effects.

In general the quality of separations obtained with proteins is similar to that produced by electrophoresis of the same mixture on strips of paper and for general use cellulose is the material of choice.

3. Particulate gels

Solid support materials of the types mentioned in (a) and (b) above are derived from naturally occurring substances and as such possess properties and structures which even after modification fall short of those required for an ideal supporting medium. They are also subject to the small variations in structure normally encountered in mixtures of naturally occurring substances. Synthetic particulate gels, such as cross-linked dextran gels, marketed under the trade name "Sephadex" (manufactured by AB Pharmacia, Uppsala, Sweden) appear to suffer from none of these drawbacks, since they consist of uniform particles having highly reproducible chemical and physical properties. It is therefore possible to construct from these granular gels, electrophoretic columns possessing a high degree of uniformity (Dose and Krause, 1962; Vendrely, Coirault, and Vanderplancke 1964). Such columns are virtually free from electro-osmotic flow and from absorption of protein although Sephadex may support bacterial and fungal growth.

Particulate gels of this type possess the important additional property of being able to discriminate between solutes on a size basis and consequently the degree of separation of a mixture attainable by electrophoresis on columns of Sephadex depends both on the size and charge of its components, i.e. two principles of separation are employed simultaneously, those of molecular sieving and of electrophoresis. It is debatable, however, whether the use of more than one separation principle is advisable since clearly in certain circumstances less satisfactory separations may be obtained than would be if each principle were to be applied separately.

The preparation of columns of Sephadex for electrophoresis follows the same principles as those prescribed for column chromatography, i.e. the gel is poured into the column as a slurry in a single operation and the excess liquid drained from the column. It is, however, essential that the gel be allowed to swell fully in the electrophoretic buffer before being packed into the column. Sephadex types G-10 to G-50 should be allowed to stand overnight in buffer at room temperature, types G-75 and G-100 should be given 3 days and type G-200, one week. With types G-10 to G-75 packing may be performed at room temperature but with types G-100 and G-200 the operating temperature should be used.

4. Miscellaneous materials

In addition to those already mentioned the following materials have received limited attention as supporting media. Glass powder prepared from glass wool (Hocevar and Northcote, 1957) was used, after careful fractionation, as the solid support in the electrophoretic separation in borate buffer of alkali-soluble polysaccharides from *Chlorella pyrenoidosa*. However, as a supporting medium, glass suffers from two serious disadvantages. It absorbs many proteins fairly strongly and produces large electro-osmotic effects. Its use has therefore not been widespread.

Granular plastic powders of various compositions have also been used, notably a copolymer of vinyl chloride and vinyl acetate, Pevikon C-870 (Müller-Eberhardt, 1960; Bocci, 1964). Nylon and methyl polymethacrylate powders (Gorkin et al., 1963) have also been employed. While satisfactory in many respects these have in general poor heat conductivity with the result that stabilization against convection in columns prepared from these materials is inferior to that in columns of starch or cellulose. Furthermore as with Sephadex a column packed with these materials contains only about half the volume of buffer contained by a column of equal size packed either with starch or cellulose.

C. Instrument design

As with density gradient electrophoresis the early instrument design of packed columns was concerned largely with scaling up the U-tube separating cell of the Tiselius type. Only one limb of the U-tube contained a supporting medium, the other being retained to provide a balance in hydrostatic pressure. Examples of this type of apparatus have been described by Kunkel (1954); Haglund and Tiselius (1950); Flodin and Porath (1954); and Carlson (1954). That described by Flodin and Porath which was based on an earlier design of Haglund and Tiselius (1950) is typical of this type and is shown diagrammatically in Fig. 2. The left

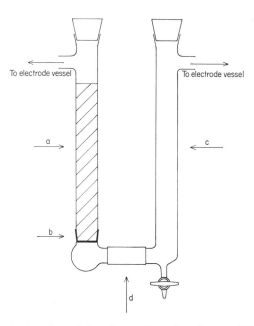

FIG. 2. Schematic drawing of the electrophoresis column of Flodin and Porath (1954); (a) glass tube 50×3 cm filled with starch; (b) glass filter; (c) glass tube; (d) rubber tubing. Reproduced with permission from Flodin and Porath (1954).

limb of the U-tube contained the solid support, in this instance potato starch. On completion of the electrophoresis this column was detached from the apparatus and the separated zones eluted with buffer. An apparatus of this type although efficient and simple in construction and in operation is limited in capacity to about 250 mg protein. The incorporation of a significantly larger column in an apparatus of this design is not feasible since the additional electrical load required increases the amount of heat generated to a level where it cannot be dissipated satisfactorily without a more efficient cooling system.

A second type of apparatus was evolved from this design (Flodin and Kupke, 1956). In this the U-tube arrangement was discarded and instead the packed column was immersed in a second wider tube containing buffer solution. Electrical contact was made between the tubes, and the buffer in the wider tube served the additional purpose of acting as a cooling fluid. With this apparatus and using a column of paper powder the maximum capacity was increased to about 450 mg protein. As with the apparatus of Flodin and Porath (1954) the separated zones were removed after completion of the electrophoresis by percolating buffer through the column.

In a third type of apparatus (Porath, 1954; Porath, Lindner, and Jerstedt,

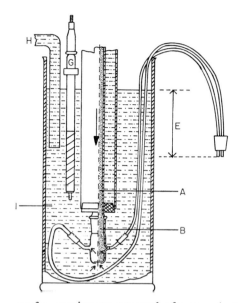

FIG. 3. Arrangement for continuous removal of zones during electrophoresis. The electrophoresis column A is filled with powder surrounded by a cooling jacket and provided with the short column B. In addition to the column the electrode G and a tube H are dipped into the electrode vessel I. The tube H connects the buffer in I with a buffer reservoir for keeping the liquid level E constant. Reproduced with permission from Porath, Lindner, and Jerstedt (1958).

1958) elution of the separated zones was performed both during and after the electrophoretic run. This was made possible by modifying the lower end of the column in the manner shown above in Fig. 3. The faster-moving components of the mixture were removed during the electrophoresis in a slow stream of buffer as they reached the lower end of the column. Those moving more slowly were eluted in the usual way after conclusion of the electrophoresis.

Using a similar principle Porath (1964) later designed a large scale apparatus. This would accommodate sample volumes exceeding 100 ml and containing more than 10 g protein. It was subsequently marketed by LKB Instruments Ltd. who have now withdrawn it in favour of their "Uniphor 7900" column electrophoresis system. Despite the fact that it is no longer available commercially its basic simplicity and elegance of design is still worthy of consideration in some detail. The apparatus consists essentially of a separation chamber formed by the annular space between two vertical coaxial transparent plastic cylinders with one electrode compartment mounted at the top and one at the foot of the column The separation column is about 80 cm in height and the internal diameter of

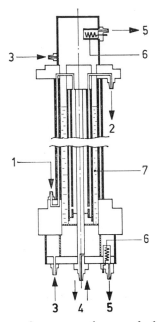

Fig. 4. The LKB column for preparative zonal electrophoresis. 1. Inlet for cooling water. 2. Outlet for cooling water. 3. Inlets for circulating electrode buffer. 4. Inlet for eluting buffer (during electrophoresis); outlet for eluate (after electrophoresis). 5. Outlets for circulating electrode buffer. 6. Platinum electrodes. 7. Separation chamber. Reproduced by permission of LKB Instruments Ltd.

the outer cylinder is 11 cm. The arrangement is shown schematically in Fig. 4. The column may be packed with cellulose, Sephadex or plastic powders of which cellulose or ethanolysed cellulose is most commonly used.

The most recently designed equipment which is commercially available is the "LKB Uniphor 7900" apparatus. Its application in zonal density gradient electrophoresis was mentioned briefly above, and it is illustrated in Fig. 5. It consists basically of a vertical separation column refrigerated by a circulating cooling fluid, and incorporating electrode compartments at its top and bottom. The column which may be any one of a range of standard LKB chromatographic columns, is mounted between two reservoirs from which buffer solution may be circulated throughout the instrument. While the instrument is undoubtedly versatile since it can be quickly adapted for use with any form of stabilization medium (Bergrahm, 1967; Bergrahm and Harlestam, 1968), it suffers from the same disadvantage when used for solid-support zonal electrophoresis as it does when used for density gradient electrophoresis, namely one of limited capacity. For

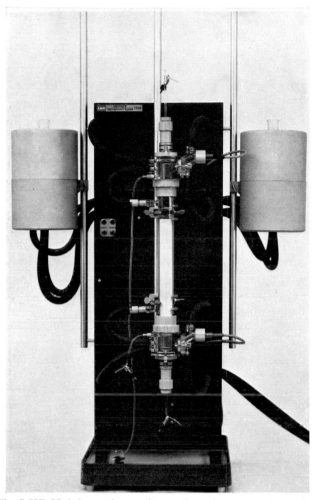

FIG. 5. The LKB Uniphor column electrophoresis system. Reproduced by permission of LKB Instruments Ltd.

protein mixtures this does not exceed 50 mg. In this respect therefore it does not provide a complete replacement for the earlier Porath column.

Finally mention should be made of a technique which uses both a solid support and a liquid density gradient as its stabilizing medium. In the simple U-tube type apparatus described by Shuster and Schrier (1967) a short column of polyacrylamide gel is layered on top of a sucrose density gradient column. During electrophoresis the migrating zones first pass down through the short gel column where some separation takes place and then through the density gradient column where further separation occurs. This technique thus seeks to couple the high resolving power of gel electrophoresis with the lower resolving power but greater ease of fractionation of a liquid column.

D. Experimental method

Whichever of the many available types of apparatus is employed the overall principles involved remain the same and there is little variation in their application. The separation column which is usually circular in cross-section but may be annular is first packed with a suspension of the stabilizing medium in buffer solution. The process varies with the medium selected as does the liquid content of the prepared column. Methods of column packing appropriate to individual support media have been outlined in Section IIIB above.

After packing, the column bed is washed extensively with the electrophoretic buffer solution and its uniformity tested. A small amount of a coloured material, (2,4-dinitrophenylethanolamine is suitable), in solution in buffer is applied to the top of the column and washed into it with further applications of buffer solution. Deformation of the coloured zone as it passes down through the column indicates irregularities in packing.

The sample to be applied to the column should be available in solution which should be dialysed against the electrophoretic buffer. It is then applied to the top of the column after careful removal of the liquid standing above the surface of the bed and displaced into the bed to a position about 3 cm from the surface by washing with buffer solution. If the sample is colourless the amount of buffer solution required for this purpose may be calculated from the behaviour under the same conditions of a coloured material of a similar type, e.g. for proteins the amount may be calculated by first determining that required to move haemoglobin, or albumin stained with bromophenol blue 3 cm into the column bed. When the sample is in the desired position buffer solution is replaced above the column, the electrodes are connected and electrophoresis is begun.

The electrophoretic conditions must of course be selected to suit particular requirements but in general the voltage applied is in the range 5–8 V per cm of column length, with a current of between 20 and 100 mA. Exceptionally with very large columns which incorporate an efficient cooling system, such as the Porath column the current may be as high as 300 mA.

At the end of the electrophoresis the column is disconnected, washed with fresh buffer and the eluate collected in fractions. The rate of elution should be about 1·5–1·7 ml/h per cm² of cross-sectional area of the column. If rates significantly greater than these are used excessive broadening of the separated zones occurs.

As an alternative, elution may be performed during the electrophoretic run itself either by using a reverse flow of buffer as described by Porath (1964) and by Porath, Lindner, and Jerstedt (1958) and mentioned in the preceding section, or by removing the faster-moving components from the foot of the column in a slow stream of buffer directed transversely across the end of the column (Vargas, Taylor, and Randle; 1960, Naughton and Taylor, 1960). Apart from the saving in time continuous elution is an advantage since it minimises zone broadening and increases the quality of the final separation.

This general method of operation is applicable to the resolution of mixtures all of whose components migrate in the same direction during electrophoresis. If both positively and negatively charged substances are present care must be taken to ensure that the starting point in the column is sufficiently far removed from the ends to prevent loss by migration out of the column. This can also be prevented by the use of Sephadex and similar gel filtration materials as the stabilizing medium. When the sample is forced into columns composed of these materials there is a tendency for electrolytes present to be retarded as the protein zone moves down the column towards its starting position. If the original sample contains a higher concentration of electrolytes than the buffer solution a layer of electrolyte will be formed in the bed above the starting position. During electrophoresis this layer will be relatively highly conducting and will thus act as a barrier to upward-moving zones. By the deliberate introduction of barriers of this kind into columns composed of substances other than gel filtration media, their capacity can be greatly increased. An excellent example of this is provided by the work of Marshall and Porath (1964) who were able, using this technique, to obtain highly purified α_1-glycoprotein from 1100 ml of human serum in a single operation on a column of the Porath type with cellulose as the supporting medium.

IV. CONCLUSION

This technique has been applied to the purification of compounds of so many types including proteins, polysaccharides, poly- and mononucleotides, enzymes, cell constituents and vitamins, that it would be impracticable to attempt to list them here, particularly as an excellent review of the subject is already available (Bloemendal, 1963).

The circumstances in which it is likely to be advantageous to apply this technique are however worthy of some consideration as are the capabilities of the technique itself. It lacks the resolving power of electrophoresis on starch and polyacrylamide gels and consequently it should not be used in analytical determinations of homogeneity. Starch and polyacrylamide gel electrophoresis on the other hand have not yet been employed successfully in other than relatively small scale purifications. In addition these techniques suffer to some extent from a limitation in the size of molecule which may be dealt with since the size of molecule able to penetrate the gel is dependent on the pore size of the gel, which, although itself dependent on gel composition, is not infinitely variable. Zonal electrophoresis in solid support columns of the type described above does not suffer from these disadvantages and is capable of producing highly purified material on a large scale. If used in combination with other methods of fractionation it can prove to be a most valuable laboratory tool.

REFERENCES

Bangham, A. D., and Dawson, R. M. C. (1962). *Biochim. biophys. Acta*, **59**, 103–115.
Berg, R. L., and Beeler, R. G. (1958). *Analyt. Chem.*, **30**, 126–129.
Bergrahm, B. (1967). *Sci. Tools*, **14**, 34–38.
Bergrahm, B., and Harlestam, R. (1968). *Sci. Tools*, **15**, 17–19.
Bernheimer, A. W. (1962). *Archs Biochem. Biophys.*, **96**, 226–232.
Bier, M. (1959). *In* "Electrophoresis; Theory, Method and Applications" (Ed. M. Bier) pp. 263–315. Academic Press, New York.
Bloemendal, H. (1963). "Zone Electrophoresis in Blocks and Columns". Elsevier Publishing Co., London.
Bocci, V. (1964). *Sci. Tools*, **11**, 7–13.
Bock, R. M., and Ling, N. S. (1954). *Analyt. Chem.*, **26**, 1543–1546.
Brakke, M. K. (1953). *Phytopathology*, **43**, 467.
Brakke, M. K. (1955). *Archs Biochem. Biophys.*, **55**, 175–190.
Carlson, L. A. (1954). *Acta chem. scand.*, **8**, 510–520.
Charlwood, P. A., and Gordon, A. H. (1958). *Biochem. J.*, **70**, 433–438.
Choules, G. L., and Ballentine, R. (1961). *Analyt. Biochem.*, **2**, 59–67.
Dierickx, L., and Ghuysen, J. M. (1962). *Biochim. biophys. Acta*, **58**, 7–18.
Dose, K., and Krause, G. (1962). *Naturwissenschaften*, **49**, 349.
Ericson, L. E., and Nihlen, H. (1953a). *Acta chem. scand.*, **7**, 980–983.
Ericson, L. E., and Nihlen, H. (1953b). *Ark. Kemi.*, **6**, 481–485.
Flodin, P., and Kupke, D. W. (1956). *Biochim. biophys. Acta*, **21**, 368–376.
Flodin, P., and Porath, J. (1954). *Biochim. biophys. Acta*, **13**, 175–182.
Gordon, A. R., and Kay, R. L. (1953). *J. chem. Phys.*, **21**, 131.
Gorkin, V. G., Avakyan, A. A., Verevkina, J. V., and Komissarova, N. V. (1963). *Fedn Proc. Fedn Am. Socs exp. Biol.*, **22**, (4 : 2) 619–622.
Haglund, H., and Tiselius, A. (1950). *Acta chem. scand.*, **4**, 957–962.
Hocevar, B. C., and Northcote, D. H. (1957). *Nature, Lond.*, **179**, 488–489.
Kobozev, G. V., and Azhits'kii, G. Yu. (1968). *Ukr. biokhem. Zh.*, **40**, 217–221.
Kolin, A. (1954). *J. chem. Phys.*, **22**, 1628–1629.
Kolin, A. (1955). *Proc. natn. Acad. Sci. U.S.A.*, **41**, 101–110.
Kolin, A. (1958). *In* "Methods of Biochemical Analysis". (Ed. D. Glick), Vol. VI, pp. 259–288. Interscience Publishers Inc., New York.
Kolin, A. (1967). *J. Chromat.*, **26**, 164–169; 180–193.
Kunkel, H. G. (1954). *In* "Methods of Biochemical Analysis" (Ed. D. Glick), Vol. I, pp. 141–170. Interscience Publishers Inc., New York.
Luner, S. J. (1968). *Analyt. Biochem.*, **23**, 357–358.

Manson, W. (1965). *Biochem. J.*, **94**, 452–457.

Marshall, W. E., and Porath, J. (1964). *J. biol. Chem.*, **240**, 209–217.

Matheka, H. D., and Geiss, E. (1965). *Arch. ges. Virusforsch.*, **15**, 301–326.

Müller-Eberhard, H. J. (1960). *Scand. J. clin. Lab. Invest.*, **12**, 33–37.

Naughton, M. A., and Taylor, K. W. (1960). *Biochem. J.*, **77**, 46–47.

Nerenberg, S. T., and Pogojeff, G. (1969). *Amer. J. clin. Path.*, **51**, 728–740.

Norris, J. R., and Manson, W. (1964). Unpublished observation.

Partridge, S. M., and Elsden, D. F. (1961). *Biochem. J.*, **79**, 26–32.

Philpot, J. S. L. (1940). *Trans. Faraday Soc.*, **36**, 38–46.

Porath, J. (1954). *Acta chem. scand.*, **8**, 1813–1826.

Porath, J. (1964). *Sci. Tools*, **11**, 21–27.

Porath, J., Lindner, E. B., and Jerstedt, S. (1958). *Nature, Lond.*, **182**, 744–745.

Shuster, L., and Schrier, B. K. (1967). *Analyt. Biochem.*, **19**, 280–293.

Sorof, S., Young, E. M., McBride, R. A., and Binder, C. L. (1964). *Sci. Tools*, **11**, 27–28.

Stein, W. H., and Moore, S. (1948). *J. biol. Chem.*, **176**, 337–365.

Svensson, H. (1954). *Iva*, **25**, 252–258.

Svensson, H. (1955). *Acta chem. scand.*, **9**, 1689–1699.

Svensson, H. (1958). *Sci. Tools*, **5**, 37–41.

Svensson, H. (1960). *In* "A Laboratory Manual of Analytical Methods in Protein Chemistry Including Polypeptides". (Eds. P. Alexander and R. J. Block), Vol. 1, pp. 193–244. Pergamon Press, London.

Svensson, H., and Valmet, E. (1955). *Sci. Tools*, **2**, 11–12.

Svensson, H., Hagdahl, L., and Lerner, K. D. (1957). *Sci. Tools*, **4**, 1–10.

Tiselius, A. (1937). *Trans. Faraday Soc.*, **33**, 524–531.

Vargas, L., Taylor, K. W., and Randle, P. J. (1960). *Biochem. J.*, **77**, 43–46.

Vendrely, R., Coirault, Y., and Vanderplancke, A. (1964). *C.R. Hebd. Seanc. Acad. Sci. Paris*, **258**, 6399–6402.

CHAPTER X

Disc Electrophoresis

KEITH E. COOKSEY

Shell Research Limited, Milstead Laboratory of Chemical Enzymology,
Broad Oak Road, Sittingbourne, Kent, England.

I.	Introduction	573
II.	Apparatus	579
	A. Description of the apparatus	579
III.	Procedure	581
	A. Preparation of sample	581
	B. Sample size	581
	C. Polymerization of the sample	582
	D. Addition of the stacking gel	582
	E. Preparation of the small-pore separation gel . . .	582
	F. Electrophoresis	583
	G. Removal of gels from the running tubes	583
	H. Staining of gels	584
	I. Variations of the above technique	584
	J. Artifacts in disc electrophoresis	586
	K. Storage of reagents and purity	587
IV.	Staining of Gels	587
	A. Proteins	587
	B. Enzymes	590
	C. Nucleic acids	591
V.	Immunological Techniques	591
	A. Immuno-disc electrophoresis	591
	B. Preparation of antisera	591
	C. Immunodiffusion techniques	591
VI.	Elution from Gels	591
VII.	Radiochemical Techniques	592
VIII.	Applications of the Disc Electrophoresis Method . . .	592
References	. . . .	593

I. INTRODUCTION

Disc electrophoresis is a comparatively new technique; the first description being published by Ornstein and Davies (1959). Kendall (1928) suggested that proteins could be separated by electrophoresis on agar gel columns, but his suggestion was never developed. As first published by Davies, the method was used to study blood proteins but many new applications have since been found. The technique of disc electrophoresis derives its name from the introduction of *dis*continuities into both the gel matrix and buffer systems.

Several supporting media have been used for electrophoretic separation of biological compounds and each has its particular advantage. The advantage of using a gel as compared with for instance, paper, is that the matrix of the gel imparts a separation based on the size of the individual components of the mixture undergoing separation. This was recognized by Smithies (1955) who suggested that the remarkable resolution obtained in starch gel electrophoresis was due to molecular sieving in the gel matrix. Starch gel is a good medium for electrophoresis, particularly in flat blocks. It does, however, suffer from the disadvantage that it is difficult to make gels of large pore size (i.e. low starch content) and difficulty is experienced in slicing and handling gels of high urea content. Starch also has a small number of ionisable groups which cause electroendosmosis during electrophoresis. Polyacrylamide (structures on p. 575) overcomes these disadvantages completely. Gels can be made and used which contain 2% to 30% (w/v) acrylamide and which are chemically more inert than starch. The approximate pore sizes obtained with gels of 7·5% and 30% acrylamide are 50 Å and 20 Å respectively (Ornstein, 1964). Table I shows the relative size of some proteins (Oncley *et al.*, 1947 and Schultz, 1958).

TABLE I

The relative size of some proteins

Protein	Molecular weight	Length Å,	Diameter Å
Albumin	69,000	150	38
Transferrin	90,000	190	37
γ-Globulin	156,000	235	44
Fibrinogen	400,000	700	38

A second advantage of the technique is the production of thin starting zones by means of discontinuities in the buffers and gels used. Two components of a mixture can only be separated if during electrophoresis, one moves more than the other. The distance one ion migrates more than the other must be greater than the distance occupied by the two ions in the original mixture, i.e. the starting zone, when the sample was applied to the matrix. Thin starting zones are arranged by the careful choice of buffer ions. The

Structure of polyacrylamide

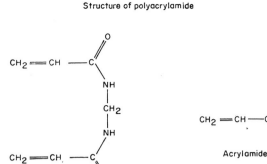

N, N'– Methylenebisacrylamide

$CH_2 = CH - CONH_2$

Acrylamide

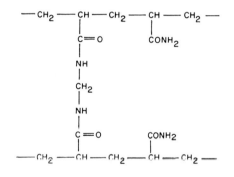

Polyacrylamide

electrode tank buffer ion must have an electrophoretic mobility slower than the slowest sample ion and the counter ion must migrate more quickly than the fastest sample ion in the sample and stacking gels. Sample ions in the large-pore gel become arranged in a series of lamina a few microns thick in order of migration velocity with the fastest ion at the bottom of the stack. The ions proceed until they reach the large pore—small pore interface. Here the relatively large sample ions are overtaken by the tank buffer ion, as the small-pore gel imposes a greater restrictive force on the large sample ions than it does on the small buffer ion. The sample ions then migrate in a constant field in the separation gel. Sample ions of similar charge are separated according to their molecular size by the sieving property of the gel. More complete discussions of factors controlling choice of buffers and pH values are given by Ornstein (1964) and Reisfeld (1964).

TABLE II
Recipes for polyacrylamide gels and electrode buffers

1. Anionic systems

System No.	% (w/v) Acrylamide in separation gel	pH of separation gel[a]	Separation gel Stock solutions[b] Reagent /100 ml water	Pts. by vol.	Sample and stacking gels Stock solutions[b] Reagent /100 ml water	Pts. by vol.	Tank buffer g/1000 ml	References	Notes
1	7·0	8·9	Acrylamide 28·0 g / Bis[c] 0·735 g	2	Acrylamide 10·0 g / Bis 2·5 g	2	Glycine 2·88 g / Tris 0·6 g pH 8·3	Davis (1964)	Originally used for proteins.
			1 N HCl 48 ml / Tris[c] 36·6 g / TEMED[c] 0·23 ml pH 8·9	1	1 M H₃PO₄ 25·6 ml / Tris 5·7 g pH 6·9	1			
			(NH₄)₂S₂O₈ 0·14 g	4	Riboflavin 0·004 g	1			
					Sucrose 40·0 g	4			
2	5·24	8·9	Acrylamide 21·2 g / Bis 0·56 g / 10 M urea 80 ml	2	Acrylamide 5·0 g / Bis 1·25 g / 10 M urea 80 ml	2	Glycine 2·88 g / Tris 0·6 g pH 8·3	Jovin, *et al.* (1964)	The final concentration of urea can be varied to suit the particular separating conditions used for proteins.
			1 N HCl 12·0 ml / Tris 9·07 g / TEMED 0·12 ml / 10 M urea 80 ml	1	1 M H₃PO₄ 12·8 ml / Tris 2·85 g / 10 M urea 80 ml	1			
			(NH₄)₂S₂O₈ 0·28 g / 8 M urea 100 ml	1	Riboflavin 0·002 g / 8 M urea 100 ml	1			
3	9·5	8·5	Acrylamide 19·0 g / Bis 1·0 g	20	Acrylamide 9·5 g / Bis 0·5 g	20	Diethyl barbituric acid 5·53 g / Tris 0·4 g pH 7·0	Richards *et. al.* (1965)	Both gels are photopolymerised. Several other gel formulae are given in the reference (pH values 5·2–9·2). The system was designed to separate soluble ribonucleic acids.
			1 N HCl 10 ml / Tris 5·88 g pH 8·5	20	1 N HCl 10 ml / Tris 1·214 g	20			
			Riboflavin 0·01 g	1	Riboflavin 0·01 g	1			

No.	pH	(As system 1)	parts	(As system 1)	parts	Electrode buffer	Reference	Used for proteins.
4	7·5	As system 1		As system 1			Williams and Reisfeld (1964)	
		1 N HCl 48 ml Tris 6·85 g TEMED 0·46 ml pH 7·5	2 / 1	1 M H_3PO_4 39 ml Tris 4·95 g TEMED 0·46 ml pH 5·5	1	Diethyl barbituric acid 5·52 g Tris 1·c g pH 7·5		
5	7·5	As system 1	1	As system 1	1	Glycine 13·7 g 2,6 Lutidine 38·2 ml pH 8·3	Taber and Sherman (1964)	Cytochromes; cathode as lower electrode (it is then strictly a cationic system).
	7·2	Acrylamide 60·0 g Bis 1·6 g $K_3Fe(CN)_6$ 0·03 g	6	Acrylamide 10·0 g Bis 2·5 g	2			
		Glycine 19·0 g TEMED 0·077 ml pH 7·2		1 N KOH 48 ml Glycine 4·8 g TEMED 0·46 ml pH 10·3	1			
		$(NH_4)_2S_2O_8$ 0·56 g	1	$(NH_4)_2S_2O_8$ 0·56 g	1			
6	9·4	Acrylamide 16·0 g Bis 0·8 g 10 M urea to 100 ml	1	Water	4	Upper electrode vessel: Tris 5·15 g Glycine 3·48 g 10 M urea 700 ml pH 8·91 Lower vessel: Tris 14·5 g 1 N HCl 60 ml pH 8·07	Reisfeld and Small (1966)	Used for the separation of immunoglobulins. Deionise urea solutions; recommended 10 M urea to be less than 3-5 μmhos.
	4·0	1 N HCl 24 ml Tris 18·15 g TEMED 0·24 ml 10 M urea to 100 ml	1	Acrylamide 8·0 g Bis 0·8 g 10 M urea to 100 ml	1			
				1 M H_3PO_4 12·8 ml Tris 2·23 g TEMED 0·1 ml 10 M urea to 100 ml	2			
		$(NH_4)_2S_2O_8$ 0·14 g 10 M urea to 100 ml	2	Riboflavin 0·001 g $(NH_4)_2S_2O_8$ 0·040 g 10 M urea to 130 ml	1			
7	5·0	Acrylamide 20·0 g Bis 0·52 g	1	No gels are used, but samples are dialysed against 0·01 M sodium phosphate buffer, pH 7·1 containing 0·1% S.D.S. and 0·1% 2-mercaptoethanol before loading on to separation gel directly		0·1 M sodium phosphate buffer pH 7·1 containing, 0·1% S.D.S.	Maizel (1966) and Shapiro et al. (1967)	Proteins are denatured and reduced in S.D.S. Mol Wt. can be determined from the mobility of the stained protein bands. All proteins are anionic because of bound S.D.S.
	7·1	TEMED 0·10 ml S.D.S.ᶜ 0·20 g 0·2 M sodium phosphate buffer pH 7·1 to 100 ml	2					
		$(NH_4)_2S_2O_3$ 0·30 g	1					

TABLE II—*continued*

Recipes for polyacrylamide gels and electrode buffers

2. *Cationic systems*

System No.	% (w/v) Acrylamide in separation gel	pH of separation gel[a]	Separation gel — Stock solutions — Reagent /100 ml water	Pts. by vol.	Sample and stacking gels[b] — Reagent /100 ml water	Pts. by vol.	Tank buffer g/1000 ml	References	Notes
8	7·5	4·3	Acrylamide 30·0 g, Bis 0·80 g	2	Acrylamide 10·0 g, Bis 2·5 g	2	β-Alanine 31·2 g, Acetic acid (glacial) 8·0 ml, pH 4·5	Reisfeld *et al.* (1962)	For basic proteins.
			1 N KOH 48 ml, Acetic acid (glacial) 17·2 ml, TEMED 4·0 ml, pH 4·3	1	1 N KOH 48 ml, Acetic acid (glacial) 2·87 ml, TEMED 0·46 ml, pH 6·8	1			
			$(NH_4)_2S_2O_8$ 0·28 g	1	Riboflavin 0·004 g	1			
			Water	4	Water	4			
9	20	2·9	Acrylamide 60·0 g, Bis 1·20 g	1·6	No gels are used. Proteins are dissolved in sample buffer: KOH 0·224 g, Sucrose 34·2 g, Acetic acid to give pH 2·9		Upper (anode) buffer Valine 35·2 g, Acetic acid to give pH 4·0. Lower (cathode) buffer. Glycine 22·5 g, Acetic acid to give pH 4·0	Shepherd and Gurley (1966)	For the separation of histones and other basic proteins.
			KOH 0·224 g, TEMED 0·4 ml, Sufficient acetic acid to give pH 2·9						
			$(NH_4)_2S_2O_8$ 100 g	0·4					
10	7·5	approx. 1·6	Acrylamide 10·0 g, Bis 0·27 g, Urea 20·0 g, Acetic acid (glacial) 46·7 ml	3	No gels are used. Proteins are dissolved in phenol: acetic acid : water, (2 : 1 : 1, w/v/v) and urea is added to give a final concentration of 2 M.		Acetic acid 100 ml	Takayama *et al.* (1966)	75% acetic acid is used for layering the separation gel in place of water. Polymerizations are carried out at 50°C for 15 min. The sample space is rinsed and filled with 75% acetic acid prior to layering the sample between this solution and the gel surface. The system has been used to separate hydrophobic proteins (membranes and mitochondrial fragments).
			TEMED (undiluted)	0·02					
			$(NH_4)_2S_2O_8$ 0·50 g, Urea 60·0 g	1					

a The pH of the separation is usually 0·5 pH units higher than the pH of the recipe buffer for an anionic system and 0·5 pH units lower for a cationic system. (Reisfeld, (1964)).

b Volumes of stock solution to be mixed to give working solution.

c Abbreviations

Bis is N,N′-methylenebisacrylamide (BDH Ltd., Poole, Dorset, U.K.);
Tris is trishydroxymethylaminomethane (Koch-Light Ltd., Colnbrook, Bucks, U.K.);
TEMED is Tetramethylethylenediamine (BDH Ltd., Poole, Dorset, U.K.);
SDS is Sodium dodecylsulphate (Sigma Chem. Co., St. Louis, Missouri, U.S.A.).

Many buffer systems are described in the literature. Table II lists a range of useful systems with comments about their applications. In general, the purest grades of chemicals obtainable should be used in order to achieve a high level of reproducibility and sensitivity.

II. APPARATUS

Disc electrophoresis apparatus is now offered by most large suppliers of laboratory equipment. The general design of commercial equipment is similar to that shown in Figs 1, 2 and 3.

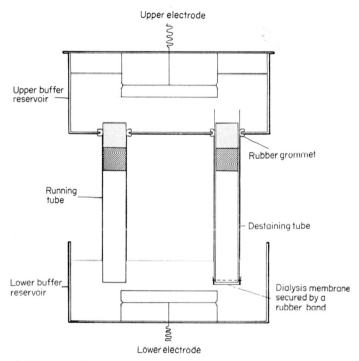

Fig. 1. Apparatus for disc electrophoresis and electrophoretic destaining.

A. Description of the apparatus

(a) The electrophoretic separation is performed on polyacrylamide gel columns contained in uniform-bore glass or acrylic tubes. The tubes most often used are 5 mm internal diameter by 7 mm external diameter and approximately 70 mm long. To contain unpolymerized gel solutions, the tubes must be capable of being stoppered. Workers have used various types of stoppers, e.g. serum caps, silicone rubber moulding paste closures or polyethylene cups (see Fig. 3).

(b) A rack for holding the tubes during polymerization should be open sided to allow the tubes to be illuminated from the side during the photopolymerization step. It should, of course, hold the tubes securely and vertically. A convenient design is shown in Fig. 2.

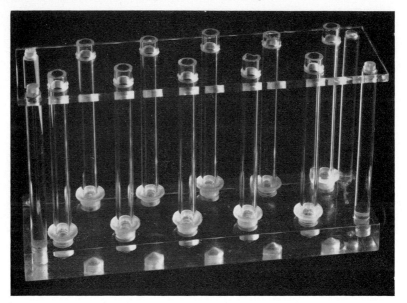

Fig. 2. A rack for holding running tubes during polymerization of the gels. The top is removable.

(c) Photopolymerization of gels requires a daylight fluorescent lamp of about 10–25 watts; more even illumination can be achieved by using several tubes of lower wattage. Ideally, gel solutions should be illuminated from both sides, especially when small-pore gels are being polymerized. The lamp needs to be constructed so that the gel tubes can be positioned about 7 cm away from the fluorescent tubes.

(d) Buffer reservoirs can be constructed from plastic refrigerator dishes. For the upper reservoir, holes are drilled in the bottom of a dish to take rubber grommets. The grommets should be of such a size that when the running tubes are in place, buffer does not leak from the reservoir (e.g. Fig. 1). Electrodes must be equidistant from each tube. They can be carbon rods cemented to the bottom of the dishes or platinum wires cemented to circular plastic formers. A convenient means of supporting the electrode in the upper reservoir is to use the reservoir lid as an electrode support.

(e) When the tubes are in place and the reservoirs filled with buffer, it only remains to connect the power supply. Most laboratory electrophoresis

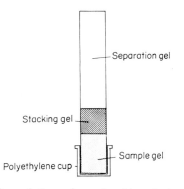

Fig. 3. Polymerization of the polyacrylamide gels in the running tube.

power supplies capable of delivering 100 milliamps at 200 volts can be used.

(f) A selection of dissecting needles and a 10 ml syringe with a number 23 gauge needle are required to remove gels from the glass running tubes after electrophoresis.

(g) Electrolytic destaining is carried out in a similar apparatus to that used for the electrophoretic separation. The same lower reservoir is used but a different upper reservoir is required. This must have larger grommets to accommodate the 10 mm (outside diameter) by 8 mm (inside diameter) by 90 mm long destaining tubes. These tubes are usually closed at one end with a plug of acrylamide gel, but I have found it quicker and more satisfactory to close the end with a piece of dialysis membrane held in place by a rubber band.

III. PROCEDURE

A. Preparation of sample

Samples containing large amounts of salt must be dialyzed before running. The tank buffer or appropriately diluted sample gel buffer can be used as dialysis solution.

B. Sample size

The weight of sample analysed depends entirely on the number of components in the mixture and the relative abundance of each one. For proteins, as little as 1 μg can be detected. If the sample is completely unknown, it is best to analyse 100 μg of total protein and adjust the sample size in the light of the result to obtain the best resolution.

C. Polymerization of the sample

In order that unpolymerized gel solutions can be covered with a layer of water, the clean running tubes are rinsed in very dilute detergent solution (a few drops of Teepol or other wetting agent in the rinsing water), allowed to drain and then dry. They are assembled in the polymerization rack with their stoppers (Fig. 2). Large-pore gel, 0·3 ml, containing the sample to be analysed is then added to the tubes by means of a Pasteur pipette. It is important not to wet the sides of the tube with sample-containing gel. This solution is then overlayered with 3–5 mm depth of water. The layering procedure is best accomplished by the use of a fine pipette control attached to a Pasteur pipette. Other workers have used a pipette drawn out to a fine tip or a Pasteur pipette threaded with a small cotton wick. I prefer the first technique. When a large sample is to be analysed it is convenient to make the large-pore gel reagent twice the concentration recommended in the recipe tables. The size of the sample can then be half the total volume of the gel, in this case 0·15 ml.

The tubes are then placed approximately 7 cm from the photopolymerization lamp. About 45 min are required for complete polymerization which is indicated by the development of opalescence in the gel.

D. Addition of the stacking gel

The water layer above the sample gel is tipped off and the last few drops in each tube removed by touching the tube with a paper tissue. The flat surface of the sample gel is rinsed once with large-pore gel reagent. Stacking gel solution, 0·2 ml, is then added to each gel tube, overlayered with water, and polymerized as described under the preparation of the sample gel. Polymerization in this case is quicker, due to the absence of sample from the gel. The time taken is 20–30 min.

E. Preparation of the small-pore separation gel

The water layer on the stacking gel is removed as described above. Small pore gel solution, 0·1 ml, is added to each tube to rinse the surface of the stacking gel. This is repeated once. It is important that the small pore—large pore interface is flat and that overlayering water (from the stacking gel preparation) does not remain to dilute the small-pore gel. Local dilution of the reagent causes irregularities in pore size which have a deterimental effect on the performance of the gel.

Small-pore gel solution is added to each tube. The tubes are filled so that a meniscus forms *above* the top of the tube. When the tubes are capped with a piece of household wrapping plastic (e.g. Mellinex, ICI Ltd., London, or Handi-Wrap, Dow Chemical Co., Midland, Michigan) the small volume of gel in the meniscus is displaced and an air-tight seal is made between the tube and the plastic film. If air bubbles are trapped beneath the plastic film they must be removed, as oxygen inhibits poly-merization of the gel. Tubes of the size recommended in the apparatus section require about 1·5 ml gel solution for rinsing and filling. Polymeri-zation takes 30 min.

F. Electrophoresis

The gel tubes are removed from the rack and the small plastic closures removed with the aid of a dissecting needle. Care must be taken not to cause a vacuum on the gel by the sudden removal of the cap, otherwise air will be sucked between the gel and the running tube. The tubes are inverted and pushed through the grommets in the upper buffer reservoir, so that the sample is uppermost. The upper and lower reservoirs are filled with tank buffer and the electrodes connected to the power supply. A hook-shaped Pasteur pipette can be used to remove air trapped beneath the gels. If the air is allowed to remain, it causes the electrical resistance of the gel column to rise and consequently sample ions will migrate less in such a tube than in others run at the same time. 0·001% Bromphenol blue, 0·5 ml/100 ml buffer, is added to the upper reservoir. Methyl green is used for cathodically migrating systems. The indicator dye migrates with the salt front in the form of a blue disc. Usually electrophoresis is allowed to take place at 4–5 milliamps per tube until the blue tracking disc has reached the end of the gel. If for some reason the tracking dye in a partic-ular tube reaches the end of the running tube before others in the same experiment, the current to that tube can be cut off by placing a sleeve of silicone rubber around the tube in the upper electrode compartment. If this is done, the tubes can be processed until all have passed the same total current.

G. Removal of gels from the running tubes

There are several ways to remove gels from the running tubes at the end of an electrophoretic separation. The method I prefer is described first, but the other methods work equally well.

(a) The gel tube is held under water in a shallow bowl and a hypodermic needle (number 23 gauge) attached to a 10 ml syringe filled with water is used to rim the sample end of the gel while water is gradually injected between the gel and the glass. The needle is pushed in carefully while rotating the glass tube. The needle is removed slowly when it has reached the top of the separation gel. The procedure is repeated from the other end of the gel. When the needle is subsequently removed, the gel will come with it. It can be stored in a test-tube for a few minutes before the next procedure, which is staining.

(b) The procedure is carried out as in (a), but the needle is attached by a long rubber tube to a water tap.

(c) A rubber teat filled with water is attached to the sample end of the gel and water is forced round the gel by squeezing the teat.

H. Staining of gels

This part of the technique depends on the nature of the sample under investigation. A detailed description of each staining method is given in a later section.

I. Variations of the above technique

These variations are described separately from the main procedure for the sake of clarity.

1. *Samples and sample gels*

Some samples, particularly coloured proteins, are known to inhibit photopolymerization of the sample gel. When this happens, it is necessary to change the technique slightly. The sample gel is replaced by 1 M sucrose or 8 M urea. The stacking gel is layered on top of this, water layered and photo-polymerized. The separation gel is added as before. When the tubes are ready for running, the dense solution in the space reserved for the sample must be removed and the surface of the stacking gel rinsed with water. The tubes are attached to the buffer reservoir as in the normal procedure, the buffer reservoirs filled and the sample, which has been made denser than the tank buffer with urea or sucrose, pipetted carefully on to the stacking gel. A final concentration of 0·25 M sucrose or 2 M urea makes the sample sufficiently dense for layering on the stacking gel. The addition should be made with an easily regulated pipette control as the sample solution has to be pipetted below the liquid level in the buffer tank. The starting current should be 1–2 milliamps per tube so that ohmic heating does not cause convective disturbances in the liquid sample. The current can be increased to 4–5 milliamps per tube when the tracking dye has entered the separation gel. The resolution of sample ions by this means is in no way inferior to that obtained with a sample gel. Some workers have dispensed with the stacking gel also (Hjerten *et al.*, 1965). They prefer to add their samples directly on to the top of the separation gel. If the conductivity of the sample is lower than that of the buffer tank, stacking of sample ions will occur as it does in a stacking gel. I have never been able to achieve the same degree of resolution without a stacking gel as with one.

2. *Internal standards*

It is sometimes necessary to run a disc electrophoresis experiment for a longer time than usual in order to separate closely migrating ions. When this happens, the tracking dye runs into the lower electrode chamber and all indication of the total current passed by individual tubes is lost. Comparison of stained gels then becomes difficult because, in spite of great care in duplicating conditions when preparing gels, more current is sometimes passed by one gel than another. This causes the same species of sample ion to migrate different distances in each gel. I have used an internal standard ion to overcome this difficulty. If a small quantity of a pure protein, for example ferritin or cytochrome C, is added to each sample, the migration of this ion can be used as a reference. These two proteins have the additional advantage that they are highly coloured and can be seen during the electrophoresis. Obviously the chosen ion will have to move at a different rate from those of ions of interest on the gel. Ferritin is used in anodically migrating systems and cytochrome C in cathodic systems (Fig. 5(g)).

3. *Split-gel technique*

For a critical comparison of two samples, the following technique is recommended. The gel tube is prepared as for a layered sample. A piece of cardboard (plastic impregnated cardboard is ideal) is cut so that it fits tightly in the top of the gel tube. The cardboard should be inserted into the top of the stacking gel. Samples are then pipetted either side of the card. Care must be taken that the samples in the two compartments do not contaminate each other. This can be checked by the addition of a standard ion to one of the samples and not to the other. The presence of a band corresponding to the sample right across the gel indicates leakage. (Fig. 5 (g)).

4. *Preparation of separation gel*

Disc electrophoresis gels can be scanned directly and the separated ions can be measured. The separation gels as prepared contain ultraviolet absorbing materials which interfere with densitometry, especially in the blue end of the spectrum. These interfering compounds can be removed by soaking the gels in appropriately diluted separation gel buffer for several days. In this case, of course, the stacking gel is added on top of the separation gel prior to its use. Separation gels can be made and kept without stacking gels for several days if they are immersed in cold buffer.

Some batches of reagents give separation gels that polymerize either too quickly or too slowly. Control of polymerization time may be obtained by varying the relative amounts of potassium ferricyanide and tetramethyl-ethylenediamine (TEMED) in the gel recipe for the separation gel. Decreasing the ferricyanide or increasing TEMED concentration will shorten the polymerization time.

If chemical polymerization is undesirable for the separation gel, the gel can be polymerized photochemically. This can be achieved by replacing the ammonium persulphate solution in the gel formulation by the ribo-flavin solution. The gel produced is usually of a larger pore size than a chemically polymerized gel so that sample ions move more quickly in it. This type of polymerization is of particular use when it is necessary to elute biologically active material from the gels, after a separation has taken place.

J. Artifacts in disc electrophoresis

1. Dissociation of complex ions

When two ions are associated in a complex, it is possible to separate them by subjecting them to a very large potential gradient. This situation occurs in the stacking and sample gels in disc electrophoresis, during the passage of the tracking dye. Complexes are separated to varying degrees depending on the length of the stacking gel. That this phenomenon is occurring in any particular system can be checked by running the same sample with long and short large-pore gels. When a long large-pore gel is used, the individual ions will be prevalent; when a short gel is used the complex ion will be prevalent.

2. The influence of ammonium persulphate on sample ions.

The effect of an oxidizing agent such as ammonium persulphate on an easily oxidized sample ion is rather obvious. Such an ion can be, for example, an enzyme requiring a free sulphydryl group for activity. Traces of persulphate left in the separation gel after polymerization are not all removed at the salt front (tracking dye) during electrophoresis. To protect such oxidisable ions from persulphate, mercaptoacetate can be added to the sample. The mercaptoacetate will migrate at the same speed as the salt front in an anodically migrating system and remove persulphate from the gel before the protein reaches it. Alternatively, the separation gel can be photopolymerized.

3. Mis-shapen bands

Bands stained after electrophoresis should be straight and at right angles to the axis of the gel. Frequent artifacts are annular zones and cup-shaped zones. The annular zones are caused by the origin (large pore-small pore interface) being dome shaped, i.e. convex upwards. When samples reach this discontinuity in the gel tube, they move down the gradient of the dome to form an annular-shaped starting zone. This is not a serious inconvenience and resolution will not be drastically impared. Incorrect water layering of the stacking gel can cause this gel to polymerize inefficiently at the water/gel interface and when the small-pore gel is added later, this takes up the shape of the large-pore gel interface. The other artifact is caused by too high an electrophoresis current causing heating in the gel and faster migration of the sample ions in the centre of the gel. Cup-shaped zones are also caused when the separation gel is allowed to polymerize too rapidly. The gel shrinks and causes the origin to be concave, thus imparting this shape to the separated ions. Both these artifacts make quantitative estimation of the separated ions by densitometry of the stained gels very difficult, if not impossible.

K. Storage of reagents and purity

All the reagents mentioned are stable for several weeks with the exception of the ammonium persulphate solutions. These are usable for one week. Only the acrylamide and Bis need to be purified, and then only for the most critical work. If separated material is to be eluted from unstained gels, then it is advisable to recrystallize the acrylamide and Bis from chloroform and acetone respectively.

IV. STAINING OF GELS

After the gels have been removed from the running tubes, the separated components of the sample need to be made visible. It is possible to scan the visible absorption of gel rods directly (Taber and Sherman, 1964) or to stain the separated bands.

A. Proteins

The most commonly used stain for proteins is Naphthalene Black (Amido Schwartz). The proteins have to be fixed before staining. When using Naphthalene Black, it is usual to include the fixative in the dye formulation. Gels are immersed in 0·5% Naphthalene black in 7% acetic acid (v/v) for 2 h. After this, the gels are removed from the stain and rinsed in tap water. Excess stain is removed by washing in several changes of 7% acetic acid (v/v). A more convenient way to remove dye is by electrolysis. The gels, after rinsing in water, are placed in tubes similar to the running tubes but a few millimetres wider. The tubes are closed at their lower end by attaching a piece of dialysis membrane with a rubber band. They are inserted

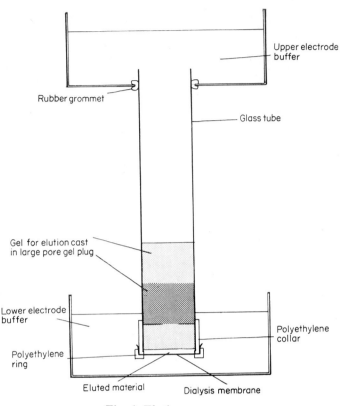

Fig. 4. Elution apparatus

into the upper reservoir of the electrophoresis bath as in the original electrophoresis. Larger grommets are needed and it is convenient to have a second upper buffer reservoir with larger grommets attached. The upper and lower baths are filled with 7% acetic acid (v/v) and current is passed at a rate of 5–10 milliamps per tube with the anode as the lower electrode. Prolonged electrolysis will remove some bound dye because it is in equilibrium with the acetic acid solution. This can be remedied by adding a very small quantity of dye to the upper electrode compartment.

Coomassie Blue R 250 (I.C.I Dyestuffs Division, Blackley, Manchester, U.K. and Colabs Laboratories Inc., Chicago, Illinois, U.S.A.) is also used for staining proteins (Chrambach *et al.*, 1967). I have found it up to four times as sensitive as Naphthalene Black for staining the toxic proteins of *Bacillus thuringiensis* but less useful than Naphthalene black for staining the vegetative cell proteins of the same organism. Gels are first fixed in 12·5% trichloroacetic acid for 30 min and then transferred to a dye solution.

The dye solution is made from 1% Coomassie Blue in water diluted 1 : 20 with 12·5% trichloroacetic acid. The gels are allowed to take up dye for 30 min to 1 h. They are then washed with water and transferred to 10%

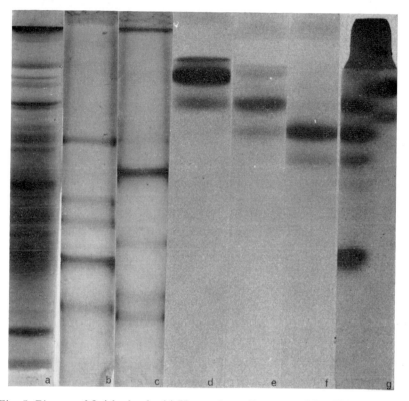

Fig. 5. Picture of finished gels. (a) Vegetative cell extract of *Bacillus thuringiensis* Mattes isolate; stained with Naphthalene black. (b) Vegetative cell extract of *Bacillus thuringiensis*, Tolworth strain, stained to show esterase activity, (see section VB 1). (c) As (a), but stained for esterase activity. (d), (e) and (f) Toxic protein fractions from a DEAE-cellulose chromatography column. The fractions were derived from *B. thuringiensis* crystal toxin. (Cooksey, 1968). The bands were stained with Naphthalene black. (g) Two samples run on the same gel. An internal control of Ferritin is present in the left-hand sample. The bands were stained with Coomassie Blue. (See section IV H and IV I (2 and 3)).

trichloroacetic acid for storage. The intensity of the staining Increases for 48 h, so that initial overstaining should be avoided. A great advantage of this method is that the polyacrylamide itself takes up little or no stain, so that the destaining step, needed with Naphthalene black, is avoided.

B. Enzymes

Several enzymically active proteins can be demonstrated on polyacryl-amide gels. The stains originally developed for histochemistry can be used. Obviously, enzymes may not survive electrophoresis for a variety of reasons. One of these, heat inactivation, can be avoided by running the experiment in a cold room and by using a low current. The removal of prosthetic groups during the stacking process can be overcome by including enzyme cofactors in the staining solution. Some of the enzyme stains do not damage the protein stained, so that elution of biologically active material from the stained band is possible.

1. Esterases

Gels are soaked in the dark in the solution given below until the bands appear. The stain consists of $0 \cdot 1$ M Tris-Maleate, pH $6 \cdot 4$, 50 ml; 1% 1-naphthylacetate in 50% acetone : water, 2 ml and Fast Blue B salt, 50 mg (Lawrence et al., 1960). The esterases appear as red-brown bands. Fast Blue B salt can be replaced by Fast Blue 2R to produce bands of a slightly different colour. The pH of the buffer can be changed to suit the requirements of the individual investigation. The gels must be removed from the stain and washed with water when the bands have appeared. (Fig. 5, (b) and (c))

2. Phosphatases

These enzymes are detected in a similar manner to esterases. The stain solution consists of buffer 50 ml; β-naphthylphosphate 100 ml, and Fast Blue B, 50 mg. The buffer can be either $0 \cdot 05$ M acetate, pH $5 \cdot 0$ (acid phosphatase) or $0 \cdot 05$ M borate, pH $9 \cdot 7$, containing $0 \cdot 01$ M Mg^{++}, (alkaline phosphatase) (Smith et al., 1967).

3. Dehydrogenases

Lactate dehydrogenase: The gels are washed in cold $0 \cdot 01$ M-Tris buffer, pH $8 \cdot 3$, and then immersed in a solution containing per 100 ml $0 \cdot 014$ M Tris-HCl buffer, pH $8 \cdot 3$, 70 ml; $0 \cdot 1$ M sodium lactate, 30 ml; Nitro-blue tetrazolium, 80 mg; N.A.D.* 30 mg; phenazine methosulphate, $1 \cdot 4$ mg, for $\frac{1}{2}$–4 h, at 37 °C (Goldberg, 1963). The substrate concentration may have to be adjusted for some applications. Other dehydrogenases can be detected with different substrates.

Other enzymes which have been demonstrated include amylases (Doane, 1965), 1 : 4 α-glucan : orthophosphate glucosyl transferase (phosphory-lase) (Frederick, 1963) and β-galactosidase (Fairbanks et al., 1965).

* N.A.D. is Nicotinamide–adeninedinucleotide.

C. Nucleic acids

The method described below is that of Richards *et al.* (1965). Gels are immersed for 4–16 h in a fixative stain solution which has the following composition:

Lanthanum acetate, 1% (w/v); acridine orange, 2% (w/v); and acetic acid, 15% (v/v). The gels are destained electrolytically or by washing with water. Nucleic acids appear as orange red bands.

V. IMMUNOLOGICAL TECHNIQUES

A. Immuno-disc electrophoresis

The high resolving power of disc electrophoresis can be used to separate antigenic substances. Gels are removed from the running tubes in the manner previously described. After blotting excess water from a gel, it is laid on a previously prepared agar-covered plate. A suitable plate is made by pipetting 12 ml of 1% diffusion agar in 0·9% (w/v) NaCl onto a lantern slide cover glass (9 cm × 9 cm). Antiserum trenches are cut about 1 cm each side of the gel rod and filled with appropriate antisera. Precipitin arcs appear after about 24 h incubation in a warm, damp atmosphere. The unstained gel can be removed and a duplicate stained gel put in its place. Precipitin arcs can then be compared with the stained bands (Cooksey, 1968). The sample load on the electrophoresis gel and the distance of the antiserum trench from the gel must be adjusted to give optimal conditions. There is very little diffusion in the polyacrylamide gels during the development of the precipitin arcs (24 h) so that the resolution of the electrophoresis system is maintained.

B. Preparation of antisera

It has been reported that it is possible to elicit antibody response by injecting homogenized gels, containing separated protein bands, directly into animals (Kunitake, 1964). This should simplify the production of monospecific antisera.

C. Immunodiffusion techniques

The gel rod can be sliced in about 3 mm transverse sections and embedded in diffusion agar in the standard patterns. Wells can be cut and filled with antiserum in the usual manner. (Oakley, this Series, Volume 5a). It is also possible to embed the gel discs in linear sequence and then cut an antiserum trench adjacent to the line of embedded discs.

VI. ELUTION FROM GELS

Unfixed protein may be eluted from polyacrylamide by homogenizing the gel in buffer or in buffer to which a trace of detergent has been added.

Homogenization of the gel can be preceded by freezing, which destroys some of the gel structure. Protein or nucleic acid may be electrophoresed from cut sections of the gel using the apparatus shown in Fig. 4. (J. R. Norris, pers. com.). The dialysis membrane prevents mixing of the sample with the lower buffer. The sample can be recovered from the elution chamber with a syringe. It is well to remember that material eluted from gels is always contaminated with gel reagents.

VII. RADIOCHEMICAL TECHNIQUES

Radioactivity associated with electrophoretically separated bands can be measured in several ways. The easiest method is to slice the gel transversely and count the slices. A freezing microtome has been used for this purpose (Scham and Roosens, 1964). The slices are then dried, placed in liquid scintillation counting fluid and counted. Provided very thin slices are cut, the error due to self absorption is not great. It is possible to exchange the water in a gel for scintillator fluid by soaking the gel slices in several changes of the fluid. Counting efficiency for C^{14} in a toluene based scintillator is reported to be $81 \pm 12\%$. (Boyd and Mitchell, 1966).

Radioautography of stained gels is possible by a method published by Fairbanks et al., (1965). These workers sliced gels longitudinally with wires while the gel rod was held in a jig. The slices were dried on a filter paper by the application of vacuum from below. After drying, radio-autographs were made in the usual manner.

VIII. APPLICATIONS OF THE DISC ELECTROPHORESIS METHOD

The following is a short list of applications of the technique.

1. There are many references in the literature where a single band on disc electrophoresis has been used as a final criterion of homogeneity e.g. lysine-ribonucleic acid synthetase was purified from Escherichia coli by Waldenström (1968). β-Galactosidase from E. coli K 12 was shown to be highly purified by Craven et al. (1965). Calender and Berg (1966) showed that highly purified tyrosyl RNA-synthetases from E. coli and Bacillus subtilis were electrophoretically distinct.

2. Henning and Yanofsky (1963) showed that mutationally altered A proteins of tryptophan synthetase from E. coli mutants have electrophoretic mobilities different from the wild type enzyme.

3. Hjerten et al. (1965) have separated 30 S and 50 S ribosomes from E. coli on both an analytical and preparative scale.

4. Zeldin and Ward (1963) followed changes in the soluble proteins of the slime mould, Physarum polycephalum during morphogenesis.

5. Leboy *et al.* (1964) showed that the ribosomal proteins of *E. coli* K 12 were different from those of other strains of *E. coli.*

6. Disc electrophoresis followed by autoradiography was used to study the synthesis of labelled protein during enzyme induction. (Fairbanks *et al.*, 1965). Newly synthesized radioactive protein was shown to be detectable by virtue of its C^{14} label at a lower concentration than was possible by staining.

7. Poliovirus protein has been fractionated into four separate components. (Maizel, 1964).

8. Several workers have noted the correlation between the molecular weight of an ion and its migration in disc electrophoresis. Shapiro *et al.* (1967) showed that the relative electrophoretic mobility of proteins in sodium dodecyl sulphate gels was linearly related to the logarithm of their molecular weight. Bishop *et al.* (1967) showed a similar relationship for nucleic acids.

9. Maizel developed a gel fractionator which depends on the extrusion of a polyacrylamide gel rod through a small aperture to disrupt the gel. The gel is propelled down a tube and through a fine hole in the end of the tube, by a motor driven piston. Here the gel fragments are mixed with a stream of sodium dodecyl sulphate solution. In the original paper (Maizel, 1966) the gel homogenate was counted for radioactivity but probably enzyme activity could also be measured in the effluent if a suitable buffer was used in place of the detergent.

10. The preparative possibilities of disc electrophoresis are obvious. Several commercial instruments are available and a discussion of their relative merits can be found in Gordon and Louis (1967) and Gordon (1968).

REFERENCES

Bishop, D. H. L., Claybrook, J. R., and Spiegelman, S. (1967). *J. molec. Biol.*, **26**, 373.

Boyd, J. B., and Mitchell, H. K. (1966). *Analyt. Biochem.*, **14**, 441.

Calendar, R., and Berg, P. (1966). *Biochemistry*, **5**, 1681.

Chrambach, A., Reisfeld, R. A., Wycoff, M., and Zaccari, J. (1967). *Analyt. Biochem.*, **20**, 150.

Cooksey, K. E. (1968). *Biochem. J.*, **106**, 445.

Craven, G. R., Steers, E. and Anfinsen, C. B. (1965). *J. biol. Chem.*, **240**, 2468.

Davies, B. J. (1964). *Ann. N. Y. Acad. Sci.*, **121**, 404.

Doane, W. W. (1965). *Amer. Zool.*, **5**, 697.

Fairbanks, G., Levinthal, C., and Reeder, R. H. (1965). *Biochem. biophys. Res. Commun.*, **20**, 393.

Frederick, J. F. (1963). *Phytochemistry*, **2**, 413.

Goldberg, E. (1963). *Science*, **139**, 602.

Gordon, A. H., and Louis, L. N. (1967). *Analyt. Biochem.*, **21**, 190.

Gordon, A. H. (1968). "Electrophoresis of Proteins in Polyacrylamide and Starch Gels". North Holland Publishing Co., Amsterdam.

Henning, U., and Yanofsky, C. (1963). *J. molec. Biol.*, **6**, 16.

Hjerten, S., Jerstedt, S., and Tiselius, A. (1965). *Analyt. Biochem.*, **11**, 219.

Jovin, T., Chrambach, A., and Naughton, M. A. (1964). *Analyt. Biochem.*, **9**, 351.

Kendall, J. (1928). *Science*, **67**, 163.

Kunitake, G. (1964). *In* "Disc Electrophoresis News Letter" number 6, Canal Industrial Corporation, Rockville, Maryland, U.S.A.

Lawrence, S. H., Melnick, P. J., and Weimer, H. E. (1960). *Proc. Soc. exp. Biol. Med.*, **105**, 572.

Leboy, P. S., Cox, E. C., and Flaks, J. G. (1964). *Proc. natn. Acad. Sci., U.S.A.* **52**, 367.

Maizel, J. V. (1964). *Ann. N.Y. Acad. Sci.*, **121**, 382.

Maizel, J. V. (1966). *Science*, **151**, 988.

Oncley, J. L., Scatchard, G. S., and Brown, A. (1947). *J. Phy. colloid. Chem.*, **51** 184.

Ornstein, L. (1964). *Ann. N.Y. Acad. Sci.*, **121**, 321.

Ornstein, L., and Davies, B. J. Unpublished work, preprinted by Distillation Products Industries, Division of Eastman Kodak Co., Rochester N.Y., U.S.A.

Reisfeld, R. A., Lewis, U. J., and Williams, D. E. (1962). *Nature, Lond.*, **195**, 287.

Reisfeld, R. A. (1964). *Ann. N.Y. Acad. Sci.*, **121**, 373.

Reisfeld, R. A., and Small, P. A. (1966). *Science*, **152**, 1253.

Richards, E. G., Coll, J. A., and Gratzer, W. B. (1965). *Analyt. Biochem.*, **12**, 452.

Schram, E., and Roosens, P. (1964). *Arch. Int. de Phys. et de Biochemie*, **72**, 695.

Schultz, H. E. (1958). *Clin. Chim. Acta.*, **3**, 24.

Shapiro, A. L., Vinuela, E., and Maizel, J. V. (1967). *Biochem. biophys. Res. Comm.*, **28**, 815.

Shepherd, G. R., and Gurley, L. R. (1966). *Analyt. Biochem.*, **14**, 356.

Smith, I., Perry, J. D., and Lightstone, P. J. (1967). *Biochem. J.*, **105**, 12P.

Smithies, O. (1955). *Biochem. J.*, **61**, 629.

Taber, H. W., and Sherman, F. (1964). *Ann. N.Y. Acad. Sci.*, **121**, 600.

Takayama, K., Maclennan, D. H., Tzagloff, A., and Stoner, C. D. (1966). *Arch. Biochem. Biophys.*, **114**, 223.

Waldenström, J. (1968). *Eur. J. Biochem.*, **3**, 483.

Williams, D. E., and Reisfeld, R. A. (1964). *Ann. N.Y. Acad. Sci.*, **121**, 373.

Zeldin, M. H., and Ward, J. M. (1963). *Nature, Lond.*, **198**, 389.

CHAPTER XI

Isoelectric Focusing and Separation of Proteins

Olof Vesterberg

Karolinska Institutet, Stockholm, Sweden

I. Introduction 595
II. Description of the theory of the Method 597
   A. Resolving power 598
   B. Determination of isoelectric point 599
   C. Separation capacity 599
III. Equipment 601
IV. Practical Techniques 603
   A. The protein solution 603
   B. Preparation of the density gradient 603
   C. Electrode solutions 605
   D. Temperature of the cooling water 606
   E. Voltage, maximum electrical load 606
   F. Electrofocusing time 607
   G. Collection of fractions 607
V Applications 607
   A. Cellulolytic and related enzymes from fungi . . . 607
   B. β-Fructofuranosidase (invertase) from yeast . . . 608
   C. Extracellular enzymes and toxins from *Staphylococcus aureus* . 609
   D. Bacteriolytic proteins 611
   E. Proteins from other bacteria 612
   F. Proteins of viral origin 613
   G. Immunoglobulins 613
References 613

I. INTRODUCTION

The technique of isoelectric focusing and separation of ampholytes, such as proteins and peptides, in stable pH gradients has found applications in several branches of natural science and especially in biochemistry. The method, which is often called electrofocusing, has two valuable main applications—

(1) For separating and purifying proteins on an analytical and a preparative scale, the method possesses a high resolving power compared to other methods.

(2) For characterizing proteins, i.e., determining their isoelectric point, pI. In a single experiment the isoelectric point of one or more proteins can be easily and exactly determined. The quantity of the protein that is needed depends entirely on the sensitivity of the method used for detecting the protein. If biological methods are used, only minimal amounts of protein, e.g., enzymes and toxins are required. The method is unique for pI determinations compared to other methods. For determining an isoelectric point at zero ionic strength, using moving boundary electrophoresis, more protein material and several experiments are needed.

Although the method of isoelectric focusing and separation of ampholytes, such as proteins, is new, the basic principles have been known for several decades. The principle is best understood if we recall the classic experiment with an electrolysis cell filled with a dilute solution of Na_2SO_4 with one electrode in each end. One of the electrodes, the anode, has a positive charge, and the other one, the cathode, a negative charge. The negatively charged sulphate ions are then attracted by the anode and the positive sodium ions by the cathode. If remixing due to convection currents is prevented, acids are collected at the anode, and bases at the cathode. This can be achieved if the electrolysis cell is divided by several membranes into many compartments as described by Williams and Waterman (1929).

It is well known that ampholytes, such as amino-acids, peptides and proteins, carry groups that upon dissociation become negatively charged (mostly carboxylic groups), or positively charged (mostly amino-groups). The net charge of the molecule depends on the number of each kind, and their respective dissociation constants; it is thus also dependent on the pH of the solution. In the acid solution at the anode, most ampholyte molecules have a positive net charge, and will thus be repelled by the anode and transported towards the cathode. But this also means going from a low to a high pH. At a certain pH the negative charges of the molecule balance the positive charges, i.e., the net charge is zero. Repulsion and attraction from both electrodes causes the migration of the ampholyte to a pH where the net charge is zero, which by definition, is the isoelectric point. This in turn depends on the type, pK values, and number of the dissociated groups in the molecule. In the case of peptides and proteins the pI is thus determined by the amino-acid composition. In the electrolysis cell there is a tendency for the most acidic ampholytes to collect close to the anode and for the others to distribute in the order of their isoelectric points towards the cathode.

The principle has been used for crude separation of various, mostly low-molecular-weight, compounds. Svensson has given a review, with references, of the achievements before 1948 (Svensson, 1948). Later attempts to improve separations gave results that were much inferior to those obtainable by electrophoresis. Difficulties were partly in an uneven field-strength distribution in the electrolysis cell, partly in a suitable pH gradient, i.e., the slowly increasing pH from the anode to the cathode proved difficult to stabilize; and partly in unstabilized convection in the cell. Svensson showed how to solve these problems with a fundamentally new approach (Svensson, 1961, 1962a, b). A description of the method as carried out at present is given in the following.

II. DESCRIPTION OF THE THEORY OF THE METHOD

Svensson has comprehensively discussed the use of density gradients in electrophoresis (Svensson, 1960). A suitable convection-free system for isoelectric focusing can be achieved by a density gradient of increasing amounts of a solute from the top to the bottom of a vertical column. Various highly water-soluble non-electrolytes making a sufficient contribution to the density of water may be used. Sucrose is often used because it is in most cases harmless to proteins. It may even have a protective effect (Hardt et al., 1946; Wadström, 1968). Sucrose has a relatively high specific density increment compared to other solutes, e.g., glycerol (Vesterberg and Berggren, 1967) and ethane diol (Ahlgren et al., 1967b) that are sometimes used. The density gradient is usually prepared from one dense and one less dense solution (see p. 600), and has two functions: it has a stabilizing effect upon thermal convections and it is a support for the focusing protein zones.

The theoretical analysis, which was necessary to overcome the earlier difficulties of the principle, was presented by Svensson (Svensson, 1961, 1962a, b) and a brief summary of this follows. Upon convection-free electrolysis of an aqueous solution of ampholytes with different isoelectric points, each ampholyte will tend to be focused at its own isoelectric point. If the ampholytes have a certain buffering capacity in the isoelectric state they will determine the pH where they are focused. Using suitable ampholytes, a stable pH gradient can be obtained with pH increasing steadily uniformly from anode to cathode. The final shape of the pH course is dictated by the pI's, the concentrations, and the buffering properties of all the protolytes in the system. Ampholytes suitable for this purpose, which must fulfil certain criteria (cf. p. 601) are called **carrier ampholytes.** If ampholytes, such as proteins, are also present in the system they will focus in narrow zones, each at its respective pI. Two proteins can be separated if they have different pI's and if carrier ampholytes are present where they are focused.

A. Resolving power

The fact that the proteins do not significantly alter the pH gradient given by the carrier ampholytes has made it easier to derive a theoretical expression for the resolving power of the method, i.e., the capacity to separate proteins having close pI values (Vesterberg and Svesson, 1966)—

$$\triangle pI = k \sqrt{\frac{D(dpH/dx)}{-E(du/dpH)}}$$

Here k is the criterion for resolution. A k value of 3 corresponds to well resolved but not completely separated protein zones. D is the diffusion coefficient, E the field strength, dpH/dx the value of the pH gradient in the electrolysis apparatus at the zone location, and (du/dpH) the (negative) mobility slope of the protein at the isoelectric point.

Some of the factors influencing the resolving power and the degree of focusing will be discussed. D and du/dpH are specific constants for each protein. The resolution can be influenced by the experimenter—

(1) By increasing the field strength E. This is, however, limited by a certain electric load above which thermal convection occurs in the density gradient. Because $E = I/\kappa$ where $I =$ current density, and κ is the conductance, it follows that the conductance should be roughly equally distributed in the column in order to obtain a fairly constant field strength. This is made possible by the new carrier ampholyte system. The conductance gradient is as important as the pH gradient.

(2) The resolution is improved by selecting carrier ampholytes having pI values in a limited pH range, whereby a low value of dpH/dx is obtained.

(3) Finally the resolution is also a result of the method used for analysis of the separation. The columns should be drained at about 2 ml/min (110 ml columns) if a good resolution is desired. The fraction volume should then be about 1 ml. The resolving power obtainable is about 0·01 pH units.

With newer separation methods having a high resolving power, many proteins have been proved to be heterogeneous, and it is necessary to analyse and characterize separated molecular forms and to prove that they are not artifacts. For many of the proteins already examined by the electrofocusing method, the heterogeneities found are also well supported by results of other methods (Flatmark and Vesterberg, 1966; Vesterberg and Svensson, 1966; Carlström and Vesterberg, 1967; Vesterberg, 1967). These papers may also be consulted for general aspects of the method.

B. Determination of isoelectric point

Svensson showed that the pH at the maximum concentration of an isoelectrically focused ampholyte is its exact pI value (Svensson, 1961). Thus it is only necessary to localize a protein focused in a pH gradient and to measure the pH in the corresponding fraction. A good degree of focusing, a low value of the pH gradient (i.e., a small pH range in the column), a proper draining procedure (even and not too rapid), and a small fraction volume will promote a good resolution as well as an accurate measurement of the isoelectric point. Reproducibility of the pI determinations is of the order of the accuracy of ordinary laboratory pH meters (Vesterberg and Svensson, 1966). (However, cf. Flatmark and Vesterberg, 1966).

Good reproducibility in the pI determinations requires that the focusing and the subsequent pH measurements should be made at the same temperature (usually $+4°C$), and it is convenient to arrange the cooling of the columns by a thermostatted water bath that is also used for temperature adjustment of fractions for pH measurements. When comparing a pI value for a protein obtained by isoelectric focusing with that determined by free electrophoresis, a discrepancy is often observed (Svensson, 1962b; Vesterberg and Svensson, 1966; Vesterberg, 1967). It is well known that pI values determined by electrophoresis depend on the kind of buffer and the ionic strength used (Brown and Timasheff, 1959). The values are higher for the lower ionic strengths of the buffer. The pI values obtained at isoelectric focusing refer to a very low ionic strength (Vesterberg and Svensson, 1966) and higher pI values are to be expected from electro-focusing than by electrophoresis.

C. Separation capacity

The focusing principle of the method offers special advantages for separating and purifying large amounts of protein. However, as mentioned before, a density gradient is used in vertical columns for stabilizing thermal convections and focused protein zones. The means that the maximum amount of protein depends on the capacity of the density gradient for carrying a focused protein zone. The protein should be soluble when focused. Because of this some proteins such as lipoproteins cannot be used in large amounts. The solubilities of some proteins have been increased by using a higher concentration of ampholytes than the usual 1% Valmet, 1968). For the same purpose urea has been used successfully in some cases (Bloemendal and Schoenmakers, 1968; Björk, 1968).

The general factors influencing the separation capacity are given below—

(1) The density gradient should have a high value. This is determined by factors such as the amount of and the density of the solute used as the dense component.

(2) The capacity is proportional to the square of the zone height. This can be influenced by the experimenter. A lower voltage will result in broader protein zones. This will, however, increase the time necessary for focusing and also decrease the resolution of closely spaced zones. The remedy is to use a narrow pH range giving a shallow pH gradient implying possibilities for improved resolution. The zones can then be allowed to have a reasonable height and a good resolution will still be obtained. Good separations have been achieved using 20 mg of protein in one zone in a 110 cm column (Vesterberg, 1967). When a very high resolution is required the amount of protein in one zone should be 5 mg (or less).

(3) It is often possible to separate rather pure protein samples into two or more zones. This means that the total amount of protein that can be applied in one experiment increases with the heterogeneity of the preparation. In work with crude protein solutions total loads of several hundred mg have been used in the 110 ml column (Vesterberg et al., 1967). Furthermore, if the contaminating proteins are focused at some distance from the protein of interest it is not necessary to restrict the amount in the corresponding zones as much as for the protein under study.

(4) The possible amount of protein in a zone is directly proportional to the cross sectional area. This can be taken advantage of in a preparative column of 440 ml capacity. As much as 5 g of a crude protein preparation has been applied in one run in the bigger column (Paleus, to be published). Some of the zones then contain almost 1 g which made them unstable, but the separation could be carried through as a preparation step.

(5) Due to the focusing principle of the method the volume of the protein solution is uncritical (cf. p. 603).

It is obvious from the above that the density gradient is a limiting factor. The use of stabilization by gels, and separation on an analytical scale are described by Wrigley (1968a, b) and Awdeh et al. (1968). Using a specially devised apparatus with membranes, the author has found it possible to further increase the capacity compared with the corresponding volume of a density gradient column. This principle also offers advantages for proteins which have a limited solubility. One instrument capable of meeting the special demands has been constructed and some of the results obtained will be published shortly.

III. EQUIPMENT

The following equipment is recommended—

(1) *Columns*

   At present two types of specially designed columns (LKB-Produkter, S-161 25 Brommal, Sweden) are available for electrofocusing (Fig. 1). The columns are equipped with platinum electrodes and cooling jackets for thermostatting. The volumes of the columns are 110 and 440 ml, respectively. (For capacity of protein, cf. pp. 599 and 600).

(2) A standard electrophoresis power supply giving a D.C. of 1200 V. (For many purposes 600 V may suffice.)

(3) A thermostatted cold water bath and a water circulating pump for cooling the columns during operation. The same bath may be used for temperature equilibration of the fractions from the columns. Usually a temperature of $+4°C$ is used. (For preliminary experiments cold tap water may be used.)

(4) A pH meter (preferably with an accuracy of about 0·01 pH units). A flow through cell for pH measurement is under development by LKB.

(6) Gradient mixer (such as the LKB type) for preparation of density gradients in the columns.

(7) Pump for filling the columns and for fractionation with a fraction collector.

(8) Fraction collector (usually the fraction volume is 1–5 ml).

(9) An analyser with a flow through cell and a recorder for estimating protein peaks by light absorption mesurements of 280 nm (LKB Vricord-II).

A column, a power supply, and a pH meter are the minimum requirements.

Chemicals

pH gradients are obtained by using *carrier ampholytes* (Ampholine, LKB-Produkter S-161 25 Brommel, Sweden), which are mixtures of various aliphatic polyaminopolycarboxylic acids having different isoelectric points (pI's) (Vesterberg, to be published). Due to their high buffering capacity and different pI values they can produce pH gradients in the pI range of most proteins. The ampholytes are available in batches giving specified pH ranges in the columns. The pH range of the carrier ampholytes should be chosen such that the pI values of the proteins under study lie well within the corresponding pH range. When making the first run with a protein sample it is often advisable to work with the pH range 3–10. Carrier ampholytes giving pH ranges of 2 pH units are also available. Still narrower pH ranges will also be offered. Usually an amount of carrier ampholyte is used which gives a final average concentration of 1% (w/v) in the columns. Higher concentrations up to 10% have been used in order to increase the solubility of some proteins (Valmet, 1968).

Other chemicals. Sucrose of a high analytical purity (cf. p. 603), phos-

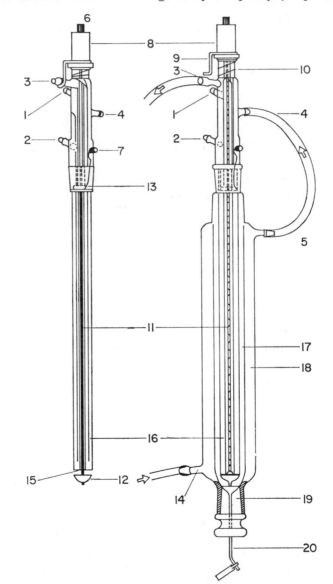

phoric- or sulphuric acid in small amounts in the anode solution (p. 605).

An amine of a high boiling point, e.g., ethylene diamine, or triethyl-amine, or diethanolamine, in small amounts in the cathode solution (p. 606).

IV. PRACTICAL TECHNIQUES

A. The protein solution

The protein solution should have only a low concentration of salt. The total amount of salt should be less than 0·5 mM. If necessary, dialyse with several changes against either a small volume of a 0·5% aqueous solution of carrier ampholytes, or against a 1% solution of glycine.

Due to the focusing character of the method, the volume of the protein solution to be applied to the column is not very critical. The proteins will migrate to their respective pI's, irrespective of where they are in the column at the start. When the density gradient is prepared, the protein solution replaces the water in either the less dense or the dense solution. In the smaller column up to 85 ml, and in the bigger column up to 340 ml, of protein solution can be applied.

Several factors determine the total load of protein allowable (cf. p. 599 and 600).

B. Preparation of the density gradient

Density gradients are used in the vertical columns. In most cases a 50% (w/v) solution of sucrose is used as the denser solution. However, other solutes such as ethylene glycol and glycerol have also been used successfully (Ahlgren *et al.*, 1967b; Vesterberg and Berggren, 1967). The latter substances are often used in 60–70% (w/v) concentrations because of their lower densities compared to sucrose.

FIG. 1. Electrofocusing column of 110 ml capacity. This is an improved version of the one described by Vesterberg *et al.*, 1967. The outer cooling jacket (18) has an inlet at (14), and an outlet at (5). From the outer jacket the water flows through a tube into the central cooling jacket at (4), and leaves the column at (3). Isoelectric focusing takes part in a compartment (17), which is filled through the tube (2). Two platinum electrodes are used. One electrode (13), in contact with the plug 7, is in the upper part of the column. The gas formed at this electrode escapes at (2). The other electrode, in contact with the plug (6), is wound on a rigid Teflon bar (11). The gas formed at this electrode escapes at (1). A piece of Teflon, (12) is connected by the rigid Teflon bar (11) to the handle (8). The central tube is open at (15), when the handle (8) is kept in the down position with the hook (9). Before draining the column the central tube is closed by lifting the plug (12), which has a rubber ring gasket on the upper surface. It is kept closed with the aid of the spring (10). At the bottom of the column there is a plug (19) with an attachment for a capillary tube for sampling fractions.

The density gradient is made of one less dense solution containing distilled water, carrier ampholytes and protein solution, and one dense solution of distilled water, sucrose and carrier ampholytes. The volume of each solution being about half that of the column.

Higher concentrations of such solutes as sucrose decrease the conductivity of water and, in order *to obtain a more uniform distribution of the field strength* at the start, 3/4 of the carrier ampholytes should be dissolved in the denser solution and the rest in the less dense solution. Furthermore, when working with a limited pH range, covering a few pH units, it is advisable to select the *electrode positions* so that well conducting ampholytes will focus in the lower part of the column. This is achieved by using the central tube (top contact on the column) as anode, for pH ranges below 6, and as

TABLE I

A volumetric aid for preparing a constant-density gradient in a 110 ml column

Fraction No.	Dense solution		Less dense solution	
	ml	running total, ml	ml	running total, ml
1	4·6	4·6	0·0	0·0
2	4·4	9·0	0·2	0·2
3	4·2	13·2	0·4	0·6
4	4·0	17·2	0·6	1·2
5	3·8	21·0	0·8	2·0
6	3·6	24·6	1·0	3·0
7	3·4	28·0	1·2	4·2
8	3·2	31·2	1·4	5·6
9	3·0	34·2	1·6	7·2
10	2·8	37·0	1·8	9·0
11	2·6	39·6	2·0	11·0
12	2·4	42·0	2·2	13·2
13	2·2	44·2	2·4	15·6
14	2·0	46·2	2·6	18·2
15	1·8	48·0	2·8	21·0
16	1·6	49·6	3·0	24·0
17	1·4	51·0	3·2	27·2
18	1·2	52·2	3·4	30·6
19	1·0	53·2	3·6	34·2
20	0·8	54·0	3·8	38·0
21	0·6	54·6	4·0	42·0
22	0·4	55·0	4·2	46·2
23	0·2	55·2	4·4	50·6
24	0·0	55·2	4·6	55·2

cathode for pH ranges above 6. This rule is not obeyed if the separation is disturbed by protein precipitates in the upper part of the column.

When working outside the pH range, 5–9, it is advisable to add carrier ampholytes of pH 6–8 in an amount of about 20% of the total.

The density gradient can be prepared with reference to Table I. The protein solution can then replace the less dense solution in some of the fractions or it may be used instead of the water when making the dense and the less dense solution. The 24 fractions are homogenized by gently turning the tubes (stoppered with a rubber bung) upside down a couple of times. They are then filled into the column, in the order of decreasing density starting with fraction number 1, using a small funnel having a capillary plastic tube (inner diameter 1–2 mm) so that the liquid slowly flows along the internal wall of the column. During this operation the central tube of the column is kept closed. The flow should be adjusted with a clip so that each fraction takes about one minute and a half to run into the column. This will give some mixing of each fraction with the preceding one in the column resulting in a smooth density gradient without steps (Vesterberg and Svensson, 1966) for the larger column multiply all volumes by four.

A gradient mixer however, is more convenient. Then the protein solution can replace a corresponding volume of the water in the less dense and/or in the dense solution.

C. Electrode solutions

To protect the carrier ampholytes from contact with the electrodes, the solution at the anode should contain acid; and the solution at the cathode base, for during electrolysis, acid and the base are attracted by the anode and the cathode respectively. Ampholytes will acquire a net positive charge in the acid solution and similarly, the base will give the ampholytes a net negative charge. The ampholytes are thus repelled from the electrodes.

Solutions

Small column 110 *ml.*

See figures below.

Larger column 440 *ml.*

For preparating solutions for this column multiply the figures below by 4·0.

Protein solution = *a* ml.

Less dense solution = 55 * *ml.*

Ordinarily 0·28*g of carrier ampholytes, i.e., 0·7*ml of a 40% (w/v) solution, is diluted with a ml protein solution and $(54-a)$ ml and distilled water.

Dense solution = 55* ml.

Ordinarily 0·8*g of carrier ampholytes, i.e., 2·0* ml of a 40% (w/v), solution is diluted with 38*ml distilled water (and/or protein solution) in which 25*g of sucrose (analytical reagent grade) is dissolved.

Anode solution

(a) Anode at the bottom of the column (in the central tube): Dilute 0·1 ml of H_3PO_4 or H_2SO_4 with 10 ml distilled water and dissolve 8 g of sucrose in this solution.
(b) Anode at the top of the column: Mix 0·05 ml of H_3PO_4 or H_2SO_4 with 6 ml of distilled water.

Cathode solution

(a) Cathode at the top of the column: Mix 0·1 ml of ethylenediamine, or ethanolamine with 8 ml of distilled water.
(b) Cathode at the bottom of the column (in the central tube): Dilute 0·2 ml of the amine with 10 ml of distilled water and dissolve 8 g of sucrose in this solution.

D. Temperature of the cooling water

It is useful to use a constant temperature during focusing. For this reason a constant temperature cooling bath at +4°C, and a pump for circulation of the water through the cooling jacket are advisable. For preliminary experiments, however tap water may be used.

E. Voltage, maximum electrical load

For most purposes a potential of 600 V is adequate†. However, when working with a narrow pH range in the interval 5–9, 900 V could be used. Initially the power imput should not exceed 5 Watt for the small column, and 20 Watt for the bigger one at the start. When there is salt left in the protein solution, it may be necessary to increase the potential by 100 V steps at intervals of some hours.

During the course of electrofocusing, the resistance of the column increases gradually.

* When using Table I the Figures should be increased by 10% due to inevitable loss in the burette and the fraction tubes.
† If too high a voltage is used the density gradient may be disturbed.

F. Electrofocusing time

After the current has apparently stabilized, the experiment should be continued for a further period of several hours. It is advantageous to use an ordinary ammeter with a range of 1–20 mA.

No blurring of the protein zones occurs if the focusing is allowed to go on longer than necessary, but usually an experiment is ready after 26–48 hours. The time is dependent on the pH range used; and longer focusing times are needed for narrow pH ranges, especially for the alkaline ones.

G. Collection of fractions

Turn off the current and *close* the central tube. Drain slowly through the bottom tubing (about 2 ml/min is recommended). In most cases a fraction volume of 2–5 ml is adequate. pH measurements on the fractions should be carried out at the temperature of the column, preferably by placing the fractions in a thermostatted water bath. For preliminary experiments a water bath with some ice may be used. Because of the low light absorption of the carrier ampholytes at 280 nm, protein concentration distribution may be measured at 280 nm. An analyser with a throughflow cell and a recorder may be used and a suitable throughflow pH measuring cell is under development. When a recorder is used it is necessary to have a constant draining speed and this can be obtained by using a peristaltic pump to suck out the contents of the column. Another alternative is to pump water into the top of the column, and to press out the contents at the bottom.

V. APPLICATIONS

Isoelectric focusing offers many possibilities for separating and purifying proteins*, as well as for characterizing them, by isoelectric-point determination. This will be illustrated with a brief summary of some applications with special reference to microbiology.

A. Cellulolytic and related enzymes from fungi†

Very little information about the physico-chemical proterties of these enzymes has been available and the first paper on isoelectric focusing of these enzymes was a result of the study of cellulase, mannase, xylanase, β-glucosidase, aryl-β-glucosidase, mannosidase and xylosidase in a crude preparation of *Aspergillus* enzymes (Ahlgren *et al.*, 1967b). In these electrofocusing experiments, ethane diol was used instead of sucrose to obtain

---

* After isoelectric focusing the carrier ampholytes can be conveniently separated from the proteins by gel chromatography on Sephadex as described by O. Vesterberg (1969).

† Cellulolytic and related enzymes from various fungi are studied in a project at the Swedish Forest Product Research Laboratory, Paper Technology Department, Stockholm, Sweden.

the density gradients, because invertase interfered with the assay of some of the enzymes.

The isoelectric points of the enzymes were in the pH range 3·9–4·7, and at least three of the enzymes were separated into two or three peaks of activity. Although the resolution in the zone electrophoresis experiments was not as good as that obtained by electrofocusing, the heterogeneity of the cellulase and aryl-β-glucosidase enzymes was demonstrated by both methods. It was possible to separate some of the enzymes on Sephadex into different peaks of activity, the latter method separating according to molecular size, and electrofocusing according to differences in pI. A run on Sephadex followed or preceded by isoelectric fractionation gave a good separation and purification.

In another paper enzymes from the wood destroying fungi *Stereum sanguinolentum*, *Fomes annosus* and *Chrysosporium lingnorum* have been studied (Ahlgren and Eriksson, 1967a). With few exceptions, the pI's of these enzymes were found to be in the pH range 3·6–4·7. Some of the enzymes were found to occur in different molecular forms. This was probably a result of the cultivation conditions (Bucht and Eriksson, 1968). The heterogeneous nature of some cellulolytic enzymes has been shown in other investigations using electrophoresis and ion exchange chromatography, and various explanations have been proposed. For review articles, cf. Whitaker (Whitaker, 1963) and Eriksson (Eriksson, 1967).

Purifying and characterizing cellulases from the fungus *C. lignorum* will be described in two forthcoming papers by Eriksson and Rezdowski (Eriksson and Rzedowski, 1969b).

The electrofocusing method has proved to be valuable for separating characterizing, and comparing enzymes from related fungi. The method will be used for preparative purification of some of the enzymes.

B. β-Fructofuranosidase (invertase) from yeast (Vesterberg and Breggen, 1967).

Preparations of invertase from brewer's yeast, isolated by autolysis and by mechanical disintegration have been studied by isoelectric focusing in density gradients of glycerol. Many peaks of activity of the enzyme were separated, and the pI's of these were essentially in the pH range 3·4–4·4. The yield after electrofocusing was close to 100% and some of the activity peaks showed pI differencies of only about 0·01 pH units. Reproducability of the isoelectric spectra was good and when a fraction of one peak was taken to electrofocusing in a new column, most of the activity was found in one peak with the same pI value.

When batches from various preparations were examined, different isoelectric spectra were obtained. This was probably caused by different conditions prevailing in the different batches before electrofocusing. Chromatography on DEAE-cellulose, and molecular sieving on Sephadex has also given results which suggest that invertase can occur in different molecular forms. One hypothesis is that the great number of enzymatically active components might arise from the action of hydrolytic, e.g., proteolytic enzymes. Lenney and others have reported a comparatively high proteolytic activity in yeast (Lenney, 1956; Lenney and Dalbec, 1967), and Kenkare et al (Kenkare et al., 1964a) have shown that the proteolytic activity in yeast can cause the conversion of yeast hexokinase into six distinct chromatographic forms. It is evident that further experiments are necessary in order to explain the observed great heterogeneity of invertase.

Electrofocusing has been used for the purification of several hundred milligrammes of invertase. The degree of purification, i.e., the specific activity, was about the same as that obtainable by carefully conducted ion-exchange chromatography.

C. Extracellular enzymes and toxins from *Staphylococcus aureus* (Vesterberg et al., 1967*)

Many methods and techniques have been used earlier for purifying and characterizing extracellular proteins from staphylococci (Elek, 1959). The difficulties encountered are partly due to the low protein concentrations obtained in the cultures.

Culture supernatants have been concentrated about one hundred times and dialysed before isoelectric focusing. Deoxyribonuclease, haemolysin, staphylokinase, protease, hyaluronate lyase and bacteriolytic activity have been assayed. In contrast to other methods, such as molecular sieving on Sephadex, and ion exchange chromatography, only minimal losses of activity were observed during electrofocusing. It was found that the carrier ampholytes and the sucrose of the density gradient have a protective effect on α-haemolysin, protease and bacteriolytic activity which are otherwise susceptible to denaturation (Wadström, 1968). This effect was specially pronounced for α-haemolysin. The pI values of the activities were distributed in the pH range 3–11. Most of the activities were found to occur in multiple molecular forms with different pI values and thus separable by electrofocusing. In later investigations more detailed studies have been made, with a higher resolution, in narrow pH ranges.

1. *Deoxyribonuclease (Wadström, 1967)*

Five components of deoxyribonuclease activity, two major and three minor peaks, have been separated. Electrofocusing was also made in

* This is a brief summary and discussion of results from Department of Bacteriology, Karolinska Institutet, Stockholm, Sweden.

6M urea and the above mentioned peaks were again found. For this reason it seems that these are not due to complex formation with other proteins. Enzymatic differences between some of the components, such as ion requirement, pH, and ionic strength for optimal activity were also noticed. The major components were isoelectric at pH 2·5 and 10·1, respectively.

2. α-Haemolysin (Wadström, 1968)

After electrofocusing the yield of the haemolytic activity, classified as α-haemolysin, varied in different experiments between 70–140%. The activity is very susceptible to inactivation when purified. The carrier ampholytes and the sucrose had a significant stabilizing effect on the toxin and three, or sometimes four, components with different pI values were separated. This can be compared with the four components with α-haemolytic activity separated in zone electrophoresis by Bernheimer (Bernheimer and Schwartz, 1963). Using Sephadex and electrofocusing a very pure product was obtained and the specific activity was of the same degree or somewhat higher than the values reported in the literature.

3. Proteolytic enzymes

After electrofocusing the yield of proteolytic activity was about 80%. Three main peaks of activity were obtained with pI's at 3·7, 5·0 and 9·7, respectively (Vesterberg et al., 1967). Further studies will be published in a forthcoming paper (Arvidsson, to be published).

4. Staphylokinase

The total activity of staphylokinase in the fractions after electrofocusing was higher than in the sample before purification. This was probably due to the removal of some inhibitory substance. In the first experiments the activity was focused in one peak with pI at 6·5 and in later experiments using a narrower pH range, it was possible to separate three peaks with closely spaced pI values. The degree of purification and the properties of the enzyme will be dealt with in a forthcoming paper (Vesterberg, K., to be published).

5. Hyaluronate lyase

Several experiments on isoelectric focusing of this enzyme have been made (Vesterberg, 1968). The recovery after electrofocusing was about 80–95%. Some interesting results on the heterogeneity of this enzyme have been obtained—

(a) The relative amount of enzyme in different peaks may show variation in different cultivation batches.

(b) Cultivation of the staphylococci in a batch for 6 h gave enzyme, most of which was isoelectric at 7·4, whereas harvest of the same culture after 14 h of cultivation gave the main enzyme peak a pI of 7·9.

(c) In concentrated-culture supernatants most of the enzyme had a pI at 5. However, if a purification on Sephadex, at 0·6 M ionic strength of salt, preceded the electrofocusing, two peaks of activity were obtained with pIs at 7·4 and 7·9. This could probably be explained by the removal of some complex-forming substance present in the component with pI at 5 (O. Vesterberg, to be published). From these experiments it can be concluded that isoelectric focusing should be made on protein material before and after another purification procedure.

(d) Storage of enzyme fractions with pI 7·4 in phosphate buffer (pH 7·0) resulted in a slow conversion to enzyme with pI 7·9 and only a small loss of activity. Carrier ampholytes and/or sucrose seemed to have a protective effect on this conversion.

6. Lipase

This enzyme was difficult to handle by isoelectric focusing, using density gradients, because of its low solubility. However, two zones of activity were obtained with pIs at 4 and 9, respectively. The solubilization and complete recovery of the activity, by dissolving the precipitates in a phosphate buffer of ionic strength 0·5 (pH 7·5), will be described (O. Vesterberg, to be published).

D. Bacteriolytic proteins

This group of proteins is of growing importance partly due to their antibiotic properties and also because of their value for studies on bacterial cell-wall composition.

1. Colicins from Escherichia coli

Strains of E. coli which carry an extrachromosomal genetic factor can produce colicins. On the basis of the existence of specific receptor sites the various colicins are grouped in 17 classes, designated alphabetically.

Class E contain the colicins E_1, E_2 and E_3 which differ in the manner in which they kill sensitive cells. Purification and characterization colicin E_2 and E_3 has been described in a paper by Herschman and Helinski (Herschman and Helinski, 1967). Colicins E_2 and E_3 have been separated by ion exchange chromatography, disc electrophoresis and isoelectric focusing and colicin E_3 was found to be homogeneous. Colicin E_2 could be separated into two components by disc electrophoresis and by isoelectric focusing and using the latter method it was shown that they are interconvertible conformal forms of a single molecule, with pI values at 7·63 and 7·41, respectively. References are given to other papers where proteins have been separated, which are supposed to differ only in conformation. For such proteins the term conformers is proposed (Kitto et al., 1966; Epstein and Schechler, 1968).

2. *Bacteriolytic activities in culture supernatants of* Staphylococcus aureus cultures

Activities that can reduce the turbidity of a suspension of *S. aureus* or *Micrococcusi lysodeckticus* and their suspensions of cell walls have been assayed (Wadström, and Hitsatsune, 1968). After electrofocusing, the yield varied probably due to the instability of the enzyme. Three peaks of activity have been separated: the main peak with a pI at 9·5, seems to be of a lysozyme-like nature. This enzyme, as well as lysozyme from hens-egg white, is less stable when stored in buffers of alkaline pH. The carrier ampholytes and sucrose increase the stability of these enzymes considerably another activity of lysis staphylocci has a pI at 10·5 (Wadstrom, to be published).

E. Proteins from other bacteria

1. *Hyaluronate lyase from streptococci*

The recovery of activity after electrofocusing was about 70% and one peak with pI at 4·5 was obtained (Vesterberg and Holmström, to be published).

2. α-*toxin from* Clostridium perfringens (*Bernheimer* et al., 1968)

The recovery after electrofocusing was about 64% and two peaks with pI's at 5·2 and 5·5 were obtained. The haemolytic activity could not be separated from the phospholipase activity, supporting the view that the enzymatic activity is also responsible for the haemolytic activity.

3. *Cereolysin*

A haemolytic product of *Bacillus cereus* (Bernheimer et al., 1968), which closely resembles streptolysin O, has been studied by electrofocusing. A partially purified preparation showed one peak of haemolytic activity with pI at 6·5.

4. *Cytochromes from* Rhodospirillum rubrum (*Sletten and Kamen*, 1967)

One major and several minor components of cytochromes c_2 and cc', respectively, were separated. In the interpretation of the heterogeneities it must be understood that the samples analysed were taken from preparations with varying histories. In cytochrome c_2, the six bands with pI on the acid side of the main component were most likely deaminated forms of the native protein, analogous to the "isocytochromes" of beefheart cytochrome c (Flatmark, 1964; Flatmark and Vesterberg, 1966). A high degree of purity of the main components was obtained.

F. Proteins of viral origin

Structural proteins from adenoviruses have been subjected to electro-focusing. In one paper dealing with purifying and characterizing adeno-virus type 2, a fibre antigen has been studied (Philipson and Petterson, 1968). In purifying this protein various methods, using different criteria for separation, have been used. Using isoelectric focusing the degree of purification was increased, and a pI of 6·1 was determined.

G. Immunoglobulins

It is well-known that it is possible to make a separation of serum proteins into different groups by using moving boundary or zone electrophoresis. Most of the immunoglobulins are found in the γ-globulin fraction, which must necessarily contain a lot of different molecular species. Using zone electrophoresis in gels it is possible to make a coarse subfractionation of the γ-globulins and by electrofocusing, the separations can be improved further (Valmet, 1968). The γ-globulins have their pI values, as determined by electrofocusing, in a wide pH range, i.e., from about pH 6 up to almost pH 10. As was mentioned previously the resolving power of the electro-focusing method, i.e. the pI difference needed for separation can be as low as about 0·01 pH units (Vesterberg and Svensson, 1966; Vesterberg and Berggren 1967). Recently a paper was published on the separation of γ-globulins by isoelectric focusing in gel (Awdeh et al., 1968). The possibility of separating antibodies by this method opens up interesting areas of study.

REFERENCES

Ahlgren, E., and Eriksson, K. E. (1967a) *Acta chem. scand.*, 21, 1193–1200.
Ahlgren, E., Eriksson, K.-E., and Vesterberg, O. (1967b). *Acta chem. scand.*, 21, 937–944.
Awdeh, Z. L., Williamson, A. R., and Askonas, B. A. (1968). *Nature*, 219, 66–67.
Berggren, B. (1966). *Ark. Kemi.*, 25, 555–565.
Berggren, B. (1967). *Ark. Kemi.*, 26, 259–266.
Bernheimer, A. W., and Schwartz, L. L. (1963). *J. gen. Microbiol.*, 30, 455–468.
Bernheimer, A. W., Grushoff, P., and Avigad, L. S. (1968). *J. Bact.*, 95, 2439–2441.
Björk, I. (1968). *Acta chem. scand.*, 22, 1355–1356
Bloemendal, H., and Schoenmakers, J. G. G. (1968). *Sci. Tools*, 15, 6–7.
Brown, R. A., and Timasheff, S. N. (1959). *In* "Electrophoresis, theory, methods, and applications" (Ed. M. Bier), pp. 322–324. Acadmic Press, New York.
Bucht, B., and Eriksson, K.-E. (1968). *Archs. Biochem. Biophys.*, 124, 135–141.
Carlström, A., and Vesterberg, O. (1967). *Acta chem. scand.*, 21, 271–278.
Elek, S. D. (1959). *In* "Staphylococcus pyogenes and its relation to disease". Livingstone Ltd., London.

Epstein, C. J., and Schechler, A. N. (1968). *Ann. New York Acad. Sci.*, **151**, 85–101

Eriksson, K.-E. (1967). *In* "Studies on cellulolytic and related enzymes". Diss, Stockholm.

Eriksson, K.-E., and Rzedowski, W. *Svensk. Kem. Tidshi.*, **79**, 660–679.

Erikson, K.-E. and Rzedowski, W. (1969a). *Archs. Biochem. Biophys.*, **129**, 683–688.

Eriksson, K.-E. and Rzedowski, W. (1969b) *Archs. Biochem. Biophys.*, **129**, 689, 695

Flatmark, T. (1964). *Acta chem. scand.*, **18**, 1656–1666.

Flatmark, T., and Vesterberg, O. (1966). *Acta chem. scand.*, **20**, 1497–1503.

Hardt, C. R., Huddleson, I. F., and Ball, C. D. (1946). *J. biol. chem.*, **163**, 211–220.

Herschman, H. R., and Helinski, D. R. (1967). *J. biol. Chem.*, **242**, 5360–5368.

Kenkare, U., Schultze, I. T., Gazith, J., and Colowick, S. P. (1964). *Sixth Intern. congr. Biochem.*, pp. 477–478. New York.

Kitto, G. B., Wassarman, P. M., and Kaplan, N. O. (1966). *Proc. natn. Acad. Sci. U.S.A.*, **56**, 578–585.

Lenney, J. F. (1956). *J. biol. Chem.*, **221**, 919–930.

Lenney, J. F., and Dalbec, J. M. (1967). *Archs. Biochem. Biophys.*, **120**, 42–48.

Philipson, L., and Petterson, V. (1968). *Virology*, **35**, 204–215.

Sletten, K., and Kamen, M. D. (1967). *In* "Preprint of symp. cytochromes", pp. 211–216, University of Tokyo Press, Osaka.

Sletten, K. and Kamen, M. D. (1968) *In* "Structures and Function of Cytochromes" (Eds. Okunuki, K., Kamen, M. D. and Sekuzu, I.) pp. 422–428, University of Tokyo Press, Tokyo; University Park Press, Baltimore, Maryland.

Svensson, H. (1948). *Adv. Protein Chem.*, **4**, 251–268.

Svensson, H. (1960). *In* "Laboratory manual of analytical methods in protein chemistry including polypeptides" (Eds. Alexander, P. and Block, R. J.), 195–244. Pergamon Press, London.

Svensson, H. (1961). *Acta chem. scand.*, **15**, 325–341.

Svensson, H. (1962a). *Acta shem. scand.*, **16**, 456–466.

Svensson, H. (1962b). *Archs Biochem. Biophys.*, **Suppl. 1,** 132–138.

Valmet, E. (1968). *Sci. Tools*, **15**, 8–11.

Vesterberg, O. (1967). *Acta chem. scand.*, **21**, 206–216.

Vesterberg, O. (1968). *Biochem. biophys. Acta.*, **168**, 218–227.

Vesterberg, O. (1969) *Science Tools*, **16**, 24–27.

Vesterberg, O., and Berggren, B. (1967). *Ark. Kemi.*, **27**, 119–127.

Vesterberg, O., and Svensson, H. (1966). *Acta chem. scand.*, **20**, 820–834.

Vesterberg, O., Wadstöm, T., Vesterberg, K., Svensson, H., and Malmgren, B. (1967). *Biochim. biophys. Acta*, **133**, 435–445.

Wadström, T. (1967). *Biochim. biophys. Acta*, **147**, 441–452.

Wadström, T. and Hitsatsune, K. (1969). *Biochim. biophys. Res. Commun.* Preliminary notes.

Wadström, T. and Hitsatsune, K. (1968). FEBS Meeting, Prague, Abstact no. 422.

Whitaker, D. R. (1963). *In* "Hydrolysis of cellulose and related materials" (Ed. E. T. Reese). Pergamon Press, London.

Williams, R. R., and Waterman, R. E. (1929). *Proc. Soc. Exptl. Biol. Med.*, **27**, 56–59.

Wrigley, C. (1968a). *Science Tools*, **15**, 17–23.

Wrigley, C. (1968b). *J. Chromatogr.*, **36**, 362, 365.

CHAPTER XII

Reflectance Spectrophotometry

F. J. Moss and Pamela A. D. Rickard

School of Biological Technology, The University of New South Wales, Kensington, N.S.W., Australia

and G. H. Roper

Department of Biological Process Engineering, School of Chemical Engineering, The University of New South Wales, Kensington, N.S.W., Australia

I.   Introduction 616

II. Theory of Band Intensity Measurement 617
    A.   Elimination of non-specific absorption 617
    B.   True band shape 618
    C.   Measured band shape 619

III. Estimation of cytochromes 621
    A.   Preparation of sample 621
    B.   Scanning the spectrum 622
    C.   Preparation of spectral data 622
    D.   Numerical analysis of spectral data 622
    E.   Comparative studies 623
    F.   Approximation method 624
    G.   Choice of reference 624
    H.   Choice of mode 624
    I.   Comparison with transmittance spectrophotometry . . 625

IV. Instrumentation 625

References 629

NOMENCLATURE

α Absorption coefficient.
α_c Absorption coefficient for a component.
α_b Background absorption coefficient.
σ Scattering coefficient.
i The intensity of incident radiation.
R Measured reflectance.
R_b Background reflectance.
R_i The contribution of the ith band to the reflectance.
$(R_i)_0$ The intensity of the reflectance of the ith band.

$\Delta \lambda_i$	The difference in wavelength between the point of measurement and the central wavelength of the ith band.
ω_i	The width of the ith band.
ω_1	Width of band in the Cauchy profile.
ω_2	Width of band in the Gauss profile.
c	Velocity of light.
Cy	Percentage of cytochrome on arbitrary scale.
C	Constant in equation (6).
A, B, C	Empirical constants.
W	Wavelength.
W_0	Central wavelength.

I. INTRODUCTION

The separation and determination of an absorbing substance from microbial material is often impracticable. A measure of concentration may be obtained from reflectance spectrophotometry using a procedure analogous to that of the analysis of transparent media by transmission spectrophotometry.

When the bands are widely separated, evaluation of the band intensity, width and shape is a relatively simple matter. The absorption spectra of biological materials, however, usually exhibit overlapping bands, and measurement of the band characteristics cannot be made separately for each band. An analytical technique which minimizes this difficulty has been used by the authors particularly for the estimation of cytochrome in micro-organisms (Moss, 1956; Moss and Tchan, 1958; Moss, Rickard, Beech and Bush, 1969; Moss, Rickard, Bush and Caiger, submitted; Rickard, Moss and Ganez, submitted; Rickard, Moss, Phillips and Mok, submitted).

Other biological studies include the following: Metzner (1965) measured the reflectance spectra of living *Chlorella* cells; Shibata (1958) examined the pigments of ivy, spinach and coleus leaves, parts of sweet pea, other flowers, carrot, radish, watermelon, squash and pole bean and of yeast; Shugarov (1965) studied the influence of nitrogen, phosphorus, potassium and sulphur deficiency on the reflection properties of sugarbeet leaves; similar studies were made by Carter and Myers (1963) who investigated the effects of Na_2SO_4, $NaCl$ and $CaCl_2$ in the soil on the reflectance properties of grapefruit leaves; the technique has provided an objective assessment of colour, and hence quality, of peanut butter (Morris *et al.*, 1953).

In the medical field, changes in tissue chemistry produced by disease have been shown to effect differences in the infrared attenuated total reflectance spectra of the tissues (Parker and Ans, 1967).

The method has been employed for the determination of oxygen saturation of human blood (see Wendlandt and Hecht, 1966). As an analytical tool it has also been applied to the determination of haem pigment concentrations in raw and cooked meats by numerous authors, including Pirko and Ayres (1957), Dean and Ball (1960), Stewart, Zipster and Watts (1965), Stewart, Hutchins, Zipster and Watts (1965), and Snyder and Armstrong (1967). The literature relative to the reflectance spectrophotometry of meat pigments was reviewed by Stewart, Zipster and Watts (1965).

The technique has been combined with fluorimetry by Chance et al. (1963) to determine the oxygenation state of haemoglobin by difference in reflectance at two wavelengths on the one hand, and the concentration of reduced pyridine nucleotide by fluorescence on the other.

Another interesting application lies in the analysis of components of mixtures resolved on thin layer chromoplates. Examples include the identification and estimation of essential oils (Jork, 1965), aspirin and salicylic acid (Frodyma et al., 1965), cations (Zaye et al., 1967, and Frei and Ryan, 1967), and nucleotides (Lieu et al., 1967).

For other applications, see Wendlandt and Hecht (1966).

II. THEORY OF BAND INTENSITY MEASUREMENT

A. Elimination of non-specific absorption

Where the technique has been applied to quantitative estimations, investigators have usually followed the theory of Kubelka (1948), Kubelka and Munk (1931), and Giovanelli (1955; 1956; 1957a, b, c). These investigators determined the ratio of the absorption coefficient to the scattering coefficient at a wavelength or wavelengths characteristic of the compound.

A comprehensive treatise of the theory of specular reflection and diffuse reflectance is to be found in Wendlandt and Hecht (1966).

A typical example of total reflectance is given in Fig. 1. The spectrum consists of a non-specific background with absorption bands centred on characteristic wavelengths.

Giovanelli (1957b) has analysed mathematically a model of randomly spaced and randomly oriented scattering particles dispersed in a matrix which may be a transparent gas or liquid.

The scattering and absorption coefficients, σ and α, are defined so that a beam of radiation of intensity i loses an amount σ_{Ids} and α_{Ids} by scattering and absorption, respectively, in a distance ds. The reflectance is a function of σ/α. The contribution of the absorbing component is α_c. The remainder of the absorption is background such that—

$$\alpha_c + \alpha_b = \alpha \tag{1}$$

A reflectance spectrum contains a number of minima representing regions where α_c has a maximum corresponding to absorption bands. The measured reflectance is—

$$R = R_b - \sum_{i=1}^{n} R_i \tag{2}$$

where n is the number of absorbing bands, R_i is the contribution of the ith band and R_b the background at the particular wavelength. The contribution of the individual bands is the function of the band intensity and of the difference between the central wavelength of the band and the wavelength of the measurement. This contribution depends upon the shape of the absorption band.

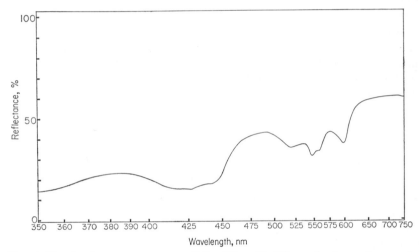

FIG. 1. Total reflectance spectrum of *Candida utilis*. The main absorption maxima of cytochromes appear as troughs at 598 nm (α-band of A complex), 558 nm (α-band of B complex), 547 nm (α-band of C complex), 518 nm (β-band of C complex), 440 nm (γ-band of A complex), and 415 nm (γ-band of C complex).

B. True band shape

Absorption or emission of radiation does not occur at a discrete wavelength or frequency. All absorption bands have a finite width. There are a number of factors which determine the linewidth. In condensed systems, collision broadening is the most important. Also effective are the Doppler effect and radiation dampening.

Collision broadening results from the perturbation of the energy levels of the absorbing vibrator by the close approach of vibrators of similar or different kinds. Lorentz (1906) and van Vleck and Weisskopf (1945) have considered electronic processes in atoms. The shape factor derived by Lorentz has been widely assumed to apply to infrared spectra of the spectra for nuclear magnetic resonance. Collision broadening results in a profile described by the Cauchy distribution which may be expressed as—

$$R_i/(R_i)_0 = 1/[1+(\Delta\lambda_i/\omega_i)^2] \qquad (3)$$

where R_i is a contribution of the ith absorbing band at a wavelength $\Delta\lambda$ distance from the central absorbing band and ω_i is its bandwidth. At $\Delta\lambda_i = \omega_i$, $R_i = \frac{1}{2}(R_i)_0$. At $\Delta\lambda_i = 0$, $R_i = (R_i)_0$, the band intensity.

Doppler broadening is the result of the thermal, or random, motion of the vibrator. Because of the random nature of this motion, at the time of absorption the vibrator will have a component of velocity in the direction of the beam. The vibrator approaching or receding with velocity v will absorb or emit at the wavelength—

$$W = W_0/(1 \pm v/c) \qquad (4)$$

where c is the velocity of light. Based on an assumed Maxwellian distribution of velocities, this broadening effect follows a Gauss distribution which may be expressed by—

$$R_i/(R_i)_0 = \exp\{-(\Delta\lambda_i/\omega_i)^2\} \qquad (5)$$

A vibrating dipole radiates energy and this energy must diminish the vibrational amplitude, resulting in damped vibrations. The radiation emitted in such a process cannot be monochromatic and the line shape has been shown to have the form of the Cauchy distribution given in equation (3).

In condensed systems of the three effects, the collision broadening is dominant.

C. Measured band shape

The measured band shape will be a function not only of the true band shape for individual vibrators, but also of (a) the presence of scattering particles, (b) the non-monochromatic nature of the light falling on the sample, and (c) the presence of background (non-specific) absorption.

Giovanelli has considered the effect of the presence of scattering particles. Assuming that the absorption of an individual vibrator follows the Cauchy distribution, he has derived a relationship—

$$R_i/(R_i)_0 = -C \ln \{1-1/[1+(\Delta\lambda_i/\omega_i)^2]\} \qquad (6)$$

for a constant background. The value of C is determined by the refractive index of the matrix and of the cover glass. The value of 0.15 is the most probable value with limits of 0.125 and 0.175.

The instrument characteristics affect the measured reflectance. The light passed to a sample in a spectrometer is non-monochromatic. Frequently it will follow a Gaussian distribution, and one characteristic of high resolution spectrometry is that the bandwidth is small. If this band-width is small compared with the bandwidth of the sample, it has little effect upon the measured curve. Additional instrument error can arise in the electronic circuits and in the graph plotter. All of these may be placed together in a term which is called "instrument function". Unfortunately, these instrument functions are difficult to determine and are not normally supplied by the manufacturers. The measured reflectance is the result from convolution of the instrument function and the true reflectance.

The convolution of a Gauss and Cauchy curve results in a profile of the shape—

$$\frac{1}{\pi\omega_1} \int_{-\infty}^{+\infty} \frac{e - (\Delta\lambda/\omega_2)^2}{1 + (\Delta\lambda/\omega_1)^2} \, dW$$

the ratio of ω_2 to ω_1 determines the proportionality between the Cauchy and Gauss characteristics of the resulting profile. If $\omega_1/\omega_2 = 0$, the curve is a pure Gauss curve, and if $\omega_1/\omega_2 = \infty$, a pure Cauchy profile is obtained.

The convolution of a Cauchy curve with a Cauchy curve results in a pure Cauchy profile, whose width is the sum of the width of each individual contributed. The convolution of a Gauss curve with a Gauss curve results in a Gauss profile whose width is similarly the sum of the width of the two contributing curves. At any wavelength the reflectance is the sum of the contributions from the various absorbing constituents. In the absence of specific absorbing vibrators, the scattering effect of the particles will result in the intensity of the reflected light being less than that of the incident. The ratio of the intensity of the reflected light to the intensity of the incident light in the absence of specific absorption bands is called the "background" reflectance. If the background reflectance were constant, the band shapes would be symmetrical. There is no clear theoretical method of allowing for the background reflectance, so it was assumed in our calculations that the background reflectance had a similar function to that of the incident light in transmittance spectrophotometry.

The validity of an assumed band profile can be tested by using it to determine band intensities in the spectra of a number of samples containing a range of known relative concentrations of pigments. This has been done with yeast spectra by Rickard et al. (1967) (see Section III for details), who determined the band intensities by each of the three assumed profiles, Giovanelli, Cauchy and Gauss, and correlated these band intensities by both linear and quadratic equations to predict the percentage of cytochrome in each sample. Each of the correlations was made using the method of least squares. The fitting of the quadratic equation—

$$Cy = A + BI + CI^2 \tag{7}$$

was found to be superior to the linear equation—

$$Cy = A + BI \qquad (8)$$

where Cy was the percentage cytochrome in the yeast, I the band intensity and A, B and C empirical constants.

Comparison of the predicted cytochrome concentrations with the true concentrations showed that for yeast spectra the Cauchy profile is least reliable. The Giovanelli profile is best, but not significantly better than the Gauss.

III. ESTIMATION OF CYTOCHROMES

The method is particularly applicable to the detection and estimation of microbial cytochromes; for example, their absorption bands in the visible and near ultraviolet are clearly defined in reflectance scans of yeast (Fig. 1). Variations in the concentration of these compounds in response to changes in environmental conditions have been of particular interest in our laboratory where reflectance spectrophotometry has been employed, as detailed below and by Rickard et al. (1967), for comparative cytochrome studies.

A. Preparation of sample

Approximately 6 g wet weight of yeast is washed in 0·15 M phosphate buffer, pH 7·0, and centrifuged. If the pH of the supernatant is less than 7·0, the procedure is repeated until this value is achieved. To ensure complete reduction of all cytochromes, the paste is treated with approximately 10 mg of sodium dithionite by mixing thoroughly with a large spatula on a flat tile. After standing for 5 min, the paste is piled into the centre of the sample well of a reflectance cell.* A sheet of optical glass, 49 mm × 49 mm × 0·5 mm is placed over the sample. It is essential that no air bubbles are trapped in the sample and this possibility is prevented by slowly and gently tamping the glass onto the yeast, rather than lowering it suddenly in one movement, until the glass rests flat on the aluminium plate and the sample fills the well. Any excess yeast which has been exuded from the well is removed and the optical cleanliness of the outside of the glass plate checked.

---

* In the authors' laboratory a reflectance cell consists of a round aluminium plate, diameter 76 mm and depth 6 mm; the sample well, diameter 45 mm and depth 2 mm, is set in one side of the plate and is concentric with it; another depression, diameter 59 mm and depth 1·5 mm, set in the opposite side, is also concentric with the plate; the second depression holds the sample clamp of the Perkin–Elmer Model 350–0150 reflectance attachment and ensures snug and rigid fitting of the sample against the rear surface port of the sample integrating sphere.

B. Scanning the spectrum

The reflectance cell is clamped into position against the rear surface port of the sample integrating sphere of a Perkin-Elmer, Model 350, recording spectrophotometer fitted with a reflectance attachment, Model 350–0150, the integrating spheres of which having previously been coated with magnesium oxide. Several sheets of filter paper are clamped against the rear surface port of the reference integrating sphere.

The total, diffuse and specular, reflectance of the sample is scanned between 350 and 700 nm employing instrumental settings which have been found to produce most satisfactory spectra, namely: Mode, 1T; resolution, 1·2; slit width at 650 nm, 0·1; scan speed, 1·2 on fast gear; response time, 3.

C. Preparation of spectral data

From the continuous spectrum thus obtained, the following data are transferred to computer punch cards: Card 1, identification; Card 2, punched according to whether cytochrome troughs read as maxima or minima; Card 3, wavelength range and interval of interest (standardized at 370–650 nm at intervals of 5 nm); Card 4, number of background points (any convenient number, not less than two)—these are the points of highest reflectance in the scan; Card 5, wavelength and percentage reflectance of the background points; Card 6, number of troughs in the scan; Card 7, wavelength and percentage reflectance of the troughs; Card 8, number of test points read around each trough; Card 9, wavelength and percentage reflectance at test points.

D. Numerical analysis of spectral data

The presence of overlapping absorption bands and of variations in non-specific absorption could present some difficulty in the numerical analysis, especially as the central wavelength of a particular absorption band is not necessarily at that point on the curve which gives a minimum in reflectance. The difficulty is diminished by a computer programme designed to resolve overlapping bands. The main absorption bands are taken at wavelengths of 415, 440, 518, 547, 558, and 598 nm. For all curves the initial bandwidth is assumed to be 10 nm. Initially, the background is estimated from points at approximately 380, 485, and 650 nm. A smooth background curve is fitted through these points and the intensity of each band estimated by the difference between the background and the total reflectance at the central wavelength. From these estimates and from an assumed Gaussian band profile, the contribution of each band may be determined. From these, an improved background curve may be obtained. From the improved background curve, a better estimate of the band intensities and bandwidths may be made. This iteration is carried out until neither the band intensity nor the bandwidth in any case changes by more than 0·1%.

The iteration is programmed in the Fortran language and used on an IBM 1620 computer, with card input–output, 40,000 digits of store, a disk file, and an on-line printer.*

E. Comparative studies

For comparison of the cytochrome concentrations of different yeasts and of the same yeast cultured under a variety of conditions, a calibration curve of band intensity versus cytochrome concentration is first prepared as follows: a range of yeast mixtures is prepared from a standard cyto-chrome-rich sample and a cytochrome-less sample, and the proportion of the former in each mixture considered to be the cytochrome concentration of that mixture relative to the standard yeast; from the continuous total reflectance spectrum of each mixture, the band intensities of the α bands of A-, B- and C-type cytochromes are calculated as described above and plotted against the relative cytochrome concentration. From the three calibration curves the cytochrome concentrations of unknown yeast samples,

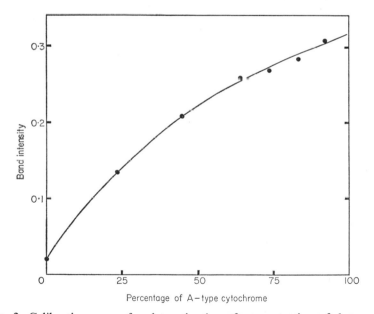

FIG. 2. Calibration curve for determination of concentration of A-type cyto-chrome in yeast relative to a standard yeast arbitrarily designated as containing 100% cytochrome. See text for method of determining band intensity.

* Copies of the programme are available from G. H. Roper.

relative to the standard yeast, can be determined after calculation of their appropriate band intensities. The calibration curve for cytochrome A is shown in Fig. 2.

F. Approximation method

Resolution of overlapping bands by programmed iteration is limited to estimation of cytochrome concentrations up to 60% of the maximum concentration recorded by us. Concentrations up to 100% of the maximum may be approximated, without resolution of overlapping bands, by the following method: a tangent, considered to approximate the background absorption of the yeast, is drawn between the reflectance maxima which occur at about 475 nm and 630 nm in the continuous spectrum; the fraction of background light reflected (I/I_B) at a given wavelength is considered as being equal to the ratio of the observed reflectance at that wavelength to the approximated background reflectance at the same wavelength; I/I_B values for the α band minima of A-, B- and C-type cytochromes are calculated at the 598 nm trough, the 558 nm shoulder and the 547 nm trough (see Fig. 1) respectively. The relative concentration of each cytochrome type is then determined from calibration curves of I/I_B vs. relative cytochrome concentration prepared from the standard yeast mixtures described in Section IIIE above.

G. Choice of reference

The double beam instruments employed for reflectance spectrophotometry require that a reference material be placed in the reference port during scanning. The reference used in scanning the spectrum shown in Fig. 1 and in preparation of the calibration curve shown in Fig. 2 was filter paper. More clearly defined absorption troughs can be obtained by employing a cytochrome-less yeast as reference. In our hands, however, the calibration curves thus obtained had no merit over those obtained with a filter paper reference. The disadvantage of a cytochrome-less yeast reference lies in the difficulty of standardizing such reference throughout a complete course of investigation.

H. Choice of mode

With some spectrophotometers it is possible to expand the reflectance scale. Although no advantage has thus far been credited to our work by this device, it nevertheless has decided potential in the detection of weak absorption bands, provided the background does not deviate far from a horizontal straight line.

I. Comparison with transmittance spectrophotometry

Similar calibration curves to that shown in Fig. 2 can be obtained by suspending yeast mixtures in glycerol and scanning their spectra by transmittance spectrophotometry. In this case either powdered cellulose suspended in glycerol or cytochromeless yeast suspended in glycerol are suitable reference materials. The calibration curves obtained with either reference offer us no advantage over reflectance spectrophotometry. In our experience, the great disadvantage suffered in the use of transmittance spectrophotometry for estimation of yeast cytochromes is its sensitivity to cell density in the yeast sample; for example, when water was added to a yeast sample so that its percentage dry weight was decreased from 19·2% to 16·3%, an apparent 20% decrease in cytochrome A concentration was found when estimated by transmittance. The apparent decrease was only 7% when estimated by reflectance.

IV. INSTRUMENTATION

Most makers of spectrophotometers provide accessories for the measurement of reflected light. Instrumental methods for measurement of reflectance and reflection are described by Wendlandt and Hecht (1966). A succinct account of most instruments is given, together with photographs and diagrams of optical paths. Microbiological samples are usually in the form of a cell paste, often somewhat fluid; horizontally placed sample cups are an advantage for this reason. It is often desirable to examine small samples so that a sensitive spectrophotometer which operates with a light beam of small diameter is ideal. The sample should be deep enough to fulfil "semi-infinite" conditions, i.e. transmitted light should be negligible.

Because of considerations of cost, most laboratories are required to use the same spectrophotometer for both reflectance and transmittance, and it is essential that the conversion be practicable. In some cases this is not so.

Two basic principles are used, reflectance and reflection. In the former case a beam of monochromatic light is diffused by the sample and integrated by a highly reflecting layer, usually magnesium oxide, on the inside of an integrating sphere of which the sample is a part. In the case of reflection, the light beam falls on the sample and is reflected to a detector which collects a selected cone of the available light diffused by the sample.

The following are among the available spectrophotometer reflectance or reflection accessories. (Information on their performance is available from manufacturers and will not be dealt with here.)

Reflection or reflectance attachments are available for Models DU, B, DB, DK-1A and DK-2A of Beckman spectrophotometers (Beckman Instruments, Inc.). Models DU and B are manually recording. The DU accessory is a reflection instrument and interchanges readily with the standard DU cell compartment. The sample, which is 1.5×2.5 in., is applied horizontally in a sliding holder.

The reflectance accessory for the Model B spectrophotometer has an incident light beam on the sample of $\frac{1}{8} \times \frac{3}{16}$ in. The sample is introduced horizontally. The wavelength range is 325–1000 nm.

A reflectance attachment is available for the DB spectrophotometer. This is a double beam instrument which covers a wavelength range of 205–770 nm.

DK-1A and DK-2A spectrophotometers are converted to reflectance instruments by replacement of the DK sample compartment by an integrating sphere reflectance unit. By using a 1P28 photomultiplier tube in combination with a lead sulphide cell as detector, the range is from 210 to 2700 nm.

Beckman attachments for infrared wavelengths are available. The microspecular attachment uses a mirror reflector from a sample area as small as 1×5 mm. The hemisphere reflectance attachment for Beckman IR-4, IR-7, IR-9 spectrophotometers uses an integrating hemisphere. The attachment on the IR-7 can be used from 250 to 1550 nm. On the IR-9 instrument the range can be extended to 2500 nm.

Three Bausch and Lomb instruments are available. The Spectronic 20 Colour Analyzer is a manual instrument which has an integrating sphere. It uses a fixed bandwidth of 20 nm. The range is 400–700 nm.

The Spectronic 505 spectrophotometer accessory, which is readily attached to the spectrophotometer, consists of an integrating sphere with sample holders and a lens system. A monochromatic light beam of 16 mm diameter is focused by the lens system onto the sample and reference surfaces. Diffuse reflected light is detected by a photomultiplier. The bandwidth of the incident beam is 5 nm and the wavelength range is 400–700 nm.

The spectroreflectometer is similar to the Spectronic 505 but employs direct illumination of the integrating sphere. The bandwidth of the incident beam is 5 nm and the wavelength range is 400–700 nm.

A diffuse reflectance attachment is supplied by the Perkin-Elmer Corporation, Norwalk, Connecticut, for the Hitachi Perkin-Elmer Model 139. Reflected monochromatic light from sample or reference material is integrated by the sphere and detected by a photomultiplier. The sample size is 24×24 mm. The wavelength range is 400–760 nm.

The reflectance attachment used routinely by the authors for the estimation of cytochromes in microbial pastes is the integrating sphere attachment, Model 350–0150, for the Perkin-Elmer, Model 350, spectrophotometer. The area of incident light at the sample surface is 25×32 mm. The wavelength range with the standard photomultiplier is 330–750 nm. This may be extended to 2500 nm by the use of lead sulphide detectors. This instrument has proved to be a highly sensitive instrument for the measurement of cytochromes *in situ*.

For Zeiss spectrophotometers, PMQ11 and RPQ20A, accessories RA2, RA3 and RA20 are available. Model RA2 is a reflection type; light normal to the sample or reference sample is focused onto the photomultiplier and controlled in amount by a shutter. RA3 can be operated in two ways: either from diffuse light from a light source placed in the integrating sphere, or from a light source outside the integrating sphere. The two modes require a change in position of light source and photomultiplier. The bandwidth varies from 0·9 to 35 nm. RA2 operates in the 200–600 nm range, RA3 from 380 to 2500 nm. The RA20 is used with the Zeiss Model RPQ20A recording spectrophotometer. Monochromatic light is reflected by an oscillating mirror through the integrating sphere onto the sample and reference. Reflected light from both sample and reference is integrated by the sphere and detected by a photomultiplier mounted on the side of the sphere. The wavelength range 200–600 nm is covered by a IP28 photomultiplier while 620–2500 nm is covered by a PbS detector.

Three types of attachments are available for Cary Models 14 and 15 spectrophotometers (available from Cary Physics Corp., 2724 South Peck Road, Monrovia, California). Model 1411 is a diffuse reflectance accessory for the Model 14 spectrophotometer. Chopped monochromatic reference and sample beams from the spectrophotometer enter the reflectance sphere at 90° to each other by a lens system. Light reflected from sample and reference is integrated by the sphere and detected by a phototube mounted at the top of the sphere. This accessory operates with a resolution of 0·25–0·7 nm.

A second type of accessory is a reflection instrument. Chopped monochromatic light reflected from sample and reference are detected by a phototube. A screen attenuator in the reference beam allows standardization of the reference. The wavelength range is 220–700 nm with a resolution of 0·13–0·7 nm.

The cell-space diffuse reflectance accessory may be used with Cary Model 14 or 15 spectrophotometers. This has two integrating chambers for the sample and reference respectively.

Unicam (York Street, Cambridge, England) offers two diffuse reflectance accessories. The SP540 reflection accessory consists of a concave annular mirror which reflects monochromatic light from sample and reference as one or other is brought into a port by means of a horizontal sliding holder.

A low-temperature attachment has been described by Symons and

Trevalion (1961) in which samples can be examined at 77°K. The development of low temperature reflectance spectrophotometry is of considerable importance, particularly for estimation of respiratory pigments *in situ*. At the temperature of liquid nitrogen overlapping cytochrome peaks are sharpened so that, for example, cytochrome c and c_1 can be recognized as separate peaks. A difficulty which needs to be overcome is interference due to condensation of water vapour on the sample surface.

The Unicam SP890 reflection accessory consists of a system of reflectors placed such that incident monochromatic light from the sample beam falls on the sample or reference and the diffuse light reflected between the angles 35° and 55° is reflected by an annular mirror onto the detector.

The reflectance attachment QV–09–00 for the Shimadzu spectrophotometer (Shimadzu Seisakusho Ltd.) type QV–50 uses an integrating sphere which collects diffuse reflected light and rejects the specular beam. Samples of 25 mm diameter are placed horizontally in a sliding holder at the bottom of the sphere.

Diffuse reflectance accessory 123–0705 is attached to the Hitachi spectrophotometer (Hitachi Ltd., Tokyo). Unlike other instruments described here, monochromatic light is transmitted through the sample which is placed against an integrating sphere opposite the port. Because the sample cell is hard up against the sphere, no light is lost by scattering. All transmitted light is integrated by the sphere and falls on the detector. A similar arrangement holds also for the reference cell. Quantitative interpretation of these measurements depends on a calibration of each type of microbial sample using a reference which is identical with the sample in all respects except for the absorbing substance of interest. An ideal example would be *Chlorella* cells containing chlorophyll as the sample and similar cells without chlorophyll as the reference.

A diffuse reflectance attachment is supplied for the Uvispec spectrophotometer MK9 (Hilger Watts Ltd.). This fits inside the cell compartment of the spectrophotometer. The wavelength range is 180–1000 nm.

Reflectance measurements can also be made by another Hilger Watts instrument, an attachment to the Spectrochem spectrophotometer. The attachment is designed to fit into the cell compartment. Light from the monochromator is directed onto the sample and that portion of the light which is reflected at 45° is collected by the photomultiplier and measured. Since reflectance is measured, a linear scale galvanometer should be used; this is connected to the photomultiplier with a cableform and jackplug, the latter automatically disengaging the internal galvanometer. The sample size is $\frac{3}{4}$ in. diameter. The bandwidth is 2–7 nm and the range 340–820 nm.

REFERENCES

Carter, D. L., and Myers, V. I. (1963). *Proc. Am. Soc. hort. Sci.*, **82**, 217–221.
Chance, B., Legallais, V., and Schoener, B. (1963). *Rev. Scient. Instrum.*, **34**, 1307–1311
Dean, R. W., and Ball, C. O. (1960). *Food Technol.*, **14**, 271–286.
Frei, R. W., and Ryan, D. E. (1967). *Analytica chim. Acta*, **37**, 187–199.
Frodyma, M. M., Lieu, V. T., and Frei, R. W. (1965). *J. Chromat.*, **18**, 520–525.
Giovanelli, R. G. (1955). *Optica Acta*, **2**, 153–162.
Giovanelli, R. G. (1956). *Optica Acta*, **3**, 127–130.
Giovanelli, R. G. (1957a). *Aust. J. Phys.*, **10**, 227–239.
Giovanelli, R. G. (1957b). *Aust. J. exp. Biol. med. Sci.*, **35**, 143–156.
Giovanelli, R. G. (1957c). *Nature, Lond.*, **179**, 621–622.
Jork, H. (1965). *Planta med.*, **13**, 489–490.
Kubelka, P. (1948). *J. opt. Soc. Am.*, **38**, 448–457.
Kubelka, P., and Munk, F. (1931). *Z. tech. Phys.*, **12**, 593–601.
Lieu, V. T., Frodyma, M. M., Higashi, L. S., and Kunimoto, L. H. (1967). *Analyt. Biochem.*, **19**, 454–460.
Lorentz, H. A. (1906). *Proc. K. ned. Akad. Wet.*, **8**, 591–611.
Metzner, H. (1965). Proc. Western-Europe Conf. Photosyn., 2nd. Conf. Centre Woudochoten, Zeist, Neth., 17–30 (published 1966).
Morris, N. J., Lohmann, I. W., O'Connor, R. T., and Freeman, A. F. (1953). *Food Technol.* **7**, 393–396.
Moss, F. (1956). *Aust. J. exp. Biol. med. Sci.*, **34**, 395–406.
Moss, F. J., and Tchan, Y. T. (1958). *Proc. Linn. Soc. N.S.W.*, **83**, 161–164.
Moss, F. J., Rickard, Pamela, A. D., Beech, G. A., and Bush, F. (1969) *Biotechnol. Bioengng*, **11**, 561–580.
Moss, F. J., Rickard, Pamela, A. D., Bush, F., and Caiger, P. C. In press.
Parker, F. S., and Ans, R. (1967). *Analyt. Biochem.*, **18**, 414–422.
Pirko, P. C., and Ayres, J. C. (1957). *Food Technol.*, **11**, 461–468.
Rickard, P. A. D., Moss, F. J., and Roper, G. H. (1967). *Biotechnol. Bioengng*, **9**, 223–233.
Rickard, Pamela A. D., Moss, F. J., Ganez, Micheline. In press.
Rickard, Pamela A. D., Moss, F. J., Phillips, D., and Mok, T. C. K. In press.
Shibata, K. (1958). *J. Biochem, Tokyo*, **45**, 599–623.
Shugarov, Yu. A. (1965(6)). *Agrokhimiya*, 141–145.
Snyder, H. E., and Armstrong, D. J. (1967). *J. Food Sci.*, **32**, 241–245.
Stewart, M. R., Hutchins, B. K., Zipster, M. W., and Watts, B. M. (1965). *J. Food Sci.*, **30**, 487–491.
Stewart, M. R., Zipster, M. W., and Watts, B. M. (1965). *J. Food Sci.*, **30**, 464–469.
Symons, M. C. R., and Trevalion, P. A. (1961). *Unicam Spectrovision*, **10**, 8–9.
van Vleck, J. H., and Weisskopf, V. F. (1945). *Rev. mod. Phys.*, **17**, 227–236.
Wendlandt, W. W., and Hecht, H. G. (1966). "Reflectance Spectroscopy", Vol. 21 of "Chemical Analysis". Interscience Publishers, New York/London/Sydney.
Zaye, D. F., Frei, R. W., and Frodyma, M. M. (1967). *Analytica chim. Acta*, **39**, 13–18.

CHAPTER XIII

Base Composition of Nucleic Acids

Wesley D. Skidmore† and Edward L. Duggan‡

Department of Biochemistry, University of California School of Medicine
San Francisco, California 94122, U.S.A.

I.	Introduction	631
II.	Methodology	632
	A. Intact DNA	632
	B. DNA or RNA altered by hydrolysis or other reactions	632
III.	Example Analyses	634
	A. A procedure applicable to DNA	634
	B. A procedure applicable to RNA or DNA	637
References		638

I. INTRODUCTION

A considerable amount of interest in the composition of nucleic acids is evident in scientific literature since the discovery of nucleic acids by Miescher in 1871. The most valuable compilation of information on this general subject is, in our opinion, the three volume series edited by Chargaff and Davidson (1955–60). We recommend that the beginning investigator of nucleic acid chemistry refer to that publication at the outset of his studies. In this Chapter, we will refer to methodology described therein as being the authoritative summary or compilation of standard procedures prior to the time of its publication. Some other reviews on the determination of base composition are now available (Rajbhandary and Stuart, 1966; Munro and Fleck, 1966).

The object of this Chapter is to guide an investigator to pertinent references in an organized manner for the purpose of determining base composition of nucleic acids. As an experimental or typical example, we describe a selected spectral method in some detail for determination of the base composition of DNA from *Bacillus subtilis*. This selection is one with which we are most familiar and have the most experience. Other recommended procedures are described briefly or given as references for the investigator to select for his specific requirements.

† Present address: Armed Forces Radiobiology Research Institute, Bethesda Md. 20014.
‡ Deceased April 1968.

The approach that we have selected to determine base composition follows: First extract DNA or RNA from the cells. Then, determine base composition of intact DNA by physical methods where applicable, or obtain a detailed chemical analysis of RNA or DNA by hydrolyzing the intact nucleic acids and either analyzing the mixture of hydrolysis products or separating the hydrolysis products by chromatography or electrophoresis. Finally, quantitate the concentrations of the isolated components spectrometrically.

II. METHODOLOGY

In general, removal of nucleic acids from cellular systems is a necessary prerequisite for analysis of base composition. Several methods are readily available for extracting, isolating or purifying nucleic acids to varying degrees of usefulness. Many of these methods are reviewed in detail elsewhere (Arnstein and Cox, 1966; Cantoni and Davies, 1966; Chargaff, 1955; Kirby, 1964; Kit, 1963; Magasanik, 1955; Munro and Fleck, 1966; Marmur et al., 1963; Stevens, 1963, and by authors in this Volume; Herbert et al., and Sutherland and Wilkinson. The extracted nucleic acids can be analyzed for base composition by selected methods which are outlined below.

After preliminary isolation of nucleic acids from other cellular components, we recommend determining base composition by the following methods:

A. Intact DNA

In recent years, methods have been described that are applicable to intact DNA, i.e., DNA that was extracted by a procedure that did not utilize hydrolysis. Reviews of these methods are in the literature (Kit, 1963; Marmur, 1963). A method applicable to partially purified DNA utilizes its buoyant density in CsCl (Schildkraut et al., 1962; Sykes, this Volume p. 55) but most methods are best applied to extensively purified nucleic acids. Another useful method directly applicable to isolated DNA utilizes its thermal denaturation temperature (Marmur, 1962). Both the CsCl and the thermal denaturation methods require other samples of DNA as controls or standards.

B. DNA or RNA altered by hydrolysis or other reactions

There are two approaches with this method. In one, intact DNA is heated with acid or is reacted with another reagent that alters its structure and it is then analysed without further separation of the hydrolytic or reaction products. Estimation of composition is made by comparison with known compounds as standards. The alternative approach requires complete hydrolysis of intact DNA or RNA followed by separation and quantitation of the components. The latter provides an accurate and comprehensive analysis of nucleic acids including those which are not base paired or which have a variety of bases.

1. Methods applicable to the direct analysis of hydrolyzed or altered DNA without separation of reaction products

(a) *Spectral.* One method (Skidmore and Duggan, 1966) provides a spectral determination for both base composition and base concentration in a single sample of DNA. Spectra of standard deoxynucleotide mixtures (adenylate with thymidylate and guanylate with cytidylate), hydrolyzed in 1 N perchloric acid, are directly proportional to spectra for DNA treated similarly. There is a linear relation between the sum of the chosen differential absorbancies on each side of an isosbestic point of the standard spectra and the base composition. The base concentration of DNA is proportional to the base concentration of the standards at the same isosbestic point. The determinations are linear for base compositions of 0–100 mole % and for base concentrations of 0–100 μM. The procedure is applicable to the four major bases paired in double-stranded DNA. Other methods show promising applicability (Hirschman and Felsenfeld, 1966; Hu *et al.*, 1964; Fredericq, *et al.*, 1961; Huang and Rosenberg, 1966 and DeLey and van Leeuwenhoek, 1967).

(b) *Chemical.* A simple and rapid method was recently proposed (Wang and Hashagen, 1964). This method is based upon the known reactivity of the commonly occurring bases and their derivatives, except adenine, with acidified N-bromosuccinimide solution. The reaction decreases absorbancy at 270 nm; the adenine content is calculated from the residual absorption. Thus, the base composition of a double stranded DNA (adenine equals thymine and guanine equals cytosine) is directly calculated.

2. Methods for RNA and DNA by quantitative hydrolysis and quantitative separation of the products

(a) *Methods of hydrolysis* These methods are reviewed in detail by others (Loring, 1955; Magasanik, 1955 and Wyatt, 1955). Those reviews describe useful methods for acid hydrolysis of DNA (Marshak and Vogel, 1951) and of RNA (Smith and Markham, 1950).

(b) *Separation of hydrolytic products of nucleic acids.* There are three chromatographic methods and one electrophoretic technique that are applicable. Some reviews are available (Saukkonen, 1964; Munro and Fleck, 1966 and Rajbhandary and Stuart, 1966).

(i) *Thin-layer chromatography* (*TLC*). TLC is a new and rapidly developing method which is explained in detail elsewhere (Randerath, 1966). Most of the information developed in paper chromatography is applicable here. Several useful methods have appeared recently in the literature (Chmielewicz and Acara, 1964; Gebicki and Freed, 1966; Harris and Warburton, 1966; Hiby and Kroger, 1967; Holdgate and Goodwin, 1964 and Pataki and Kunz, 1966, 1967).

(ii) *Paper chromatography* has found wide application in many fields, including specialized applicability for separating hydrolytic products of nucleic acids (Wyatt, 1955 and Niemierko and Krzyzanowska, 1967).

(iii) *Column chromatography* is a technique that provides quantitative separations and recoveries of nucleic acid components. Reviews describing applicable procedures are available (Cohn, 1955; Katz and Comb, 1963; Green, *et al.*, 1966 and Blattner and Erickos, 1967).

(iv) *Electrophoretic techniques* can be useful for separations also. Descriptions of applicable procedures are available (Smith, 1955; Edstrom, 1960 and 1964; Click, 1966 and Hiby and Kroger, 1967).

3. *Spectrometric quantitation of isolated bases by ultraviolet absorption*

This analysis is readily performed on isolated constituents containing the aromatic structure of the bases. The spectra and molar absorbancy or extinction coefficients for nucleic acid bases, nucleosides, and nucleotides at specified pH values are available elsewhere (Beaven *et al.*, 1955 and Voet *et al.*, 1963). Detailed information for calculations is also available in these two references.

III. EXAMPLE ANALYSES

A typical example of determining base composition of bacterial nucleic acids is outlined below.

A. A Procedure applicable to DNA

1. Isolate DNA from *B. subtilis* by procedures described by Marmur, 1961, or modifications of that procedure as reviewed by Arnstein and Cox, 1966 and other authors in this Series.

2. Determine base composition and concentration by perchloric acid hydrolysis of intact DNA and direct spectral analysis of the reaction mixture (Skidmore and Duggan, 1966). A detailed procedure, including a figure illustrating the method, follows:

(a) *Reagents*

(1) *Perchloric acid* (70% w/v, 12 N) reagent grade.
(2) *Standard deoxynucleotides.*

Deoxynucleotide monophosphates of adenylate (A), thymidylate (T), guanylate (G), and cytidylate (C) were obtained from Calbiochem Corp., Los Angeles, California (lot nos. 30475, 30476, 30477 and 30474 respectively). Standard solutions were prepared in water (50 μM each). The base concentration of each was determined spectrometrically with the molar absorbancy coefficient (E) at the appropriate pH value and wavelength using the maximal absorbancy of each (for A, 15,200, pH 7, 259 nm; for T, 9600, pH 7, 267 nm; for G, 13,800, pH 7, 253 nm and for C, 13,200, pH 2, 280 nm).

(3) Standard deoxynucleotide mixtures

Standard mixtures of AT and GC were prepared by mixing equal volumes of the working standards of A with T and then G with C. The total base concentration of each standard mixture is 50 μM. The base compositions are 100 mole % AT for the AT standard mixture and 100 mole % GC for the GC standard mixture. The base concentrations of these mixtures were verified by the molar absorbancy coefficient at the isosbestic point (273 nm) of the AT, GC spectra ($E_{273}^{AT,GC}$ equals 9450) as obtained by the procedure described below.

(b) *Procedure.* Add dilute samples (3·0 ml) of DNA and of two standard deoxynucleotide mixtures of AT and GC to individual test tubes (18 × 150 mm). Next, mix perchloric acid (0·30 ml) with each sample. Now cap the tubes with clean glass marbles and place them in any convenient form of tube heater (pre-heated to 110–112°C) for 15 min. Remove the tubes and cool them to room temperature in cold tap water. Determine the spectra of each sample between 315 and 240 nm with a recording spectrophotometer (e.g. Beckman DK-2 or Cary 14). Record absorbancy values at three wavelengths from spectra of AT, GC and DNA samples for calculations. The wavelengths are defined as follows: an isosbestic point of the standard spectra defines one wavelength (273 nm). The maximal absorbancy of the AT spectrum defines another (264 nm). The third wavelength (286 nm) is the one whose absorbancy value of the GC spectrum is equal to the absorbancy value of the same spectrum at the wavelength previously defined for the AT spectrum (Fig. 1).

(c) *Calculation of base composition and base concentration.* Linear equations are useful for calculating base concentration and base composition. First, the equation for base concentration stems directly from Beer's Law. After substituting the data from the spectra (Fig. 1) for standard mixtures of AT and of GC and for *B. subtilis* DNA, the equation appears in the following easily recognizable form:

$$h/g \times 50 \ \mu M = \chi \mu M \text{ base} \tag{1}$$

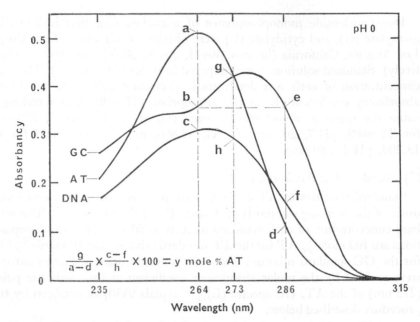

FIG. 1. *Illustration of the method for determining base composition and base concentration of DNA.* The following samples, treated as in the procedure described in the text, yield the following spectra: standard deoxynucleotide mixture of 100 mole % AT at 50 μM base concentration; a similar standard mixture of 100 mole% GC at 50 μM base concentration and a sample of *B. subtilis* DNA with x μM base concentration, y mole % AT, and z mole % GC base composition. The points designated by letters are: a, absorbancy of AT spectrum at 264 nm; b, absorbancy of GC spectrum at 264 nm; c, absorbancy of DNA spectrum at 264 nm; d, absorbancy of AT spectrum at 286 nm; e, absorbancy of GC spectrum at 286 nm; f, absorbancy of DNA spectrum at 286 nm; g, absorbancy of isosbestic point at 273 nm; h, absorbancy of DNA spectrum at 273 nm. Calculations to determine x μM base concentration, y mole % AT, and z mole % GC of DNA are given in text.

Secondly, the equation for base composition (Fig. 1) also stems from proportionality. The total absorbancy difference for the DNA (c–f) is normalized for concentration by multiplying by g/h. The resulting number is arithmetically compared with the total absorbancy difference for the standards $(a-b)+(e-d)$. Since b equals e (Fig. 1), this value reduces to $(a-d)$. The resulting calculation is:

$$g/h \, (c-f) \times 1/(a-d) \times 100 \text{ mole } \% \text{ AT} = y \text{ mole } \% \text{ AT} \qquad (2)$$

After replacing letters with abbreviations and rearranging the equation so that standard values are in one fraction, the equation follows:

$$[A_{273}^{\text{AT, GC}}/(A_{264}^{\text{AT}} - A_{286}^{\text{AT}})] \times [(A_{264}^{\text{DNA}} - A_{286}^{\text{DNA}})/A_{273}^{\text{DNA}}] \times 100 = y \text{ mole } \% \text{ AT} \quad (3)$$

(Note; $A_{273}^{AT,GC}$ is the absorbancy value for the AT and GC spectra at 273 nm and the other abbreviations in this equation for A are also absorbancy values for different spectra at the designated wavelengths.)

In our experience, the value for the standard fraction

$$[A_{273}^{AT, GC}/(A_{264}^{AT} - A_{286}^{AT})]$$

is 1·064. Thus, the DNA spectrum provides the data for the calculations. And lastly, the equation for mole % GC of DNA follows:

$$100 - y \text{ mole } \% \text{ AT } = z \text{ mole } \% \text{ GC} \tag{4}$$

3. Compare calculated values for base composition with published values (Belozersky and Spirin, 1960; Schildkraut et al., 1962 or Marmur et al., 1963).

There are several advantages of the method reported here over others. One advantage is that calculations are linear for base composition between 0–100 mole % AT or GC, and for base concentrations between 0–100 μM. Also, DNA fractions as obtained from some special methods like column chromatography are ready for analysis. A heating period in 1 N perchloric acid virtually eliminates uncertainties associated with the hypochromic effect in DNA. The pH value of the DNA solution does not interfere so long as the buffering capacity does not appreciably neutralize 1 N acid. Calculations are not based on "native" DNA as a standard but on commercially available deoxynucleotides. Various contaminants, such as sodium chloride, phosphate and small amounts of protein (less than 5%) do not interfere. Also, the procedure is experimentally simple. Special equipment is not required (a recording spectrophotometer is convenient but not necessary).

One limitation of this method is that impurities (RNA, aromatic amino-acids, and phenol) which absorb ultraviolet light interfere, unless they are removed by dialysis. Another limitation is that analyses are restricted to DNA containing the four usual bases, unless the unusual bases are incorporated stoichiometrically in the standard mixtures. Spectral curves of the standard mixtures have rather steep slopes at 286 nm which will cause variability in absorbancy readings at this wavelength. The differential absorbancy at 286 nm added to that at 264 nm allows high sensitivity, but it also adds variability. Using the determination at one wavelength (264 nm) removes much of the variability, however it decreases the sensitivity by more than one half. For some purposes this might be worthwhile for higher accuracy.

B. A procedure applicable to RNA or DNA

1. Isolate ribosomal, nuclear, messenger, transfer or viral RNA as appropriate (Arnstein and Cox review, 1966), or if a comprehensive

physical and chemical analysis of major and minor bases of a DNA sample is desired, use the DNA sample isolated above and continue with the following procedure for RNA or DNA.

2. Hydrolyze the nucleic acids with perchloric acid (12 N, 1 h, 100°C) as previously described for RNA (Smith and Markham, 1950) or DNA (Marshak and Vogel, 1951).

3. Isolate purine bases and pyrimidine nucleotides from the RNA hydrolysis mixture quantitatively, or total bases from the DNA mixture, utilizing thin-layer chromatography (Randerath, 1966; Pataki, 1967). Recover the hydrolytic products quantitatively from TLC by elution. Dilute eluted samples appropriately with a diluent of proper pH value for subsequent spectral quantitation.

4. Quantitate the isolated bases or nucleotides spectrophotometrically (Beaven et al., 1955; Voet et al., 1963). Briefly, this is accomplished by recording absorbancy values at wavelengths of maximal absorption at the appropriate pH value for each sample. Molar concentrations are calculated from Beer's Law by dividing an absorbancy value with the required molar absorbancy coefficient. Base ratios are estimated directly from the molar concentrations of the bases.

REFERENCES

Arnstein, H. R. V., and Cox, R. A. (1966). *Br. med. Bull.*, **22**, 158.

Beaven, G. H., Holiday, E. R., and Johnson, E. A. (1955). *In* "The Nucleic Acids", Vol. 1, p. 493. Academic Press, New York and London.

Belozersky, A. N., and Spirin, A. S. (1960). *In* "The Nucleic Acids", Vol. **3**, **p. 147.** Academic Press, New York and London.

Blattner, F. R., and Erickos, H. P. (1967). *Analyt. Biochem.*, **18**, 220.

Cantoni, G. L., and Davies, D. R., Eds. (1966). "Procedures in Nucleic Acid Research."

Chargaff, E. (1955). *In* "The Nucleic Acids", Vol. 1, p. 307. Academic Press, New York and London.

Chmielewicz, Z. F., and Acara, M. (1964). *Analyt. Biochem.*, **9**, 94.

Click, R. E. (1966). *Biochim. biophys. Acta*, **129**, 424.

Cohn, W. E. (1955). *In* "The Nucleic Acids", Vol. 1, p. 211. Academic Press, New York and London.

De Ley, J., and van Leeuwenhoek, A. (1967). *J. Microbiol. Serol.*, **33**, 203 (Eng.).

Edstrom, J. E. (1964). *Biochim. biophys. Acta*, **80**, 399.

Edstrom, J. E. (1960). *J. biophys. biochem. Cytol.*, **8**, 39.

Fredericq, E., Oth, A., and Fontaine, F. (1961). *J. molec. Biol.*, **3**, 11.

Gebicki, J. M., and Freed, S. (1966). *Analyt. Biochem.*, **14**, 253.

Green, J. G., Nunley, C. E., and Anderson, N. G. (1966). *Natn. Cancer Inst. Monograph No.* 21, 431.

Harris, A. B., and Warburton, R. (1966). *Nature, Lond.*, **212**, 1359.

Hiby, W., and Kroger, H. (1967). *J. Chromat.*, **26**, 545.

Hirschman, S. Z., and Felsenfeld, G. (1966). *J. molec. Biol.*, **16**, 347.

Holdgate, D. P., and Goodwin, T. W. (1964). *Biochim. biophys. Acta*, **91**, 328.

Hu, P. C., Liu, C. P., Wu, H., and Liang, C. C. (1964). *Scient. Sinica* (Peking), **13**, 761.

Huang, P. C., and Rosenberg, E. (1966). *Analyt. Biochem.*, **16**, 107.

Katz, S., and Comb, D. G. (1963). *J. biol. Chem.*, **238**, 3065.

Kirby, K. S. (1964). *Prog. Nucleic Acid Res.*, **3**, 1.

Kit, S. (1963). *A. Rev. Biochem.*, **32**, 43.

Loring, H. S. (1955). *In* "The Nucleic Acids", Vol. 1, p. 191. Academic Press, New York and London.

Magasanik, B. (1955). *In* "The Nucleic Acids", Vol. 1, p. 373. Academic Press, New York and London.

Marmur J. (1961) *J. molec. Biol.*, **3**, 208.

Marmur, J., and Doty, P. (1962). *J. molec. Biol.*, **5**, 109.

Marmur, J., Falkow, S., and Mandel, M. (1963). *A. Rev. Microbiol.*, **17**, 329.

Marshak, A., and Vogel, H. (1951). *J. biol. Chem.*, **189**, 597.

Munro, H. N., and Fleck, A. (1966). "Methods of Biochemical Analysis" XIV, p. 113.

Niemierko, W., and Krzyzanowska, M. (1967). *J. Chromat.*, **26**, 424.

Pataki, G. (1967). *J. Chromat.*, **29**, 126.

Pataki, G., and Kunz, A. (1966). *J. Chromat.*, **23**, 465.

Rajbhandary, U. L., and Stuart, A. (1966). *A. Rev. Biochem.*, **35**, 759.

Randerath, K. (1966). "Thin-layer Chromatography", Academic Press, New York and London.

Saukkonen, J. J., *In* M. Lederer (Ed.). (1964). *Chromat. Rev.*, **6**, 53.

Schildkraut, C. L., Marmur, J., and Doty, P. (1962). *J. molec. Biol.*, **4**, 430.

Skidmore, W. D., and Duggan, E. L. (1966). *Analyt. Biochem.*, **14**, 223.

Smith, J. D. (1955). *In* "The Nucleic Acids", Vol. 1, p. 267. Academic Press, New York and London.

Smith, J. D., and Markham, R. (1950). *Biochem. J.*, **46**, 509.

Stevens, A. (1963). *A. Rev. Biochem.*, **32**, 15.

Voet, D., Gratzer, W. B., Cox, R. A., and Doty, P. (1963). *Biopolymers*, **1**, 193.

Wang, S. Y., and Hashagen, J. M. (1964). *J. molec. Biol.*, **8**, 333.

Wyatt, G. R. (1955). *In* "The Nucleic Acids", Vol. 1, p. 243. Academic Press, New York and London.

Author Index

Numbers in *italics* refer to the pages on which references are listed at the end of each chapter

A

Aaronson, S., 179, *200*
Aasmundrud, O., 350, *382*
Abrams, A., 430, *450*
Abul-Fadl, M. A. M., 233, *336*
Acara, M., 634, *638*
Acs, G., 507, *512*
Afonso, E., 486, *509*
Ahlgren, E., 597, 603, 607, 608, *613*
Ainsworth, G. C., 46, *52*
Aitkhozhin, M. A., 134, *205*
Akabori, S., 447, *450*
Alan, R., 535, *547*
Alan, S., 535, *547*
Albaum, H. G., 286, 290, *336*
Alberts, A. W., 437, *454*
Alberts, B., 401, 419, *421*
Alberts, B. M., 401, *421*
Albertsson, P–A., 386, 388, 392, 393, 394, 395, 396, 398, 399, 400, 401, 402, *403, 404, 405, 406, 407, 408, 410,* 411, 412, 413, 415, 416, 417, 418, 420, *421, 422, 423*
Albro, P. W., 307, *340*
Alburn, H. E., 507, *510*
Alderton, G., 405, *422*
Alexander, P., 412, *421*, 442, 448, *453,* 551, 552, 554, *571*, 597, 599, *614*
Ali, D., 47, *48*
Allen, F. W., 371, *381*, 467, *510*
Allen, J. M., 467, *509*
Allen, R. H., 447, *452*
Allen, R. J. L., 226, 228, *336*
Allfrey, V., 507, *509*
Alper, R., 191, *197, 202*
Amano, T., 45, *48*
Amaral, D., 283, *336*
Amaral, G., 283, *336*
Ambrose, A. J., 514, 515, 544, *546, 547,* 548
Amelunxen, F., 524, 530, 533, 538, *547*

Ames, B. N., 98, 99, 194, *197, 202*
Amodio, F. J., 401, *421*
Anderson, I. C., 177, *197*
Anderson, L., 230, *336*
Anderson, N. G., 56, 64, 83, 84, 88, 89, 91, 92, 126, 128, *197, 198, 199,* 634, *638*
Andreus, P., 444, *450*
Anfinsen, C. B., 477, *511*, 592, *593*
Anno, K., 507, *512*
Ans, R., 616, *629*
Apgar, J., 376, *382*
Archibald, A. R., 329, *336, 337*
Archibald, W. J., 107, 110, 111, *198*
Argaman, M., 157, 159, *204*
Armstrong, D. J., 617, *629*
Armstrong, J. J., 329, *336*
Arnstein, H. R. V., 632, 634, 637, *638*
Aronson, A. I., 140, *197*
Aronsson, T., 485, *509*
Artmann, M., 328, *336*
Aschaffenburg, R., *509*
Ascoli, I., 303, *338*
Ascensiò, C., 283, *336*
Ashton, G. C., 493, 499, *509*
Askonas, B. A., 600, 613, *613*
Asselineau, J., 303, *336*, 354, 355, *381*
Atfield, G. N., 473, *509*
Atno, J., 277, *343*
August, J. T., 400, 407, *422, 423*
Aurisicchio, A., 114, *206*
Avakyan, A. A., 563, *570*
Avi-Dor, Y., 216, *341*
Avigad, G., 195, 196, *200*, 283, *336*
Avigad, L. S., 612, *613*
Avigan, J., 157, 159, *204*
Awdeh, Z. L., 600, 613, *613*
Ayad, S. R., 88, *197*
Ayres, J. C., 617, *629*
Azhits'kii, G. Yu, 558, *570*

B

Babinet, C., 400, *421*
Bachmann, B. J., 46, *48*
Bachrach, H. L., 125, *198*
Bacon, E., 45, *49*
Bacon, J. S. D., 45, *49*
Baddiley, J., 329, *336, 337, 339*
Bagdasarian, M., 468, *509*
Bailey, K., 426, *454*
Bailey, L., 398. *421*
Baird, G. D., 402, 403, 404, 405, 415, 416, *421, 422*
Bakay, B., 150, 152, *204*
Balazs, E. A., 283, *343*
Balcavage, W. A., 30, *50*
Baldwin, R. L., 77, 103, *198, 207*
Bale, W. F., 507, *511*
Ball, C. D., 597, *614*
Ball, C. O., 617, *629*
Ballentine, R., 552, 559, *570*
Baltscheffsky, H., 171, *201, 421*
Bangham, A. D., 559, *570*
Barbu, E., 323, *340*
Bardawill, C. J., 245, *339*
Barker, S. A., 362, *381*
Barnard, F. J., 25, *48*
Barney, J. E., 230, *337*
Barollier, J., *546*
Baron, D. N., 434, *451*
Barrenscheen, H. K., 286, *336*
Barret, J., 349, *381*
Bartha, R., von, 189, 190, *204*
Bartolotti, T. R., 236, *337*
Bauer, G., 531, 542, *546*
Bauer, K., 42, *50*
Baumann-Grace, J. B., 44, *52*
Bayer, M. E., 156, *198*
Bayley, S. T., 137, 161, *198, 201*, 304, *343*
Beal, D., 231, *340*
Beams, J. W., 81, 110, *201, 206*
Beaufay, H., 70, 71, 84, 89, *199*
Beaven, G. H., 634, 638, *638*
Bechmann, H., 146, *203*
Becka, J., 235, *336*
Beckman, H., 137, 149, *206*
Beech, G. A., 616, *629*
Beeler, R. G., 552, 555, 558, *570*
Beet, A. E., 217, *336*
Begg, D., 322, *338*

Beijerinck, M. W., 386, *421*
Beljanski, M., 286, *336*
Bell, J. C., 277, *336*
Bell, R. D., 224, *336*
Bellows, S. L., 354, *382*
Belozersky, A. N., 309, 310, *336*, 637, 638
Be Miller, J. N., 302, *344*
Benedict, S. R., 230, *336*
Bengtsson, S., 405, 406, 417, 418, *421*
Bennich, H., 445, *453*
Bentley, R., 302, *343*
Berg, P., 592, *593*
Berg, R. L., 552, 555, 558, *570*
Berger, L., 433, *450*
Bergeron, J. A., 187, *198*
Berggren, B., 597, 603, 608, 613, *613, 614*
Bergkvist, R., 333, *336, 337*
Berglund, F., 230, *337*
Bergstrom, J., 149, *206*
Bergstrom, L., 152, 155, *206*
Bergrahm, B., 552, 566, *570*
Bernard, Claude, 278, *337*
Bernheimer, A. W., 559, *570*, 610, 612, *613*
Bernstein, L., 283, *336*
Berthet, J., 65, 70, 71, 84, 89, *199*
Bertolacini, R. J., 230, *337*
Besman, L., 256, *343*
Bevan, G. H., 262, 263, 320, *337*
Bidwell, E., 527, 532, *546*
Biederman, M., 171, 172, 185, *203*
Bier, M., 514, *546*, 551, *570*, 599, *614*
Biggins, J., 179, 181, 182, *198*
Bigh, E., 544, *548*
Binder, C. L., 562, *571*
Birdsell, D. C., 157, *198*
Birnie, G. D., 88, *198*
Bishop, D. H. L., 593, *593*
Bister, F., 354, 364, *383*
Bitter, T., 292, *337*
Björk, I., 599, *613*
Black, C. C., 179, *198*
Blanchar, R. W., 229, *337*
Blatt, W. F., 449, *450, 451*
Blattner, F. R., 634, *638*
Blau, F., 239, *337*
Bligh, E. G., 351, *381*
Bloch, K., 303, *336*, 354, 355, *381*

Block, R. J., 333, *337*, 412, *421*, 442, 448, *453*, 462, 465, *509*, 551, 552, 554, 571, 597, 599, *614*
Bloemendal, H., 493, 499, 507, *510*, 551, 569, *570*, 599, *613*
Bloor, W. R., 303, *337*
Boardman, N. K., 442, *451*
Boatman, E. S., 166, *198*
Bocci, V., 499, *510*, 563, *570*
Bock, R. M., 85, 134, 143, *198*, *205*, 442, *451*, 554, 556, *570*
Bodansky, A., 224, *337*
Bodman, J., 487, 497, 502, *510*
Budo, G., 448, *451*
Bogdanov, A. A., 136, *205*
Boivin, A., 366, *382*
Bollum, F. J., 419, *421*
Bolognani, L., 317, *337*
Boman, H. G., 442, *451*
Bond, H. E., 126, 127, *199*
Bonner, D. M., 46, *48*
Bonsall, R. W., 88, *198*
Booth, V. E., 24, *48*
Borenfreund, E., 285, 291, 295, 317, *338*
Borgstrom, B., 307, *337*
Boring, J., 161, *201*
Bosch, L., 507, *510*
Bottiger, L. E., 507, *510*
Bourke, M. E., 46, *51*
Bourne, E. J., 362, *381*
Bourne, G. H., 60, 65. 128, *199*, *201*
Bowen, C. C., 178, *203*
Bowen, T. J., 134, *198*
Boyd, J. B., 592, *593*
Boyer, J., 434, *451*
Boyer-Kawenoki, F., 386, *422*
Brakke, M. K., 91, 92, 94, *198*, 550, 551, 552, 557, 559, *570*
Brattstein, L., 468, *510*
Brattsten, I., 516, 524, *548*
Bräumer, K., 524, 535, *546*
Braunitzer, G., 529, 543, *546*
Breese, S., 125, *198*
Brenner, S., 139, *198*, *202*
Bridgman, P. W., 12, *49*
Briggs, A. P., 224, *337*
Bril, C., 184, 185, 188, *198*
Brishammar, S., 388, *423*
Britten, R. J., 85, 86, *198*
Brønsted, J. N., *421*

Brown, A., 574, *594*
Brown, A. D., 160, 161, *198*
Brown, A. H., 286, 287, *337*
Brown, R. A., 599, *614*
Brüel, D., 217, *337*
Bruening, G., 143, *198*
Brumm, A. F., 333, 334, *340*
Bruner, R., 83, 90, 101, 102, *206*
Bryson, V., 136, 140, 146, *203*
Bublitz, C., 434, *451*
Buchanan, J. G., 329, *336*
Buchanan, P. J., 30, *49*
Buchner, E., 1, *49*
Bucht, B., 608, *614*
Buckley, E. S., 236, *337*
Buetow, D. E., 30, *49*
Bull, M. J., 165, 172, *198*
Bullen, W. A., 431, 437, *451*
Bulmer, G. S., 138, 158, *203*
Bungenberg, de Jong, H. G., 387, *421*
Burger, M., 45, *49*
Burges, H. D., 497, 498, *511*
Burr, H. E., 126, 127, *199*
Burris, R. H., 47, *52*, 430, 432, *451*, *453*
Burrous, S. E., 158, *198*
Burton, A. L., 294, *341*
Burton, K., 142, *198*, 314, 317, 318, 319, 320, 323, 324, 325, 326, *337*
Bush, F., 616, *629*
Buus, O., 467, 507, *511*
Buzzard, Z., 495, *510*

C

Caiger, P. C., 616, *629*
Caldwell, A. C., 229, *337*
Caldwell, P. C., 286, *337*
Calendar, R., 592, *593*
Calvin, M., 180, *198*
Cammack, K. A., 134, 135, 136, *198*
Campbell, J-J. R., 330, *341*
Campbell, P. N., 489, *510*
Campillo, A., del, 31, *50*, 438, *452*
Candler, E. L., 89, *198*
Cantoni, G-L., 632, *638*
Capecchi, M. R., 400, *421*
Carlson, L. A., 507, *510*, 561, 563, *570*
Carlstrom, A., 598, *614*
Carnahan, J. E., 431, *451*
Carroll, W. R., 426, *454*

Carter, C. E., 286, 288, 290, *341*
Carter, D. L., 616, *629*
Casassa, E. F., 116, *198*
Caselli, P., 467, *510*
Cass, B., 329, *336*
Castle, J. E., 431, *451*
Ceriotti, G., 286, 288, 290, 320, 322, *337*
Chaikoff, I. L., 20, *49*
Chaix, P., 45, *52*
Chambers, L. A., 27, *49*
Chan, B., 157, *203*
Chance, B., 617, *629*
Chao, F-C., 138, 145, 147, *198*
Chapeville, F., 399, *421*
Chapman, J. A., 149, 153, *204*
Chargaff, E., 286, 308, 309, 310, 311, 317, 320, *336, 337, 338*, 632, *638*
Charlwood, P. A., 552, 554, *570*
Charm, S., 11, *49*
Charman, R. C., 488, *510*
Chen, P. S., 225, *337*
Chermann, J. C., 405, *421*
Chervenka, C. H., 80, 105, *198*
Chibnall, A. C., 218, *337*
Chmielewicz, Z. F., 634, *638*
Cho, Y. S., 433, *451*
Choules, G. L., 552, 559, *570*
Chrambach, A., 576, 588, *593, 594*
Christian, W., 142, *206*, 263, 264, 322, *344*, 447, *454*
Chun, P. W. L., 147, *201*
Chung-Yen, W., 507, *511*
Church, B. D., 42, *50*
Ciocalteu, V., 249, 250, *338*
Civen, M., 433, *451*
Claybrook, J. R., 593, *593*
Clayton, R. K., 188, *198*
Cleland, W. W., 430, *451*
Click, R. E., 634, *638*
Cline, G. B., 89, 126, *197, 199*
Cocking, E. C., 254, 262, 334, *344*
Cohen, H. R., 224, *340*
Cohen, P. P., 437, *451*
Cohen, S. S., 138, *199*, 320, *337*, 507, *510*
Cohen-Bazire, G., 148, 150, 151, 162, 163, 164, 166, 169, 170, 171, 173, 174, 175, 183, *198*
Cohn, E. J., 147, *199*, 447, *451*

Cohn, M., 428, *451*
Cohn, W. E., 308, 328, *339, 340, 344*, 634, *638*
Coirault, Y., 562, *571*
Colbert, L., 45, *49*
Cole, J. A., 48, *49*, 194, 195, *199*
Coleman, G., 96, 130, 133, 137, *199*
Coll, J. A., 576, 591, *594*
Colowick, S. P., 22, 41, 42, *49, 51*, 81, 111, 142, 147, *204*, 400, 401, 407, *421, 422*, 427, 433, 438, 440, 441, *450, 451, 452*, 462, *511*, 609, *614*
Comb, D. G., 634, *639*
Commerford, S. L., 38, *50*
Comsa, J., 524, 535, *547*
Connolly, T. N., 435, *451*
Conover, M. S., 150, 152, *204*
Consden, R., 472, 499, *510*
Conti, S. F., 150, 151, 164, *201*
Contopoulou, R., 190, *205*
Conway, E. J., 222, 256, *337*
Cooksey, K. E., 589, 591, *593*
Coon, M. J., 438, *453*
Cooper, C., 434, *452*
Cooper, P. D., 25, *49*
Coppi, G., 317, *337*
Cori, C. F., 433, 436, *450, 452*
Cori, G. T., 433, *454*
Corrigal, J. J., 467, *512*
Cost, K., 168, *201*
Cota Robles, E. H., 4, *50*, 157, *198*
Cote, W. A., 191, *197*
Cowan, C. I., 330, *341*
Cowan, R. M., 39, *49*
Cox, E. C., 146, *202*, 593, *594*
Cox, R. A., 632, 634, 637, 638, *638, 639*
Craig, D., 412, *421*
Craig, L. C., 412, *421*, 448, *451*
Crane, F. L., 358, *382*
Craven, G. R., 592, *593*
Creeth, J. M., 57, 143, *199*
Crespi, H. L., 179, 182, *199*
Crestfield, A. M., 467, *510*
Crestofield, A. M., 442, *451*
Critchley, P., 329, *337*
Croft, D. N., 319, *337*
Croson, M., 304, *340*
Cruft, H. J., 378, *382*
Cullum, D. C., 230, *337*
Cummings, D. J., 117, *199*

Cunningham, V. C., 22, *50*
Cusanovitch, M. A., 175, 176, 177, *199*
Cuzin, F., 399, *421*
Czarnetzky, E. J., 31, *51*

D

Dagley, S., 133, 134, 140, *198*, *199*
Dalbec, J. M., 609, *614*
Dalton, A. J., 89, *201*
Daly, M. M., 507, *509*
Danielli, J. F., 60, 65, 128, *199*, *201*
Dankert, M., 351, 355, *382*, *383*
Dark, F. A., 42, *49*, *51*, *52*, 213, 214, 222, 258, 260, 261, 279, 301, *343*
Datta, A., 429, 430, *453*
Davern, C. I., 115, 139, *199*, *202*
David, M. M., 245, *339*
Davidson, A. L., 329, *336*
Davidson, E. A., 284, 292, *337*
Davidson, J. M., 469, *510*
Davidson, J. N., 286, 308, 309, 310, 311, 317, 320, 324, 328, *336*, *337*, *338*, *339*, *340*, *341*
Davidson, S. J., 434, *451*
Davies, B. J., 573, 576, *593*, *594*
Davies, D. A. L., 366, *382*
Davies, D. R., 632, *638*
Davies, R., 28, *49*
Davison, P. F., 11, *50*
Dawes, E. A., 190, 193, *199*, *204*, 315, *337*
Dawson, R. M. C., 559, *570*
Dayton, S., 267, *343*
Dean, R. W., 617, *629*
Dehm, P., 537, *547*
Deichmann, W., 292, *337*
Delacourt, A., 467, *510*
Delacourt, R., 467, *510*
Delaparte, B., 304, *340*
De Ley, J., 132, 133, *199*, 283, *337*, 405, *421*, 633, *638*
Delivs, H., 146, 147, *203*
Delorey, G. E., 225, 231, *337*, *340*
Denis, W., 235, *337*
Denneny, J., 322, *344*
Denson, K. W. E., 527, 532, *546*
Depinto, J. A., 351, *382*

Deriaz, R. E., 317, *337*
De Rover, W., 179, *206*
Devi, A., 296, *338*
Dicks, J. W., 137, *205*, 233, 238, 329, 330, *338*, *343*
Dierickx, L., 559, *570*
Digeon, M., 405, *421*
Dike, G. W. R., 527, 532, *546*
Di Lallo, J., 488, *512*
Dillon, R. T., 255, 256, 259, 260, *344*
Dilworth, M. J., 432, *453*
Dische, Z., 266, 277, 278, 285, 286, 287, 291, 292, 293, 295, 296, 297, 312, 314, 317, 319, 320, *338*
Dittmer, J. C., 307, *339*, *344*
Dixon, M., 426, 440, 441, *451*
Doane, W. W., 590, *593*
Dobry, A., 386, *422*
Doctor, B. P., 376, *382*
Dodgson, K. S., 229, 231, *338*
Doherty, D. G., 507, *510*
Doisy, E. A., 224, *336*
Dolin, M. I., 31, *49*
Donikan, R., 466, *510*
Donninger, C., 497, *512*
Dose, K., 562, *570*
Doty, P., 124, *204*, 401, *421*, 632, 634, 637, 638, *639*
Doudoroff, M., 190, 191, 193, *202*, *205*, 283, 305, *337*, *338*, 430, *454*
Douglas, H. C., 166, *198*
Downing, 373
Drevon, B., 466, *510*
Drews, G., 163, 166, 167, 171, 172, 185, 186, *200*, *203*
Dreyer, W. J., 477, *511*
Dreywood, R., 266, *338*
Drury, H. F., 286, 288, *338*
Dua, R. D., 430, *451*
Duane, W. C., 181, *199*
Dubois, M., 272, 274, 275, 276, 302, *338*
Duell, E. A., 45, *49*
Duggan, E. L., 633, 634, *639*
Duguid, J. P., 330, *343*, *344*, 359, *382*
Duke, E. J., 499, *510*
Dundas, I. E. D., 435, *452*
Dunn, F., 10, *49*, *50*
Durrum, E. L., 333, *337*, 462, 463, 465, *509*, *510*

Duve, C., de, 65, 70, 71, 84, 89, 129, 199
Dvonch, W., 507, *510*
Dyer, W. J., 351, *381*

E

Ebergagen, D., 526, 532, *548*
Echlin, P., 178, *199*
Eddy, A. A., 45, *49*
Edebo, L., 12, 31, *49*
Edelstein, S. J., 78, 80, 81, 104, *199*, *204*
Edsall, J. T., 147, *199*
Edstrom, J. E., 634, *638*
Edwards, M. R., 148, *202*
Efron, M. L., 473, *510*
Ehtisham-ud-Din, A. F. M., 152, 157, *204*
Eimhjellen, K. E., 350, 357, *382*
Eisenberg, H., 116, *198*, *199*
Ekholm, K., 467, *511*
Elder, S. A., 9, *49*
Eldjarn, L., 234, *338*
Elek, S. D., 609, *614*
Ellar, D., 191, *199*
Ellis, S. J., 507, *510*
Ellwood, D. C., 329, 330, *343*
Elorza, M. V., 45, *49*
El'Piner, E. E., 9, 12, *49*
Elsden, D. F., 551, 552, 553, 555, *571*
Elson, D., 141, *199*, 507, *510*
Elson, H. E., 156, *203*
Elson, L. A., 284, *338*
Elvehjem, C. A., 312, *342*
Elwood, D. C., 330, *338*
Emanuel, C. F., 20, *49*
Emerson, M. R., 46, *49*
Emerson, S., 46, *49*
Engelberg, H., 328, *336*
Eoyang, L., 400, 407, *422*
Epstein, C. J., 612, *614*
Epstein, J. H., 295, *342*
Epstein, R. L., 42, *50*
Epstein, W., 213, *338*
Erdman, V. A., 141, *203*
Erickos, H. P., 634, *638*
Ericson, L. E., 550, *570*
Eriksson, K-E., 597, 603, 607, 608, *613*, *614*
Esche, R., 9, *49*

Euler, H. U., 317, *338*
Evered, D. F., 468, *510*
Everett, M. R., 271, *343*

F

Fahian, L. A., 437, *451*
Fairbanks, G., 590, 592, 593, *593*
Falaschi, A., *422*, *451*
Fales, F. W., 267, 268, *338*
Falkow, S., 103, 114, *202*, 632, 637, *639*
Farr, A. L., 3, *50*, 245, 249, 251, 252, *341*, 398, *422*
Farrow, J., 376, *382*
Farrow, R. P. N., 232, *338*
Favre, J., 401, *422*
Fawcett, J-K., 218, *338*
Feinberg, M. P., 449, *450*
Feldberg, N., 527, *547*
Felix, F., 435, *451*
Felsenfeld, G., 135, *199*, 321, *338*, 633, *638*
Fennessey, P., 351, *383*
Ferris, F. L., 401, *421*
Fessler, J. H., 71, *199*
Few, A. V., 152, *200*
Fewson, C. A., 179, *198*
Field, C. W., 499, *510*
Filmer, D., 189, *203*
Fincham, J. R. S., 429, *451*
Fischer, E. H., 447, *451*, 463, *512*
Fischer, W. D., 126, *199*
Fiske, C. H., 224, 225, 235, *338*
Fitz-James, P. C., 149, 153, *199*, 323, 327, *338*
Fiver, E. M., 449, *454*
Flaks, J. G., 146, *202*, 593, *594*
Flamm, W. G., 126, 127, *199*
Flatmark, T., 598, 599, 613, *614*
Fleck, A., 308, 309, 312, 313, 314, 315, 321, 322, 325, *338*, *341*, 631, 632, 633, *639*
Fleming, A., 43, *49*
Flin, O., 249, 250, 255, *338*
Flodin, F., 444, *453*
Flodin, P., 448, 449, *451*, 561, 563, 564, *570*
Florkin, M., 360, *382*
Flory, P. J., 388, *422*
Flosdorf, E. W., 27, 31, *49*, *51*
Flynn, F. V., 485, *510*

Flynn, H. G., 9, *49*
Folch, J., 155, *199*, 217, 303, *338, 344,*
Fontaine, F., 633, *638*
Foons-Bech, P., 507, *510*
Forrester, I. A., 515, *546*
Forsyth, W. G., 189, *200*, 304, 305, *339*
Foster, J. V., 271, *343*
Foster, J. W., 39, *49*
Fowler, N., 254, *341*
Fox, C. F., 434, *453*
Frame, E. G., 255, *339*
Francois, C. I., 526, *546*
Fraser, D., 38, 44, *49, 50*
Frederick, J. F., 590, *593*
Frederick's W. W., 179, 180, *200*
Fredericq, E., 633, *638*
Freed, S., 634, *638*
Freeman, A. F., 616, *629*
Freeman, G. G., 366, *382*
Freer, J. H., 150, 152, 155, *204*
Frei, R. W., 617, *629*
Freimer, E. H., 152, *200*
Freiser, H., 236, *340*
French, G. S., 13, *51*
Frenkel, A. W., 150, 151, 167, 168, 169, 183, *200, 201*
Frick, G., 407, 410, 411, *422*
Friedberg, I., 195, 196, *200*
Friedman, H. S., 234, *339*
Friedrich, A., 217, 221, *339*
Frodyma, M. M., 617, *629*
Fröhlich, Ch., 526, 532, *546*
Fry, M., 328, *336*
Fuerst, R., 333, *339*
Fuhs, G. W., 179, *200*
Fujii, T., 526, 536, 537, *546*
Fujikawa, K., 45, *48*
Fujisawa, H., 428, 447, 448, *451*
Fujita, A., 298, 299, *339*
Fujita, H., 57, 103, *200, 207*
Fujita, Y., 180, *200*
Fulco, L., 179, *200*
Fuller, R. C., 150, 151, 164, 177, *197, 201*, 357, *382*
Furano, A. V., 135, *200*
Furness, G., 25, *49*
Furuya, E., 434, *454*

G

Gaebler, O., 65, *201*
Galli, E., 42, *50*
Ganez, Micheline, 616, *629*
Ganser, M., 531, 542, *546*
Garcia, A., 167, 171, 177, 183, 186, 188, *200*
Garcia, A. F., 184, *206*
Gardell, S., 281, 285, 302, *339*
Garen, A., 428, *451*
Garner, R. J., 235, *339*
Gavrilova, L. P., 72, 125, 136, *200, 202*
Gazith, J., 609, *614*
Gebicki, J. M., 45, *49*, 634, *638*
Gefter, M., 436, *452*
Gehred, G. A., 437, *451*
Geiduschek, P., 114, *206*
Geiss, E., 559, *571*
Gelotte, B., 449, *451*
Gesteland, R. F., 136, *200*
Ghuysen, J. M., 15, *19*, 559, *570*
Giaja, J., 45, *49*
Gibbon, J. A., 130, 150, 163, 164, 165, 166, 172, 193, 194, 195, *205*
Gibbons, M. N., 298, *339*
Gibbons, N. E., 161, *201*, 304, *343*
Gibbs, M., 179, *198*
Gibbs, S. P., 167, *200*
Gibian, H., *546*
Gibor, A., 328, *339*
Gibson, D. T., 438, *451*
Gibson, J. G., 236, *337*
Gibson, K. D., 169, 173, 174, 185, 186, *200*
Giesbrecht, P., 163, 166, 167, 186, *200*
Gilby, A. R., 152, *200*
Giles, K. W., 319, *339*
Gilles, K. A., 272, 274, 275, 276, 302, *338*
Ginsburg, V., 194, *197*, 284, 292, *337, 343, 344*
Giovanelli, R. G., 617, *629*
Giuntini, J., 405, *421*
Glaser, L., 41, *49*
Glass, B., 180, *203*, 295, *338*
Glick, D., 448, *451*, 550, 558, 563, *570*
Glimcher, M. I., 526, *546*
Godson, G. N., 133, *200*
Goldberg, A., 135, 137, *200*
Goldberg, E., 590, *593*

Goldsworthy, P. D., 507, *510*
Goldzieher, J. W., 231, *343*
Golov, V. F., 72, 125, 136, *202*
Golstein, N. P., 295, *342*
Good, C. A., 278, *339*
Good, N. E., 435, *451*
Gooder, H., 129, 151, *200, 201*
Goodwin, T. W., 634, *638*
Gordon, A. H., 499, *510*, 552, 554, *570*, 593, *593*
Gordon, A. R., 553, *570*
Gordon, J., 400, *422*
Gorini, L., 435, *451*
Gorkin, V. G., 563, *570*
Gornall, A. G., 245, *339*
Got, C., *511*
Gottschalk, G., 189, 190, *204*
Gould, H. J., 89, *201*
Goulian, M., 115, *200*
Grabmann, W., 516, 527, *546*
Grabar, P., 499, 502, *510*
Granick, S., 328, *339*
Grant, C. T., *422*
Gratzer, W. B., 576, 591, *594*, 634, 638, *639*
Graves, D. J., 430, 434, *451*
Gray, G. W., 158, *200*
Graziosi, F., 114, *206*
Green, A. A., 433, 440, *451, 454*
Green, D. E., 24, *48*
Green, J. G., 634, *638*
Greenawalt, J. W., 30, *49*, 157, *204*
Greenberg, D. M., 435, *452*
Greenberg, G. R., 329, *336*
Greenberg, R. E., 399, *421*
Greengard, O., 489, *510*
Greenstein, J. P., 255, *339*
Griebel, R., 191, *200*, 405, *422*
Grieff, D., 434, *451*
Griffith, O. M., 91, 110, *200*
Grimm, F. L., 507, *510*
Grindley, G. B., 307, *340*
Grisolia, S., 430, 434, *453*
Gronwall, A., 485, *509*
Gropper, L., 91, *200*
Gross, D., 472, 473, *510*
Grossman, L., 118, 119, 120, *205*, 426, *452*

Grushoff, P., 612, *613*
Guest, H., 44, *50*, 150, 163, 168, 169, 175, 189, *198, 201, 203, 206*
Guest, J. F., 524, *546*
Gunsalus, I. C., 31, 41, 44, *49, 50, 51*, 149, 151, 157, *202*, 438, *452*
Gurley, L. R., 577, *594*
Guthrie, C., 140, 141, *200*
Gutman, E. D., 401, *421*
Guze, L. B., 149, 151, 152, 153, *199, 200, 202, 206*

H

Hagdahl, L., 91, 92, *205*, 555, 557, 558, *571*
Hagedorn, J. C., 298, *339*
Hagihara, B., 46, *51*, 435, 436, 442, 447, 448, *450, 451, 452, 453*
Haglund, H., 563, *570*
Hagopian, H., 477, *511*
Hahn, L., 317, *338*
Haley, H. B., 488, *512*
Hall, B. D., 114, *205*
Hall, C. E., 147, *200*
Hall, D. O., 30, *49*
Halvorson, H. O., 190, *201*, 333, *339*, 435, *452*
Hamilton, J. K., 272, 274, 275, 276, 302, *338*
Hamilton, P., 255, 256, 259, 260, *344*
Hanahan, D. J., 307, *339*
Hancock, R., 332, 333, *339*, 349, *382*
Hanes, C. S., 469, *510*
Hannig, K., 516, 518, 519, 521, 522, 524, 526, 527, 528, 529, 530, 531, 532, 535, 538, 542, 543, *546, 547, 548*
Hanson, S. W. F., 292, *339*
Hanzon, V., *422*
Harbers, E., 535, *547*
Harden, A., 2, *49*
Hardt, C. R., 597, *614*
Harlestam, R., 552, 566, *570*
Harold, F. M., 195, 196, *200*, 330, 331, *339*, 368, 370, 371, *382*
Harrington, W. F., 66, *202*
Harris, A. B., 634, *638*
Harris, D. L., 70, *205*
Harris, E., 507, *510*
Harris, J. I., 145, *206*

Harrison, G. A., 231, *339*
Harrison, J. S., 267, 272, 279, 286, 316, 323, 324, 327, *344*
Hartley, B. S., 66, *202*
Hartley, R. W., Jr., 426, *454*
Harvey, A. E., 236, *339*
Harvey, D. R., 88, *198*
Harvey, E. N., 27, *49*
Hashagen, J. M., 633, *639*
Haslewood, G. A. D., 231, *340*
Hastings, J. R. B., 85, *203*
Haugaard, G., 277, *343*
Hawke, P. B., 260, *339*
Hawkes, J. G., 497, *511*
Hawley, S. A., 10, *49*
Hay, J. B., 329, *339*
Hayaishi, O., 427, 428, 430, 431, 434, 436, 437, 447, 448, *451, 452, 553, 454*
Hayashi, H., 400, *422*
Hayward, A. C., 189, *200*, 304, 305, *339*
Hearst, J. E., 114, 117, 119, *206*
Hecht, H. G., 616, 617, 625, *629*
Hecker, E., 412, *422*
Heden, C. G., *49, 52*
Heidrich, H. G., 530, 538, *547*
Helinski, D. R., 611, *614*
Heller, G., 524, 530, 533, 538, *547*
Helmreich, E., 436, *452*
Henderson, A. R., 88, 99, *200*
Hendler, R. W., 130, *200*
Hennig, S., 537, *547*
Henning, U., 592, *594*
Heppel, L. A., 40, 43, *49, 51*
Herbert, D., 129, *201*, 286, 287, 288, 289, 290, *339*
Herbst, E. J., 141, *203*
Herriott, R. M., 249, *339*
Herschman, H. R., 611, *614*
Hershey, A. D., 142, *202*
Hess, E. L., 147, *201*
Hewlett, R. T., 25, *48*
Hexner, P. E., 110, *201*
Heyman-Blanchet, T., 45, *52*
Hibbard, P. L., 230, *339*
Hickman, D. D., 150, 151, 167, 168, 169, 183, *200, 201*
Higashi, L. S., 617, *629*
Higashi, Y., 351, *382*
Hilby, W., 634, *638*

Hill, A. G., 232, *338*
Hill, L. R., 103, 114, *201*
Hill, R., 239, *339*
Hinshelwood, C., 286, *337*
Hirata, M., 436, 437, *452*
Hirata, T., *422*
Hirsch, J. G., 45, *49*
Hirschfelder, A. D., 235, *339*
Hirschman, A., 256, *343*
Hirschman, S. Z., 633, *638*
Hitsatune, K., 612, *614*
Hjerten, H., 515, *547*
Hjerten, S., *422*, 441, 444, *452, 454*, 584, 592, *594*
Hoare, D. S., 130, 163, 164, 165, 172, 189, *203, 205*
Hobom, G., 529, 543, *546*
Hocevar, B. C., 563, *570*
Hodge, J. E., 298, *339*
Hoffenberg, H. B., 449, *450*
Hoffman, Ph., 524, 536, *547*
Hoffman, W. S., 233, *340*
Hofreiter, B. T., 298, *339*
Hofsten, B., 403, 404, 405, *421, 422*
Hogden, C. G., 245, *342*
Hogeboom, G. H., 5, *50*, 60, 65, 89, 128, *201*, 312, *342*
Hogg, J. F., 8, 27, *50*
Hohl, M. C., 181, *199*
Holde, K. E., van, 80, 103, 105, 108, 109, 110, *206, 207*
Holdgate, D. P., 634, *638*
Holiday, E. R., 262, 263, 320, *337*, 634, 638, *638*
Holley, R. W., 376, *382*
Hollingsworth, B. R., 134, 139, 142, 147, *206*
Holmes, P. K., 435, *452*
Holt, S. C., 150, 151, 163, 164, 169, 170, 171, *201*
Holt, S. J., 5, *50*
Holter, H., 217, *337*
Honegar, C. G., 481, *510*
Hoppel, C., 434, *452*
Horecker, B. L., 283, 294, *336, 339*, 360, *382*
Horikoshi, K., 45, *50*

Horio, T., 171, *201*, 442, 448, *452, 453, 454*
Horn, R., 147, *201*
Horne, R. W., 279, 280, 323, *342*
Hosoda, J., 42, *51*
Hosokawa, K., 140, *201*
Hoste, J., 317, *341*
Hough, L. 302, *339, 340*
Hsias, S. H., 493, *510*
Hu, P. C., 633, *639*
Huang, P. C., 633, *639*
Huddleson, I. F., 597, *614*
Hugget, A. St-G., 282, *340*
Hughes, D. E., 4, 8, 9, 10, 11, 12, 22, 25, 27, 28, 29, 31, 41, 42, 47, 48, *49, 50, 51,* 194, 195, *199*
Hughes, R. C., 367, *382*
Hughes, W. L., 440, *451*
Hughes, W. L., Jr., 447, *451*
Hugo, W. B., 25, 31, 41, *50*
Hunt, A. L., 48, *50*
Hunt, S., 88, *198*
Hunter, G. D., 303, *340*
Hunter, J. R., 132, 137, *205, 206,* 233, 238, 329, *343*
Hunter, M. J., 38, *50*
Hunter, R. L., 497, *511*
Hurlbert, R. B., 333, 334, *340*
Hurwitz, R., 399, *421*
Huston, C. K., 307, *340*
Hutchins, B. K., 617, *629*
Hutchison, W. C., 286, 308, 313, *340, 342*
Hyatt, E. A., 31, *51*

I

Ichikawa, S., 45, *48*
Ift, J. B., 117, *201*
Iida, S., 45, *50*
Ikawa, M., 303, *340*
Ikenaka, T., 447, *450*
Immers, J., 284, 285, *340*
Inagaki, A., 436, *452*
Ingles, O. G., 294, *340*
Ingraham, J. L., 351, *382*
Ingram, M., 40, *50*
Ingram, V. M., 478, *512*
Inoue, S., 45, *49*
Isherwood, F. A., 469, *510*
Ishida, M., 148, 154, 155, *202*

Israel, G. C., 294, *340*
Itada, N., 431, *452*
Itoh, T., 139, *201*
Ivanov, D. A., 136, *200*
Iwatake, D., 298, 299, *339*
Izawa, S., 435, *451*

J

Jackson, A. P., 124, *207*
Jacob, F., 139, *198,* 428, *452*
Jacobs, H. R. D., 233, *340*
Jacobs, S., 217, *340*
Jaffé, W. G., 527, *547*
Jagendorf, A. T., 179, 180, *200*
Jahn, W. F., 537, *547*
Jakoby, W. B., 447, *452*
James, A. M., 45, *49*
James, A. T., 303, *340*
James, H. M., *457,* 515, 544
Jankowski, S. J., 236, *340*
Jann, K., 360, *382*
Jarabak, J., 434, *452*
Jatzkewitz, H., 525, 537, *547*
Jeanloz, R. W., 283, *343*
Jeng, D. Y., 431, 437, 438, *452*
Jenkins, M., 401, *421*
Jenkins, W. T., 439, *452*
Jensen, B. N., 298, *339*
Jensen, S. L., 350, 357, *382*
Jermyn, M. A., 467, *511*
Jerstedt, S., 564, 565, 569, *571,* 584, 592, *594*
Joffe, S., 7, *50*
John, S., 527, *547*
Johns, E. W., 493, *511*
Johnson, C. R., 239, *340*
Johnson, E. A., 320, *337,* 634, 638, *638*
Johnston, J. P., 75, *201*
Jones, A. S., 374, *382*
Jones, J. K. N., 302, *340*
Jork, H., 617, *629*
Joubert, F. J., 388, *422*
Jovin, T., 576, *594*
Julien, J., 328, *342*
Juni, E., 322, 330, *340*

K

Kaempfer, R. O. R., 115, *201*
Kagamiyama, H., 431, 434, *453*
Kahler, H., 64, 65, *201*

Kaiser, A. D., 399, *422*
Kallio, R. E., 438, *451*
Kalnitsky, G., 30, *50*
Kamen, M. D., 175, 176, 177, *199*, 322, 330, *340*, 612, *614*
Kaplan, A., 544, *548*
Kaplan, N. O., 22, 41, 42, *49*, *51*, 81, 111, 142, 147, *204*, 376, *382*, 400, 401, 407, *421*, *422*, 427, 438, 440, 441, *451*, *452*, 462, *511*, 612, *614*
Karfunkel, P., 179, *200*
Karunairatnam, M. C., 44, *50*, 163, 169, *201*
Kasai, N., 307, *340*
Kashiba, S., 45, *48*
Katagiri, M., 434, *454*
Katchalsky, A., 101, 102, *204*
Kates, M., 350, 351, *382*
Katz, A. M., 477, *511*
Katz, J. J., 179, 182, *199*
Katz, S., 634, *639*
Katze, J. R., 400, *422*
Kaufmann, W., 42, *50*
Kauzmann, W., 430, *452*
Kawaguchi, K., 46, *51*
Kay, L. M., 256, *343*
Kay, R. L., 553, *570*
Ke, B., 167, 186, *200*
Keech, D. B., 437, *454*
Kegeles, G., 110, 111, *201*
Kelemen, M. V., 329, *336*
Kellenberger, E., 148, *204*
Kelley, W. S., 355, *382*
Kendall, J., 573, *594*
Kenkare, U., 609, *614*
Kennedy, E. P., 434, *451*
Kennedy, I. R., 437, *452*
Kent, L. H., 285, *340*
Kent, R., 83, 90, 101, 102, *206*
Kernot, B. A., 489, *510*
Kerr, S. E., 317, *340*
Kim, K. H., 138, *203*
King, E. J., 218, 222, 225, 226, 231, *340*
King, H. K., 31, *52*
King, T. P., 448, *451*
Kinsky, S. C., 46, *50*
Kirby, K. S., 85, *203*, 328, *340*, 375, *382*, 632, *639*
Kircher, H. W., 302, *340*
Kissane, J. Q., 488, *511*

Kisselev, N. A., 136, *205*
Kit, S., 632, *639*
Kitahara, K., 154, 155, *202*
Kitai, R., 475, *512*
Kitto, G. B., 612, *614*
Klainer, S. M., 110, 111, *201*
Klaschka, F., 532, *548*
Klett, R. P., 426, *452*
Kling, H., 524, *548*
Klingenberg, M., 5, *51*
Klofat, 531, 545, *547*
Klucis, E. S., 22, 39, *50*, 89, *201*
Knight, C. G., 143, *199*
Knowles, J. R., 400, *422*
Knox, W. E., 433, *451*
Kobozev, G. V., 558, *570*
Koch, J. R., 438, *451*
Kodicek, E., 472, *511*
Koehler, H. L., 267, 269, *340*
Koenig, R. A., 239, *340*
Kohlhage, H., 508, *512*
Kohn, J., 377, *382*, 462, 483, 484, 487, 489, *511*
Kohn, L. D., 447, *452*
Kojima, Y., 431, *452*
Kolb, J. J., 150, 152, *204*, 302, *340*
Kolin, A., 445, *452*, 515, *547*, 550, 558, 559, *570*
Kolthoff, I. M., 230, 235, *340*
Komarmy, J. M., 236, *339*
Kominek, L. A., 190, *201*
Komissarova, N. V., 563, *570*
Koningsveld, R., *422*
Konrad, M. W., 72, 115, *201*
Koprowski, H., 81, 89, 103, 111, *197*, *202*, 405, 407, 408, *421*, *423*
Korkes, S., 31, *50*, 438, *452*
Korn, E. D., 363, 368, *382*
Kornberg, A., 115, *200*, 399, *422*, *451*, *453*
Kornberg, H. L., 8, 27, *50*
Kornberg, S. R., 193, 194, *201*
Kotani, S., 434, 436, *453*
Kouns, D. M., 87, *204*
Kraemer, E. O., 106, *201*
Kramer, B., 234, 236, *340*
Kramer, H., 278, *339*
Krause, G., 562, *570*
Krebs, E. G., 447, *451*

Kretchmer, N., 399, *421*
Kroger, H., 634, *638*
Krogman, D. W., 179, 181, *199, 205*
Kronish, D. P., 42, *50*
Krüsmann, W. F., 531, 542, *546, 547*
Kruyt, H. R., 387, *421*
Krzyzanowska, M., 634, *639*
Kubelka, P., 617, *629*
Kubinski, H., 114, *205*
Kübler, H., 524, 530, 533, 538, *547*
Kubowitz, F., 447, *452*
Kucaynski, M., 216, *341*
Kuendig, W., 281, *342*
Kuff, E. L., 5, *50*, 60, 65, 89, 128, *201*
Kühn, K., 524, 535, *546*
Kunimoto, L. H., 617, *629*
Kunisawa, R., 150, 163, 164, 166, 169, 170, 171, 173, 174, 183, 190, *198, 205*
Kunitake, G., 591, *594*
Kunitz, M., 447, *452*
Kunkel, H. G., 507, *511*, 563, *570*
Kunz, A., 634, *639*
Kupke, D. W., 81, *206*, 561, 564, *570*
Kurland, C. G., 140, 143, *201*
Kurland, G. S., 11, *49*
Kushner, D. J., 161, *201, 203*
Kusunose, E., 438, *453*
Kusnose, M., 438, *453*
Kuttner, T., 224, *340*
Kuyama, S., 508, *511*
Kuzovleva, O. B., 507, *511*

L

Laipis, P., *423*
Lamanna, C., 25, *50*
Lampen, J. O., 136, 140, 146, *203*
Landelout, H., 189, *206*
Landowne, R. A., 488, *510*
Lang, N. J., 178, *201*
Langen, P., 132, 136, *205*
Lansine, W. D., 106, *201*
Lanz, H., 81, *202*
Lapointe, J., 400, *422*
Largier, J. F., 388, *422*
Larsen, H., 160, 161, *203*
Lascelles, J., 148, 150, 151, 162, 165, 172, 174, *198, 201*
Laskowski, M., Sr., 525, 533, 534, 535, *548*

Laurell, A. H. F., 462, *511*
Law, J. H., 193, *201*, 305, 306, 307, *343*
Lawrence, N. S., 13, *51*
Lawrence, S. H., 590, *594*
Laybourn, R. L., 359, 369, *382*
Layne, E., 3, *50*, 438, *452*
Lea, C. H., 307, *340*
Leach, A. A., 444, *452*
Leaver, J. L., 378, *382*
Le Bar, F. E., 106, *201*
Le Baron, F. N., 303, *338*
Leberman, R., 407, *422*
Lebowitz, J., *423*
Le Boy, P. S., 146, *202*, 593, *594*
Le Comte, J. R., 431, 437, *451*
Lederer, E., 303, *336*, 354, 355, *381*
Lederer, M., 462, *511*, 633, *639*
Lee, K. W., 323, *340*
Lee, R. T., 526, *547*
Lee, Y. C., 275, 285, *340*
Leene, W., 148, *206*
Lees, M., 155, *199*, 303, *338*
Legallais, V., 617, *629*
Lehman, J. R., 399, *422*
Lemieux, R. U., *340*
Lemoigne, M., 304, *340*
Lenhoff, H. M., 376, *382*
Lenney, J. F., 609, *614*
Lerche, Ch. H., 544, *547*
Lerman, M. I., 72, 125, 136, *202*
Lerner, K. D., 91, 92, *205*, 555, 557, 558, *571*
Lester, R. L., 358, *382*
Leeuwenhoek, A. van, 633, *638*
Levin, O., 441, *452, 454*
Levine, L., 187, *203*
Levinthal, C., 11, *50*, 428, *451*, 478, *512*, 590, 593, *593*
Levy, H. B., 308, *344*, 434, *452*
Lewis, U. J., 146, *204*, 577, *594*
Lezius, A. G., 537, *547*
Li, C. H., 507, *510, 512*
Liang, C. C., 633, *639*
Lichtenstein, J., 138, *199*, 507, *510*
Liener, I. E., 507, *511*
Liu, C. P., 633, *639*
Lieu, V. T., 617, *629*

Lightstone, P. J., 590, *594*
Lin, F. M., 302, *341*
Lindegren, C. C., 330, *341*
Linderstrøm-Lang, K., 81, *202*, 217, *337*
Lindigkeit, R., 132, 136, *205*
Lindner, E. B., 564, 565, 569, *571*
Ling, N. S., 442, *451*, 554, 556, *570*
Ling, Nan-Sing, 85, *198*
Linnane, A. W., 25, *52*
Lipmann, F., 180, *203*, 507, *512*
Lis, H., 471, *512*
Lissenden, A., 9 ,*50*
Little, J. W., 399, *422*
Llaurado, G. J., 234, *341*
Lloyd, A. G., 229, 230, *338, 341*
Lloyd, B. J., 64, 65, *201*
Lloyd, D., 22, 30, 39, *50*
Loeb, L. A., 400, *422*
Loening, U. E., 328, *341*
Logan, J. E., 322, *341*
Lohmann, I. W., 616, *629*
Loomis, A. L., 27, *49*
Lopez, J. A., 225, *341*
Lorentz, II. A., 618, *629*
Loring, H. S., 633, *639*
Louis, L. N., 593, *593*
Lovern, J. A., 303, *341*
Lowry, O. H., 3, 50, 225, 245, 249, 251, 252, *341*, 398, *422*
Lubran, M., 319, *337*
Luderitz, O., 354, 360, 364, *382, 383*
Ludlum, D. B., 117, *202*
Lundgren, D. G., 148, 189, 190, 191, *197, 199, 202, 204, 206*
Lundquist, F., 467, 507, *511*
Luner, S. J., 558, *570*
Lusena, C. V., 286, *341*
Lutwak, L., 230, *342*
Lynch, V., 180, *198*

M

Ma, T. S., 218, *341*
Maag, T. A., 39, *49*
McBride, R. A., 562, *571*
McCarty, M., 45, *50*
McCleod, R. M., 10, *49, 50*
MacDonald, H. J., 488, *511*
McElroy, W. D., 180, *203*, 295, *338*

MeEwen, C. R., 98, 101, *202*
Macfadyen, D. A., 254, 255, 256, 259, 260, *341, 344*
Macfarlane, M. G., 303, *341*
Macfayden, A., 2, *50*
Macheboeur, M., 286, *336*
McIlreavy, D. J., 140, 144, *202*
McIlwain, H., 30, *50*
McIndoe, W. M., 324, *341*
Maclennan, D. H., 577, *594*
MacLeod, M., 294, *341*
McQuillen, K., 44, *51*, 149, 151, 152, 157, *200, 202*, 544, *547*
Macrae, R. M., 191, *202*, 358, *382*
McRary, W. L., 286, 287, *341*
Magasanik, B., 632, 633, *639*
Mager, J., 216, *341*
Mahler, H. R., 44, *49, 50*
Mahoney, R. P., 148, *202*
Maier, K. P., 530, 538, *547, 548*
Maizel, J. V., 577, 593, *594*
Mallette, M. F., 25, 47, *50, 51*, 330, *341*
Malmgren, B., 600, 603, 609, 610, *614*
Malmstrom, B., 467, *511*
Mandel, M., 103, 114, *202*, 632, 637, *639*
Mandelkern, L., 143, *204*
Mandell, J. D., 142, *202*
Mandell, N., 33, *50*
Mandelstam, J., 331, 333, *341*
Mandeveille, S. E., 179, 182, *199*
Mangiarotti, D., 133, 141, *202*
Mangold, H. K., 333, *341*
Mann, F., 292, *341*
Mann, T., 330, *341*
Manncll, W. A., 322, *341*
Manson, W., 552, 553, 555, 558, *571*
Maramorosch, K., 81, 89, 103, 111, *197, 202*, 405, 407, 408, *421, 423*
Marchessault, R., 191, *197, 199, 202*
Marinetta, G. V., 352, *383*
Marini, M. A., 376, *382*
Markert, C. L., 497, *511*
Markham, R., 81, *202*, 217, 286, 288, 333, *341*, 470, 471, *511*, 633, 638, *639*
Markov, G. G., 322, *344*
Marmur, J., 103, 114, 124, *202, 204*, 371, 372, *382*, 632, 634, 637, *639*

Marr, A. G., 4, 40, *50, 51,* 149, 150, 151, 159, 163, 164, 169, 170, 171, *201, 202, 203, 204,* 351, *382*
Marr, H. G., 5, *50*
Marsh, G. E., 374, *382*
Marshak, A., 633, 638, *639*
Marshall, V. C., 6, *50*
Marshall, W. E., 569, *571*
Marston, J. H., 151, *202*
Martin, A. J. P., 463, 499, *510, 511*
Martin, D. B., 437, *454*
Martin, R. G., 98, 99, *202*
Martinez, R. J., 195, 196, *202*
Mashburn, T. A., Jr., 524, 536, *547*
Mason, S. G., 11, *51*
Mason, W. P., 9, *49*
Massart, L., 317, *341*
Massey, V., 66, *202*
Masters, M., 229, *341*
Matheka, H. D., 559, *571*
Matsubara, H., 435, 436, *452, 453*
Matthaei, J. H., 134, *203,* 524, 530, 533, 538, *547*
Mattoon, J. R., 30, *50*
Maxwell, 352
Mayo, P. de, 485, *510*
Mazzone, H. M., 103, 111, *202*
Mears, G. E. F., 388, *422*
Meath, J. A., 303, *338*
Mehl, E., 525, 537, *547*
Mehl, J. W., 245, *341,* 507, *510*
Mejbaum, W., 286, *341*
Mel, H. C., 515, *547*
Melamed, M. B., 494, *511*
Melnick, P. J., 590, *594*
Melo, A., 41, *49*
Melvin, E. H., 267, *343*
Menzel, C., 537, *547*
Merrick, J. M., 191, 193, *200, 203,* 405, *422*
Merrill, S. H., 376, *382*
Meselson, M., 114, 115, 116, 135, 139, *198, 199, 201, 202, 206*
Mesrobeanu, I., 366, *382*
Mesrobeanu, L., 366, *382*
Metz, E., 537, *547*
Metzner, H., 616, *629*
Meynell, G. G., 129, *201*
Michl, H., 472, 475, *511*
Mickle, H., 25, *51,* 279, 323, 324, *341*

Midgley, J. E. M., 140, 142, 144, *202*
Miles, T., 321, *338*
Militzer, W. E., 286, 288, *341*
Miller, G. L., 294, *341*
Miller, L. L., 507, *511*
Miller, P. S., 230, *342*
Mills, G. T., 292, *339*
Mills, R. S., 256, *343*
Milner, H. W., 13, *51*
Mirsky, A. E., 507, *509*
Mitchell, H. K., 307, *344,* 592, *593*
Mitchell, P., 42, 44, *51,* 214, *341*
Miura, T., 148, 154, 155, *202*
Mizushima, H., 448, *453, 454*
Mizushima, S., 141, 146, 148, 154, 155, *202, 203*
Mohamed, M. S., 435, *452*
Mohan, R. R., 42, *50*
Mohr, V., 160, 161, *203*
Mok, C. C., *422*
Mok, T. C. K., 616, *629*
Moldave, K., 118, 119, 120, *205,* 426, *452*
Mollenhauer, H., 167, 171, 177, 183, 184, 186, 188, *200*
Moller, M. L., 138, *203*
Moller, W., 146, *203*
Monier, R., 142, *204,* 328, *342*
Monod, J., 428, *452*
Montgomery, R., 275, 276, 285, *340, 341*
Moore, D., 283, *336*
Moore, P. B., 146, 147, *203*
Moore, S., 254, 333, *341, 343,* 442, *451,* 561, *571*
Morawetz, H., 388, *422*
Morgan, C., 157, *203*
Morgan, D. M., 471, *512*
Morgan, W. T. J., 284, *338, 342*
Morino, Y., 428, 435, *453*
Morioka, T., 45, *48*
Morita, K., 430, *454*
Morris, C. J. O. R., 473, *509*
Morris, D. L., 267, 269, *341*
Morris, I., 178, *199*
Morris, J. A., 431, 437, 438, *452*
Morris, N. J., 616, *629*
Morse, M. L., 286, 288, 290, *341*
Mortensen, L. E., 377, *382,* 431, 437, 438, *451, 452*

Morton, R. A., 357, *382*
Mortin, R. K., 41, *51*
Mosbach, R., 444, *452*
Mosch, 407,
Moskowitz, M., 439, *452*
Moss, F. J., 616, 620, 621, *629*
Mower, H. F., 431, *451*
Moyle, J., 42, 44, *51*, 214, *341*
Mudd, S., 31, *51*
Muhhammed, A., 4, *51*
Muir, H. M., 292, *337*
Muller, F., *512*
Muller-Eberhard, H. J., 507, *511*, 563, *571*
Munk, F., 617, *629*
Munoz-Ruiz, E., 45, *49*
Munro, A. L., 190, 195, *206*
Munro, H. N., 286, 308, 309, 312, 313, 314, 315, 321, 325, *340, 341, 342*, 631, 632, 633, *639*
Munson, T. O., 432, *453*
Muntwyler, E., 267, *343*
Murray, K., 499, *511*
Murray, R. G. E., 148, 156, 177, 178, *203, 206*
Murrell, W. G., 233, *342*
Muys, G. T., 25, *51*
Myers, A., 319, *339*
Myers, J., 180, *200*
Myers, M., 434, *451*
Myers, V. I., 616, *629*

N

Nagasaki, K., 45, *51*
Nakai, H., 407, *422*
Nakai, M., 436, *452*
Nakamoto, T., 434, *453*
Nakayama, T., *422*, 435, 436, *453*
Nakazawa, A., 430, *453*
Nakazawa, T., 434, 436, *453*
Naughton, M. A., 477, *511*, 569, *571*, 576, *594*
Necas, O., 42, *51*
Nelson, N., 299, 301, *342*
Némethy, G., 430, *453*
Neppiràs, E. A., 9, 29, *51*
Nerenberg, S. T., 560, *571*
Ness, A. G., 213, 214, 222, 258, 260, 261, 279, 301, *343*

Neth, R., 524, 530, 533, 538, *547*
Netschey, A., 156, *204*
Neu, H. C., 40, *51*
Neufeld, E. F., 194, *197*, 284, 292, *337, 343, 344*
Nehaus, F. C., 329, *336*
Neukom, H., 281, *342*
Neurath, H., 426, *454*
Newman, D. W., 57, *206*
Newton, G. A., 174, 175, 176, 187, *203*
Newton, J. W., 162, 172, 174, 175, 176, 183, 187, *203*
Newton, W. A., 428, 435, *453*
Nicholas, D. J. D., 376, *382*
Nickerson, W. J., *51*
Nielson, F. J., 286, *342*
Niemierko, W., 634, *639*
Nihlen, H., 550, *570*
Nikkila, E., 467, *511*
Nirenberg, M. W., 134, *203*
Nixon, D. A., 282, *340*
Noll, H., 88, 99, *203*
Noller, H., 146, 147, *203*
Noltingk, B. E., 9, *51*
Nomura, M., 42, *51*, 136, 137, 139, 140, 141, 146, *200, 201, 203, 206*
Nordwig, A., 527, *546*
Nordwig, H., 537, *547*
Norrby, E., 407, 410, *422*
Norris, J. R., 42, *51*, 497, 498, *511*, 530, 538, 539, 542, *547*, 558, *571*
Norris, K. P., 216, *342*
Northcote, D. H., 279, 280, 281, 323, *342*, 363, 368, *382*, 563, *570*
Northrop, J. H., 447, *453*
Norton, J. E., 138, 158, *203*
Norton, J. W., 141, *203*
Nossal, N. C., 40, *51*
Nossal, P. M., 25, *51*
Novic, B., 267, *343*
Novoa, W. B., 434, *453*
Noyons, E. C., 231, *342*
Nozaki, M., 431, 434, 436, 442, 448, *452, 453, 454*
Numa, S., 430, *453*
Nunley, C. E., 89, *198*, 634, *638*
Nyborg, W. L., 8, 9, 10, 11, *50, 51*
Nygaard, O., 234, *338*
Nyns, J., 393, 394, *421*

O

Öberg, B., *423*
Ochoa, S., 31, *50*, 438, *452*
O'Connor, R. T., 616, *629*
Oelze, J., 171, 172, 185, *203*
Ogston, A. G., 75, *201*
Ogur, M., 313, 314, 327, *342*, 373, *382*
Ohnishi, T., 46, *51*
Okamura, K., 191, *199*
Okazaki, T., 399, *422*, *453*
Okunuki, K., 435, 436, 442, 448, *452*, *453*, *454*
Oldershausen, H. F., von, 532, *548*
O'Leary, W. M., 351, *382*
Olson, J. M., 189, *203*
Oncley, J. L., 574, *594*
Onishi, H., 161, *203*
Ono, K., 434, 436, *453*
Ornstein, L., 573, 574, 575, *594*
Osawa, S., 95, 139, *201*, *203*
Osborne, M. J., 296, *342*
Oser, B. L., 260, *339*
O'Shea, D. C., 444, *452*
Oster, G., 64, *197*
O'Sullivan, D. G., 5, *50*
Otaka, E., 139, *201*
Oth, A., 633, *638*
Ott, P., 447, *452*
Ouchterlony, O., 499, 502, *511*
Oumi, T., 95, *203*
Owne, J. A., 484, *511*
Oxenburgh, M. S., 439, *453*
Ozaki, M., 141, 146, *203*

P

Packham, P. M., 447, *452*
Pacovsk, E., 245, *341*
Paech, K., 286, 288, *341*
Paigen, K., 114, *206*, 507, *511*
Pain, R. H., 57, *199*
Paleus, S., 467, *511*
Palmstierna, H., 278, *342*
Pangborn, J., 149, 159, *203*
Pankratz, H. S., 178, *203*
Pardee, A. B., 130, 150, 163, 169, 179, *204*, 428, *453*, *454*, 507, *511*
Parish, J. M., 85, *203*
Parker, F. S., 616, *629*
Parkes, A. S., 434, *453*
Parmeggiani, A., 524, 530, 533, 538, *547*

Partridge, S. M., 442, *451*, 551, 552, 553, 555, *571*
Pascher, G., 531, 542, *546*
Pastuska, G., *511*
Pataki, G., 634, 638, *639*
Patterson, J. B., 401, *422*
Paul, R., 41, *49*
Pavolec, A., 134, *203*
Peacocke, A. R., 11, *51*
Pearson, P., 146, 147, *203*
Pechmann, E., 467, *512*
Pedersen, K. O., 69, 72, *203*, *205*
Peham, A., 286, *336*
Penefsky, H. S., 429, 430, 434, *453*
Perry, J. D., 590, *594*
Persson, K. O. U., 331, 333, 334, *342*
Petermann, M. L., 132, 134, 142, *203*
Peters, J. P., 260, *342*
Peterson, E. A., 426, 442, *453*, *454*
Peterson, J. A., 438, *453*
Petrack, B., 180, *203*
Petterson, V., 613, *614*
Pettijohn, D. E., 394, 398, 401, *422*
Pfeiderer, G., 467, 507, *512*
Pfennig, N., 150, 163, 164, 166, *198*, *203*
Pfister, R. M., 191, *202*
Philippi, K., 277, *343*
Philipson, L., 405, 406, 407, 410, 411, 417, 418, 419, *421*, *422*, *423*, 613, *614*
Phillips, D., 616, *629*
Phillips, G. O., 47, *50*
Philpot, J. S. L., 550, *571*
Phipps, P. J., 286, 287, 288, 289, 290, *339*
Pianotti, R. S., 42, *50*
Pickels, E. G., 65, *203*
Picou, D., 230, *342*
Pirie, N. W., 318, *342*
Pirko, P. C., 617, *629*
Pitot, H. C., 433, *451*
Plazin, J., 217, *344*
Pogojeff, G., 560, *571*
Polge, C., 434, *453*
Pollister, A. W., 64, *197*
Polson, A., 388, *422*
Pomeranz, Y., 302, *314*
Pope, C. G., 256, *342*
Pope, L. M., 189, *203*
Porath, J., 442, 443, 444, 445, 449, *451*, *453*, 563, 564, 565, 569, *570*, *571*

Postel, S., 507, *511*
Postigate, J. R., 376, *382*
Potgieter, G. M., 388, *422*
Potter, V. R., 22, *51*, 312, 333, 334, *340*, *342*
Poulik, M. D., 493, 499, *511*, *512*
Powell, E. O., 190, 195, *206*, 216, *342*
Powell, F. J. N., 234, *342*
Powell, J. F., 233, 239, *342*
Pramer, D., 508, *511*
Preeze-Mendez, G., 526, *547*
Prestidge, L. S., 507, *511*
Price, C. A., 46, *51*
Prins, J , 442, *453*
Pritchard, M. J., 11, *51*
Privitera, C. A., 434, *451*
Prokowa, D., 530, 538, *547*
Przyrembel, L. R., 544, *547*
Pullman, M. E., 429, 430, 434, *453*
Putnam, F. W., 493, *510*

R

Raacke, I. D., 507, *512*
Racker, E., 429, 430, *453*
Radding, C. M., 399, *422*
Radford, L. E., 110, *201*
Radin, M. A., 234, *339*
Radloff, R., 189, *203*, *423*
Raijman, L., 430, *453*
Rajbhandary, U. L., 631, 633, *639*
Ramaiah, T. R., 286, *342*
Ramsay, W. N. M., 239, *342*
Randall, R. J., 3, *50*, 245, 249, 251, 252, *341*, 398, *422*
Randerath, K., 333, *342*, 634, 638, *639*
Randle, P. J., 569, *571*
Rankin, C. T., 92, 99, *205*
Raper, J. R., 31, *51*
Raskas, H. J., 115, *201*
Raskas, H. S., 136, *203*
Ray, J., 25, *51*
Raynaud, M., 405, *421*
Razin, S., 157, 159, *203*, *204*
Rebers, P. A., 272, 274, 275, 276, 302, *338*
Reddi, K. K., 472, *511*
Reeder, R. H., 590, 593, *593*
26–5*b*

Rees, N. W., 218, *337*
Reese, E. T., 608, *614*
Rehm, G., 229, *337*
Reiner, J. M., 322, 330, *340*
Reisfeld, R. A., 146, *204*, 575, 576, 577, 588, *593*, *594*
Reith, A., 5, *51*
Remsen, C., 148, 177, 178, *204*
Rettig, H. L., 507, *512*
Rezdowski, 608, *614*
Rhodes, D. N., 307, *340*
Rhodes, J. M., 507, *512*
Ribbons, D. W., 190, *199*, 315, *337*
Richards, E. G., 80, 103, 105, 107, 110, *204*, 576, 591, *594*
Richmond, M. H., 42, *51*
Richter, J., 132, 136, *205*
Rickard, Pamela A. D., 616, 620, 621, *629*
Ribopoulus, N., 357, *382*
Rinde, H., 77, *205*
Ringelmann, E., 430, *453*
Risse, H. J., 364, *382*
Ritchie, G. A. F., 193, *204*
Ritschard, W. J., 481, 482, *512*
Rizui, S. B. H., 374, *382*
Roach, G. I., 22, 39, *50*
Robbins, F. M., 449, *451*
Robbins, P. W., 351, 354, 355, *382*, *383*
Roberts, C. F , 33, *50*
Roberts, J. B., 189, *200*, 304, 305, *339*
Roberts, R. B., 85, 86, 134, 140, *198*, *199*, *204*
Robinson, C. G., 277, *336*
Robinson, H. K., 290, 309, 311, 316, 328, *342*, *344*
Robinson, H. W., 245, *342*
Robinson, S. M., 449, *451*
Robison, R., 294, *341*
Robrish, S. A., 40, *51*, 149, 159, 170, *203*, *204*
Rodgers, A., 8, 25, 27, 48, *50*, *51*
Roe, D. A., 230, *342*
Roe, J. H., 295, *342*
Romano, A. H., *51*
Romano, C. A., 189, *203*
Rondle, C. J. M., 284, *342*
Roodyn, D. B., 141, *199*
Roosens, P., 592, *594*
Roper, G. H., 620, 621, *629*
Rose, H. M., 157, *203*

Rosebrough, N. J., 245, 249, 251, 252, 341, 398, 422
Rosen, G., 313, 314, 327, 342, 373, 382
Rosen, H., 254, 342
Rosenberg, E., 633, 639
Rosenbloom, J., 100, 101, 204
Rosenbrough, M. I., 3, 50
Rosenkranz, H. S., 157, 203
Ross, H. E., 312, 342
Rosset, R., 142, 204, 328, 342
Rossiter, R. J., 322, 341
Rotman, B., 507, 512
Rowland, S., 2, 50
Rozits, K., 217, 337
Rubin, A. L., 527, 547
Rubin, M. M., 101. 102, 204
Rudin, L., 398, 400, 401, 422
Rueff, F., 515, 547
Ruhenstroth-Bauer, G., 531, 542, 546
Ruhland, W., 303, 341
Runscheidt, F. D., 11, 51
Russell, J. A., 255, 339, 342
Ruttenberg, E., 88, 197
Ryan, D. E., 617, 629
Ryle, A. P., 475, 512
Ryter, A., 148, 204

S

Sacks, L. E., 405, 422
Saidel, L. J., 256, 342
Sakami, W., 432, 453
Salo, T., 87, 204
Salton, M. R. J., 149, 150, 152, 153, 155, 156, 157, 204, 253, 302, 304, 342
Samis, H. V., 87, 204
Sander, G., 524, 530, 533, 538, 547
Sandman, R. P., 526, 547
Sanger, F., 475, 512
San Pietro, A., 150, 168, 175, 189, 198, 201, 203
Santer, M., 140, 204
Saravis, C. A., 449, 450, 451
Sargent, J. R., 480, 482, 512
Sarker, S., 528, 534, 536, 547
Sarnat, M., 114, 206
Sattler, L., 267, 342
Saukkonen, J. J., 633, 639
Saxton, T., 78, 205
Sayre, F. E., 526, 547

Scatchard, G. S., 574, 594
Schachman, H. K., 57, 70, 78, 80, 81, 103, 104, 105, 106, 107, 110, 111, 130, 147, 150, 163, 169, 179, 198, 199, 204, 205
Schaitman, C., 157, 204
Schatz, G., 430, 453
Schatzberg, G., 216, 341
Schechler, A. N., 612, 614
Scheraga, H. A., 143, 204, 430, 453
Scherr, G. H., 488, 512
Schildkraut, C. L., 124, 204, 632, 637, 639
Schlegel, H. G., 189, 190, 204
Schlenk, F., 286, 288, 342
Schlessinger, D., 130, 133, 134, 139, 141, 142, 147, 202, 204, 206
Schliselfeld, L. H., 526, 537, 547
Schmid, H. H. O., 333, 341
Schmidt, G., 265, 310, 342, 372, 382
Schmitz, H., 333, 334, 340
Schnaitman, C., 191, 202
Schneider, W. C., 5, 50, 60, 128, 142, 201, 204, 265, 312, 315, 327, 342
Schnell, G. W., 295, 343
Schöch, G., 524, 530, 533, 538, 547
Schockman, G. D., 150, 152, 204
Schoener, B., 617, 629
Schoenmakers, J. G. G., 599, 613
Schöniger, W., 230, 342
Schooler, J. M., 437, 451
Schram, E., 592, 594
Schrier, B. K., 567, 571
Schroeder, W. A., 256, 343
Schudel, G., 293, 344
Schuler, B., 5, 51
Schulte, K. E., 512
Schulte-Holthausen, H., 364, 382
Schultz, H. E., 574, 594
Schultz, J., 527, 547
Schultz, I. T., 609, 614
Schumacher, H., 467, 510
Schumaker, V., 78, 100, 101, 204
Schwart, Kz., 286, 287, 338
Schwartz, L. L., 610, 613
Schweiger, A., 528, 530, 535, 538, 547, 548
Schweigert, 373
Schwert, G. W., 430, 453

Scott, T. A., 267, *343*
Scrutton, M. C., 430, 437, *453*, *454*
Sealock, R. W., 430, 434, *451*
Sedat, J., 407, *422*
Seeds, A. E., Jr., 434, *452*
Seegmiller, J. E., 294, *339*
Seely, G. R., 148, 150, 151, 162, *198*
Seibert, F. B., 277, *343*
Seifter, S., 267, *343*
Seitz, W., 526, 532, *548*
Seki, Y., 45, *48*
Sekuze, I., 448, *452*
Seligson, D., 234, *344*
Sells, B. H., 140, *204*
Semenza, G., 442, *453*
Seno, V., 507, *512*
Seraidarian, K., 317, *340*
Serles, E. R., 235, *339*
Sevag, M. G., 361, 366, *382*
Shakulov, R. S., 134, 136, *205*
Shapiro, A. L., 577, 593, *594*
Sharon, N., 283, *343*
Shatkin, A. J., 180, *205*
Shaw, C. H., 31, *51*
Sheldrick, P., 114, *205*
Shepherd, G. R., 577, 587, *594*
Shepherdson, M., 428, *453*
Sher, I. H., 47, *51*
Sherman, F., 576, *594*
Sherratt, H. S. A., 323, *343*
Shetlar, M. R., 267, 271, *343*
Shettles, L. B., 295, 297, *338*
Shibata, K., 616, *629*
Shizuta, Y., 430, *453*
Shorey, C. D., 160, 161, *198*
Shugarov, Yu. A., 616, *629*
Shukuya, R., 430, *453*
Shultz, S. G., 213, *338*
Shuster, C. W., 430, *454*
Shuster, L., 567, *571*
Siakotos, A. N., 89, *205*
Siegelman, H. W., 441, 449, *454*
Signer, R., 412, *422*
Silberman, H. J., *511*
Siliprandi, D., 471, *512*
Siliprandi, N., 471, *512*
Silverman, M., 434, *453*
Singh, R. M. M., 435, *451*
Sinsheimer, R. L., 115, 133, *200*, 407, *422*

Sistrom, W. R., 148, 150, 151, 156, 162, 167, 168, 170, 172, 173, 174, 176, *198*, 200, *205*, *207*
Sizer, I. W., 439, *452*
Skidmore, W. D., 633, 634, *639*
Skilna, L., 140, *204*
Slack, H. G. B., 229, *343*
Slater, E. C., 434, *453*
Slater, R. J., 507, *511*
Slattery, M. C., 286, 287, *341*
Slayter, H. S., 147, *200*
Slein, M. W., 295, *343*, 433, *450*
Slepecky, R. A., 193, *201*, 305, 306, 307, *340*, *343*
Sletten, K., 612, *614*
Sloane-Stanley, G. H., 155, 303, *199*, *338*
Small, P. A., 576, *594*
Smellie, R. H. S., 496, *510*
Smillie, R. M., 357, *382*
Smith, A. J., 189, *203*, 236, *343*
Smith, A. U., 434, *453*
Smith, F., 272, 274, 275, 276, 302, *338*
Smith, I., 462, 484, 489, 497, 502, *510*, *511*, *512*, 574, 590, *594*
Smith, I. W., 330, *343*
Smith, J. D., 333, *341*, 469, 470, 471, *511*, *512*, 633, 634, 638, *639*
Smith, J. E., 24, *52*
Smith, K. R., 149, 157, *205*
Smith, L. F., 475, *512*
Smith, M., 426, *452*
Smith, Z., 191, *200*, 405, *422*
Smithies, O., 489, 499, *512*, 574, *594*
Smithies, W. R., 304, *343*
Smith–Kielland, I., 332, 333, *343*
Smyrniotis, P. Z., 294, *339*
Snell, E. E., 428, 435, *453*
Snoswell, A. M., 439, *453*
Snyder, H. E., 617, *629*
Sobel, A. E., 256, *343*
Sober, H. A., 426, 442, *453*, *454*
Sohler, A., *51*
Sokatch, J. R., 138, 158, *203*
Soll, D., 137, *206*, 400, *422*
Solvok, H., 467, *511*
Somogyi, M., 278, 299, 300, *339*, *343*
Sørensen, M., 277, *343*
Sørbo, B., 230, *337*
Sorof, S., 562, *571*

Soyenoff, B. C., 225, *343*
Spahr, P. F., 145, *205*
Spencer, B., 229, 230, *338*, *343*
Spencer, E. W., 478, *512*
Spicer, C. C., 40, *50*
Spiegelman, S., 114, *205*, 322, 330, 333, *339*, *340*, 507, *512*, 593, *593*
Spies, J. R., 256, *343*
Spirin, A. S., 72, 125, 134, 136, 144, *200*, *202*, *205*, 309, 310, *336*, 637, *638*
Spiro, N., 284, *343*
Spitnik-Elson, P., 145, 146, *205*
Spizizen, J., 42, 44, *50*, *51*, 163, 169, *201*
Spooner, E. T. C., 152, *206*
Spragg, S. P., 78, 92, 99, *205*
Stacey, M., 317, *337*, 362, *381*
Staehlin, T., 136, *203*
Stafford, D. W., 401, *422*
Stahl, E., 333, *341*, 481, *512*
Stahl, F. W., 114, 115, 116, 135, *202*
Stahn, R., 530, 538, *547*, *548*
Stanier, W. H., 472, *510*
Stanier, R. Y., 44, *51*, 130, 149, 150, 151, 157, 163, 169, 179, 190, *202*, *204*, *205*, 305, *338*
Stanley, W. M., 134, 143, *205*
Starr, M. P., 151, 166, *206*
Stauff, J. F., 47, *52*
Stedingk, L. V. von, 171, *201*
Steed, P., 156, *203*
Steers, E., 592, *593*
Stein, W. H., 254, 333, *341*, *343*, 442, *451*, 561, *571*
Steinberg, I. Z., 430, *453*
Stellwagen, E., 106, *205*
Stengar, V. A., 230, *340*
Stenti, G., 72, 115, *201*
Stelzel, K. H., 527, *547*
Stephenson, M., 42, *51*, 129, *205*
Stevens, A., 632, *639*
Stevens, M. F., 256, *342*
Stewart, M. R., 617, *629*
Stickland, H. L., 245, *343*
Stocker, B. A. D., 152, *206*
Stodolsky, M., 114, *206*
Stoll, R. D., 307, *340*
Stone, G. C. H., 231, *343*
Stoner, C. D., 577, *594*
Storck, R., 43, *52*, 114, *205*

Stotz, E. H., 360, *382*
Strange, R. E., 42, *49*, *51*, *52*, 137, *206*, 213, 214, 222, 233, 235, 236, 237, 239, 258, 260, 261, 279, 285, 301, *340*, *342*, *343*
Straub, E., 515, *548*
Strauch, L., 527, *546*
Strickler, A., 521, 522, 544, *548*
Strominger, J. L., 43, *51*, 351, 364, *382*
Stuart, A., 631, 633, *639*
Stumpf, P., 320, *343*
Stuy, J. H., 286, *343*
Subba Row, Y., 224, 225, 235, *338*
Sueoka, N., 124, *205*
Sugawara, S., 46, *52*
Sulkowski, E., 525, 533, 534, 535, *548*
Summers, W. C., 401, *422*
Summerson, W. H., 260, *339*
Sumner, J. B., 225, *343*
Susor, W. A., 179, 181, *205*
Sussman, A. S., 46, *52*
Sutherland, I. W., 360, 364, 376, *382* *383*
Sutton, C. R., 31, *52*
Suyama, Y., 24, *52*
Suzue, G., 356, *383*
Suzuki, H., 434, *454*
Svedberg, T., 69, 77, *205*
Svensson, H., 91, 92, *205*, 445, 446, *454*, 516, *548*, 551, 552, 554, 555, 557, 558, *571*, 597, 598, 599, 600, 603, 605, 609, 610, 613, *614*
Sviensson, S. L., 234, *338*
Sweeley, C. C., 302, *343*, 351, *382*
Sybesma, C., 189, *203*
Sykes, J., 96, 130, 132, 133, 134, 135, 140, 141, 142, 150, 152, 163, 164, 165, 166, 172, 193, 194, 195, *198*, *199*, *206*, *207*
Sylvan, S., 330, *339*
Symons, M. C. R., 628, *629*
Synge, R. L. M., 463, 468, *510*, *511*, 524, *546*
Szilagyi, J. F., 151, *201*
Szybalski, W., 113, 114, 117, 118, 119, 120, 122, *205*, 401, *422*

T

Taber, H. W., 576, 587, *594*
Taborsky, G., *422*

Tagawa, K., 448, *452*
Takayama, K., 577, *594*
Takemori, S., 434, *454*
Talalay, P., 434, *451, 452*
Tanaba, S., 356, *383*
Tashiro, Y., 148, *205*
Tavel, P. v., 412, *422*
Taylor, G. I., 11, *52*
Taylor, J. F., 426, 433, *454*
Taylor, K. W., 569, *571*
Tchan, Y. T., 616, *629*
Teagre, O., 499, *510*
Tedeschi, H., 70, *205*
Teece, E. G., 317, *337*
Teller, D. C., 80, 103, 105, 110, 140, *204*
Tempest, D. W., 132, 137, 140, 152, *205, 206*, 213, 233, 235, 236, 237, 238, 329, 330, *338, 343*
Thannhauser, S. J., 265, 310, *342*, 372, *382*
Thiele, O. W., 307, *344*
Thomas, A. J., 323, *343*
Thomas, D. B., 230, *337*
Thomas, J. B., 179, *206*
Thomas, R., 467, *511*
Thorp, R. F., 239, *343*
Thorsteinsson, T., 467, 507, *511*
Thurman, P. F., 158, *200*
Tillmans, J., 277, *343*
Timasheff, S. N., 599, *614*
Tisdall, F. J., 234, 236, *340*
Tiselius, A., 441, *454*, 457, *512*, 550, 553, 563, *570, 571*, 584, 592, *594*
Tissieres, A., 134, 139, 142, 147, *206*
Tobback, P., 189, *206*
Tocchini-Valenti, G. P., 114, *206*
Toennies, G., 302, *340*
Toennis, G., 150, 152, *204*
Tokushige, M., 430, 436, *452, 453, 454*
Tollens, B., 292, *341*
Tomcsik, J., 44, *52*
Tompa, H., 388, *422*
Toribara, T. Y., 225, *337*
Torriani, A., 428, *454*
Toschi, G., 467, *512*
Toukada, K., 356, *383*
Touster, O., 526, 537, *547*
Tracey, M. V., 286, 288, 292, *341, 343*
Traub, P., 136, 137, 139, 140, 146, *203, 206*

Traut, R. R., 146, 147, *203, 206*
Trautman, R., 57, 75, 76, 111, 125, *198, 206*, 507, *511*
Travers, S., 78, *205*
Travis, P. M., 8, *52*
Trentini, W. C., 151, 166, *206*
Trevalion, P. A., 628, *629*
Trevelyan, W. E., 267, 272, 279, 286, 316, 323, 324, 327, *344*
Treybal, R. E., 392, *423*
Trinks, H., *511*
Tsanev, R., 322, *344*
Turner, B. C., 441, *454*
Tuttle, A. L., 163, 169, *206*
Tzagloff, A., 577, *594*

U

Ulrich, D., 81, *206*
Umbreit, W. W., 47, *52*, 286, 290, *336*
Urbaschek, B., 307, *344*
Utsunomiya, T., 147, *201*
Utter, M. F., 30, *50*, 45, *49*, 430, 437, *453, 454*

V

Vadlamudi, B. V., 482, *512*
Vagelos, R. R., 437, *454*
Valmet, E., *571*, 599, 602, 613, *614*
Valois, F., 148, 177, 178, *204*
Vanderplancke, A., 562, *571*
Vanderwinckel, E., 43, *52*, 148, *206*
Van Eys, J., 526, 537, *547*
Van Iterson, W., 148, *206*
Van Slyke, D. D., 217, 255, 256, 259, 260, *342, 344*
Vargas, L., 569, *571*
Vasseur, E., 284, 285, *340*
Vatter, A. E., 169, *206*
Vendrely, R., 562, *571*
Venekamp, 407
Verevkina, J. V., 563, *570*
Vernon, L. P., 148, 150, 151, 162, 167, 168, 171, 175, 177, 183, 184, 186, 188, 189, *198, 200, 201, 203, 206*
Verwoerd, D. W., 508, *512*
Vesterberg K., 446, *454*, 600, 603, 609, 610, *614*

Vesterberg, O., 445, 446, *454*, 597, 598, 599, 600, 603, 605, 607, 608, 609, 610, 613, *613*, *614*

Villanueva, J. R., 45, 46, *49*, *52*

Vinograd, J., 71, 83, 90, 101, 102, 114, 116, 117, 119, 135, *199*, *201*, *202*, *206*, *423*

Vinter, V., 229, *344*

Vinuela, E., 577, 593, *594*

Viswanata, T. W., 507, *511*

Vitols, E., 25, *52*

Vleck, J. H. van, 618, *629*

Voet, D., 117, *201*, 634, 638, *639*

Vogel, H., 136, 140, 146, *203*, 633, 638, *639*

Vogell, W., 5, *51*

Vogt, M., 535, *547*

Voigt, H. P., 524, 530, 533, 538, *547*

Volkin, E., 308, *344*

Volwiler, W., 507, *510*

Vorbreck, M. L., 353, *383*

W

Waddell, W. J., 398, *423*

Wade, H. E., 134, 135, 136, *198*, 213, 290, 309, 311, 316, 324, 328, *342*, *343*, *344*, 471, *512*

Wadström, T., 446, *454*, 597, 600, 603, 609, 610, 612, *614*

Wagner, R. P., 333, *339*

Wahl, R., 323, *340*

Waldenström, J., 592, *594*

Waldo, A. L., 267, *344*

Waldschmidt-Leitz, E., 524, *548*

Waldvogel, M. J., 286, 288, *342*

Wallenfels, K., 467, *512*

Waller, J. P., 145, *206*

Wallis, O. C., 30, *49*

Walter, H., 396, *423*

Wang, J. H., 430, 434, *451*

Wang, S. Y., 633, *639*

Wang, W. S., 189, 190 ,*206*

Warashina, E., 307, *339*

Warburg, O., 142, *206*, 263, 264, 322, *344*, 447, *454*

Warburton, R., 634, *638*

Ward, J. M., 592, *594*

Wargon, M., 317, *340*

Warner, H., 225, *337*

Warner, R. C., 117, *202*, 430, *453*

Warth, A. D., 233, *342*

Wassarman, P. M., 612, *614*

Watanabe, M., *423*

Waterlow, J. C., 230, *342*

Waterman, R. E., 596, *614*

Watson, J. D., 134, 139, 140, 142, 147, *203*, *206*

Watson, K., 24, *52*

Watson, R., *423*

Watson, S. W., 148, 177, 178, *203*, *204*

Watt, W. B., 468, *510*, 524, *546*

Watts, B. M., 617, *629*

Watts, D. C., 497, *512*

Watzke, E., *546*

Waymouth, C., 317, *337*

Weare, J. H., 447, *451*

Webb, E. C., 426, 440, 441, *451*

Webb, J. M., 308, 317, *344*

Weber, G., 437, *454*

Weibull, C., 43, *52*, 149, 151, 152, 155, *206*, 302, *344*

Weidel, W., 42, *52*

Weidner, M. A., 302, *340*

Weigle, J., 83, 90, 101, 102, 114, *206*

Weimer, H. E., 590, *594*

Weisiger, J. R., 217, *344*

Weiss, B., 46, *52*

Weiss, J. B., *512*

Weiss, S. B., 114, *206*, 434, *453*, 507, *512*

Weissbach, H., 436, *452*

Weissberger, A., 412, *421*

Weisskopf, V. F., 618, *629*

Wells, M. A., 307, *344*

Wells, W. W., 302, *343*

Wendlandt, W. W., 616, 617, 625, *629*

Werkman, C. H., 30, *50*

Wesslén, T., 407, *423*

Westley, J., 307, *344*

Westphal, O., 354, 364, *382*, *383*

Wetter, L. R., 467, *512*

Wheat, R., 284, *344*, 360, *382*

Whistler, R. L., 266, 278, 281, 285, 298, 302, *338*, *339*, *340*, *342*, *344*

Whitaker, D. R., 608, *614*

Whitaker, J. R., 444, *454*

White, A. E., 140, *199*

Wiame, J. M., 43, *52*, 194, *206*, 330, *344*

Wicken, A. J., 329, *339*
Widdowson, J., 146, *203*
Widra, A., 194, *206*
Wieck, K., 532, *548*
Wieczorek, G. A., 441, *454*
Wieland, T., 463, 467, 507, *512*
Wiggins, L. F., 317, *337*
Wild, D. G., 140, *199*
Wilhelmi, A. E., 255, *339*
Wilkinson, J. F., 190, 191, 195, *202*, *206*, *207*, 304, 305, 330, *343*, *344*, 358, 359, 360, *382*, *383*
Wilkinson, R. H., 234, *344*
Williams, 352
Williams, C. A. 499, 502, *510*
Williams, D. E., 146, *204*, 576, 577, *594*
Williams, E. F., 218, *337*
Williams, J. W., 103, *206*
Williams, R. E. O., 40, *50*
Williams, R. R., 596, *614*
Williams, R. T., 292, *339*
Williamson, A. R., 600, 613, *613*
Williamson, D. H., 41, 45, *49*, *50*, 190, 195, *207*, 304, 305, *344*, 359, *383*
Williamson, M. B., 488, *512*
Willstätter, R., 293, *344*
Winder, F. G., 322, *344*
Winget, G. D., 435, *451*
Winitz, M., 255, *339*
Winter, W., 435, *451*
Winzler, R. J., 245, *341*
Winzor, D. J., 99, *207*
Wirth, H., 521, *547*
Wirth, M. E., 89, *205*
Wober, W., 307, *344*
Woiwod, A. J., 256, *344*
Wolfe, R. S., 169, *206*
Wolfrom, M. L., 266, 278, 281, 285, 298, 302, *338*, *339*, *340*, *342*, *344*
Wood, W. A., 158, *198*
Worden, P. B., 156, 167, 168, 170, 172, 173, 174, 176, *200*, *207*
Wrba, H., 542, *547*

Wren, J. J., 307, *344*
Wrigley, C., 600, *614*
Wright, A., 351, 354, 355, *382*, *383*
Wu, H., 633, *639*
Wyatt, G. M., 236, *339*
Wyatt, G. R., 633, 634, *639*
Wycoff, M., 588, *593*

Y

Yamanaka, T., 448, *453*, *454*
Yamano, A., 307, *340*
Yamashita, J., 442, *452*
Yanofsky, C., 524, *546*, 592, *594*
Yates, R. A., 428, *454*
Yemm, E. W., 254, 262, 334, *344*
Yoneda, M., 448, *452*
Yoshida, A., *52*, 194, *207*
Young, E. M., 562, *571*
Young, F. E., 124, *207*
Young, T. W., 132, 135, 140, 141, 142, *205*, *207*
Young, W. J., 2, *49*
Yount, V., 357, *382*
Yphantis, D. A., 80, 105, 108, 109, 148, *205*, *207*, 439, *452*

Z

Zaccari, J., 588, *593*
Zajdela, F., 45, *52*
Zambotti, V., 317, *337*
Zaye, D. F., 617, *629*
Zciler, K., 531, 543, *547*
Zeldin, M. H., 592, *594*
Zeppezauer, M., 388, *423*
Zerban, F. W., 267, *342*
Zernike, J., 392, *423*
Zettner, A., 234, *344*
Zillig, W., 508, *512*
Zipf, R. E., 267, *344*
Zipster, M. W., 617, *629*
Zuazaga, G., 218, *341*
Zweig, G., 333, *337*, 462, 465, *509*

Subject Index

A

Abrasives, 30

Absorption band width,
factors influencing, 618–621

Absorption coefficients,
in reflectance spectrophotometry, 617

Absorption optical systems,
in ultracentrifugation, 73, 78–80

Acetate,
electrophoresis buffers, in, 459, 524–525
protein extraction with, 376–377

Acetic acid,
polysaccharide extraction, in, 366
protein stains, in, 586
ribosomal protein analysis, in 145
stain solvent, as, 464–465

Acetone,
enzyme crystallization by, 447
enzyme protection by, 434
extraction with,
carotenoid, of, 357
ferredoxin, of, 377–378
lipid, of, 352
quinone, of, 357
fractional precipitation with, 439–440

Acetyl coenzyme A,
and pyruvate carboxylase, 437

Acetyl coenzyme A carboxylase,
activation of, 437
cold-labile, 430

Acetyl glutamate, 437

N-acetyl hexosaminidase, 559

Acetylated Sudan black B, 465

Acid extraction,
acetone and, 353
basic proteins, for, 378

Acid "locks",
in zonal electrophoresis, 554

Acid proteins,
electrophoresis buffers for, 528

Aconitase, 19–21

Acrylamide, 575

Adenosine diphosphate (ADP),
separation of, 471
threonine deaminase activation by, 436

Adenosine monophosphate (AMP),
separation of, 469, 471
enzyme activation by, 436

Adenosine triphosphatase,
separation of, 467
stabilization of, 434
temperature effects on, 429–430

Adenosine triphosphate (ATP),
separation of, 471

Adenine nucleotides,
separation of, 533

Adenovirus proteins,
separation of, 613

Adenyl cyclase,
activation of, 437

Aerobacter spp.,
disintegration of, 44–45

Aerobacter aerogenes,
ATP content of, 213
diphenylamine reagent with, 319
disintegration of, 41
free amino-acids in, 335
glycogen from, 279
ionic balance in, 137
magnesium content of, 235
nucleic acid extraction from, 326
nucleic acid pool in, 336
orcinol reagent with, 291
polyphosphate from, 194–196, 331
protein determination in, 244–248, 256–258
sodium content of, 232
total carbohydrate determination in, 270–277
total nitrogen content of, 222, 243

Agar-gel electrophoresis, 499–503
apparatus for, 500
buffers for, 500–501
gel preparation for, 501
sample application in, 501–502

Agar-gel electrophoresis—*cont.*
  immunoelectrophoresis by, 502–503
Agarose, 444
Aggregate formation,
  in iso-osmotic buffers, 537
Albumin,
  electrophoresis marker, as, 568
  molecular dimensions of, 574
  protein protection, in, 433
  serum protein electrophoresis, in,
    464–467, 482, 489
Alcohol dehydrogenase, 434–435
Aldo-pentoses, 291
Aldose determination, 293–295
Algae,
  biphasic separation of, 402, 404
  polyphosphates in, 369
Alkali extraction, 362–363
  capsular polysaccharides, of, 362
  glucan, of, 363
  glycogen, of, 362–363
  mannan, of, 363
Alkali "locks",
  in zonal electrophoresis, 554
Alkali soluble polysaccharides,
  separation of, 563
Alkaline copper methods,
  for reducing sugars, 299–300
Alkaline ferricyanide methods,
  for reducing sugars, 298–299
Alkaline hypochlorite,
  extraction with,
    phosphates, for, 371
    poly-β-hydroxybutyrate, for, 359
Alkylated cellulose, 561
Allosteric enzymes, 436
Alumina,
  enzyme purification, for, 441
  microbial disintegration, in, 30
Amberlite IRC-50, 442
Aminco French press, 13–15, 53
Amine separation, 481
Amino-acids,
  determination of, 253–262
    copper complex method, by, 255–
      258
      β-naphthoquinone sulphonate,
        by, 255
      ninhydrin methods, by, 253–256,
        259–260

    nitrous acid, by, 255
  extraction of, 348–349
  pool, estimation of, 331–335
  separation of, 468, 478, 481
Aminoethyl cellulose, 443
Amino group dissociation, 458
2-amino-2 methyl-1,3-propanediol-
    phosphate buffer, 530
α-Amino nitrogen (*see also* amino-acids),
  determination of, 253–262
Amino sugars,
  phenol-sulphuric acid reagent on, 275
Ammonium acetate buffers,
  in electrophoresis, 470, 525
Ammonium formate buffers,
  in electrophoresis, 470
Ammonium persulphate,
  in gel polymerization, 576–577, 585
Ammonium phosphomolybdate, 469,
    470
Ammonium sulphate, 435, 440, 447
Ampholine, 446, 601
Ampholyte separation,
  by isoelectric focusing, 595–613
α-Amylase,
  crystallization of, 447
  proteinases, on, 435
  separation of,
    disc electrophoresis, by, 589
    starch block electrophoresis, by, 505
  stabilization of, 435
Anabaena cylindrica, 180
Anabaena variabilis, 179–181
Anacystis nidulans, 178–180
Anaerobic environment, 432–433
Analytical cell electrophoresis, 515
Analytical centrifugation, 58–60, 63
Angle rotors, 60–61, 83, 125–127
Ankistrodesmus sp., 558
Annular zones,
  in disc electrophoresis, 585
Anode solution,
  in isoelectric focusing, 606
Anthrone reagent, 266–272
Antibiotic,
  cell surface effects of, 544
Antigen–antibody complexes,
  separation of, 418–419
Antigen isolation, 405
Apparent diffusion coefficient, 76–78

Aqueous extraction,
  microbial cells, of, 349
  slime polysaccharide, of, 360–362
Aqueous polymer solutions, 388
Arabinose,
  orcinol reagent, with, 288–290
  phenol-sulphuric acid reagent, with,
    276
Araldite, 553
Archibald method, 110–112
Arginase,
  stabilization of, 435
Aryl-β-glucosidase,
  isoelectric focusing of, 607
Ascites tumour cells,
  disintegration of, 22
  separation of, 542
Aspartate transcarbamylase, 428
Aspergillus spp.,
  enzyme separation in, 607
  proteinase purification in, 537
Aspergillus niger, 24
Aspergillus oryzae,
  α-amylase from, 447
  disintegration of, 45
Athiorhodaceae,
  cytomembranes from, 150, 166–174
  poly-β-hydroxybutyrate granules in,
    189
  sub-chromatophore particles from,
    183–7
Atomic absorption spectrophotometry,
  magnesium estimation by, 238
  sulphate estimation by, 230
Autoanalysers, 211–212
Autolysis, 42
Azocarmine B.,
  in electrophoresis, 464, 487
Azotobacter spp.,
  disintegration of, 40
  nitrogen-fixing systems from, 437
Azotobacter agilis,
  cytomembranes from, 149, 159
Az. beijerinckii,
  poly-β-hydroxy butyrate granules in,
    189, 193
Az. chroococcum,
  poly-β-hydroxy butyric acid in, 304
Az. vinelandii,
  disintegration of, 44

B

Babes–Ernst granules, 193
Bacillaceae,
  poly-β-hydroxy butyrate granules in,
    189
Bacillus spp.,
  autolysis of, 42
  calcium content of, 233
  DNA extraction from, 372
  disintegration of, 45
  poly-β-hydroxy butyrate in, 304
Bacillus amyloliquefaciens,
  intracellular ionic balance in, 137
  polyribosomes from, 133
  ribosomes in, 130
B. cereus,
  cereolysin from, 612
  iron content of, 239
  poly-β-hydroxy butyrate in, 190–1
  spore resistance of, 323
B. globigii,
  electrophoretic separation of, 544
B. lichiniformis,
  cytomembranes from, 152
B. megaterium,
  cytomembranes from, 148–149, 152–
    155, 405
    composition of, 148
    lipid from, 155
    membrane fragments from, 154
    mesosomal elements from, 149,
      153
    plasma membrane from, 149
  disintegration of,
    egg-white lysozyme, by, 43–44
    French press, by, 15
    Sonomec shaker, by, 26
  poly-β-hydroxy butyrate from, 189–
    192, 304, 358–359
  polyphosphate from, 195
B. stearothermophilus,
  cytomembranes from, 152
B. subtilis,
  α-amylase from, 447
  cytomembranes of, 148, 155
  D.N.A. from, 400, 634–637
  disintegration of,
    freeze-pressing, by, 23
    lysozyme, by, 43–44
  magnesium in, 238

B. subtilis—cont.
  ribosomes from, 97
  teichoic acids from, 329–330
B. thuringiensis,
  esterases from, 498
  metabolite separation in, 498
  protoxin from, 538–539
Background reflectance, 620
Bacterial cell number, 216
Bacterial cytomembranes, 148–189
  Athiorhodaceae, from, 166–174
  cell envelopes in, 157–161
  Chlorobacteriaceae, from, 163–166
  "chromatophores", in, 148, 150, 162–
    163, 169–177, 183
  components of, 148–151
  "ghosts" from, 149, 152, 157–158
  Gram-negative bacteria, from, 156–
    161
  Gram-positive bacteria, from, 151–
    156
  intra-cytoplasmic membrane in, 148,
    157
  mesosomal elements in, 148–150, 157
  Nitrobacteriaceae, from, 177–178
  photosynthetic vesicles in, 164–166
  plasma membrane in, 148–150, 156–
    157
  sub-chromataphore particles in, 183–
    139
  Thiorhodaceae, from, 174–177
Bacterial enzymes,
  electrophoretic separation of, 559
Bacterial homogenates,
  centrifugal fractionation of, 128–129
Bacterial mixtures,
  separation of,
    countercurrent distribution, by,
      415–416
    electrophoresis by, 529, 531, 544–
      546
Bacterial oxidosomes,
  separation of, 405
Bacterial particles,
  biphasic distribution of, 403–404
Bacterial photosynthetic apparatus, 150
Bacterial "pool" constituents, 331–336
  amino-acid determination in, 334–335
  extraction of, 332–333
  identification of, 333–334

nucleic acid in, 335–336
Bacterial polyribosomes,
  centrifugal isolation of, 129–148
  preservation of, 133
  ribonuclease on, 134
Bacterial ribosomes,
  centrifugal isolation of, 129–132,
    138–140
  characterization of, 142–145
  disassembly sequences for, 136–137
  EDTA on, 133
  growth medium on, 132
  ionic environment on, 134–138
  location of, 130
  molecular weight of, 143–145, 147
  polyamines on, 138
  protein, in, 145–147
  purity of, 140–141
  release of, 130
  ribonuclease on, 134
  sedimentation constants for, 142
  ultraviolet absorbance of, 142
Bacterial spores,
  isolation of, 405
Bacterial sub-cellular components,
  isolation and characterization of,
    55–207
Bacteriolytic proteins,
  isoelectric focusing of, 609
Bacteriophage (*see also* phage),
  isolation of, 47
  purification of, 407–412
Bacteriophage T_2,
  purification of, 408–412
Ballast ions, 523
Ballotini beads, 8, 25
Banard and Howlett mill, 25
Band capacity,
  in rate-zonal centrifugation, 90–92
Band forming caps,
  swinging bucket rotors for, 90
Band sedimentation studies, 83, 90, 92
Barbitone buffer,
  in electrophoresis, 459, 463, 485
Barbitone-acetate buffer,
  in electrophoresis, 484, 500
Barbitone-calcium lactate buffer,
  in electrophoresis, 500
Base composition,
  of nucleic acids, 631–639

Base concentration,
of DNA, 635–637
Basic proteins,
acid extraction of, 378
Batch harvesting, 57
Beckman Analytrol, 118
Benzene,
metabolism of, 438
Benzyl alcohol dehydrogenase, 434
Bio-Gel P–2,
protein purification by, 449
Biphasic separation,
microbial particles of, 385–423
counter current distribution, by,
412–419
polymer phase systems for, 386–
393
nucleic acids, of, 393–401
proteins, of, 393–401
viruses, of, 405, 407–412
Biuret method,
absorption spectra of, 246
interference in, 248
microbial cells, for, 245–248
ribosomal proteins, for, 145
standard curves for, 244
Blood cells,
separation of,
buffers, by, 531
electrophoresis, by, 542–543
Block electrophoresis, 503–508
apparatus for, 504
applications of, 507–508
glass powder in, 506
polyvinyl chloride in, 506
preparation of, 503–504
protein elution in, 505–506
sample application in, 504–505
Sephadex in, 506–507
starch in, 503–506
Blue-green algae, *see also under specific
names*,
cytomembranes from, 178–183
lysozyme on, 179
photosynthetic apparatus of, 150
pigmentation of, 151
Booth Green mill, 24
Borate buffer,
in electrophoresis, 459, 555
Borate-glycine buffer,

in electrophoresis, 527
Borate ions,
with carbohydrates, 472
Boric acid–sodium chloride buffer,
in carbohydrate separation, 472
Boric acid–sodium hydroxide buffer,
in starch gel electrophoresis, 493
Bound lipid,
extraction of, 351, 353
Bovine serum albumin (*see also* albumin),
sedimentation constant for, 72
total nitrogen determination on, 217–
224
Braun shaker, 25, 53
Brain protein,
electrophoresis buffers for, 527
Brevibacterium liquefaciens,
adenyl cyclase from, 437
Bromophenol blue,
electrophoresis in, 464, 486, 495, 501
Brucella abortus,
disintegration of, 40
lipid analysis in, 307
Buffer capacity, 523, 555
Buffer effects, 458–459
Buffers,
enzyme stability in, 435
electrophoresis, in, 524–531, 559
Buoyant density, 103, 114, 116, 120–124
determination of, 120–124
Burton's diphenylamine method,
absorption spectra for, 319
microbial DNA, for, 317–320, 325–
326
Butanol extraction, 355
n-Butyric acid–sodium hydroxide buffer
in electrophoresis, 471

C

Caesium acetate, 118
Caesium chloride,
DNA analysis, in, 113–114, 632
sedimentation equilibrium in, 112–
113
Caesium formate, 118
Caesium ions,
on bacterial ribosomes, 137
Caesium sulphate, 117–118, 121

Calcium ions,
   determination of, 233–234
   enzymes, on, 435, 447
   ribosomes, on, 138
Calcium phosphate gel, 441
Calf thymus,
   nucleic acids from, 311, 319
Candida albicans,
   disintegration of, 46
Candida utilis,
   cytochromes of, 618
Cane sugar,
   in electrophoresis, 529
Capsular polysaccharide,
   extraction of, 362–364
Carbamyl phosphate synthetase,
   activity of, 434, 436–437
Carbohydrates,
   analysis of, 265–302
      anthrone reagent, by, 266–272
      chromatography, by, 301–302
      Molisch test, by, 266
      phenol method, by, 272–277
      polysaccharide content, by, 278–281
      sugar content, by, 282–302,
   cytomembranes, in, 155
   electrophoretic separation of, 472, 507
Carbon determination, 261–217
Carbon dioxide,
   on cavitation, 29
Carbon rods,
   as electrodes, 579
Carbon tetrachloride,
   in electrophoresis, 470, 476
Carbowax polyethylene glycol 6000,
   392, 408, 411
Carboxyl group dissociation, 458
Carboxymethyl cellulose, 442–443, 378
Carboxymethyl dextran, 388
Carotenoids,
   extraction of, 356–357
Carrier ampholytes, 597
Cartilage extracts,
   fractionation of, 536
Catalase,
   proteinases on, 435
Cathode solution,
   in isoelectric focusing, 606
Cauchy distribution, 619–621

Cavitation, 9–10, 28–29
Cell disintegration, 6–40
Cell envelopes,
   from Gram-negative bacteria, 157–161
Cell particles,
   separation of,
      biphasic, 401–405
      electrophoretic, 537–540
Cell wall,
   disintegration of,
      enzymes, by 43–46
      Hughes press, by, 34–37
   isolation of, 405
Cellulolytic enzymes,
   isoelectric focusing of, 607–608
Cellulose,
   polysaccharide fractionation, in, 281
   zonal electrophoresis, in, 561–562
Cellulose acetate electrophoresis, 482–489
   buffers for, 484–485
   protein stains in, 486–489
Cellulose ion exchangers, 442–443
Cellulose nitrate tubes, 125–127
Cellulose powder,
   chromatography on, 307
   electrophoresis on, 507
Centrepieces,
   analytical ultracentrifuge for, 59
Centrifugation,
   of sub-cellular components, 55–128
Centrifuge shaker, 53
Cerebroside sulphatase,
   purification of, 525, 537
Cereolysin,
   from *B. cereus*, 612
Cetylpyridinium bromide, 468
Chaikoff press, 20–22, 24
Chemical extraction methods,
   for microbial cells, 346–381
Chlorella cells,
   separation of,
      biphasic, 402, 404, 415
      electrophoretic, 558
Chlorella pyrenoidosa,
   disintegration of, 15
   separation of, 563
Chloride buffer, 555

Chlorobacteriaceae,
cytomembranes from, 162–166
sub-chromatophore particles from, 189
Chlorobenzene, 471, 476
Chlorobium limicola,
cytomembranes from, 163–164
Chlorobium thiosulphatophilum,
cytomembranes from, 163–164
mesosomal elements in, 150
photosynthetic apparatus of, 150
photosynthetic vesicles from, 165–166
Chloroform, 484
Chloroform-methanol extraction, 351–352
Chlorophylls,
extraction of, 350
Chloropseudomonas ethylicum,
photosynthetic apparatus for, 150
photosynthetic vesicles from, 164–165
sub-chromatophore particles from, 189
Chondroitin sulphate,
electrophoretic separation of, 535
Chromatium sp.,
cytomembranes from, 162, 174
Chromatium okenii,
cytomembranes from, 174
Chromatium strain D,
chromatophores from, 174–177
sub-chromatophore particles from, 187–188
Chromatium weissei,
cytomembranes in, 174
Chromatophores,
Athiorhodaceae, from, 150, 169–174
bacterial cytomembranes, in, 148–150, 162–163
fragmentation of, 183
photosynthetic bacteria from, 162–163
Thiorhodaceae, from, 174–177
Chrombacterium violaceum,
electrophoretic separation of, 544
poly-β-hydroxybutyrate in, 189
Chryoosporium lingnorum, 608
Citrate,
electrophoresis buffers, in, 459, 469
enzyme activation by, 437

extraction with,
haemoproteins, of, 377
mannan, of, 368
Cleland's reagent, 430
Clostridium pasteurianum,
ferredoxin from, 378
nitrogen-fixing systems from, 430–431, 437
Clostridium perfringens,
nucleic acid from, 311
toxin from, 559, 612
Clostridium tetanomorphum,
threonine deaminase from, 436
Clostridium welchii, extracts of, 4, 23
Coenzymes,
extraction of, 347–350
electrophoresis of, 468 473, 532
Coenzyme Q,
extraction of, 357–358
Cold-labile enzymes, 429–430, 434
"Cold shock", 213
Colicin,
from *Escherichia coli*, 611–612
Collagen,
electrophoresis buffers for, 526
Collagenase,
electrophoresis buffers for, 527
Collision broadening, 618–619
Colloid mills, 8, 54
Column chromatography,
of ribosomal protein, 147
Column design,
in isoelectric focusing, 600–602
Column resistance, 606
Complex coacervates, 387
Complex ion dissociation, 585
Computer analysis,
reflectance spectra of, 622–623
Condensed systems,
absorption band width and, 618–619
Conductance gradient, 598
Contact staining,
in electrophoresis, 496, 501
Continuous action rotors, 57
Continuous analysis,
by free flow electrophoresis, 513–546
Continuous culture, 132
Continuous deflection electrophoresis, 516, 521
Continuous electrophoresis, 565

Continuous flow automatic analysers, 211–212
Continuous flow rotors, 128
Controlled overflow method, 95–98
Convection current effects, 596–597
Convex gradients, 91
Cooled metal plates,
    in electrophoresis, 473–475
Cooling water,
    in iso-electric focusing, 601–602, 606
Coomassie Blue R. 250, 587
Copper complex method,
    amino-acids by, 255–258
Corynebacterium sp.,
    polyphosphate synthesis in, 4
Corynebacterium diphtheriae,
    polyphosphate in, 330
C. xerosis,
    disintegration of, 26
    polyphosphate in, 194, 330
Cotton seed oil, 484
Coulter counter, 3
Counter-current distribution (CCD),
    antigen–antibody complexes, of, 418–419
    apparatus for, 412–413
    bacteria, of, 415–417
    virus, of, 417–418
Craig glass machine, 413
Critical point, 390–391
Cross flow electrophoresis, 499
Cross linked dextran gels, 562–563
Cup-shaped zones, 585
Current stabilization,
    in electrophoresis, 522
Current strength,
    in electrophoresis, 457–458, 464, 517
Cysteine-sulphuric acid reagent, 296
Cytochrome *c*,
    biphasic separation of, 395
    crystallization of, 448
    electrophoretic separation of, 467, 584
Cytochromes,
    Candida utilis, from, 618, 621–625
    extraction of, 376–378
    iso-electric focusing of, 612
Cytomembranes (*see also* bacterial cytomembranes), 148–189
    blue-green algae from, 178–183

Cytosine monophosphate,
    separation of, 469

D

Dactylium dendroides,
    galactose oxidase from, 283
Decanol, 447
Dehydrogenases,
    separation of, 467
Deionized agar, 501
Dekalin, 484
Density gradients,
    automatic analyser for, 96–97
    fractionator for, 95
    free flow electrophoresis, in, 575
    iso-density sedimentation, in, 112–115
    iso-electric focusing, in, 597–605
        stabilization of, 597–600
        preparation of, 603–605
    mixers for, 85–90
    rate zonal centrifugation, in, 83–98
        formation of, 85–90
        materials for, 83–85
        sampling of, 92–98
    zonal electrophoresis in, 550–559
        apparatus for, 555–556
        applications of, 558–559
        columns for, 554–555
        experimental method in, 555–558
        gravitational stability in, 550
Density inversions, 90–92
Density shelf, 557
Density transfer experiment, 115
6-Deoxyhexose determination, 297
Deoxymononucleotides,
    separation of, 533
2-Deoxypentose determination, 298
Deoxyribonuclease,
    DNA, on, 152
    isoelectric focusing of, 446, 609–610
Deoxyribonucleic acid (DNA),
    anthrone reaction with, 270–272
    base composition of, 309–310, 631–639
    biphasic distribution of, 393–401
    chemical analysis of, 310–328
        Burton's diphenylamine reagent, by, 317–320

Deoxyribonucleic acid (DNA)—*cont.*
　cysteine sulphuric acid reagent, by, 320
　micro-organisms, in, 322–328
　Ogur and Rosen method, by, 313, 315–316
　Schmidt and Thannhauser method, by, 310–311, 313–315
　Schneider method, by, 312–314
　ultraviolet absorption, by, 320–322
chemical composition of, 124, 311
equilibrium centrifugation of, 113–115, 118–121
extraction of, 326, 371–374
　detergent, by, 371–372
　perchlorate, by, 373–374
　trichloroacetic acid, by, 372–373
hybrids of, 114–115
hydrolysate, determination of, 469, 632–639
isolation of, 395, 399–401, 634
molecular weight of, 328
orcinol reaction with, 288, 290–291
phenol-sulphuric acid reagent with, 276
replication of, 115
Deoxyribonucleic acid phage,
　isolation of, 545
Deoxyribonucleic acid polymerase,
　purification of, 399, 537
Deoxyribonucleotides,
　separation of, 553–554
Deoxyribose,
　as endo-osmotic standard, 460
Desalting, 448
Detergent extraction, 371, 374–375
Dextran,
　endo-osmotic standard, as, 400
　phase-separation, in, 387, 392
Dextran-polyethylene glycol phase system, 389–405, 415–417
　antigen isolation by, 405
　bacterial distribution in, 403–404
　cell wall isolation by, 405
　Chlorella distribution in, 402
　counter current distribution, in, 415, 417
　DNA isolation by, 400–401
　DNA polymerase isolation by, 399

nucleic acid partition in, 394–399
partition in, 389
poly-β-hydroxybutyrate isolation by, 405
protein partition in, 394
protoplast isolation by, 405
spore isolation by, 405
virus distribution in, 405
Dextran-polyethylene glycol-water system,
　phase diagram of, 391
Dextran sulphate,
　polymer phase systems in, 387–388, 391–392, 405, 408, 410
Dichloromethane, 483
2,6-dichlorophenol-indophenol, 497
Diethanolamine, 603
Diethylaminoethyl cellulose (DEAE),
　ferredoxin extraction, in 378
　β-fructofuranoside preparation on, 609
　lipid separation on, 355–356
　properties of, 442–443
Diffuse reflectance, 622
Diffusion coefficient, 99, 100, 102
3,4-dihydroxyphenol alanine (DOPA), 497
Dimethyl sulphoxide, 434
2,4-dinitrophenylethanolamine, 568
6-dinitrophenyl lysine, 480
Dinucleotides,
　separation of, 470–471
Diphenyl alanine, 466
Disassembly sequences, 136
Disc electrophoresis, 573–592
　apparatus for, 578–580
　applications of, 591–592
　artefacts in, 585
　definition of, 591–592
　gel elution in, 590–591
　gel staining in, 586–590
　immunological technique in, 590
　principles of, 575
　procedure for, 580–586
　radiochemical techniques in, 591
　ribosomal protein separation by, 146
Discrete automatic analysers, 211–212
Doppler effect, 618–619
Dry mills, 6–7

Drying,
    microbial disintegration by, 41
Dumas elemental analysis, 216–217
Dye elution,
    in electrophoresis, 586

E

Eberthella typhi,
    disintegration of, 45
ECHO virus,
    concentration of, 410–412
ECTEOLA (mixed amines) cellulose,
    443
Effective density gradient, 116
Egg-white lysozyme, 43–45
Eimhjellen and Jensen method, 357
Elastase,
    purification of, 526, 536–537
Electrical capacity,
    in electrophoresis, 553
Electrical field strength,
    in electrophoresis, 457, 520
Electrical load,
    iso-electric focusing, in, 606
    zonal electrophoresis, in, 558
Electrode buffer,
    electrophoresis, in, 521
    iso-electric focusing, in, 605–606
Electrode design,
    electrophoresis, in, 461, 552–554, 559
    iso-electric focusing, in, 604
Electrodecantation, 492
Electroendosmosis,
    in electrophoresis, 459–460, 467, 500,
        574
Electrofocusing (*see under* iso-electric
        focusing)
Electrolysis cell,
    iso-electric focusing, in, 596
    dye elution, in, 586
Electrolysis products,
    in electrophoresis, 461, 521
Electromagnetic rotation, 559
Electro-osmotic flow,
    electrophoresis, in,
        free flow, 522
        starch gel, 561
Electrophoresis,
    agar gel, 499–503
    block, 503–508

cellulose acetate, 482–489
disc, 571–592
free-flow, 513–546
moving boundary, 456–457, 530
paper, 463–480
    high voltage, 472–480
    low voltage, 463–472
preparative zonal, 549–570
    density gradient, 550–559
    solid support, 559–569
starch gel, 489–499
theory of, 456–463, 513–516
thin layer, 481–482
Electrophoretic band shape, 585
Electrophoretic mobility, 514, 517, 523,
    575
Elemental analysis,
    of microbial cells, 216–242
Elson–Morgan reaction, 284
Elution apparatus,
    for disc electrophoretograms, 587, 591
Enolase,
    separation of, 467
Enrichment culture techniques, 427
Enteric bacteria,
    lipid analysis in, 308
Enterobacteriaceae,
    DNA extraction from, 372
    lipid in, 351
    polysaccharide production by, 360
Enzymes,
    activators, for, 436–438
    concentration of, 449
    crystallization of, 446–448
    denaturation of, 435–436
    desalting of, 448
    detection of, 589–590
    dialysis of, 448–449
    disintegrated cells, from, 4–5, 19–21
    fractionation of,
        acetone, by, 439–440, 447
        alumina, by, 441
        amberlite IRC-50, by, 442
        ammonium sulphate, by, 440, 447
        calcium phosphate gel, by, 441
        cellulose ion exchangers, by, 442–
            443
        electrophoresis, by, 467, 497, 499,
            507, 536, 591
        ethanol, by, 439–440, 447

Enzymes—*cont.*
hydroxyapatite, by, 441
iso-electric focusing, by, 445–446, 595–613
molecular sieve chromatography, by, 444–445
Sephadex ion exchanger, by, 443–444, 449
sequence of methods for, 449–450
inactivation of, 436–438
induction of, 427–428
location of, 5–6
nucleic acid removal from, 438–439
protectors for, 433–435
reactivation of, 430
stability of, 429–433
stabilization of, 433–435
susceptibility to hydrolysis of, 435–436
ultrafiltration of, 449
Epon centrepieces, 115, 118
Equilibrium sedimentation methods, 58–60, 102–128
(*see under* sedimentation equilibrium methods *and* iso-density sedimentation equilibrium method)
Erythrocytes,
separation of, 542
Escherichia coli
alkaline phosphatase from, 428
amino-acid separation in, 333
analytical ultracentrifugation of, 130–131
aspartate transcarboxylase from, 428
autolysis of, 42
colicins from, 611–612
counter-current distribution of, 415–416
cytomembranes from, 148, 156–157
disintegration of,
explosive decompression, by, 38
French press, by, 15–21
Hughes press, by, 23
ionizing radiation, by, 47
leucozyme, by, 45
lysozyme, by, 44
osmosis, by, 40
phagocytin, by, 45
ultrasonics, by, 21
wet mills, by, 25

X press, by, 23
enzyme extracts of, 19–21
enzyme separation in, 591
β-galactasidase from, 428
glutamic decarboxylase from, 430
glycogen from, 278–279
lipids from, 304, 350–351
nucleic acids in, 311, 373, 401
nucleotide extraction from, 332–333
polyribosomes from, 133
ribosomes from, 135–138, 143–147, 530, 538
cations on, 135–138
molecular weights of, 143–144, 147
mutant, 545–546
protein from, 145, 147
RNA in, 143–144
threonine deaminase from, 430, 436
tryptophanase from, 428
Esterases,
electrophoresis of, 497, 589
Ethanediol (ethylene glycol),
in density gradients, 554, 597, 603, 607
Ethanol,
density gradient, in, 550
enzyme crystallization, in, 447
enzyme protection by, 434
fractional precipitation, in, 437–440
lipid extraction by, 354–355
metabolite extraction by, 348–349
Ethanolysed cellulose, 561
Ethylene diamine, 603
Ethylene-diamine tetra-acetate (EDTA),
electrophoretic buffers, in, 459, 528
ribosomes, on, 133
Ethylene glycol (*see* ethanediol),
Euchromatin, 535
Euglena gracilis,
disintegration of,
glass beads, by, 30
homogenization, by, 23
proteolytic enzymes, by, 46
Explosive decompression,
cell disintegration by, 38–40, 53
Extinction coefficients,
for nucleic acid bases, 634
Extracellular proteins,
separation of, 559

"Extractable" lipid,
extraction of, 351–353
Extrusion methods,
for density gradient sampling, 93–95

F

Fast blue dyes, 589
Fatty acid hydroxylation,
in *Pseudomonas oleovorans*, 438
Ferredoxin, 377
Ferritin, 584
Ferrobacillus ferro-oxidans,
poly-β-hydroxybutyrate in, 189–190
Fibrinogen, 574
Ficoll,
density gradient material, as, 84
phase separation, in, 387
Filters,
microbial disintegration by, 27
Finger printing,
electrophoresis in, 478, 482
Flavine,
extraction of, 348
nucleotides, separation of, 471
Fluorescent lamp,
photopolymerization by, 579
Folin–Ciocalteu method, 249–252
Fomis anosus,
enzyme separation in, 608
Foot and Mouth disease virus,
preparation of, 412
Formaldehyde,
anthrone reaction with, 272
electrophoresis, in, 466
Formamide extraction,
for polysaccharides, 367–368
Formic acid,
in electrophoresis buffers, 493, 524
Fraction collection,
in iso-electric focusing, 607
Free amino-acids,
estimation of, 334–335
identification of, 333
Free-flow electrophoresis, 516–543
apparatus for, 516–523
bacteria of, 544–546
buffers, for, 524–531
phages, of, 543–544
separation media for, 523–532
soluble biological materials, of, 532–537
subcellular particles, of, 537–540
whole cells, of, 540–543
Freeze pressure,
microbial disintegration by, 12–24
Fremyella diplosiphon,
lysozyme on, 179
French press,
cell disintegration in, 15–22, 53
enzyme activity, on, 15, 19–20
Frictional coefficient, 68–69
Frictional force 68
Frictional resistance, 514
Friedrich–Kjeldahl titration method,
221–222
β-fructofuranosidase (invertase),
purification of, 467, 608–609
D-fructose,
anthrone reaction with, 269–271
orcinol reaction with, 287–288
phenol-sulphuric acid reagent with,
276
L-Fucose,
anthrone reaction with, 269–271
phenol sulphuric acid reagent on, 276
Fujita's equation, 76
Fumarase,
in *E. coli* extracts, 19–20
Fungal enzymes,
iso-electric focusing of, 607–608
Fuchsin sulphite, 465

G

Galactosamine determination, 283–285
Galactose,
anthrone reaction with, 269–271
determination of, 283
orcinol reagent with, 287–288
phenol-sulphuric acid reagent with,
276
β-Galactosidase,
from *E. coli*, 428
β-Galactosides,
electrophoresis of, 589
"Galactostat", 283
Galacturonic acid,
determination of, 292–293
phenol-sulphuric acid reagent on, 276

Gas chromatography,
 in lipid analysis, 309
Gauss distribution,
 in reflectance spectrophotometry, 620
Gel cutter, 493
Gel-filtration media, 569
Gel pore size, 574
Gel slicing, 496
Geons, 506
"Ghosts",
 bacterial cytomembranes from, 149,
  152, 157–158
Giovanelli distribution,
 in reflectance spectrophotometry,
  619–620
Glass beads,
 microbial disintegration by, 25, 30
Glass powder,
 electrophoresis in, 506, 563
Gleocapsa sp.,
 cytomembranes in, 178
Globulins,
 molecular dimensions of, 574
 separation of, 460, 466–467, 482,
  532–533, 613
Glucans,
 determination of, 279–281
 extraction of, 363
Glucosamine,
 determination of, 284–285
Glucose,
 determination of,
  anthrone reagent, by, 267–271
  glucose oxidase, by, 282
  phenol-sulphuric acid method, by,
   272–276
 electrophoresis in, 529
 orcinol reagent, with, 287–288
Glucose-6-phosphate dehydrogenase,
 purification of, 434
β-glucosidase,
 iso-electric focusing of, 607
glucuronic acid,
 anthrone reaction with, 269–271
 determination of, 292–293
 phenol-sulphuric acid reagent on, 276
 orcinol reagent, with, 288
Glutamate decarboxylase,
 Clostridium welchii, in, 4
 Escherichia coli, in, 436

Glutamic aspartic transaminase,
 stability of, 439
Glutamic dehydrogenase,
 from Neurospora crassa, 429
Glutaminase,
 in Clostridium welchii, 4
D-Glycero-D-guloheptose,
 anthrone reaction with, 269–271
 phenol-sulphuric acid reagent on, 276
Glycerokinase,
 purification of, 434
Glycerol,
 density gradients, in, 550, 597, 603–
  605, 608–609
 electrophoresis, in, 486
 enzyme protection by, 433–434
 rate-zonal centrifugation, in, 85
Glycine-NaCl-NaOH buffer,
 in electrophoresis, 528
Glycogen,
 alkali extraction of, 362–363
 determination of, 278–281
Glycogen phosphorylase,
 rabbit muscle, from, 430
 activity of, 436
Glycine,
 density gradient, in, 554
 electrophoretic buffer, in, 500, 528
 iso-electric focusing, in, 603
Glycoprotein,
 electrophoretic separation of, 465–
  466, 487–488, 502, 506
Gradients (see under density gradients)
Gram-negative bacteria,
 cytomembranes from, 156–161
 extraction of, 364–367
 lipid A, in, 351, 353
Gram-positive bacteria,
 cytomembranes from, 151–156
 extraction of, 367
Granulocytes,
 electrophoretic separation of, 542
Grinding,
 microbial disintegration by, 30–31
Guanosine monophosphate (GMP), 469
Guanidoethyl cellulose, 443

H

Haemins,
 extraction of, 349–350

Haemoglobins,
electrophoresis of, 489, 568
α-Haemolysin,
separation of, 609–610
Haemoproteins,
extraction of, 376–368
Halobacterium sp.,
cell envelopes from, 160–161
Halobacterium cutirubrum,
ribosomes from, 137
H. salinarium,
enzymes from, 435
Hartmanella castellanii,
disintegration of, 22–23, 27
Heat evolution,
in electrophoresis, 472–473
Heavy metal ions,
in electrophoresis, 523
Helix pomatia,
cell disintegration by, digestive juices
of, 45–46
Heparin,
separation of, 535
Hepatic ribosomes,
molecular weight of, 148
Heptane,
as electrophoresis coolant, 476
Heptose,
determination of, 296–297
phloroglucinol reagent, with, 292
Heterochromatin, 535
Hexosamine,
determination of, 283–285
Hexose,
determination of, 267–276
orcinol reagent, with, 288, 290
phloroglucinol reagent, with, 292
Hexuronic acids,
determination of, 292–293
orcinol reagent, with, 288, 290
phloroglucinol reagent, with, 291
High voltage electrophoresis (*see also*
paper electrophoresis), 472–
480
Histochemical tests,
enzyme detection by, 467
Homogenizers, 20, 22–24, 53
Homogentisicase,
from bovine liver, 434
Horizontal electrophoresis, 460, 490, 492

Hormone separation, 507
Hughes press, 12–13, 19, 23, 31–37, 52
cell wall membranes, on, 34–37
description of, 31–32
efficiency of, 23, 33–34
weight dropping device for, 34
Hyaluronate lyase,
separation of, 609–612
Hyaluronic acid,
separation of, 535
Hydration,
of macromolecules, 71
Hydraulic press, 53
Hydraulic shear,
cell disintegration by, 10–12
Hydrochloric acid,
in electrophoresis, 493
Hydrogen,
determination of, 216
Hydrogenomonas sp.,
poly-β-hydroxybutyrate in, 189–
190
Hydrophobic ladder, 388
Hydrosulphite,
on enzyme crystallization, 447
Hydroxapatite, 441
β-Hydroxybutyrate,
determination of, 193
β-Hydroxybutyric acid dehydrogenase,
from *Rhodospirillum rubrum*, 430
N-2-hydroxyethylpiperazine-N′-2-
ethane sulphonic acid,
enzyme stability in, 435
N-β-hydroxy-myristyl glucosamine,
353
Hydroxy-propyl dextran,
in phase separation systems, 387–
388
Hydroxy-steroid dehydrogenases, 434
Hyperchromic effect, 321
Hypobromite oxidation,
aldose determination by, 294–295
Hypoiodite oxidation,
aldose determination by, 293–294

I

"Ideal" particles, 56
Immuno-diffusion (*see* immuno-electrophoresis)

Immuno-electrophoresis,
  agar gels for, 502–503
  disc electrophoresis for, 591
  starch gels for, 499–500
Immunoglobulins,
  separation of, 613
Infinite dilution,
  electrophoretic theory and, 514
Inorganic polyphosphates,
  acid extraction of, 369–370
  hypochlorite digestion of, 371
  micro-organisms, in, 368–371
Inorganic salts,
  in micro-organisms, 349
Integrating sphere,
  in reflectance spectrophotometry,
    621–622, 625–628
Interference fringe pattern, 74, 80
Intermediary metabolites,
  extraction of, 347–350
Internal standards,
  in disc electrophoresis, 584
Intra-cytoplasmic membrane, 148, 157,
    162–163
Ion-exchange,
  cellulose,
    sRNA purification by, 376
  chromatography,
    haemoprotein fractionation by, 377
    polysaccharide fractionation by,
      281
  membranes,
    electrophoresis in, 521–522
  resins,
    amino-acid analysis by, 333
Ion-focusing electrophoresis, 456
Ionic charge,
  electrophoresis and, 457
Ionic environment,
  effect on bacterial ribosomes, 134–138
Ionic strength,
  electrophoretic effects in, 458–459,
    514, 555
  iso-electric points, effects on, 599
Ionising radiation,
  microbial disintegration by, 47
Iron determination, 239–242
Isco gradient analyser, 94
Isocitrate dehydrogenase,
  in E. coli extracts, 19–21

"Iso-cytochromes", 613
Iso-density sedimentation equilibrium
    methods, 58–60, 103, 112–128
  analytical, 115, 118–120
  buoyant densities by, 120–124
  DNA analysis by, 114–115, 118–119,
    121, 124
  molecular weights by, 114, 116–117
  practical aspects of, 116–128
  preparative, 124–128
  RNA analysis by, 114–115, 118
  solvents for, 112–113, 117–118
  virus, preparation by, 115, 117
Iso-electric focusing,
  carrier ampholytes, in, 445–446
  deoxyribonuclease, of, 446
  protein separation, for, 445–446,
    595–613
    applications of, 607–613
    chemicals for, 601–603
    electrolysis in, 596
    equipment for, 601–603
    practical techniques in, 603–607
    theory of, 597–600
Iso-electric point, (pI),
  bacterial proteins, of, 612–613
  bacteriolytic proteins, of, 611–612
  cellulolytic enzymes, of, 607
  definition of, 596
  determination of, 599
  β-fructofuranosidase, of, 608–609
  immunoglubulins, of, 613
  iso-electric focusing, by, 595–613
Iso-electric range,
  electrophoresis and, 523
Iso-electric spectra,
  of β-fructofuranosidase, 608–609
Iso-kinetic gradients, 88, 99
Iso-octane extraction, 357
Iso-osmotic medium,
  electrophoresis in, 528
Iso-prenoid compounds, 351
Iso-pycnic banding, 128
Iso-pycnic centrifugation, 58–60

J

Johnston–Ogston effect, 75
"Joule effect", 520
Joyce–Loebl double beam micro-den-
    sitometer, 118

Joyce–Loebl "Meccolab" autoanalyser, 212

K

Kel F centre pieces, 115, 118
Ketohexose determination, 295–296
Kidney bean proteins, 527
Kjeldahl method, for total nitrogen analysis, 217–224
   distillation, by, 218–221
   modifications of, 218–219
   Nesslerization, by, 222–224

L

Lactic dehydrogenase,
   crystallization of, 447
   electrophoresis of, 497, 589
   proteinases on, 435
Lactobacillus arabinosus,
   disintegration of, 23
   teichoic acids from, 329
L. plantarum,
   protein purification from, 439
Law and Slepecky method, 305–307
Leptospira pomona,
   disintegration of, 40
Leuco-malachite green, 487
Leucozyme,
   E. coli disintegration by, 45
Light green dye, 465
Lipase,
   electrophoretic spearation of, 467
   iso-electric focusing of, 611
Lipid,
   content, micro-organisms of, 350–351
   cytomembranes, in, 155
   determination, 302–308
     chromatography, by, 307–308
     Law and Slepecky method, by, 305–307
     total, 304
   extraction of, 303–304, 350–358
     acetone, by, 352–353, 357
     butanol, by, 355–356
     chloroform-methanol, by, 351–352
     ethanol-ether, by, 354–355
     iso-octane, by, 357–358
     methanol, by, 356
     petroleum, by, 352–353

phenol-water, by, 354
   Soxhlet technique, by, 353
"Lipid A",
   extraction of, 353–354
   Gram-negative bacteria, in, 351
   ultracentrifugation of, 364–366
Lipid crimson dye, 502
Lipid granules,
   extraction of, 359
   micro-organisms, in, 189
Lipid-oligosaccharides,
   extraction of, 354–356
Lipopolysaccharides,
   extraction of, 364–366
   purification of, 365
Lipoproteins,
   separation of, 488, 507, 599
   stains for, 465, 489, 497, 502
Liquid cooling,
   in high voltage electrophoresis, 475–477
Liquid–liquid distribution, 386, 414
Liquid-interface distribution, 414–415
Liquid paraffin,
   cellulose acetate clearing by, 484
Lithium chloride, 145
Lymph node cells,
   electrophoresis of, 531–542
Lysine-ribonucleic acid synthetase,
   separation of, 591
Lysosomes,
   separation of, 530, 538
Lysozyme,
   bacterial cell wall, on, 133, 151–156
   biphasic separation of, 394
   blue-green algae, on, 179
   cell lysis by, 354
   iso-electric focusing of, 612
   stabilization of, 435

M

Macromolecule finger prints, 478
Magnesium acetate-acetic acid buffer,
   in electrophoresis, 530
Magnesium acetate-citric acid buffer,
   in electrophoresis, 530
Magnesium ions,
   cytomembranes, on, 152
   determination of, 235–239
   enzyme crystallization, in, 447

Magnesium oxide,
  in integrating spheres, 622, 625–628
Magnetostrictors, 9
Malate dehydrogenase,
  in *E. coli* extracts, 19–21
Mammalian tissue,
  disintegration of, 38–39
Manganese ions,
  enzyme stabilization by, 435
Mannan,
  determination of, 279–281
  extraction of, 363, 368
Mannase,
  iso-electric focusing of, 607
Mannose,
  anthrone reagent with, 269–271
  orcinol reagent with, 287–288
  phenol-sulphuric acid reagent with, 276
Mannosidase,
  iso-electric focusing of, 607
Manometric ninhydrin-carbon dioxide method, 256, 259–260
Manton Goring homogenizer, 20, 53
"Melinex" sheet, 494
Membrane fraction,
  from Gram-positive bacteria, 151–156
Membrane proteins,
  separation of, 465
Mercaptoacetate, 585
Mercaptoethanol,
  enzyme crystallization, in, 447–448
  toxin separation, in, 539
Mercuric chloride,
  in protein electrophoresis, 464
Mercuric ions,
  in enzyme crystallization, 447
Merismopedia glauca,
  cytomembranes in, 178
Mesosomal elements, 148–150, 153, 157, 162
  B. megaterium, from, 153
  Chlorobacteriaceae, in, 162
  E. coli, in, 157
  photosynthetic bacteria, in, 150
Metachromatic granules, 193–197
Metalloproteins,
  extraction of, 376
Metapyrocatechase, 430–431, 434

Methanol extraction,
  for carotenoids, 356
Methyl cellulose,
  phase separations in, 388
Methyl ethyl ketone, 472
5-methyl pentose determination, 297
Methylated albumin,
  in RNA fractionation, 142
N,N'-methylenebis-acrylamide, 575
Michl's apparatus, 475
Mickle shaker, 25, 53
Micro mill, 30
Microbial disintegration, 1–53
  autolysis, by, 42
  bacteriophage, by, 47
  cavitation microstreaming in, 9–10
  Chaikoff press, by, 20–22
  chemical methods, by, 40–47
  choice of methods for, 47–48
  crushing and grinding, by, 6–7, 30–31
  digestive enzymes, by, 45–46
  drying and extraction, by, 41–42
  enzyme location after, 5–6
  enzymic, 43–46
  explosive decompression, by, 38–40
  filters, by, 27
  French press, by, 13–22
  homogenizers, by, 22–24
  Hughes press, by, 12–13, 31–37
  hydraulic shear, by, 10 12
  ionizing radiation, by, 47
  leucozyme, by, 45
  liquid shear, by, 13–30
  lysozyme, by, 43–45
  measurement of, 3–4
  mechanical, 6–13
  osmosis, by, 40–41
  phagocytin, by, 45
  proteolytic enzymes, by, 46
  solid shear, by, 30–40
  Streptomyces enzymes, by, 45
  ultrasonics, by, 9–10, 27–30
  vibration mills, by, 25–26
  wet mills, by, 7–9, 24–25
Microbial powders, 6–7
Micrococcus aureus,
  disintegration of, 23
M. halodentrificans,
  poly-β-hydroxybutyric acid in, 304

M. lysodeikticus,
 bacteriolytic activity from, 612
 cytomembranes from, 152–153
 disintegration of, 45
 lipids of, 155
 nucleic acids from, 311
 polyphosphate in, 194, 196–197
 RNA,
  content of, 155
  polymerase from, 434
M. pyogenes, mutants of, 544
Micropipette, 464, 486, 544
Microsomes,
 separation of, 538
Migration velocity, 574–575
Miscibility,
 of polymers, 386
Mitochondria,
 DNA from, 328
 purification of, 530, 538
Mobility, 514
Molecular charge, 596
Molecular sieves,
 chromatography, in, 444–445
 electrophoresis, in, 457, 489, 562
 iso-electric focusing and, 609
Molecular weight determinations,
 ribosomal RNA, of, 143–145
 sedimentation equilibrium methods,
  by, 103–112, 114, 116–117
 sedimentation velocity methods, by,
  80–81, 99
Molisch test, 266
Monocyte,
 separation of, 542
Mononucleotides,
 separation of, 469–471, 478
Morphogenesis,
 disc electrophoresis and, 591
Mouse serum proteins, 482
Moving   boundary   centrifugation
  methods, 60–81
 analytical, 63, 72–81
  absorption optical system for, 72,
   78–80
  apparent diffusion coefficient by,
   76–78
  boundary position in, 73
  cell for, 63
  molecular weights by, 80–81

 refractometric methods in, 74–78
 rotor for, 63
 sedimentation coefficient by, 75
 preparative, 60–72
  angle rotor for, 60–61
  frictional coefficient in, 68–69
  medium in, 69–71
  performance index, 65, 67
  sedimentation coefficient in, 65–67,
   70
  sedimentation constants in, 66–
   67
  swinging bucket rotor for, 62, 64
  wall effects in, 64–65
  zonal rotor for, 62, 64
Moving boundary electrophoresis, 456–
  457, 530, 550
Moving zone methods (*see under* rate
  zonal centrifugation methods)
Mucopolysaccharides,
 separation of, 524, 535–536
Multiphase polymer systems, 387
Mutant enzymes,
 electrophoretic detection of, 591
Mycobacterium sp.,
 disintegration of, 15, 23
 glycolipids in, 351, 354–355
Mycobacterium phlei, 23
M. smegmatis, 23
M. tuberculosis var. *bovis*, B.C.G., 15–
  16
Mycoplasma sp.,
 plasma membrane from, 157–159
Mycoplasma bovigenitalum, 159
M. gallinarum,
 disintegration of, 40
M. laidlawii, 159
Mycloperoxidase, 527
Myosin digest,
 separation of, 482
Myxomycetes,
 disintegration of, 22
Muys mill, 25

                    N

Naphthalene black,
 as protein stain, 465, 487, 496, 586
β-naphthoquinone sulphonate, 255
1-Naphthylacetate, 589

1-Naphythyl disodium orthophosphate, 497

β-Naphthyl phosphate, 589

Nelson's arsenomolybdate reagent, 301

Nepton, 522

Neurospora sp.,
 disintegration of, 30, 46

Neurospora crassa,
 amino-acid pool in, 333
 disintegration of, 46
 glutamic dehydrogenase from, 429

Nicotinamide coenzymes,
 separation of, 471–472, 497

Nigrosine,
 as protein stain, 487, 496, 502

Ninhydrin reaction, 253–255

Nitrobacter agilis,
 poly-β-hydroxybutyrate in, 189

Nitrobacteriaceae,
 cytomembranes from, 177–178
 poly-β-hydroxybutyrate in, 189

Nitro blue tetrazolium,
 as dehydrogenase stain, 497, 589

Nitrocystis oceanus,
 cytomembranes from, 177–178

Nitrogen,
 determination of, 217–224, 243
 fixation of, 430–431, 437
 release, in cell disintegration, 15, 22

Nitrosomonas sp.,
 cytomembranes in, 177

Nitrous acid,
 amino-acid determination by, 255

Nodes,
 in phase diagrams, 390

Non-buffering salts,
 electrophoresis and, 523

Non-ionic polymers,
 in phase separation systems, 387

Non-monochromatic light,
 reflectance spectrophotometry and, 619–620

Non-specific absorption,
 in reflectance spectrophotometry, 617

Nossal shaker, 25

Nostoc muscorum,
 extract of, 179

Nuclear DNA,
 separation of, 328

Nucleic acid,
 anthrone reaction with, 270–272
 bases,
  composition of, 631–639
  identification of, 333–334
 composition of, 309, 311
 determination of, 308–328
  difficulties in, 313–316
  diphenylamine method, by, 317–320, 325–326
  micro-organisms in, 322–328
  Ogur and Rosen method, by, 313, 315–316
  orcinol method, by, 316–317, 325
  Schmidt–Thannhauser method, by, 310–311, 313–315, 322, 324
  Schneider method, by, 312, 314, 323
  sugar analysis, by, 316–320
  ultraviolet absorption, by, 320–322
 extraction of, 371–376, 632
  detergent, by, 371–372, 374–375
  perchlorate, by, 373–374
  phenol, by, 375–376
  trichloroacetate, by, 372–373
 fractions, determination of, 328
 hydrolysis of, 633
 hyperchromic effect in, 321
 microbial proteins, effect on, 263
 phenol-sulphuric acid reagent, effect on, 276
 polymer-phase distribution of, 393–401
 pool, 331–336
 separation of,
  electrophoresis, by, 500, 507, 533–535
  proteins, from, 438–439
 stains for, 590
 zonal centrifugation of, 84–85

Nucleohistone,
 separation of, 535

Nucleosides,
 electrophoresis of, 524–525
 extraction of, 348
 extinction coefficient of, 634
 identification of, 537

Nucleotides,
 enzyme activators, as, 436–437
 extraction of, 348–349

Nucleotides—*cont.*
  extinction coefficients of, 634
  identification of, 333–334
  pyrophosphatase, purification of, 537
  separation of, 468–470, 520, 524–525,
    533–535
  sugar peptides in, 348
  sugars in, 348
Numerical analysis,
  in reflectance spectrum, 622–623
Nutrients,
  electrophoresis in, 529
Nylon powder,
  as electrophoresis support, 563, 566

O

Ogur and Rosen method, 313, 315–316
Ohms Law, 458
Oil red dyes, 497
Oligonucleotides,
  electrophoresis of, 478, 533
Orcinol method, 285–291, 316–317
  interference in, 287–288, 290
  microbial extracts, for, 291
  pentose, for, 285–291
  RNA, for, 316–317
  specificity of, 287–288
Organic phosphate,
  extraction of, 349
Orthophosphate,
  determination, 331
Oscillatoria tenuis,
  lysozyme on, 179
Osmosis,
  cell disintegration by, 40–41
Overlapping reflectance bands, 622–623
Oxidosomes,
  from ribosomes, 405
Oxygen labile enzymes,
  isolation of, 430–433

P

Paper chromatography,
  analysis by,
    amino-acid, 333
    lipid, 307
    nucleic acid, 634

Paper electrophoresis, 463–481
  high voltage, 472–480
    aparatus for, 473–477
    applications of, 478–480
    cooled metal plate, 473–475
    current source in, 477–478
    liquid cooled, 475–477
  low voltage, 463–472
    amino-acid separation by, 468
    carbohydrate separation by, 472
    nucleotide separation by, 468–472
    protein separation by, 463–468
Paper powder,
  electrophoresis in, 564
Para-amino benzyl cellulose, 443
Partial specific volume, 116
Particulate cell components,
  bacterial extracts, in, 56
  separation of, 558
Particulate gels, 562–563
Partition coefficient,
  in polymer phase systems, 395–398
Peltier batteries, 517, 519–520
Penicillin, 43
Penicillium sp.,
  nucleotides in, 331
Pentose determination,
  orcinol, by, 285–291
  phloroglucinol, by, 291–292
Pepsin,
  separation of, 467
Peptides,
  extraction of, 348–349
  separation of, 468, 478, 481, 499, 524,
    595–613
Perchloric acid,
  extraction with, 348, 369–370, 373–
    374
    nucleic acid, for, 373–374
    polyphosphates, for, 369–370
  nucleic acid hydrolysis by, 633
Peristaltic pump, 88, 519, 607
Peroxidase separation, 467
Petroleum extraction, 352
Pevikon C 870, 499, 506, 563, 566
Pfluger's method, 278
pH,
  electrophoresis in, 458–459, 523
  gradients, 597, 601–602
  meters, 601, 607

Phage,
exonuclease from, 399
RNA polymerase from, 399
separation of, 89, 529, 544–545
Phagocytin, 45
Phase diagrams, 390–392
Phase polymers, 386–387, 419
Phase separation,
in polymer mixtures, 386–388
Phenol,
anthrone reaction with, 272
electrophoresis buffer, in, 524
extraction with, 354, 363–366
Phenol-sulphuric acid method,
carbohydrate for, 272–277
specificity of, 276
Phloroglucinol method,
pentose by, 291–292
RNA by, 317
Phormidium luridum,
disintegration of, 179
protoplasts from, 182
Phosphatases,
detection of, 589
Phosphate,
detection of, 469, 497
determination of, 224–229
electrophoretic buffers, in, 459, 493,
500, 529, 585
separation of, 471
Phosphate-polyethylene glycol phase
system, 405
Phosphate-sucrose buffer, 531
Phospholipase C,
separation of, 559
Phospholipid,
in cytomembranes, 148, 156
Phosphomethyl cellulose, 443
Phosphoprotein,
separation of, 558
Phosphopyruvate kinase,
crystallization of, 447
Phosphoric acid,
in iso-electric focusing, 606
Phosphorus-deficient medium, 369
Phosphorylase,
crystallization of, 447
electrophoresis of, 589
stabilization of, 434
Phosphorylated cellulose, 442–443

Photopolymerization,
of electrophoresis gels, 579
Photosynthetic bacteria,
cytomembranes from, 150, 161–177
sub-chromatophore particles from,
183–189
Photosynthetic vesicles,
from Chlorobacteriaceae, 164–166
Phycoerythrin, 395
Piezo-crystalline transducers, 9
Pigment separation, 529
Plant material,
fractionation of, 524
Plasma albumins (*see also* albumins),
crystallization of, 447
Plasma membrane, 148–150, 156–159
B. megaterium, in, 149
Gram-negative bacteria, in, 156–157
Mycoplasma sp., from, 159
Strep. faecalis, in, 150
Platinum electrodes, 553–554, 579
Plectonema calothricoides,
disintegration of, 179
Plysarum polycephalum,
enzyme separation in, 591
Polio virus,
antibody complex, 419
concentration of, 406, 410–412, 417–
418
protein separation from, 592
Polyacrylamide gel (*see also* disc electrophoresis),
chromatography, in, 444
electrophoresis in, 456–457, 468, 497,
567
polymerization of, 578–581
recipes for, 576–577
structures of, 575
Polyamines,
effect on ribosomes, 138
Polyaminopolycarboxylic acids,
as carrier ampholytes, 597, 601–
602
Polyelectrolytes,
in phase separation systems, 387–
388
Polyethylene cups, 578
Polyethylene glycol,
in phase separation system, 387–388,
392

Polyethyleneimine cellulose, 443

Poly-β-hydroxybutyrate,
  depolymerase for, 192
  determination of, 304–307, 358
  extraction of, 191–192, 359
  purification of, 405

Polymer phase separations, 385–421
  cell particles for, 401–405
  counter current distribution, in, 412–419
  nucleic acid, of, 393–401
  partition coefficient for, 396–398
  phase diagrams for, 390–392
  polymers for, 387, 392
  protein, of, 393–401
  virus purification by, 407–412

Polyphosphate,
  determination of, 330–331
  extraction of, 193–197
  overplus reaction for, 369
  synthesis of, 4

Polypropylene,
  in polymer phase systems, 388

Polyribosomes,
  bacterial, *see under* bacterial poly-ribosomes.

Polysaccharides,
  determination of, 278–281
  extraction of, 359–368
    acetic acid, by, 366
    alkali, by, 362–363
    citrate, by, 368
    ethylene diamine, by, 368
    formamide, by, 367–368
    phenol, by, 363, 366
    trichloroacetic acid, by, 366–367
    water, by, 360–362

Polystyrene,
  cell disintegration and, 3
  moving boundary sedimentation, in, 65

Polytomella sp.,
  disintegration of, 23

Polytomella caeca, 22–23, 27

Polyvinyl alcohol,
  in polymer phase systems, 388

Polyvinyl chloride,
  in electrophoresis, 566

Polyvinyl pyrrolidone,
  as density gradient, 85

Polyvinyl sulphate,
  in electrophoresis, 538

Ponceau S, 487

Porath column, 565

Potassium,
  determination of, 233
  ribosomes, effect on, 135, 137

Potassium bromide,
  as density gradient, 117

Potassium ferricyanide,
  in electrophoresis gels, 585

Potassium iodide, 465

Potato starch, 503, 560–561

Potter–Elvehjem homogenizer,
  cell disintegration, in, 22
  electrophoresis gels and, 489

Power packs,
  electrophoresis, for, 460, 477–478
  iso-electric focusing, for, 601

Precipitin arcs, 590

Precipitation time, 67, 69

Preparative electrophoresis, *see also* zonal electrophoresis,
  advantages of, 569–570
  free-flow, 516–521
  gel, in, 592
  zonal density gradient, 550–559
    experimental method in, 555–559
    theory of, 550–555
  zonal solid support, 559–560
    experimental method in, 568–569
    instruments for, 563–568
    media in, 559–563

Primary charge effect, 72

Procion blue, 487

Prolamine, 524

Protamine sulphate, 438–439

Proteinases (proteases),
  enzyme susceptibility to, 436
  separation of, 609–610
  stabilization of, 435

Proteins,
  analyser for, 607
  cytomembranes, from, 148, 155
  denaturation of, 426
  determination of, 242–265, 304, 398
    biuret, by, 244–249
    copper complex, by, 255–258
    Folin–Ciocalteu, by, 249–252
    Kjeldahl, by, 243–244

Proteins—*cont.*
β-naphthoquinone sulphonate, by, 255
ninhydrin-carbon dioxide, by, 253–255
nitrous acid, by, 255
ultraviolet absorption, by, 262–264
digest, fractionation of, 524
dimensions of, 574
extraction, 264–265, 348, 376–378
acetate, by, 376–377
acetone, by, 377–378
acid, by, 378
citrate, by, 377
urea, by, 377
heterogeneities in, 598
oxidation of, 430
polymer phase distribution of, 387–388, 393–401
purification of, *see under* separation of
reactivation of, 430
ribosomes, in, 137, 145–147
separation of,
acetone, by, 439, 447
ammonium sulphate, by, 440, 447
chromatography, by, 441–445
concentration in, 449
crystallization, by, 446–448
desalting, by, 448
dialysis, by, 448–449
electrophoresis, by,
agar gel, 499, 501, 502
block, 507
cellulose acetate, 482, 486, 487, 489
disc, 573–592
density gradient, 555–559
elution in, 499
free-flow, 520, 524, 532–533, 536–537
paper, 462–468
polymers in, 499
solid support, 559–570
starch gel, 497–499
sub-units, in, 499
ethanol, by, 439, 447
fractional absorption, by, 440–441
fractional denaturation, by, 439
ion-exchange resins, by, 442–444

iso-electric focusing, by, 445–446
molecular sieve chromatography, by, 445–446
nucleic acid removal, in, 438–439
polysaccharide complex, in, 524
preparative, 520–524
sequence of methods in, 449–450
ultracentrifugation, by, 449
stains for, 426, 496–497, 502, 586, 589
structure of, 426
Proteolytic enzymes,
cell disintegration by, 46
β-fructofuranosidase, effect on, 609
Proteus sp.,
disintegration of, 44
Proteus morganii,
disintegration of, 23
Protocatechuate 3,4-dioxygenase,
crystallization of, 447–448
Pseudomonas aeruginosa, from, 427–428
Protoplasts, 153
Prototheca zopfii,
disintegration of, 23, 30
Pro-toxin, 533
Pseudomonadaceae,
poly-β-hydroxybutyrate in, 189
DNA extraction in, 372
Pseudomonas sp.,
benzyl alcohol dehydrogenase from, 434
disintegration of, 44
enzyme crystallization from, 447
metapyrocatechase from, 434
poly-β-hydroxybutyrate in, 304
viability loss in, 213
Pseudomonas aeruginosa,
cell envelopes from, 158
protocatechuate dioxygenase from, 427–428
Ps. fluorescens,
cell envelopes from, 158–159
cytochromes in, 376
electrophoretic separation of, 545
protein determination in, 250–251
Ps. NCMB 845,
cell envelopes from, 160
Ps. oleovorans,
fatty acid hydroxylation in, 438

Ps. putida,
 benzene metabolism in, 437–438
 lysis of, 152
Ps. saccharophilia,
 galactose dehydrogenase from, 283
 poly-β-hydroxybutyrate in, 189
Ps. solanacearum,
 poly-β-hydroxybutyrate in, 189
Ps. testosteroni,
 3-α-hydroxy steroid dehydrogenase
   from, 434
Pyrex powder, 30
Pyridine,
 in electrophoresis buffers, 524
Pyridoxal derivates,
 separation of, 471–472
Pyrophosphatase,
 electrophoresis buffers for, 526
Pyrophosphate,
 separation of, 471
Pyruvate,
 adenylyl cyclase activation by, 437
Pyruvate carboxylase, 430, 437

Q

Quinones,
 extraction of, 357–358

R

Radial dilution correction, 77
Radiation dampening, 618
Radioautographic estimation, 496
Radio-chemical techniques,
 in disc electrophoresis, 591
Rat liver mitochondria,
 velocity sedimentation of, 70
Rat serum proteins,
 separation of, 482
Rate-zonal centrifugation methods, 81–
   102
 analytical, 83
 band sedimentation in, 90–93, 101–
   102
 density inversions in, 90–92
 diffusion coefficient by, 99–100
 gradients for, 83–85, 89, 91
 gradient forming devices for, 85–90
 molecular weights by, 99

preparative, 83
 resolution in, 83–85, 97
 zone location in, 91–98
Rayleigh interferometer, 80, 105–108
Reducing sugar,
 determination of, 298–301
Reflectance,
 cell, 621
 definition of, 617
Reflectance spectrophotometry, 617–
   625
 absorption coefficient in, 617
 cytochromes, estimation of, 621–625
 instruments for, 625–628
 introduction to, 616–617
 measured band shape in, 619–621
 non-specific absorption in, 617–618
 scattering coefficient in, 617
 theory of, 617–619
 true band shape in, 618–619
Reflection,
 in reflectance spectrophotometers,
   625
Refractive index,
 electrophoresis and, 456
 velocity sedimentation in, 74–78
Resolving power,
 free flow electrophoresis, in, 522
 iso-electric focusing, of, 598
 zonal electrophoresis, of, 558
Resorcinol method, 295
Rhamnose,
 anthrone reaction with, 269–271
 phenol-sulphuric acid reagent on,
   276
Rhizobium sp.,
 poly-β-hydroxybutyrate in, 189, 191
Rhodomicrobium vanielli,
 cytomembranes from, 166
Rhodopseudomonas sp. *N.H.T.C.*,
 sub-chromatophore particles from,
   186
Rhodopseudomonas capsulata,
 cytomembranes from, 172–174
Rh. palustris,
 cytomembranes from, 167
 light intensity on, 151
 sub-chromatophore particles from,
   187
 lipids from, 351–352

Rh. spheroides,
cell free extract of, 130–131
cytomembranes from, 162, 172–174
sub-chromatophore particles from, 185–186
Rh. viridis,
cytomembranes from, 163, 166–167
Rhodospirillum fulvum,
cytomembranes from, 167–169
R. molischianum,
cytomembranes from, 167–169
photosynthetic apparatus from, 150
R. photometricum,
cytomembranes from, 167–169
R. rubrum,
chromatophores from, 150, 163
cytomembranes from, 169–172
disintegration of, 23, 40, 44
β-hydroxybutyric acid dehydrogenase from, 430
light intensity, effect on, 151
photosynthetic apparatus for, 150
poly-β-hydroxybutyrate in, 189–190, 193
RNA, in, 156
spheroplast formation in, 163
sub-chromatophore particles from, 183–185
Riboflavin,
separation of, 471–472, 576–577, 585
Ribonuclease,
inhibitors of, 134
polymer phase distribution of, 394
polyribosomes, on, 134
ribosomes, on, 134
ribosomal protein analysis, in, 145
separation of, 457
Ribonucleic acid (RNA),
anthrone reaction with, 270–272
base composition of, 309–310, 631–634, 638
chromatography of, 142
composition of, 142, 311
cytomembranes, in, 155–156
determination of,
micro-organisms, in, 322–328
Ogur and Rosen method, by, 313, 315–316
orcinol, by, 286, 288, 290–291, 316–317

phloroglucinol, by, 317
Schmidt–Thannhauser method, by, 310–311, 313–315
Schneider method, by, 312, 314
sugar analysis, by, 316–320
ultraviolet absorption, by, 320–322
extraction of, 326, 328, 374–376
detergent, by, 374
perchlorate, by, 373–374
phenol, by, 375–376
sodium chlorate, in, 374–375
hydrolysate of,
determination of, 632–634, 638
separation of, 328, 469
iso-density sedimentation analysis of, 114–115, 118
molecular weight of, 143–145
phenol-sulphuric acid reagent with, 276
polymerase, purification of, 400, 434
polymer phase distribution of, 396
protein complex with, 528
ribosomes, in, 142–145
separation of, 478, 534
sedimentation constants for, 143
sedimentation velocity centrifugation of, 74, 79
soluble, extraction of, 376
Ribose,
anthrone reaction with, 269–271
electrophoresis, in, 529
orcinol method for, 287–290
phenol-sulphuric acid reagent on, 276
Ribosomes (*see also* bacterial ribosomes),
oxidosomes from, 405
separation of, 530, 538, 591–592
sub-units, extracts in, 132–133
Rotors,
analytical, 63
angle, 61
iso-density equilibrium techniques, in, 115
moving boundary techniques, in, 60–64
performance indices of, 65, 67
sedimentation equilibrium techniques in, 109–110
swinging bucket, 62
zonal, 62, 89

Rubidium chloride,
   in density gradients, 117
Ryles apparatus, 475–476

S

Saccharomyces cerevisiae,
   amino-acid pool from, 333
   iron determination in, 242
   nucleic acids from, 242
   polysaccharide determination in, 279–281
   ribosomes from, 138, 147
Salmonella para-B,
   disintegration of, 45
Salmonella typhi,
   antigens from, 403
S. typhimurium,
   disintegration of, 41
Salt gradients (*see also* density gradients),
   sedimentation studies in, 112–115, 117–118
Salting out, 440
Sample,
   application of, 473, 522, 580
   electrophoresis gels for, 575, 581, 583
   layering of, 90–92
Saponification,
   in lipid extraction, 353, 358
Sarcina sp.,
   disintegration of, 45
S. lutea
   cytomembranes from, 152
   disintegration of, 15, 25, 44
   lipid analysis in, 307
   RNA in, 155
Scattering coefficient, 617
Scheraga–Mandelkern equation, 143
Schiff-periodic acid stain, 465, 487, 502
Schlieren optical system, 73–78, 131
Schmidt–Thannhauser method,
   nucleic acid in,
      determination of, 310–311, 313–315
      extraction of, 372–373
Schneider method,
   nucleic acid determination by, 312, 314, 323, 324
Sedimentation coefficient, 65–67, 70, 75, 99–102, 130, 143

Sedimentation constant,
   bovine serum albumin, of, 72
   determination of, 66–67
   RNA, of, 143
   ribosomes, of, 132, 147
Sedimentation equilibrium methods, 103–112
   diffusion and, 102
   high speed, 105–106
   low speed, 106–108
   molecular weights by, 103–112
   optical systems for, 104–108
   practical aspects of, 108–110
   rotors for, 109–110
   theory of, 103–108
Sedimentation rate, 70–72
Separation gel, 581, 584
Separation medium,
   electrophoresis in, 523–532
Sephadex,
   electrophoresis in, 495, 506–507
   iso-electric focusing and, 609
   lipid separation on, 356
   molecular sieve chromatography, in, 444
   protein purification by, 443–444, 449
Sepharose, 444
Sequential extraction methods, 378–381
Serratia marcescens,
   disintegration of, 39, 41
   electrophoretic separation of, 545
S. plymuthica,
   disintegration of, 40
Serum caps, 578
Serum proteins,
   electrophoresis of, 463–467, 482, 489, 507, 526, 527, 532, 558
Shakers,
   for microbial disintegration, 7
Shape factor
   in velocity sedimentation, 69
Sharples supercentifuge, 57
Shigella sonnei,
   disintegration of, 41
Silica gel G, 482
Silicic acid chromatography, 307
Silicone oil,
   as electrophoresis coolant, 476
Silicone rubber,
   disc electrophoresis in, 578

Silver–silver chloride electrodes, 553
Simple coacervates, 387
Slime moulds,
  electrophoretic separation in, 591
Slime polysaccharide
  aqueous extraction of, 360–362
Slot former,
  in starch gel electrophoresis, 492
Small pore gel,
  in disc electrophoresis, 581
Sodium acetate,
  in electrophoresis buffers, 471, 495,
    500, 524, 529
Sodium azide,
  in cellulose electrophoretic column,
    562
Sodium chlorate,
  in RNA detergent extraction, 374–
    375
Sodium dithionate,
  in cytochrome estimation, 621
Sodium ions,
  determination of, 231–233
  ribosomes, effect on, 135–137
Sodium lauryl sulphate,
  in nucleic acid extraction, 372–374
Sodium phosphate buffers, see phos-
    phate buffers
Sodium tetraborate buffer, see also
    borate, 463
Solid support electrophoresis, 559–
    507
Soluble polysaccharide,
  separation of, 370
Solvent spectrum, 388–390
Somatic antigen, 366–367
Sonicators, 52
Sonomec shaker, 8, 26
Soxhlet extraction,
  for lipid, 353
Specific conductivity, 517, 523
Specific protectors,
  for enzymes, 433
Specular reflectance, 622
Spheroplasts,
  formation of, 156–157
  Ps. fluorescens, from, 158
  R. rubrum, from, 163
Spirillum itersonii,
  polyphosphate in, 195

Split beam photoelectric scanning sys-
    tem, 78–80
Split gels,
  in disc electrophoresis, 584
Spores,
  electrophoretic separation of, 530
  harvesting of, 213
  nucleic acid determination in, 327–
    328
Stabilized moving-boundary centrifuga-
    tion, 65
Stacking gel, 575, 581, 583
Staphylococcus sp.,
  disintegration of, 45
S. aureus,
  amino-acid pool in, 332–333, 349
  bacteriolytic activity in, 612
  disintegration of,
    autolysis, by, 42
    explosive decompression, by, 40
    French press, by, 15
    wet mill, by, 25
  electrophoresis of, 544
  extracellular enzymes from, 609–612
  toxins from, 609
S. epidermis,
  amino-acid pool in, 333
Staphylokinase separation, 609–610
Starch,
  electrophoresis in,
    block, 503–508
      apparatus for, 504
      block preparation in, 503–506
      protein elution in, 506
      sample application in, 504–505
      starch grains in, 503
    gel,
      apparatus for, 489–493
      applications of, 499
      buffers for, 493–494
      gel preparation in, 494–495
      procedure for, 495–496
      protein staining in, 496–499
      two dimensional, 499
    gel,
      disc electrophoresis in, 574
    phenol-sulphuric acid reagent on, 276
    slurry, 503–504
Stereum sanguinolentum,
  enzyme separation in, 608

Stokes' law, 69, 514
Streaming velocity, 517
Streptococci sp.,
   hyaluronate lyase from, 612
   protoplasts from, 153
S. fecalis
   ATP-ases from, 430
   cytomembranes from, 152–153
   disintegration of, 23
   plasma membrane of, 150
S. pyogenes
   disintegration of, 23
   ribosomes from, 147
Streptolysin O., 45, 612
Streptomyces albus G.,
   N-acetyl hexosaminidase of, 559
Streptomycin,
   nucleic acid precipitation by, 438–439
Sub-cellular components (*see also* bacterial sub-cellular components)
   electrophoretic separation of, 537–540
Sub-cellular organelles,
   velocity sedimentation of, 70
Sub-chromatophore particles, 183–189
Sucrose,
   density gradients, in, 84, 115, 139, 154, 550, 554, 555, 597–605
   disc electrophoresis, in, 583
   phenol-sulphuric acid reagent on, 276
Sudan black B.
   lipoprotein stain, for, 469, 488, 497
   poly-β-hydroxybutyrate detection with, 304
Sugar determination,
   chemical methods, by, 283–301
   chromatography, by, 301–302
   enzymatic methods, by, 282–283
Sugar phosphates,
   extraction of, 348
Sulphate determination, 229–231
Sulphoethyl cellulose, 442–443
Sulpholane, 84–85
Sulphomethyl cellulose, 443
Sulphur-deficient medium, 369
Sulphydryl reagents,
   enzyme reactivation by, 430
Surface charge,
   cell separation by, 544–546

Svedberg unit, 66
Swinging bucket rotors,
   analytical centrifugation, in,
      moving boundary, 62, 64
      rate-zonal, 90
      sedimentation equilibrium, 125–127
Synechococcus cedrorum,
   cytomembranes from, 179–180
Szent-Gyorgyi-Blum rotor, 57

T

Takadiastase, 467
Taxonomic classification,
   electrophoresis in, 497
Technicon "Autoanalyser" 212
Teichoic acid,
   analysis of, 329–330
   extraction of, 367
Temperature activation,
Temperature control,
   in free-flow electrophoresis, 522
Terminal velocity, 67–68
Tetrahymena sp.,
   disintegration of, 22, 24
T. pyriformis, 27
Tetramethylethylene diamine (TEMED), 585
Thermal denaturation,
   DNA analysis by, 632
Thermal diffusion,
   in free-flow electrophoresis, 522
Thin layers,
   chromatography, in,
      amino-acids by, 333
      lipids by, 307, 308
      nucleic acids by, 633–634
      counter-current distribution in, 421–414
   electrophoresis, in, 481–482
Thiocapsa floridiana,
   cytomembranes in, 174
Thiococcus sp.,
   cytomembranes in, 174
Thiorhodaceae,
   cytomembranes from, 174–177
   sub-chromatophore particles from, 187–188

Thiospirillum jenense,
  cytomembranes in, 174
Thorium oxide, 85
Threonine deaminase, 430, 436
Tie lines,
  in phase diagrams, 390
Titan Yellow,
  magnesium estimation by, 236–238
Titanium zonal rotors, 127–128
Tobacco mosaic virus protein,
  electrophoretic purification of, 528,
    534
Toluene,
  as electrophoresis coolant, 476
Tolypothrix tenuis,
  extract of, 179
Torula utilis,
  determination in,
    DNA, of, 319
    iron, of, 242
    magnesium, of, 238
    pentose, of, 291
    total carbohydrate, of, 268–269,
      273–275, 277
Total lipid extraction, 353
Total reflectance,
  definition of, 622
  spectrum, *Candida utilis* of, 618
α-Toxin,
  Clostridium perfringens from, 595–
    613
Toxin separation, 595–613
Transfer RNA,
  electrophoresis of, 468
Transferrin, 574
Tricarboxylic acid cycle intermediates,
  extraction of, 348
Trichloroacetic acid,
  extraction by,
    DNA, of, 372–373
    microbial cells, of, 347–348, 366–
      367
    polyphosphate, of, 370
    somatic antigen, of, 366–367
    teichoic acids, of, 367
  protein stains, in, 587
Triethanolamine buffers, 530, 531
Triethylamine,
  in iso-electric focusing, 603
Triethylamine cellulose, 443

Trimethylacetic acid
  in electrophoresis buffers, 526
Trinucleotides,
  separation of, 470–471
Tris, buffers, 354
  in electrophoresis, 459, 493, 500,
    526–528, 530–531, 555
N-tris(hydroxymethyl)-methylamino-
    ethanesulphonic acid, 435
Trypsin, 467
Tryptophan,
  absorption spectrum of, 262
  anthrone reaction with, 270–271,
    273
  synthetase, separation of, 591
Tryptophanase, 428, 435
Tube-slicer,
  centrifugation and, 93
Tyrosinase, 497
Tyrosine, 262
Tyrosyl-RNA synthetase, 591

U

Ultrasonic disintegration, 9–10, 27–30
  apparatus for, 27, 29
  cavitation in, 9, 28–29
  efficiency of, 28
  factors influencing, 28–29
  procedure for, 27–28
Ultrasonic generator, 27–30
Ultraviolet absorbance,
  bacterial ribosomes of, 142
  protein determination by, 262–264
  nucleic acid determination by, 320–
    322
  nucleotide determination by, 469
"Uniphor" column, 551–552
Urea,
  electrophoresis in, 494, 583
  endo-osmotic standard, as, 400
  extraction with, 377
  iso-electric focusing in, 599, 610
Urea-tris-glycine buffer, 527
Uridine monophosphate, 469
U-tube,
  in preparative electrophoresis, 551,
    555, 563, 564, 567
Uvicord flow monitor, 96

V

Vacuum drying boxes, 432
Valmet, 599
Van-Slyke-Neill manometer, 216–217
Vapour saturation, 462
Versol, *see* white spirit.
Velocity profiles,
    free-flow electrophoresis in, 522
Velocity sedimentation methods, (*see also* moving-boundary centrifugation *and* rate-zonal centrifugation)
    for sub-cellular components, 60–102
Veronal-acetate buffer, 527
Vertical electrophoresis, 460, 491–492
Vibration mills,
    microbial disintegration by, 8, 25–26
Vidikon, 521
Vinyl chloride, (*see also* Pevicon), 495, 563, 566
Virus, (*see also* bacteriophage, phage)
    precipitation of, 387
    protein, separation of, 613
    purification of, 401–403, 405–412, 417–418
    separation of, 89, 115, 559
Viscosity effects, 514, 554
Vitamin K, 357–358
Volatile buffers, 480
Voltage effects,
    in electrophoresis, 457–458
Volutin granules,
    micro-organisms, in, 193, 194
    polyphosphates in, 330

W

Wall effects, 64–65
Weight dropping device, 34
Wet mills,
    microbial disintegration by, 7–9, 24–25
    colloid mills in, 8
    Booth and Green Mill in, 24–25
    Banard and Howlett Mill in, 25
    Muys mill in, 25
Wettex sponge, 462, 484, 505
Whitemor oil, 484
White spirit,
    as electrophoretic coolant, 471, 476, 477

Whole cell separation,
    free-flow electrophoresis by, 540–543
Wood destroying fungi,
    enzyme separation in, 608
Wool hydrolysates,
    electrophoresis of, 499

X

X-press, 12, 31–33, 52
    Hughes press and, 23
    description of, 31, 33
Xylanase, 607
Xylose,
    estimation of,
        orcinol reagent, by, 288, 290
        phenol-sulphuric acid reagent, by, 276
Xylosidase, 607

Y

Yeast,
    ATP-ases from, 430
    counter-current distribution of, 415
    cytochrome estimation in, 621–625
    disintegration of,
        autolysis, by, 42
        explosive decompression, by, 42
        French press, by, 15, 22
        homogenization, by, 23
        Hughes press, by, 37
        snail digestive juices, by, 45–46
        Streptomyces enzymes, by, 45
        vibration mill, by, 25–26
        wet mill, by, 25
    extraction of,
        alkali, by, 363
        citrate, by, 368
        ethylene diamine, by 368
    glycoproteins from, 368
    hexokinase from,
        proteolytic enzymes on, 609
    polyphosphates in, 369
    polysaccharide determination in, 279–281
    reflectance spectrum of, 620–625
    RNA determination in, 227
    total carbohydrate determination in, 266–267
Yeda press, 53

Z

Zeo-karb 225, 494
Zeta potential effects, 460
Zone capacity,
  in rate-zonal centrifugation, 91–92
Zonal electrophoresis, (see also preparative electrophoresis), 456–509
  agar gel, 499–503
  block, 503–508
  cellulose acetate, 482–489
  density gradient,
    buffers in, 555
    electromagnetic rotation in, 559

    theory of, 550–559
  paper, 463–481
    high voltage, 472–481
    low voltage, 463–472
  polysaccharide separation in, 281
  solid media,
    instruments in, 563–568
    method of, 568–669
    supporting media for, 559–563
  starch gel, 489–499
  theory of, 456–463
  thin layer, 481–482
Zonal rotor, 62, 64, 83, 88, 89
Zwitterionic buffers, 435, 458